Sedimentology of Paralic Reservoirs: Recent Advances

Geological Society books refereeing procedures

The Society makes every effort to ensure that the scientific and production quality of its books matches that of its journals. Since 1997, all book proposals have been refereed by specialist reviewers as well as by the Society's Books Editorial Committee. If the referees identify weaknesses in the proposal, these must be addressed before the proposal is accepted.

Once the book is accepted, the Society Book Editors ensure that the volume editors follow strict guidelines on refereeing and quality control. We insist that individual papers can only be accepted after satisfactory review by two independent referees. The questions on the review forms are similar to those for *Journal of the Geological Society*. The referees' forms and comments must be available to the Society's Book Editors on request.

Although many of the books result from meetings, the editors are expected to commission papers that were not presented at the meeting to ensure that the book provides a balanced coverage of the subject. Being accepted for presentation at the meeting does not guarantee inclusion in the book.

More information about submitting a proposal and producing a book for the Society can be found on its website: www.geolsoc.org.uk.

It is recommended that reference to all or part of this book should be made in one of the following ways:

Hampson, G. J., Reynolds, A. D., Kostic, B. & Wells, M. R. (eds) 2017. *Sedimentology of Paralic Reservoirs: Recent Advances*. Geological Society, London, Special Publications, **444**.

Lambiase, J. J., Riadi, R. S., Nirsal, N. & Husein, S. 2017. Transgressive successions of the Mahakam Delta Province, Indonesia. *In:* Hampson, G. J., Reynolds, A. D., Kostic, B. & Wells, M. R. (eds) *Sedimentology of Paralic Reservoirs: Recent Advances*. Geological Society, London, Special Publications, **444**, 335–348. First published online June 29, 2016, https://doi.org/10.1144/SP444.2

GEOLOGICAL SOCIETY SPECIAL PUBLICATION NO. 444

Sedimentology of Paralic Reservoirs: Recent Advances

EDITED BY

G. J. HAMPSON
Imperial College London, UK

A. D. REYNOLDS
BP Exploration Operating Co. Ltd, UK

B. KOSTIC
Badley Ashton & Associates, UK

and

M. R. WELLS
BP Exploration Operating Co. Ltd, UK

2017
Published by
The Geological Society
London

THE GEOLOGICAL SOCIETY

The Geological Society of London (GSL) was founded in 1807. It is the oldest national geological society in the world and the largest in Europe. It was incorporated under Royal Charter in 1825 and is Registered Charity 210161.

The Society is the UK national learned and professional society for geology with a worldwide Fellowship (FGS) of over 10 000. The Society has the power to confer Chartered status on suitably qualified Fellows, and about 2000 of the Fellowship carry the title (CGeol). Chartered Geologists may also obtain the equivalent European title, European Geologist (EurGeol). One fifth of the Society's fellowship resides outside the UK. To find out more about the Society, log on to www.geolsoc.org.uk.

The Geological Society Publishing House (Bath, UK) produces the Society's international journals and books, and acts as European distributor for selected publications of the American Association of Petroleum Geologists (AAPG), the Indonesian Petroleum Association (IPA), the Geological Society of America (GSA), the Society for Sedimentary Geology (SEPM) and the Geologists' Association (GA). Joint marketing agreements ensure that GSL Fellows may purchase these societies' publications at a discount. The Society's online bookshop (accessible from www.geolsoc.org.uk) offers secure book purchasing with your credit or debit card.

To find out about joining the Society and benefiting from substantial discounts on publications of GSL and other societies worldwide, consult www.geolsoc.org.uk, or contact the Fellowship Department at: The Geological Society, Burlington House, Piccadilly, London W1J 0BG: Tel. +44 (0)20 7434 9944; Fax +44 (0)20 7439 8975; E-mail: enquiries@geolsoc.org.uk.

For information about the Society's meetings, consult *Events* on www.geolsoc.org.uk. To find out more about the Society's Corporate Affiliates Scheme, write to enquiries@geolsoc.org.uk.

Published by The Geological Society from:
The Geological Society Publishing House, Unit 7, Brassmill Enterprise Centre, Brassmill Lane, Bath BA1 3JN, UK

The Lyell Collection: www.lyellcollection.org
Online bookshop: www.geolsoc.org.uk/bookshop
Orders: Tel. +44 (0)1225 445046, Fax +44 (0)1225 442836

British Library Cataloguing in Publication Data

A catalogue record for this book is available from the British Library.
ISBN 978-1-78620-274-1
ISSN 0305-8719

Distributors
For details of international agents and distributors see:
www.geolsoc.org.uk/agentsdistributors

Typeset by Nova Techset Private Limited, Bengaluru & Chennai, India
Printed and bound by CPI Group (UK) Ltd, Croydon CR0 4YY

Contents

Acknowledgements

The Geological Society thanks the following companies for their generous sponsorship of the Petroleum Group and the conference that led to this volume. The Petroleum Group is thanked for its contribution towards the colour printing in this volume.

Petroleum Group corporate sponsors at the time of the conference:

Sponsors of Sedimentology of Paralic Reservoirs: Recent Advances and their Applications conference:

Introduction to the sedimentology of paralic reservoirs: recent advances

GARY J. HAMPSON[1]*, ANTONY D. REYNOLDS[2], BORIS KOSTIC[3] & MARTIN R. WELLS[2]

[1]*Department of Earth Science and Engineering, Imperial College London, London SW7 2AZ, UK*

[2]*BP Exploration Operating Co. Ltd, Sunbury-Upon-Thames, Middlesex TW16 7LN, UK*

[3]*Badley Ashton & Associates, Winceby House, Winceby, Horncastle, Lincolnshire LN9 6PB, UK*

**Correspondence: g.j.hampson@imperial.ac.uk*

Paralic reservoirs reflect a range of clastic depositional environments developed along or near coastlines, including deltas, shoreline–shelf systems and estuaries. Such reservoirs provide the backbone of production in many mature basins around the world, and contribute significantly to global conventional hydrocarbon production. Strata that host these reservoirs are shaped by a wide variety of depositional processes and controls which reflect the upstream supply of sediment and water, the characteristics of the receiving basin, relative sea level, tectonic setting, and the internal dynamics of depositional systems. Consequently, they exhibit much variability in their stratigraphic architecture and sedimentological heterogeneity, which translates into complex patterns of reservoir distribution and reservoir performances that are challenging to predict, optimize and manage. This Special Publication presents new research and developments in established approaches to exploration and production of paralic reservoirs. It arises from the conference and associated core workshop titled 'Sedimentology of Paralic Reservoirs: Recent Advances and their Applications', which was organized by the Petroleum Group of the Geological Society of London and held in London from 18 to 21 May 2015.

Volume organization

The papers in this Special Publication are arranged into three sections, each with a distinct theme. The aim of this introduction is to provide a broader context for the papers in these three sections.

Subsurface characterization of paralic reservoirs

The first group of papers explores the sedimentological character of paralic reservoirs using subsurface data. The papers consider the subdivision of paralic reservoirs into stratigraphic layers; the dimensions, geometries and distributions of depositional elements (cf. geobodies) within such layers; and facies characteristics within depositional elements. These aspects of reservoir sedimentology all influence fluid flow and hydrocarbon recovery. Stratigraphic layering of paralic reservoirs commonly defines the position of laterally extensive barriers to vertical flow (e.g. mudstone intervals) and the thickness of intervening flow units (*sensu* Ebanks 1987) that are bounded by such barriers. The spatial arrangement of depositional elements controls the lateral extent, continuity and connectivity of fluid-flow pathways within stratigraphic layers (e.g. Weber & Van Guens 1990), and influences the volumetric proportion of hydrocarbon-bearing rock that is contacted by fluids moving through the reservoir layers (i.e. the areal and vertical sweep efficiency). Facies characteristics determine the grain sizes, textures and rock properties (e.g. permeability, porosity) of depositional elements, and they influence in turn the efficiency with which hydrocarbons are displaced from swept volumes of the reservoir at the pore scale (i.e. displacement efficiency) (e.g. Weber 1986).

Reynolds (2016) provides an overview and synthesis of paralic reservoirs, which explains their global significance and identifies their distinctive features across a range of scales. The impact of sedimentology on reservoir characterization and associated field-development strategy is emphasized throughout. This paper explains the overall motivation for previous and current research on paralic reservoirs, including the examples presented in the volume. The sedimentological character and stratigraphic architecture of the Lower Jurassic Cook Formation reservoir in the Knarr Field, offshore Norway, are documented by **Churchill *et al.* (2016)**. In their study, the depositional environments

From: HAMPSON, G. J., REYNOLDS, A. D., KOSTIC, B. & WELLS, M. R. (eds) 2017. *Sedimentology of Paralic Reservoirs: Recent Advances*. Geological Society, London, Special Publications, **444**, 1–6.
First published online May 9, 2017, https://doi.org/10.1144/SP444.14

and stratigraphic zonation of the reservoir are constrained by integrated sedimentological, micropalaeontological and palynological analyses. The reservoir is interpreted to record a temporal evolution in the depositional-process regime, from net-regressive and tide-dominated in its lower part to net-transgressive and wave-dominated in its upper part. **Heldreich *et al.* (2017)** use a combination of core, wireline-log and 3D seismic data to develop a depositional model for the fluvio-deltaic, Upper Triassic Mungaroo Formation, a major exploration and production target in the North Carnarvon Basin, offshore Australia. Their analysis characterizes the geometries, dimensions and vertical facies successions of depositional elements to aid reservoir characterization and modelling, and discusses uncertainties in interpreting allogenic (external) and autogenic (internal) controls on the architecture of strata that contain these geobodies. Characterization of stratigraphic architectures that may have resulted from autogenic behaviours, such as avulsion of distributary channels in fluvio-deltaic strata, is attempted by **Flood & Hampson (2016)** using a combination of core-based facies and palaeosol analyses, and the application of spatial statistical tools to channelized sandbody distributions in upper coastal plain strata in the Middle Jurassic Brent Group reservoir, Brent Field, offshore UK.

Tidal heterogeneity in paralic systems

Heterogeneity in lithological character is a recurring theme in paralic reservoirs. Arguably, heterogeneity is at its greatest in reservoirs that were deposited under the influence of tidal currents, where intercalation of sandstones and mudstones across a wide range of scales is common (e.g. Nio & Yang 1991). Such intercalations have repeatedly been shown to have a significant impact on fluid flow (e.g. Jackson *et al.* 2005; Ringrose *et al.* 2005) and remain an area of active research (Massart *et al.* 2016). Accordingly, the second group of papers in the Special Publication reviews and extends the literature that describes tide-influenced paralic reservoirs and their outcrop analogues.

Longhitano & Steel (2016) present a generic depositional model for deltas that advanced into seaways or straits in which tidal currents were dominant, based on two well-documented outcrop examples. The model identifies diagnostic aspects of the plan-view geometries, three-dimensional stratal stacking patterns, and vertical successions of facies, palaeocurents and depositional-process indicators in such deltas. The model may allow reinterpretation and improved characterization of hitherto-enigmatic sandstone reservoirs along the margins of elongate marine basins. **Mellere *et al.* (2016)** document the facies characteristics, stratigraphic architecture and palaeogeography of a series of stacked tide-dominated and tidally influenced deltas and estuaries in the Middle Jurassic Bryne and Sandnes formations, Søgne Basin, offshore Norway and Denmark, using core, wireline-log and 3D seismic data. These strata contain complex spatial patterns of reservoir distribution and character, which reflects their deposition in an evolving salt-influenced rift basin. A facies analysis and stratigraphic architectural study of net-transgressive, early Miocene tide-dominated deltaic reservoirs from the Balingian province, offshore Sarawak, Malaysia is presented by **Amir Hassan *et al.* (2016)**. Increased wave influence is characteristic of transgressive intervals in these reservoirs. Vertical and lateral stacking patterns of sequence stratigraphic units reflect tectonic controls on hinterland denudation and subsidence variations, respectively. **Johnson *et al.* (2016)** document the stratigraphic context and internal architecture of an estuarine outcrop analogue (from the Cretaceous Straight Cliffs Formation, Utah, USA) that exhibits variable tide and wave influence. The detailed bed-scale architecture of inclined heterolithic strata from a bayhead delta in the inner estuary is characterized in order to develop diagnostic criteria for bayhead delta depositional elements in subsurface reservoir data. **McLean & Wilson (2016)** use a quantitative approach to outcrop data from the Plio-Pleistocene Erin Formation, Trinidad to document both tidal and river flood signatures in thinly bedded mudstones and sandstones – an approach that is readily, but rarely, transferred to subsurface core data.

Analogue studies

Understanding the distribution and character of paralic reservoirs often relies on insights from a range of analogues, which provide appropriate context or additional detail to aid interpretation of sparse subsurface data. Analogues take many forms: modern sedimentary systems, in which processes can be observed and linked directly to depositional product; sub-modern depositional systems that are imaged at high resolution and in three dimensions using shallow seismic data; analogous reservoirs rich in static and dynamic data; outcrops of ancient strata, which allow preserved stratigraphic architectures and facies characteristics to be observed and integrated across a wide range of spatial scales; and numerical and physical (e.g. flume tank) models that allow the underlying assumptions of our interpretations to be investigated experimentally. The final group of papers in this Special Publication presents a range of paralic reservoir analogues from studies of modern sediments and ancient outcrops.

Lane *et al.* (2016) unravel temporal and spatial patterns of river-channel avulsion in a

Holocene–modern fluvial megafan and associated mixed-influence delta (Mitchell River, Queensland, Australia) using a combination of mapping, vibracoring, trenching and topographical surveying. The resulting patterns of stratigraphic architecture and channelized sediment-body distribution are integrated with vertical facies successions and intra-sediment-body architectural panels in order to develop a generic model of depositional-element distribution within a regressive megafan delta. **Lambiase *et al.* (2016)** characterize sedimentation patterns in the modern Mahakam Delta, Borneo, which is undergoing transgression via retrogradational stacking and switching of delta lobes. Diagnostic core-based sedimentological criteria are presented for three sharp-based sandstone depositional elements (regressive fluvial distributary channels; transgressively back-filled fluvio-estuarine distributary channels; transgressive wave-dominated shorefaces) that appear similar in wireline-log data but have distinctly different geometries, spatial distributions and sequence stratigraphic implications. The stratigraphic architecture of a thick, net-aggradational interval of paralic strata, characterized by high accommodation and high sediment supply, is described from outcrops of the Cretaceous Straight Cliffs Formation, Utah, USA by **Mulhern & Johnson (2016)**. Transgressive–regressive cycles composed of regressive, wave-dominated deltaic and shoreface deposits and transgressive, wave-dominated estuarine deposits can be correlated between successions of different thickness that underwent variable local tectonic subsidence and early compaction. **Gomis-Cartesio *et al.* (2016)** document the architecture of ancient wave-dominated and mixed-influence shelf-edge deltas and underlying gullies and slope channels at outcrop (Permian Waterford Formation, South Africa). The facies character, geometry and dimensions of depositional elements associated with the shelf edge are documented and synthesized into a model for sediment bypass from deltaic distributary channels onto the slope via gullies that become wider and shallower down depositional dip.

Recurring themes and future directions

Heterogeneity in paralic reservoirs is increasingly well understood at all scales. **Reynolds (2016)** highlights that (1) advances in imaging due to improvements in seismic data acquisition and processing, (2) widespread application of sequence stratigraphic concepts, (3) availability of robust facies models, (4) improved use and organization of analogue databases, and (5) increased integration of available data types have all played key roles in allowing heterogeneity from pore scale to reservoir scale to be described in paralic reservoirs. Consequently, reserve volumes and their distributions are estimated earlier in field life, allowing better and more timely development decisions. Despite this undoubted progress, the conference and the papers presented here amply demonstrate that progress continues in the sedimentological description of paralic reservoirs and that several recurring themes represent fruitful directions for future work:

- The tectonostratigraphic context of paralic successions influences their subsidence patterns and preserved thickness, sediment input and routing, and potentially also the process regime (e.g. via suppression or amplification of waves and tides: Stride 1982; Yoshida *et al.* 2007; Ainsworth *et al.* 2008; Wells *et al.* 2010). Spatial and temporal variations in tectonostratigraphic context may result in the development of subtle stratigraphic trapping configurations, and complex spatial patterns of reservoir distribution and character. Integrated analysis of seismic, well-log, core and biostratigraphic data is essential to unravel the stratigraphic architecture of such paralic successions, and thereby to identify and de-risk paralic reservoir targets, even in mature basins (e.g. **Amir Hassan *et al.* 2016**; **Churchill *et al.* 2016**; **Mellere *et al.* 2016**). Analogues have an important role to play in demonstrating the viability of concepts and models that are unproven in a subsurface hydrocarbon-play context (e.g. **Gomis-Cartesio *et al.* 2016**; **Longhitano & Steel 2016**; **Mulhern & Johnson 2016**).
- Theoretical and experimental studies have demonstrated that the internal dynamics of sediment-routing systems may influence stratigraphic architectures across a range of spatial and temporal scales (e.g. Muto & Steel 1992; Kim *et al.* 2006; Muto *et al.* 2007). This key theoretical contribution has yet to be routinely incorporated into the sequence stratigraphic interpretation of subsurface and outcrop data. However, distinguishing allogenic (external) and autogenic (internal) controls on sedimentation is important in predicting the extent, continuity and degree of order in paralic stratigraphic architectures, particularly in subsurface settings where data are sparse (e.g. Wang *et al.* 2011; see also discussions in **Flood & Hampson 2016**; **Lane *et al.* 2016**; **Heldreich *et al.* 2017**). The theory needs to be tested against more ‘real world’ datasets in order to further develop diagnostic tools and predictive models.
- Facies analysis of paralic deposits has traditionally emphasized the use of ‘end-member’ examples, dominated by riverine, wave or tidal processes, at the expense of mixed-influence

depositional systems in which two or more of these processes are significant. In recent years, the stratigraphic architecture of mixed-influence paralic deposits, and the geometry and distribution of the depositional elements that they contain, have been addressed in a new classification scheme (Ainsworth *et al.* 2011), and in case studies of modern and ancient reservoir analogues (e.g. Legler *et al.* 2014; Ainsworth *et al.* 2016; **Johnson *et al.* 2016**; **Lambiase *et al.* 2016**; **Lane *et al.* 2016**). Numerical models may provide a key role in extending this approach by exploring the range of possible stratigraphic architectures and facies distributions that lie between those documented in high-quality outcrop and subsurface case studies (e.g. Edmonds & Slingerland 2010; Geleynse *et al.* 2011).

- Lithological heterogeneity within depositional elements and their constituent facies typically occurs at spatial scales smaller than reservoir-model grid cells. This heterogeneity is arranged in a hierarchical manner (e.g. laminae, laminasets, beds and bedsets: Campbell 1967), and may be architecturally complex due to a combination of cross-stratification, erosional scour, lapout, dewatering and bioturbation. Deriving values of effective permeability, for example, that are applicable at the scale of facies in a reservoir-model grid cell can be far from straightforward, because laboratory core measurements may not sample a sufficiently large volume of the reservoir to represent the heterogeneity present (i.e. Representative Element Volume *sensu* Bear 1972). Similarly, upscaling of core measurements using simple averaging methods may not be appropriate if the measurements do not capture the spatial structure of the heterogeneity. Characterizing the architecture of lamina-scale to bedset-scale heterogeneity is therefore an essential step in deriving values of effective rock properties for a particular facies (e.g. Willis & White 2000; Jackson *et al.* 2005; **Johnson *et al.* 2016**; **McLean & Wilson 2016**; **Reynolds 2016**), although detailed architectural studies at these scales remain relatively rare.
- The recent explosive growth in production from shale gas and shale oil reservoirs, particularly in onshore North America, requires robust sedimentological characterization of paralic mudstone successions. Although not addressed by the papers in this volume, it is clear that paralic facies models must develop to incorporate the geometrical, textural and mineralogical characteristics of mudstones in addition to those of sandstones that are typically the focus of such models (e.g. MacQuaker *et al.* 2007; Schieber *et al.* 2010; Bohacs *et al.* 2014). As well as unconventional shale reservoirs, there is potential to reassess hydrocarbons stranded in the shaly 'waste zones' of paralic sandstone reservoirs as reserves in 'halo plays' that surround these sandstones.

We thank the Geological Society, in particular its Petroleum Group, and the SEPM (Society for Sedimentary Geology) for their support of the conference, the related core workshop and this Special Publication. The success of the conference and core workshop owed much to the generous financial help of sponsors BP, Statoil and Chevron, and to the support of Weatherford Laboratories, DEA and the British Geological Survey in releasing and displaying cores. The Petroleum Group of the Geological Society is thanked for supporting the reproduction of colour figures in this volume.

We are indebted to the authors of the 34 talks, 20 posters and four core displays presented at the conference and core workshop, and to the many attendees for their lively contributions to discussions. Bruce Ainsworth, Janok Bhattacharya and Ron Steel are thanked for their input as conference co-convenors, in addition to the four editors of this volume. We are grateful to the authors who have contributed papers to this volume, and to the reviewers who provided insightful and constructive comments that shaped the final form of these papers. Finally, we thank the staff of the Geological Society's Conference Office and Publishing House for their administrative and logistical support throughout, in particular Tamzin Anderson, Jo Armstrong, Laura Griffiths, Angharad Hills and Annie Sewell.

References

Ainsworth, R.B., Flint, S.S. & Howell, J.A. 2008. Predicting coastal depositional style: influence of basin morphology and accommodation to sediment supply ratio within a sequence stratigraphic framework. *In*: Hampson, G.J., Steel, R.J., Burgess, P.M. & Dalrymple, R.W. (eds) *Recent Advances in Models of Siliciclastic Shallow-Marine Stratigraphy*. Society for Sedimentary Geology (SEPM), Special Publications, **90**, 237–263.

Ainsworth, R.B., Vakarelov, B.K. & Nanson, R.A. 2011. Dynamic spatial and temporal prediction of changes in depositional processes on clastic shorelines: toward improved subsurface uncertainty reduction and management. *American Association of Petroleum Geologists Bulletin*, **95**, 267–297.

Ainsworth, R.B., Vakarelov, B.K., MacEachern, J.A., Nanson, R.A., Lane, T.I., Rarity, F. & Dashtgard, R.A. 2016. Process-driven architectural variability in mouth-bar deposits: a case study from a mixed-process mouth-bar complex, Drumheller, Alberta, Canada. *Journal of Sedimentary Research*, **86**, 512–541.

Amir Hassan, M.H., Johnson, H.D., Allison, P.A. & Abdullah, W.H. 2016. Sedimentology and stratigraphic architecture of a Miocene retrogradational, tide-dominated delta system: Balingian Province, offshore Sarawak, Malaysia. *In*: HAMPSON, G.J., REYNOLDS, A.D., KOSTIC, B. & WELLS, M.R. (eds) *Sedimentology of Paralic Reservoirs: Recent Advances*. Geological Society, London, Special Publications, **444**. First published online August 30, 2016, https://doi.org/10.1144/SP444.12

BEAR, J. 1972. *Dynamics of Fluids in Porous Media*. American Elsevier, New York.

BOHACS, K.M., LAZAR, O.R. & DEMKO, T.M. 2014. Parasequence types in shelfal mudstone strata – quantitative observations of lithofacies and stacking patterns, and conceptual link to modern depositional analogues. *Geology*, **42**, 131–134.

CAMPBELL, C.V. 1967. Lamina, laminaset, bed and bedset. *Sedimentology*, **8**, 7–26.

CHURCHILL, J.M., POOLE, M.T., SKARPEID, S.S. & WAKEFIELD, M.I. 2016. Stratigraphic architecture of the Knarr Field, Norwegian North Sea: sedimentology and biostratigraphy of an evolving tide- to wave-dominated shoreline system. *In*: HAMPSON, G.J., REYNOLDS, A.D., KOSTIC, B. & WELLS, M.R. (eds) *Sedimentology of Paralic Reservoirs: Recent Advances*. Geological Society, London, Special Publications, **444**. First published online August 12, 2016, https://doi.org/10.1144/SP444.4

EBANKS, W.J., JR. 1987. Flow unit concept – integrated approach to reservoir description for engineering projects. *American Association of Petroleum Geologists Bulletin*, **71**, 551–552.

EDMONDS, D.A. & SLINGERLAND, R.L. 2010. Significant effect of sediment cohesion on delta morphology. *Nature Geoscience*, **3**, 105–109.

FLOOD, Y.S. & HAMPSON, G.J. 2016. Analysis of floodplain sedimentation, avulsion style and channelized fluvial sandbody distribution in an upper coastal plain reservoir: Middle Jurassic Ness Formation, Brent Field, UK North Sea. *In*: HAMPSON, G.J., REYNOLDS, A.D., KOSTIC, B. & WELLS, M.R. (eds) *Sedimentology of Paralic Reservoirs: Recent Advances*. Geological Society, London, Special Publications, **444**. First published online June 29, 2016, https://doi.org/10.1144/SP444.3

GELEYNSE, N.L., STORMS, J.E.A., WALSTRA, D.J.R., JAGERS, H.R.A., WANG, Z.B. & STIVE, M.J.F. 2011. Controls on river delta formation; insights from numerical modelling. *Earth and Planetary Science Letters*, **302**, 217–226.

GOMIS-CARTESIO, L.E., POYATOS-MORÉ, M., FLINT, S.S., HODGSON, D.M., BRUNT, R.L. & WICKENS, H.DeV. 2016. Anatomy of a mixed-influence shelf edge delta, Karoo Basin, South Africa. *In*: HAMPSON, G.J., REYNOLDS, A.D., KOSTIC, B. & WELLS, M.R. (eds) *Sedimentology of Paralic Reservoirs: Recent Advances*. Geological Society, London, Special Publications, **444**. First published online July 14, 2016, https://doi.org/10.1144/SP444.5

HELDREICH,G., REDFERN, J., LEGLER, B., GERDES, K. & WILLIAMS, B.P.J. 2017. Challenges in characterizing subsurface paralic reservoir geometries: a detailed case study of the Mungaroo Formation, North West Shelf, Australia. *In*: HAMPSON, G.J., REYNOLDS, A.D., KOSTIC, B. & WELLS, M.R. (eds) *Sedimentology of Paralic Reservoirs: Recent Advances*. Geological Society, London, Special Publications, **444**. First published online May 8, 2017, https://doi.org/10.1144/SP444.13

JACKSON, M.D., YOSHIDA, S., MUGGERIDGE, A.H. & JOHNSON, H.D. 2005. Three-dimensional reservoir characterisation and flow simulation of heterolithic tidal sandstones. *American Association of Petroleum Geologists Bulletin*, **89**, 507–528.

JOHNSON, C.L., STRIGHT, L., PURCELL, R. & DURKIN, P. 2016. Stratigraphic evolution of an estuarine fill succession and the reservoir characterization of inclined heterolithic strata, Cretaceous of southern Utah, USA. *In*: HAMPSON, G.J., REYNOLDS, A.D., KOSTIC, B. & WELLS, M.R. (eds) *Sedimentology of Paralic Reservoirs: Recent Advances*. Geological Society, London, Special Publications, **444**. First published online June 29, 2016, https://doi.org/10.1144/SP444.1

KIM, W., PAOLA, C., SWENSON, J.B. & VOLLER, V.R. 2006. Shoreline response to autogenic processes of sediment storage and release in the fluvial system. *Journal of Geophysical Research*, **111**, F04013.

LAMBIASE, J.J., RIADI, R.S., NIRSAL, N. & HUSEIN, S. 2016. Transgressive successions of the Mahakam Delta Province, Indonesia. *In*: HAMPSON, G.J., REYNOLDS, A.D., KOSTIC, B. & WELLS, M.R. (eds) *Sedimentology of Paralic Reservoirs: Recent Advances*. Geological Society, London, Special Publications, **444**. First published online June 29, 2016, https://doi.org/10.1144/SP444.2

LANE, T.I., NANSON, R.A., VAKARELOV, B.K., AINSWORTH, R.B. & DASHTGARD, S.E. 2016. Evolution and architectural styles of a forced-regressive Holocene delta and megafan, Mitchell River, Gulf of Carpentaria, Australia. *In*: HAMPSON, G.J., REYNOLDS, A.D., KOSTIC, B. & WELLS, M.R. (eds) *Sedimentology of Paralic Reservoirs: Recent Advances*. Geological Society, London, Special Publications, **444**. First published online July 7, 2016, https://doi.org/10.1144/SP444.9

LEGLER, B., HAMPSON, G.J., JACKSON, C.A.-L., JOHNSON, H.D., MASSART, B.Y.G., SARGINSON, M. & RAVNAS, R. 2014. Facies relationships and stratigraphic architecture of distal, mixed tide-and-wave-influenced deltaic deposits: lower Sego Sandstone, western Colorado, USA. *Journal of Sedimentary Research*, **84**, 605–625.

LONGHITANO, S.G. & STEEL, R.J. 2016. Deflection of the progradational axis and asymmetry in tidal seaway and strait deltas: insights from two outcrop case studies. *In*: HAMPSON, G.J., REYNOLDS, A.D., KOSTIC, B. & WELLS, M.R. (eds) *Sedimentology of Paralic Reservoirs: Recent Advances*. Geological Society, London, Special Publications, **444**. First published online July 13, 2016, https://doi.org/10.1144/SP444.8

MACQUAKER, J.H., TAYLOR, K.G. & GAWTHORPE, R.L. 2007. High-resolution facies analysis of mudstones: implications for paleoenvironmental and sequence stratigraphic interpretations of offshore ancient mud-dominated successions. *Journal of Sedimentary Research*, **77**, 324–339.

MASSART, B.Y.G., JACKSON, M.D., HAMPSON, G.J. & JOHNSON, H.D. 2016. Effective flow properties of heterolithic, cross-bedded tidal sandstones, part 2: flow simulation. *American Association of Petroleum Geologists Bulletin*, **100**, 723–742.

McLEAN, A.Q. & WILSON, B. 2016. Recognizing seasonal fluvial influence in ancient tidal deposits. *In*: HAMPSON, G.J., REYNOLDS, A.D., KOSTIC, B. & WELLS, M.R. (eds) *Sedimentology of Paralic Reservoirs: Recent Advances*. Geological Society, London, Special Publications, **444**. First published online July 13, 2016, https://doi.org/10.1144/SP444.6

MELLERE, D., MANNIE, A., LONGHITANO, S., MAZUR, M., KULAUSA, H., BROUGH, S. & COTTON, J. 2016. Tidally influenced shoal water delta and estuary in the Middle Jurassic of the Søgne Basin, Norwegian North Sea: sedimentary response to rift initiation and salt tectonics. *In*: HAMPSON, G.J., REYNOLDS, A.D., KOSTIC, B. & WELLS, M.R. (eds) *Sedimentology of Paralic Reservoirs: Recent Advances*. Geological Society, London, Special Publications, **444**. First published online September 21, 2016, https://doi.org/10.1144/SP444.11

MULHERN, J.S. & JOHNSON, C.L. 2016. Time–space variability of paralic strata deposited in a high accommodation, high sediment supply setting: example from the Cretaceous of Utah. *In*: HAMPSON, G.J., REYNOLDS, A.D., KOSTIC, B. & WELLS, M.R. (eds) *Sedimentology of Paralic Reservoirs: Recent Advances*. Geological Society, London, Special Publications, **444**. First published online June 29, 2016, https://doi.org/10.1144/SP444.7

MUTO, T. & STEEL, R.J. 1992. Retreat of the front in a prograding delta. *Geology*, **20**, 967–970.

MUTO, T., STEEL, R.J. & SWENSON, J.B. 2007. Autostratigraphy: a framework norm for genetic stratigraphy. *Journal of Sedimentary Research*, **77**, 2–12.

NIO, S.-D. & YANG, C.-S. 1991. Diagnostic attributes of clastic tidal deposits: a review. *In*: SMITH, D.G., REINSON, G.E., ZAITLIN, B.A. & RAHMANI, R.A. (eds) *Clastic Tidal Sedimentology*. Canadian Society of Petroleum Geologists, Memoirs, **16**, 3–28.

REYNOLDS, A.D. 2016. Paralic reservoirs. *In*: HAMPSON, G.J., REYNOLDS, A.D., KOSTIC, B. & WELLS, M.R. (eds) *Sedimentology of Paralic Reservoirs: Recent Advances*. Geological Society, London, Special Publications, **444**. First published online July 20, 2016, https://doi.org/10.1144/SP444.10

RINGROSE, P.S., NORDHAL, K. & WEN, R. 2005. Vertical permeability estimation of heterolithic tidal deltaic sandstones. *Petroleum Geoscience*, **11**, 29–36, https://doi.org/10.1144/1354-079303-614

SCHIEBER, J., SOUTHARD, J.B. & SCHIMMELMANN, A. 2010. Lenticular shale fabrics resulting from intermittent erosion of water-rich muds – interpreting the rock record in the light of recent flume experiments. *Journal of Sedimentary Research*, **80**, 119–128.

STRIDE, A.H. 1982. *Offshore Tidal Sands*. Chapman & Hall, London.

WANG, Y., STRAUB, K.M. & HAJEK, E.A. 2011. Scale-dependent compensational stacking: an estimate of autogenic time scales in channelised sedimentary deposits. *Geology*, **39**, 811–814.

WEBER, K.J. 1986. How heterogeneity affects oil recovery. *In*: LAKE, L.W. & CARROLL, H.B. (eds) *Reservoir Characterization*. Academic Press, New York, 487–544.

WEBER, K.J. & VAN GUENS, L.C. 1990. Framework for constructing clastic reservoir simulation models. *Journal of Petroleum Technology*, **43**, 1248–1253.

WELLS, M.R., ALLISON, P.A., PIGGOTT, M.D., HAMPSON, G.J., PAIN, C.C. & GORMAN, G.J. 2010. Tidal modeling of an ancient tide-dominated seaway, part 2: the Aptian Lower Greensand seaway of northwest Europe. *Journal of Sedimentary Research*, **80**, 411–439.

WILLIS, B.J. & WHITE, C.D. 2000. Quantitative outcrop data for flow simulation. *Journal of Sedimentary Research*, **70**, 788–802.

YOSHIDA, S., STEEL, R.J. & DALRYMPLE, R.W. 2007. Changes in depositional processes – an ingredient in a new generation of sequence-stratigraphic models. *Journal of Sedimentary Research*, **77**, 447–460.

Paralic reservoirs

ANTONY D. REYNOLDS

BP Exploration, Sunbury on Thames, Middlesex TW16 7LN, UK
tony.reynolds@bp.com

Abstract: Paralic environments, deltas, estuaries and shoreline-shelf systems respond sensitively to changes in sea-level and sediment supply, commonly by shifts in facies belts across sequence stratigraphic surfaces. As a result, they are often interleaved in highly layered successions. Layering and a wide range of associated sandbody types control the distinctive nature of paralic oil and gas fields. Layering results in stacked reservoirs. Thick extensive sandstones have structurally defined spill points, large aquifers and good reservoir properties. Such sands have high recovery, good reservoir performance and form primary completion targets. By contrast, thin, laterally restricted, isolated sandstones have low recovery factors and poor performance. Each isolated sand may have its own hydrocarbon contacts and a distinct outline that reflects both structural and stratigraphic trapping. Such sands form secondary completion targets. Fault compartmentalization is a recurring risk. Sealing faults increase the number of reservoir compartments and may decrease performance and recovery. They can also generate traps. Sedimentology is pivotal in describing heterogeneity in paralic reservoirs at all scales. Established techniques are likely to be complemented in the future by forward stratigraphic modelling, integration with palynology and more studies of surveillance data to show how sedimentological variability controls fluid movement in real reservoirs.

Paralic reservoirs are globally important, but vary in nature from small isolated sands with limited connectivity, low reservoir quality and poor recovery to laterally extensive sheets with good reservoir properties and high recovery. Sedimentology plays a pivotal role in unravelling this complexity. It provides the language and concepts that allow complex successions of sediment bodies, both sandstones and mudstones, to be described, differentiated and quantified. This paper aims to illustrate that process and how, if done well and early in field life, it can result in optimal development decisions, higher recovery and improved economics.

The paper commences by briefly reviewing the meaning and value of the term paralic and the global importance of paralic reservoirs. The bulk of the discussion considers paralic reservoirs at a series of scales. Discussion at the field and reservoir scale focuses on layering and the importance of the early identification of discrete reservoirs and flow units. At the sandbody scale, sandbody shape, size, stacking and connectivity are emphasized and their interaction with structure to form off-structure hydrocarbon traps and complex, amalgamated field footprints is illustrated. The complexity that characterizes the internal nature of many paralic sandbodies is then considered at the facies and core plug scale. The paper concludes with summaries of development and surveillance strategies that have proved successful in paralic reservoirs. Throughout, an attempt is made to illustrate how sedimentology can help in the development of paralic oil and gas reservoirs and, to this end, the final section considers fruitful avenues for further work.

Paralic depositional environments

Paralic environments consist of siliciclastic settings that occur at or close to sea-level and include deltas, coastal plain to shoreline-shelf systems and estuaries. These distinct settings are commonly described separately, but, for several reasons, it is also useful to consider them together. Firstly, many authors have shown how paralic environments respond similarly and sensitively to changes in sea-level and sediment supply and, as a result, a suite of concepts, terminology and stratigraphic models has been developed that is applied across them (Van Wagoner *et al.* 1990). Secondly, paralic environments form lateral equivalents with, for example, deltas passing laterally into shoreline-shelf systems. Thirdly, distinct environments are commonly interleaved, with deltas and coastal plain to shoreline-shelf systems being favoured during regression, and estuaries and lagoons favoured during transgression (Boyd *et al.* 1992; Fig. 1). Finally, it is often helpful to consider these environments together as they can be hard to distinguish.

Each environment can be described as displaying a number of sub-environments, generally linked to the degree of marine influence which they display (Table 1). In turn, the sub-environments are composed of a range of sediment bodies, of which

From: Hampson, G. J., Reynolds, A. D., Kostic, B. & Wells, M. R. (eds) 2017. *Sedimentology of Paralic Reservoirs: Recent Advances*. Geological Society, London, Special Publications, **444**, 7–34.
First published online July 20, 2016, https://doi.org/10.1144/SP444.10

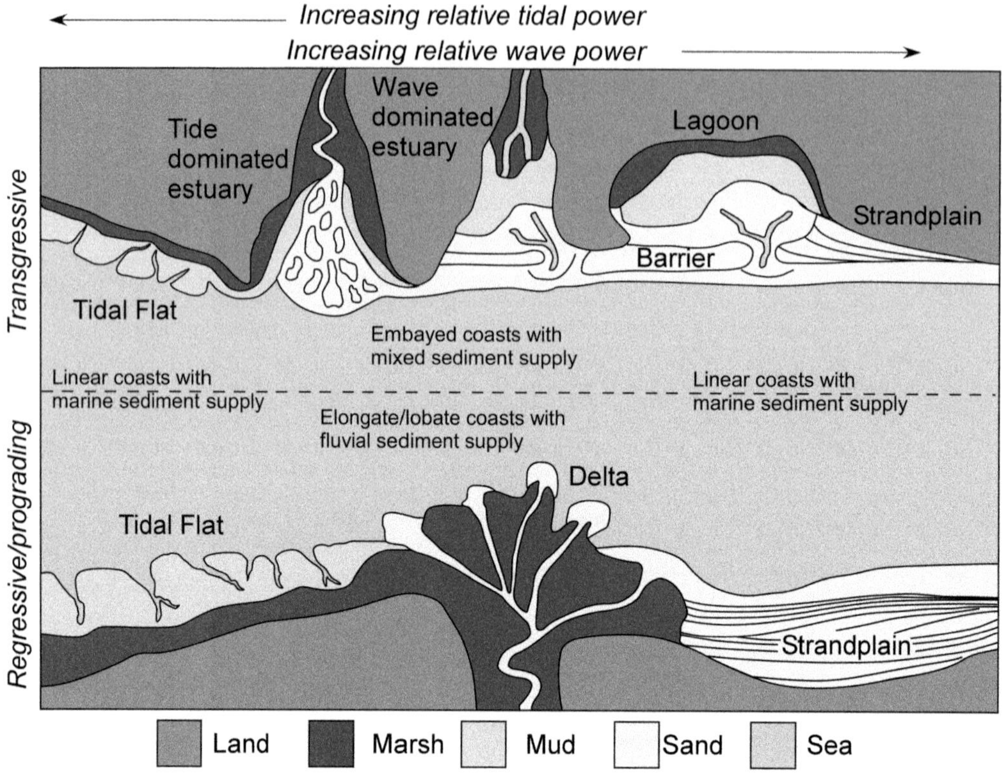

Fig. 1. Paralic depositional environments from Boyd *et al.* (1992).

sandstone bodies have received the most focus. In terms of their relative contributions to the rock record, these settings are unequal. Deltas dominate because they are the fundamental source (via the fluvial systems that feed them) of sediment in the paralic realm. They feed some of that sediment along

Table 1. *Paralic environments, sub-environments and sandbodies*

Paralic environment	Sub-environment and *sandbodies*		
	Fluvially dominated	Mixed processes	Marine dominated
Deltas	Upper delta plain *mobile channel belts* *distributary channels* *crevasse channels* *crevasse splays*	Lower delta plain *tidal creeks* *distributary channels* *crevasse channels* *crevasse splays* *levees*	Delta front *mouth bars* *shoreline-shelf sands* *cheniers*
Coastal plain to shoreline-shelf systems	Upper coastal plain *minor fluvial channels* *crevasse channels* *crevasse splays*	Lower coastal plain *bayhead deltas* *lagoons*	Shoreline-shelf *barrier islands* *flood tidal deltas* *ebb tidal deltas* *shoreline-shelf sands* *tidal flats*
Estuaries	Bayhead delta	Central estuary *tidal flats* *tidal–fluvial channels*	Estuary mouth shoal *tidal ridges* *barriers* *flood and ebb tidal delta*

Italic font denotes sandbodies.

shore to coastal plain to shoreline-shelf systems, and portions of them are reworked into estuaries during abandonment and transgression.

The global importance of paralic reservoirs

There is no readily available, complete, compilation of oil and gas fields that describes depositional environment. However, the global importance of paralic reservoirs can be inferred by comparing data presented in a statistical review of world energy (BP Statistical Review of World Energy 2015) with a second and separate BP database: Reservoirs in Paralic Environments (RIPE). The RIPE database is not exhaustive – for example, it excludes the USA and Canada – but it is large, containing 2500 fields with a total reserve of 900 billion barrels, and reports data from most of the key paralic basins around the world, with the Arabian and West Siberian basins dominating. In detail, the databases are not directly comparable as the paralic database reports estimates of initially recoverable reserves and the world energy review totals the proved reserves remaining in 2015. Despite this, the current and future importance of paralic reservoirs is clear as, when the initial reserves for paralic fields are totalled by country, they result in numbers of similar magnitude to those reported on a country basis (Fig. 2).

Analysis of the RIPE database suggests that most of the hydrocarbon is in a few large fields. Of these fields, 125 (5%) exceed 1 billion barrels in size and together contain 700 billion barrels, almost 80% of the reserves captured in the database. Some of the fields are very large indeed (Fig. 3). However, the small fields should not be ignored as 60% of the fields contain <50 million barrels, but have a total reserve of 20 billion barrels.

Reported recovery factors for paralic oil reservoirs range steadily from less than 10% to over 50% (Fig. 4). Through accurate and detailed description of reservoir architecture, extent and quality, sedimentology can help to assess where an individual field is likely to sit within this spectrum. If these descriptions are developed and used to influence development decisions early in the field life, then recovery can be optimized through, for example, the implementation of appropriate pressure support schemes and well locations. Conversely, if the

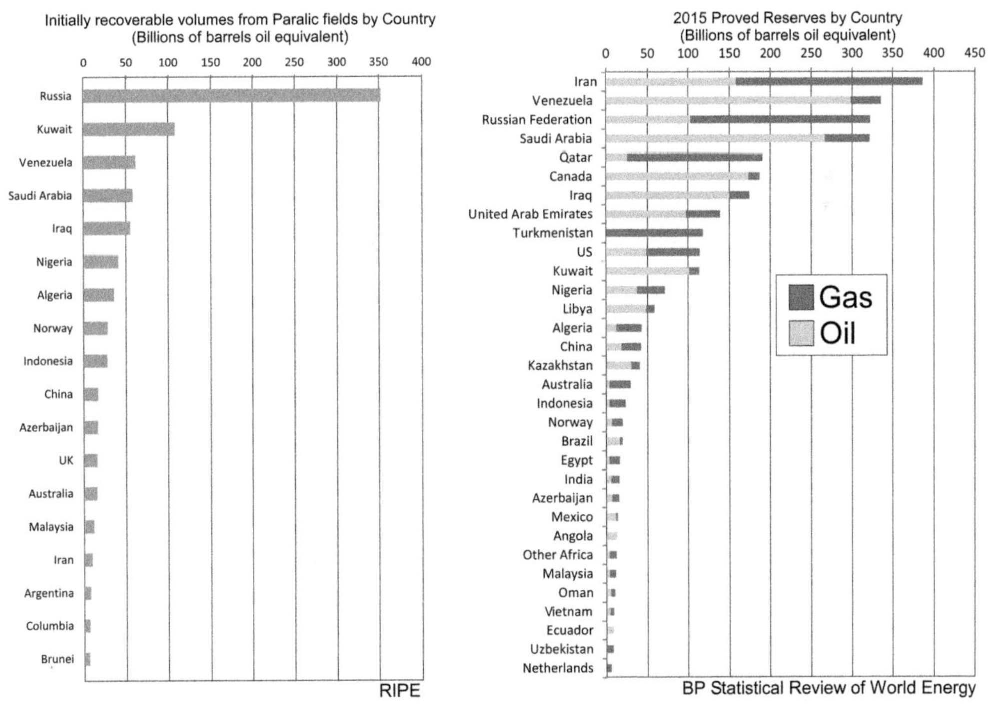

Fig. 2. The figure shows data from two separate databases. On the left, data from the BP RIPE database captures information from paralic fields, whereas, on the right, data from the BP Statistical Review of World Energy summarizes data from fields from all environments. The similar scales of (1) total hydrocarbons initially recoverable in paralic oil and gas fields and (2) proved 2015 country reserves for all reservoir types suggest that paralic reservoirs have global significance. Only countries with totals in excess of 5 billion barrels are shown here.

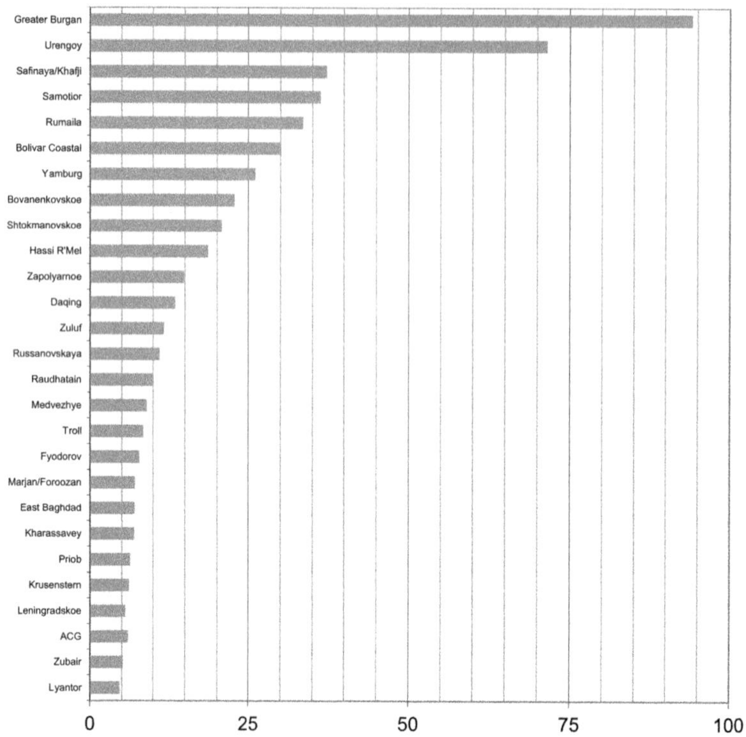

Fig. 3. Initially recoverable volumes in billions of barrels of oil equivalent for paralic oil and gas fields in excess of 5 billion barrels from the BP RIPE database.

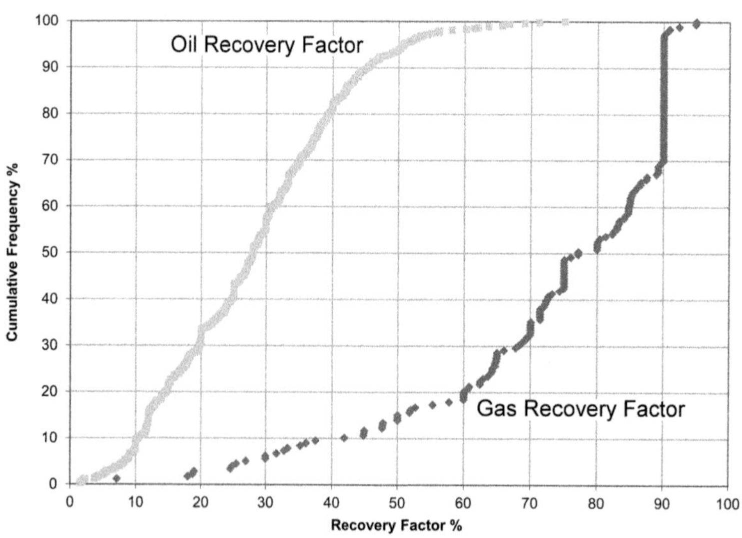

Fig. 4. Reported oil and gas recovery factors for paralic reservoirs from the BP RIPE database. Oil recovery factors typically range from 10 to 50%. There are fewer data points for the gas recovery factor and many of these are related to gas produced as a by-product of oil production. Despite this, the higher recovery factors in gas reservoirs are clear.

descriptions are poor, if they do not affect development decisions, or if they arrive late in field life, then, at best, they may offer remedial benefit and, at worst, they may destroy value.

Paralic reservoirs at the field and reservoir scale

A reservoir is a petroleum accumulation separated from other reservoirs by either pressure isolation, across for example sealing mudstones, or by physical separation from other hydrocarbon pools that share a common aquifer. Reservoirs are important entities that form the basis for (1) the reporting of reserves and (2) developing depletion plans – that is, descriptions of how hydrocarbons will be produced, including, for example, the number of wells and any necessary pressure support scheme. Paralic reservoirs are highly variable in nature. At their best, they are laterally extensive sheets with excellent reservoir quality. Such reservoirs form targets for early field development and can deliver high recovery factors and world-class well performance. The best individual wells can produce dry oil under natural flow conditions for decades at rates of tens of thousands of barrels per day and cumulative volumes of hundreds of millions of barrels for an individual well. By contrast, reservoirs formed by small sands, ribbons and dots typically have low recovery factors and poor well performance. In some cases reservoirs can be too small to justify drilling a single well. In such examples development may proceed by exploiting the multi-layered nature of paralic oil and gas fields. Paralic reservoirs rarely exist in isolation; instead, they typically occur in stacked successions and an individual well can exploit several reservoirs over time.

A schematic section through the Handil field on the Mahakam Delta in Indonesia provides a type example of how paralic reservoirs stack to form paralic fields (Fig. 5). The field is composed of around 150 discrete reservoirs extending across a vertical interval of over 2 km (Verdier *et al.* 1980). Most stratigraphic intervals below a depth of 500 m contain some hydrocarbons, but not all do so and the arrangement of oil, water and gas might initially have appeared bewildering. Compare, for example, well HL1, which drilled a crestal location in the southern compartment and found largely oil-bearing sands from 500 to 2500 m, with well HE1, which did not encounter hydrocarbons until it had drilled a 1500 m deep wet segment and then proceeded to penetrate almost exclusively gas columns. With such complexity the development of correlation panels akin to that shown in Figure 5 is far from trivial, but it is aided by noting that the reservoir patterns are far from random. Instead, they occur in discrete stacks, defined by a combination of structural elements and sealing mudstones. For example, the crest lines of many of the reservoirs are vertically superposed one above the other, simply because discrete reservoir sands have been folded together. Others occur in inclined stacks reflecting through-going sealing faults that form trap elements for a number of reservoirs. While structure is important in controlling the lateral positions of reservoir stacks, sealing mudstones impart a measure of vertical consistency, with the thickest mudstones providing the largest seals and separating reservoirs with broadly similar pressures and comparable fluids. The fluids may change progressively, for example, from more oil-prone fluids in lower reservoirs to more gas-prone fluids in reservoirs towards the top of the reservoir stack. In the Handil example, markers R6 and R4-3, R1, R0-09 and R0-02 mudstones appear to have this role. Such mudstones commonly extend between adjacent fields, giving them similar reservoir stacking patterns.

Handil reservoirs were delineated by marker correlation and MDT style pressure analysis (Verdier *et al.* 1980). Over 70 marker beds were identified with an average vertical separation of 35 m and, in general, a field-wide extent. Such marker beds are invaluable, the more distinctive the better, with coals, limestones and mudstones being prominent in many correlation schemes. Discrete reservoirs can also be indicated by the vertical pattern of fluids with, as a first pass, separate reservoirs being suggested when, for example, oil is found above gas, or water is encountered above oil. A visual analysis of any of the wells illustrated in Figure 5 shows the power of this approach. Initial models of this type can be tested by many data sets. For example, within a discrete reservoir fluids will be similar in nature, initially be in pressure equilibrium, and as production proceeds they will be demonstrably in pressure communication (Brown 2003; Reynolds 2006). Individual reservoirs may themselves be internally complex, as illustrated by the 19-7 reservoir in Handil (Fig. 6) which lies between two field-wide marker beds and is composed of up to three distinct ‘bars’ with complex outlines and variable reservoir sand quality, and more laterally restricted channel sands of generally good reservoir quality. This assemblage can be reinterpreted to record four parasequences connected by channels that hang below flooding surfaces FS3 and FS4 and cut through mudstones above FS2 and FS3, which would otherwise compartmentalize the reservoir.

The layers are important in that they not only control the trapping of hydrocarbons into discrete reservoirs, but they also strongly influence fluid flow during production. A water flood simulation based on an outcrop of the Ametlla Formation in the

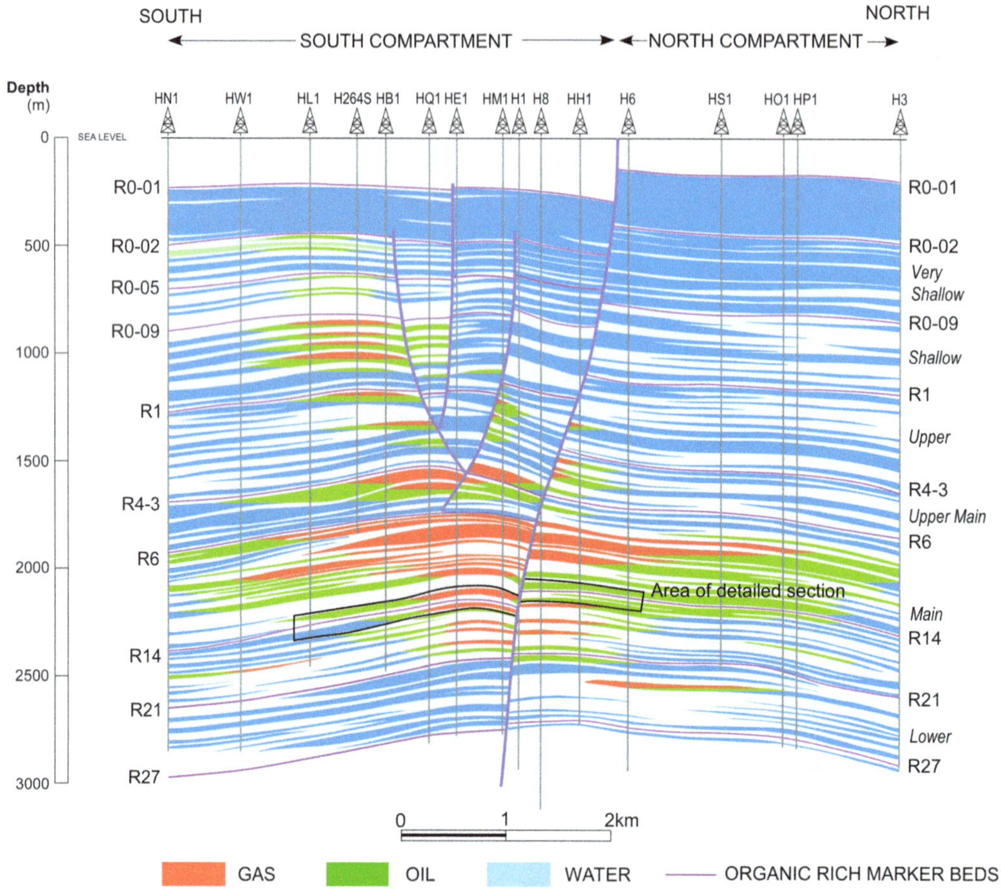

Fig. 5. The Handil Field, Mahakam delta, Indonesia, shows a multi-layered reservoir configuration typical of many paralic oil and gas fields, with multiple reservoirs in numerous sands compartmentalized by sealing faults. The 'area of detailed section' is shown in Figure 6. Modified from Verdier *et al.* (1980) and used with permission from the American Association of Petroleum Geologists.

Spanish Pyrenees provides an illustration (Fig. 7). In this example water is injected on the right-hand side of the model into all layers and is moving towards a producer on the left-hand side that is similarly completed in all zones. However, the layer performance is variable and higher permeability (and/or lower 'storage', the product of the net-to-gross ratio and porosity) results in water arriving first, 'breaking through', in the central layer, while in other layers breakthrough is delayed by reduced permeability or higher storage, and some sands, unconnected to the injector, are not swept at all. Water handling facilities commonly have limited capacity and well performance is generally severely reduced by water production. So an ability to predict these behaviours is invaluable as it enables choices to be made to manage the flood front, delaying water breakthrough by selective staged completion and injection of distinct layers. Strategies might include producing the lowest and uppermost sands together and not initially producing the central layer, or producing one layer at a time from the bottom upwards and abandoning (plugging off) each layer as it waters out.

To make optimum development choices, it is necessary to develop robust layering schemes and predictions of the inter-well variability early in field life. Commonly, however, the impact of layering is only fully revealed after significant production, as illustrated by three wells drilled into the Pereriv unit of the ACG field in Azerbaijan (Fig. 8). The pressures shown are datum pressures calculated by extrapolating the measured pressures to a reference (mid-reservoir) depth along an assumed oil gradient. On such graphs, the pressure points measured in vertical wells in oil columns in pressure

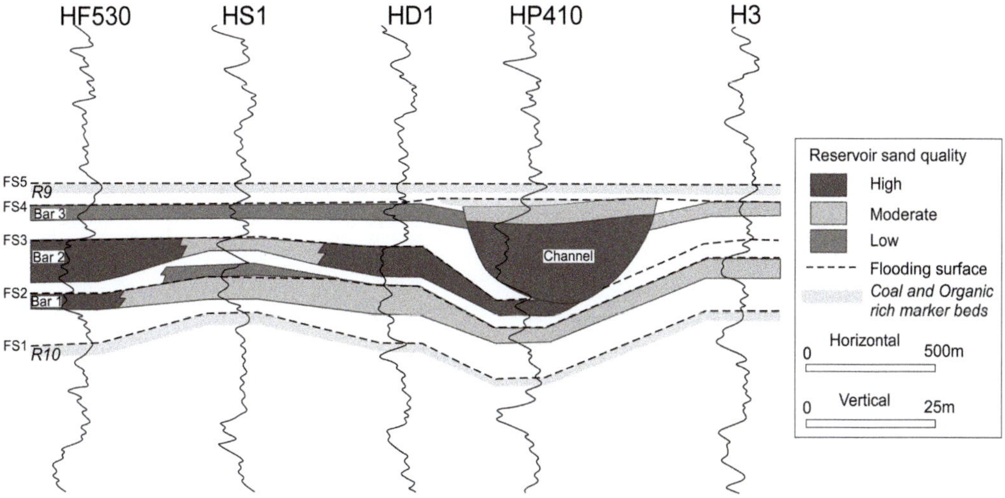

Fig. 6. Detailed correlation of the 19-7 reservoir between the R9 and R10 markers in the Handil Field, Mahakam delta, Indonesia. The panel is located on Figure 5. Modified from Verdier *et al.* (1980) and used with permission from American Association of Petroleum Geologists.

equilibrium plot as vertical lines (Reynolds 2006). Progressively lower pressures, more pressure breaks and a wider spread of pressures are encountered as production proceeds. Early in the life of this field, the A, D and E layers were revealed to be at least baffled from the B and C layers. However, there

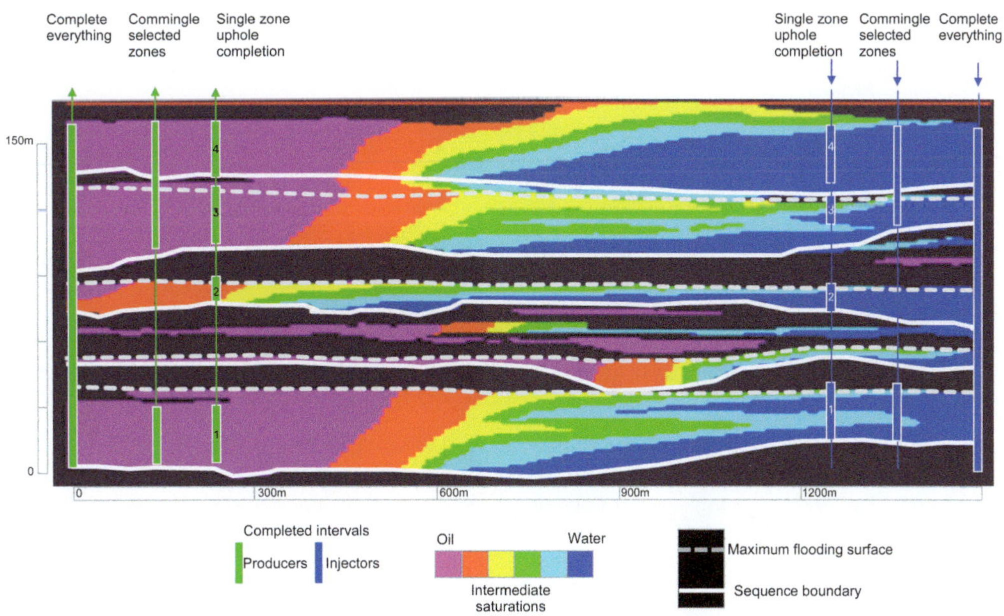

Fig. 7. A water flood simulation of the Ametlla Formation, Spain, based on an outcrop panel in Dreyer & Falt (1993). Models of this kind allow different development options to be compared. The example shown here assumes that water is injected into all sands on the right-hand side of the model (the 'Complete everything' case). Two alternative completion schemes give options to manage early water breakthrough in the central layer. Such simulation models are clearly founded on robust correlation and prediction of the reservoir properties between wells. Simulation model from Martin (1999) used with permission of Imperial College London.

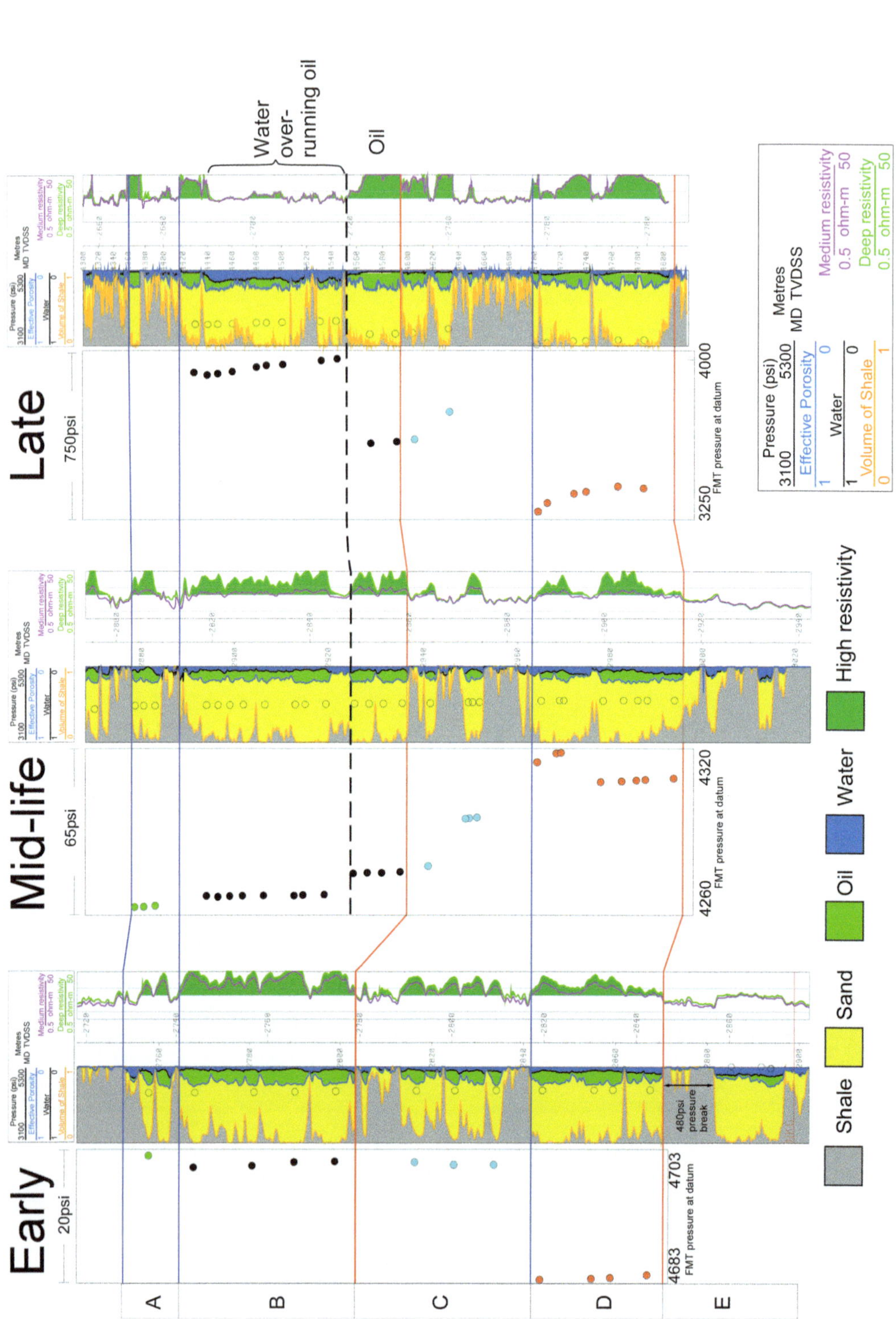
Early
20psi
4683
4703
FMT pressure at datum
480psi pressure break
Mid-life
65psi
4260
4320
FMT pressure at datum
Late
750psi
3250
4000
FMT pressure at datum
Water over-running oil
Oil
A
B
C
D
E
Shale
Sand
Oil
Water
High resistivity
Pressure (psi)
3100 5300
Effective Porosity
1 0
Water
1 0
Volume of Shale
0 1
Metres
MD TVDSS
Medium resistivity
0.5 ohm-m 50
Deep resistivity
0.5 ohm-m 50

was little hint of the internal pressure breaks seen within these layers in a mid-life well, nor that the thin mudstones in layer B would strongly influence local flood patterns in late life.

How can sedimentology help in the early assessment of layering?

Layering schemes in paralic reservoirs are commonly based on wireline log correlation and typically proceed by:

(1) developing true stratigraphic thickness displays for the wells;

(2) delineating marker intervals that can be recognized with confidence in each well and using them to datum the different wells to produce correlation panels;

(3) correlating laterally extensive mudstones that are likely to subdivide the succession into distinct reservoirs;

(4) checking the correlation panels with ancillary data, such as pressure measurements, seismic data and any independent dating information; and

(5) reviewing the panels and attempting to correlate the flow units (Ebanks *et al.* 1993) between sealing mudstones.

Such an approach may well produce a layering scheme. However, if it is combined with (1) detailed sedimentological analysis and (2) careful consideration of plausible stratigraphic models, a better layering scheme is likely to be achieved, both sooner and with more limited data.

High-resolution sequence stratigraphy

Perhaps the key sedimentological contribution to the process of reservoir layering is to correlate paralic successions by using high-resolution sequence stratigraphic principles. Such an approach would begin at the very largest scale by broadly matching sequence sets (Mitchum & Van Wagoner 1991) between wells and then picking low-order maximum flooding surfaces to be used as datums. Once the logs have been datumed and low-order correlations have been made, higher order flooding surfaces can be correlated, typically at first maximum flooding surfaces, then parasequence sets and, in turn, flooding surfaces (Van Wagoner *et al.* 1990). With a flooding surface framework in place, sequence boundaries can be recognized. This approach is well established and has been used successfully in the subsurface and at outcrop across the globe. It provides a predictive framework at several scales and goes far beyond the correlation of easily recognized marker beds. By placing such beds in their depositional and stratigraphic context, their lateral persistence can be explained and it can be predicted how and when they may peter out or be truncated.

From a reservoir point of view, the value of this approach is crystallized around an association between sequence stratigraphic surfaces of different order and connectivity within the subsurface. Mudstones associated with low-order flooding surfaces typically bound reservoir stacks (Figs 5 & 9) whereas maximum flooding surfaces and well-developed flooding surfaces may form seals to individual reservoirs (Fig. 6) and minor flooding surfaces may provide local baffles that become significant during depletion (Fig. 8).

The relationships between the order of a given sequence stratigraphic surface and its impact on trapping and fluid movement can be very close, but they are not exact – classifying surface order is an interpretative step and mudstones may be breached by faults. More interestingly, the impact of a mudstone can depend on the relative scales of the mudstone, the trap geometry and the hydrocarbon column, with, for example, previously distinct reservoirs merging as traps fill down to spill points defined by the termination of flooding surface mudstones (Fig. 10). In different scenarios the same flooding surface may either provide a seal to a reservoir or a key intra-reservoir baffle.

A related theme that also benefits from a high-resolution sequence stratigraphic context is the potential of channels to provide vertical connectivity. Valleys, typically recognized as being over-deepened by comparison with other channels, provide a pre-eminent example of this. By using a high-resolution sequence stratigraphic template, the presence and stratigraphic position of valleys can be predicted, lying in type examples between

Fig. 8. Three wells from the Pereriv Formation in the ACG Field, Azerbaijan, illustrate how the impact of layering is commonly only confirmed later in field life. At discovery there will be no pressure breaks within a reservoir. As production proceeds, differential off-take and injection above and below laterally extensive thin mudstones typically results in lower pressures, more and larger pressure breaks, and a wider spread of pressures. In this example, a large pressure break is revealed at the top of the D subunit early in field life, but there is little hint of the intra-subunit pressure breaks seen in later wells. The late life well shows how a laterally extensive thin mudstone near the base of the B subunit (dashed line) indicated in the mid-life well can control fluid flow with, in this example, injected water running over the mudstone, leaving unswept dry oil below. Surveillance data of this type allow injection and production strategies to be modified and recovery maximized.

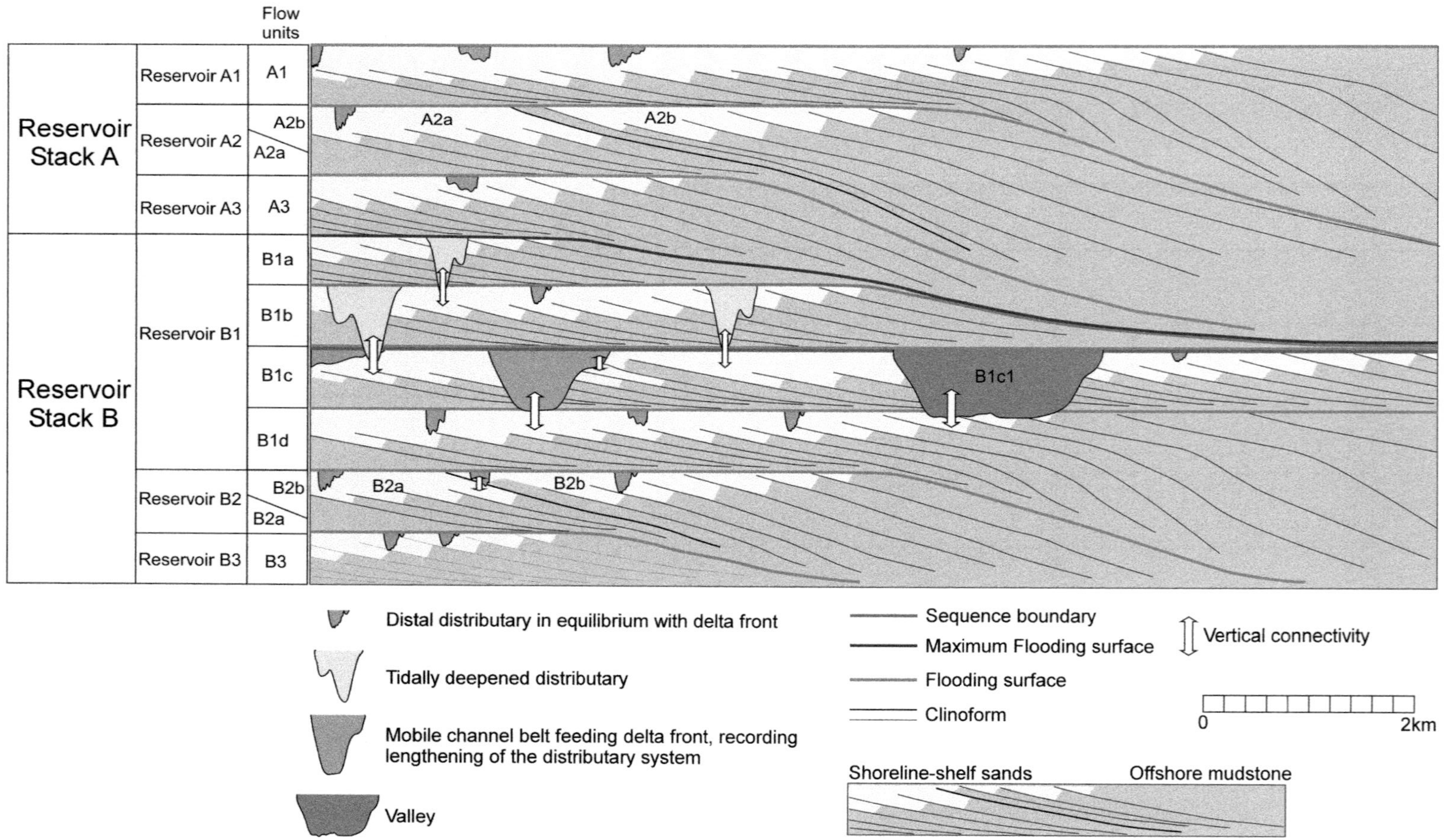

Fig. 9. A schematic illustration of relationships between key sequence stratigraphic surfaces and paralic reservoirs showing two reservoir stacks, separated in this example by a maximum flooding surface. Individual reservoirs are bounded by flooding surfaces, but some parasequences are connected by valley and channel incision.

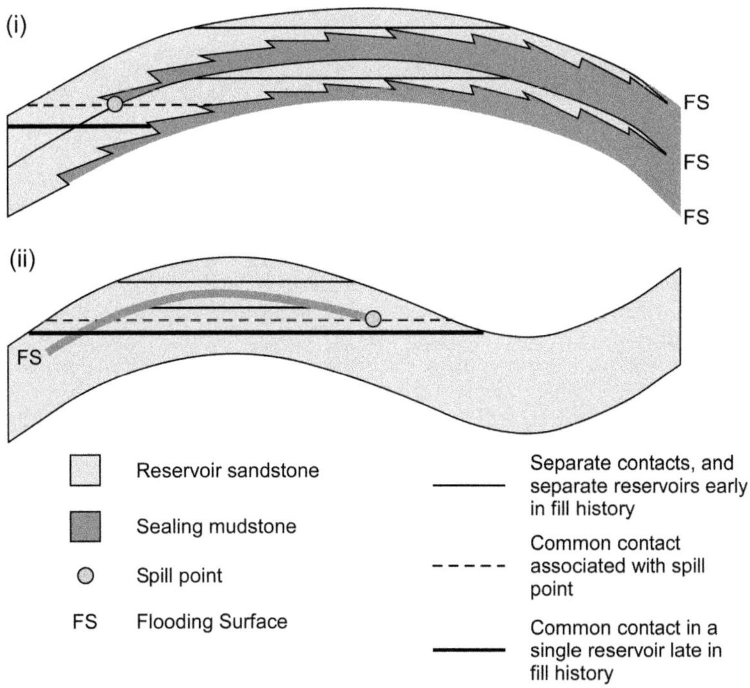

Fig. 10. The influence of mudstones in forming separate reservoirs will change as traps fill down to spill points and connect hydrocarbon columns merging previously distinct reservoirs.

an underlying highstand, progradational parasequence set and an overlying transgressive, retrogradational parasequence set (Fig. 9).

In addition to valleys, overly deepened channels can also be associated with the tidal reworking of abandoned distributaries (Dalrymple *et al.* 2003; Legler *et al.* 2013). Such reworking may have been favoured in transgressive systems tracts when shelves were wide and tides were enhanced (Stride 1982; Reynolds 1996; Ainsworth *et al.* 2011).

At the parasequence scale, one model is to assume that distributary channels are thinner than the parasequence they feed and that, as a result, they do not provide vertical connectivity between parasequences. This model is supported by the widespread recognition of parasequence-scale flooding surface mudstones acting as reservoir seals and by descriptions of strongly distributive systems where the delta front is fed by numerous shallow terminal channels. In other cases, flooding surface mudstones may be naturally breached during normal progradation. For example, as a system progrades, the distributary system lengthens and a restricted number of large distributaries and mobile channel belts (Gibling 2006) are expected to locally replace the terminal distributaries; their greater depth may result in vertical connectivity between shoreline sands. This process seems likely to be favoured on wave- and storm-dominated deltas, which are typically fed by a restricted number of presumably deeper distributaries. Connectivity through channel incision is also likely to be enhanced when parasequences are thinner, as tends to occur in late highstand and lowstand sequence sets and systems tracts – intervals that are also likely to be affected by mobile channel belt extension. By contrast, parasequences tend to be thicker in transgressive systems tracts and are far less likely to be affected by subsequent valley incision, though they may be connected by over-deepened tidally reworked distributaries.

In all cases, with valleys, tidally deepened distributaries and distributary channels, the vertical connectivity naturally depends on the nature of the channel-fills and this may vary both between channels and along their lengths.

Further applications of sedimentology to layering

Beyond the application of high-resolution sequence stratigraphic principles, sedimentology can aid the development of layering schemes by encouraging correlation panels to be constructed parallel and perpendicular to strike, so that variations in sandbody continuity with orientation can be investigated. In a linked manner, an appreciation of sandbody scale can help to answer the question of whether a sand encountered in one well is likely to be seen in an adjacent penetration, while a knowledge of depositional environment will inform what may occur adjacent to a laterally restricted sandstone body. Similarly, a sedimentological approach would carefully consider the generally low depositional gradients encountered in paralic systems and draw sub-horizontal and sub-parallel correlation lines accordingly. Important exceptions related to erosive channel and valley margins and to clinoforms would also be considered. Although the latter are commonly seen at outcrop and can be shown to have important influences on fluid flow, they are difficult to recognize in the subsurface and, in some cases, have to be forced into the layering framework (Wehr & Brasher 1996; Graham *et al.* 2015 and references cited therein).

Paralic reservoirs at the sandbody scale

Paralic sandbodies display a wide range of dimensions, variable shapes and differing internal lithological and textural trends. It is helpful to be able to predict these parameters, not least because they are a cornerstone of geocellular modelling.

In many studies, sandbody thickness has been measured from core and wireline log data and analogue data have been used to predict body widths. Reynolds (1999) showed that widths can vary by three orders of magnitude for a given thickness and argued that the best way of differentiating the data was by recognizing distinct sandbodies within depositional sub-environments differentiated by sedimentary processes (Fig. 11). Each sandbody type, for example, crevasse channel or shoreline-shelf sandstone, displayed a restricted range of widths and thicknesses.

Paralic sandbody widths range from 10 m to 100 km, but their thicknesses only extend from around a metre to a few tens of metres. The restricted thicknesses are thought to reflect the increments in accommodation space generated in paralic settings and recorded by parasequences. As discussed above, following transgression, sandbodies deposited in the subsequent progradation tend to fill the available accommodation space rather than exceed it by cutting into the underlying units.

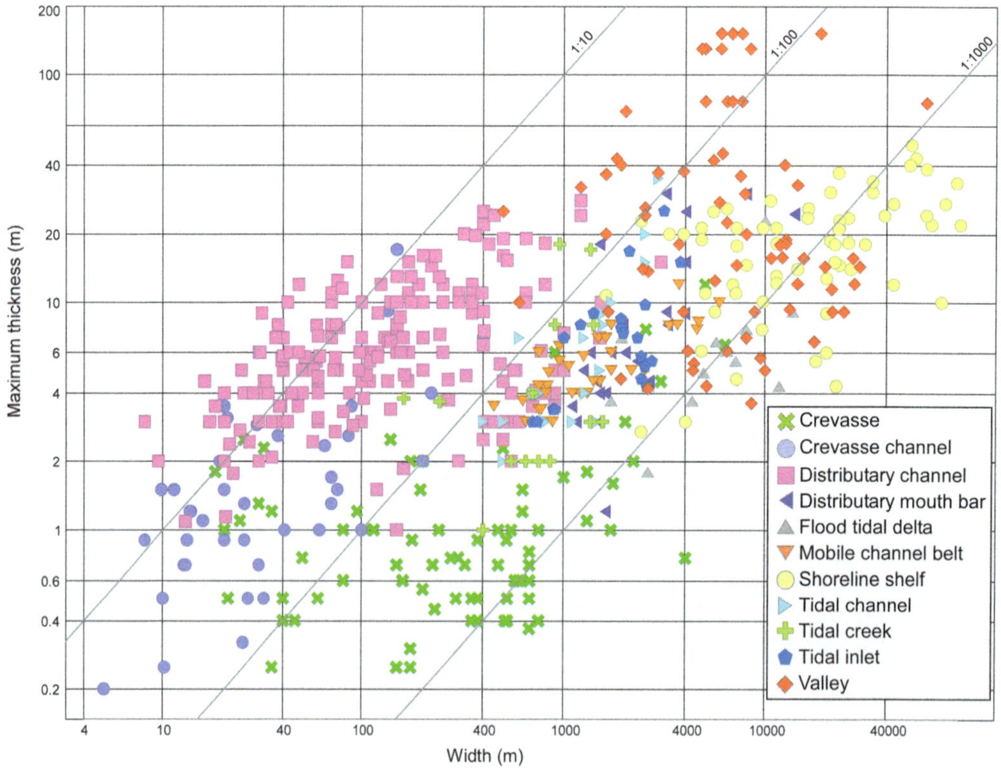

Fig. 11. Dimensions of paralic sandstone bodies. Modified from Reynolds (1999).

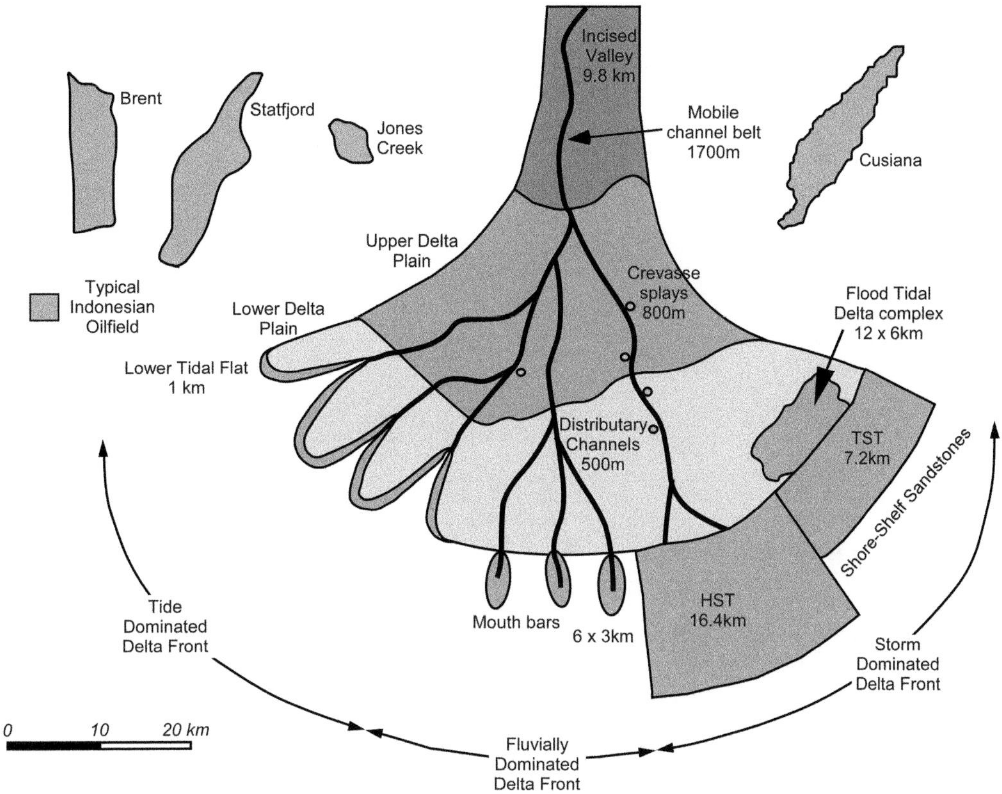

Fig. 12. A schematic delta showing a range of sandbody types at their average dimensions, together with a number of paralic oil and gas fields at the same scale. The delta front is divided into three segments that are storm, fluvial and tide-dominated, respectively. Modified from Reynolds (1999). HST, highstand systems tract; TST, transgressive systems tract.

Valleys form a clear exception to this and provide the thickest sandbodies in the database reported by Reynolds (1999).

The largest sandbodies occur at the scale of individual reservoirs or field segments and are likely to be correlated between wells and recognized as deterministic elements in layering schemes (Fig. 12). By contrast, small sandbodies are likely to be penetrated only once, if at all, and for these sandbodies to become effective reservoirs they need to stack together into composite bodies, or link to adjacent sands of larger dimensions. Figure 13 shows schematic cross-sections that emphasize these relationships, together with the increased complexity introduced by faulting. With large sandbodies, higher net-to-gross and an absence of faulting, correlation can be expected to be relatively easy, strong aquifer support is likely and recovery may be high. By contrast, as faulting increases and sandbody dimensions and net-to-gross reduce, connectivity between sands will decrease and become increasingly difficult to predict and pressure support will rely on expansion and gas solution drive rather than on aquifer support or injection, so that recovery reduces. In many examples, sandbody stacking is systematic with relationships between sandbody type and systems tract reflecting changing rates of generation of accommodation space and, as noted above, different shoreline processes operating as shelf width changes. The Niger delta provides an example of how sandbodies may stack in response to growth faulting with more and thicker channel sandstones together with improved vertical communication in the growth depocentre (Fig. 14; Weber 1971).

These relationships can be quantified through geocellular modelling (e.g. Larue & Hovadik 2006) and flow simulation. They can also be illustrated directly by field examples, as sandbody variability can have a direct impact on reservoir outlines. Outlines tend to be simple when the extent of the sandbody exceeds that of the structural closures. Shoreline-shelf sandstones in the Niger delta are generally mapped in this way, resulting in simple

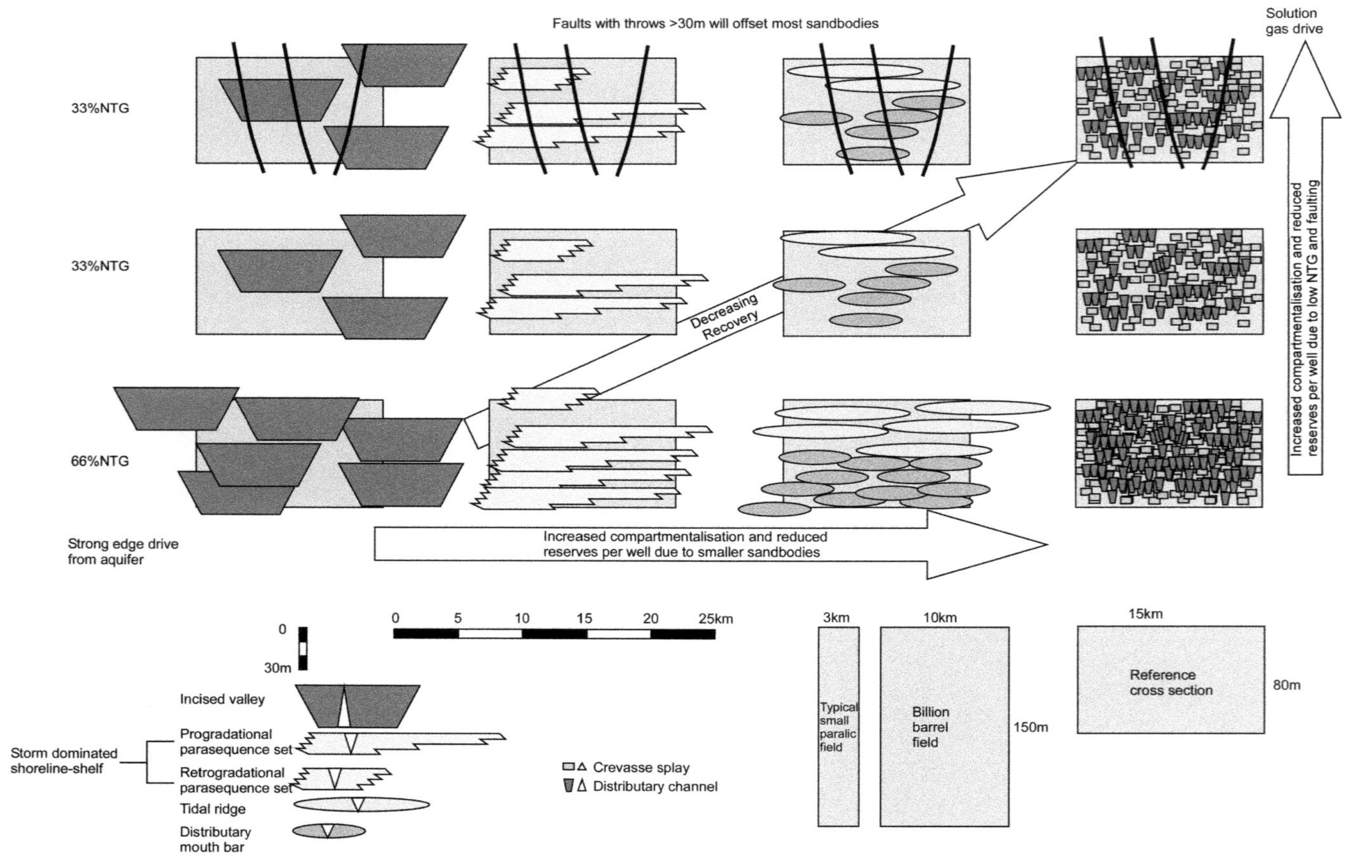

Fig. 13. Sandbody dimensions play a key part in determining reservoir architectural complexity. Here, a range of sandbody types is compared, at their average size, with a schematic field cross-section. Two net-to-gross (NTG) levels and a suite of faults are illustrated. As the sandbodies decrease in size and the NTG level reduces, recovery is expected to fall. This effect is likely to be compounded by faulting as even small fault throws will offset most paralic sandbodies. Dimensions based on Reynolds (1999).

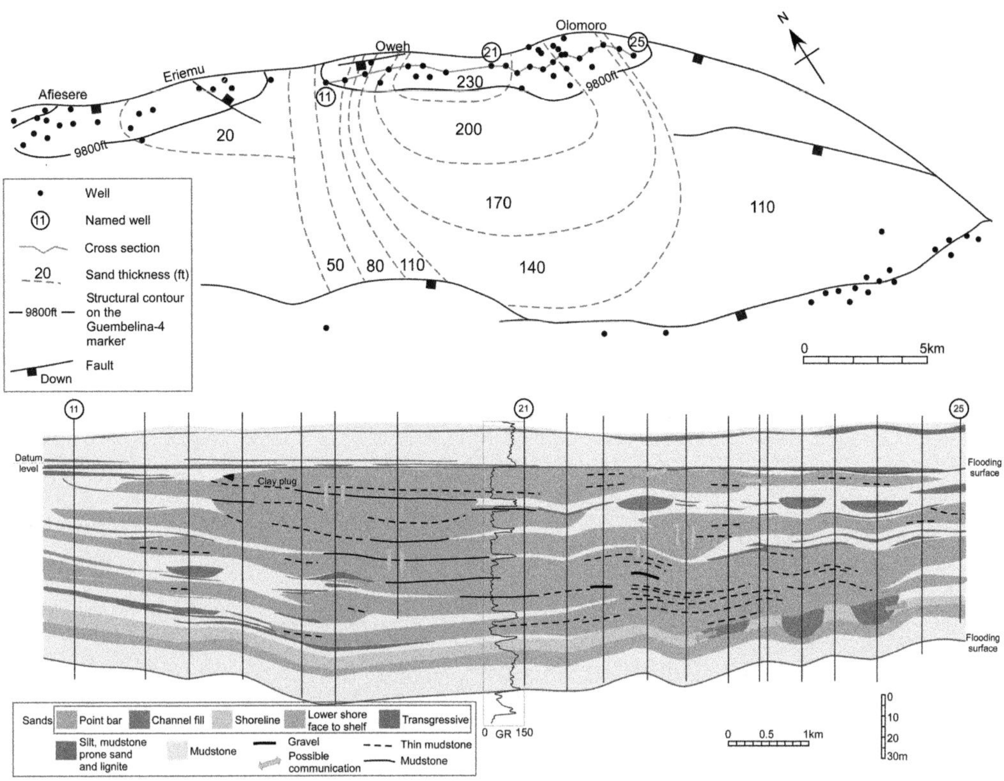

Fig. 14. A map and section through a growth fault depocentre on the Niger Delta (modified from Weber 1971). The section is bounded by biostratigraphically defined marine flooding surfaces. In the depocentre core sands are thicker, more numerous and better connected than in marginal locations where mudstones are thicker and more correlateable. As might be expected from the dimensional data shown on Figure 11 the shoreline sand at the base of the interval is the most correlateable.

reservoir outlines that reflect more or less faulted rollover anticlines (Fig. 15; Doust & Omatsola 1989). By contrast, the L1 sand in the Oficina Formation was mapped by Hedberg *et al.* (1947) as having a complex channel outline, dissected by sealing faults to form five distinct reservoirs (Fig. 16). Faults commonly increase the complexity in paralic reservoirs as (1) the sandbodies are generally thin (Fig. 11) and easily offset by small fault displacements and (2) interbedded successions result in sand–mudstone cross-fault juxtaposition and low-permeability gouge (Bouvier *et al.* 1989; Doust & Omatsola 1989). As shown by the Oficina example, they can also generate traps. In assessing the probable impact of faults, cross-fault juxtaposition can be assessed using Allan diagrams (Allan 1989) whereas the probable nature of fault rock can be assessed using shale gouge ratios calibrated to pressure and fluid data (Gibson 1994; Yielding *et al.* 1997).

Taken together, complex sandbody outlines, faulting and the stacking of reservoirs can make field outlines – that is, the composite footprint of all reservoirs in any given field – highly complex. With continued drilling, areas that were initially considered as separate fields can merge to produce large accumulations with amoeboid shapes and satellite fields. Mencher *et al.* (1953) documented an early example on the eastern margin of Lake Maracaibo in Venezuela, where at least seven different giant fields amalgamated during step-out drilling into the lake to form one of the world's largest accumulations, the Bolivar coastal field (Fig. 17).

Determination of sandbody type

Sedimentology has a key role in determining sandbody type in paralic reservoirs, ideally through the analysis of continuous core and typically via the description and interpretation of facies and facies associations, founded on an analysis of sedimentary structures, bedding contacts and vertical trends. In some examples, this approach can quickly constrain the depositional setting with, for

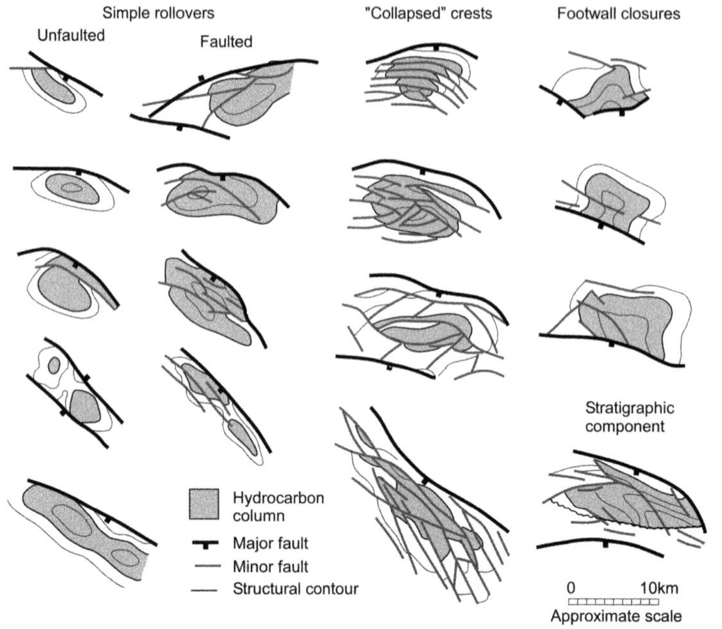

Fig. 15. Field outlines for laterally extensive sheet sandstones reflect structural geometries. These examples from the Niger delta were presented by Doust & Omatsola (1989), whereas the scale bar is estimated based on examples in Weber (1987) and Orife & Avbovbo (1982). Modified figure used with permission from American Association of Petroleum Geologists.

example, periodically spaced, double mud drapes on decimetre- to metre-scale cross-beds recording sub-tidal dunes. In other cases the interpretation may be equivocal or controversial and, in most instances, it can be improved by the integration of trace fossil and palynological analyses. For example, thin beds in the Productive Series of the ACG field in Azerbaijan display loaded bases, coarsening- and fining-upwards profiles and distinctive successions of sedimentary structures, plane laminae, ripples, climbing ripples, ripples and then back in to plane laminae (Fig. 18). By comparison with bedform phase diagrams (Allen 1985), flume studies (Ashley *et al.* 1982) and literature reviews, these beds have been interpreted as hyperpycnites deposited by waxing and waning hyperpycnal flows in a delta front setting (Nemec 1995; Mulder *et al.* 2003). This interpretation is not immediately compatible with the sharp-based fining-upwards profiles of the associated sandbodies that have been interpreted as fluvial channels. However, an extensive palynological database unequivocally records a lacustrine setting and the sandstones have accordingly been interpreted as a mixture of distributary channel sands, sharp-based mouth bars and crevasse sands.

This approach of interpreting the depositional environment and using it to constrain the likely range of sandbody types is well established and is aided by comparison with type examples and classification schemes. Pre-eminent of these in paralic environments is the ternary classification of deltas introduced by Galloway (1975) which considers the relative influence of the three processes that move sand in paralic shoreline settings, rivers, waves and tides. Another seminal paper, by Orton & Reading (1993), extended this classification by emphasizing the important role that grain-size has in controlling the nature of deltas. Both papers placed a range of modern deltas on their summary diagrams, so that if we can determine from an ancient example the relative importance of fluvial, tide and wave processes and the dominant coarse grain-size, then it is possible to infer likely modern analogues and, in turn, the range of possible sandstone bodies. Building on work by Bhattacharya & Giosan (2003), Ainsworth *et al.* (2011) suggested several refinements to this approach. In particular, they argued that individual sandbodies can be placed on the ternary diagram by assessing the relative dominance of wave, tide and fluvially generated sedimentary structures. If numerous sandstone bodies can be plotted in this way, then a picture of the entire depositional system can be constructed. If only a limited number can be described, then a series of models can suggest a range of possible

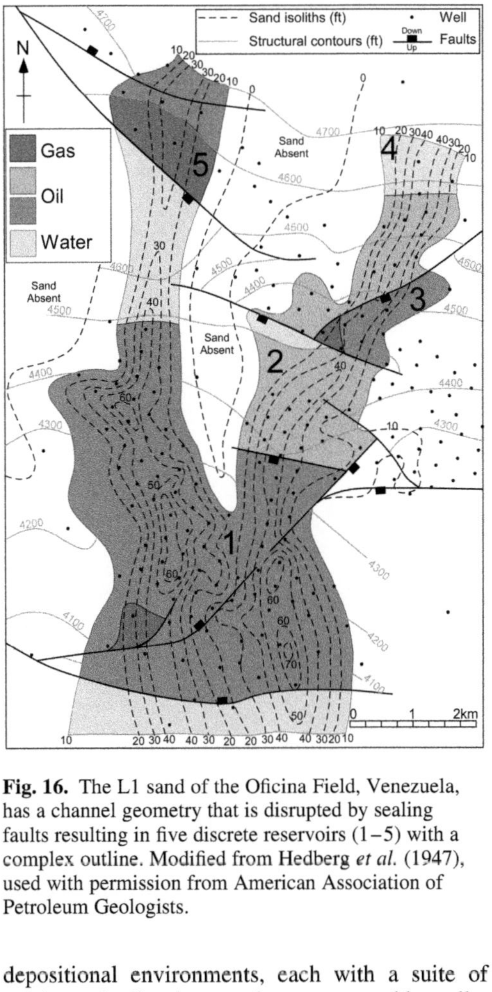

Fig. 16. The L1 sand of the Oficina Field, Venezuela, has a channel geometry that is disrupted by sealing faults resulting in five discrete reservoirs (1–5) with a complex outline. Modified from Hedberg *et al.* (1947), used with permission from American Association of Petroleum Geologists.

depositional environments, each with a suite of sandstone bodies that may be encountered laterally.

Paralic reservoirs at the facies and lamination scales

Complexity at the facies scale is common in paralic reservoirs. This is perhaps most easily illustrated in tidal sandstones, where low-permeability mud drapes, deposited during tidal slack water, result in complex flow pathways in ripples and dune-scale cross-bedding. However, the complexity is not restricted to mud drapes. Most paralic sandstones are laminated, with the laminae reflecting exquisite sorting of grains by grain-size, by density and by shape (Fig. 19).

Studies of reservoir quality (i.e. permeability and porosity) reveal that grain-size and grain composition are key controls on permeability and, indeed, can be used to predict it (Bryant *et al.* 1993; Cade *et al.* 1994). As a result, we would expect each lamination and lamina set to have distinct flow properties, and Miniperm maps of cross-bedded sands, core flood experiments and simulation models confirm this (Corbett *et al.* 1992; Ringrose & Corbett 1994; Huysmans *et al.* 2008). The predominance of laminated sands and the scarcity of structureless sands deposited from high-density suspensions sets paralic reservoirs apart from their deep water cousins. At the small scale this difference matters, not least because many core analysis experiments either assume or benefit from homogeneous core plugs (1.5 inches across and 3–4 inches long).

The nature of these laminae, and that of mud drapes, their thickness, shape and extent, can be described and modelled and, as a result, their impact can also be assessed at the facies scale.

The origins of cross-laminae have been investigated by experimental physicists who place them amongst more general concerns relating to the mixing of granular material (Makse *et al.* 1998; Ball 2009). Lamination descriptions are most complete in the sedimentological literature where grain-flow (avalanche) and grain-fall (from suspension) processes are distinguished (Jopling 1965; Hunter 1985; Hunter & Kocurek 1986). Buck (1985) excavated tidal dunes and showed (1) how lamination extent and lamination dip varied systematically with variations in peak flow velocity, and (2) systematic shape and grain-size sorting with respect to position on a cross-lamination (Fig. 20). Larger, flakier and lighter grains tend to travel further down the slope and to rise to the top and margins of grain flows, resulting in inversely graded laminae and a tendency for the lower parts of cross-beds to be slightly coarser grained. Similarly, the grain-flow deposits themselves have distinct shapes parallel to dune crest lines. When the rate of sand supply to the crest line is low the avalanches have a tongue-like geometry. As the rate of sand supply to the crest is increased (and the bedform migrates more quickly) the grain flows widen, become sheet-like and, at the highest migration rates, they become diffuse, reflecting almost continuous avalanching.

The impact of laminae at the facies scale can be modelled by accounting for (1) longer wavelength variations in grain-size, as, for example, occur between finer grained neap tidal bundles and coarser grained thicker spring tidal bundles (Allen & Homewood 1984) and (2) lamination termination and truncation associated with bedform growth and migration (Rubin 1987; Cornish 1901).

Two examples are worthy of more detailed consideration and both concern mud-draped tidal current ripples. In the first example Nordahl *et al.* (2005) and Ringrose *et al.* (2005) used commercially available software (SBED) to mimic sediments deposited by the migration of bipolar

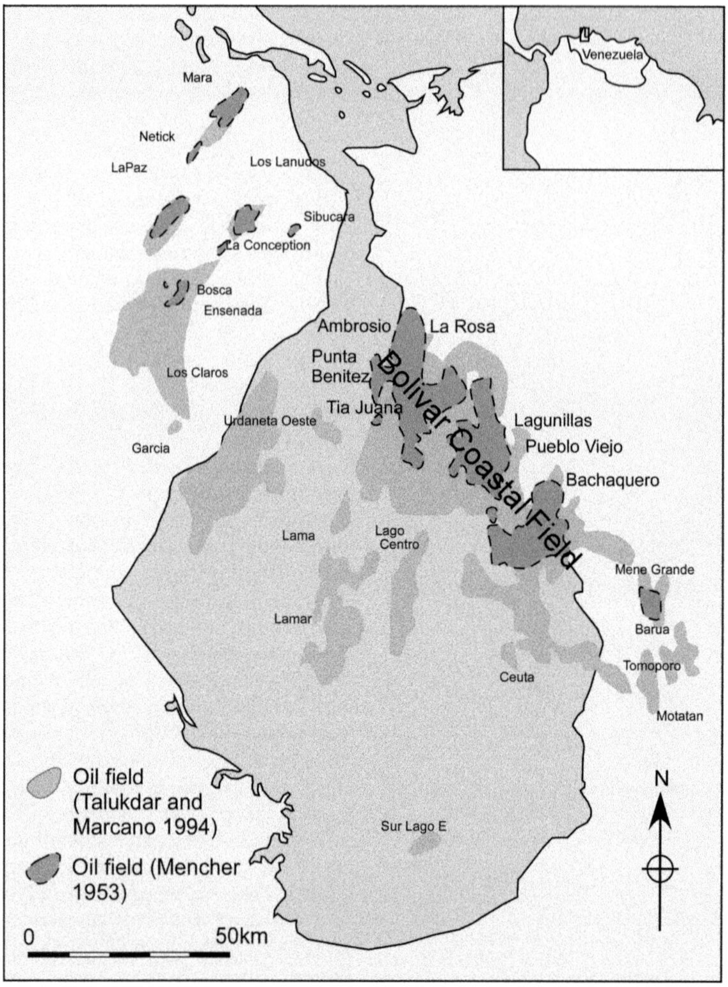

Fig. 17. Mencher *et al.* (1953) describe how the amoeboid shape of the Bolivar Coastal Field, on the eastern side of Lake Maracaibo, was defined by step-out drilling of seven regions (Ambrosio, La Rosa, Punta Benitez, Tia Juana, Lagunillas, Pueblo Viejo and Bachaquero) previously considered as separate giant fields. Field amalgamation is a common feature of paralic reservoirs as a result of stratigraphic trapping and combination traps forming 'off-structure' hydrocarbons. Modified from Mencher *et al.* (1953) and Talukdar & Marcano (1994) and used with permission from American Association of Petroleum Geologists.

opposed current ripples and mud drapes (Fig. 21a). In their models they varied the proportions, properties and relative positions of sand and mud and then calculated the impact on effective properties. Figure 21b shows some key results. At very low sand volumes the effective permeability approaches the permeability of mudstone and at high sand volumes it approaches the permeability of sand. Intermediate cases result in intermediate effective properties that differ depending on the flow direction. It is always easier to flow horizontally than vertically, because the only heterogeneity in the model, the mudstones, are sub-horizontal. However, the mud drapes dip preferentially in the direction of the ripple migration because they drape ripple foresets as well as filling ripple troughs. As a result, at less than 35% sand volume it would be harder to flow fluids parallel to the palaeoflow than perpendicular to it. The impact of the variation in grain-size between and within sand laminae was not considered.

A second, closely comparable, example is from the work of Jackson *et al.* (2005), who constructed a two-dimensional simulation model of mud-draped tidal ripples based on serial sections of a sandstone block. The model was 57% sand, with a permeability

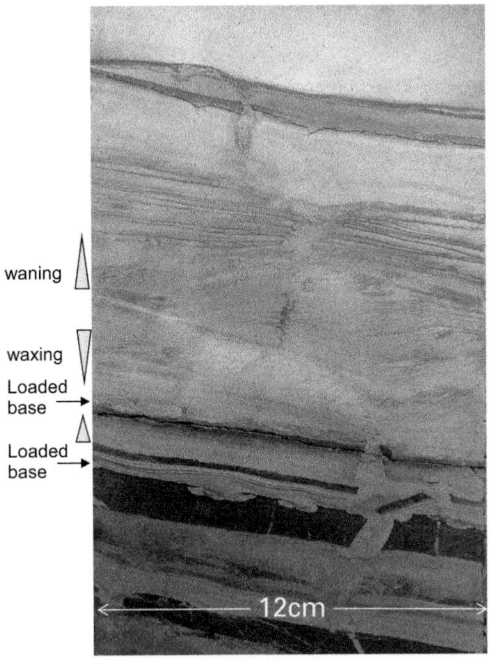

Fig. 18. Hyperpycnites from the Productive Series on the ACG field in Azerbaijan.

of 1000 mD, and 43% silt, with a permeability of 0.1 mD. Figure 22 shows three simulations, each using the same permeability and porosity arrays, but with different fluid properties. In one model (Fig. 22a) displacement is favourable, so that a high proportion of the oil is swept from each sand cell before the next sand cell is flooded. In a second model displacement is unfavourable, far lower proportions of oil are swept, and water breakthrough is rapid (Fig. 22b). In a third model (Fig. 22c), strong capillary pressure is introduced so that the injected water is sucked into the silts, and the sands have high residual oil saturations at water breakthrough.

Both examples illustrate that heterogeneity at the facies scale can influence fluid flow, and can do so in surprising ways. The importance of fluid–rock interactions cannot be underestimated. However, models of this nature are typically only built when complex facies form large volumes of large oil reservoirs. They are difficult to build, requiring properties, skill sets and software that are not widely available. They are also difficult to upscale, particularly when multiphase flow (oil, water and gas) is considered.

Despite this, sedimentological guidance can still have a part to play at the facies scale. Figure 23 shows three cells from a geocellular grid penetrated by a well. Each cell has a distinct arrangement of low- and high-permeability lithologies and each represents a distinct sedimentary facies. By chance, the arrangement of low-permeability lithologies at the well is similar in each layer, resulting in comparable log-derived permeability values. In geocellular modelling, the log-derived permeabilities in each cell are averaged, or 'blocked' to derive single values, effective permeabilities, that can be used to model flow across the cell in three dimensions. In Figure 23 (i) and 23 (ii), the low-permeability lithologies lie parallel to the grid and only partially

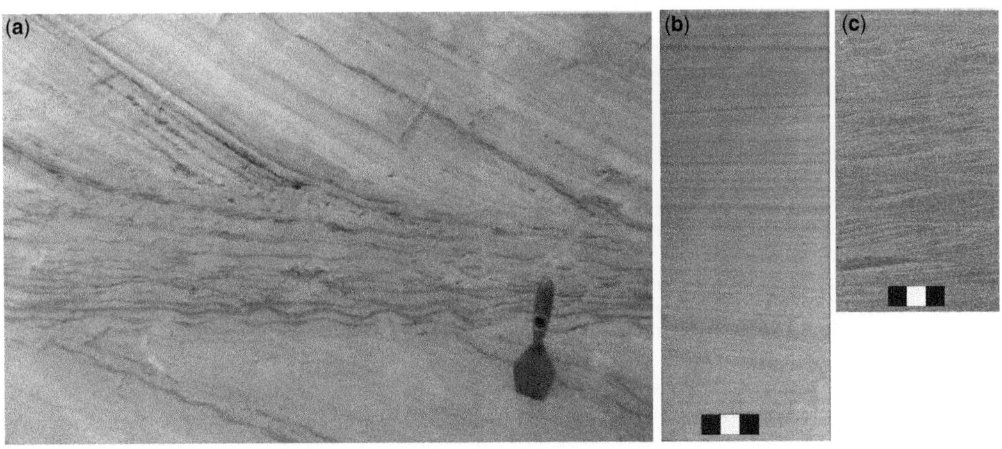

Fig. 19. Paralic sandstones are typically laminated. (**a**) Mud-draped ripples, mud drapes along dune foresets and sand laminae in dune foresets from a subtidal shoal setting, as described by Visser (1980); (**b**) gently fanning laminae in very fine-grained sands interpreted as hummocky cross-stratification, Viking Formation, Alberta Canada; (**c**) climbing current ripple laminations defined by variations in grain-size and grain composition, Productive Series of Azerbaijan. Part (a) has a hand trowel for scale, whereas (b) and (c) have 3 cm long scale bars.

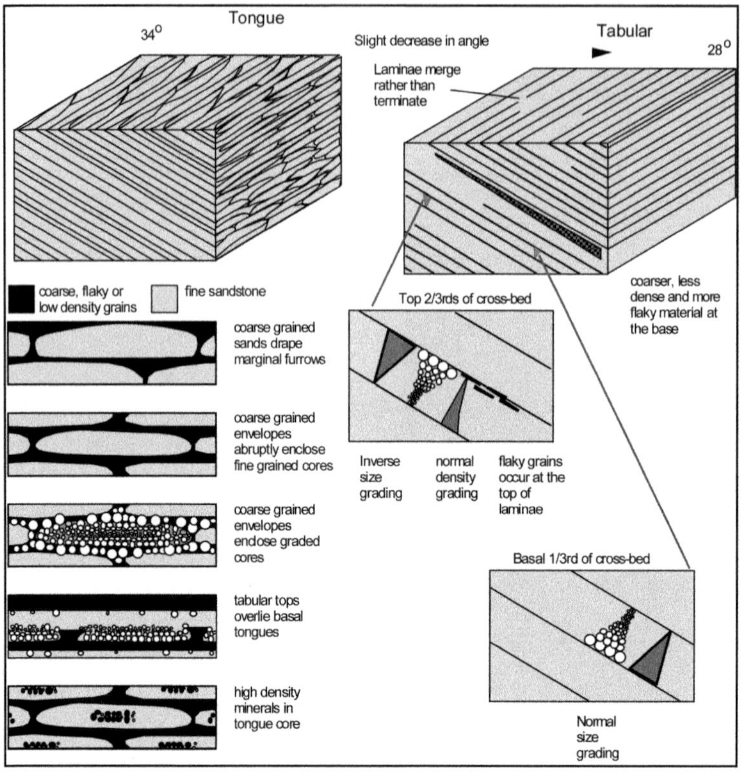

Fig. 20. The shapes and properties of tidal cross-laminae after Buck (1985).

impede the lateral flow. By contrast, their lateral extent is variable. In Figure 23 (i) the lengths are short, and it is possible to flow from the top to the base of the cell. In Figure 23 (ii) one low-permeability layer extends across the whole of the cell and will have a significant influence on vertical flow. In Figure 23 (iii) two thin, low-permeability layers have not been resolved by the wireline log data and some of the low-permeability layers are dipping in a manner that will impede horizontal as well as vertical flow. The way in which the log-derived values are averaged is ideally guided by sketches of this type, resulting in different averaging techniques being used for different facies during blocking or, as illustrated above, by the construction of small test models to investigate the likely effective properties. In both cases sedimentologists have a role in describing, in three dimensions, mudstone continuity and abundance, as well as intersection relationships (point c in Fig. 23 (iii)) and the properties of sand laminae.

Grain-size

Most sedimentological studies focus on trends in grain-size and are concerned with crucial questions for facies and environmental analyses, such as does a given sandbody coarsen or fine upwards, or is there a large jump in grain-size across a given surface that may suggest a key stratigraphic surface, for example the base of a valley-fill? As a result, published sedimentological descriptions generally use simplified grain-size classifications. For example, of 21 sedimentary logs in Reading & Collinson (1996), 9 use simplified grain-size classifications and 12 have no grain-size scale, but the vertical trends are clear and, together with the sedimentary structures shown, they are sufficient to develop significant insights into depositional environments and the origins of sandbodies. In reservoirs, however, this understanding is only a step along the route to describing the three-dimensional architecture of reservoir quality, porosity and permeability and, in that context, a detailed description of grain-size can be of great value as grain-size is commonly closely linked to permeability. Figure 24 illustrates this by considering a moderately well-sorted quartz arenite with 20% porosity and showing how permeability can be modelled (Bryant *et al.* 1993; Cade *et al.* 1994) as doubling between Wentworth grain-size classes (Wentworth 1922), with an order of magnitude change occurring over four classes, say from

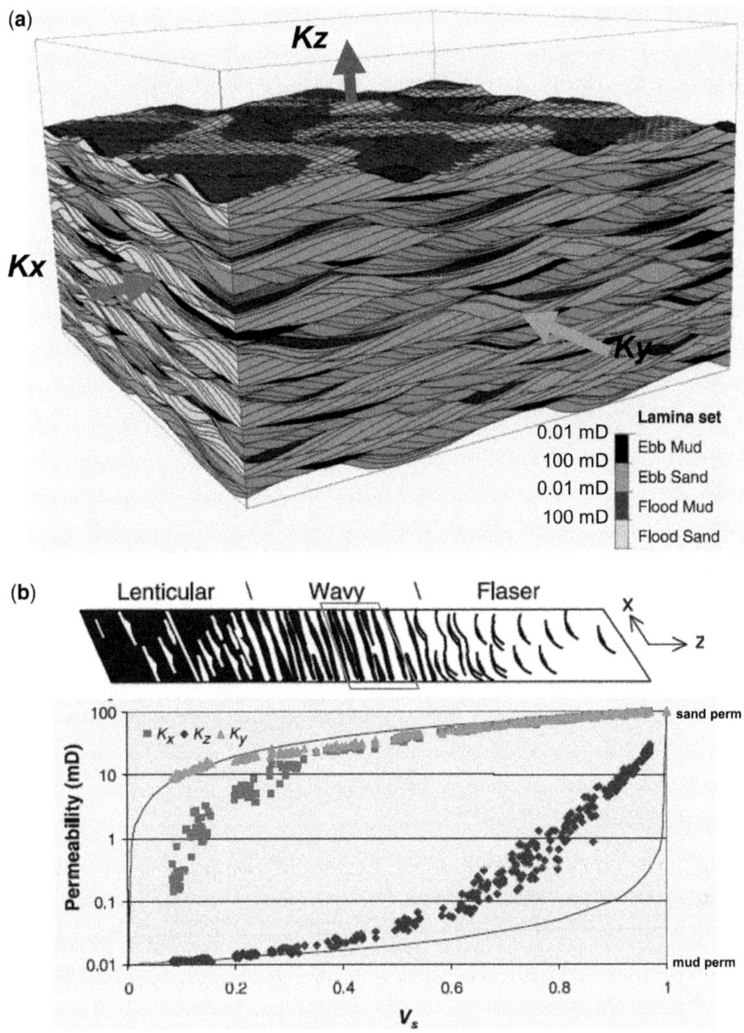

Fig. 21. (**a**) An SBED model of flaser bedding from Ringrose *et al.* (2005). Note the bidirectional sand lamina sets (lines) and the partially preserved mud drapes (dark tones). The cuboid has dimensions of 0.1 × 0.5 × 0.5 m and contains 90% sand. (**b**) Effective permeabilities derived from a suite of models similar to that shown in (a) with variable sand content (V_s, volume of sand).

vfL sand to fU sand. Grain-size is not the only depositional control on reservoir quality – ductile content and sorting also play a part, but often these too can be linked to grain-size. In addition to its influence on reservoir quality, grain-size also has a close link with sedimentary structures. Hummocky cross-stratification is invariably developed in very fine sands. Current ripples occur in silts and very fine and fine sands, but are typically not seen in coarse sands, whereas the converse is broadly true for dunes and lower stage plane beds (Allen 1985). In a related way, distinct transport mechanisms are linked to certain grain-size populations. Plumes move silts and muds, which are held in suspension by slow settling velocities and minor turbulence, whereas jets, turbulent flows with significant inertia, can, in addition, transport finer sands. High-density turbulent flows (Lowe 1982) are rare in paralic successions and medium and coarse sands generally move as bedload. Finally, it is worth noting that the grain-size range for many paralic reservoirs is quite limited and that significant grain-size breaks are commonly associated with key sequence stratigraphic surfaces.

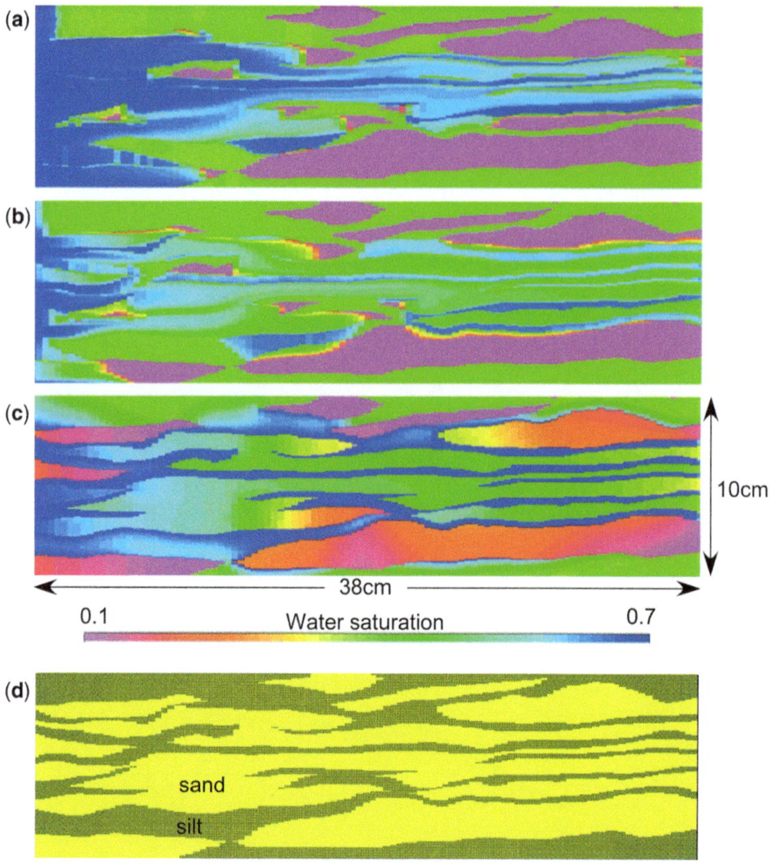

Fig. 22. (**a–c**) Waterflood simulations through a selected two-dimensional slice of the three-dimensional models reported by Jackson *et al.* (2005). Reproduced with permission of Imperial College London. Water is injected on the left-hand side of the model and oil and water are produced on the right. The geology (**d**) is the same in each simulation model, but changes in fluid properties, described in the text, result in very different flood performance.

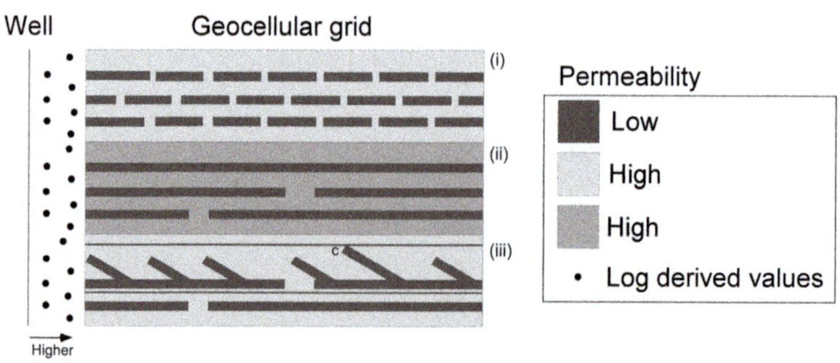

Fig. 23. Three different sedimentary facies fill three model grid cells (i) to (iii) and are represented at a well by log-derived permeability values. Sedimentology has a key role in describing the three-dimensional distribution of high and low-permeability values and thereby in determining appropriate averaging of the log values to derive effective permeabilities, vertical and horizontal, for the entire cell.

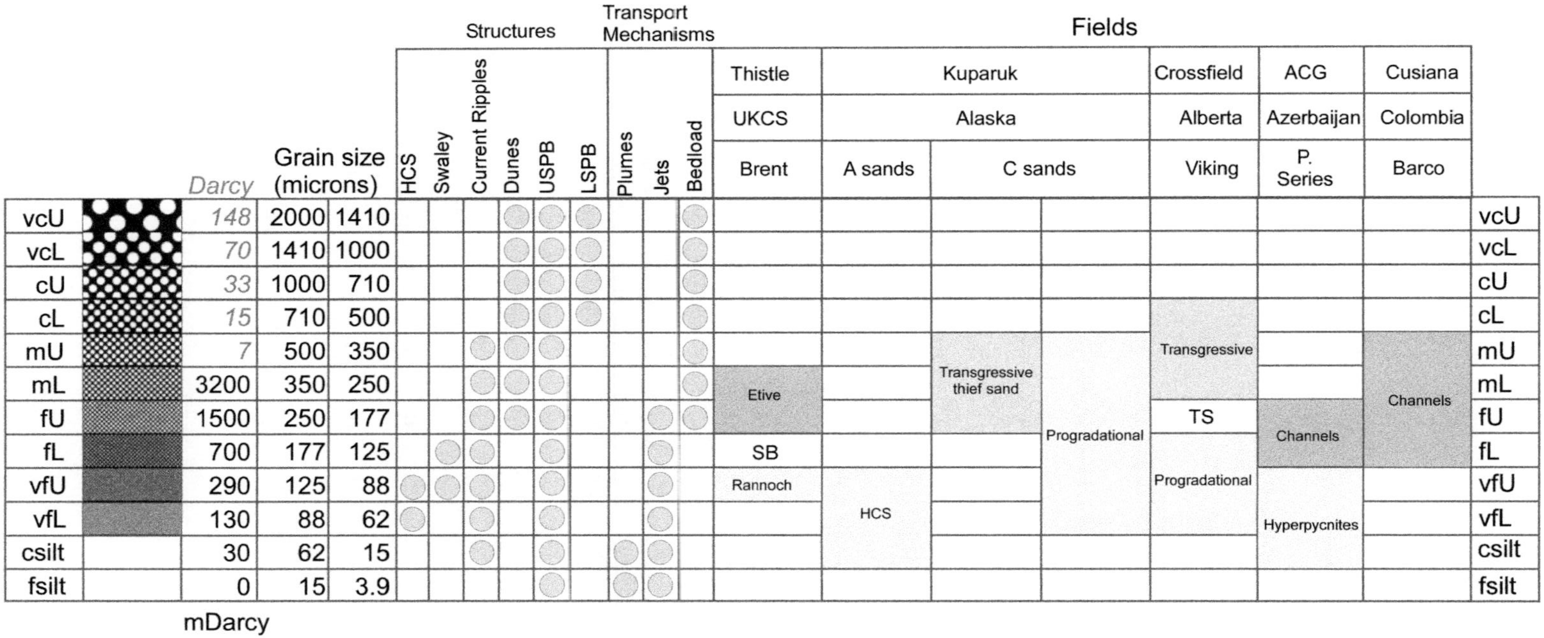

Fig. 24. Relationships amongst grain-size, sedimentary structures, transport mechanisms and example paralic fields. Dots show the typical range of grain-sizes seen associated with sedimentary structures and processes likely to be encountered in paralic successions. Permeability values are based on a moderately well-sorted quartz arenite with 20% porosity (Bryant *et al.* 1993; Cade *et al.* 1994). HCS, hummocky cross-stratification; USPB, upper stage plane bedding; LSPB, lower stage plane bedding.

Development strategies in paralic reservoirs

As in all reservoir types, the development strategies in paralic reservoirs vary depending on the key drivers. If the drivers are largely economic, then it is typical for development to proceed via early exploitation of the largest, simplest reservoirs: sheet sands. These are generally relatively easy to appraise and are likely to deliver high-rate wells, larger volumes per well and greater recovery. The high rates and volumes justify capital expenditure and result in shorter pay-back times. In fields with reservoirs of this type, the deepest high-quality reservoir is generally developed first and shallower reservoirs are appraised by collecting data as development wells are drilled. In fields where maximizing recovery is paramount, or where reservoirs are relatively complex, then a bottom-up completion strategy may be used. In such a scheme each well produces from each layer, with wells being recompleted upwards when production dies or waters out in any given layer. Such an approach may result in lower initial field rates, but may maximize the number of off-take points in any given reservoir, thereby increasing recovery in complex reservoirs. To aid recovery, injection is often required to maintain rates through pressure support and to sweep oil towards producers. With numerous layers, and usually limited off-take points, a detailed understanding of the layers and likely flow paths between the injectors and producers becomes of paramount importance and the part that sedimentology can play in this activity is clear. As discussed above, the description of structure can also play a key part and it may influence the drilling strategy, with slant wells being targeted down the backs of faults to appraise multiple reservoirs trapped against sealing faults. Consider, for example, the added appraisal value of accessing numerous hydrocarbon columns by drilling a well from the H6 location on the Handil field parallel to the major fault 100 m south of the well head (Fig. 5). In some cases, however, complexity can be very high, too high to be described in detail, and it may be necessary to drill cheap wells on a grid pattern and exploit any encountered hydrocarbons by pressure depletion alone.

Surveillance in paralic reservoirs

Because paralic oil and gas fields are commonly strongly layered, tracking fluid movements and pressure changes within them can be complex, particularly when several reservoirs are being developed at once, and more than one fluid type must be monitored. Table 2 notes some of the techniques that may be employed and provides selected examples where they have been applied in paralic reservoirs – given that these datasets form the basis for production geology, the number of published studies that demonstrate their value is remarkably limited.

Discussion

Paralic reservoirs are increasingly well understood at all scales. Applications of sequence stratigraphic concepts and ichnology, and the use of analogue databases and increased integration have all played key parts. However, at least as much as in the past, the present is an exciting time to study paralic reservoirs. The idea that the genesis and shape of sandbodies can be usefully described and differentiated by the proportions of the wave-, fluvial- and tide-generated sedimentary structures that they display (Ainsworth *et al.* 2011) is powerful, but is largely untested in the published literature. The extent to which these processes can be distinguished, and the capacity of the proportions determined to be predictive, are fundamental sedimentological questions for paralic reservoirs that are likely to be best addressed by high-quality outcrop studies. The ability to quantify outcrops through three-dimensional models built using LIDAR and photogrammetry will aid this process and increasingly set outcrop studies on a more robust quantitative footing that goes beyond remaining dimensional questions (in particular the paucity of channel length data) to develop databases on, for example, depositional element proportions, lateral transitions and bedding dips. In doing this, it will be helpful to ensure that grain-size is described in detail to allow links to be developed to the reservoir quality of depositionally analogous reservoirs.

Three-dimensional outcrop models may also increasingly provide data to constrain forward stratigraphic models that have the potential to test or at least illustrate reservoir depositional models, and to suggest alternative scenarios beyond a single most-likely case. Similar advances may follow from studies of time-lapse satellite imagery. Coastal regions with significant tides can be rather confusing because coastline movements are a key feature of time-lapse images, but on delta and coastal plains and on shorelines where the tidal range is small, or in lakes where the lake level is broadly constant or changing steadily, time-lapse satellite imagery can, in combination with field studies, provide an opportunity to quantify and describe paralic successions in unparalleled detail. Discharge, sediment load variations, physiography and basin processes could be linked to sediment body geometry and size, sediment calibre, facies and time. As part of any such study, it will undoubtedly be fruitful for sedimentologists to actively link with

Table 2. *Tools and techniques used to monitor fluid and pressure movements and examples where they have been applied in paralic reservoirs*

Tool		Open hole	Cased hole	Fluids	4D seismic
Vertical resolution	Perfect	Core			
	Very high	Image logs	Fibre optic distributed temperature surveys (DTS)		
	High		Spinner surveys Saturation logging		
	Good	Deep resistivity MDT style pressure data			
	Low		Downhole pressure gauges	Tracers Oil water and gas production streams	Low
Lateral coverage		Reflects well density		drainage areas swept inter-well volumes	Excellent
Measurement Repeatability		Requires new wells	DTS measurements are continuous. Other cased hole tools often require access to flowing wells which invariably reduces well rates, which is unpalatable even if only for a short period		Seismic reshoot, processing, interpretation and application may take many months
Key Limitations				Tracers require long term careful monitoring	
Examples		ACG (Reynolds 2006)			
		Endicott, (Reynolds *et al.* 1997)			
			Cusiana (Pulham *et al.* 1997)		Pembina (Chopra *et al.* 2015)
			Brent (Quint 1999)		Brent (James *et al.* 1999)
		Fulmar (Kuhn 2003)			
		Scott (Brook *et al.* 2010)			
			Worthington (2010)		

geomorphologists, who are well versed in the analysis of modern environments.

Progress is also likely to be seen through the integration of sedimentological descriptions with quantitative analyses of the palynology of modern settings and outcrops. It is difficult to overstate the value of the palynological analysis of core data in describing paralic reservoirs, but similar detailed study of outcrops is almost completely lacking. Diverse palynofacies in cores cannot be compared with a rich suite of analogous studies of outcrops or modern environments and much insight may be being lost as a result.

There is also a significant opportunity for further and more detailed studies of fluid movement in complex paralic reservoirs. Far more effort has been expended on describing and predicting likely reservoir performance than in showing how fluid movement in real reservoirs has actually been controlled by sedimentological variability.

Conclusions

Paralic oil and gas fields are globally important for both current and future production. They are typically composed of a stacked series of individual reservoirs and correctly identifying these separate hydrocarbon accumulations is crucial, as is recognizing the layering within them. High-resolution sequence stratigraphy and sedimentological insights can result in an improved, and more rapid, understanding of the layering. Individual reservoirs range from thick extensive sheets to thin ribbons and dots. Recovery and performance vary accordingly. Sheet sands tend to benefit from aquifer support, have high reservoir quality, and display high recovery and high well rates. By contrast, sands of restricted size have lower rates and lower recoveries, particularly if they occur in low net-to-gross intervals. Analogue data help to constrain the dimensions of sandstone bodies and these vary most clearly with depositional environment. Interactions between sandbodies and trap geometry and size give paralic reservoirs much of their distinctive character. Large sheet sands may extend beyond trap boundaries and display simple outlines. Smaller sands exhibit stratigraphic pinch-out and result in complex reservoir outlines. This is particularly likely when faulting is an issue as most paralic sandstones are thin (<20 m) and easily offset. Sealing faults increase the number of compartments and may decrease performance and recovery. They can also generate traps.

At the plug scale laminated sands predominate, while at the facies scale complexity generated by bedforms and facies juxtapositions is the norm. Heterogeneity can be modelled at all scales, but plug and facies scale models are relatively demanding and are generally only constructed when the heterogeneity is complex and volumes are large.

Depletion strategies of paralic oil and gas fields are commonly driven by the range of encountered reservoir types, high-value sheet sands or ribbons and dots, their vertical arrangement and the ease of access to any high-value oil rims. In most cases the highest quality reservoirs are depleted first and other reservoirs, certainly the shallower ones, form targets for appraisal during development drilling and later production.

This paper resulted from collaboration with many colleagues at BP over many years. Key among them was Mark Eller, with whom I researched paralic reservoirs in the early 1990s. The insights on field outlines presented here are, for example, his. Mark's input and the help of reviewers Gary Hampson and Simon Lang are gratefully acknowledged. BP, Chevron, Exxon Azerbaijan Limited, Inpex, Itochu, ONGC, SOCAR, Statoil, and TPAO are thanked for allowing me to publish Figures 8, 18 and 19c.

References

AINSWORTH, R.B., VAKARELOV, B.K. & NANSON, R.A. 2011. Dynamic spatial and temporal prediction of changes in depositional processes on clastic shorelines: toward improved subsurface uncertainty reduction and management. *American Association of Petroleum Geologists Bulletin*, **95**, 267–297.

ALLAN, U.S. 1989. Model for hydrocarbon migration and entrapment within faulted structures. *American Association of Petroleum Geologists Bulletin*, **73**, 803–811.

ALLEN, J.R.L. 1985. *Principles of Physical Sedimentology*. Allen and Unwin, London.

ALLEN, P.A. & HOMEWOOD, P. 1984. Evolution and mechanics of a Miocene tidal sandwave. *Sedimentology*, **31**, 63–82.

ASHLEY, G.M., SOUTHARD, J.B. & BOOTHROYD, J.C. 1982. Deposition of climbing-ripple beds: a flume simulation. *Sedimentology*, **29**, 67–79.

BALL, P. 2009. *Flow: Nature's Patterns: a Tapestry in Three Parts*. Oxford University Press, Oxford.

BHATTACHARYA, J.P. & GIOSAN, L. 2003. Wave-influenced deltas: geomorphological implications for facies reconstruction. *Sedimentology*, **50**, 187–210.

BOYD, R., DALRYMPLE, R. & ZAITLIN, B.A. 1992. Classification of clastic coastal depositional environments. *Sedimentary Geology*, **80**, 139–150.

BOUVIER, J.D., KAARS-SIJPESTEIJN, C.H., KLUESNER, D.F., ONYEJEKWE, C.C. & VAN DER PAL, R.C. 1989. Three-dimensional seismic interpretation and fault sealing investigations, Nun River Field, Nigeria. *American Association of Petroleum Geologists Bulletin*, **73**, 1397–1414.

BP STATISTICAL REVIEW OF WORLD ENERGY 2015, https://bp.com/statisticalreview

BROOK, G.R., WARDELL, J.R., FLANAGAN, S.F. & REGAN, T.P. 2010. The Scott Field: revitalization of a mature field. *In*: VINING, B.A. & PICKERING, S.C. (eds) *Petroleum Geology: From Mature Basins to New*

Frontiers–Proceedings of the 7th Petroleum Geology Conference. Geological Society, London, Petroleum Geology Conference Series, **7**, 387–403, https://doi.org/10.1144/0070387

Brown, A. 2003. Improved interpretation of wireline pressure data. *American Association of Petroleum Geologists Bulletin*, **87**, 295–311.

Bryant, S., Cade, C. & Mellor, D. 1993. Permeability prediction from geologic models. *American Association of Petroleum Geologists Bulletin*, **77**, 1338–1350.

Buck, S.G. 1985. Sand-flow cross strata in tidal sands of the Lower Greensand (Early Cretaceous), southern England. *Journal of Sedimentary Research*, **55**, 895–906.

Cade, C.A., Evans, I.J. & Bryant, S.L. 1994. Analysis of permeability controls: a new approach. *Clay Minerals*, **29**, 491–501.

Chopra, S., Kumar Sharma, R., Hossein Nemati, M., Perz, M. & Cary, P. 2015. Impedance inversion: big aid in interpretation. *American Association of Petroleum Geologists Explorer*, **July**, 34–35.

Corbett, P.M.W., Ringrose, P.S., Jensen, J.L. & Sorbie, K.S. 1992. Laminated clastic reservoirs: the interplay of capillary pressure and sedimentary architecture. Paper SPE 24699 presented at the SPE 67th Annual Technical Conference, 4–7 October, Washington, DC, https://doi.org/10.2118/24699-MS

Cornish, V. 1901. On sand-waves in tidal currents. *The Geographical Journal*, **18**, 170–200.

Dalrymple, R.W., Baker, E.K., Harris, P.T. & Hughes, M.G. 2003. Sedimentology and stratigraphy of a tide-dominated, foreland-basin delta (Fly River, Papua New Guinea). *In*: Sidi, F.H., Nummedal, D., Imbert, P., Darman, H. & Posamentier, H.W. (eds) *Tropical Deltas of Southeast Asia: Sedimentology, Stratigraphy, and Petroleum Geology*. SEPM Special Publications, **76**, 147–173.

Doust, H. & Omatsola, E. 1989. Niger delta. *In*: Edwards, J.D. & Santogrossi, P.A. (eds) *Divergent/Passive Margin Basins*. American Association of Petroleum Geologists Memoirs, **48**, 201–238.

Dreyer, T. & Falt, L.M. 1993. Facies analysis and high-resolution sequence stratigraphy of the Lower Eocene shallow marine Ametlla Formation, Spanish Pyrenees. *Sedimentology*, **40**, 667–697.

Ebanks, W.J., Jr, Scheihing, M.H. & Atkinson, C.D. 1993. Flow units for reservoir characterization. *In*: Morton-Thompson, D. & Woods, A.M. (eds) *Development Geology Reference Manual*. American Association of Petroleum Geologists Methods in Exploration Series, **10**, 282–285.

Galloway, W.E. 1975. Process framework for describing the morphologic and stratigraphic evolution of deltaic depositional systems. *In*: Broussard, M.L. (ed.) *Deltas*. Houston Geological Society, Houston, TX, 87–98.

Gibson, R.G. 1994. Fault-zone seals in siliciclastic strata of the Columbus Basin, offshore Trinidad. *American Association of Petroleum Geologists Bulletin*, **78**, 1372–1385.

Gibling, M.R. 2006. Width and thickness of fluvial channel bodies and valley fills in the geological record: a literature compilation and classification. *Journal of Sedimentary Research*, **76**, 731–770.

Graham, G.H., Jackson, M.D. & Hampson, G.J. 2015. Three-dimensional modeling of clinoforms in shallow-marine reservoirs: part 2. Impact on fluid flow and hydrocarbon recovery in fluvial-dominated deltaic reservoirs. *American Association of Petroleum Geologists Bulletin*, **99**, 1049–1080.

Hedberg, H.D., Sass, L.C. & Funkhouser, H.J. 1947. Oil fields of Greater Oficina area Central Anzoategui, Venezuela. *American Association of Petroleum Geologists Bulletin*, **31**, 2089–2169.

Hunter, R.E. 1985. Subaqueous sand-flow cross strata. *Journal of Sedimentary Research*, **55**, 886–894.

Hunter, R.E. & Kocurek, G. 1986. An experimental study of subaqueous slipface deposition. *Journal of Sedimentary Research*, **56**, 387–394.

Huysmans, M., Peeters, L., Moermans, G. & Dassargues, A. 2008. Relating small-scale sedimentary structures and permeability in a cross-bedded aquifer. *Journal of Hydrology*, **361**, 41–51.

Jackson, M.D., Yoshida, S., Muggeridge, A.H. & Johnson, H.D. 2005. Three-dimensional reservoir characterisation and flow simulation of heterolithic tidal sandstones. *American Association of Petroleum Geologists Bulletin*, **89**, 507–528.

James, S., Pronk, D., Abbots, F., Ward, V., Van Dierendonck, A. & Steves, D. 1999. The Brent Field: improving subsurface characterization for late field life management. *In*: Fleet, A.J. & Boldy, S.A.R. (eds) *Petroleum Geology of Northwest Europe: Proceedings of the 5th Conference*. Geological Society, London, 1039–1049, https://doi.org/10.1144/0051039

Jopling, A.V. 1965. Hydraulic factors controlling the shape and laminae in laboratory deltas. *Journal of Sedimentary Petrology*, **35**, 777–791.

Kuhn, O., Smith, S.W., Van Noort, K. & Loiseau, B. 2003. The Fulmar Field, Blocks 30/16, 30/11b, UK North Sea. *In*: Gluyas, J.G. & Hichens, H.M. (eds) *United Kingdom Oil and Gas Fields, Commemorative Millennium Volume*. Geological Society, London, Memoirs, **20**, 563–585, https://doi.org/10.1144/GSL.MEM.2003.020.01.46

Larue, D.K. & Hovadik, J. 2006. Connectivity of channelized reservoirs: a modelling approach. *Petroleum Geoscience*, **12**, 291–308, https://doi.org/10.1144/1354-079306-699

Legler, B., Johnson, H.D., Hampson, G.J., Massart, B.Y., Jackson, C.A.L., Jackson, M.D. & Ravnas, R. 2013. Facies model of a fine-grained, tide-dominated delta: lower Dir Abu Lifa Member (Eocene), Western Desert, Egypt. *Sedimentology*, **60**, 1313–1356.

Lowe, D.R. 1982. Sediment gravity flows: II. Depositional models with special reference to the deposits of high-density turbidity currents. *Journal of Sedimentary Petrology*, **52**, 279–297.

Makse, H.A., Havlin, S., King, P.R. & Stanley, H.E. 1998. Spontaneous stratification in granular mixtures. *Nature*, **386**, 379–381.

Martin, T. 1999. *The effects of large scale heterogeneity representation on waterflood performance – a simulation study of the Ametlla Formation, Ager Basin, Spanish Pyrenees*. MSc thesis, Imperial College London.

MITCHUM, R.M. & VAN WAGONER, J.C. 1991. High-frequency sequences and their stacking patterns: sequence-stratigraphic evidence of high-frequency eustatic cycles. *Sedimentary Geology*, **70**, 131–160.

MENCHER, E., FICHTER, H.J., RENZ, H.H., WALLIS, W.E., RENZ, H.H., PATTERSON, J.M. & ROBIE, R.H. 1953. Geology of Venezuela and its oil fields. *American Association of Petroleum Geologists Bulletin*, **37**, 690–777.

MULDER, T., SYVITSKI, J.P., MIGEON, S., FAUGERES, J.C. & SAVOYE, B. 2003. Marine hyperpycnal flows: initiation, behavior and related deposits. A review. *Marine and Petroleum Geology*, **20**, 861–882.

NEMEC, W. 1995. The dynamics of deltaic suspension plumes. *In*: POSTMA, G. & OTI, M.N. (eds) *Geology of Deltas*. Balkema, Amsterdam, 31–93.

NORDAHL, K., RINGROSE, P.S. & WEN, R. 2005. Petrophysical characterization of a heterolithic tidal reservoir interval using a process-based modelling tool. *Petroleum Geoscience*, **11**, 17–28, https://doi.org/10.1144/1354-079303-613

ORTON, G.J. & READING, H.G. 1993. Variability of deltaic processes in terms of sediment supply, with particular emphasis on grain size. *Sedimentology*, **40**, 475–512.

ORIFE, J.M. & AVBOVBO, A.A. 1982. Stratigraphic and unconformity traps in the Niger Delta. *In*: *The Deliberate Search for the Subtle Trap*. American Association of Petroleum Geologists, Memoirs, **32**, 251–265.

PULHAM, A.J., MITCHELL, A., MACDONALD, D. & DALEY, C. 1997. Sequence stratigraphic evaluation and production characteristics of the Eocene Mirador Formation, Cusiana Field, Eastern Colombia. *In*: SHANLEY, K.W. & PERKINS, B.F. (eds) *Shallow Marine and Non Marine Reservoirs*. GCSSEPM Foundation 18th Annual Research Conference Proceedings, 225–233.

QUINT, E. 1999. Monitoring contact movement in depressurizing Brent reservoirs. Society of Petroleum Engineers Paper 56951.

READING, H.G. & COLLINSON, J.D. 1996. Clastic coasts. *In*: READING, H.G. (ed.) *Sedimentary Environments: Processes, Facies and Stratigraphy*. 3rd edn. Blackwell Science, Oxford, 154–231.

REYNOLDS, A.D. 1996. Paralic successions. *In*: EMERY, D. & MYERS, K. (eds) *Sequence Stratigraphy*. Blackwell Publishing, Oxford, 134–177.

REYNOLDS, A.D. 1999. Dimensions of paralic sandstone bodies. *American Association of Petroleum Geologists Bulletin*, **83**, 211–229.

REYNOLDS, A.D. 2006. The geological use of pressure data: examples from the development of the giant ACG Oilfield, Azerbaijan. Paper SPE-101175-MS presented at the Abu Dhabi International Petroleum Exhibition and Conference, Society of Petroleum Engineers, 5–8 November, https://doi.org/10.2118/101175-MS

REYNOLDS, T., SHAW, A. & WARREN, E. 1997. Integrated description of a complex low net/gross sandstone reservoir: upper subzones of the Endicott field, North Slope of Alaska. *In*: SHANLEY, K.W. & PERKINS, B.F. (eds) *Shallow Marine and Non Marine Reservoirs*. GCSSEPM Foundation 18th Annual Research Conference Proceedings, 245–257.

RINGROSE, P.S. & CORBETT, P.W.M. 1994. Controls on two-phase fluid flow in heterogeneous sandstones. *In*: PARNELL, J. (ed.) *Geofluids: Origin, Migration and Evolution of Fluids in Sedimentary Basins*. Geological Society, London, Special Publications, **78**, 141–150, https://doi.org/10.1144/GSL.SP.1994.078.01.12

RINGROSE, P., NORDAHL, K. & WEN, R. 2005. Vertical permeability estimation in heterolithic tidal deltaic sandstones. *Petroleum Geoscience*, **11**, 29–36, https://doi.org/10.1144/1354-079303-614

RUBIN, D.M. 1987. *Cross-Bedding, Bedforms, and Palaeocurrents*. SEPM Concepts in Sedimentology and Paleontology, **1**.

STRIDE, A.H. 1982. *Offshore Tidal Sands*. Chapman and Hall, London.

TALUKDAR, S.C. & MARCANO, F. 1994. Petroleum systems of the Maracaibo basin, Venezuela. *In*: MAGOON, L.B. & DOW, W.G. (eds) *The Petroleum System – from Source to Trap*. American Association of Petroleum Geologists, Memoirs, **60**, 463–463.

VAN WAGONER, J.C., MITCHUM, R.M., CAMPION, K.M. & RAHMANIAN, V.D. 1990. *Siliciclastic Sequence Stratigraphy in Well Logs, Cores, and Outcrops*. American Association of Petroleum Geologists, Methods in Exploration Series 7.

VERDIER, A.C., OKI, T. & SUARDY, A. 1980. Geology of the Handil field (East Kalimantan – Indonesia). *In*: HALBOULTY, M.T. (ed.) *Giant Oil and Gas Fields of the Decade: 1968–1978*. American Association of Petroleum Geologists, Memoirs, **30**, 399–421.

VISSER, M.J. 1980. Neap-spring cycles reflected in Holocene subtidal large-scale bedform deposits: a preliminary note. *Geology*, **8**, 543–546.

WEBER, K.J. 1971. Sedimentological aspects of oil fields in the Niger Delta. *Geologie en Mijnbouw*, **50**, 559–576.

WEBER, K.J. 1987. Hydrocarbon distribution patterns in Nigerian growth fault structures controlled by structural style and stratigraphy. *Journal of Petroleum Science and Engineering*, **1**, 91–104.

WEHR, F.L. & BRASHER, L.D. 1996. Impact of sequence-based correlation style on reservoir model behaviour, lower Brent Group, North Cormorant Field, UK North Sea. *In*: HOWELL, J.A. & AITKEN, J.F. (eds) *High Resolution Sequence Stratigraphy: Innovations and Applications*. Geological Society, London, Special Publications, **104**, 115–128, https://doi.org/10.1144/GSL.SP.1996.104.01.08

WENTWORTH, C.K. 1922. A scale of grade and class terms for clastic sediments. *The Journal of Geology*, **30**, 377–392.

WORTHINGTON, P.F. 2010. A road map for the identification and recovery of by-passed pay. *In*: VINING, B.A. & PICKERING, S.C. (eds) *Petroleum Geology: From Mature Basins to New Frontiers–Proceedings of the 7th Petroleum Geology Conference*. Geological Society, London, Petroleum Geology Conference Series, **7**, 453–462, https://doi.org/10.1144/0070453

YIELDING, G., FREEMAN, B. & NEEDHAM, D.T. 1997. Quantitative fault seal prediction. *American Association of Petroleum Geologists Bulletin*, **81**, 897–917.

Stratigraphic architecture of the Knarr Field, Norwegian North Sea: sedimentology and biostratigraphy of an evolving tide- to wave-dominated shoreline system

JAMES M. CHURCHILL[1]*, MATTHEW T. POOLE[1], SILJE S. SKARPEID[2] & MATTHEW I. WAKEFIELD[1]

[1]*BG Group, Thames Valley Park, Reading, Berkshire RG6 1PT, UK*

[2]*BG Norge, Løkkeveien 111, 4007 Stavanger, Norway*

**Correspondence: james.churchill@bg-group.com*

Abstract: The Knarr Field is located on the Tampen Spur, Norwegian continental shelf and was discovered in 2008 by the Jordbær well (34/3-1S), with additional resources later added to the field by the Jordbær Vest well (34/3-3S) in 2011. Within the Knarr Field, the Cook Formation is informally divided into the Lower Cook and Upper Cook successions and appears to have prograded from east to west. The Lower Cook consists of Sands 1, 2 and 3 and the Upper Cook consists of Sands 4 and 5, with the sands separated by intraformational mudstones that are commonly chronostratigraphically constrained; the J15 maximum flooding surface separates the Lower and Upper Cook. The tide-dominated Lower Cook is notably heterolithic, with intricate intercalations of sandstone and mudrock lithologies representing tidal channel, tidal bar and intertidal bar facies. The Upper Cook represents a series of coarsening-upwards cycles that displays the systematic changes in facies and ichnology expected for a shoreface succession, consisting of offshore, offshore transition zone and shoreface facies. Palynomorphs confirm these observations and suggest that the Lower Cook was deposited in brackish-water conditions, whereas the presence of more marine fossils in the Upper Cook suggests an increase in marine influence. The integration of the sedimentology and biostratigraphy described herein enabled the establishment of a robust reservoir zonation that has been utilized during the development and ongoing exploitation of the Knarr Field.

The Knarr Field lies on the eastern fringes of the Tampen Spur, Norwegian continental shelf (Fig. 1) and is divided into Knarr Sentral and Knarr Vest accumulations. It was discovered in 2008 by wildcat well 34/3-1S (Jordbær Sentral) and subsequently appraised by wells 34/3-1A, 34/3-3S (Jordbær Vest discovery well) and 34/3-3A. Prior to sanction, the Knarr Field was called the Jordbær prospect and then the Jordbær Field, with the discovery and appraisal wells termed Jordbær Sentral (34/3-1S), Jordbær Øst (34/3-2S) and Jordbær Vest (34/3-3S), respectively. BG Group has operated the field from discovery, through appraisal and on to development via a floating production storage and offloading vessel. The field is currently producing oil from three wells (34/3-A-1 H, 34/3-A-2 H and 34/3-A-4 H), which are supported by three paired water injection wells (34/3-B-1 H, 34/3-B-2 H and 34/3-B-4 H). Although not within the Knarr Field, well 34/3-2S (Jordbær Øst) represents a key data point and is included herein.

Hydrocarbons are reservoired within the Lower Jurassic Cook Formation (Fig. 2) and the stratigraphic architecture of the reservoir section represents an overall transgressive succession, with a tide-dominated lower interval, referred to as the Lower Cook, and a wave-dominated upper section referred to as the Upper Cook, where the succession is bounded and punctuated by key sequence stratigraphic surfaces that are chronostratigraphically constrained. Detailed integration of stratigraphic and sedimentological data revealed a complex stratigraphy and intricate facies mosaic; it is only through the integration of data that a complete picture of the reservoir has been achieved.

In this paper, we present the key stratigraphic and sedimentological aspects of the Cook Formation and how these data have been interpreted to produce a detailed understanding of the reservoir interval.

Geological background

The Knarr Field is located on the Tampen Spur and lies *c.* 50 km NE of the giant Snorre Field (Fig. 1) in 410 m of water. The Tampen Spur area is located in the northern part of the Norwegian North Sea and represents a continuation of the East Shetland Platform. For discussions of the evolution and

From: Hampson, G. J., Reynolds, A. D., Kostic, B. & Wells, M. R. (eds) 2017. *Sedimentology of Paralic Reservoirs: Recent Advances*. Geological Society, London, Special Publications, **444**, 35–58.
First published online August 12, 2016, https://doi.org/10.1144/SP444.4

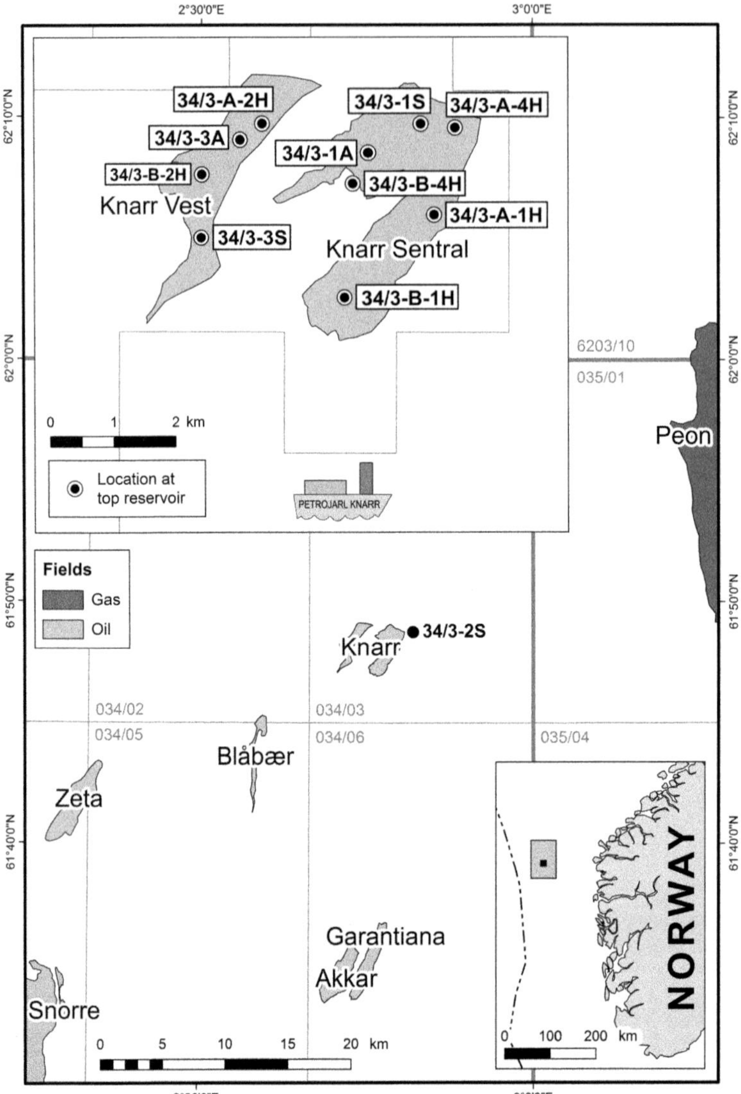

Fig. 1. Field and well location map.

petroleum geology of the Tampen Spur, readers are referred to Kirk (1980), Eynon (1981), Karlsson (1986), Graue *et al.* (1987), Erichsen *et al.* (1987), Olaussen *et al.* (1993) and Horstad *et al.* (1995).

Stratigraphy

The reservoirs in the Knarr Field consist of the marginal to shallow-marine sandstones of the Cook Formation, which is part of the Dunlin Group (Fig. 2), as defined by Doré *et al.* (1984). In stratigraphic order, the Dunlin Group consists of the Amundsen, Johansen, Burton, Cook and Drake formations. The key formations within the Knarr Field are the Cook Formation reservoir, the Drake Formation mudstones, which represent the principal seal overlying the Cook Formation (Fig. 3), and the underlying mudstones and thin sandstones of the Amundsen Formation. The Dunlin Group overlies the Triassic to early Jurassic Statfjord Formation of the Hegre Group of Vollset & Doré (1984) and Lervik (2006) and is overlain by the Brent Group. Charnock *et al.* (2001) undertook a detailed sequence stratigraphic study of the Dunlin Group lithostratigraphic units in conjunction with high-resolution biostratigraphy, which defined a series of maximum flooding

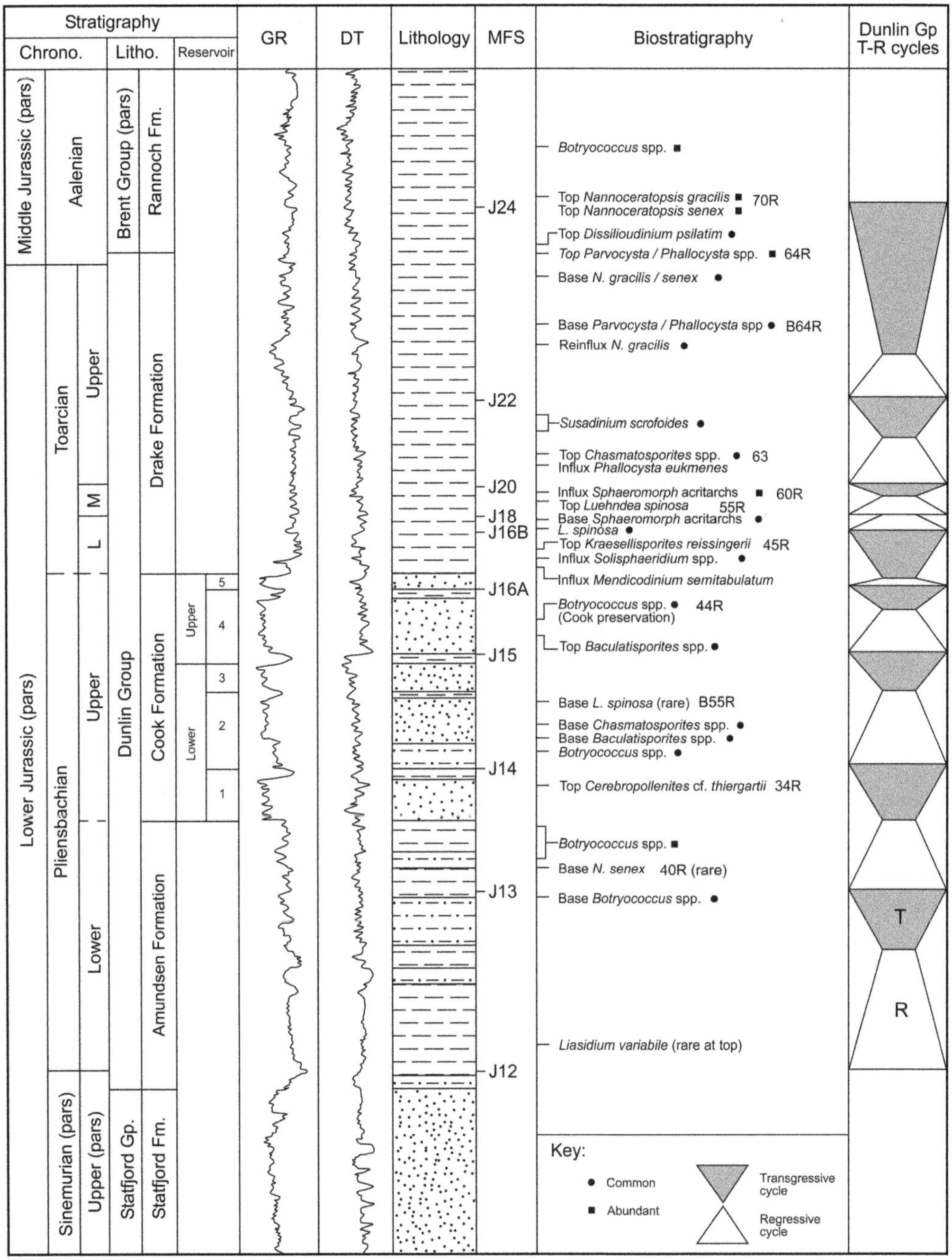

Fig. 2. Stratigraphy of the Knarr Field area and reservoir layering of the Cook Formation. The biostratigraphy is based on Charnock *et al.* (2001) with their coded events labelled.

surfaces (MFSs) (equivalent definition to Galloway 1989). The stratigraphy of Charnock *et al.* (2001) is used herein, assigning the Statfjord Formation to the Statfjord Group (Fig. 2).

The top of the Cook Formation reservoir within the Knarr Field is interpreted to coincide with the base of the Toarcian and is capped by the J16B MFS, whereas the base of the Cook Formation is

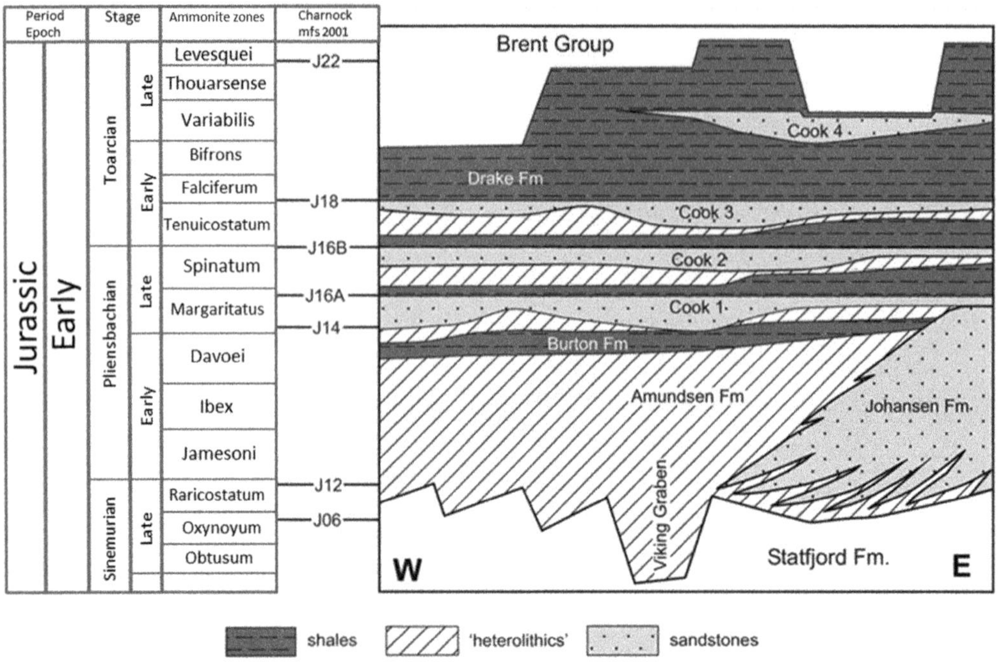

Fig. 3. Revised stratigraphic architecture of the Dunlin Group across the northern North Sea. Redrawn from Marjanac & Steel (1997). J sequence maximum flooding surface (MFS) after Charnock *et al.* (2001).

interpreted to coincide with the base of the Upper Pliensbachian (Fig. 2). Charnock *et al.* (2001) and Marjanac & Steel (1997) showed that, within the North Viking Graben area, the Cook Formation can be much younger, ranging into the Middle Toarcian, where it is capped by the J20 MFS (Fig. 3).

Five sand units are identified within the Cook Formation and informally subdivide the Cook Formation into the Lower Cook and Upper Cook, which are separated by the J15 MFS (Fig. 2). Sand 1 occurs at the base of the reservoir and is overlain by a well-developed mudstone interval that represents the J14 MFS; the remainder of the Lower Cook consists of Sands 2 and 3, which are subdivided by a thin mudstone and are seen in most wells. Sand 4 overlies the J15 MFS, which is well developed in all the study wells and is separated from Sand 5 by the J16A MFS.

Regional palaeography

The Cook Formation is regionally interpreted as a series of basin-wide regressive–transgressive cycles that represent a mixture of tidal- and wave-influenced sediments overlying the distal shelfal sediments of the Burton and Amundsen formations (Marjanac & Steel 1997; Fig. 3) and underlie the offshore sediments of the Drake Formation (Doré *et al.* 1984). Marjanac & Steel (1997) interpreted four regressive–transgressive cycles from the Late Pleinsbachian to Middle Toarcian that comprise the Cook Formation. MFS J16A, J16B and J18 cap three regressive–transgressive cycles observed by Marjanac & Steel (1997; Fig. 3) and the fourth transgressive–regressive cycle occurs between the J18 and J22 MFS. In the Knarr area only the two earliest regressive–transgressive cycles of Marjanac & Steel (1997) are present as well-developed sandstones, with the majority of Knarr area Cook Formation sandstones being deposited in the first regressive–transgressive cycle between the J14 and J16A MFSs, spanning the earliest Late Pliensbachian. An additional maximum flooding shale, J15, was described by Charnock *et al.* (2001) and corresponds to another regressive–transgressive cycle previously undescribed by Marjanac & Steel (1997). The J15 MFS is present across the Knarr Field and caps Sand 3 (Fig. 2).

Dreyer & Wiig (1995) presented a detailed sequence stratigraphic framework and palaeogeography of the Cook Formation to the south of the Knarr Field, focused around the Gullfaks Field. The palaeography presented by Dreyer & Wiig (1995) describes the Cook Formation as being dominated by shoreline sediments in the lower successions and as a tide-dominated system in the upper parts of the Cook Formation, the reverse of that seen in the Knarr Field.

Database and methodology

All exploration and appraisal wells of the Knarr Field have an extensive suite of open hole wireline logs, while development wells have measured or logged while drilling logs. The discovery well 34/3-1S (Jordbær Sentral) includes 60.5 m of conventional core that covers a significant part of the Upper Cook and 80.4 m of conventional core was cut within Jordbær Vest (34/3-3S), covering much of the Upper Cook and a significant part of the Lower Cook. Additional core was taken in offset wells 34/3-2S (26.9 m) and 34/5-1S (Blåbær; 48.4 m). Post-well biostratigraphic analysis was carried out on all 11 Jordbær and Knarr wells, generating a rich dataset consisting of *c.* 1200 samples, which are a mixture of microfossil and palynological analyses, although microfossil recovery in the reservoir interval was limited. In addition, petrographic analysis was conducted on 103 samples to help understand the reservoir properties of the field.

Detailed core descriptions were obtained at a 1:50 scale using routine core description techniques with particular attention paid to the ichnology of the succession as this provided important palaeoenvironmental information; the bioturbation scheme applied here is that of Taylor & Goldring (1993). The cored intervals were depth-shifted relative to wireline logs and subjected to a rigorous interpretation workflow with objective description and depositional process interpretation before integration with complementary data and the assignment of the most likely depositional setting. Individual lithofacies do not have direct environmental implications, but provide the building blocks for sedimentological process interpretation and combine to form facies associations; these are inferred to relate to likely environments of deposition and form the basis for depositional modelling. This approach allows for the transparent and robust interpretation of cores, with all possibilities considered, before a final depositional environment is assigned and applied to the uncored intervals using the well logs as a semi-quantitative guide. The interpretation of depositional settings for the succession was very much an iterative process as new wells and data became available. For brevity, the details of each lithofacies are not described in detail; however, the sedimentological character of each facies association is fully described in the following text and summarized in Table 1.

Although beyond the scope of this paper, complementary techniques, such as chemical stratigraphy, palaeomagnetometry and S_r residual salt isotope analysis, have been carried out to refine the intra-reservoir correlation of the succession.

Sedimentology

Four facies associations have been assigned within the Lower Cook: tidal channel complex, tidal bar complex, intertidal bar complex and proximal intertidal bar complex. Six facies associations have been assigned within the Upper Cook and represent the constituent parts of a shoreface succession: upper shoreface, proximal lower shoreface, distal lower shoreface, proximal transition zone, distal transition zone and offshore.

Lower Cook. Sediments interpreted as tidal channel complex deposits are cored in wells 34/3-1S, 34/3-2S and 34/3-3S. Tidal channel complex sediments here are dominated by current–ripple cross-laminated sandstones with abundant rip-up clasts (Figs 4c & 5b) and high-angle cross-stratification is locally recognized (34/3-2S), with highly variable grain sizes and sorting. Intervals locally display upwards-increasing inclined heterolithic stratification, where mudclast-rich intervals include disarticulated bivalve shell debris and coated grains. The intensity of bioturbation is variable, although locally intense, and includes *Asterosoma*, *Skolithos*, *Palaeophycus* and *Planolites* traces, with individual beds locally exhibiting intense *Taenidium* bioturbation. The dominance of low-relief current–ripple cross-lamination records tractional sand transport and locally draped toesets may simply reflect hydrodynamic sorting; however, when observed with local reactivation and bidirectional current-ripple foresets they suggest a degree of tidal modulation. Although not mutually exclusive, the presence of inclined heterolithic stratification is consistent with tidal deposition (Mutti *et al.* 1985; Thomas *et al.* 1987; Choi *et al.* 2004; Johnson & Dashtgard 2014). Common mudstone rip-up clasts are suggestive of currents that were capable of eroding coeval mud-prone depositional facies. The bulk of the trace fossil assemblage is consistent with marginal marine deposition, e.g. *Asterosoma* has been described within brackish coastal lagoons in association with *Palaeophycus* (Martino 1979). However, locally intense *Taenidium* bioturbation suggests at least short-lived freshwater conditions or a depositional setting between fully aquatic and terrestrial (cf. Buatois & Mángano 1995; Savrda *et al.* 2000). Short-lived intervals of strong fluvial input are consistent with influxes of the predominantly freshwater alga *Botryococcus* within Sand 2 and indeed throughout the Lower Cook. Modern *Botryococcus* generally lives in freshwater pools, lakes and marshes, but can also tolerate some brackish habitats (Batten & Grenfell 1996). The term complex

Table 1. *Facies associations assigned to the cored intervals and extrapolated to uncored intervals*

	Description	Facies association
Upper Cook	Organized into metre-scale beds that are cross-stratified, planar-bedded and locally current–ripple cross-laminated. The sandstones are clean with negligible detrital silt and clay. Bioturbation is sparse with local *Skolithos*, *Planolites* and burrow mottling (Figs 4a & 5a)	Upper shoreface
	Well-sorted, upper to medium lower grained sandstones that include a minor silt/detrital clay component. Dominantly swaley cross-stratification and hummocky cross-stratified. Bioturbation tends to be sparse with bivalve burrows, resting traces, *Planolites* and *Palaeophycus*	Proximal lower shoreface
	These upper fine- to medium-grade silty sandstones exhibit weak bioturbation, including *Planolites*, *Asterosoma*, *Arenicolites*, bivalve burrows, resting traces and escape traces. Hummocky cross-stratification is locally observed	Distal lower shoreface
	Highly silty and highly to intensely bioturbated with *Palaeophycus* and *Asterosoma* with subordinate *Rosselia*, *Planolites* and *Phycosiphon* burrows. Storm-generated structures are locally preserved	Proximal transition zone
	Silty sandstones that locally preserved storm-generated hummocky cross-stratification and pinstripe lamination recording minor low-density turbidity flows. Highly bioturbated with a diverse trace fossil assemblage including *Teichichnus*, *Rosselia*, *Phycosiphon*, *Planolites*, *Rhizocorallium*, *Arenicolites*, bivalve burrows, *Skolithos* and *Chondrites*	Distal transition zone
	Offshore sediments only comprise a small proportion of the Cook Formation. They are composed of intensely bioturbated micaceous siltstones with a dominantly *Phycosiphon* ichnofabric and subordinate *Chondrites*, *Teichichnus* and *Schaubcylindrichnus*	Offshore
Lower Cook	Tidal channel complex sediments here are dominated by current–ripple cross-lamination with abundant rip-up clasts (Figs 4b & 5b). Highly variable grain size, sorting and bioturbation, including *Asterosoma*, *Skolithos*, *Palaeophycus* and *Planolites* traces. Current–ripple locally display silt drapes on foresets and bidirectionality	Tidal channel complex
	These sandstones are composed of metre-scale coarsening-upwards intervals with current–ripple cross-lamination at the base and dune-scale cross-stratification at the top of cycles. Bioturbation intensity is highly variable (Fig. 4c) and characterized by distinctive *Rosselia* traces (Fig. 5c)	Tidal bar complex
	Variably bioturbated mudrocks with *Palaeophycus*, *Asterosoma*, *Schaubcylindrichnus*, *Rosselia*, *Planolites*, *Teichichnus* and *Phycosiphon* burrows (Figs 4d & 5d). Locally preserved lamination and pinstripes. Local echinoid debris	Intertidal bar
	Not cored. Assigned to relatively high gamma ray intervals recognized towards the base of Sand 2 in wells 34/3-1A, 34/3-B-1 H and 34/3-B-4 H. Ditch cutting descriptions suggest these intervals are composed of heterolithic, bioturbated, argillaceous sandstones	Proximal intertidal bar

is used here as the sediments are highly heterolithic, with beds organized at a decimetre to metre scale, which suggests multiple channels within a channel complex, rather than a single channel body.

Tidal bar complex facies are best expressed in core from Sand 3, where they exhibit strong progradation. They are interpreted to form much of Sand 1, although they are not cored here and are inferred from well log expression. These sandstones form metre-scale coarsening-upwards intervals (Fig. 4b) that exhibit a corresponding increase in grain size and bed thickness. Heterolithic intervals are common towards the base of individual bed sets that are current–ripple cross-laminated. Larger-scale bed forms commonly exhibit drapes and locally include minor proportions of mudstone rip-up clasts. Bioturbation is present throughout, ranging between sparse and intense at the top of bed sets, and is characterized by distinctive *Rosselia* traces (Fig. 5c); other notable traces include bivalve

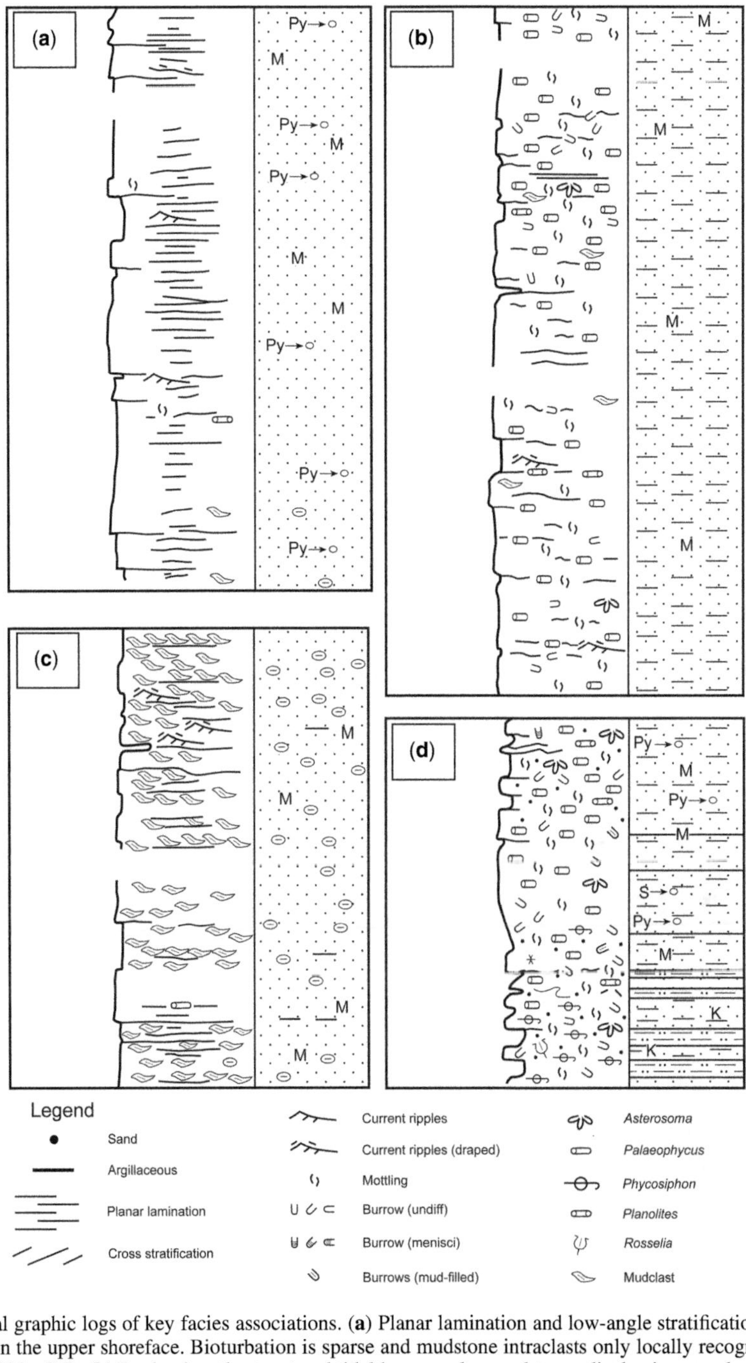

Fig. 4. Typical graphic logs of key facies associations. (**a**) Planar lamination and low-angle stratification are common within the upper shoreface. Bioturbation is sparse and mudstone intraclasts only locally recognized (34/3-1S, 3876–3880 mDD. (**b**) Predominantly structured tidal bar complex sandstones display increased amounts of bioturbation with common *Palaeophycus* and locally notable *Astersoma* (34/3-1S, 3902–3906 mDD). (**c**) Tidal channel complex deposits include abundant mudstone intraclasts with current ripples locally preserving mud drapes (34/3-2S, 3931–3934 mDD). (**d**) Intertidal bar complex sediments are highly bioturbated with a diverse trace fossil assemblage including *Rosselia*, *Asterosoma*, *Palaeophycus* and *Phycosiphon* locally (34/3-1S, 3908–3910 mDD). K, kaolin; M, mica; Py, pyrite; S, siderite. Graphic log images courtesy of Robertson, part of CGG GeoConsulting.

Fig. 5. Selected core photographs. (**a**) Upper shoreface facies characterized by cross-stratified sandstones. Common trough cross-stratification and beds developed on a decimetre scale (34/3-3S, 3918.7 mDD). (**b**) Common deformed mudstone intraclasts (arrowed) are indicative of tidal channel complex sandstones. Mudstone intraclasts are locally sideritized (34/3-3S, 3959.7 mDD). (**c**) Distinctive *Rosselia* burrows characteristic of tidal bar complex deposits (34/3-3S, 3951.7 mDD). (**d**) Highly bioturbated intertidal bar complex sediments with a diverse trace fossil assemblage including *Rosselia* (Rs), *Teichichnus* (Te) and *Planolies* (Pl) (34/3-3S, 3942.7 mDD). Core photographs courtesy of Ichron, part of the RPS Group.

burrows, *Asterosoma* and *Palaeophycus*. The common occurrence of cross-stratification suggests these sediments were deposited by moderate to high-energy tractional currents. Repeated reworking is also evident by local mudstone rip-up clasts and may suggest some local channelization towards the top of bars. The trace fossil assemblage is consistent with persistent marginal-marine deposition, with *Rosselia* traces indicating a degree of tidal influence where their clay-rich fill reflects silt incorporated into the burrow during periods of low depositional energy. The facies arrangement indicates deposition by tidal bars, rather than channels, as a result of the strongly progradational profile across wells and less abundant mudstone rip-up clasts.

Sands 1, 2 and 3, which comprise the Lower Cook within the Knarr Field, are divided from one another by field-wide flooding surfaces and corresponding mudrock-rich intervals composed of variably bioturbated mudstones and heteroliths. Commonly structureless or weakly laminated with local pinstripes, the mudstones are weakly to intensely bioturbated with *Palaeophycus*, *Asterosoma*, *Schaubcylindrichnus*, *Rosselia*, *Planolites*, *Teichichnus* and *Phycosiphon* burrows (Figs 4d & 5d), with echinoid debris recognized locally. These sediments resemble offshore transition zone deposits of a shoreline system; however, within the context of the surrounding sandstones they are interpreted as intertidal bar deposits. Biofacies analysis of these intervals is ambiguous regarding the degree of marine influence, however; sediments below the J15 MFS (Lower Cook) have a marine influence index between 5 and 10%, compared with *c.* 40% at the J16A MFS, suggesting relatively less marine deposition. The lithology, sedimentary structures and ichnology of this facies is suggestive of relatively low-energy deposition, with low rates of deposition. The fluctuating amount of sand input is consistent with offshore deposition, although the vertical succession suggests an overall tide-dominated rather than wave-dominated setting.

Proximal intertidal bar complex deposits are not cored and are tentatively assigned to relatively high gamma ray intervals recognized towards the base of Sand 2 in wells 34/3-1A, 34/3-B-1 H and 34/3-B-4 H. These intervals are composed of sandstone that is less mud-prone than the underlying intertidal bar complex mudstones, but more silty than the immediately overlying tidal channel complex sandstones.

Upper Cook. Within the Upper Cook, cleaning and coarsening-upwards intervals are commonly capped by clean, cross-stratified sandstones interpreted as upper shoreface deposits. High-angle cross-stratification and increased heterogeneity may suggest local channel preservation within the shoreface (Fig. 5a). These sandstones are composed of metre-scale, cross- and planar-bedded clean sandstones that are moderately to well-sorted and locally structureless (Fig. 4a). Some cross-sets exhibit sigmoidal geometries and/or scour and drape features reminiscent of swaley cross-stratification, although these features are ambiguous. Bioturbation is sparse and characterized by *Skolithos*. Pervasively cemented intervals are locally focused around concentrations of calcareous shelly debris and may indicate minor ravinement surfaces. These sediments record the deposition of moderate to high-energy tractional currents that formed dune-scale bed forms and the local structureless nature of the sandstones may attest to rapid deposition or dewatering due to wave action because there is no suggestion of complete levels of bioturbation. The sparse, low-diversity dwelling ichnofauna is also consistent with high-energy deposition. The local fining-upwards grain-size motifs may suggest a degree of channelization, where the package at the top of Sand 4 in well 34/3-1A may represent a channel cutting the shoreface as it exhibits a distinct inorganic chemical character.

Proximal lower shoreface sediments are commonly recognized below upper shoreface facies and these sandstones are characterized as moderately to well-sorted, upper fine to medium lower grained, micaceous sandstones with a minor silt component. Sedimentary structures are dominated by swaley and hummocky cross-stratification, with lesser amounts of low-angle cross-stratification and current ripples. Bioturbation tends to be sparse, with bivalve burrows, resting traces, *Planolites* and *Palaeophycus*. These sediments represent the deposition of moderate to high-energy oscillatory currents recording storm activity, with subordinate fair-wave processes. The slight increase in silt content and bioturbation compared with the upper shoreface deposits suggests a slightly lower depositional energy.

Underlying the proximal lower shoreface sediments are distal lower shoreface deposits, which form a continuum of lower shoreface deposits similar to those described by Gowland (1996). They are characterized by moderately to well-sorted, upper fine- to medium-grade, silty sandstones with weak biotutrbation, including *Planolites*, *Asterosoma*, *Arenicolites*, bivalve burrows, resting traces and escape traces. Hummocky cross-stratification and pebble-grade mudstone clasts are locally observed. These sediments represent a similar environment of deposition to the proximal lower shoreface deposits; however, greater proportions of silt suggest a lower depositional energy.

The proximal transition zone sediments observed in core display many of the same characteristics as the distal lower shoreface deposits, but with a greater proportion of detrital silt and an increasingly diverse trace fossil assemblage. Proximal transition zone sediments consist of moderately to well-sorted, lower fine-grained, silty, micaceous sandstones with locally preserved hummocky cross-stratification. Bioturbation is commonly high to intense and dominated by *Palaeophycus* and *Asterosoma,* with subordinate *Rosselia*, *Planolites* and *Phycosiphon.* A significant proportion of detrital silt, preserved storm-generated sedimentary structures and a diverse trace fossil assemblage suggest an offshore transition zone depositional environment, but with a notable sand input, probably from offshore currents.

Distal transition zone deposits consist of highly bioturbated siltstones. Pinstripe and centimetre-scale current–ripple cross-lamination is locally preserved where bioturbation is less intense. Trace fossils include *Teichichnus*, *Rosselia*, *Phycosiphon*, *Planolites*, *Rhizocorallium*, *Arenicolites*, bivalve burrows, *Skolithos* and *Chondrites*.

Offshore sediments only make up a small proportion of the Cook Formation within the Knarr Field and are characterized by intensely bioturbated micaceous siltstones. The trace fossil assemblage is dominated by *Phycosiphon*, with *Chondrites*, *Teichichnus* and *Schaubcylindrichnus* also observed, attesting to offshore deposition.

In combination, these facies associations are interpreted to represent a complex tide-dominated system for the Lower Cook and a shoreline-to-shelf system for the Upper Cook. Tidal depositional systems are characteristically complex and have been extensively discussed by Dalrymple & Choi (2007) and Smith *et al.* (1991). In contrast, linear shoreline systems can be generalized as relatively organized with a predictable manner. For a detailed description and discussion of shoreface deposits, see Clifton (2006), Hampson & Howell (2005) and Reading & Collinson (1996); for tidal-influenced shoreface deposits, see Dashtgard *et al.* (2009).

Biostratigraphy

The biostratigraphic zonation in the Dunlin Group largely follows that of Charnock *et al.* (2001), who used both palynological (miospores, pollen, algae, acritarchs and dinoflagellate cysts) and micropalaeontological (benthonic forams and ostracods) data. Some modifications to the published scheme are noted in Appendix A and enable correlation of the Knarr Field wells. Within the Knarr Field area, palynological data have proved the most biostratigraphically robust. The micropalaeontological bioevents of Charnock *et al.* (2001), although present in the upper Drake and upper Amundsen formations, were poorly developed, if present at all, in the lower Drake and Cook formations. Therefore only the palynological bioevents are discussed here; the Charnock *et al.* (2001) bioevent codes are augmented by additional bioevents identified during study of the five Jordbær exploration and appraisal wells and the six Knarr development wells.

Quantitative palynology. The eight wells of palynological data discussed here represent a robust set of consistent results. The other three wells within the Knarr Field area (34/3-1S, 34/3-1A and 34/3-2S) are not included in the statistical analysis to avoid possible taxonomic and numerical inconsistencies complicating the interpretation as they were analysed by a different palynologist who used a different data collection and counting methodology.

The sample material was entirely from ditch cuttings ranging from 2.5 to 5 m spacing. The development wells were drilled at an angle of 37–47° from the vertical, giving a true vertical sampling resolution of *c.* 2–3.5 m. All samples were prepared using standard palynological techniques, including initial hydrochloric acid treatment, followed by the hydrofluoric digestion of silicates. Oxidation with warm nitric acid further lightened palynomorphs and removed pyrite. Residues were sieved using a 10 μm mesh to ensure that the small dinoflagellate cysts (*Parvocysta* and *Phallocysta*) were not lost, which would have happened if the standard 20 μm mesh had been used. The resulting residue was strew-mounted on glass slides with coverslips and the palynomorphs were counted, including dinoflagellate cysts, acritarchs, miospores, pollen and algae. Counts were performed quantitatively and an initial count of 300 specimens was made to emphasize more subtle events. This was augmented by a concurrent count of 100 dinocysts where possible.

Only the miospore and pollen census data were used during the biofacies analysis. The dinoflagellate cyst and acritarch census data were removed because they were so rare in the Cook Formation. In addition, the freshwater algae, predominantly *Botryococcus*, were treated separately.

Biofacies analysis. Fuzzy c-means (FCM) clustering is a data analysis/classification method that identifies groups of samples with similar compositions. Fuzzy set theory allows for the possibility that a sample can have an association with multiple classes or clusters (Zadeh 1965), unlike the hard clustering methods commonly used by palaeontologists. This is an important feature for biostratigraphic data analysis because such datasets often include samples that are transitional between two or more 'pure' faunal or floral assemblages. This is a novel technique that has not been commonly applied to palaeontological data; see Gary *et al.* (2009) for a full review of the biostratigraphic application of FCM and Da Gama *et al.* (2014), Daly *et al.* (2011), Erbs-Hansen *et al.* (2011, 2012) and Martin *et al.* (2013) for examples of its application to palaeontological data.

FCM was implemented using the biostratigraphic data analysis software package TACS-Works, which was developed by the Technical Alliance for Computational Stratigraphy at the

University of Utah and was used to accomplish the following tasks:

1. determine the appropriate number of clusters;
2. resolve cluster 'centres' and memberships;
3. rank species based on correlation with each cluster to evaluate the suitability and utility of index species;
4. visualize cluster 'centres' as faunal/floral assemblage bar graphs; and
5. visualize cluster memberships as downhole assemblage curves.

Determining the appropriate number of clusters is an iterative process. The fuzzy exponent was varied between 1.1 (a hard cluster) and 3.0 (a soft cluster where memberships to clusters tend to be highly gradational) to understand the data structure and better constrain the FCM parameters. Gary *et al.* (2009) described two criteria for determining the appropriate number of clusters: compaction and separation. Compaction refers to a high sample density around each of the cluster centres. Compact clusters are preferred, as samples associated with a cluster will be well represented by the characteristic biota. Separation is a measure of the degree to which cluster centres are unique. In other words, it is preferable that the distance between cluster centres is maximized.

An initial model with six clusters was calculated, from which correlation coefficients (r) for each taxon to each cluster were calculated (Gary *et al.* 2009), where $r = +1$ indicates identical samples and $r = -1.0$ indicates totally dissimilar samples. Examination of these data indicated a large number of taxa that had little, if any, significance in many, if not all, of the clusters. To develop a simpler and more robust model, taxa that had a correlation coefficient between -0.1 and $+0.1$ against four or more of the resulting clusters were removed from the data and the model was re-run. The resulting model structure is shown in Figure 6. This clearly shows that six compact clusters could be calculated from the revised dataset with a fuzzy exponent of 2.0. If more clusters were calculated, then their integrity would be lost (Fig. 6a). The similarity measure of the clusters indicates that if more than six clusters were calculated, then their similarity would be too high. The resulting six-cluster model was not perfect in that an ideal centroid similarity trend was not present and two of the clusters appear to be fairly similar (Fig. 6b). Further data mining did not reveal a different model structure. The six FCM clusters are generally named after the taxon with the highest correlation coefficient to that cluster (Fig. 7).

The palaeoenvironmental significance of each FCM cluster was determined by identifying the most important taxa in each. The known palaeoenvironmental significance of those taxa was then

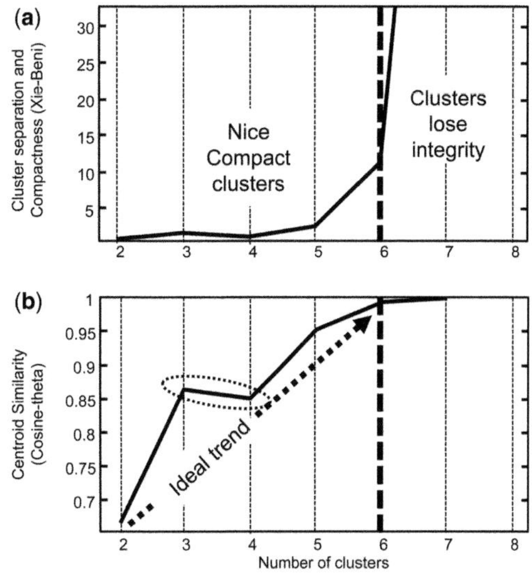

Fig. 6. Fuzzy c-means clustering model structure: (**a**) cluster separation and compactness (Xie-Beni); (**b**) centroid similarity (cosine-theta). See text for detailed explanation.

assigned to that FCM cluster. Known palaeoecological controls on the spores and pollen that were used to determine the palaeoenvironment within the Knarr Field succession are summarized in Table 2 to save repetitious referencing and to improve readability. The most important taxa were arbitrarily considered to be those with a correlation coefficient of $r = +0.5$ to $+1$ and were considered to be good candidates for proxy indicators for that biofacies cluster. Those taxa with a strongly negative correlation coefficient values ($r = -0.5$ to -1) were also noted, as these should be diametrically opposed to those with strongly positive correlation coefficient values. All the nominated taxa show a good positive correlation with FCM clusters, i.e. those with a high percentage occurrence of the taxon have high membership values (Fig. 7). Refinements to the palaeoenvironmental and palaeoclimatological interpretation for each FCM cluster were made using information from the accessory taxa (Tables 2 & 3).

- The *Cerebropollenites* cf. *thiergartii* cluster. *Cerebropollenites* cf. *thiergartii* is the pollen from a conifer considered to be similar to the modern *Tsuga* and dominates this cluster ($r = 0.87$; *c.* 52% of the cluster, Table 3). *Tsuga* prefers moist, but not mire, conditions. Negatively correlated taxa including *Classopollis* spp. ($r = -0.43$) and *Cerebropollenites mesozoicus* ($r = -0.31$) are considered to be more coastal in position and may have preferred slightly

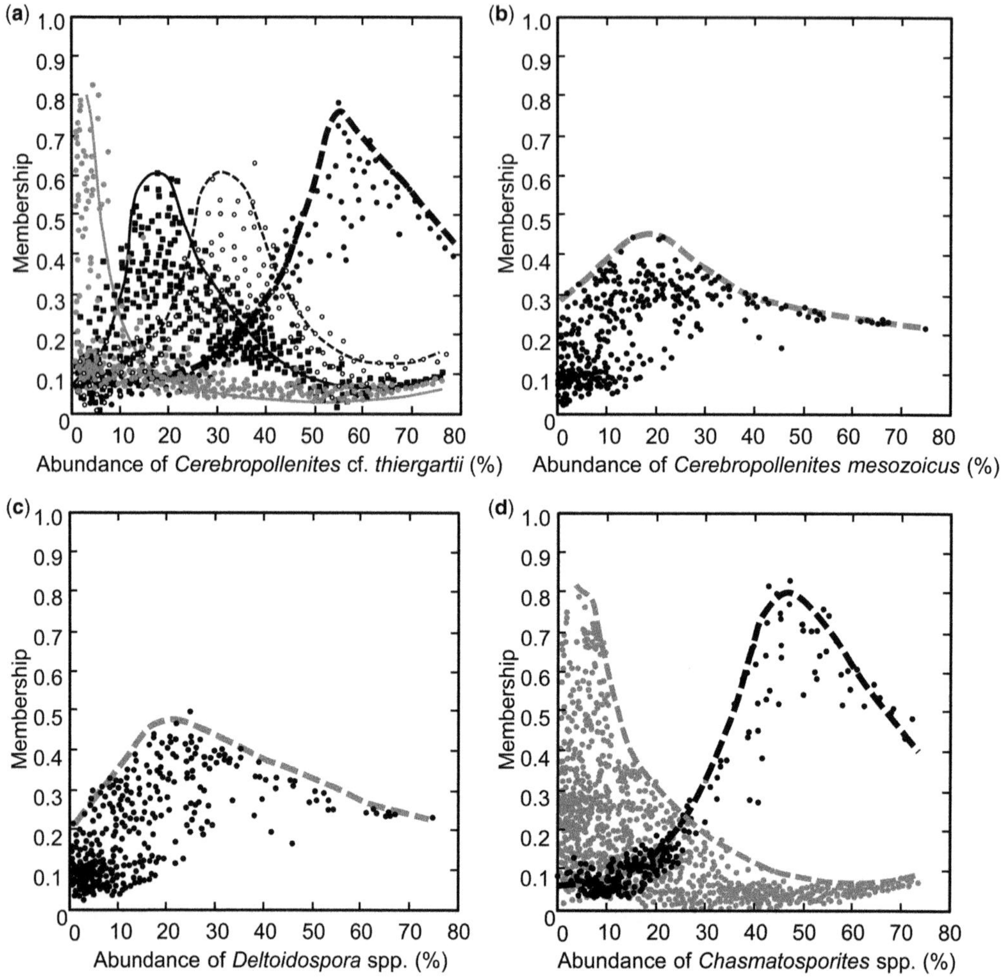

Fig. 7. (B2). Membership curves (% taxon occurrence). (**a**) *Cerebropollenites* cf. *thiergartii* against its eponymous cluster (black), the *Baculatisporites*/*Cerebropollenites* cluster (open circles), the *Baculatisporites* spp. cluster (black squares) and the *Chasmatospotites* spp. cluster (grey circles). Note that the gradation between these clusters is probably related to a palaeoenvironmental gradient (see text). (**b**) *Cerebropollenites mesozoicus* against its eponymous cluster. (**c**) *Deltoidospora* spp. against its eponymous cluster. (**d**) *Chasmatosporites* spp. against its eponymous cluster (black) and all other clusters (grey).

drier conditions. This cluster is related to the following two FCM clusters (Fig. 7a).

- The *Baculatisporites* cluster. The Osmundacean fern spore *Baculatisporites* spp. ($r = 0.77$; 28.5% of the cluster, Table 3) indicates warm and wet lowland marsh conditions. *Baculatisporites* are accompanied by spores from the filicopsid marsh/swamp ferns *Biretisporites* spp. ($r = 0.44$) and *Dictyophyllidites* spp. ($r = 0.35$) and the swamp cycad spore *Cycadopites* spp. ($r = 0.28$). Spores from the bryophyte *Stereisporites* spp. ($r = 0.23$) indicate the presence of some freshwater marshes.
- The *Baculatisporites*/*Cerebropollenites* cluster. *Cerebropollenites* cf. *thiergartii* has an r value of 0.49 and *Baculatisporites* spp. an r value of 0.52. This cluster is clearly transitional between the *C.* cf. *thiergartii* and *Baculatisporites* clusters (Fig. 7a) with an increase in marsh conditions, e.g. *Stereisporites* spp. has an r value of 0.11 compared with an r value of -0.02 in the *C.* cf. *thiergartii* cluster.
- The *Cerebropollenites mesozoicus* cluster. The lowland plain conifer pollen *C. mesozoicus* dominates this cluster ($r = 0.66$, Fig. 7b; *c.* 21% of the cluster, Table 3) with the beach/back

mangrove arborescent filicopsid fern spore *Deltoidopsora minor* ($r = 0.48$; *c.* 23% of the assemblage, Table 3), the forest mire inhabiting club moss spore *Retitriletes* spp. ($r = 0.44$; *c.* 11% of the assemblage, Table 3) and the coastal conifer pollen *Classopollis* ($r = 0.36$; 14% of the assemblage). The coastal conifer pollen *Callialasporites turbatus* has a reasonable positive correlation with the cluster ($r = 0.34$), but is only a minor component of the biofacies (<1%, Table 3), as is the case with the lowland fern spore *Striatella* spp. ($r = 0.33$ and 1.2% of the assemblage). Negatively correlated taxa are *C.* cf. *thiergartii* ($r = -0.68$) and *Baculatisporites* spp. ($r = -0.47$). This cluster is interpreted as being from a coastal, possibly beach front, forest with some understorey marshes and was probably climatically drier than the *Baculatisporites* cluster. Table 3 clearly reveals this change in dominant taxa compositions between these three FCM clusters (Fig. 7a).

- The *Deltoidospora* cluster. This is similar to the *C. mesozoicus* cluster in that it is also dominated by *C. mesozoicus* ($r = 0.61$; *c.* 21% of the assemblage) and the fern spore *D. minor* ($r = 0.40$ Fig. 7c; 20% of the assemblage). Accessory taxa also include *Retitriletes* spp. ($r = 0.45$), *Classopollis* ($r = 0.45$), *C. turbatus* ($r = 0.35$), *Striatella* spp. ($r = 0.35$). Negatively correlated taxa are *C.* cf. *thiergartii* ($r = -0.71$) and *Baculatisporites* spp. ($r = -0.43$). The cluster is interpreted as being a slightly drier part of the coastal forest than the *C. mesozoicus* cluster due to the increase in the conifer pollen *Classopollis* spp. and the spore *Striatella* spp. from a lowland fern preferring warm dry conditions.
- The *Chasmatosporites* cluster. The cycad spore *Chasmatosporites* spp. dominates this cluster ($r = 0.90$; *c.* 45% of the assemblage, Table 3) and is negatively correlated with the other five clusters (Fig. 7d). The parent cycad of *Chasmatosporites* probably formed the upper canopy of a mire setting and dominated the palynoflora at the top of the Cook Formation and the base of the Drake Formation. Modern cycads dominate in subtropical to tropical conditions and so a warm wet humid palaeoenvironment is suggested. Accessory taxa include *Classopollis* spp. ($r = 0.16$). Although *D. minor* does not appear to be significant ($r = 0.00$), it does make up *c.* 15% of the cluster assemblage with the Chleirolepidacean conifer pollen *Classopollis* at *c.* 12%, indicating some coastal forest component to this biofacies (Table 3). The marsh fern spore *Baculatisporites* spp. and conifer pollen *C.* cf. *thiergartii* are moderately strongly negatively correlated ($r = -0.35$ and -0.40, respectively).

The six clusters have been arranged to reflect the relationships between the clusters and to enable palaeoenvironmental trends to be elucidated (Table 3, Fig. 8). The six clusters also reflect the stratigraphy of the succession being studied. The *C.* cf. *theiragtii* cluster dominates at the base of the Cook Formation (the 34R bioevent of Charnock *et al.* 2001), while the *Chasmatosporites* cluster dominates in the lowermost Drake and uppermost Cook formations (the 63 bioevent of Charnock *et al.* 2001); see Figure 2.

There are some clear differences in the stratigraphic development of the FCM clusters in a geographical sense (Fig. 9). The *C.* cf. *thiergartii* cluster is well developed in the Amundsen Formation and the lowermost part of Sand 1 in the Lower Cook, apart from 34/3-A-1 H, 34/3-B-4 H and 34/3-B-1 H. It is variably developed in the remainder of Sand 1 in all the study wells and is dominant throughout Sand 1 in 34/3-3A and 34/3-A-4 H and, to a lesser extent, in 34/3-3S. Numbers of *Botryococcus* vary greatly within this interval, indicating a variable riverine input. Overall, Sand 1 appears to be highly variable both stratigraphically and geographically, although the greatest dominance of the *C.* cf. *thiergartii* cluster is in the two most northerly study wells 34/3-3A and 34/3-A-4 H. The *Chasmatosporites* cluster is well developed in the lowermost Drake Formation in the Knarr Vest segment, with only a weak presence within the Cook Formation (e.g. 34/3-A-2 H). However, in the Knarr Sentral area the cluster is also evident immediately above the J15 MFS and is dominant in 34/3-B-1 H and 34/3-B-4 H. This is interpreted as a proximal (Sentral) to distal (Vest) trend that was probably enhanced by wave action in the Upper Cook, restricting spore distribution and keeping it closer to shore. This distribution also indicates that an increase in humid conditions started just below the J15 MFS and peaked at around the J18 MFS (e.g. 34/3-B-1 H) (Fig. 9).

Synthesis

In the Knarr Field the Cook Formation is informally divided into the Lower Cook and Upper Cook, with the Lower Cook consisting of Sands 1, 2 and 3 and the Upper Cook consisting of Sands 4 and 5. Sands are separated by intraformational mudstones, which display remarkably consistent thicknesses across the field. Individual sands also display a notably consistent thickness (Fig. 10) and are described in the following sections.

Sand 1

The base of Sand 1 is generally sharp (Fig. 10) and is interpreted to represent erosion onto the underlying

Table 2. *Botanical affinity and palaeoecological interpretations of the taxa used in the final iteration of the FCM model*

Taxon	Botanical affinity and palaeoenvironmental information
Auritulinasporites scanicus	Filicopsid fern spore. Lowland swamps
Baculatisporites spp.	Osmundacean fern spore; lowland marsh/river fern (Hubbard & Boulter 1997). Osmundacean ferns are considered to prefer generally drier conditions than lycopods and mosses and often form part of the understorey ground cover
Biretisporites spp.	Filicopsid fern spore. Lowland swamps
Callialasporites turbatus	Gymnosperm pollen from an Araucariacean coastal conifer. Indicative of a seasonal climate and able to withstand extended cold and dry periods (Dettmann *et al.* 1992; Abbink 1998) as well as the influence of salt wind; probably grew in coast-fringing forests (Abbink 1998; Vakhrameev 1991)
Cerebropollenites cf. *thiergartii*	Gymnosperm conifer pollen of Taxodiaceaen or possibly Pinaceaen affinity. Possibly related to the coniferophyte genus *Tsuga* (see Nielsen *et al.* 2010). Modern *Tsuga* prefers moist seasonal conditions with high rainfall, but not necessarily mire conditions. These conifers formed part of the upper canopy of a forest and may have been tolerant of dry conditions (Barrón *et al.* 2006)
Cerebropollenites mesozoicus	Gymnosperm pollen. Conifer (Taxodiaceae or Pinaceae?). Abbink (1998) viewed the parent plant as being a pioneer form that inhabited post-regression lowland plains
Chasmatosporites spp.	Gymnosperm pollen. Possible cycad/ginkgo parent plant. Upper canopy in mire settings, although most authors consider it to be indicative of somewhat drier climatic conditions.
Classopollis spp.	Gymnosperm pollen from a Cheirolepidiaceaen conifer superficially similar to the extant family Cupressaceae. The genus has been reported in high proportions from near-coastal settings such as sandy bars or coastal islands with well-drained soils (Abbink 1998). Batten & MacLennan (1984) concluded that the parent plants of *Classopollis* were salt marsh or mangrove shrubs/trees, an argument supported and expanded upon by Riding *et al.* (2013). Vakhrameev (1981) considered the Cheirolepidiaceae to be drought-resistant, thermophilous, xerophytic shrubs and trees that resembled modern juniper bushes. It is likely that they lived in a wide variety of habitats and that they could also survive in semi-arid to arid conditions
Concavisporites spp.	Filicopsid fern spore. Possibly tolerant of brackish waters (Frederiksen 1992)
Cycadopites spp.	Gymnosperm pollen. Cycads from lowland freshwater swamps, possibly drought tolerant (Abbink 1998; Barrón *et al.* 2006)
Deltoidospora australis	Spores from filicopsid arborescent ferns that were probably several metres tall. Probably has an affinity to the genera *Dicksonia* and *Coniopteris.* Similar to the living beach fern *Acrostichum aureum*. Found in mangroves, inner/proximal elevated areas and palm sands with reduced salinity brackish water
Deltoidospora minor	As above. Dominant in more proximal parts of the mires
Densoisporites crassus	Known *in situ* spore from a Pleuromeiaceae heterosporous lycopsid with a close relationship to modern Isoetaceae (Raine *et al.* 1988). The parent plants appear to have been salt tolerant and probably formed the groundcover in coastal, tidally influenced mires (Abbink 1998). Lycopsids often favour humid acidic soils
Dictyophyllidites spp.	Filicopsid fern spore from forest swamps
Exesipollenites scabratus	Gymnoperm pollen. Conifer, Cupressaceae/Taxodiaceae. Forms upper canopy in mires
Kraeuselisporites reissingeri	Spore. Lycophytes, Selaginellaceae. Coastal swamps or ponds with humid soils (Barrón *et al.* 2006)
Lycopodiacidites rugulatus	Spore. Lycophyte, club mosses. Lowland freshwater marsh close to rivers (Hubbard & Boulter 1997)
Quadraeculina anellaeformis	Gymnosperm pollen from a Podocarpaceaen conifer. Possibly xerophilous and restricted to (sub)tropical areas. Generally placed in a dry upland habit (Vakhrameev 1991) so they have probably been blown into the Knarr area

(*Continued*)

Table 2. *Continued*

Taxon	Botanical affinity and palaeoenvironmental information
Retitriletes spp.	Spore. Lycopodiaceae. Club mosses that inhabited forest mires (Nielsen *et al.* 2010). Lycophytes require humid conditions and their high water requirements meant that they grew on perennially or seasonally wet substrates in abandoned channels and wet flood-basin settings
Spheripollenites spp.	Gymnosperm conifer pollen. Taxodiaceaen = cypresses. Swamp cypress trees can grow in standing water (Nielsen *et al.* 2010). Inhabit coastal forests and parts of environments comparable with mangroves in a humid environment (Barrón *et al.* 2010)
Stereisporites spp.	Spore. Sphagnaceae, Bryophyte. Sphagnum swamp mosses. Lowland freshwater marsh, swamp mosses, near river banks or coastal swamps (Barrón *et al.* 2006)
Striatella spp.	Filicopsid fern spore. Lowland ground cover. Warm and dry conditions (Abbink 1998; Nielsen *et al.* 2010)
Todisporites major	Filicopsida, Osmundaceae, fern spore. Coastal swamps or ponds with humid soils (Barrón *et al.* 2006)

Predominantly from Abbink (1998) and references cited therein, but see also Vakhrameev (1981, 1991), Batten & Maclennan (1984), Balme (1995), Hubbard & Boulter (1997), Ulicý *et al.* (1997), Shrank & Mahmoud (2002), Oosting (2004), Barrón *et al.* (2006, 2010), Stefanowicz (2008), Nielsen *et al.* (2010), Wagstaff *et al.* (2013) and Krupnik *et al.* (2014).

fine-grained shallow shelf sediments of the Amundsen Formation. The Amundsen Formation is dominated by the *C.* cf. *thiergartii* and *Baculatisporites/Cerebropollenites* clusters, with intervals where the freshwater alga *Botryococcus* is also abundant (Fig. 8). Small numbers of brackish-marine prasinophyte algae are recorded, along with a marine benthonic assemblage including forams (e.g. *Nodosaria metensis*, *Lenticulina varians* and *Reophax horridus*) and ostracods (*Ogmoconchella danica*, *Ogmoconcha amalthei* and *Ogmochoncha contractula*). The Amundsen Formation was deposited in shallow-marine waters with riverine input close to the Knarr Field location.

No core has been taken within Sand 1; however, it is interpreted to be dominated by tidal channel complex sediments based on its well log character and association with the overlying cored sands. The eastern part of the area (Knarr Sentral) is dominantly composed of tidal channel complex sediments; these transition to subtidal bar complex sediments to the west, which represents a basinwards direction (Fig. 11). Further west the tidal sandstones grade into muddy deposits and are interpreted from well logs as distal shelf areas. Intertidal to supratidal deposits are not observed in the Knarr area; however, they are predicted further to the east towards the hinterland. Sediments at the base of the Cook Formation record a rapid and consistent decrease in the presence of the *C.* cf. *thiergartii* cluster in the middle of Sand 1 and an increase in the *Baculatisporites* spp. cluster, although the transitional *Baculatisporites/Cerebropollenites* biofacies dominates. This is interpreted as indicating a wetter, marshier hinterland to the marginal-marine depositional setting of the Cook Formation. Small amounts of the coastal forest biofacies are present, as are small numbers of brackish-marine actritarchs and prasinophyte algae. The top of Sand 1 is marked by a laterally continuous mudrock interval that corresponds to the J14 MFS; see discussion in Appendix A.

Sand 2

Sand 2 overlies J14 MFS and consists of a relatively thick, high net-to-gross interval that is interpreted as tidal channel complex sediments grading westwards into subtidal tidal sand bars (Fig. 11) and is extensively cored in wells 34/3-1S and 34/3-2S. The lower sections of Sand 2 are locally composed of tidal bar complex sediments and proximal intertidal bar complex facies, where relatively high gamma ray intervals towards the base of the interval are interpreted as intertidal bars, although this interval is not cored (Fig. 10). The distal mudstones west of the Knarr Field (Blåbær) are interpreted as extending closer to the Knarr area than in Sand 1 due to a slightly greater portion of subtidal tidal sand bar facies in Sand 2 than in Sand 1; this was potentially in response to a decrease in the amount of sediment supply. Across the field Sand 2 displays notable heterogeneity, with individual sands appearing to pinch-out between wells (Fig. 10), reflecting the complex mosaic of facies observed and the dynamic nature of the depositional system. Conditions appear to have become wetter and more humid during Sand 2 deposition compared with Sand 1; the dominance of the *Baculatisporites* spp. cluster over the *Baculatisporites/Cerebropollenites* cluster occurs at the top of Sand 2. There is

Table 3. *Centroid percentage compositions for the six fuzzy c-means (FCM) clusters*

Taxa	*Chasmatosporites*		*Cerebropollenites mesozoicus*		*Deltoidospora*		*Baculatisporites*		*Baculatisporites* and *Cerebropollenites*		*Cerebropollenites* cf. *thiergartii*
Auritulinasporites scanicus	0.20		0.20		0.20		0.04		0.07		0.12
Baculatisporites spp.	6.57	→	8.93	→	10.34	→	28.51	←	22.80	←	14.96
Biretisporites spp.	0.00		0.02		0.03		0.32		0.06		0.01
Callialasporites turbatus	0.04		0.83		0.82		0.09		0.05		0.01
Cerebropollenites cf. *thiergartii*	3.97	→	4.44	→	5.47	→	18.82	→	30.53	→	52.36
Cerebropollenites mesozoicus	6.97	→	20.74		18.50	←	5.33	←	3.94	←	2.25
Chasmatosporites spp.	45.46	←	8.28	←	9.41	←	7.65	←	5.82	←	5.01
Classopollis spp.	12.64	→	13.66	→	14.39	←	10.45	←	7.70	←	3.63
Concavisporites spp.	0.00		0.00		0.00		0.00		0.00		0.00
Cycadopites spp.	0.35		0.24		0.30		0.76		0.46		0.25
Deltoidospora australis	0.27		0.49		0.63		2.25		0.51		0.16
Deltoidospora minor	14.76	→	22.62		20.29	←	11.34	←	14.68	←	10.76
Densoisporites crassus	0.06		0.11		0.13		0.30		0.41		0.44
Dictyophyllidites spp.	0.01		0.02		0.04		0.25		0.03		0.00
Exesipollenites scabratus	0.04		0.20		0.20		0.35		0.06		0.01
Kraeuselisporites reissingeri	0.02		0.04		0.07		0.27		0.18		0.15
Lycopodiacidites rugulatus	0.16		0.16		0.18		0.37		0.51		0.74
Quadraeculina anellaeformis	1.08		2.07		2.09		1.14		1.16		1.43
Retitriletes spp.	3.16	→	10.84		10.51	←	6.85	←	7.00	←	4.98
Spheripollenites spp.	3.13		4.36		4.61		3.32		2.72		1.86
Stereisporites spp.	0.49		0.53		0.58		1.01		0.78		0.59
Striatella spp.	0.63		1.21		1.19		0.45		0.44		0.28
Todisporites major	0.01		0.01		0.02		0.14		0.08		0.01

Trend lines are shown indicating interpreted palaeoenvironmental gradients. The order of the clusters also represents the approximate stratigraphic position within the Knarr Field. The *C.* cf. *thiergartii* cluster dominates in the upper Amundsen and lowermost Cook formations, while the *Chasmatosporites* spp. cluster dominates in the Upper Cook and lower Drake formations.

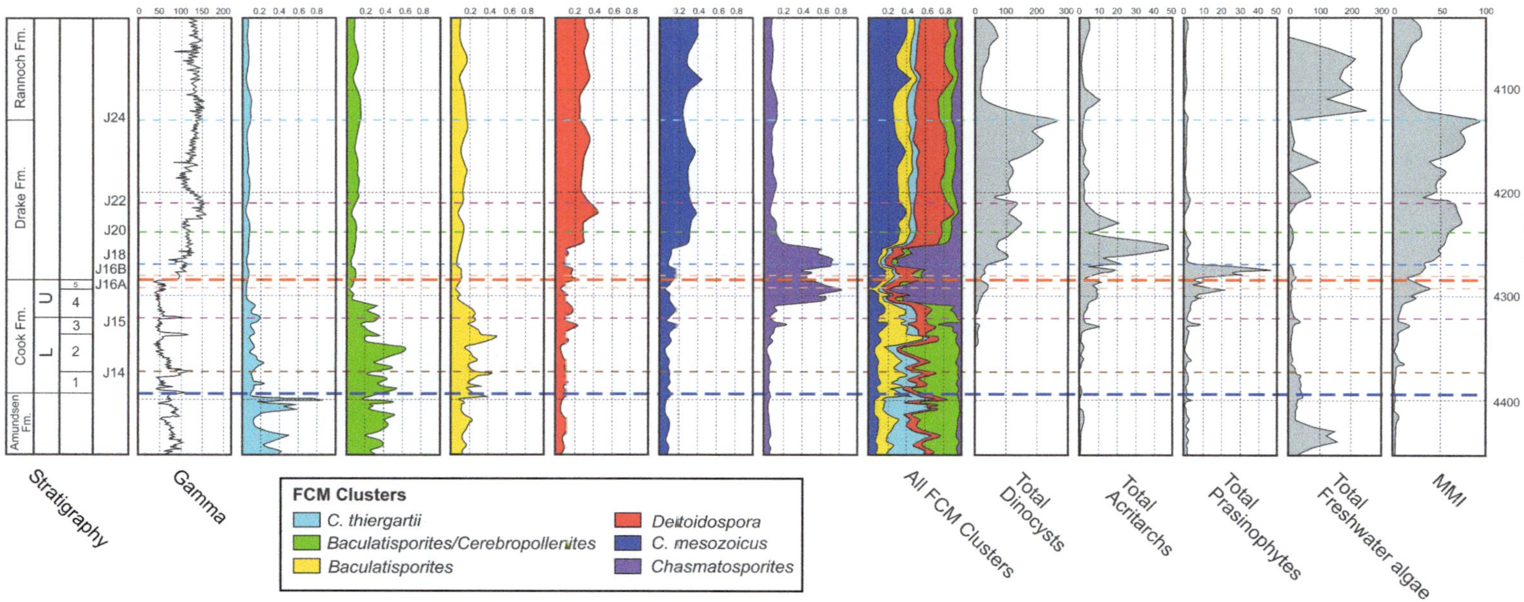

Fig. 8. Plot showing the fuzzy c-means (FCM) clusters and additional key palaeoenvironmental signals for the 34/3-B-1 H well. This is used as a standard well for the area. The marine influence index (MII) is calculated as the total marine specimens divided by the total palynoflora, but without the buoyant easily transported bisaccate pollen. This takes into account the Neves effect of Chaloner & Muir (1968), whereby the maximum occurrence of bisaccate pollen is bivariate; maxima occur both in nearshore and deep basin positions, where it is related to sea-level and marine transgression. Depths shown are measured depths below Kelly Bushing.

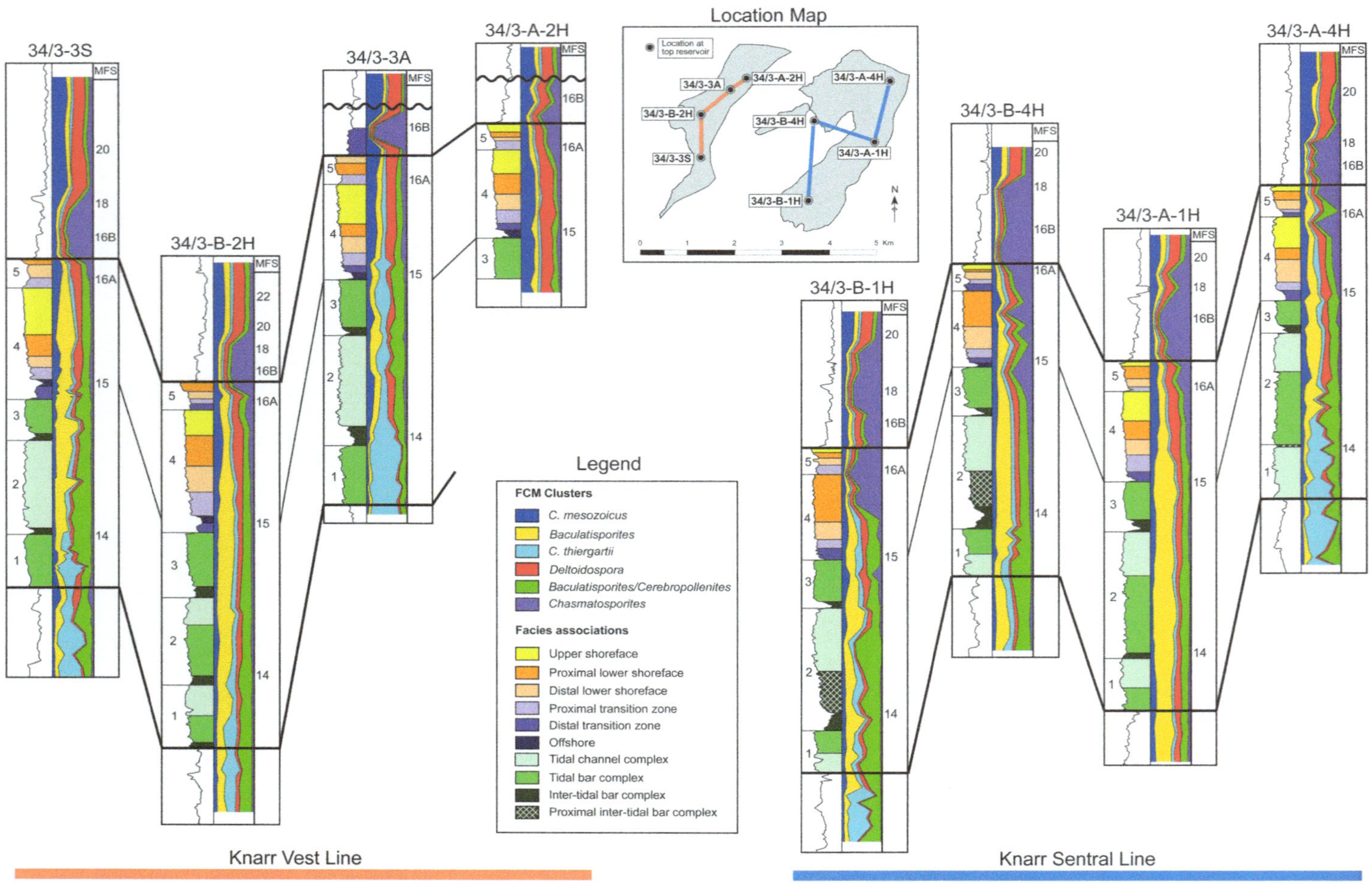

Fig. 9. Fuzzy c-means (FCM) clusters for each Knarr Field well plotted geographically to elucidate palaeoenvironmental trends and groupings. Two correlation lines are shown over Knarr Vest and Knarr Sentral. Depths shown are measured depths below Kelly Bushing.

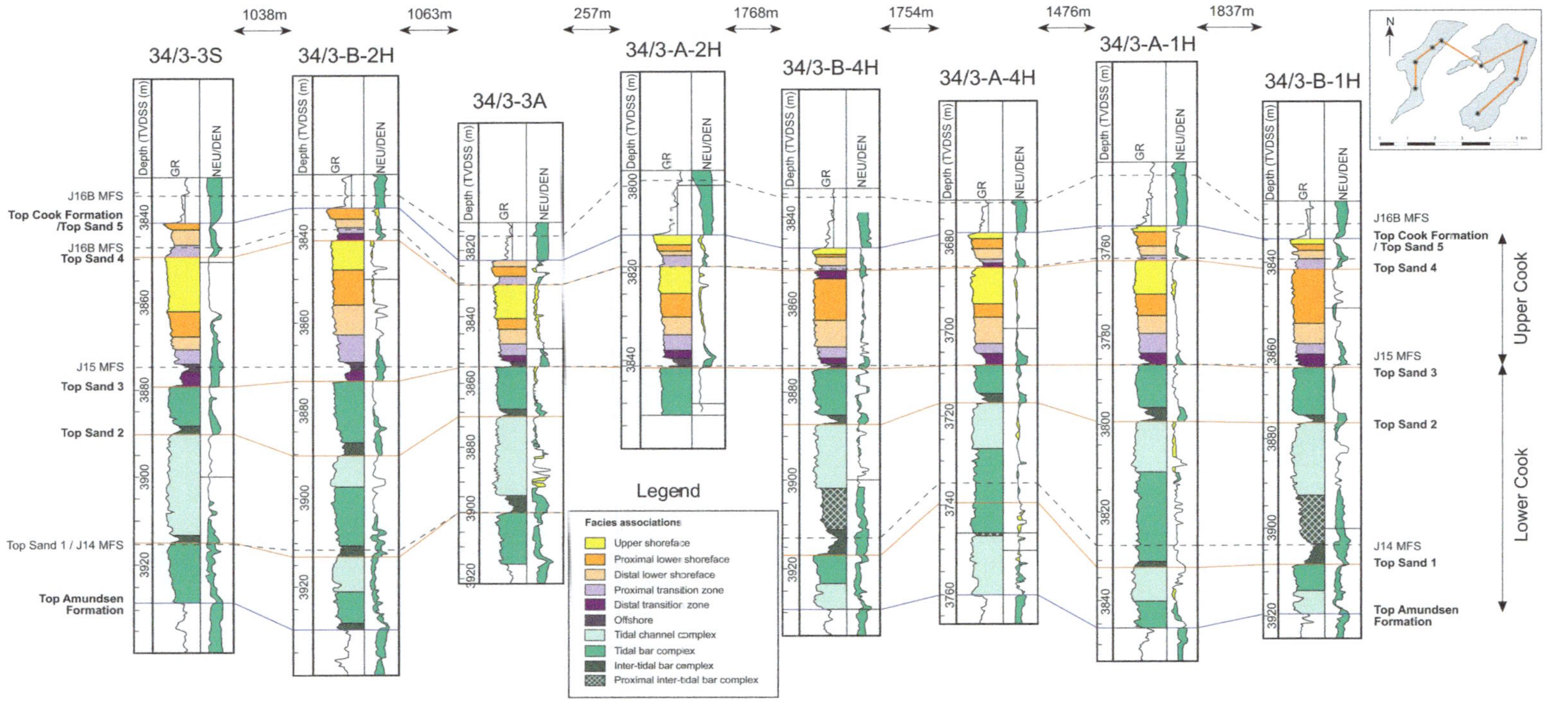

Fig. 10. The architecture of the Cook Formation within the Knarr Field is defined by a series of field-wide flooding surfaces that display notable consistency and where the flooding surface capping Sand 3 corresponds to the J15 MFS, which is recognized regionally. The Lower Cook Formation is dominated by tidally influenced sediments, which contrasts with the Upper Cook Formation that represents a wave-dominated shoreline.

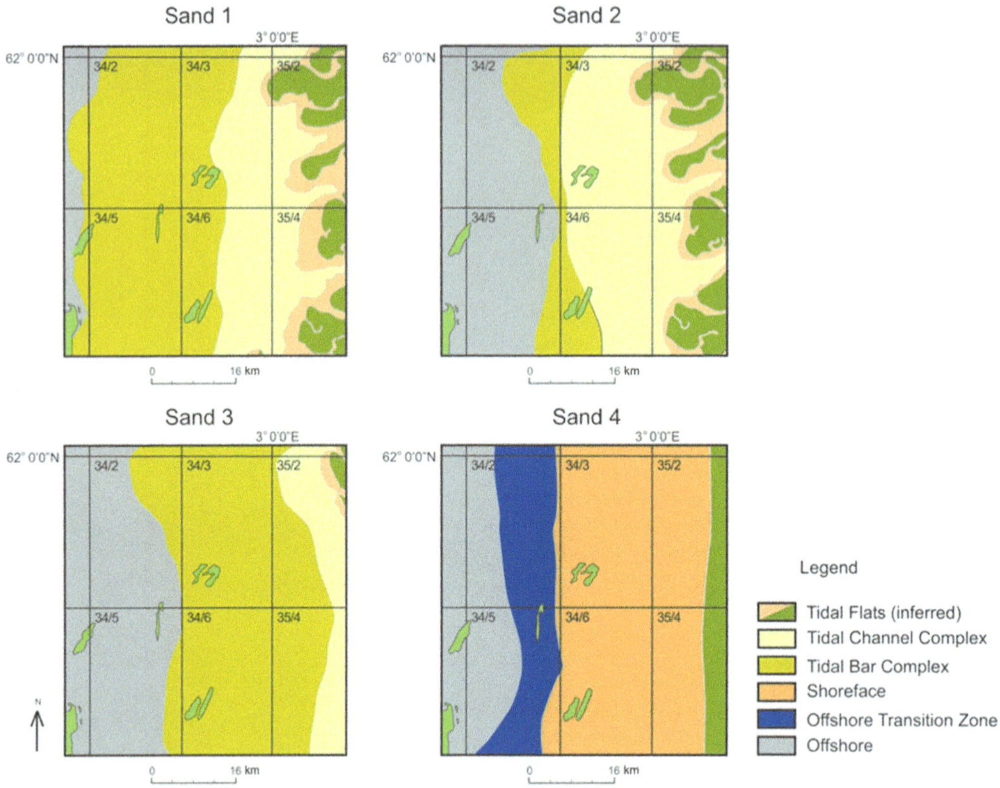

Fig. 11. During the deposition of Sand 1, the Knarr Field area was dominated by tidal sand bars. The deposition of Sand 2 represents a slight regression with the progradation of tidal channel facies belts into the field area. The relatively offshore tidal sand bar facies belts appear relatively thin at this stage, with subtidal/offshore facies recognized within the Blåbær discovery. Sand 3 then records a transgressive event with a return to tidal sand bar deposition. The J15 MFS marks a significant change in deposition with the establishment of a shoreline system that records the onset of transgression, which culminates at the J16B maximum flooding surface.

a minor increase in the coastal forest biofacies to >20% of the assemblage as compared, in places, with 5–10% at the base of the Cook Formation. However, the depositional system remains highly variable, with locally intense *Taenidium* bioturbation and *Botryococcus* influxes reflecting significant fresh water input.

Sand 3

This sand displays the most consistent character within the Lower Cook across the field with a strong cleaning-upwards profile (Fig. 10) and coarsening-upwards succession in core and is dominated by tidal bar complex facies. The sand is locally variable; it thins to the east and SE of the field and is more heterolithic in Knarr Sentral than in Knarr Vest. An increasing marine signature is expressed within the depositional facies of Sand 3. Tidal bar complex facies form a consistent package across the field, but variations in bioturbation reflecting varying salinity and tidal influence. The top of Sand 3 is marked by the J15 MFS, which is consistently recognized across the field (Fig. 10).

Sand 4

A major change in deposition occurs at the J15 MFS in the Knarr Field. Depositional cycles older than the J15 MFS are characterized by tide-dominated sediments, whereas younger sediments represent a shoreline-to-shelf depositional system. Sand 4 is interpreted, from extensive core, as a coarsening-upwards shoreface succession, although wave-dominated local evidence of channels cutting the upper shoreface may suggest a degree of reworking. The transition zone and lower shoreface sediments are notably heterolithic compared with the clean and medium-grained sandstones of the upper shoreface deposits.

The presence of dinoflagellate cysts increases towards the top of the Cook Formation. These include *Luehndea spinosa*, *Mancodinium semitabulatum* and small numbers of *Nannoceratopsis* taxa, all of which can tolerate marginal-marine settings, although the latter could have been caved downhole from the Drake Formation. The marine influence index increases from 5 to 10% just below the J15 MFS to *c.* 40% at the J16A MFS (Fig. 8). The number of brackish-marine to marine-tolerant acritarchs and prasinophyte algae mirrors the increase in dinoflagellates, indicating a far more marine-dominated depositional environment, although variable short-lived abundances of *Botryococcus* with Cook Formation preservation style (see Appendix A) indicate that small rivers were still emptying into the area.

The *Chasmatosporites* cluster becomes more numerous immediately above the J15 MFS and is dominant in wells 34/3-B-4-H, 34/3-B-1 H and 34/3-B-4 H, but in none of the Knarr Vest wells. This is interpreted as a proximal (Knarr Sentral) to distal (Knarr Vest) trend that was probably enhanced by wave action in the Upper Cook, restricting the spore distribution and keeping it closer to shore. This distribution also indicates that an increase in humid conditions started just below the J15 MFS and peaked around the J18 MFS, e.g. 34/3-B-1 H (Fig. 8). Sand 4 is capped by the J16A MFS and represents the end of the second Cook regressive–transgressive cycle (Marjanac & Steel 1997; Fig. 3).

Sand 5

The next regressive–transgressive cycle (cf.Marjanac & Steel 1997; Fig. 3) is composed of Sand 5 in the Knarr Field and is interpreted as lower to upper shoreface sediments. It is relatively thin and well logs indicate a fining-upwards character into the J16B MFS (Fig. 10), suggesting the main depocentre had shifted away from the Knarr Field. Indeed, upper shoreface deposits are only recognized in the eastern parts of the field, suggesting a continued east to west depositional gradient. The top of the Cook Formation and the base of the Drake Formation are marked by a major marine transgression, although there is still a marked hinterland palynological signature. Local brackish conditions are indicated by the short-lived appearance of abundant *Solisphaeridium* spp., whereas the *Chasmatosporites* spp. cluster dominates this interval across the entire Knarr area, suggesting a far more humid palaeoenvironment.

It is also interesting to note that, immediately above the J22 MFS in the Knarr area, there is a field-wide siltier package clearly evident on the gamma ray log that records a regressive–transgressive cycle younger than those discussed by Marjanac & Steel (1997). No sand is seen in the Knarr Field over this interval, although the palaeontological data indicate a freshwater influx into the area at this time (Fig. 8).

The integration of the sedimentology and biostratigraphy described here enabled the establishment of a robust reservoir zonation that has been utilized during the development and ongoing exploitation of the Knarr Field. Although the use of routine and novel techniques creates a detailed picture of the subsurface architecture, the intricate mosaic of facies is likely to produce unforeseen complexity and the subsurface view of the Knarr Field will be continually reviewed and updated.

Conclusions

The Cook Formation within the Knarr Field records an evolving Late Pliensbachian depositional system. The formation is informally divided into the Lower Cook and Upper Cook by the J15 MFS and a field-wide interval of more offshore deposits. The Lower Cook is divided into Sands 1, 2 and 3 and consists of a complex mosaic of tide-dominated facies; the Upper Cook is divided into Sands 4 and 5, which are composed of shoreline-to-shelf deposits. Consequently, the J15 MFS reflects a major change in depositional system and reflects the gradual transgression of the system, which is finally drowned at the J16B MFS in the Knarr area. Close integration of sedimentological and biostratigraphic data has provided valuable insights into the architecture of the reservoir and allowed the production of a robust zonation and layering scheme, which is currently used to guide the development of the field.

The authors thank Knarr Field partners BG Group, Idemitsu, DEA and Wintershall for giving permission to publish this work. A special thank you is extended to Michael Oates for his guidance in understanding the well data within the context of the field. We acknowledge the contribution of Nick Butler, John Gregory (PetroStrat Ltd) and Stan Duxbury (Duxbury Stratigraphic Consultants) for their consistent biostratigraphic data and discussion. We also thank Ichron, part of the RPS Group, and Robertson, part of CGG GeoConsulting, for their sedimentological support. The constructive comments of Mike Martin (MMGeos) are also acknowledged. BG Group is a subsidiary of Royal Dutch Shell plc.

Appendix A

The J24 MFS and bioevent 70R of Charnock *et al.* (2001) have been placed within the Aalenian stage (probable Murchisonae Ammonite Zone) following Butler *et al.* (2005). This is contrary to Charnock *et al.* (2001), who placed this bioevent and MFS at the top of the Aalenian.

The top of abundant and diverse Phallocystaceae event (bioevent 64R) has been broadly used by Charnock *et al.*

(2001) to define the basal lower Aalenian/top upper Toarcian. However, Riding (1984) and Butler *et al.* (2005) noted that this event occurs throughout the Concavum Ammonite Zone of the lower Aalenian in Yorkshire, UK.

The influx of common/abundant *M. semitabulatum* occurs between the J16A and J16B MFSs. It is then rare until numbers increase again below the J22 MFS. This reduction in numbers appears to be Europe-wide and is controlled by the increased availability of nutrients (Bucefalo Palliani & Riding 1999*a*, *b*; Bucefalo Palliani *et al.* 2002). This is expressed by the dominance of prasinophyte algae and Sphaeromorph acritarchs (the 60R bioevent of Charnock *et al.* 2001) in the Lower Toarcian, culminating with the development of anoxic conditions at the J18 MFS.

The influx of common/abundant *Solisphaeridium* spp. occurs immediately above the top of the Cook Formation in the Knarr Field and is almost certainly palaeoenvironmentally mediated; it prefers brackish waters.

Within the Cook Formation biostratigraphic control in a regional sense is poor. Charnock *et al.* (2001) noted common *Botryococcus* spp. (the 44R bioevent), although its presence in the Jordbær and Knarr wells was variable within Sand 4. The specimens of *Botryococcus* spp. encountered in the Cook Formation are generally more poorly preserved (abraded) than those seen in the Rannoch Formation (fresh and cauliflower-like). This distinction enables caved specimens to be excluded from counts within the Cook and Amundsen formations.

The base of the dinocyst *L. spinosa* (B55R) occurs just below the J15 MFS in Charnock *et al.* (2001). The record of this taxon within the Knarr Field reservoir sands is relatively patchy, often as single specimens, but it is generally seen just above the base of common/abundant *Chasmatosporites* spp. and the top of common/abundant *Baculatisporites* spp. at the top of Sand 3. It is used to help constrain the position of the J15 MFS. In other wells, e.g. 34/3-A-1H, the record of *L. spinosa* is very good, indicating probable variations in local abundance.

The base of the Cook Formation is constrained in Charnock *et al.* (2001) by an increase in *Botryococcus* spp. and the further increase in *C.* cf. *thiergartii* and common *Baculatisporites*/*Osmundacidites* spp. (bioevent 34R), in conjunction with the general base of *Nannoceratopsis senex* (Event 40R). In the Knarr Field, the downhole increase in *Botryococcus* spp. occurs within the basal Cook Formation (within Sand 1) and is overlain by a brackish-marine interval. Only in the basal part of Sand 1 does *C.* cf. *thiergartii* and *Baculatisporites*/*Osmundacidites* spp. increase. For this reason, we place the J14 MFS within the basal Cook Formation rather than below it as in Charnock *et al.* (2001).

Abundant *Botryococcus* spp. are always present immediately below the Cook Formation in the Knarr Field area. A local *Botryococcus* ACME is developed in the uppermost Amundsen Formation and then numbers decline downhole to the base common *Botryococcus* spp. event noted by Charnock *et al.* (2001), who place it coincident with the Pliensbachian/Sinemurian boundary. This bioevent has been used to constrain the position of the J13 MFS. The ACME of *Botryococcus* described here is separate from the original concept of the 34R bioevent.

The absence of the Upper Sinemurian index marker *Liasidinium variablie* (bioevent 30R) from this area is interpreted to be due to more marginal-marine facies that are not conducive to the proliferation of dinoflagellate cyst taxa.

References

ABBINK, O.A. 1998. *Palynological investigations in the Jurassic of the North Sea region*. PhD Thesis, University of Utrecht, LPP Contributions Series No. 8.

BALME, B.E. 1995. Fossil in situ spores and pollen grains: an annotated catalogue. *Review of Palaeobotany and Palynology*, **87**, 81–323.

BARRÓN, E., GÓMEZ, J.J., GOY, A. & PIEREN, A.P. 2006. The Triassic–Jurassic boundary in Asturias (northern Spain): palynological characterisation and facies. *Review of Palaeobotany and Palynology*, **138**, 187–208.

BARRÓN, E., SOLEDAD, U., GOY, A. & LASSALETTA, L. 2010. Palynology of the Toarcian–Aalenian Boundary Stratotype Section and Point (GSSP) at Fuentelaz (Lower–Middle Jurassic, Iberian Range, Spain). *Review of Palaeobotany and Palynology*, **162**, 11–28.

BATTEN, D.J. & GRENFELL, H.R. 1996. *Botryococcus*. *In*: JANSONIUS, J. & MCGREGOR, D.C. (eds) *Palynology: Principles and Applications*. Vol. **1**. American Association of Stratigraphic Palynologists Foundation, Dallas, Texas, 205–214.

BATTEN, D.J. & MACLENNAN, A.M. 1984. The paleoenvironmental significance of the conifer family Cheirolepidiaceae in the Cretaceous of Portugal. *In*: REIF, W.-E. & WESTPHAL, F. (eds) *Third Symposium on Mesozoic Terrestrial Ecosystems, Germany*. Attempto Verlag, Tübingen, 7–12.

BUATOIS, L.A. & MÁNGANO, M.G. 1995. The palaeoenvironmental and paleoecological significance of the lacustrine Mermia ichnofacies: an archetypical subaqueous nonmarine trace fossil assemblage. *Ichnos*, **4**, 151–161.

BUCEFALO PALLIANI, R. & RIDING, J.B. 1999*a*. Early Jurassic (Pliensbachian–Toarcian) dinoflagellate migrations and cyst paleoecology in the Boreal and Tethyan realms. *Micropaleontology*, **45**, 201–214.

BUCEFALO PALLIANI, R. & RIDING, J.B. 1999*b*. Relationships between the early Toarcian anoxic event and organic-walled phytoplankton in central Italy. *Marine Micropaleontology*, **37**, 101–116.

BUCEFALO PALLIANI, R., MATTIOLI, E. & RIDING, J.B. 2002. The response of marine microplankton and sedimentary organic matter to the early Toracian (Lower Jurassic) oceanic anoxic event in Northern England. *Marine Micropaleontology*, **46**, 223–245.

BUTLER, N., CHARNOCK, M.A., HAGER, K.O. & WATKINS, C.A. 2005. The Ravenscar Group: a coeval analogue for the Middle Jurassic reservoirs of the North Sea and offshore Mid-Norway. *In*: POWELL, A.J. & RIDING, J.B. (eds) *Recent Developments in Applied Biostratigraphy*. Micropalaeontological Society, Special Publications. Geological Society, London, 43–53.

Chaloner, W.G. & Muir, M. 1968. Spores and floras. *In*: Murchison, D. & Westall, T.S. (eds) *Coal and Coal Bearing Strata*. Elsevier, New York, 127–146.

Charnock, M.A., Kristiansen, I.L., Ryseth, A. & Fenton, J.P.G. 2001. Sequence stratigraphy of the Lower Jurassic Dunlin Group, northern North Sea. *In*: Martinsen, O.J. & Dreyer, T. (eds) *Sedimentary Environments Offshore Norway – Palaeozoic to Recent*. Norwegian Petroleum Society, Special Publications, **10**, 145–174.

Choi, K.S., Dalrymple, R.W., Chun, S.S. & Kim, S. 2004. Sedimentology of modern, inclined heterolithic stratification (ihs) in the macrotidal Han River Delta, Korea. *Journal of Sedimentary Research*, **74**, 677–689.

Clifton, E.H. 2006. A re-examination of facies models for clastic shorelines. *In*: Posamentier, H.W. & Walker, R.G. (eds) *Facies Models Revisited*. SEPM Special Publications, **84**, 293–337.

Da Gama, R.O.B.P., Lutz, B., Desjardins, P., Thompson, M., Prince, I. & Espejo, I. 2014. Integrated palaeoenvironmental analysis of the Niobrara Formation: Cretaceous Western Interior Seaway, northern Colorado. *Palaeogeography, Palaeoclimatology, Palaeoecology*, **413**, 66–80.

Dalrymple, R.W. & Choi, K.S. 2007. Morphologic and facies trends through the fluvial–marine transition in tide-dominated depositional systems: a schematic framework for environmental and sequence-stratigraphic interpretation. *Earth-Science Reviews*, **81**, 135–174.

Daly, R.J., Jolley, D.W., Spicer, R.A. & Ahlberg, A. 2011. A palynological study of an extinct arctic ecosystem from the Palaeocene of Northern Alaska. *Review of Palaeobotany and Palynology*, **166**, 107–116.

Dashtgard, S.E., Gingras, M.K. & MacEachern, J.A. 2009. Tidally-modulated shorefaces. *Journal of Sedimentary Research*, **79**, 793–807, https://doi.org/10.2110/jsr.2009.084

Dettmann, M.E., Molnar, R.E. *et al.* 1992. Australian Cretaceous terrestrial faunas and floras: biostratigraphic and biogeographic implications. *Cretaceous Research*, **13**, 207–262.

Doré, A.G., Hamar, G.P., Lilleng, T., Shaw, N.D., Skarpnes, O. & Vollset, J. 1984. Revised Jurassic lithostratigraphy of the Norwegian North Sea, northern area. *In*: Vollset, J. & Dore, A.G. (eds) *A Revised Triassic and Jurassic Lithostratigraphic Nomenclature for the Norwegian North Sea*. Norwegian Petroleum Society, Special Publications, **3**, 2–53.

Dreyer, T. & Wiig, M. 1995. Reservoir architecture of the Cook Formation on the Gullfaks field based on sequence stratigraphic concepts. *In*: Steel, R.J., Felt, V.L. *et al.* (eds) *Sequence Stratigraphy on the Northwest European Margin*. Norwegian Petroleum Society, Special Publications, **5**, 109–142.

Erbs-Hansen, D.R., Knudsen, K.L., Gary, A.C., Jansen, E., Gyllencreutz, R., Scao, V. & Lambeck, K. 2011. Late Younger Dryas and early Holocene palaeoenvironments in the Skagerrak, eastern North Atlantic: a multiproxy study. *Boreas*, **40**, 660–680.

Erbs-Hansen, D.R., Knudsen, K.L., Gary, A.C., Gyllencreutz, R. & Jansen, E. 2012. Holocene climatic development in Skagerrak, eastern North Atlantic: foraminiferal and stable isotopic evidence. *The Holocene*, **22**, 301–312.

Erichsen, T., Helle, M. & Rognebakke, A. 1987. Gullfaks. *In*: Spencer, A.M. (ed.) *Geology of Norwegian Oil and Gas Fields*. Graham & Trotman, London, 273–286.

Eynon, G. 1981. Basin development and sedimentation in the Middle Jurassic of the northern North Sea. *In*: Illing, L.V. & Hobson, G.D. (eds) *Petroleum Geology of the Continental Shelf of North Western Europe*. Heyden, London, 196–204.

Frederiksen, N.O. 1992. *Palynology of Mesozoic and Lower Tertiary Samples from Northern, Central, and Southern Pakistan*. US Geological Survey Open-File Report, 92–215.

Galloway, W.E. 1989. Genetic stratigraphic sequences in basin analysis, I. Architecture and genesis of flooding surface bounded depositional units. *American Association of Petroleum Geologists Bulletin*, **73**, 125–142.

Gary, A.C., Wakefield, M.I., Johnson, G.W. & Ekart, D.D. 2009. Application of Fuzzy C-means clustering to palaeoenvironmental analysis: example from the Jurassic, Central North Sea, UK. *In*: Demchuk, T.D. & Gary, A.C. (eds) *Geologic Problem Solving with Microfossils: a Volume in Honor of Garry D. Jones*. SEPM Special Publications, **93**, 9–20.

Gowland, S. 1996. Facies characteristics and depositional models of highly bioturbated shallow marine siliciclastic strata: an example from the Fulmar Formation (Late Jurassic), UK Central Graben. *In*: Hurst, A., Johnson, H.D. *et al.* (eds) *Geology of the Humber Group: Central Graben and Moray Firth, UKCS*. Geological Society, London, Special Publications, **114**, 185–214, https://doi.org/10.1144/GSL.SP.1996.114.01.09

Graue, E., Helland-Hansen, W. *et al.* 1987. Advance and retreat of Brent delta system, Norwegian North Sea. *In*: Brooks, J. & Glennie, K. (eds) *Petroleum Geology of North West Europe*. Graham & Trotman, London, 915–938.

Hampson, G.J. & Howell, J.A. 2005. Sedimentologic and geomorphic characterization of ancient wave-dominated shorelines: examples from the Late Cretaceous Blackhawk Formation, Book Cliffs, Utah. *In*: Bhattacharya, J.P. & Giosan, L. (eds) *River Deltas: Concepts, Models, and Examples*. SEPM Special Publications, **83**, 133–154.

Horstad, I., Larter, S.R. & Mills, N. 1995. Migration of hydrocarbons in the Tampen Spur area, Norwegian North Sea: a reservoir geochemical evaluation. *In*: Cubitt, J.M. & England, W.A. (eds) *The Geochemistry of Reservoirs*. Geological Society, London, Special Publications, **86**, 159–183, https://doi.org/10.1144/GSL.SP.1995.086.01.12

Hubbard, R.N.L.B. & Boulter, M.C. 1997. Mid Mesozoic floras and climates. *Palaeontology*, **40**, 43–70.

Johnson, S.M. & Dashtgard, S.E. 2014. Inclined heterolithic stratification in a mixed tidal–fluvial channel: differentiating tidal v. fluvial controls on sedimentation. *Sedimentary Geology*, **301**, 41–53.

Karlsson, W. 1986. The Snorre, Statfjord and Gullfaks oilfields and the habitat of hydrocarbons on the Tampen Spur, offshore Norway. *In*: Spencer, A.M. (ed.) *Habitat of Hydrocarbons on the Norwegian Continental Shelf*. Norwegian Petroleum Society. Graham & Trotman, London, 77–85.

Kirk, R.H. 1980. Statfjord field: a North Sea giant. *In*: Halbouty, M.T. (ed.) *Giant Oil and Gas Fields of the Decade: 1968–1978*. American Association of Petroleum Geologists, Memoir, **30**, 95–116.

Krupnik, J., Ziaja, J., Barbacka, M., Feldman-Olszewska, A. & Jarzynk, A. 2014. A palaeoenvironmental reconstruction based on palynological analyses of Upper Triassic and Lower Jurassic sediments from the Holy Cross Mountains region. *Acta Palaeobotanica*, **54**, 35–65.

Lervik, K.-S. 2006. Triassic lithostratigraphy of the Northern North Sea Basin. *Norwegian Journal of Geology*, **86**, 93–116.

Marjanac, T. & Steel, R.J. 1997. Dunlin Group sequence stratigraphy in the northern North Sea: a model for Cook sandstone deposition. *American Association of Petroleum Geologists Bulletin*, **81**, 276–292.

Martin, M.A., Wakefield, M.I., Macphail, M.K., Pearce, T. & Edwards, H.E. 2013. Sedimentology and stratigraphy of an intra-cratonic basin coal seam gas play: Walloon Subgroup of the Surat Basin, eastern Australia. *Petroleum Geoscience*, **19**, 21–38, https://doi.org/10.1144/petgeo2011-043

Martino, R.L. 1979. Trace fossils from marginal marine facies of the Kanawha Formation (Middle Pennsylvanian), West Virgnia. *Journal of Palaeontology*, **63**, 389–403.

Mutti, E., Rosell, J., Allen, G.P., Fonnesu, F. & Sgavetti, M. 1985. Eocene Baronia tide dominated delta-shelf system in the Ager Basin. *In*: Mila, M.D. & Rosell, J. (eds) *Excursion Guidebook. 6th European Regional Meeting*, 15–18 September 1985. International Association of Sedimentologists, Lleida, Spain, 579–600.

Nielsen, L.H., Petersen, H.I., Dybkjær, K. & Surlyk, F. 2010. Lake-mire deposition, earthquakes and wildfires along a basin margin fault: Rønne Graben, Middle Jurassic, Denmark. *Palaeogeography, Palaeoclimatology, Palaeoecology*, **292**, 103–126.

Olaussen, S., Beck, L., Fält, L.M., Graue, E., Jacobsen, K.G., Malm, O.A. & South, D. 1993. Gullfaks field – Norway East Shetland Basin, northern North Sea. *In*: Foster, N.H. & Beaumont, E.A. (eds) *Structural Traps VI*. American Association of Petroleum Geologists, Tulsa, OK, 55–83.

Oosting, A.M. 2004. *Palaeoenvironmental and climatic changes in Australia during the Early Cretaceous*. PhD Thesis, LPP Contributions Series No. 17, University of Utrecht.

Raine, J.L., De Jersey, N.J. & Ryan, K.G. 1988. Ultrastructure and lycopsid affinity of *Densoisporites psilatus* (de Jersey) comb. nov. from the Triassic of New Zealand and Queensland. *In*: Jell, P.A. & Playford, G. (eds) *Palynological and Palaeobotanical Studies in Honour of Basil E. Balme*. Association of Australian Palaeontologists Memoir, **5**, 79–88.

Reading, H.G. & Collinson, J.D. 1996. Clastic coasts. *In*: Reading, H.G. (ed.) *Sedimentary Environments; Processes, Facies and Stratigraphy*. 3rd edn. Wiley-Blackwell, London.

Riding, J.B. 1984. A palynological investigation of Toarcian to early Aalenian strata from the Blea Wyke area, Ravenscar, North Yorkshire. *Proceedings of the Yorkshire Geological Society*, **45**, 109–122, https://doi.org/10.1144/pygs.45.1-2.109

Riding, J.B., Leng, M.J., Kender, S., Hesselbo, S.P. & Feist-Burkhardt, S. 2013. Isotopic and palynological evidence for a new Early Jurassic environmental perturbation. *Palaeogeography, Palaeoclimatology, Palaeoecology*, **374**, 16–27.

Savrda, C.E., Blanton-Hooks, A.D. *et al*. 2000. *Taenidium* and associated ichnofossils in fluvial deposits, Cretaceous Tuscaloosa Formation, Eastern Alabama, Southeastern USA. *Ichnos*, **7**, 223–242.

Shrank, E. & Mahmoud, M.S. 2002. Barremian angiosperm pollen and associated palynomorphs from the Dakhla Oasis area, Egypt. *Palaeontology*, **45**, 35–56.

Smith, D.G., Reinson, G.E., Zaitlin, B.A. & Rahmani, R.A. (eds) 1991. *Clastic Tidal Sedimentology*. Canadian Society of Petroleum Geologists Memoir, **16**.

Stefanowicz, S. 2008. *Palyno-stratigraphy and palaeoclimatic analysis of the Lower–Middle Jurassic (Pliensbachian–Bathonian) of the Inner Hebrides, NW Scotland*. Master's thesis. Lund University.

Taylor, A.M. & Goldring, R. 1993. Description and analysis of bioturbation and ichnofabric. *Journal of the Geological Society, London*, **150**, 141–148, https://doi.org/10.1144/gsjgs.150.1.0141

Thomas, R.G., Smith, D.G., Wood, J.M., Visser, J., Calverley-Range, E.A. & Koster, E.H. 1987. Inclined heterolithic stratification – terminology, description, interpretation and significance. *Sedimentary Geology*, **53**, 123–179.

Ulicý, D., Kvacek, J., Svobodová, M. & Spicáková, L. 1997. High-frequency sea-level fluctuations and plant habitats in Cenomanian fluvial to estuarine succession: Pecínov quarry, Bohemia. *Palaeogeography, Palaeoclimatology, Palaeoecology*, **136**, 165–197.

Vakhrameev, V.A. 1981. Pollen *Classopollis*: indicator of Jurassic and Cretaceous climates. *The Palaeobotanist*, **28–29**, 301–307.

Vakhrameev, V.A. 1991. *Jurassic and Cretaceous Floras and Climates of the Earth*. Cambridge University Press, Cambridge.

Vollset, J. & Doré, A.G. 1984. *A revised Triassic and Jurassic lithostratigraphic nomenclature for the Norwegian North Sea*. Norwegian Petroleum Directorate Bulletin, **3**.

Wagstaff, B.E., Gallagher, S.J., Norvick, M.S., Cantrill, D.J. & Wallace, M.W. 2013. High latitude Albian climate variability: palynological evidence for long-term drying in a greenhouse world. *Palaeogeography, Palaeoclimatology, Palaeoecology*, **386**, 501–511.

Zadeh, L.A. 1965. Fuzzy sets. *Information and Control*, **8**, 338–352.

Challenges in characterizing subsurface paralic reservoir geometries: a detailed case study of the Mungaroo Formation, North West Shelf, Australia

G. HELDREICH[1]*, J. REDFERN[1], B. LEGLER[2], K. GERDES[3,4] & B. P. J. WILLIAMS[1]

[1]*School of Earth and Environmental Sciences, University of Manchester, Oxford Road, Manchester M13 9PL, UK*

[2]*Wintershall Holding GmbH, Friedrich-Ebert-Strasse 160, 34119 Kassel, Germany*

[3]*Shell International E&P, 40 Bank Street, London E14 5NR, UK*

[4]*Shell Centre for Exploration Geoscience, Applied Geosciences Unit, School of Energy, Geoscience, Infrastructure & Society, Heriot-Watt University, Edinburgh EH14, 4AS, UK*

**Correspondence: georgina_heldreich@hotmail.co.uk*

Abstract: Prediction of sandstone body dimensions within paralic depositional systems is crucial for the development of predictive 3D reservoir models. Continental-scale paralic reservoir targets have complex architectures, with interpretation often further compounded because they are often located at subsurface depths ≥2 km in pre- and synrift basinal settings, typified by poor seismic resolution. As such, in many cases analysis relies on core and wireline-log data, from which depositional facies are interpreted and thicknesses of sandstone reservoir units measured. Estimation of width:thickness (W:T) ratios for different reservoir elements relies on analogue data. However, the inherent uncertainty in the initial interpretation of core/log data and the wide W:T ranges for different sandstone bodies within published analogue datasets hinders prediction of accurate reservoir geometry/dimensions. Without access to quality seismic data, constraining the evolution and dimensions of reservoirs in these depositional systems is challenging.

This paper presents a detailed case study of the Triassic Mungaroo Formation, and assesses the uncertainty and limitation of interpretations that can be made about reservoir architecture and the evolution of a paralic depositional system, if only wireline and core data are available.

The Mungaroo Formation is characterized by upper and lower delta-plain channel sandstones, swamps and restricted embayments, through to delta-front and pro-delta heterolithics. Core-to-wireline-log calibration allowed identification of key marine intervals that enabled well correlations to be established across the study area, based on candidate flooding surfaces. Applying classic sequence stratigraphic models to low gradient fluvio-deltaic systems is difficult, due to a lack of preservation and/or confident identification of laterally continuous chronostratigraphic markers. The large-scale temporal change in reservoir architecture was likely to have been controlled by eustatic sea level causing overall transgression of the depositional system. However, the observed complex spatial facies variability is most likely controlled by climatic-induced changes in sediment supply/fluvial discharge and autocyclic processes.

Four classes of sandstone bodies/reservoir elements have been identified from core and wireline data based on their thickness distributions. Whereas core to wireline to seismic calibration has enabled large-, medium- and small-scale geobodies to be identified representing fluvial channel belt complexes; multistorey channel belts; and single-storey channel belts, respectively. To predict channel body widths and sinuosity from thickness data extracted from wells, typically, analogue data are used. However, for each geobody type there is a large range of possible sandstone body dimensions based on published literature. The largest-scale sand bodies, with most significance as reservoir targets, have possible interpretations as either incised valleys or amalgamated channel belts, based on their thickness ranges alone. This poses significant uncertainty for understanding the evolution of the depositional system and input into predictive reservoir models.

This study emphasizes the importance of understanding the range of uncertainty of interpretation and the need for refined analogue data to better constrain reservoir element dimensions when relying solely on well-log datasets. Where seismic attribute analysis from high-quality 3D seismic data are available, W:T dimensions for reservoir elements can be constrained more accurately and correlated to core and log data. The existing global database is limited, often poorly constrained due to the use of variable terminology and potential for misinterpretation. Many studies lack statistical rigour. We conclude that further high-resolution studies are required to build more robust and quantitative analogue datasets.

From: Hampson, G. J., Reynolds, A. D., Kostic, B. & Wells, M. R. (eds) 2017. *Sedimentology of Paralic Reservoirs: Recent Advances*. Geological Society, London, Special Publications, **444**, 59–108.
First published online May 8, 2017, https://doi.org/10.1144/SP444.13

The stratigraphically complex fluvio-deltaic Mungaroo Formation forms one of the most important reservoirs on the North West Shelf of Australia, and remains a major exploration and production target. Understanding the controls on sediment distribution and dimensions of the sandstone reservoir units are of paramount importance when it comes to developing 3D depositional and reservoir models for this succession. This study is located within the Northern Carnarvon Basin, which is approximately 500 000 km^2. We utilize a multidisciplinary dataset comprising wireline logs from 21 wells, approximately 830 m of core from eight of these wells, and 10 000 km^2 of 3D seismic data situated over an area NW of the Exmouth and Barrow sub-basins, located within the Kangaroo Syncline and part of the Exmouth Plateau. The interval of well control penetrates up to 1.4 km of the Mungaroo Formation, with core coverage restricted to discrete intervals of interest (typically focused on sand intervals for reservoir characterization and plug porosity–permeability measurements). This dataset offers a valuable opportunity to investigate the characteristic facies associations and sedimentary architecture of the subsurface Triassic Mungaroo Formation over a large area, in order to improve our understanding of the spatial and temporal trends within a low-relief fluvio-deltaic depositional system and to evaluate the active controls.

Many studies of ancient and modern fluvial systems have been focused on internal geometry and identification of architectural elements of channel bodies and associated bounding surfaces (Miall 1996; Holbrook 2001). A number of authors have also addressed the challenge of quantifying channel sandstone dimensions from core/wireline and outcrop studies (Lorenz *et al.* 1985; Robinson & McCabe 1997; Reynolds 1999; Bridge & Tye 2000; Pranter *et al.* 2009). With limited subsurface spatial resolution and inherently sparse well control, studies of channel deposits depend upon the application of suitable analogues to estimate the range of dimensions: however, there remains a considerable amount of uncertainty. Estimates of channel widths from published literature, when only sand body thickness is known from subsurface well data, have significant variability (Reynolds 1999; Gibling 2006). Bridge & Tye (2000) emphasized the drawbacks of techniques used to relate channel-bar thickness to maximum palaeochannel depth, channel width and channel-belt width.

Another area of uncertainty relates to our understanding of the relative control of eustatic, local and regional tectonism and climate on the evolution of large-scale fluvio-deltaic systems. This challenge stems from the difficulty in identifying chronostratigraphic surfaces that can be correlated confidently over hundreds of kilometres (Bhattacharya 2011). Difficulties also arise when applying sequence stratigraphic concepts and terminology to more proximal, fluvial-dominated strata which cannot be easily correlated to corresponding flooding surfaces in distal settings (Shanley *et al.* 1992). Shanley & McCabe (1994) suggested relative sea level as a key base-level control, whilst Wright & Marriott (1993) proposed the importance of water-table fluctuations. Dip-orientated models for large-scale depositional systems have focused upon stream profile as a function of the interplay between relative sea level, tectonics and sediment supply/discharge for base-level control (Holbrook *et al.* 2006; Wood 2013).

The Mungaroo Formation provides an opportunity to test the effectiveness of current sequence stratigraphic models on a low-gradient depositional system. In low-relief coastal plains, small fluctuations in relative sea level (a few metres) are likely to cause marine incursions that rapidly transgress and affect facies distributions over very large distances (of the order of tens to hundreds of kilometres). High-frequency climatic signatures are unlikely to be preserved, due to shredding of climatic signals within a large-scale sediment-transfer system (Jerolmack & Paola 2010).

Identification of incised valleys (created during relative sea-level fall), which contain a characteristic transgressive fill during relative sea-level rise, can be effective in assessing eustatic controls (Shanley & McCabe 1994; Zaitlin *et al.* 1994). However, as low-gradient depositional profiles lack a significant break in slope, the shelf will not necessarily undergo major incision, and instead may develop a series of complex diachronous terraces without preservation of clearly recognizable valley walls (Blum & Tornqvist 2000). Current sequence stratigraphic models are based on systems of $\leq$300 km in length and consider a near linear relationship between source to sink, whereby the stream gradient profile can be used to predict facies distributions and changes (Shanley *et al.* 1992; Holbrook *et al.* 2006). However application of these models to larger-scale low-gradient fluvio-deltaic systems becomes problematic. This issue is augmented by limited studies of large-scale depositional systems either in the subsurface or outcrop (Wood 2013).

Whilst core and wireline data remain invaluable in terms of understanding depositional environments, reservoir architecture and spatial/temporal evolution of depositional systems, the key uncertainties of reservoir dimensions and the relative controls on the depositional system must be recognized and understood as they have significant implications for prediction of sandstone body connectivity and in building reservoir models.

Geological setting

The Northern Carnarvon Basin is one of four major sedimentary basins which make up the North West Shelf of Australia (Longley *et al.* 2002) (Fig. 1). Covering an area approximately 500 000 km^2; the basin is dominated by NE–SW-trending sub-basins and platforms, which were major depocentres for mainly Mesozoic–Cenozoic siliciclastic sedimentary successions (reaching up to 15 km in thickness in the deepest sub-basins) (Fig. 2). During the Triassic, the Northern Carnarvon Basin was located on the NE margin of Gondwana, with siliciclastic sediments sourced from the Pilbara craton and potentially from a huge internal catchment to the north and south, being deposited in a broad and relatively stable intra-cratonic sag basin (Bradshaw *et al.* 1988; Hocking 1988; Longley *et al.* 2002). Later, more intense extension resulted in polyphase rifting and, ultimately, sea-floor spreading related to the final break-up of Gondwana (separating Australia from Antarctica and India) during the Jurassic–Cretaceous (Hocking *et al.* 1987; Hocking 1988; Metcalfe 1996, 1999; Longley *et al.* 2002). The Triassic stratigraphic succession (composed of the Chinty Formation, Locker Shale, Mungaroo Formation and Brigadier Formation) is approximately 4 km thick in inboard areas (to the SE and east) and up to 6 km outboard on the NW margin of the North Carnarvon Basin, of which up to 3 km comprises the Mungaroo Formation alone (Hocking 1988; Adamson *et al.* 2013) (Fig. 2).

The palaeoclimate during the Late Triassic (Carnian–Norian) is interpreted to have been temperate to warm, humid and monsoonal with wet and dry periods (Bradshaw *et al.* 1988; Ratcliffe *et al.* 2010; Payenberg *et al.* 2013). Adamson *et al.* (2013) and Heldreich *et al.* (2013) documented broadly east–west channel orientations, suggesting a dominant westwards transport direction of sediment within the fluvial system of the Mungaroo Formation. Heavy mineral analysis and age dating of detrital zircons from the Mungaroo Formation supports the presence of continental-scale drainage for the Triassic river systems, sourced not only from the relatively proximal Pilbara and Yilgarn cratons, onshore Western Australia, but also may have included much larger river systems that flowed along the western margin of the Australian continent, and through the infant Perth Basin, fed by systems from Greater India (Lewis & Sircombe 2013) or as far away as central/eastern Australia (Jablonski 1997). Sediment pathways may have also drained through the southern and central Australian Officer and Canning basins during the Triassic (Adamson *et al.* 2013; Lewis & Sircombe 2013).

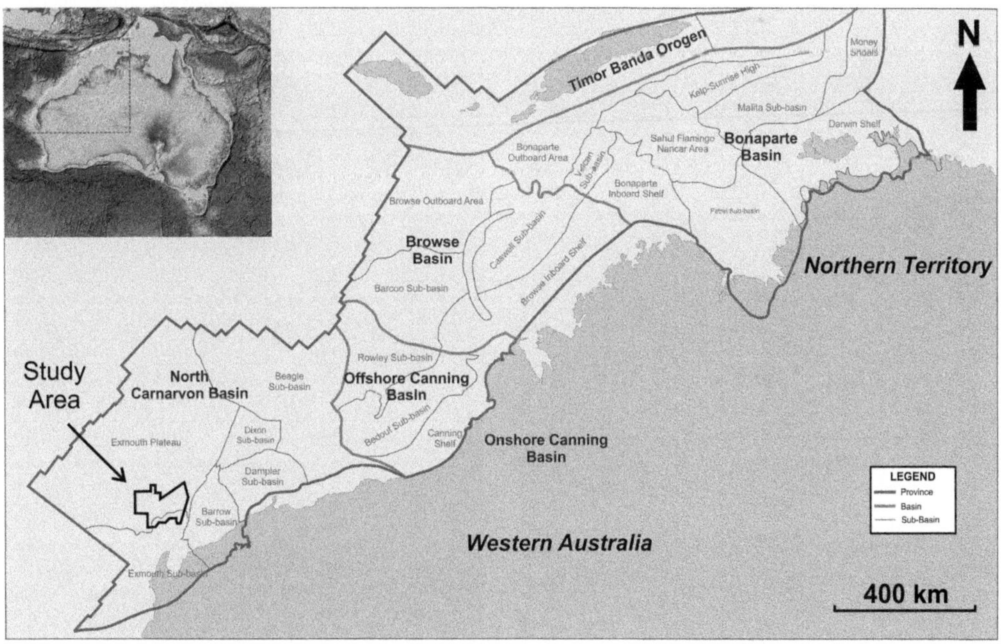

Fig. 1. Location map of the North West Shelf of Australia and its major sedimentary basins (after Longley *et al.* 2002). The study area (outlined in black) is situated within the Northern Carnarvon Basin, NW of the Exmouth and Barrow sub-basins.

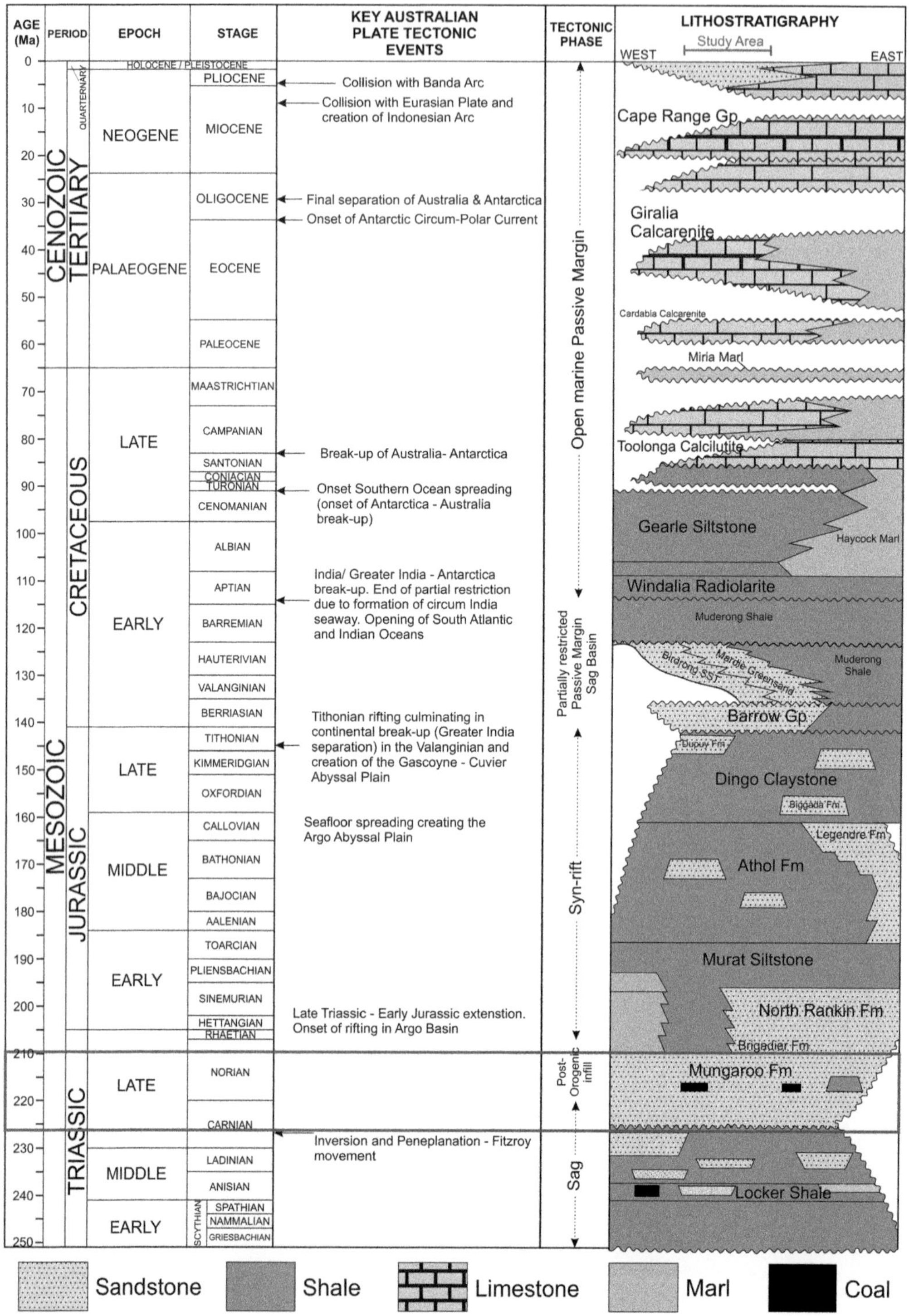

Fig. 2. Tectonostratigraphic diagram of the Northern Carnarvon Basin (after Longley *et al.* 2002; Ghori 2011). The Mungaroo Formation interval of interest is outlined.

The Northern Carnarvon Basin contains the thickest and most extensive Triassic sequence on the North West Shelf (AGSO North West Shelf Study Group 1994), with estimated sedimentation rates of $<0.2\ \mathrm{m\ ka^{-1}}$ (Adamson *et al.* 2013).

The Mungaroo Formation is dominated by upper delta-plain to alluvial-plain fluvial channel sandstones of varying types, floodplain fines with crevasse splays, through to lower delta-plain distributary channels, soils and coals, mouth bar and interdistributary bay deposits (Adamson *et al.* 2013; Heldreich *et al.* 2013). Palynological assemblages (spores, pollen and dinoflagellates) reported by Backhouse & Balme (2002), Backhouse *et al.* (2002) and Nicoll (2002) supports a fluvio-deltaic depositional setting.

Methodology

The multidisciplinary dataset comprises wireline logs from 21 wells, approximately 830 m of core from eight of these wells and 3D seismic situated over an area of approximately 10 000 $\mathrm{km^2}$ (Figs 3 & 4). All the data used in this study have been released for public access and are available from the Western Australia Department of Mines and Petroleum and Geological Survey of Western Australia (GSWA). The interval of well control penetrates up to 1.4 km of the Mungaroo Formation, with core coverage restricted to discrete intervals of interest (typically focused on sand intervals for reservoir characterization and plug porosity–permeability measurements) (Fig. 3). Image log data were unavailable to use during this study: however, an interpretive well completion report for Clio-2 provides discussions on palaeosediment dispersal directions. The Clio-2 well is located within the 3D dataset, and provides the key control point where core, wireline and seismic can be confidently calibrated.

Detailed sedimentary descriptions and interpretations of conventional core have been produced (Figs 3 & 4). The cored sections sample the mid-upper part of the Mungaroo Formation. Core taken from the Clio-2 and Chandon-2 wells provide the longest and most continuous record through two separate parts of the stratigraphy (Fig. 4). Detailed ichnological observations have also been made to aid interpretation of depositional environments. Sedimentary facies identified in core have been calibrated with their corresponding wireline-log response to define a series of characteristic well-log facies, allowing interpretation of possible

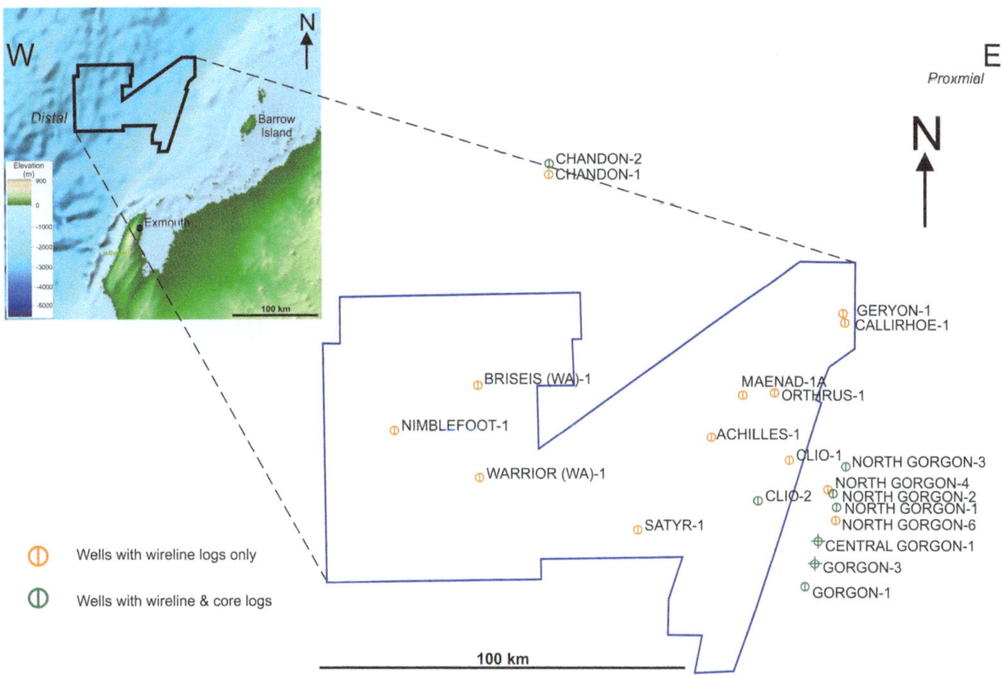

Fig. 3. Location and extent of the 3D seismic volume showing all the well locations used in this study. Orange wells contain only wireline-log data. Green wells contain core and wireline-log data and were used for core-to-wireline calibration.

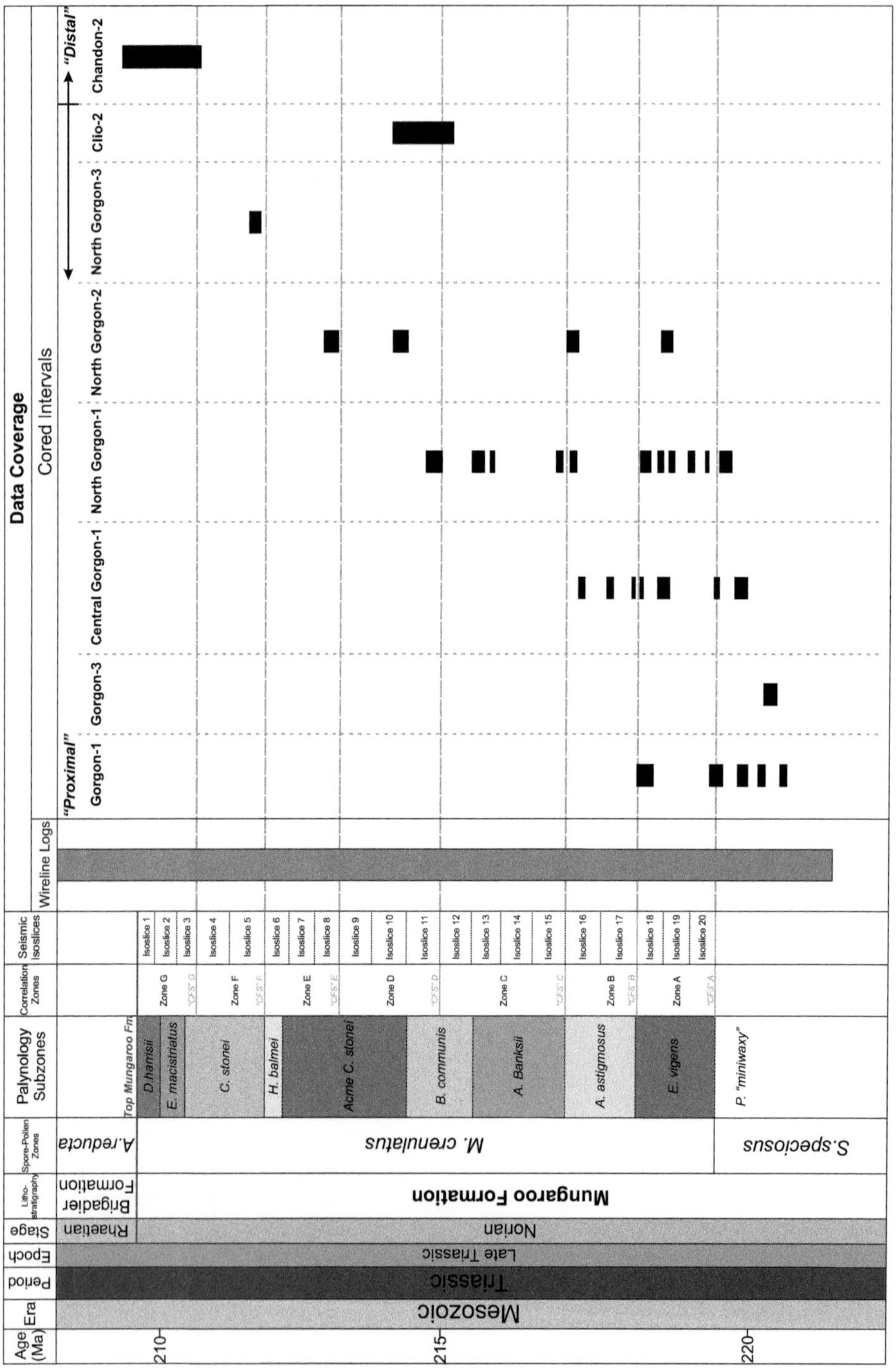

Fig. 4. Data coverage across the dataset within the mid–upper part of the Mungaroo Formation. Wireline logs penetrate the entire *M. crenulatus* Zone, whereas cored intervals typically target sandstone bodies within the formation for porosity–permeability tests. Palynology provided by Dixon *et al.* (2012) has enabled the upper part of the Mungaroo Formation to be subdivided into seven stratigraphic zones (A–G) based on identification of candidate flooding surfaces. These surfaces are used for well correlations.

sedimentary facies over intervals without core. The main logs utilized were gamma-ray, neutron-porosity, density and sonic logs.

Well correlations were developed across the study area, based on integrating vertical well-log trends with palynological studies undertaken by Dixon *et al.* (2012) which identified a series of semi-regional marine incursions. Spatial and temporal facies trends were identified across the interval of well control. Subsequent correlation of potential flooding surfaces across the dataset enabled the upper Mungaroo Formation to be divided into seven subzones in relation to a sequence stratigraphic framework: within which, detailed spatial and temporal changes in sand net:gross (N:G) and sandstone body thicknesses can be examined.

Seismic attribute analysis of a 10 000 km^2 3D seismic volume was interpreted to identify the spatial and temporal distribution of reservoir units and depositional facies within the Mungaroo deltaic system (Fig. 3). A subset of log-correlated potential flooding events could be identified on the seismic data and were used to divide the upper Mungaroo Formation into a suite of isoproportional horizon slices/sculpts for attribute analysis with sequence stratigraphic significance. Twenty isoslices, 40 ms thick (which lie approximately parallel to stratigraphy) were generated through the seismic volume and within the interval of well control. Root mean square (RMS) energy seismic attributes were calibrated to core and wireline-log interpretations, and used to differentiate between sandstone and mudstone lithologies: therefore enabling geomorphic features such as channel and channel-belt edges to be imaged within the 3D seismic volume. This paper focuses on wireline-log and core data interpretation, and the implications of predicting geobody geometries without seismic control: however, some of the seismic control data and interpretation is also presented for completeness.

Results and interpretations

Sedimentary facies and facies associations

Adamson *et al.* (2013) documented a comprehensive range of facies observed within the Mungaroo Formation across the entire Northern Carnarvon Basin. In this paper, we present a high-resolution study of part of the basin. Eleven depositional facies associations (FA) have been identified from sedimentological analysis of the core based on lithology, grain size, sedimentary structures and ichnofacies (where present). Many of these are comparable to those described by Adamson *et al.* (2013). Sedimentary characteristics and the organization/stacking of individual lithofacies, together with ichnofacies, have been calibrated with their corresponding wireline-log signatures to define these facies associations. In this study, all sedimentary logs and descriptions are fully documented, and can be linked to specific wells, cores and depths to provide future reference sections for further work.

FA 1: fluvial channel bars

Description. FA 1 typically comprises ≥5–10 m-thick sandstone intervals, which often display fining-upwards trends, comprised of 10 cm–1 m-thick sandstone units with sharp or erosive bases (Fig. 4). Commonly grey- to buff-coloured and visually comprised of well to moderately sorted, subrounded to rounded quartz grains, which typify quartz arenites and arkoses. Individual units/co-sets show fining-upwards trends from very-coarse- to coarse-grained planar/trough cross-bedded sandstone (foreset dips $>10°$) up to medium- to fine-grained low-angle cross-bedding (foreset dips $<10°$) and ripple to wavy/horizontal lamination at the top. Massive sandstones and dewatered, slumped/overturned foresets are also present. Individual cross-bed foresets (1–10 cm thick) fine up from coarse- to medium-grained sandstone and are often capped by a muddy/carbonaceous parting. In some cases, there are low-angle discordances between finer-grained/carbonaceous layers and coarser sandstone beds below. Matrix and clast-supported pebble lags, a few centimetres thick, containing subrounded quartz and extraformational clasts, mud rip-up intraclasts and wispy carbonaceous material, often line coarser-grained bed and co-set bases. Thin (millimetre-scale) micaceous mud layers/caps are often found at the top of fining-upwards units/co-sets.

Thick, blocky, low gamma-ray (GR) log motifs, characterized by sharp bases and either sharp tops or, more commonly, a fining-upwards (muddying-upwards) log response, correlate to thick, medium-to very- coarse-grained and cross-bedded sandstone. These blocky sandstones observed from wireline logs can range from 10 to 70 m thick (Fig. 5).

Interpretation. Cross-bedded sandstones and sharp-based erosional scour surfaces of FA 1 represent high-energy erosion and deposition by fluvial channels (Tucker 2001). Pebble lags above erosive scour surfaces indicate rapid deposition of poorly sorted bedload (Reineck & Singh 1980). A change from trough/planar cross-bedded coarse- to medium-grained sandstone to ripple-laminated fine-grained sandstone within each unit records the decrease in flow regime within active fluvial channels (Reineck & Singh 1980). Mudstone/carbonaceous drapes along the foresets are interpreted to have formed as a result of slackwater periods in river discharge rather than representing tidal channels due to a lack of distinct tidal bundles,

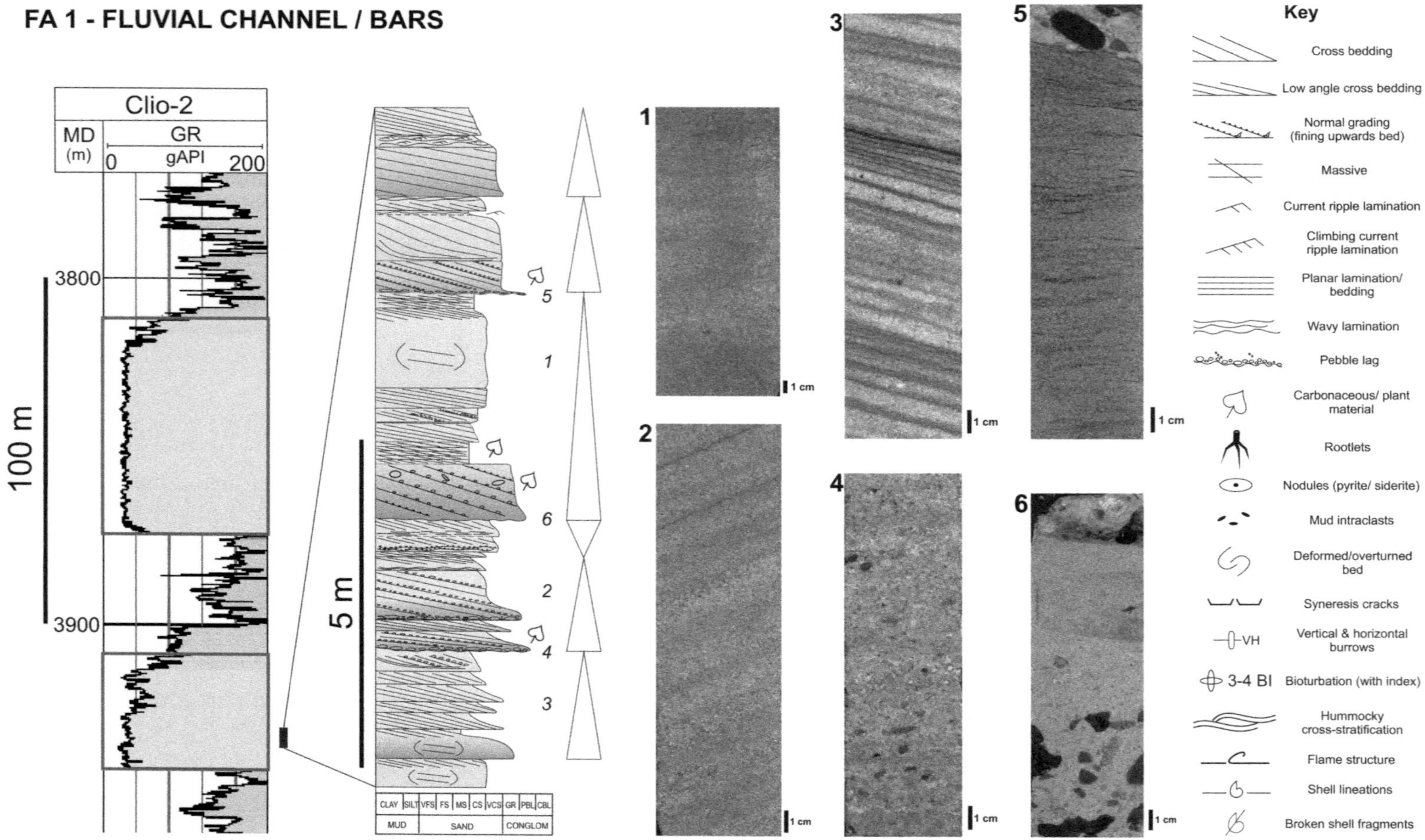

Fig. 5. Type section and key characteristics of Facies Association 1 from Clio-2 well. The bold outlined boxes highlight this facies on wireline-log motifs. The black line indicates where the sedimentary core type section is taken in relation to the wireline-log data. Vertical trend arrows are highlighted on the sedimentary core section. Numbers 1–6 relate the core photos to their position within the type section.

reactivation surfaces, bi-directional foresets or bioturbation. A low-angle discordance between finer-grained layers and coarser beds below may also represent toesets of the next migrating dune. Slumping and infraformational soft-sediment deformation is likely to be due to the high-energy shear forces exerted by the unidirectional current within these channel systems and or bank collapse (Tucker 2001). Upper and lower channel bars can be identified from stacking patterns of individual cross-bedded sandstone units (Bridge & Tye 2000; Bridge 2006). It is difficult to identify hierarchical surfaces/individual fluvial elements described by Holbrook (2001) from 1D, core or wireline data alone.

Single ≥5 m-thick fining-upwards units likely represent individual channel fills. Stacked units >10 m thick, lacking mudstone interbeds, represent amalgamated fluvial channel bars deposited in a single or multistorey channel-belt system (depending on the thickness of the interval and level of amalgamation) as a result of vertical aggradation and lateral migration (Allen 1983; Miall 1996; Bridge 2003). Multistorey fluvial channel belts are often capped by channel-fill abandonment facies (FA 3). It is likely that these thicker, blocky sandstone bodies represent large-scale fluvial channel belts, which may vertically/laterally stack and amalgamate to form channel-belt complexes (Heldreich *et al.* 2013).

FA 2: distributary channels

Description. FA 2 comprises 1–10 m-thick overall fining upwards sandstone packages that are composed of sharp- to erosive-based 50–25 cm-thick coarse-/medium- to fine-grained sandstone bars, often capped by the channel-fill abandonment facies (FA 3). Typically, light-grey coloured (but can also be buff-brown- to grey-coloured) and visually composed of well- to very-well-sorted sandstone (cleaner looking than FA 1) (Fig. 6). Sedimentary structures are dominated by scour surfaces, trough and/or planar cross-bedding occurring at the base of individual units, and ripple to wavy/planar laminations at the top. Rounded lithic pebble clasts suspended in a matrix of subrounded to rounded quartz grains are also commonly observed at the base of individual units. Individual cross-beds/forests often fine upwards from medium- to fine-grained sandstone, whilst disseminated carbonaceous material line foreset laminae. There is a higher percentage of carbonaceous material dispersed throughout individual units as small wispy flecks, and an increased occurrence of mud/carbonaceous-draped sedimentary features and mud intraclasts in comparison to FA 1. Mud interbeds between sandstone units range from 2 mm to 15 cm thick and are either massive or planar laminated. They often contain siderite and pyrite nodules, silt to very-fine-grained sandstone laminae, syneresis cracks and rare *Planolites* ichnofauna (Fig. 6).

The corresponding wireline-log response displays a blocky or fining-upwards 'bell-shaped' GR log motif often isolated within overall serrated (interbedded sands and shales), low N:G intervals (Fig. 6).

Interpretation. The thinner (up to 10 m thick) sandstone bodies of FA 2 are interpreted to represent erosion, subsequent infill and abandonment of either single-storey fluvial channels or distributary channels (Heldreich *et al.* 2013). The presence of mud/carbonaceous laminae is a key differentiator from FA1. Although sedimentary structures reflect unidirectional flow conditions, current velocity and direction were likely to be variable within fluvial distributary channels in response to some tidal modulation, resulting in the deposition and preservation of clay layers during low river stages (Reineck & Singh 1980). Conversely, structureless mudstone layers (>0.5 cm thick) may represent fluid mud deposits as a result of a tidal influence, commonly found within the fluvial–tidal transition zone of many modern estuaries: this can be greatly affected by river flood periods, and accumulates as pools and blanket deposits within these tidally influenced depositional settings (Ichaso & Dalrymple 2009).

Given the stratigraphic position and contextual association within lower delta-plain facies and larger-scale fining upwards trends identified within FA 2, the preferred interpretation is that of fluvially dominated, tidally influenced distributary channels, representing progressive channel abandonment as distributary channels bifurcate downstream across the lower delta plain (Bhattacharya 2010).

FA 3: channel abandonment

Description. FA 3 is composed of 25 cm–1 m-thick grey/brown- to black-coloured mudstone–siltstone units with interbeds of siltstone and buff-coloured very-fine-grained sandstone (Fig. 7). Individual interbeds decrease in thickness from 4 cm at the base of the unit to ≤1 cm at the top. Planar to wavy lamination is preserved within some interbeds: however, most sedimentary structures are mud draped and disturbed by rootlets. Mudstones/siltstones are commonly massive or planar laminated. A high abundance of carbonaceous plant material and rootlets/rhizoliths are observed, which often gives rise to colour mottling at the very top of FA 3. Swelling clays such as smectite are often present within these colour-mottled intervals based on visual observations (Fig. 7).

FA 3 is represented in wireline logs by a high GR motif that increases upwards, overlying the thick

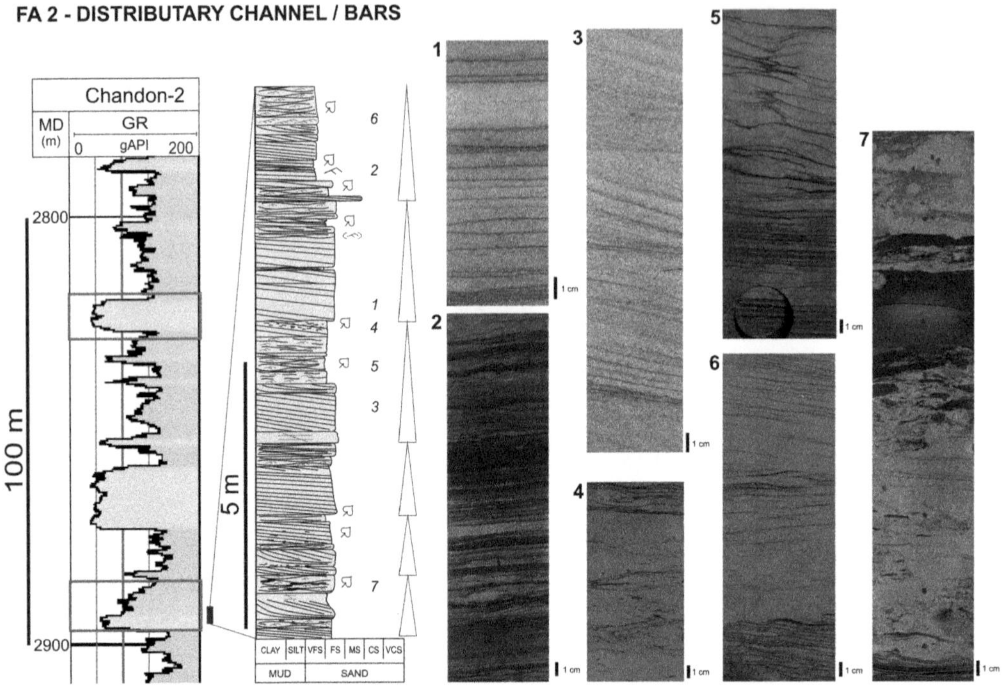

Fig. 6. Type section and key characteristics of Facies Association 2 from Chandon-2 well. The bold outlined boxes highlight this facies on wireline-log motifs. The black line indicates where the sedimentary core type section is taken in relation to the wireline-log data. Vertical trend arrows are highlighted on the sedimentary core section. Numbers 1–7 relate the core photos to their position within the type section. (See Fig. 5 for the key to the sedimentary structures and symbols.)

fining-upwards sandstone units of FA 1 and FA 2 (Figs 5 & 6).

Interpretation. The fine-grained sediments were deposited via low-energy tractional currents and suspension settling, and are interpreted to represent channel abandonment (either fluvial or distributary channel) or the formation of an oxbow lake mud plug, as a result of chute/neck cut-off or channel avulsion (Reineck & Singh 1980; Bridge 2006; Miall 2010). Rootlets and colour mottling suggests post-depositional pedogenic alteration and incipient palaeosol formation (Miall 2010; Adamson *et al.* 2013).

FA 3 is very similar to FA 5 in composition and wireline-log character, but forms a gradational, thinner interval from genetically related channel-fill facies such as FA 1 and FA 2 below. There is, however, significant uncertainty in distinguishing between FA 3 and FA 5 from wireline-log responses alone.

FA 4: crevasse splays

Description. FA 4 is composed of very-fine- to coarse-grained, well- to very-well-sorted, light-grey sandstone and dark grey to black mudstone, which can form packages up to 4 m thick (Fig. 8). Typically dominated by rounded, quartz, feldspar and lithic grains, sedimentary structures include massive sandstones, high- and low-angle cross-bedding, ripple lamination, climbing ripple lamination, horizontal to wavy lamination and parallel-laminated mudstone layers. Both fining- and coarsening-upwards vertical stacking patterns are observed. Fining-upwards trends are composed of cross-bedded and massive sandstones, which commonly pass upwards into ripple/climbing ripple and horizontal lamination. Fine-grained rippled to low-angle cross-bedded sandstones which pass upwards into higher-angle planar and/or cross-bedded coarse-grained sandstones typify coarsening-upwards trends. Interbedded very-fine-grained sandstone, siltstone and mud layers are common, as are wavy, lenticular mudstone laminae and wispy lenses of carbonaceous material. Siderite and rare pyrite nodules are occasionally present within FA 4. Rootlets can be very pervasive, disturbing original sedimentary structures. These units stack up into decimetre- to metre-scale upwards-fining or -coarsening trends (Fig. 8).

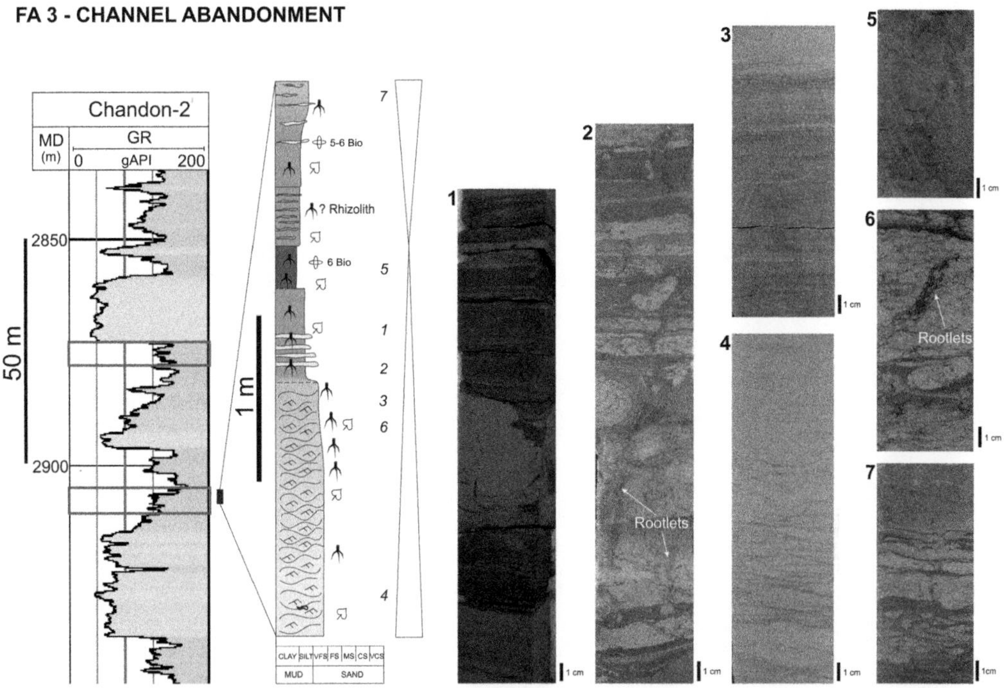

Fig. 7. Type section and key characteristics of Facies Association 3 from Chandon-2 well. The bold outlined boxes highlight this facies on wireline-log motifs. The black line indicates where the sedimentary core type section is taken in relation to the wireline-log data. Vertical trend arrows are highlighted on the sedimentary core section. Numbers 1–7 relate the core photos to their position within the type section. (See Fig. 5 for the key to the sedimentary structures and symbols.)

FA 4 corresponds to very-thin (1–3 m thick), low GR responses, displaying a coarsening- or fining-upwards trend. GR values are usually slightly higher than those of FA 1 and FA 2 (Fig. 8). These are interpreted to record thin sandstone intervals that are encased within either dominantly high GR (muddy) sections or overall ratty/interbedded sandstone and shale units (Fig. 8).

Interpretation. FA 4 is interpreted to represent crevasse splays adjacent to fluvial and distributary channel bodies. Individual flood units on the floodplain are likely to be sheet-like, which usually thin in the flow direction (Bridge 1984). Overbank crevasse-splay deposits closest to the main channel can be confused with upper channel-bar deposits displaying a progressive decrease in grain size with increasing distance from the channel margin. High- and low-angle cross-bedding is common in sandy crevasse-splay deposits: however, there is also evidence for the cessation of current discharge, in the form of mud layers and root casts/rootlets. Gradational coarsening- and fining-upwards sequences within overbank environments may reflect progradation and abandonment of individual crevasse splay, migration or cut-off of channels within the active channel belt or avulsion of the entire channel belt (Bridge 2006).

FA 5: floodplain heterolithics and palaeosols

Description. FA 5 is dominated by homogenous mudstones to siltstones with thin (10–25 cm thick) very-fine- to fine-grained sandstone intervals, forming sequences 9 m thick or more. Individual sandstone units are typically sharp/erosively based and show a fining-upwards trend. They contain low-angle to wavy laminations and ripple laminations, whereas mudstone and siltstone intervals are generally homogeneous and massive or parallel laminated. Siderite nodules are often observed in sandstone units, whereas pyrite nodules are commonly found in mudstone/siltstone intervals. Lenticular and flaser lamination are occasionally present, whilst thinly interbedded very-fine sandstone, siltstone and mudstone units are common. Irregular and lenticular laminae dominate within these heterolithic intervals, and swelling clays are also often present. There is a very high proportion of carbonaceous material distributed within the mudstones and sandstones of FA 5, along with thin (5–15 cm thick) coaly horizons. *Planolites* ichnofauna and rootlets are abundant and commonly

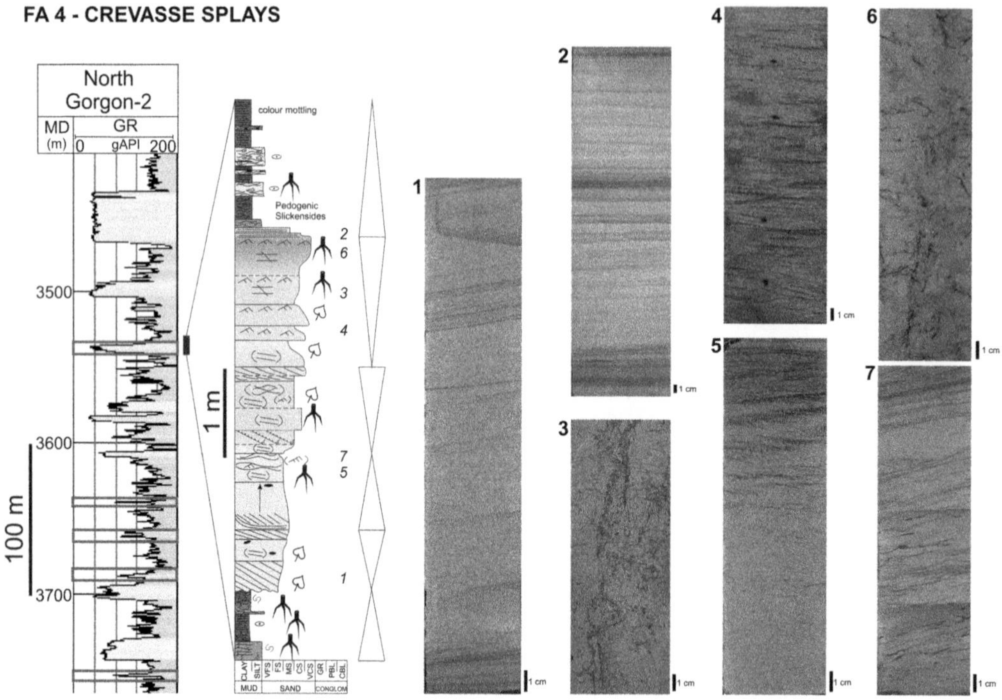

Fig. 8. Type section and key characteristics of Facies Association 4 from North Gorgon-2 well. The bold outlined boxes highlight this facies on wireline-log motifs. The black line indicates where the sedimentary core type section is taken in relation to the wireline-log data. Vertical trend arrows are highlighted on the sedimentary core section. Numbers 1–7 relate the core photos to their position within the type section. (See Fig. 5 for the key to the sedimentary structures and symbols.)

destroy original sedimentary structures. Intensely rooted intervals are associated with colour mottling (in both mudstone and at the top of sandstone units), and pedogenic slickensides occur within mudstones and siltstones. Black homogenous siltstones pass upwards into colour-mottled dark-grey to green-grey mudstone (Fig. 9).

FA 5 is characterized on wireline logs by a ratty and often increasing (fining-upwards) motif of high GR values (Fig. 9). Thin sheet sandstones are often recorded as very thin, low GR peaks. The abundance of carbonaceous material and the presence of coaly horizons correspond to high peaks in the sonic (DT), high neutron porosity (NEU) and low bulk density (DEN) peaks (Fig. 9). These well-log responses are typical for coaly horizons: however, they are not completely diagnostic for every palaeosol interval (although the high NEU and low DEN spikes usually indicate coaly horizons).

Interpretation. The presence of rootlets and deposition of mudstone and sandstone sheets/lenses from suspension settling and low-energy traction currents suggest that FA 5 represents floodplain basins, floodplain lakes and associated palaeosol horizons (Bridge 2006). Fine-grained sandstone and siltstone are interpreted to represent thin sheet sands and distal portions of splays deposited during river flood conditions. The abundance of carbonaceous material, pedogenic slickensides, disturbed bedding and colour mottling are indicative of palaeosols/soil horizons typically formed in a humid climate with thick vegetation cover (Reineck & Singh 1980). Poorly drained areas of the floodplain are swampy and constantly waterlogged, enabling the accumulation and preservation of organic/plant matter and the precipitation of calcium carbonate and iron oxide nodules, such as pyrite and siderite, under chemically reducing or anoxic conditions at and below the sediment surface (Reineck & Singh 1980; Bridge 2006).

These wet, poorly drained flood-basin soils, typically located on the lower delta plain, are often gleyed giving rise to the grey-green coloration observed in FA 5 (Bridge 2006).

FA 6: distributary mouth bars/distal mouth bars

Description. FA 6 is dominated by interbedded, very-fine- to fine-grained sandstone, siltstone and mudstone, which form trendless to coarsening-

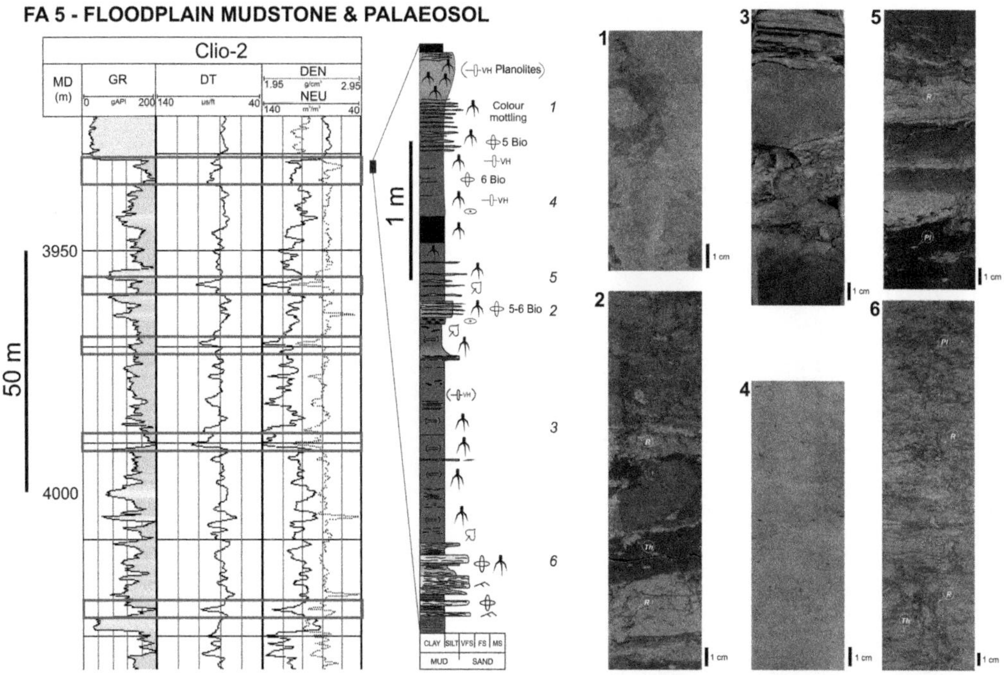

Fig. 9. Type section and key characteristics of Facies Association 5 from Clio-2 well. The bold outlined boxes highlight this facies on wireline-log motifs. The black line indicates where the sedimentary core type section is taken in relation to the wireline-log data. Vertical trend arrows are highlighted on the sedimentary core section. Numbers 1–6 relate the core photos to their position within the type section. Trace fossils include rootlets (*R*), *Thalassinoides* (*Th*) and *Planolites* (*Pl*). (See Fig. 5 for the key to the sedimentary structures and symbols.)

upwards packages 3–4 m thick. Individual sandstone and siltstone interbeds range from 5 to 50 cm thick and contain low-angle cross-bedding, wavy to ripple lamination, hummocky cross-stratification (HCS) and planar laminations. Mudstone interbeds are typically 5–15 cm thick and are structureless or planar laminated. Heterolithic intervals contain well-developed wavy, flaser and lenticular bedding. Some soft-sediment deformation structures can be observed, whilst shallow- to open-marine ichnofacies are abundant and locally intense within some beds. Ichnofacies include *Planolites*, *Skolithos*, *Ophiomorpha nodosa*, *Diplocraterion parallelum*, *Palaeophycus*, *Teichichnus* and *Rosselia* (Fig. 10). A high proportion of carbonaceous material is present within sandstone and mudstone intervals, and is concentrated along individual laminae. Possible rootlets appear locally. Some sedimentary structures appear to show bi-directional current flow, whilst 5–10 cm-thick alternations of mudstone, siltstone and sandstone, containing planar and wavy laminae, are often mud draped in a rhythmic pattern (Fig. 10).

FA 6 corresponds on the GR to intervals of very-thin, typically 1–3 m thick, mid to high GR responses, which often show a reducing-upwards (coarsening-upwards) trend (Fig. 10). The GR values are usually slightly higher than those of FA 1 and FA 2. These low GR intervals (that can be calibrated to thin sandstones on the mud logs) are encased within either dominantly high GR (muddy) sections or overall ratty/interbedded sandstone and mudstone units (Fig. 10).

Interpretation. Alternating moderate- to low-energy tractional currents, density currents/hyperpycnal flows and suspension settling suggest deposition within distributary mouth bars and distal mouth bars (Bhattacharya 2010). Wave/storm effects and possible tidal influence highlight the continuous reworking of sediment at the delta front (Reineck & Singh 1980; Bhattacharya 2010). Muddy heterolithic intervals are likely to represent alternations between low and high river flood discharge (Ichaso & Dalrymple 2009). Rhythmically mud-draped structures suggest that river flow was modulated by tidal currents (Ichaso & Dalrymple 2009). Structureless mud layers are likely to represent deposition of fluid mud during river flood periods, whereby sediment concentrations were higher during times of high river discharge or that the turbidity maximum was pushed further basinwards at this time (Ichaso & Dalrymple 2009). Ichnofacies indicate

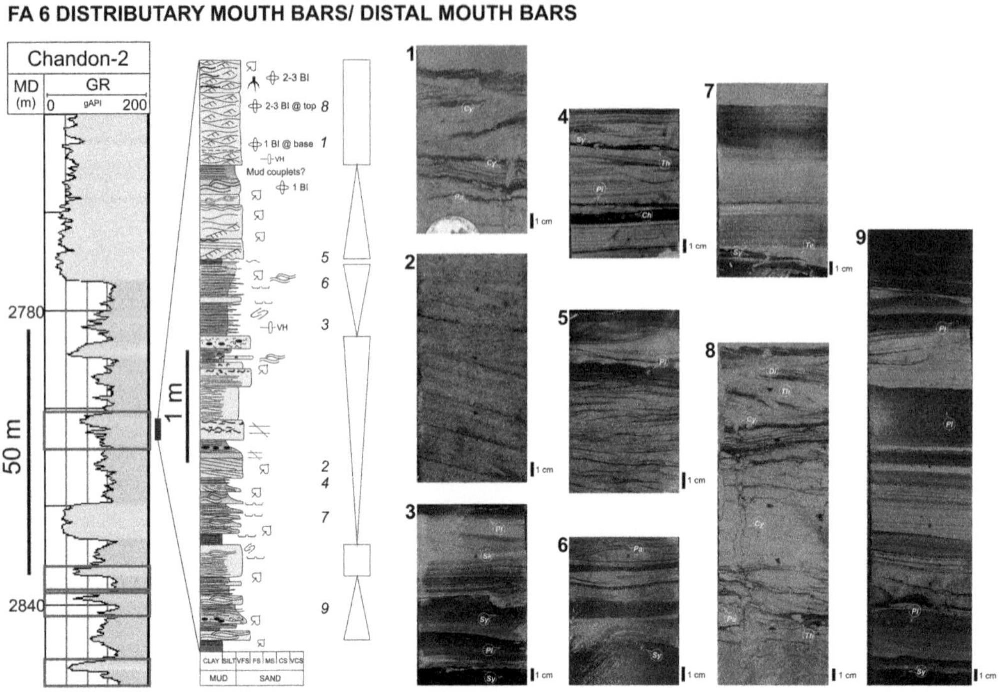

Fig. 10. Type section and key characteristics of Facies Association 6 from Chandon-2 well. The bold outlined boxes highlight this facies on wireline-log motifs. The black line indicates where the sedimentary core type section is taken in relation to the wireline-log data. Vertical trend arrows are highlighted on the sedimentary core section. Numbers 1–9 relate the core photos to their position within the type section.Trace fossils include *Cylindrichnus* (*Cy*), *Palaeophycus* (*Pa*), *Teichichnus* (*Te*), syneresis cracks (*Sy*), *Planolites* (*Pl*), *Chondrites* (*Ch*), *Diplocraterion parallelum* (*Di*) and *Skolithos* (*Sk*). (See Fig. 5 for the key to the sedimentary structures and symbols.)

an 'open marine–mid shoreface' setting, which corresponds to the proximal–mid delta-front depositional environment with often stressed assemblages associated with periods of freshwater output from river flood events (Gerard & Bromley 2008; Adamson *et al.* 2013).

There is inherent uncertainty in subsurface interpretations and, based on lithofacies characteristics alone, FA 6 is very similar to FA 7 and FA 9. Both FA 6 and FA 9 encompass delta-front settings: however, FA 6 has lower diversity and abundance of ichnofauna, which is likely to reflect its proximity to fluvial, freshwater input from distributary channels.

FA 7: crevasse delta

Description. FA 7 is characterized by a coarsening-upwards succession, typically 3 m thick, from interbedded siltstone and very-fine- to fine-grained sandstone containing starved ripples, wavy to ripple laminae, parallel laminations and HCS, up to fine-grained, massive to indistinctly cross-bedded sandstone. Syneresis cracks, lenticular laminae and ichnofacies including *Planolites*, *Thalassinoides* and *Teichichnus* are common and locally intense within thinly interbedded siltstone and sandstone beds. Sparse intervals of homogeneous mud a few centimetres thick are also present within FA 7. A high abundance of carbonaceous material is common, and is clearly seen within massive sandstones at the top of the unit. Rootlets, irregular mudstone laminae and a thin (millimetres to centimetres thick) coaly horizon are present at the very top of this succession (Fig. 11).

The wireline-log response of this facies association is characterized by small-scale reducing-upwards (coarsening-upwards) motifs within an overall high GR (muddy) or ratty/interbedded sandstone and mudstone sequence (Fig. 11).

Interpretation. FA 7 is interpreted to represent the progradation of a crevasse delta into a marginal-marine restricted embayment. Interbedded mudstone, siltstone and sandstone with syneresis cracks, wavy to ripple lamination and low-angle, probable HCS beds, reflect brackish-water conditions and wave action at the coastline (Boyd 2010). However, the dominance of fine-grained sediment and the

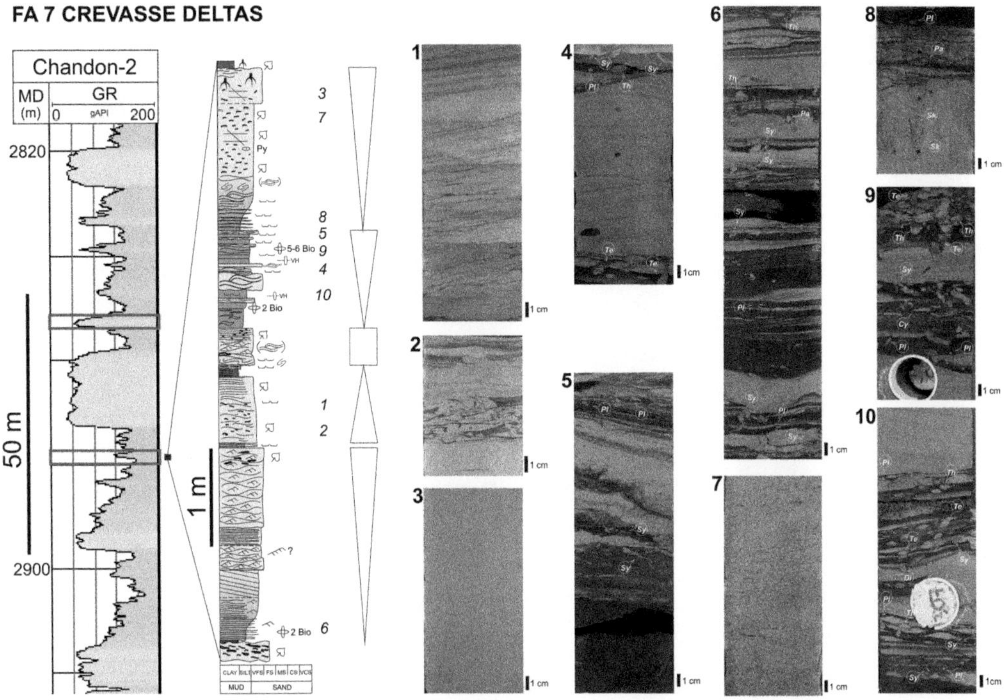

Fig. 11. Type section and key characteristics of Facies Association 7 from Chandon-2 well. The bold outlined boxes highlight this facies on wireline-log motifs. The black line indicates where the sedimentary core type section is taken in relation to the wireline-log data. Vertical trend arrows are highlighted on the sedimentary core section. Numbers 1–10 relate the core photos to their position within the type section. Trace fossil include *Planolites* (*Pl*), syneresis cracks (*Sy*), *Thalassinoides* (*Th*), *Palaeophycus* (*Pa*), *Teichichnus* (*Te*), *Cylindrichnus* (*Cy*), *Diplocraterion parallelum* (*Di*) and *Skolithos* (*Sk*). (See Fig. 5 for the key to the sedimentary structures and symbols.)

presence of some homogeneous mud intervals may indicate tidal effects, and deposition of fluid mud within an overall low-energy, protected embayment (Boyd 2010). The *Planolites* and *Teichichnus* ichnofacies assemblage is also indicative of a brackish, restricted embayment setting (Gerard & Bromley 2008). The overall progression up to massive to weakly cross-bedded fine-grained sandstone at the top of FA7 with abundant carbonaceous material and evidence for subaerial exposure in the form of rootlets and coaly horizons suggests the progradation of a crevasse channel forming a small delta into a low-energy embayment (Boyd 2010). Gugliotta *et al.* (2015) also identified river-dominated but tidal-influenced crevasse sub-deltas (on a similar scale to FA7) from the Lajas Formation, Argentina.

FA 7 displays similar characteristics to FA 6; nevertheless, based on their stratigraphic position, contextual association within interdistributary bay facies associations (FA 8), lower diversity and abundance of 'environmentally stressed' ichnofauna, and high concentrations of carbonaceous material and rootlets, FA 7 is interpreted to represent crevasse deltas within a brackish embayment rather than mouth bars in a strictly open-marine delta-front setting. However, differentiating between mouth bars and crevasse deltas without core is impossible.

FA 8: lower delta-plain heterolithics

Description. Thinly interbedded mudstone, siltstone and very fine-grained sandstone sequences, which form metre-thick sand- or mud-prone heterolithic intervals. These sequences often stack up into upwards-coarsening sequences (*c.* 3 m thick), associated with increasing carbonaceous material, and well developed *in situ* rootlets and pedogenic colour mottling (Fig. 12). The uppermost part of these sequences contains an increase in pyrite and siderite nodules/iron oxide staining. A thin (5–10 cm thick) carbonaceous/coaly cap is only occasionally preserved at the top of FA 8. Individual sandstone interbeds are sharp based, commonly erosive into underlying mudstone/siltstone units and show normal and reverse grading. Wavy to planar lamination and current and wave ripples are common within siltstone–sandstone units, whilst some streaky/lenticular siltstone–sandstone laminae are also present.

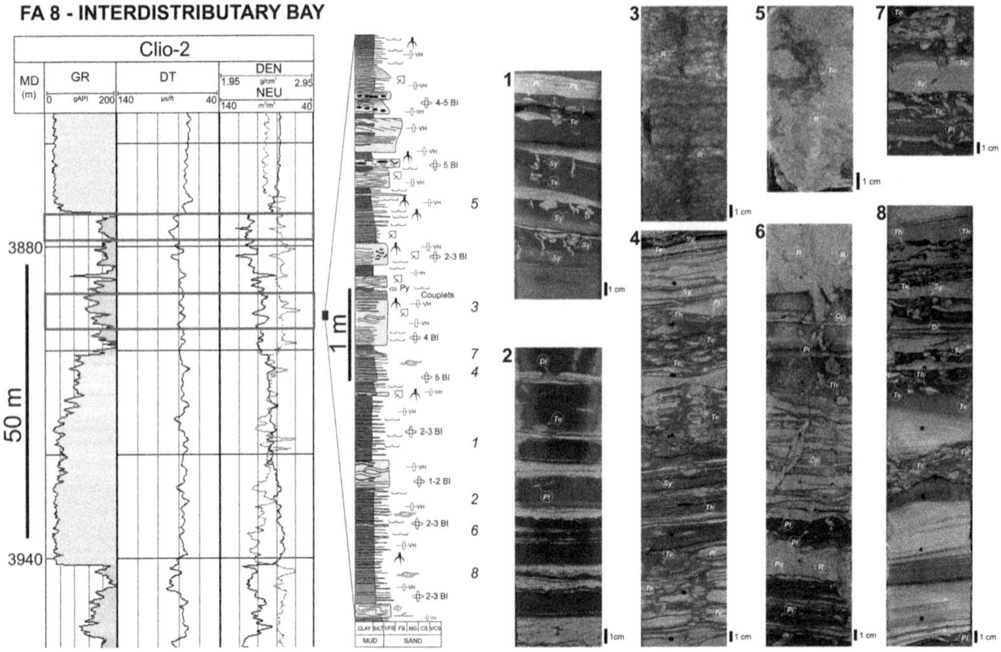

Fig. 12. Type section and key characteristics of Facies Association 8 from Clio-2 well. The bold outlined boxes highlight this facies on wireline-log motifs. The black line indicates where the sedimentary core type section is taken in relation to the wireline-log data. Vertical trend arrows are highlighted on the sedimentary core section. Numbers 1–8 relate the core photos to their position within the type section. Trace fossils include *Teichichnus* (*Te*), *Planolites* (*Pl*), syneresis cracks (*Sy*), *Palaeophycus* (*Pa*), *Thalassinoides* (*Th*), Rootlets (*R*), *Ophiomorpha nodosa* (*Op*) and *Diplocraterion Parallelum* (*Di*). (See Fig. 5 for the key to the sedimentary structures and symbols.)

HCS is dominant within thicker (10–25 cm thick) very-fine-grained sandstone units. Some sedimentary structures are also rhythmically mud draped. Syneresis cracks are abundant within thinly (millimetre- to centimetre-scale) interbedded intervals (Fig. 12). *Planolites*, *Teichichnus*, *Palaeophycus*, *Diplocraterion parallelum* and *Thalassinoides* are all common within FA 8 and locally intense within some mudstone–siltstone beds (Fig. 12).

The typical wireline-log response for FA 8 is of a ratty (interbedded mudstone and sandstone) high GR motif, which often decreases (coarsens) up to a crevasse splay or distributary channel sandstone facies on top (Fig. 12). However, this trend is not diagnostic and some sequences also show an upwards-increasing (fining) motif. FA 8 also stacks vertically into a thicker interval (10–20 m thick) composed of a series of small-scale coarsening-upwards packages observed on wireline logs (Fig. 12) in association with crevasse delta, distributary channel and/or floodplain mudstone facies. Coaly horizons are identified from high peaks in the DT and NEU logs, and a low spike in the DEN log. However, the wireline-log motif of FA 8 is very similar to that of FA 5.

Interpretation. Heterolithic sequences with individual intercalations of sandstones to mudstones recording low- to moderate-energy tractional currents, low-energy density currents/hyperpycnal flows and suspension settling are characteristic of interdistributary bay-fill sequences (Elliott 1974). Sedimentation within these bays is dominated by flood-generated incursions from the fluvial distributaries, including overbank flooding, crevassing and avulsion (Elliott 1974). Fine-grained sediment is deposited from suspension due to the quiet, ponded waters under normal conditions and the absence of strong wave currents; whilst coarser-grained sediment is likely to be sourced from crevasse channels and splays entering the embayment (Elliott 1974; Reineck & Singh 1980). The ichnofauna assemblage within FA 8 represents restricted to semi-open marine embayments, suggesting a sheltered/protected environment, dominated by brackish to marine water conditions giving rise to restricted/stressed ichnofacies and the prevalence of syneresis cracks (Gerard & Bromley 2008; Buatois & Mangano 2011). The presence of wave ripples and HCS indicates the effects of weak waves and storm events within an embayment; whereas scoured sandstone beds and current ripple laminations are likely to have been produced by tidal currents or flood periods from the distributaries (Reineck & Singh 1980). Mud-draped structures or rhythmic bedding within some intervals may reflect a periodic weak tidal influence into the embayment, depending on the relative influence of tides (Reineck & Singh 1980; Bhattacharya 2010).

Small-scale coarsening-upwards sequences may represent infilling of these shallow embayments, due to delta progradation or channel avulsion, leading to delta-lobe switching and eventually to the development of lagoons or delta-plain lakes, marshes/swamps and levees with extensive vegetation (Allen 1965; Elliott 1974; Bhattacharya 2010). As a result, sedimentary structures within the upper part of the bay-fill sequence are often obliterated by plant activity and pedogenesis, often capped by a peat layer/coal, carbonaceous shale, palaeosol or intertidal flat (Elliott 1974; Bhattacharya 2010).

FA 9: distal delta-front heterolithics

Description. Very-fine- to fine-grained sandstone units with thin interbeds of siltstone and mudstone; these units form trendless to slightly upwards-coarsening packages 3–4 m thick. Individual sandstone beds range from 10 to 40 cm thick, and contain low-angle cross-bedding, planar to wavy lamination, current ripple laminae and HCS. Small-scale scour surfaces and mud intraclasts often line the base of sharp or erosive sandstone beds. Carbonaceous material is abundant along with the presence of syneresis cracks and deformed beds. Ripple structures are often mud draped and some cross-laminae often show evidence for rhythmic mud drapes. There is also occasional evidence for reverse current directions in some cross-bedding structures, whilst other sandstone beds can be argillaceous and highly cemented. Heterolithic intervals commonly show very small-scale (millimetre–centimetre) fining-upwards/normal grading from fine-grained sandstone to mudstone or inverse grading and are frequently bioturbated (Fig. 13). Bioturbation varies between low–moderate diversity and moderate (locally intense) abundance, and occurs within both heterolithic and thicker sandstone beds. Ichnofauna include *Planolites*, *Teichichnus*, *Terebellina* (*Schaubcylindrichnus*), *Thalassinoides*, *Skolithos*, *Cylindrichnus*, *Scolicia*, *Rhizocorallium*, *Palaeophycus*, *Taenidium* and *Diplocraterion parallelum*

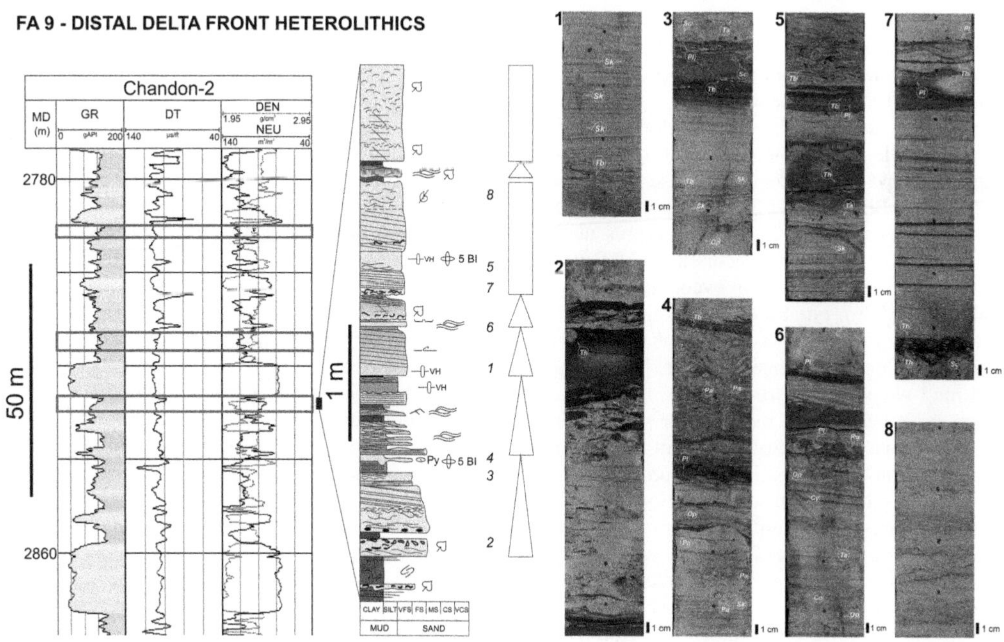

Fig. 13. Type section and key characteristics of Facies Association 9 from Chandon-2 well. The bold outlined boxes highlight this facies on wireline-log motifs. The black line indicates where the sedimentary core type section is taken in relation to the wireline-log data. Vertical trend arrows are highlighted on the sedimentary core section. Numbers 1–8 relate the core photos to their position within the type section. Trace fossils include *Planolites* (*Pl*), *Scolicia* (*Sc*), *Terebellina* (*Schaubcylindrichnus*) (*Tb*), *Taenidium* (*Ta*), *Skolithos* (*Sk*), *Ophiomorpha nodosa* (*Op*), *Palaeophycus* (*Pa*), *Thalassinoides* (*Th*), *Cylindrichnus* (*Cy*), *Rosselia* (*Ro*) and *Conichnus* (*Co*). (See Fig. 5 for the key to the sedimentary structures and symbols.)

(Fig. 13). Within low-angle cross-bedded fine-grained sandstone intervals, rare *Conichnus*, *Rosselia* and *Ophiomorpha nodosa* are preserved. *In situ* pyrite and siderite nodules are also common within these heterolithic units.

FA 9 corresponds to 1–4 m-thick, ratty (interbedded heterolithics) mid to high GR responses, which often either show a trendless or upwards-decreasing (coarsening) trend to a potential distributary channel sandstone on top (Fig. 13). This well-log motif is very similar to FA 6, FA 7 and FA 8, although distinguished by the lack of a significant palaeosol or peat/coaly signature from the DT, DEN and NEU logs (Fig. 13).

Interpretation. Relatively fine-grained heterolithic intervals including a mix of unidirectional (and potentially bi-directional) current-, wave- and tidal-influenced sedimentary structures are representative of marginal-marine, mid-distal delta-front deposits (Bhattacharya 2010). The lack of coarser-grained clean sandstone, indicative of coalesced distributary mouth bars and distributary channels, places this facies association in a slightly more medial to distal setting dominated by interbedded delta-front sands, silts and muds (Reineck & Singh 1980; Bhattacharya 2010); although variations in fluvial discharge may produce an irregular coarsening-upwards succession with interbedded mudstones and sandstones throughout (Bhattacharya 2010). The presence of HCS within sandstone–siltstone intervals reflects storm influence in the middle–lower shoreface settings, typically situated above the storm wave base (Reineck & Singh 1980; Bhattacharya 2010). Massive to weakly laminated mudstone intervals may represent fluid mud deposits, which often occur on top of wave rippled to HCS sandstone storm event beds (Ichaso & Dalrymple 2009). Fluid mud within the delta-front environment may form as a result of intense wave action, causing re-suspension of previously deposited mud which is then re-deposited on top of the storm event bed created by the same storm (Ichaso & Dalrymple 2009). Soft-sediment deformation features are also common in river-dominated delta-front settings due to high sedimentation rates which cause overpressure in underlying pro-delta muds (Reineck & Singh 1980; Bhattacharya 2010). Gas heave structures formed within delta-front sandstones and siltstones also may have formed the deformed beds observed within FA 9 (Reineck & Singh 1980).

Olariu *et al.* (2010) identified that the delta-front setting of the Cretaceous fluvial-dominated Panther Tongue delta, cropping out in Utah, USA, was formed by highly sediment-concentrated hyperpycnal flow deposits. Small-scale inverse- or normal-graded bedding observed in FA 9 may represent hyperpycnal and density flow deposits dominated by parallel lamination and containing significant amounts of organic matter shed from terminal distributary channels (Olariu *et al.* 2010).

Salinity changes within the delta-front area occur as a result of freshwater and saltwater mixing, leading to the formation and preservation of syneresis cracks and siderite nodules (Bhattacharya 2010), and alternations of mudstone and siltstone beds with higher and lower diversity/abundance of ichnofauna. Individual beds containing higher diversity and abundance of ichnofauna represent periods of saline, open-marine conditions, whereas beds with lower diversity (stressed)/lower abundance of ichnofauna and syneresis cracks are interpreted to indicate a freshwater/brackish influence during periods of river flood into the delta-front setting. Although FA 9 is similar to FA 6 and FA 7, ichnofacies assemblages within FA 9, and, in particular, *Terebellina* (*Schaubcylindrichnus*) found within argillaceous sandstone units, are diagnostic of open-marine distal delta-front settings, which further supports this interpretation of depositional environment (Gerard & Bromley 2008).

FA 10: pro-delta

Description. FA 10 is dominated by mudstone, siltstone and thinly interbedded muddy heterolithic intervals, which stack into metre-thick trendless packages (Fig. 14). Individual very-fine-grained sandstone and siltstone interbeds are commonly light grey in colour and 1–5 cm thick. The base of these sandstone interbeds are often sharp to erosive, and occasionally contain broken shell material and crinoid fragments as basal lags. Mudstone intervals are massive to planar laminated, whilst very thin sandstone and siltstone interbeds (up to 1 cm thick) are commonly streaky, lenticular and irregular. Thicker, very-fine-grained sandstone interbeds (2–5 cm thick) contain planar to wavy laminations, starved and current ripple laminae, swaley lamination, and HCS. Siltstone and sandstone interbeds often form small-scale fining-upwards units 5–10 cm thick, showing normal grading and an associated colour change from light-grey siltstone/sandstone to black mudstone. Some intervals also exhibit reverse grading. Several sandstone interbeds exhibit orange iron oxide/siderite staining, whilst pyrite clusters are very common within mudstone intervals. Syneresis cracks and carbonaceous material are rare. Soft-sediment deformation is also present within some sandstone interbeds. Bioturbation is often monospecific and only locally intense within muddy intervals (Fig. 14). The bioturbation index is generally lower in comparison to FA 8. *Phycosiphon* is the dominant ichnofauna, whilst *Teichichnus*, *Thalassinoides*, *Zoophycos*, *Planolites*, *Chondrites*, *Taenidium*, *Cylindrichnus*, *Scolicia*,

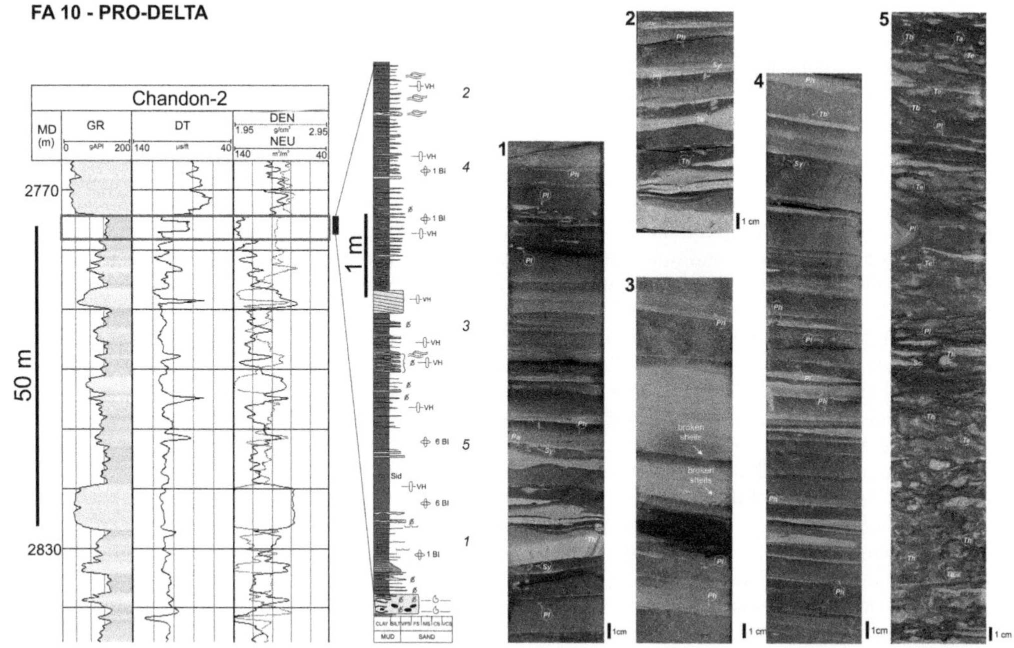

Fig. 14. Type section and key characteristics of Facies Association 10 from Chandon-2 well. The bold outlined boxes highlight this facies on wireline-log motifs. The black line indicates where the sedimentary core type section is taken in relation to the wireline-log data. Vertical trend arrows are highlighted on the sedimentary core section. Numbers 1–5 relate the core photos to their position within the type section. Trace fossils include *Planolites* (*Pl*), *Palaeophycus* (*Pa*), *Phycosiphon* (*Ph*), syneresis cracks (*Sy*), *Thalassinoides* (*Th*), *Teichichnus* (*Te*), *Terebellina* (*Schaubcylindrichnus*) (*Tb*), *Taenidium* (*Ta*) and *Scolicia* (*Sc*). (See Fig. 5 for the key to the sedimentary structures and symbols.)

Asterosoma and *Terebellina* (*Schaubcylindrichnus*) are present in very low abundance (Fig. 14).

The wireline-log response for this facies association is dominantly a high value, ratty (interbedded) GR motif. A cleaning-upwards (sandier-upwards) motif can only occasionally be observed, whereas a muddying-upwards or trendless high GR response is typical (Fig. 14). This wireline motif is similar to FA 5 and FA 8; although there is a distinct lack of any palaeosol or peat/coaly signature from the DT, DEN and NEU logs.

Interpretation. Thinly laminated siltstone and mudstone deposited by density currents, hyperpycnal flows and suspension settling, along with thin storm beds, suggest a distal delta-front to pro-delta environment of deposition (Reineck & Singh 1980; Bhattacharya 2010). Suspension settling within the pro-delta setting can be more rapid than previously thought as a result of hyperpycnal fluid muds and rapidly collapsing hypopycnal plumes (Bhattacharya 2010). Wavy laminations, HCS and shelly lags within sandstone–siltstone interbeds indicate deposition from wave currents and during storm events, whereas rhythmically laminated/mud-draped interbedded muds and silts may signify some tidal influence during deposition (Reineck & Singh 1980; Bhattacharya 2010). Inverse- and normal-graded silty and sandy beds indicate deposition by river floods, and reflect proximity to the distributary channel mouths. Centimetre-thick normally to inversely graded siltstone and very-fine-grained sandstone beds with internal scours may also represent hyperpycnal deposits formed during waxing, as well as waning, flow conditions, which are more common in river-dominated fluvio-deltaic systems than historically believed (Bhattacharya & MacEachern (2009). These muddy, pro-delta hyperpycnal deposits can be differentiated from anoxic laminites by a higher bioturbation index and higher ichnological diversity, a presence of wave ripples and HCS (Bhattacharya & MacEachern 2009). Wave cross-lamination and HCS within sandstone interbeds suggest a linkage with major storms in a setting above storm wave base (Bhattacharya & MacEachern 2009).

A result of periodic freshwater input into the pro-delta setting also initiates the formation of rare syneresis cracks, and greatly affects the abundance and intensity of bioturbation. The bioturbation index can be higher along bedding surfaces when normal salinity and slow sedimentation conditions resume in-between flood periods (Bhattacharya 2010). Discrete intervals which show high ichnological stress (monospecific and stunted/small-sized ichnofacies) may be a direct indicator of a river plume nearby that induces brackish-water or high sedimentation stress (Hampson & Howell 2005; Bhattacharya 2010). Conversely, river systems may alternate between hyperpycnal and hypopycnal conditions, resulting in alternations of more or less bioturbated intervals (Bhattacharya & MacEachern 2009). The ichnofacies present within FA 10 is indicative of distal delta-front and pro-delta settings, which further support our interpretations (Gerard & Bromley 2008).

Within cores, interdistributary bay and proximal pro-delta deposits may be indistinguishable as they reflect intergradational depositional environments (Bhattacharya 2010). Likewise, the pro-delta setting can often be difficult to distinguish from shelf–mud environments (Reineck & Singh 1980). However, biostratigraphic data (Dixon *et al.* 2012) and regional well correlations (Adamson *et al.* 2013; Marshall & Lang 2013; Payenberg *et al.* 2013) can help to identify the lateral extent of these facies associations and highlight laterally extensive marine intervals to build conceptual depositional models.

FA 11: transgressive lags

Description. FA 11 is characterized by 25–30 cm-thick units of light-grey-/buff-coloured fine- to medium-grained sandstone or siltstone, containing mud intraclasts and abundant broken shell (bioclast) material. No sedimentary structures are preserved and the intervals are often conglomeratic. Wavy to irregular muddy layers/lenses are also common along with siderite and pyrite clusters. Calcite cement is dominant within these units (Fig. 15).

Interpretation. FA 11 is interpreted to represent a transgressive lag deposited in response to marine erosion during transgression, producing a ravinement surface. Tidal currents and storm waves rework and erode older material on the shelf, and deposit a thin conglomeratic lag of reworked extra-basinal clasts of mudstone, sandstone and bioclasts (Plint 2010).

FA11 could also be interpreted as a storm bed/tempestite: however, given the contextual association with overlying marls, transgressive lags are the preferred interpretation.

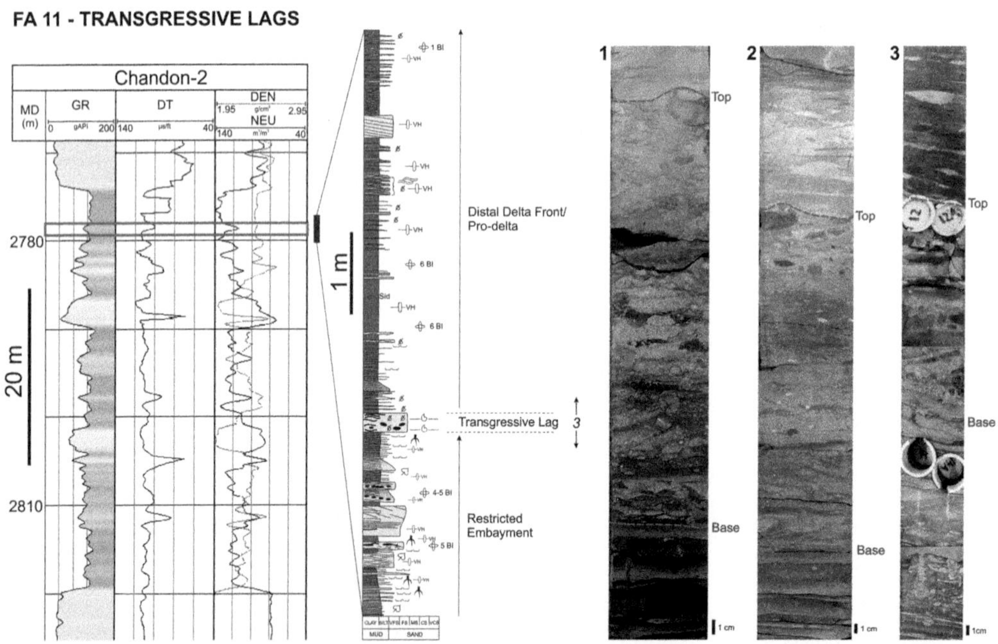

Fig. 15. Type section and key characteristics of Facies Association 11 from Chandon-2 well. The bold outlined box highlight this facies on wireline-log motifs. The black line indicates where the sedimentary core type section is taken in relation to the wireline-log data. Vertical trend arrows are highlighted on the sedimentary core section. Numbers 1 and 2 are examples of this facies from other intervals of the stratigraphy encountered in Chandon-2. Number 3 relates the core photo to the type section sedimentary core log. (See Fig. 5 for the key to the sedimentary structures and symbols.)

Spatial and temporal facies variability

The finer-grained FAs (FA 5, 8, 9 and 10) are commonly indistinguishable on logs, reducing the ability to confidently identify them for well correlation. As such, five major architectural associations have been defined based on core-to-wireline log calibration and their wireline-log signature/trends. These architectural associations have been used to develop well-log correlations in order to understand the spatial and temporal facies trends within the Mungaroo Formation.

Association A: multistorey channel sandstones. Association A comprises sandstone units >20–70 m thick, with blocky, low GR log motifs characterized by sharp bases and either sharp tops or, more commonly, a fining-/muddying-upwards log response. They are interpreted to represent vertically stacked and amalgamated multistorey fluvial channel belts, incised valleys or fortuitously stacked, but genetically unrelated, channel-belt sandstones. These channel belts or valleys are likely to range from large single/discrete geobodies through to laterally amalgamated 'multi-valley sheets' (Adamson *et al.* 2013). They extend over long enough distances to have a clear and distinct seismic response (Heldreich *et al.* 2013).

Association B: single-storey channel sandstones. Individual fluvial and distributary channel-fill and abandonment facies are characterized by thinner (*c.* 10–20 m thick) blocky or fining-upwards 'bell-shaped' GR log motif and often isolated within overall high GR/low N:G or serrated (thinly interbedded shale and sandstone) intervals (Heldreich *et al.* 2013).

Association C: thin sheet sandstones. Thin (<5 m and usually 1–3 m thick) low GR peaks within serrated well-log motifs are interpreted to represent either sheet-like crevasse-splay sandstones or mouth bars underlain and overlain by heterolithic and mudstone deposits (Heldreich *et al.* 2013). These facies are often part of a larger-scale coarsening- or fining-upwards trend, reflecting progradation (and successive abandonment) of the crevasse or mouth bar. Coarsening-upwards crevasse-splay, crevasse-deltas and mouth-bar successions are impossible to differentiate from wireline log without core.

Association D: overbank mudstone and heterolithics. Serrated or consistently high GR responses forming overall low N:G intervals are interpreted to represent mudstone and heterolithic facies (Heldreich *et al.* 2013). Non-marine facies commonly exhibit a serrated, trendless or often increasing-upwards (fining-upwards) GR log motif. Coaly horizons are common, identified by high sonic (DT), high neutron (NEU) and low density (DEN) peaks.

Association E: delta-front heterolithics. Within core, mudstone facies can be further subdivided into overbank (FA 5), interdistributary bays (FA 8), delta-front (FA 9) and pro-delta (FA 10) mudstones. However, it is difficult to differentiate these distinct facies associations from wireline-log response alone.

Marine-influenced facies such as interdistributary bays also show a serrated GR log motif with abundant coaly well-log signatures, similar to Association D; but GR responses often display an overall decreasing-upwards (coarsening-upwards) trend over 1–4 m, which are interpreted to reflect small-scale interdistributary bay-fill sequences. Conversely, larger-scale (10–30 m thick) coarsening-upwards GR log trends capped by Association B and C are likely to represent delta-front progradation cycles.

Marine pro-delta facies are commonly several metres thick, dominated by consistently high GR, high NEU and low DEN values, with no evidence from peaks in the NEU, DEN or DT well logs for carbonaceous/coaly horizons. Increasing-upwards (fining-upwards) trends in GR are also common: however, differentiation of pro-delta facies from delta front or Association D is impossible in the absence of core, which poses significant uncertainties for constructing well-log correlations and sequence stratigraphic models.

Well-log correlations

Semi-regional correlation. Robust, detailed correlations within paralic systems over long distances are typically very difficult due to the paucity of reliable biostratigraphic datums, repetitive well-log signatures and marked lateral facies variability. Correlations between wells based confidently on marine incursions are commonly restricted to distal localities. Palaeosols or interfluves have been used in published studies of similar depositional environments to aid well correlations (Dalrymple 1997): however, over distances of tens to hundreds of kilometres, it is unlikely that confident correlations can be made as there is a significant local control on palaeosol development (Wright & Marriott 1993).

In this study, we present a dip-section well correlation, approximately 130 km in length (Fig. 16), controlled by moderate- to high-resolution biostratigraphy/paylnostratigraphy (Backhouse *et al.* 2002; Dixon *et al.* 2012) that offers a coarse framework in which to constrain correlations.

The main datum for this semi-regional well correlation is the *Heibergella balmei* (*H. balmei*) Zone,

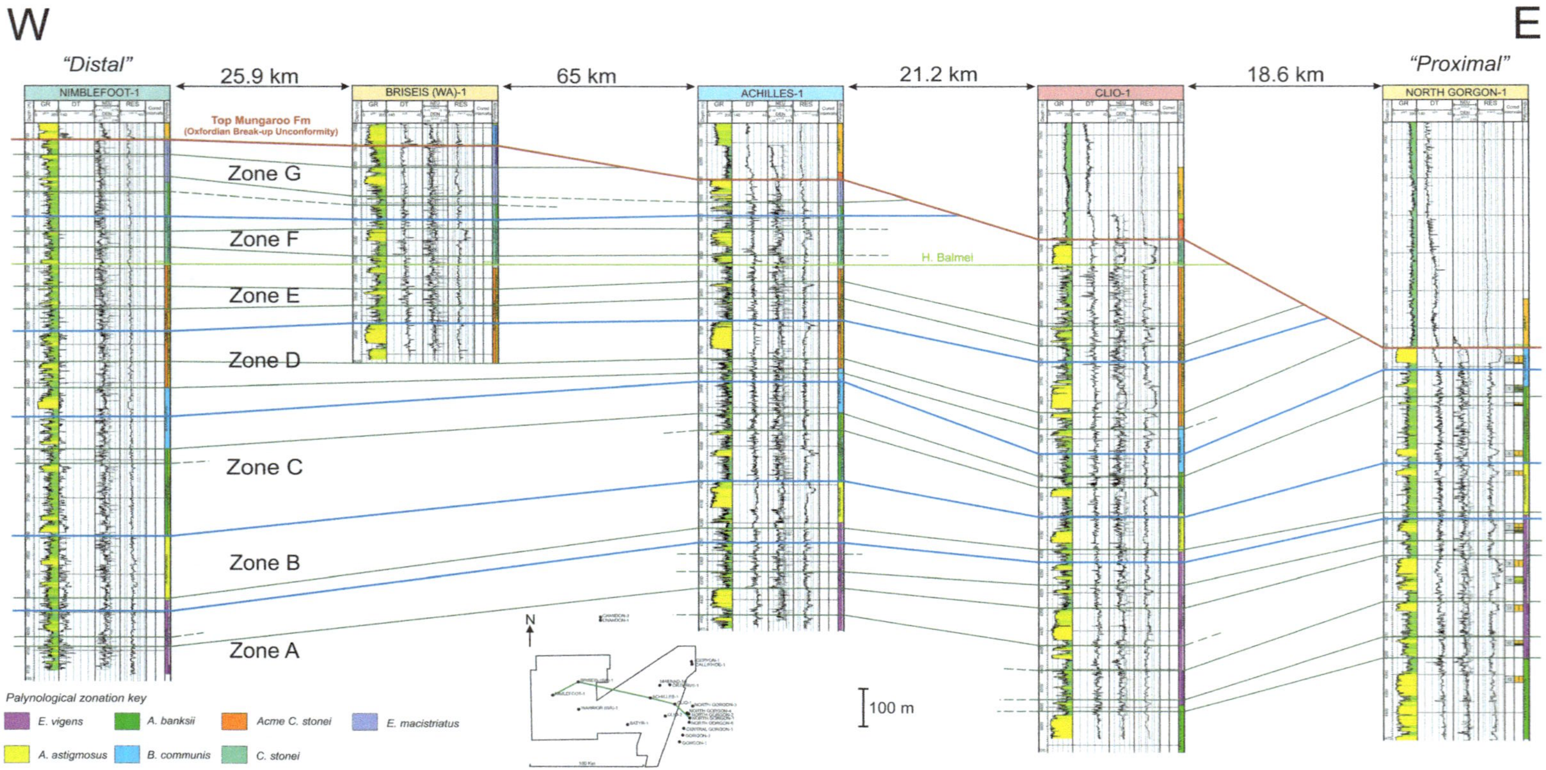

Fig. 16. Semi-regional well-correlation panel across the study area (the *H. Balmei* marine incursion surface is used as the regional datum). Blue lines represent regional flooding surfaces and green lines represent semi-regional flooding/marshy signals from palynology (Dixon *et al.* 2012). Candidate flooding surfaces (blue lines) subdivide the upper Mungaroo Formation into seven zones (A–G).

one of the most widespread marine incursions observed across the study area. Integration with the identification of brackish-marine intervals from core-to-wireline calibration, often occurring at the base of coarsening-upwards packages, and marginal to open-marine dinoflagellates and acritarchs, identified by Dixon *et al.* (2012), have allowed candidate flooding surfaces to be correlated (blue surfaces in Fig. 16). Revised higher-resolution palynostratigraphy by Dixon *et al.* (2012) identified elevated coastal, swamp and lacustrine signals, which often coincide with marine incursions (possibly reflecting a rising water-table associated with the downing of sedimentary systems), allowing these signals to be traced further inboard than marine incursions alone (green surfaces in Fig. 16). Consequently, the upper Mungaroo Formation has been subdivided into a series of seven zones (A–G) in this study: within which, detailed spatial and temporal variability in N:G and sandstone body thickness can be examined (Figs 16 & 17). Fining- (muddying-) and coarsening-upwards (sandier-upwards) trends on the scale of 10–70 m thick can also be observed on wireline logs within biozones. This semi-regional proximal to distal 'trend' sits within (and is consistent with) the regional-scale models that exist within the basin (Longley *et al.* 2002; Adamson *et al.* 2013; Marshall & Lang 2013; Payenberg *et al.* 2013).

From North Gorgon-1 (relatively 'proximal') to Nimblefoot-1 (relatively 'distal'), there is an overall decrease in N:G in a basinwards direction (a decrease in sand N:G from 0.7 to 0.35, respectively), along with a decrease in the occurrence of thicker, blocky sandstone bodies within each individual zone (Fig. 17). There is, however, significant lateral variability in this basinwards trend, which is likely to be a result of autocyclicity within the depositional system. Thicker, amalgamated sandstone bodies (ranging from 20 up to 70 m thick) are present in a 'proximal' setting, whilst thinner (5–35 m thick) sandstone bodies dominate in a 'distal' setting (Figs 16 & 17). There is also a basinwards decrease in the thickness of coarsening-upwards packages from Clio-1/Achilles-1 to Nimblefoot-1 (Fig. 16). The observed spatial variability may reflect

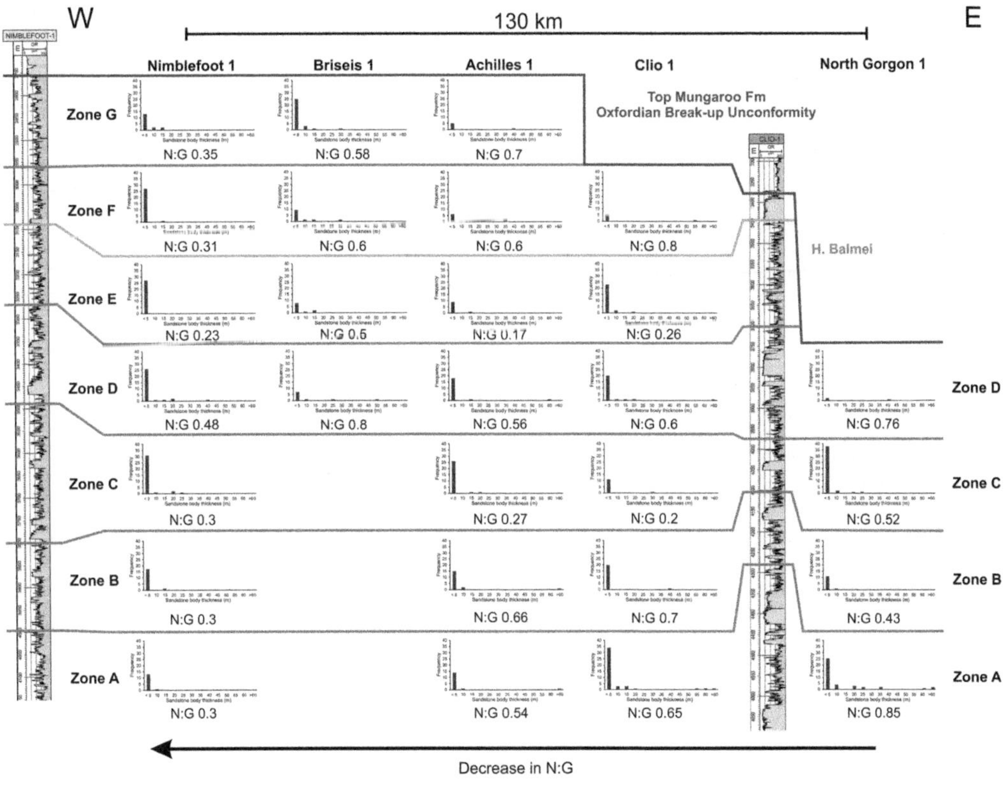

Fig. 17. Frequency histograms of sandstone body thickness and sand N:G calculated within individual correlation zones from each well of the regional well-correlation panel (Fig. 15). Clio-1 and Nimblefoot-1 wireline well logs represent typical 'proximal' v. 'distal' positions of the correlation panel and highlight the decrease in N:G down depositional dip both overall and within individual zones A–G.

proximal to distal changes in reservoir dimensions and distribution: however, an alternative interpretation is that these trends represent axial v. non-axial locations with regard to channel belts.

Sand N:G also varies temporally between individual zones, illustrating an overall distinct change to a dominance of thinner (>5–10 m thick) sandstone bodies in the uppermost younger stratigraphy, which likely reflects the overall backstepping and transgression of the deltaic system (Figs 16 & 17). Superimposed upon this long-term relative sea-level rise are a series of higher-frequency, lower-magnitude cycles. These third- or fourth-order cycles often generate 10–70 m-thick coarsening-upwards sequences which are likely to be progradational in character, and can be related to either sea-level-driven variations in accommodation, climatic-induced changes in fluvial discharge and sediment supply or to autocyclic lobe switching (Fig. 16).

Field-scale correlation. Higher-resolution spatial and temporal variability can be identified from more closely spaced well correlations over a smaller 'field scale'. Alternating zones of relatively high and low N:G can be observed in the eight well strike-section correlation over the Gorgon Field (which is 45 km long and 8 km wide) in Figure 18.

High N:G zones are sandstone prone, and contain thick, stacked and amalgamated fluvial channel sandstones (Association A and B), which range from 10 to 55 m in thickness, as well as crevasse-splay sandstones and minor overbank mudstone (Fig. 19). These amalgamated units can also stack vertically into large-scale amalgamated complexes up to 100 m thick, as seen in North Gorgon-1, Gorgon-3 and Central Gorgon-1.

Low N:G zones are mudstone prone, dominated by overbank shale and siltstone, crevasse splays (Association C and D), coal and only minor fluvial channels/channel belts (Association B). Channel facies commonly occur as isolated, single-storey channels and range in thickness from 5 to 15 m (Fig. 19).

Isolated sandstone bodies within low N:G intervals commonly occurring as discrete fining-upwards units have traditionally been interpreted as representing deposition by meandering fluvial systems. However, Bridge & Tye (2000) and Miall (2010) both state that channel styles (braided v. meandering) cannot be interpreted from vertical lithofacies profiles alone as a result of variable degrees of preservation of fluvial deposits and their associated floodplain fines.

These high and low N:G alternations are likely to be controlled by changes in accommodation and/or sediment supply (or A/S ratio). The amount of accommodation significantly alters the degree of preservation, resulting in potential misinterpretations in fluvial style (Miall 2010). Under low accommodation conditions, channels and channel belts may scour back and forth, effectively cannibalizing earlier deposits and resulting in the removal of channel and overbank fines (Miall 1980). This scenario may represent the depositional processes occurring during deposition of high N:G intervals recording low accommodation conditions (Figs 18 & 19). In addition, during high accommodation conditions, channel systems are able to aggrade and migrate laterally without removing or eroding into earlier deposits, resulting in a more complete preservation of channel and floodplain fines, giving rise to low N:G intervals (Miall 1980). As such, high and low accommodation conditions can arise due to periods of transgression and regression, respectively, caused by relative sea-level changes.

Alternatively, fluctuations in sedimentation rate/supply caused by climatic-induced changes in fluvial discharge and/or large-scale system avulsions can change sediment supply (into the study area), leading to periods of increased channel/channel-belt amalgamation (Figs 18 & 19), which can also be responsible for the observed variability and stacking patterns over the Gorgon Field. Confident differentiation between accommodation and sediment supply as the overriding controlling factor on facies variability and architecture is inherently challenging. The stratigraphic succession is more likely to represent a composite record of changes in accommodation (sea-level driven) and sediment supply (climatically driven) (Schlager 1993). Therefore, it is interpreted that changes within the A/S ratio may have controlled the observed temporal and spatial variability and stratigraphic architecture both on a field and semi-regional scale.

Reservoir element dimensions

Integration of core and wireline-log data allows four classes of sandstone bodies to be defined, based on thickness and log character:

- thick amalgamated channel sandstones (ranging from ≥30 to 75 m thick) interpreted as vertically amalgamated channel belts or incised valleys;
- blocky to fining-upwards sandstone bodies (typically 20–30 m thick) interpreted to represent multistorey fluvial channel belts;
- thin sandstone bodies (10–20 m thick) deposited as single-storey channels and distributary channels;
- very thin sheet-like sandstones (≤5 m thick), which are most likely crevasse-splay deposits and mouth-bar sandstones.

Estimation of channel-body dimensions using sandstone-body thicknesses observed in core and

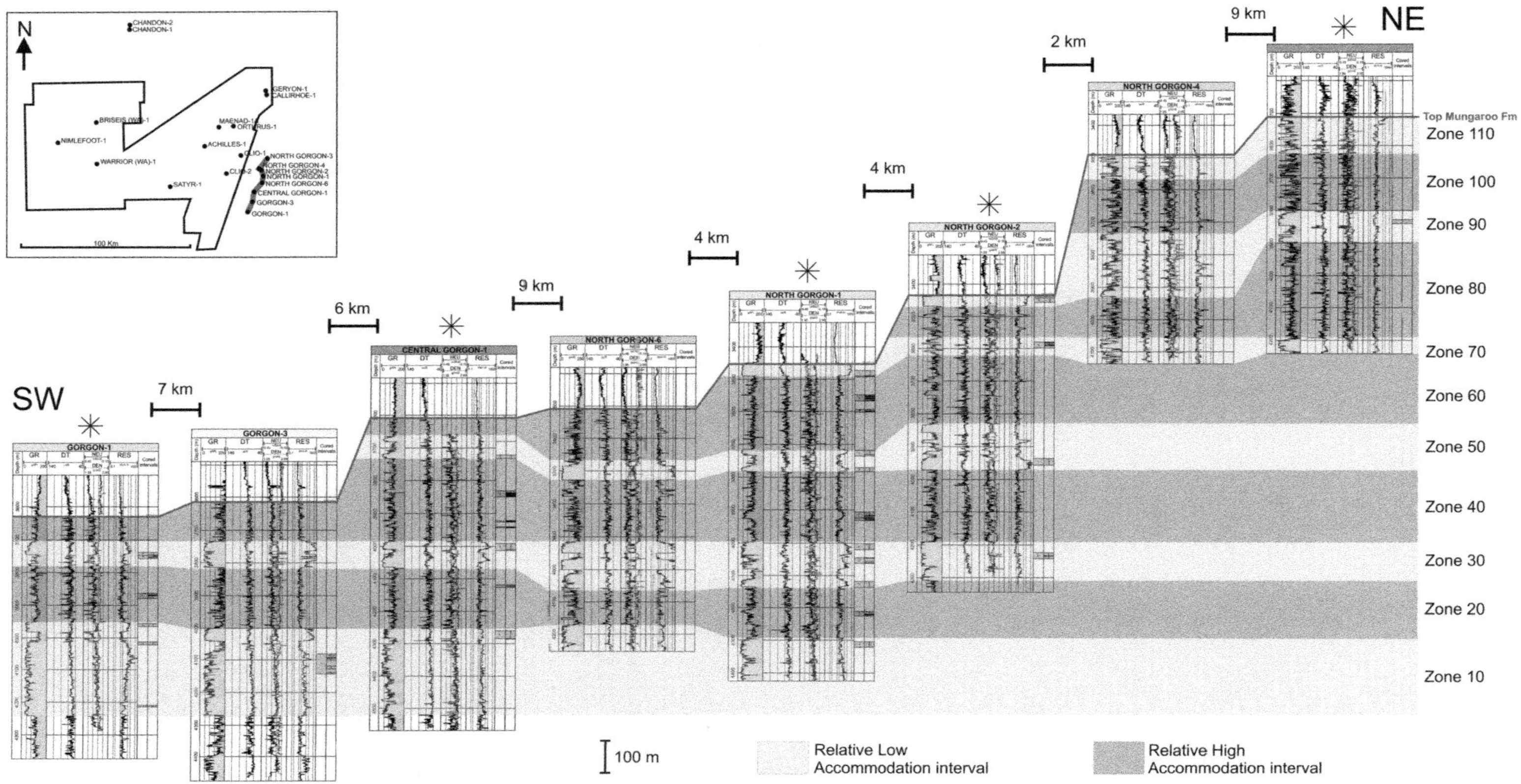

Fig. 18. Well correlation panel across the Gorgon Field showing relative low and high accommodation zones based on the Brenneman *et al.* (1998) model. The thick line on the location map shows the location and orientation of the well-correlation panel. (Note: wells which have an asterisk are the wells that were used in the Brenneman *et al.* (1998) model.)

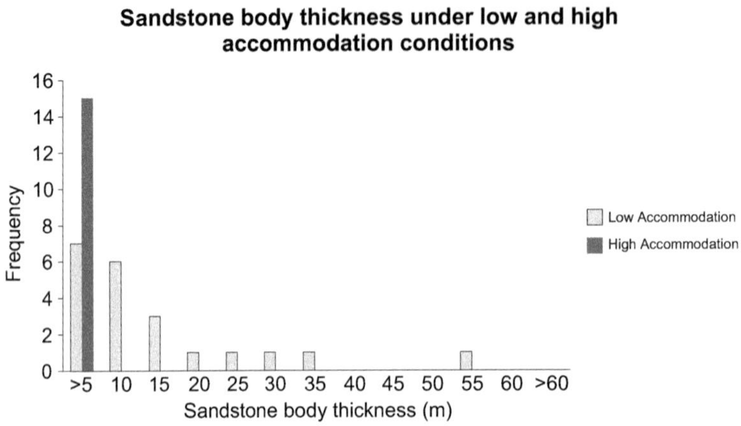

Fig. 19. Frequency histogram to show sandstone body thicknesses under high and low accommodation zones from the Gorgon Field well-correlation panel (Fig. 18).

wireline logs typically relies on comparison with published empirical width-to-thickness relationships based on outcrop analogue studies. However, a critical analysis of these methods highlights the uncertainties, especially when dealing with subsurface data (Bridge & Tye 2000). Channel and channel-belt widths can be estimated as a function of maximum bankfull channel depth: however, maximum bankfull channel depth requires confident estimation of complete, untruncated, channel-bar or channel-fill thickness interpreted from core and wireline logs (Bridge & Tye 2000). Identification of individual unit bars and stacked channel bars is often difficult, if not impossible, without core. In addition, in many cases the presence of sandy-muddy upper bar deposits, used to recognize the top of a bar form, are either not present (in the case of bar amalgamation within multistorey channel belts) or can be confused with near-channel overbank deposits, meaning that accurate identification of palaeo-bankfull level is problematic (Bridge & Tye 2000). The relationship of maximum channel depth to bar thickness can also vary spatially by a factor of 2; thus, limited data from a few wells may not be representative (Bridge 1993; Bridge & Tye 2000). Palaeochannel patterns cannot be reliably interpreted from wireline and core data alone as there is only a weak correlation between sandstone-body dimensions and palaeochannel pattern. For example, the channel-belt width/maximum bankfull depth for the meandering Mississippi River is similar to that of the braided Brahmaputra River (Bridge & Tye 2000).

Lorenz *et al.* (1985) used empirical equations to calculate palaeochannel depths and widths from the thicknesses of fining-upwards trends in core and wireline logs of the non-marine Mesaverde Group, Piceance Creek Basin, NW Colorado. However, recognition of individual point-bar deposits within stacked and amalgamated composite sandstone bodies observed in wireline log alone is very difficult without core in order to distinguish basal erosion surfaces and diagnostic characteristic sedimentary structures between point bars. Despite these recognized pitfalls, Lorenz *et al.* (1985) suggested that channel depths of 3–4 m have channel widths of 45–67 m and correspond to meander belt widths of 350–520 m. The empirical equations used for their study assumed that sandstone bodies in outcrop were deposited from highly sinuous meandering fluvial systems with sinuosities greater than 1.7. They suggested that there was little relationship between the width and depth of low-sinuosity rivers (Lorenz *et al.* 1985).

Bridge & Tye (2000) presented a method whereby channel belt widths can be estimated using well-to-well correlation of channel-belt sandstones. This method, however, relies on many assumptions, such as sandstones at the same stratigraphic level being connected, or that palaeochannel pattern and channel-belt geometries can be inferred from vertical sequences of channel deposits, and, thus, sandstone-body widths can be estimated from palaeochannel patterns. The spatial resolution and density of well control is critical for well–well correlation methods. In areas with widely spaced wells, sand-body connectivity at similar stratigraphic intervals is likely to be highly variable, if we accept the considerable width ranges for sandstone bodies from published literature (Reynolds 1999; Gibling 2006). Furthermore, the sinuosity of reservoir elements is a key parameter for reservoir modelling, which is impossible to extract from 1D borehole data.

The most commonly used method, where spatial data are sparse in subsurface investigations, estimates reservoir dimensions from a database of suitable analogues. Published channel-body width to thickness ranges are used to help constrain and predict the width of sandstone bodies from thickness data extracted from wireline-log data or core studies (Fig. 20).

Extensive databases of widths, lengths and thicknesses of fluvial–paralic sandstone bodies have been compiled by Reynolds (1999), Gibling (2006) and others (Table 1). Outcrop studies have further enhanced width–thickness relationships that can be developed for various fluvial architectural elements (Robinson & McCabe 1997; Pranter *et al.* 2009) (Table 1). Figure 20 highlights the range of published terminology and nomenclature for fluvio-deltaic depositional systems. There is a vast array of terminologies, which produces significant uncertainty and ambiguity when collating individual depositional elements between different studies in the literature, and also in understanding their dimensions and geometries.

The study by Gibling (2006) distinguished channel bodies based on their geomorphic setting, geometry and internal structure, whilst recognizing the large amount of uncertainty in the data. The range of data observed from Gibling's study suggests that selection of appropriate outcrop analogues is paramount, requiring detailed interpretation of both the outcrop and the subsurface depositional setting, which can be difficult given that many depositional models are qualitative and outcrops are rarely extensive enough to allow confident determination of 3D reservoir geometries and dimensions (Bridge & Tye 2000). An outcrop analogue is only as good as the exposure and it is unlikely that data from a few large exposures is representative of all fluvio-deltaic channel belts (Bridge & Tye 2000).

On a field scale, where wells are spaced a few kilometres apart (Fig. 18), connectivity may exist between thicker sandstone bodies within high N:G zones. However, analysis of potential connectivity between channel sandstones in outcrop suggests that a large proportion of them are smaller than a well spacing of 402 and 201 m (Pranter *et al.* 2009), which further supports the understanding that subsurface well correlations and prediction of sandstone-body connectivity, even on a field scale, are complex and enigmatic.

An objective view of the large number of studies would suggest that, at present, there is no reliable method to accurately determine, with any degree of confidence, the width to thickness ratio of channel bodies from core and wireline data alone. For a given thickness, the potential range of widths is very large and, whilst some very broad classes may be distinguished, the significant overlap of width values between different reservoir bodies is problematic: making it challenging to use these data to usefully constrain reservoir geometry (Fig. 21). For example, thick amalgamated sandstones $\geq$30–75 m in thickness may range from 20 m up to 105 km wide, while 20–30 m-thick blocky, fining-upwards sandstones could range in width from 16 m to 15 km. Thinner sandstone bodies identified in wireline logs, which are commonly approximately 10–20 m thick, could range from 5 m up to 5.9 km wide, whereas thin ($\leq$5 m thick) sheet-like crevasse-splay/channel sandstones could range from 5 m to 7.7 km wide (Fig. 21).

In part, this problem is due to our limited ability to accurately classify geobody types and the variety of terminology and methodologies that proliferate in the literature (Fig. 20). Gibling (2006) noted this in his review, and the problem is exemplified in the data plots produced and ranges suggested. In addition, the level of difficulty in identifying bar forms or composite bar forms from subsurface data with confidence, even when extensively cored, adds a significant level of uncertainty.

Seismic attribute analysis

In this study, seismic attribute analysis, calibrated with core and wireline-log studies, has been used to corroborate the range of depositional facies encountered with the Mungaroo Formation. Furthermore, the dimensions of some of the larger-scale sandstone geobodies can be constrained by the seismic data.

RMS attribute was chosen to optimally image the range of sandstone bodies across the 10 000 km^2 sample of the Mungaroo delta system (Figs 22 & 23). Large- through to small-scale seismic geobodies were recognized, representing fluvial channels, channel belts and distributary channels of varying scales (Figs 22 & 23), whilst diffuse high-amplitude areas are interpreted to represent remnants of fluvial channel meander scroll bars or amalgamated crevasse splays, and low-amplitude areas characterize inter-channel, overbank or swamp/embayment facies (Fig. 24). Planform geometries of these geobodies integrated with interpretative proprietary reports of image log data from Clio-2 suggest that the largest-scale geobodies are straight to low sinuosity, containing dominantly downstream accreting mid-channel bars, indicative of braided channel belts (Sixsmith 2009; Ellis *et al.* 2010). Medium- and small-scale geobodies commonly display low–moderate sinuosities in planform, and it is possible that these represent more mixed transitional braided–meandering systems containing downstream accreting side-attached and mid-channel bars (such as patterns 3, 4, 9 and 10 described by Schumm 1981). Given the resolution of the seismic

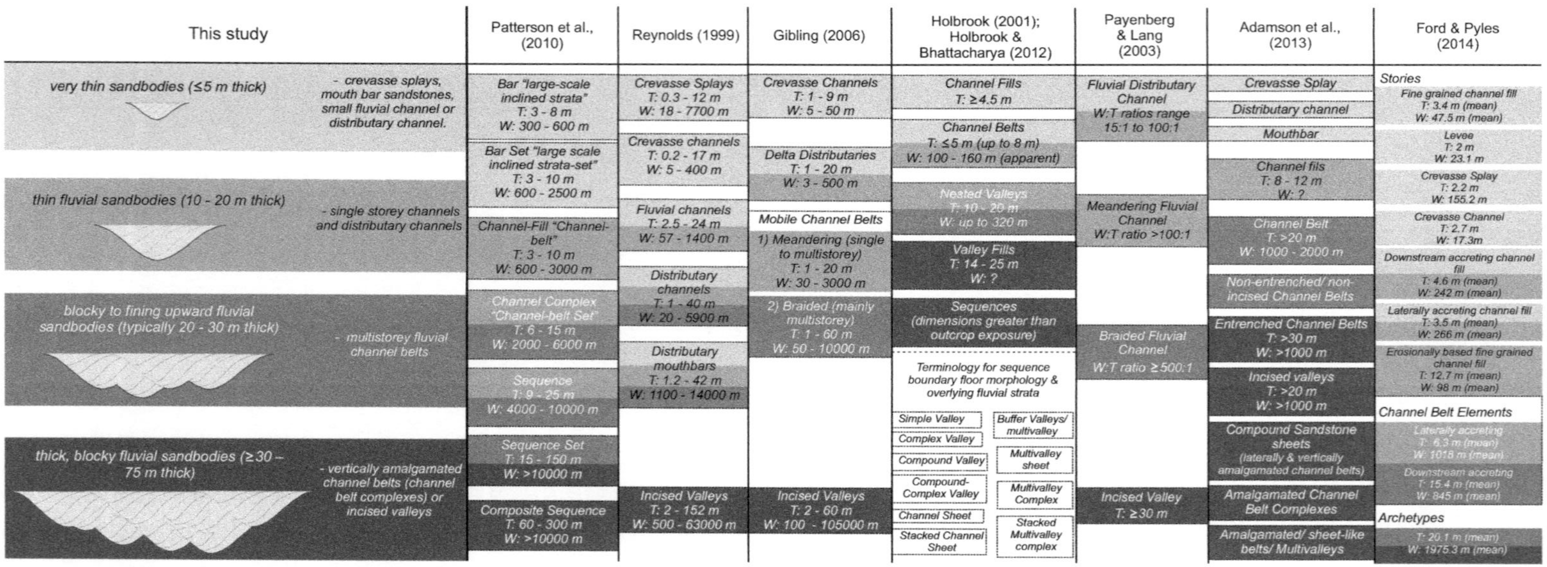

Fig. 20. Comparison of fluvial architectural elements identified from this study, and how they relate to the range of terminology and nomenclature from published literature. Where possible, width (W) and thickness (T) data for the individual architectural elements documented in the published literature are also presented. The greyscale shading relates to the classification scheme interpreted from this study and how the dimensions of these architectural elements correspond to different terminologies used in the literature (Reynolds 1999; Holbrook 2001; Payenberg & Lang 2003; Gibling 2006; Patterson *et al.* 2010; Holbrook & Bhattacharya 2012; Adamson *et al.* 2013; Ford & Pyles 2014).

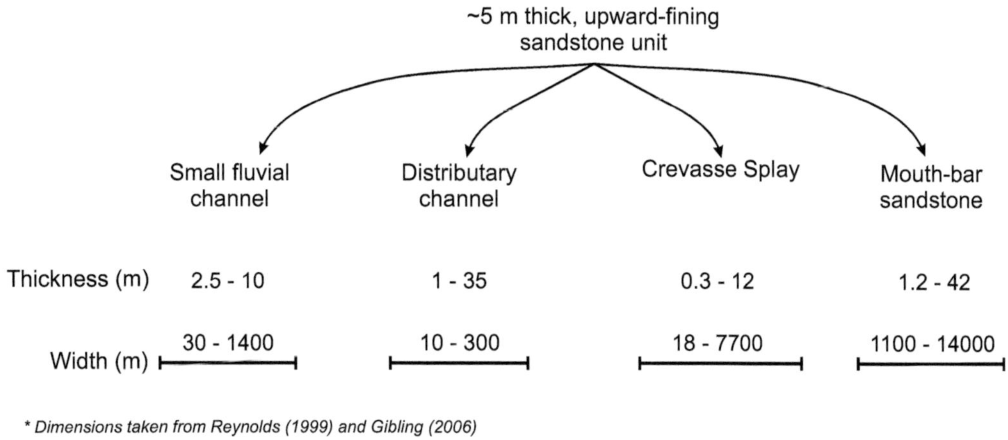

Fig. 21. Schematic diagram to highlight the range of uncertainty in interpretation of a single 5 m-thick, upwards-fining, sandstone unit identified from core and wireline data. The thickness of the sand body is within the ranges identified for four very different architectural elements, each with a very different range of width distributions. The inherent uncertainty and potential for misinterpretation of this sand body have major implications for understanding and predicting sand-body connectivity within 3D static reservoir models (dimensional data derived from Reynolds 1999 and Gibling 2006).

dataset used in this study, it is likely that channel belts are dominantly imaged. Payenberg *et al.* (2014) used channel-belt rugosity (smoothness) to assess the dominant style of bar-form migration (lateral v. downstream) within channel belts. Using this method, it is possible that straight to low-sinuosity large- to medium-scale geobodies may contain dominantly downstream accreting barforms associated with braided channel morphologies within the channel belt: whereas relatively higher sinuosity, small-scale geobodies may represent deposition from more mixed braided–meandering systems or single-thread meandering channel morphologies.

Imaged fluvial seismic geobodies trend in an east–west and NE–SW orientation, further supporting interpretations made by Adamson *et al.* (2013) and Lewis & Sircombe (2013) of a west- to NW-directed palaeodrainage and sediment transport direction for the Mungaroo delta system (Fig. 23). Furthermore, interpretative image log reports corroborate an overall west to NW sediment dispersal direction for both fluvial sandstone bodies and heterolithics/mudstones based on cross-bedding trends. Spatially, there is a dominance of large- and medium-scale geobodies present in the 'proximal' setting, whilst smaller-scale geobodies are more common in the 'distal' setting, which supports the trends observed from wireline-log correlation (Figs 22 & 23). Through time there is also a general evolution from large- and medium-scale geobodies that dominates in the older stratigraphy (zones A–D) into mostly small- and medium-scale geobodies in the younger stratigraphy (zones E–G) (Figs 22 & 23), which substantiates interpretations made from well correlations of overall transgression and backstepping of the deltaic system through time.

Given the range of uncertainty in predicting the dimensions of reservoir sand bodies, imaging them in planform from 3D seismic is the only way to accurately constrain W:T ranges. The seismic interpretation indicates that the large-scale geobodies representing channel-belt complexes are >900 m wide (typically 1–2 km wide). Medium-scale geobodies range from 400 to 900 m wide and are interpreted to represent multistorey channel belts: whereas small-scale geobodies are typically ≤400 m wide and are likely to represent single-storey channel belts or distributary channels (Table 1). The seismic geobodies imaged in Figures 22 and 23 display much narrower widths in comparison to the huge dimension ranges provided by published literature, such as Reynolds (1999) and Gibling (2006).

Where available, seismic data can also be used to corroborate potential controls on the depositional system. Based on core and the wireline-log signature, the thickest sand bodies can be interpreted as incised valleys or multistorey amalgamated channel belts. However, imaging geobodies in plan-view enables cross-cutting geometries to be identified, which are impossible to interpret based on 1D core and wireline data alone. Well locations are often targeted to encounter the maximum potential net sand thickness, which may be at the intersection of cross-cutting geobodies (Fig. 23). Therefore, anomalously thick, amalgamated sand bodies (>30 m thick), identified from core and wireline analysis, may in

(a)

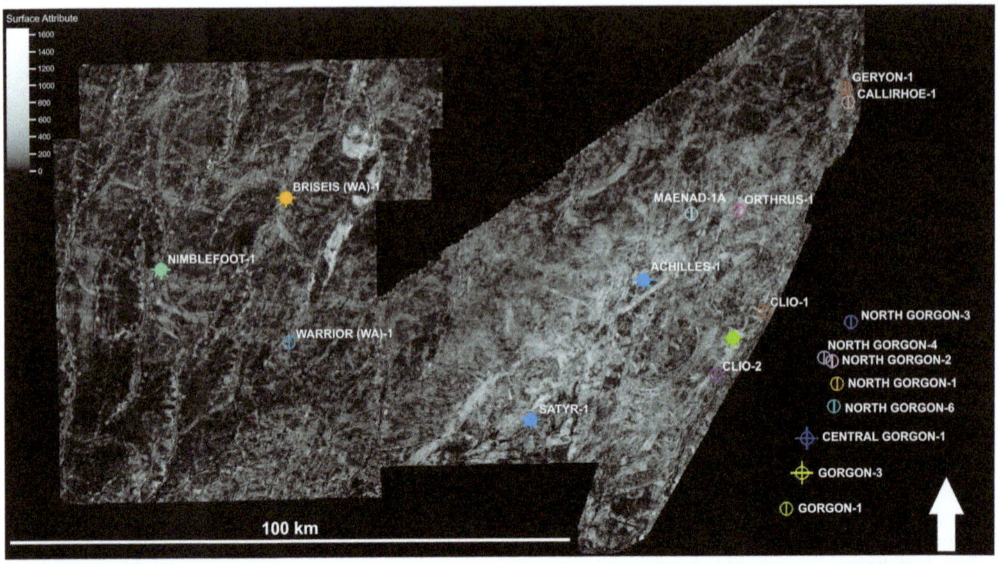

(b)

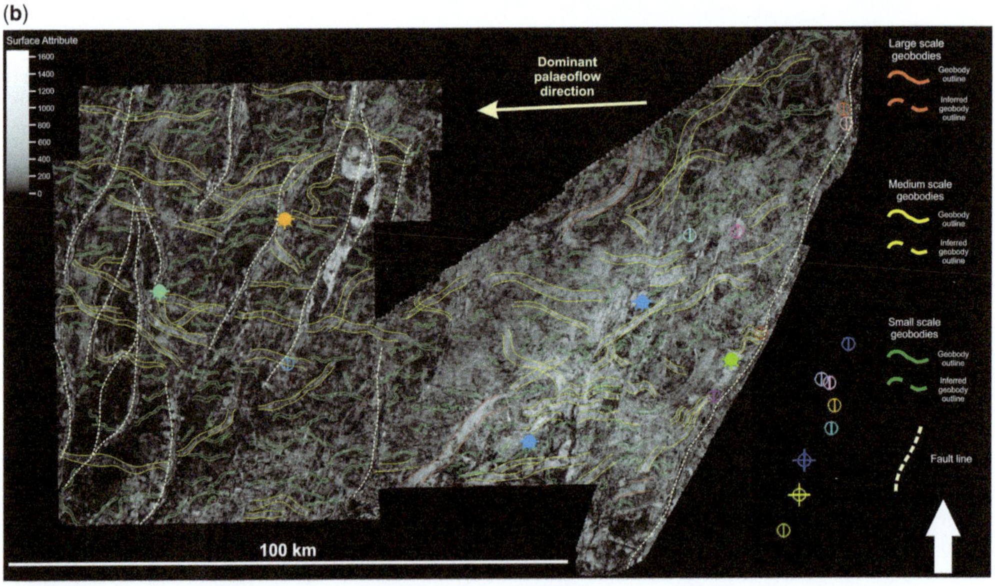

Fig. 22. Seismic attribute (RMS) maps of a 40 ms sculpt from the shallow (relatively younger) part of the stratigraphy (Zone F). (**a**) Uninterpreted RMS sculpt. (**b**) Interpreted RMS sculpt showing the range of geobodies present.

fact be the product of two or more cross-cutting genetically unrelated channel belts, rather than a single incised valley fill. This is the case in the Clio-2 well. The potential for significant misinterpretation and uncertainty of interpretation is further compounded where only one well location (in this study, Clio-2) allows confident core–wireline–seismic calibration.

Discussion

Mungaroo conceptual depositional model

Overall, the upper Mungaroo Formation records the progressive, yet punctuated, drowning and backstepping (second-order transgressive cycle) of a basin-scale, low-relief fluvio-deltaic system. This is interpreted from core by the increased occurrence

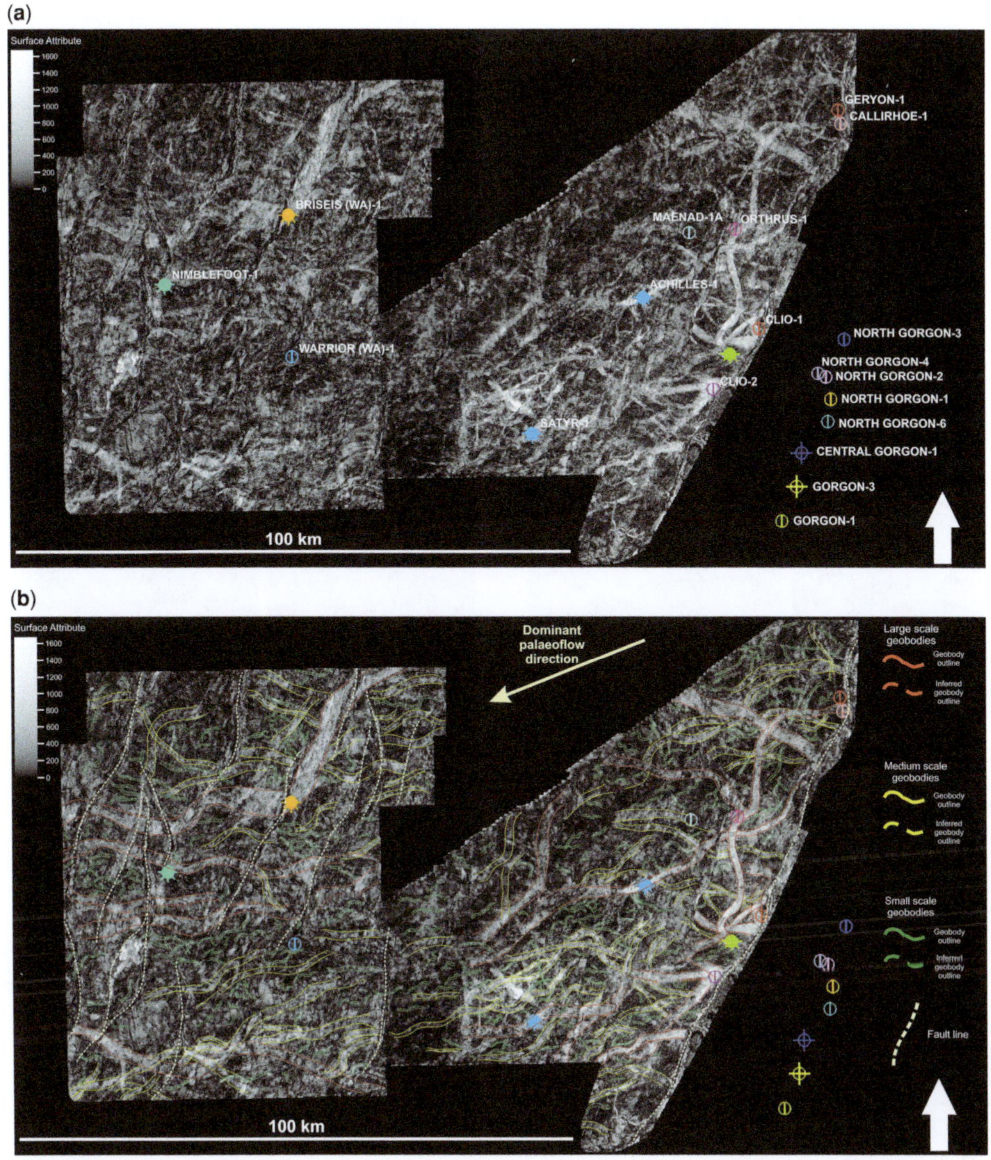

Fig. 23. Seismic attribute (RMS) maps of a 40 ms sculpt from the deeper (relatively older) part of the stratigraphy (Zone D). (**a**) Uninterpreted RMS sculpt. (**b**) Interpreted RMS sculpt showing the range of geobodies present.

of tidal- and wave-influenced intervals vertically through the stratigraphy. Floodplain and soil fabrics observed within the upper Mungaroo Formation are dominantly gleyed. Regional models presented by Adamson *et al.* (2013) and Seggie (pers. comm. 2013) identify moderately to very-well-drained reddened mature soils present in parts of the lower-mid Mungaroo Formation. This regional trend supports the interpretation for an overall-upwards wetting of the depositional system, related to either climatic changes or, perhaps, changing water-table as a result of base-level fluctuations and overall transgression (Adamson *et al.* 2013). Analysis of 830 m of core within the stratigraphic succession suggests that the deltaic system was dominated by fluvial processes with varying degrees of wave/storm and tidal influence throughout. As such, the study area was located predominantly within the upper to lower delta-plain setting of the Mungaroo deltaic system (Fig. 25).

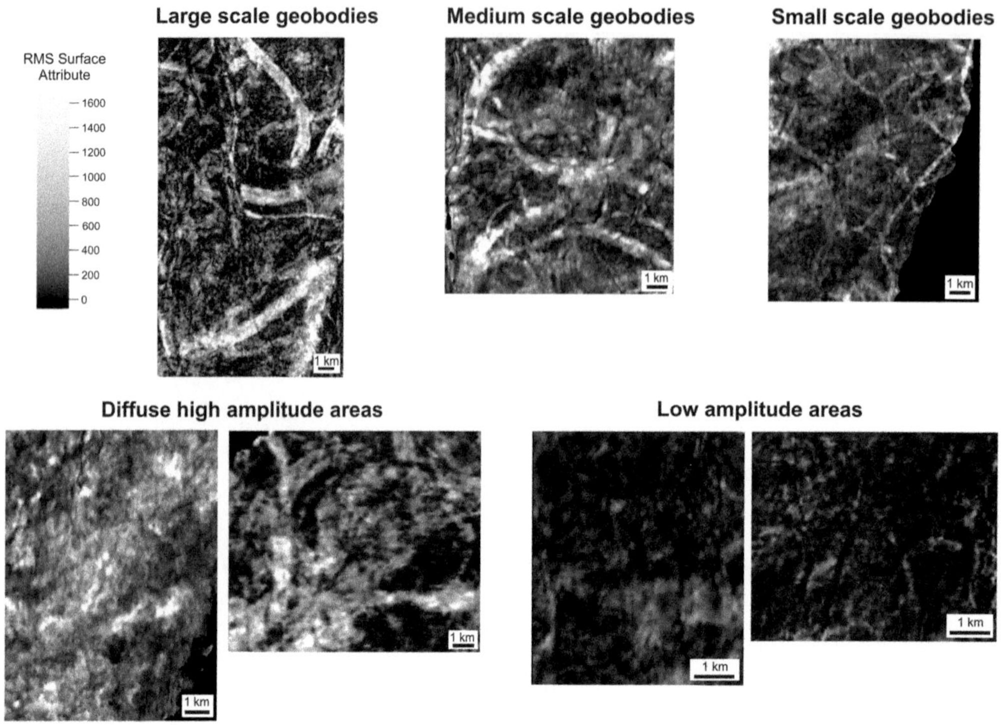

Fig. 24. Seismic facies observed from seismic attribute analysis. Large-, medium- and small-scale geobodies represent fluvial channel belts on varying scales. Diffuse high-amplitude areas are likely to contain remnants of sinuous channel elements, scroll bars and coals. Dominantly low-amplitude areas are likely to represent mudstone-dominated overbank settings.

Relatively ‘proximal’ locations have a dominance of thick (30–70 m), sharp-based, stacked and amalgamated channel-belt sandstones, which often fine up into floodplain mudstones with interbedded crevasse-splay sandstones, reflecting deposition within a predominantly alluvial–upper delta-plain setting (Fig. 25). These features are typically observed over the Gorgon and Clio fields

Table 1. *Sandstone-body thickness distributions from this study (regional well-correlation panel) compared to published literature*

Fluvial architectural element	Data source					
	This study		Gibling (2006)		Reynolds (1999)	
	Width (m)	Thickness (m)	Width (m)	Thickness (m)	Width (m)	Thickness (m)
Incised valley	>900	≥30	100–105 000	2–210	500–63 000	2–152
Channel-belt complex	>900	≥30	–	–	–	–
Fluvial channel	–	–	–	–	57–1400	2.5–24
Multistorey channel belt	400–900	20–30	–	–	–	–
Single-storey channel belt	≤400	10–20	–	–	–	–
Braided low- sinuosity river	–	–	50–13 000	1–1200	–	–
Meandering river	–	–	30–15 000	1–38	–	–
Distributary channel	≤400	5–10	3–1000	1–35	20–5900	1–40
Crevasse channel	–	–	5–400	1–9	5–400	0.2–17
Crevasse splay	–	≤5	–	–	18–7700	0.3–12

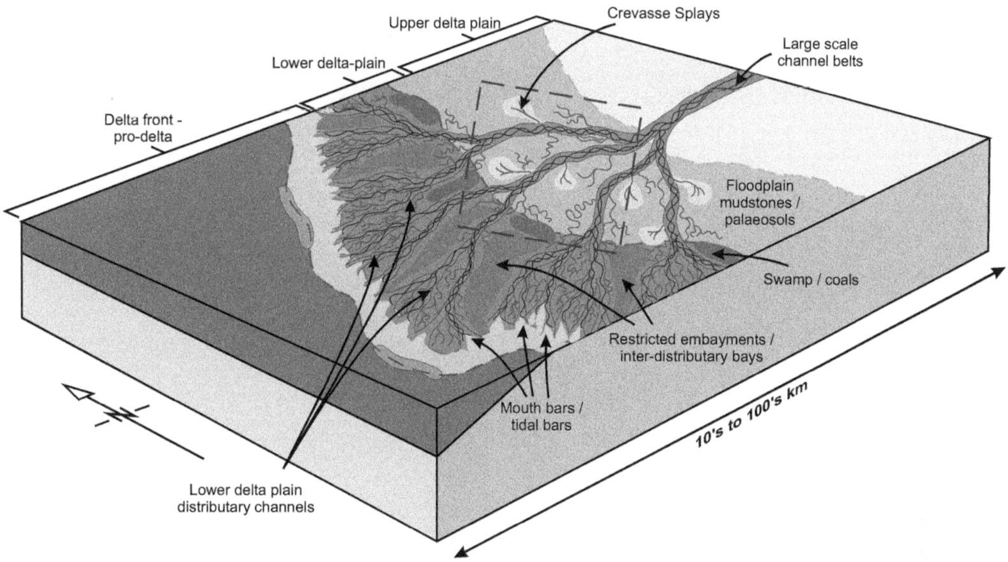

Fig. 25. Conceptual depositional model of the Mungaroo Formation, showing the range of depositional environments encountered from core analysis (the dominant relative position of study area within the depositional system is highlighted by the dashed outlined box). Given the size of the study area in relation to the entire Northern Carnarvon Basin, it is probable that the Mungaroo fluvio-delatic system was composed of a multi-sourced amalgamated coastal plain rather than a single-sourced delta.

(North Gorgon-1; Clio-1 and Clio-2) and deeper in the stratigraphic succession (zone A). In medial to relatively 'distal' positions (Warrior, Briseis and Nimblefoot), the sequence has a different character, with a dominance of 15–50 m-thick, upwards-cleaning (increasingly sandy) packages and typically 5–10 m-thick sandstone bodies. This is interpreted to reflect deposition of distributary channels and progradation of delta-front mouth bars, characteristic of the lower delta-plain to delta-front setting, respectively.

Although thicker sandstone bodies become proportionally less common down depositional dip (basinwards), their characteristics (>25 m-thick stacked and amalgamated channel-fill sandstones) remain broadly similar and are likely to reflect the low gradient of the depositional system. Transgressions recorded over distances of at least 100 km or more also suggest that deposition occurred over a very-low-gradient, broad, low-relief basin, with even small-scale fluctuations in relative sea level causing widespread lateral shifts in depositional environments/facies belts (Adamson *et al.* 2013). Empirical relationships and equations from Bridge & Tye (2000), LeClair & Bridge (2001) and Holbrook & Wanas (2014) were applied to this dataset to calculate a palaeo-shelf angle of approximately 0.02° and a palaeo-depositional slope of 3.12×10^{-4} gradient or 0.017° for the Mungaroo delta.

The upper Mungaroo Formation can be subdivided based on the identification of candidate maximum flooding surfaces (MFS) into sequence stratigraphic systems tracts (Fig. 26b) (Posamentier & Allen 1999; Adamson *et al.* 2013). Core observations and biostratigraphy support the identification of these phases of transgression and marine influence. Weak marine influences are encountered within proximal areas; the most significant and extensive being the *H. balmei* maximum flooding surface (MFS), associated with deposition of proximal delta-front to pro-delta facies across much of the Northern Carnarvon Basin (Adamson *et al.* 2013).

High N:G zones, consisting of stacked and amalgamated channel belts (Fig. 26a, b), can be defined by an erosive sequence boundary at their base and a flooding surface at the top (Posamentier & Allen 1999). Vertically amalgamated sandstone bodies are likely to represent entrenched channel belts (or potentially incised valleys), and are considered to represent deposition during periods of low accommodation and low A/S ratios, generating lowstand systems tract (LST) to early transgressive (TST) stratigraphic intervals (Fig. 27) (Brenneman *et al.* 1998; Adamson *et al.* 2013). Erosional sequence-boundary surfaces may be comparable with diachronous 'regional composite scour' surfaces of Holbrook & Bhattacharya (2012), whereby amalgamated and/or entrenched channel belts or valleys

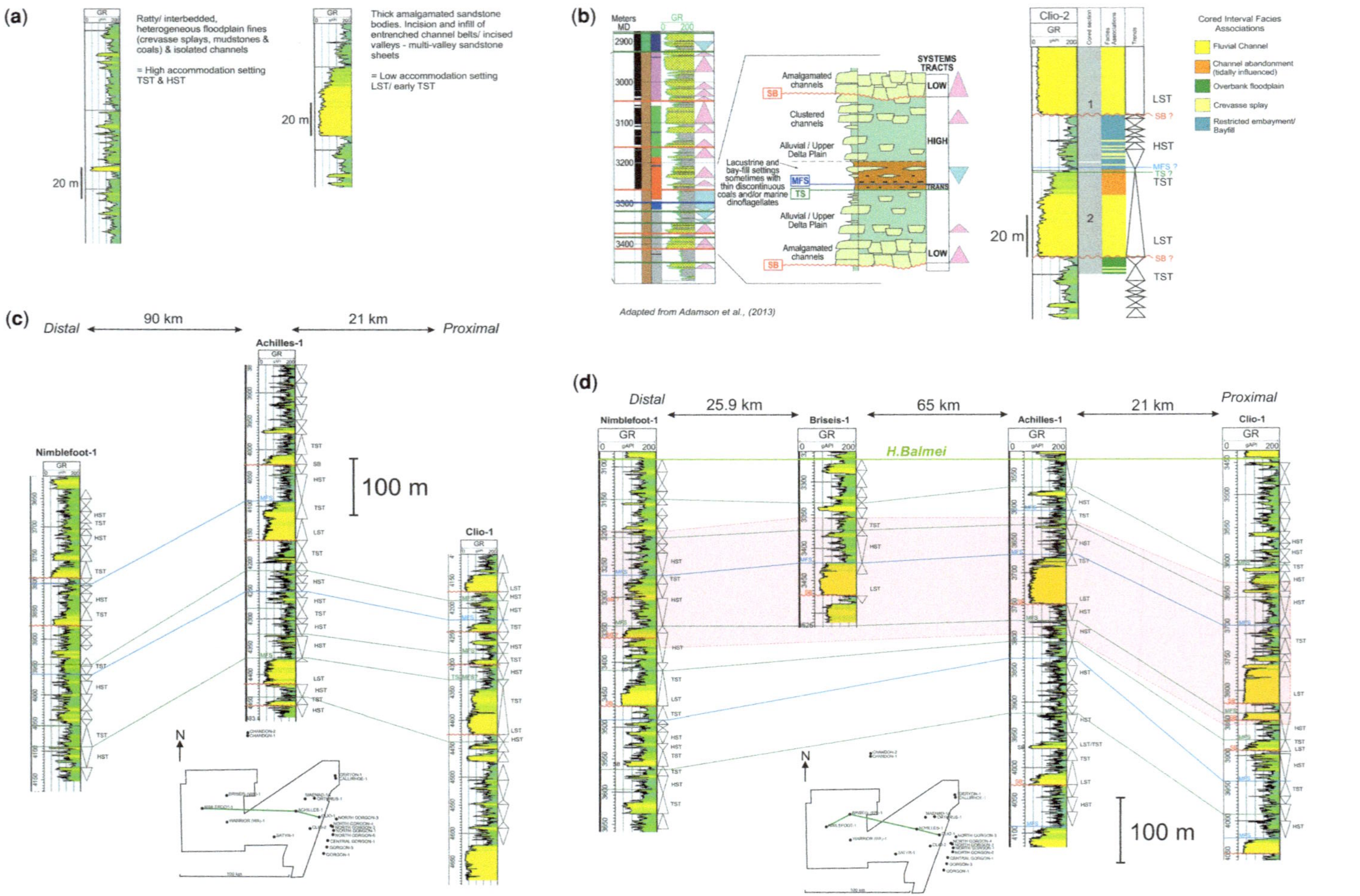

Fig. 26. **(a)** Well-log motifs interpreted to represent deposition under high and low accommodation conditions. **(b)** Potential systems tracts observed from core and wireline studies (adapted from Adamson *et al.* 2013). **(c)** & **(d)** Well-correlation panels over the study area to highlight potential systems tracts in relation to correlation techniques within fluvio-deltaic depositional systems.

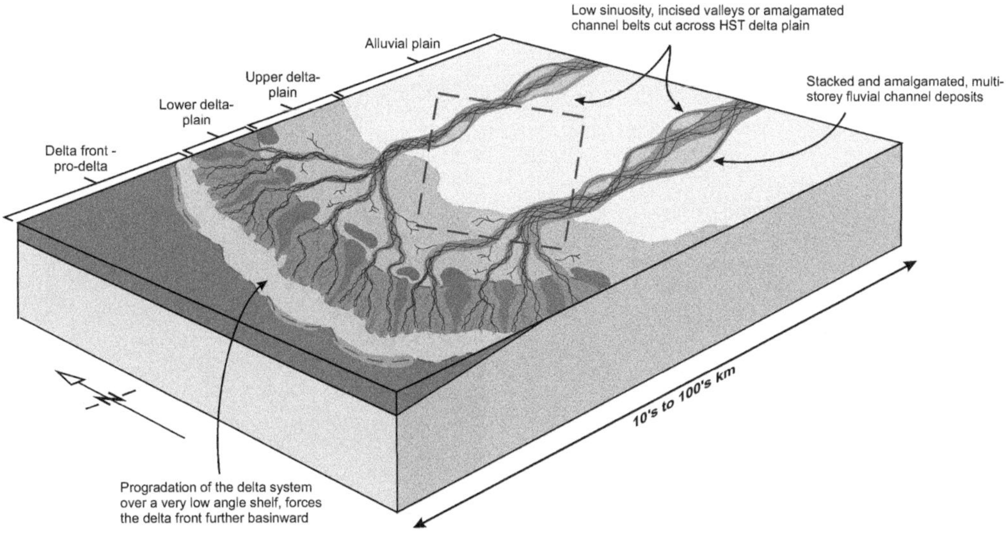

Fig. 27. Depositional model of the Mungaroo Formation under relative regressive phase/LST (the relative position of the study area within the depositional system is highlighted by the dashed outlined box).

were deposited under consistently low accommodation conditions. Recognition of time-equivalent interfluves is required to test whether amalgamated sandstone intervals represent the classic Dalrymple *et al.* (1992) fluvio-estuarine incised valley-fill model. However, due to the well spacing of field and semi-regional well correlations in this dataset, they cannot be differentiated confidently from normal floodplain palaeosol profiles/well-log motifs.

Mudstone-dominated or heterogeneous, low N:G, intervals consisting of floodplain fines (crevasse splays, mudstones and coals), embayment facies and isolated channels/channel belts may reflect deposition under periods of higher accommodation (Adamson *et al.* 2013) or high A/S ratios, and are potentially characterized by a flooding surface at their base and a sequence boundary at the top (Figs 26a, b & 28) (Posamentier & Allen 1999). Brenneman *et al.* (1998) and Adamson *et al.* (2013) interpreted these low N:G intervals as representing deposition under high relative base-level conditions, resulting in relative transgressive (TST) to highstand systems tracts (HST) (Fig. 28).

In a medial–distal position within the dataset, well correlations, based on the identification of flooding surfaces from well-log signatures and palynology, identify 10–30 m-thick, increasing-upwards (coarsening-upwards) packages, with candidate flooding surfaces at their base (Fig. 16). These coarsening-upwards packages contain several smaller-scale coarsening-upwards units (*c.* 5 m thick) and are more common in a medial–distal position within the dataset. These stacking patterns are likely to represent individual parasequences (5 m thick) (Adamson *et al.* 2013) within larger-scale sequences (10–30 m thick), which are likely to be progradational in character.

However, characteristic vertical stacking patterns (Fig. 26d) can change with location across this dataset within a certain time equivalent zone, making identification and correlation of potential sequence stratigraphic packages very challenging. For example, a thick, amalgamated, blocky sandstone unit representing low A/S ratio conditions (possibly associated with a LST/early TST sequence) is not always overlain by a fining-upwards/'transgressive' package; and where relative coarsening-upwards sequences can be observed in a 'proximal–medial' area, a relative fining-upwards sequence is observed in 'distal' areas within the time-equivalent zone (Fig. 26d). These observations suggest that accommodation (driven by changing relative sea level) was not the overriding control on the depositional system, and therefore it is likely that fluctuations in sediment supply caused by climatic/environmental changes in fluvial discharge may also have had a significant effect on reservoir architecture.

Comparison of well-log signatures from relatively 'proximal' to 'distal' settings also emphasize the difficulty and uncertainty in correlating and applying sequence stratigraphic concepts to fluvial-dominated systems over 100 km or more (Fig. 26c). Whilst thick blocky sandstone units in a proximal setting may possess an erosional sequence boundary at their base, correlation into relative distal regions

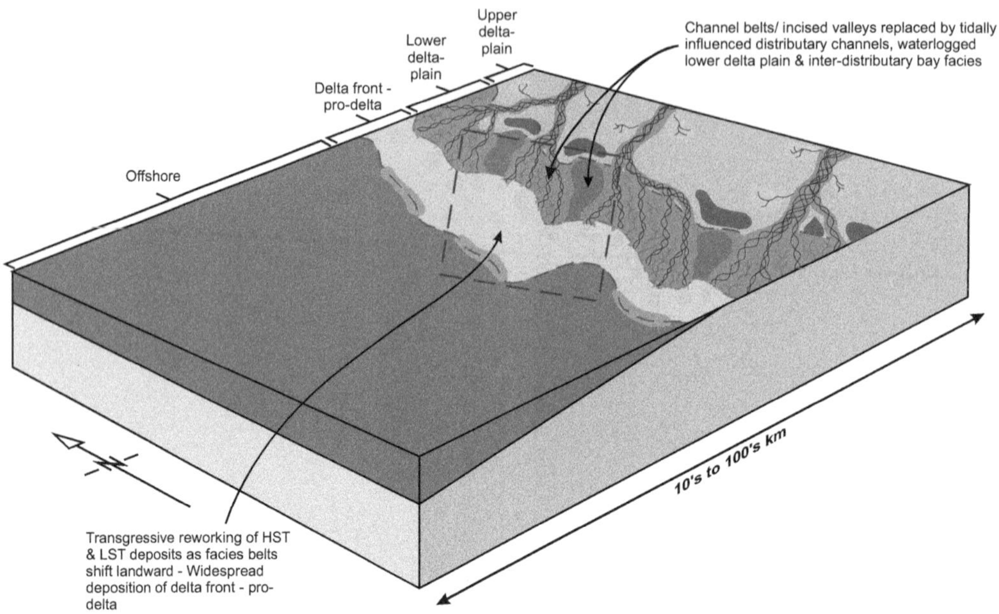

Fig. 28. Depositional model of the Mungaroo Formation under relative transgressive phase/TST (the relative position of study area within the depositional system is highlighted by the dashed outlined box). Under transgression, there is potential for the development of estuaries in place of the channel belts/incised valleys formed under regressive conditions.

is problematic when attempting to identify condensed sections and correlative unconformities (Fig. 26c, d).

Consequently, the lateral facies variability highlights the complex interplay of allocyclic and autocyclic controls on the depositional system at a location where fluvial processes dominate, and emphasizes the challenges in identifying which controls are responsible for the observed stacking patterns.

Controls on depositional evolution

Allocyclic controls

Tectonic. The highly uniform seismic stratigraphy displays no major thickness changes of individual depositional packages, and a lack of syndepositional faulting. Well data also record a uniform isopachous stratigraphy, indicating deposition of the fluvio-deltaic sediments of the Mungaroo Formation in a broad, slowly subsiding, pre-rift sag tectonostratigraphic setting. The main phase of rifting and tectonic activity, related to the onset of Eastern Gondwana fragmentation, was not initiated until the Jurassic and early Cretaceous (Longley *et al.* 2002). As a result, it is unlikely that tectonic activity played any significant role in the location and stacking patterns of reservoir sandstone units during deposition of the Mungaroo Formation throughout the Norian–Carnian.

Climate. Changes in climate affect vegetation and rainfall, which consequently influences fluvial erosion, sediment supply and discharge both within the sedimentary basin and at the source of fluvial drainage systems (Shanley & McCabe 1994). Regional studies suggest that the Northern Carnarvon Basin was located within a temperate to warm, humid and monsoonal climate which was subject to both wet and dryer periods during the late Triassic (Bradshaw *et al.* 1988; Adamson *et al.* 2013). Seasonality is recorded within some palaeosol horizons and floodplain facies, which record an overall upwards-wetting climate: with well-drained reddened mature palaeosols present in older stratigraphy of the Mungaroo Formation, and gleyed/waterlogged gleysols dominating in younger stratigraphy (Adamson *et al.* 2013). Ratcliffe *et al.* (2010) also suggested an overall wetting-upwards of climate from heavy mineral and geochemical data over the Rankin Platform and the Gorgon Field. Adamson *et al.* (2013) speculated that highly seasonal variations in discharge (fluctuating precipitation) during deposition of the Mungaroo Formation might explain the co-association of sinuous planform geometries of channel belts/geobodies within seismic data and unidirectional bed/barform migration observed from image log data. Pebble lags containing extra-basinal clasts within sandstone units observed in core from this dataset may record fluctuations in river discharge and/or precipitation (Reineck & Singh 1980; Tucker 2001), supporting Adamson *et al.* (2013) interpretations.

Furthermore, it is possible that reservoir stacking patterns observed within this dataset represent aggradation and degradation cycles related to climatic, environmental and hydrological changes within source drainage areas as opposed to changes in relative sea level. This would be comparable to sections analysed by Blum (1990) within the Colorado River, Texas.

Eustatic. Tracing eustatically generated sequences and regionally extensive erosion surfaces/sequence boundaries (especially over distances >100 km) updip into entirely non-marine (fluvial-dominated) strata and discriminating them from channel scours is often difficult and largely impractical, because fluvial aggradation and incision can form in both updip and downdip areas simultaneously, independent of sea-level fluctuations and accommodation space (Shanley & McCabe 1994; Holbrook 2001). Moreover, both architectural element analysis and sequence stratigraphy focus on the recognition of bounding surfaces which exist within a bounding-surface hierarchy continuum (Holbrook 2001): however, identification of such surfaces is problematic when using 1D wireline and core data.

Identification of true sequence boundaries in alluvial strata, especially from subsurface datasets, is intrinsically difficult due to a paucity of extensive marker horizons, lack of robust biostratigraphic indicators and the discontinuous nature of alluvial strata (Shanley & McCabe 1994). The amalgamated channel deposit cored in Clio-2 could be interpreted as a classic incised valley fill described by Dalrymple *et al.* (1994) or Hampson *et al.* (1999), based on observed vertical facies changes alone (Figs 26b & 29). However, confident identification of a sequence boundary at its base cannot be made, as not enough of the underlying mudstone succession is cored to identify marine biostratigraphic indicators to infer that the thick amalgamated sandstone package represents a basinwards shift in facies. It is also likely that during regression of the fluvial systems after base-level fall, late transgressive strata (and potential marine indicators) are eroded. Thus, in proximal settings, marine biostratigraphic markers are likely to be relatively rarely preserved, which could easily result in misinterpretation of depositional setting and base-level control.

The formation of a regionally extensive sequence boundary incision, as a result of relative sea-level fall, also depends on the angle of slope between a shallow-marine shelf and an alluvial plain. In low-relief ramp settings, where the fluvial profiles near the coast are similar to those of bathymetric profiles of the shallow shelf, a base-level fall may result in only minor fluvial incision, extension of the fluvial profile and, possibly, a change in channel patterns (Shanley & McCabe 1994). Likewise, Posamentier (2001) identified 'unincised alluvial bypass systems' that developed over low-gradient shelves in response to eustatic sea-level fall. Therefore, it is likely that the development and classic fill of incised valleys (Van Wagoner *et al.* 1990) occurs within relatively high-gradient systems (Ferron Delta and Castlegate Sandstone) (McLaurin & Steel 2000; Yoshida 2000; Moiola *et al.* 2004; Ryer & Anderson 2004). Consequently, quantifying the effects of sea-level fluctuations in lower-gradient systems such as the Mungaroo Formation are problematic when channel incision/entrenchment may not be deep enough to develop well-defined incised valley fills. As such, it is likely that intervals of high N:G and vertically/laterally amalgamated channel-belt sandstones reflect periods of low A/S and enable diffuse and diachronous 'sequence boundaries' or regional composite scour (RCS) surfaces (as described by Holbrook & Bhattacharya 2012) to develop instead of true incised valleys. This interpretation was also made by Adamson *et al.* (2013).

The Holbrook *et al.* (2006) base-level 'buffers and buttresses' model can be employed to help quantify the differences in fluvial architecture and stacking patterns over the study area between relative 'proximal' and 'distal' locations (*c.* 130 km dip distance). The increased dominance of amalgamated fluvial channel deposits in 'proximal' areas and spatial architectural complexity may record higher-frequency variability in upstream base-level controls (climate and discharge) between buffers updip, gaining increasing influence over lower-frequency downstream controls (sea level) (Fig. 30) (Holbrook *et al.* 2006).

Can we distinguish allocyclic v. autocyclic controls?. Spatial and temporal facies distributions within the upper Mungaroo Formation are interpreted to be strongly controlled by the rate of accommodation change, which, in the absence of significant tectonic effects, is likely to be controlled by a complex interplay of allocyclic (eustatic sea level and climate) induced base-level variations (the A/S ratio) and autocyclic controls. Over a 130 km area of this study, the updip dominance and increased thickness of amalgamated sandstone packages, and downdip increase and thickening of floodplain and marginal-marine facies along with the presence of periodic, intercalated marine intervals and flooding surfaces in a relatively proximal

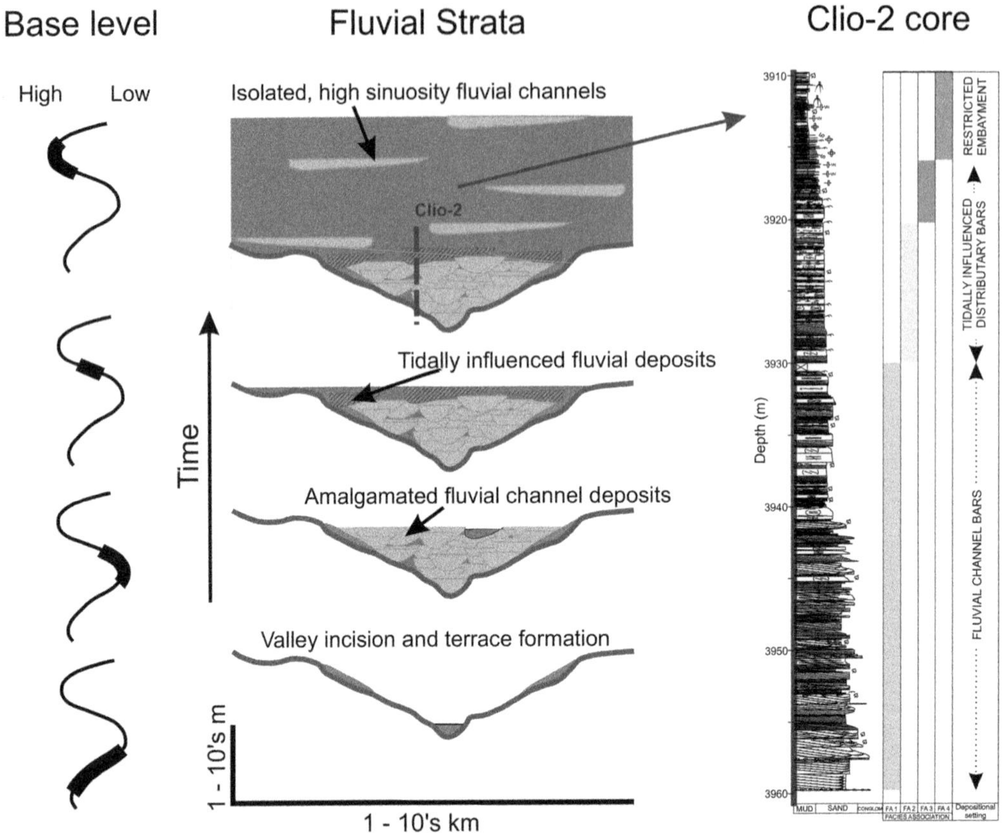

Fig. 29. The vertical association of facies observed from core studies alone (e.g. in Clio-2 well) could be interpreted as a classic incised valley fill. The schematic model of incised valley formation and successive transgressive fill (modified from Shanley & McCabe 1994) could be applied to this part of the cored stratigraphic interval within the dataset.

position, suggest that eustatic sea level had a significant control during deposition.

The combined effect of an allocyclic control on facies distributions and deposition on a large, very-low-angle platform is interpreted to limit accommodation availability, with even small fluctuations in eustatic sea level shifting the depositional system and associated facies belts hunderds of kilometres across the shelf. The presence of intervals with stressed marine ichnofauna, weak-marine biostratigraphic markers, transgressive surfaces and the *H. Balmei* flooding surface within the upper Mungaroo succession infer that eustatic sea-level fluctuations primarily controlled accommodation generation during these periods. The observed vertical change from non-marine fluvial channel sandstone and floodplain mudstones to lower delta-plain, delta-front and pro-delta deposits identified from core in the uppermost Mungaroo Formation also suggest a high-magnitude sea-level control on the depositional system. This vertical trend is also observed by regional studies of Adamson *et al.* (2013) and Payenberg *et al.* (2013).

Adamson *et al.* (2013, p. 22) identified '3rd and 4th order high magnitude transgressive–regressive depositional cycles as recording higher frequency variations in accommodation primarily driven by sea-level change together with sediment supply and/or climate'. Their study also recognized repeated cycles of entrenchment/incision within entrenched channel belts and buttress-type 'valleys' related to periods of lowstand (LST), followed by floodplain mudstone dominated intervals reflecting increased accommodation during transgressive (TST) and highstand (HST) periods (Adamson *et al.* 2013).

Given the relatively small areal extent of this study in relation to published regional studies across the entire Northern Carnarvon Basin, it is important to understand how far inland the sea-level

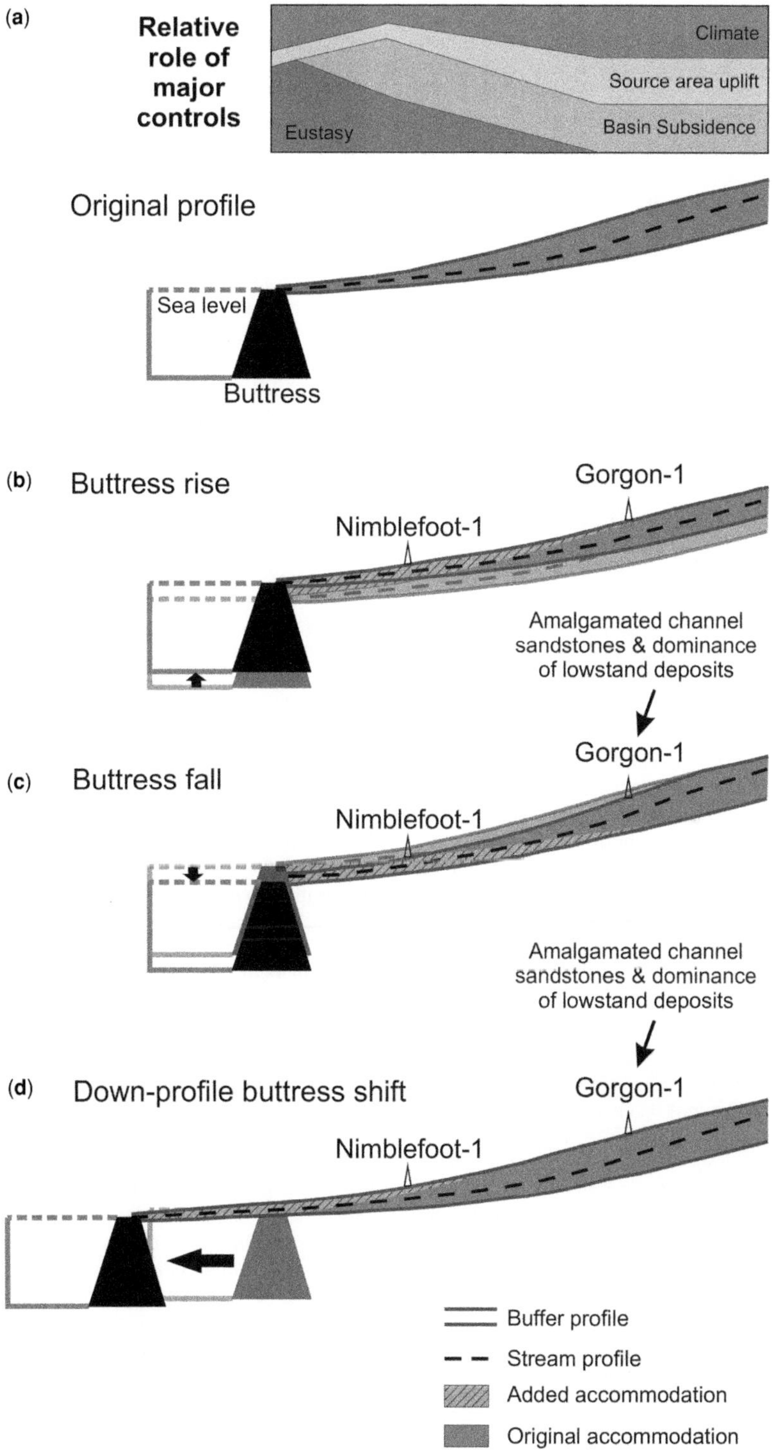

Fig. 30. Schematic diagram to show the effects of changing relative base level on the coastal and inland locations of the depositional system (adapted from Holbrook 2001). The relative importance of controls on the depositional system in a basinwards and landwards direction are also highlighted (adapted from Shanley & McCabe 1994).

fluctuations actually control fluvial incision, aggradation and stacking patterns, and the applicability of Posamentier & Vail (1988) sequence stratigraphic concepts even within relatively proximal areas of a continental-scale, very low gradient, fluvial-dominated depositional system. Changes in relative base level (sea level or subsidence) significantly influence the fluvial graded profile and equilibrium surfaces in the lower coastal plain, and will be the major factor in controlling the stratal architecture of fluvial systems in this region (Shanley & McCabe 1994). However, in 'proximal' areas, the effect of relative sea-level change becomes diminished, and climate, tectonic uplift of source area, sediment supply and discharge become increasingly important controls on stratigraphic base level and fluvial architecture (Shanley & McCabe 1994; Blum & Tornqvist 2000; Catuneanu 2002).

Autocylic controls. Muto & Steel (2004) and Muto & Swenson (2005) proposed that the autogenic response of fluvial deltas to steady sea-level fall is that the deltaic system behaves aggradationally (without major incision) during initial sea-level fall until the 'auto-incision threshold' of the delta system is reached, after which, incision occurs. As such, the base of thick amalgamated sandstone units would represent the delayed response to steady sea-level fall rather than sudden allocyclic events/changes in the rate of base-level movement. Conversely, with a constant rise in relative sea level, initially the deltaic shoreline continues to migrate and prograde basinwards, but later begins to 'autoretreat' (Muto *et al.* 2007). The identification and correlation of time-equivalent, chronostratigraphic surfaces across the basin and integration with existing sequence stratigraphic models of the Mungaroo Formation would be problematic.

Autocyclic processes, such as delta-lobe switching, were observed by Kosters & Suter (1993) and Kulp *et al.* (2005)from the Mississippi delta, and the data suggest this may be an active process within parts of the upper Mungaroo Formation. The cyclic juxtaposition of embayment and delta-front facies above fluvial channel and distributary channel sandstone bodies observed in the upper part of the Chandon-2 core may either represent a marine flooding surface related to eustatic sea-level rise or delta-lobe switching associated with autocyclic controls. Intervals of thick, stacked and amalgamated fluvial-dominated sandstones may represent periods of regression/delta-lobe progradation (Kulp *et al.* 2005), which are often abruptly overlain by swamp and restricted embayment facies, which reflect delta-lobe abandonment, long-term compaction, subsidence and relative sea-level rise.

The backwater effect described by Nittrouer *et al.* (2011, 2012), Chatanantavet *et al.* (2012) and Jerolmack & Swenson (2007) may have also influenced the spatial and temporal trends in fluvial architecture observed within the Mungaroo Formation. Backwater effects can extend for tens to hundreds of kilometres upstream in low-gradient river systems (Nittrouer *et al.* 2012). As rivers enter the backwater segment, non-uniform flow conditions cause bedload in the channel to be deposited, causing aggradation and sedimentation in the upper reaches of the backwater zone (Nittrouer *et al.* 2011). The zone of net aggradation and sedimentation in the backwater segment coincides with the location of increased distributary channel avulsions and a change in channel morphology (Jerolmack & Swenson 2007; Chatanantavet *et al.* 2012), which results in localized concentrations of channel and channel-belt deposits (Blum *et al.* 2013). The observed increase in occurrence and amalgamated thickness of channel belts in relatively 'proximal' locations, and the downdip decrease in geobody thickness and N:G within individual stratigraphic zones, may have been controlled by backwater lengths extending hundreds of kilometres upstream (as observed in the lowermost 150–600 km above the modern-day Mississipppi River outlet: Nittrouer *et al.* 2011, 2012). However, distinguishing between backwater processes or climatically induced variations in fluvial discharge and sediment supply as the major control on the observed changes in fluvial architecture is likely to be very challenging.

Depositional patterns and key sequence stratigraphic surfaces which reflect allocyclic or autocyclic controls within updip and downdip locations is inherently difficult to differentiate within fluvial-dominated settings, as flooding surfaces are likely to amalgamate with autocyclic erosion surfaces upstream that may not represent any allocyclic forcing mechanism. Correlations between fluvial and coeval marine deposits rely on the identification of systems tracts and major stratal surfaces, such as sequence boundaries and landwards equivalents of maximum flooding surfaces (Posamentier & Vail 1988; Shanley & McCabe 1994; Posamentier & Allen 1999). There is evidence, however, for widespread flooding surfaces and coeval transgressive surfaces within the Mungaroo Formation, which can be used to understand the stratigraphic framework over a distance of approximately 130 km. This framework and the range of uncertainties and alternative interpretations can be integrated into larger-scale regional studies within the basin.

Allocyclic changes can cause autocyclic process changes: for example, changing eustatic sea level can affect the nature/degree of delta-lobe switching (Cecil 2003). As such, deciphering between the products of allocyclic and autocyclic processes is challenging given that both processes are inherently interrelated. In this study, the regional correlations indicate that eustatic sea level is likely to have controlled the large-scale, overall transgressive nature of the depositional system through time and was the dominant control during certain intervals within the upper Mungaroo Formation. The low gradient of the depositional system facilitated the incursion of marine transgression events into proximal locations within the basin. These candidate flooding surfaces can be used for semi-regional and regional well correlations.

However, between these maximum flood intervals, the lack of cyclic high and low N:G packages (Fig. 16) suggests that sea level is not the overriding control. Given the medial position of the study area in relation to the whole Mungaroo delta, the observed complex spatial variability in facies and N:G is most likely to be due to an interplay of climatic controls affecting sediment supply/fluvial discharge and autocyclic processes.

Limitations and uncertainties of interpretation

This study utilizes a range of high-resolution datasets over an area of 10 000 km^2. Despite the large scale of this study extending 130 km in length across the medial portion of the Mungaroo delta, it still represents only approximately 5% of the entire Northern Carnarvon Basin. Results from this study can be compared and integrated with regional studies to refine depositional models and assess the range of potential controls on the system.

Cored intervals within this dataset are typically exploration well cores that target reservoir intervals: in this case, fluvial sandstone bodies (Fig. 4). As such, representative core data through coeval delta-front facies are rare. There is always a level of uncertainty of facies interpretation, and this increases for some facies with only sporadic or no core control and those with non-unique wireline-log signatures. In addition, if there is a lack of robust data, such as high-resolution biostratigraphy, identification of true sequence boundaries at the base of thick stacked sandstone units is impossible and confidence in differentiating vertically amalgamated channel belts from incised valleys is low. The implications of this uncertainty and potential for misinterpretation can considerably affect models of sand-body connectivity. In this study, to address this uncertainty, the 11 sedimentological facies associations have been condensed into five well-log facies associations/motifs that have higher levels of confidence. This study addresses the impact of scale of interpretation from core to wireline datasets and the uncertainties of making confident well correlations over hundreds of kilometres.

Characterization of channel pattern (braided v. single-thread low- to moderate-sinuosity channel belts) from core, wireline and image log reports is subject to significant uncertainty. As identified by Schumm (1981), channel patterns form a continuum, and low-sinuosity channel belts dominated by downstream accreting side-attached and mid-channel bars can be formed by purely braided systems through to mixed transitional braided–meandering systems. Changes in channel morphology related to critical threshold values of stream power, gradient and sediment load can take place along any longitudinal position of the river system (Schumm 1981). Therefore, where image log data may infer a dominance of downstream accreting mid-channel bars associated with braided channel patterns within a channel belt at one borehole site, it is possible that channel morphology can change to become composite braided–meandering systems a short distance from that well location.

In subsurface studies, core and wireline-log data are the key data for interpreting the depositional settings and, in particular, understanding reservoir facies and their hydrocarbon/fluid flow potential. These data provide good vertical control, but lack the critical element of lateral control. It can be used to determine sandstone-body thickness, although determining whether units are vertically stacked or laterally connected is still difficult. Channel-body widths and architecture can be estimated from channel-body thicknesses by comparison to published literature, based on outcrop or modern studies (e.g. Reynolds 1999; Gibling 2006); however, the compiled data for different fluvial geobodies have a significant range of dimensions for a given thickness. In most cases, the results have an order of magnitude range, and this level of uncertainty poses significant problems for predicting channel-body connectivity and correlation in the subsurface, with major implications for the development of 3D reservoir models. As input for reservoir modelling, they are widely regarded as being far from satisfactory (Bridge & Tye 2000).

Core and wireline data can only take us so far in our understanding of subsurface geobody dimensions and their potential connectivity. Analysis of high-resolution 3D seismic data (calibrated to core and wireline logs) provides the missing link: enabling channel belts to be imaged at a range of scales, and, where drilled, widths/thickness dimensions of reservoir units can be constrained much more accurately. For areas where seismic data quality is not good enough to image channel/channel-belt morphologies, analogue data are the only option. It is clear that further high-resolution quantitative studies are required, with more precise terminology, to build robust ranges of geobody dimensions for well-constrained depositional settings, and to generate significantly larger meaningful statistical datasets.

Conclusions

- Detailed analysis of wireline-log data and approximately 830 m of conventional core provides a number of type sections for the Mungaroo Formation, with facies that record deposition predominantly within upper delta-plain (fluvial channel, levees, crevasse splays, overbank, palaeosols and swamp/marsh) to lower delta-plain distributary channels, restricted embayments and, to a lesser extent, delta-front to pro-delta settings.
- This paper documents the core and log data, all available as open files through the GSWA, and provides the first published descriptions that can be linked to specific wells and core depths for future reference.
- Semi-regional well correlations are presented that define vertical and spatial facies variations within the interval of well control:
  - a gross decrease in N:G downdip, which reflects a typical facies change basinwards within the deltaic system;
  - the proportion of thick fluvial-dominated amalgamated sandstone bodies also decreases through time, which reflects the overall drowning and retrogradation of the Mungaroo delta system throughout the Norian.
- The uncertainty of interpretation of facies using wireline-log characteristics is highlighted in this study. With only sporadic core control, alternative facies interpretations are possible for some facies. The implications of this uncertainty needs to be recognized and can be significant for developing reservoir models. Where there is insufficient, robust data to confidently differentiate amalgamated channel belts from incised valleys, the misinterpretation can considerably affect modelled sand-body connectivity.
- The application of sequence stratigraphic models to continental, low-gradient fluvio-deltaic systems has been reviewed. Accommodation availability governs the generation and preservation of laterally continuous chronostratigraphic markers such as marine flooding surfaces. Correlation of these stratigraphic surfaces becomes highly uncertain in a landwards direction due to the progressive decrease in accommodation created in response to relative sea level.
- The study area is located within a dominantly medial part of the Mungaroo delta system. As such, spatial facies variability is interpreted to be caused by the interplay of autocyclic and allocyclic controls on accommodation at different scales of observation. Eustatic sea level is likely to have controlled the large-scale, overall transgression of the depositional system through time, and the low gradient of the depositional system facilitated marine-transgression events extending into 'proximal' locations within the basin during certain intervals within the upper Mungaroo Formation.
- However, between these maximum flood intervals, the observed complex spatial facies variability is most likely controlled by the combination of climatic-induced changes in sediment supply/fluvial discharge and autocyclic processes.
- Three classes of sandstone bodies/reservoir elements have been identified from core–wireline–seismic data integration and calibration based on their thickness distributions. Large-, medium- and small-scale geobodies represent fluvial channel-belt complexes, multistorey channel belts and single-storey channel belts, respectively.
- If seismic data were poor, analogue data would typically be used to predict channel-body widths and sinuosity from thickness data extracted from wells. We critically assess this and show how the range of possible sandstone-body dimensions based on published literature can have a range that is an order of magnitude, and as such results in significant uncertainty for input into reservoir models.
- The results emphasize the need for the collection (globally) of additional high-resolution and well-constrained analogue data to more precisely characterize geobody type and provide improved statistical data to improve our understanding of reservoir element dimensions, if relying on well-log datasets.
- If high-quality 3D seismic data are available, seismic attribute analysis can enable W:T dimensions for reservoir elements to be constrained more accurately, with significant implications for understanding sand-body connectivity and improving static 3D reservoir models.

This research is sponsored by Shell International. Many thanks to Prof. Keith Gerdes for initiating the project and Stuart Taylor (Shell Development, Australia) for valuable and insightful discussions during time spent in Perth, Australia. Also many thanks go to the whole team at Shell Development Australia for their valuable discussions and support. Luz Gomis-Cartesio, Marcello Gugliotta, Miguel Poyatos-More and Brian Burnham are also gratefully acknowledged for their constructive discussions during core analysis. In addition, much appreciation is expressed to Prof. Annette George for her support at UWA and the core store team at the Geological Survey of Western Australia, Perth Core Library for their support and allowing access to the core material in an excellent facility. Schlumberger are gratefully acknowledged for software academic licences. The authors would also like to thank Keith Adamson and Tobi Payenberg for their detailed reviews and constructive comments for improving the manuscript.

Appendix A

Table A1. *Lithofacies*

Code	Lithofacies	Sedimentary description	Interpretation and depositional environment
Gms	Matrix-supported conglomerate	Rounded to well-rounded quartz, feldspar and lithic granules to pebbles. Often poorly sorted clasts suspended in a medium- to very-coarse-grained, quartz-rich sandstone matrix. Forms a 5–10 cm-thick layer with a sharp basal contact	High-energy flow conditions. Forming a pebble lags at the base of individual bed sets and fluvial channel storeys
Gcs	Clast-supported (sandy) conglomerate	Rounded to well-rounded quartz, feldspar and lithic granules to pebbles. Forming a poorly sorted 1–6 cm-thick clast layer with a sharp basal contact	High-energy flow conditions. Concentrated at the base of bed sets and forming pebble lags at the base of individual channel storeys
Gm	Intra-formational mud conglomerate	Clay and carbonaceous intraclasts suspended in a fine- to coarse-grained sandstone matrix. 0.1–5 cm diameter mud clasts often present along with Gcs and Gms lithofacies at sharp basal contacts	High erosive power to erode underlying mud during river flood periods. Mudstone rip-up clasts from the underlying bed
Sm	Massive sandstone	Structureless grey-, buff- to fawn-coloured, fine- to coarse-grained sandstone. Rounded to subrounded well-sorted quartz grains	Rapid deposition within a fluvial channel, from high- density sediment gravity flow deposits and/or post-depositional modification (recrystallization–replacement)
St	Trough cross-bedded sandstone	Grey- to buff-coloured trough cross-bedded, fine- to coarse-grained sandstone. Moderate- to well-sorted, rounded quartz grains are dominant. Angle of individual cross-beds often increases or decreases vertically though a single continuous bed set unit: i.e. from 20° to 25° at the base to 5°–10° at the top of a unit and also becoming more argillaceous/darker coloured at the top	Unidirectional migration of 3D sinuous to linguoid-crested dunes in high- to moderate-energy fluvial channels under lower flow regime
Sp	Planar cross-bedded sandstone	Light-grey- to buff-coloured, fine- to coarse-grained sandstone. Tabular cross-bedding with individual foreset angles ranging between 15° and 20°. Coarse-grained cross-bedded units usually poorly to moderately sorted with subangular to subrounded quartz grains. Medium- to very-fine-grained cross-bedded units, usually well to very-well sorted with rounded to subrounded quartz grains. Individual cross-beds range in thickness from 0.5 to 8 cm and often show a fining-upwards trend within each foreset laminae from very-coarse-/coarse- to medium-/fine-grained-sandstone. Cross-bed sets typically range from 10 cm – 1 m in thickness	Unidirectional migration of 2D straight-crested dunes and thalweg bars in high-energy fluvial channels, under lower flow regime
Sl	Low-angle cross-bedded sandstone	Similar to Sp, but with cross-bed angles of >15°. Dominantly fine- to medium-grained sandstone	Washed-out dunes and anti-dunes that occur between subcritical and supercritical flow regimes

(*Continued*)

Table A1. *Lithofacies* (*Continued*)

Code	Lithofacies	Sedimentary description	Interpretation and depositional environment
Sh	Horizontally bedded/laminated sandstone	Buff-coloured, very-well-sorted quartz-rich, parallel-laminated sandstone. Dominantly very-fine- to medium-grained sandstone	Planar bed flow under lower and upper (flash flood) flow regimes. Parallel-laminated medium sandstone deposited under high flow velocities in shallow water depths: whereas parallel lamination in very-fine- to fine-grained sandstone can be deposited from suspension, slow-moving sediment clouds/low-density turbidity currents
Sr	Ripple cross-laminated sandstone	Light-grey- to fawn-, buff-coloured, very-fine- to medium-grained sandstone with asymmetrical erosive and climbing unidirectional ripple laminations. Very- well- to moderately sorted sandstone and often associated with wavy laminated structures. Usually contains abundant carbonaceous flecks and rootlets. Commonly occurs at the top of individual bed sets/bar units and often depicted by >1 mm-thick micaceous mud partings/drapes	Mutually erosive ripples represent migration of ripples with a low–moderate rate of sedimentation, whilst climbing ripples are formed during ripple migration under high suspended sedimentation rates. May represent shallow stable flow conditions on top of levees or crevasse channels/splays as the flood energy subsides or deposited on barform tops such as point bars during waning/low flow regime conditions
Swr	Wave ripple-laminated sandstone	Similar to Sr, but contains symmetrical bidirectional ripples	Bidirectional, oscillatory wave motion creates straight-crested symmetrical wave ripples. Represents deposition under wave action or in shallow water with stable flow conditions
Sw	Wavy-laminated sandstone	Wavy-laminated silt and fine- to medium-grained, grey- to buff-coloured sandstone. Often associated with Sr and Fl facies	Associated with vertical facies changes from parallel-laminated to wavy-laminated to climbing rippled sandstone. Wavy laminae may be the product of previously rippled surfaces
Shcs	Hummocky cross-stratified sandstone	Grey- to buff-coloured, very-fine- to fine-grained, hummocky cross-stratified sandstone. Also identified present in interbedded very fine sandstone, siltstone and mudstone intervals (heterolithic units)	Represents deposition during high-energy oscillatory or combined flows (waves and currents) under storm wave conditions (below fair-weather wave base). Storm event beds
Se	Erosional scour surfaces	Associated with Gms, Gcs, Gm, St and Sp facies. Sharp and undulating surfaces cut into underlying sediments	High-energy, erosive flow conditions. Rapid deposition of poorly sorted bedload in fluvial scour fills
Sb	Bioturbated sandstones	Low- to high-intensity bioturbation in very-fine- to fine-grained sandstone, silt–mudstone units and heterolithic intervals. Often associated with Sro. Ichnofauna include *Planolites*, *Teichichnus*, *Palaeophycus*, *Phycosiphon*, *Thalassinoides*, *Ophiomorpha nodosa*, *Diplocraterion parallelum*, *Skolithos* and *Cylindrichnus*	Represents colonization of fine-grained sediment by organisms within quiet, low-energy settings such as overbank floodplains, brackish restricted embayments, swamp/marshes and marginal-marine to pro-delta environments

Sro	Rooted sandstone	Very-well-sorted, very-fine- to medium-grained grey sandstone with large-scale root casts, small rootlets and carbonaceous material. Root casts range from 5 to 20 cm long. Often associated with Sw, Sr and Sc intervals	Plant colonization of fine-grained sediment within quiet, low energy settings such as overbank floodplain, floodplain lakes and shallow-water swamps/marshes.
Sd	Deformed/convolute laminated sandstone	Contorted and overturned laminae in very-fine- to medium-grained, grey- to buff-coloured, well-sorted sandstone. Also present within very-fine sandstones of heterolithic intervals.	Post-depositional slumping and soft-sediment deformation associated with channel-bank collapse, dewatering processes as a result of rapid deposition and/or shearing of the sediment surface by currents
Ssh	Shell-rich/bioclastic sandstone	Broken shells, bioclast material and belemnites suspended in a matrix of light-grey, fine- to medium-grained, poorly sorted sandstone. Calcite-cemented interval. Irregular mudstone lenses and intraclasts also present. Pyrite clusters common	High-energy shoreface transgressive lag/ravinement surface. A discontinuity surface produced by transgressive coastal erosion
Fm	Massive mudstones and siltstones	Structureless dark-grey- to black- coloured mudstone and siltstone. Carbonaceous material, rootlets, and pyrite and siderite nodules/clusters common	Deposition from high-viscosity sediment–water flows under low-energy conditions or in standing bodies of water (mudflows). Indicative of channel abandonment/oxbow lakes and floodplain fines
Fl	Horizontally laminated mudstone, siltstone and very-fine sandstone	Dark-grey- to black-coloured planar-laminated mudstone, siltstone and sandstone. Often present at the top of fining-upwards Sr facies and also within interbedded very-fine-grained sandstone, siltstone and mudstone (heterolithic) units. Associated with Fc, Fh and Fb	Deposition from suspension settling of clay and silt-grade sediment in very-low-energy conditions. Indicative of overbank floodplain (waning flood deposits), floodplain lake, interdistributary embayment and pro-delta environments
Fn	Nodular and mottled siltstone and mudstone	Pyrite and siderite nodules/clusters within dark-grey- to black-coloured siltstone and mudstone. Often colour- mottled light green to grey. Associated with Fro and Sro facies	Indicative of overbank floodplain deposits under reducing conditions. Associated with the development of immature and 'gleyed' waterlogged palaeosols (entisols or inceptisols)
Fw	Wavy-laminated very-fine sandstone and siltstone	Wavy-laminated, dark to light grey siltstone and sandstone. Associated with interbedded sandstone, siltstone and mudstone units (Fh)	Wavy laminae may be the product of previously rippled surfaces under low-flow regimes
Fro	Rooted siltstone and mudstone	Rootlets present within dark-grey to black siltstone and mudstone. Often colour mottled light green, yellow and red to grey. Pedogenic slickensides and carbonaceous material common. Associated with Fn, Fb and Fh facies	Colour mottling formed as a result of pedogenic processes causing irregular distributions of iron oxides/hydroxides within the sediment. Plant colonization of fine-grained sediment under very-low-energy conditions such as overbank floodplain, floodplain lakes, swamp/marshes and brackish embayments

(Continued)

Table A1. *Lithofacies* (*Continued*)

Code	Lithofacies	Sedimentary description	Interpretation and depositional environment
Fb	Bioturbated very-fine-grained sandstone, siltstone and mudstone	Moderate- to high-intensity bioturbated dark-grey to black heterolithic intervals. Burrows commonly siltstone and sandstone filled in mudstone interbeds. Common ichnofauna include *Teichichnus*, *Palaeophycus*, *Cylindrichnus*, *Planolites*, *Phycosiphon*, *Diplocraterion parallelum*, *Asterosoma*, *Rhizocorallium* and *Taenidium*	Represents colonization of fine-grained sediment by organisms within quiet, low-energy settings such as brackish restricted embayments and marginal-marine–pro-delta environments
Fc	Carbonaceous very-fine sandstone, siltstone and mudstone	High concentration of carbonaceous material and broken plant fragments suspended within fine-grained heterolithics. Associated with Fro facies and can occur as wispy flecks/lenses or plant fragments	Terrestrial source of sediment and associated with river flood conditions. Indicative of floodplain lake, overbank and swamp/marsh settings and even marginal-marine environments
Fh	Heterolithic units	Thinly interbedded mudstone, siltstone and very-fine sandstone. Graded bedding (normal and inverse) and syneresis cracks are very common. Often planar to wavy-laminated siltstone and sandstone. Streaky sandstone/siltstone laminae. Small-scale ripples and HCS in some sandstone beds/lenses. Can be carbonaceous and rooted/bioturbated. Individual sandstone/mudstone beds typically range from 2 to 70 cm thick. Associated with Fc, Fb and Fro facies	Graded bedding reflects changes in flow conditions during sedimentation. Syneresis cracks are formed as a result of sediment contraction through the loss of porewater in response to changing salinity: inducing volume changes in clay minerals. Tidal- and wave-influenced intervals represent deposition in deltaic/interdistributary bay settings and pro-delta environments

References

ADAMSON, K.R., LANG, S.C., MARSHALL, N.G., SEGGIE, R.J., ADAMSON, N.J. & BANN, K.L. 2013. Understanding the Late Triassic Mungaroo and Brigadier Deltas of the Northern Carnarvon Basin, North West Shelf, Australia. *In*: KEEP, M. & MOSS, S.J. (eds) *The Sedimentary Basins of Western Australia IV. Proceedings of the Petroleum Exploration Society of Australia Symposium*. Petroleum Exploration Society of Australia, Perth, Australia, 1–29.

AGSO NORTH WEST SHELF STUDY GROUP 1994. Deep reflections on the North West Shelf: changing perceptions of basin formation. *In*: PURCELL, P.G. & PURCELL, R.R. (eds) *The Sedimentary Basins of Western Australia. Proceedings of the Petroleum Exploration Society of Australia Symposium*. Petroleum Exploration Society of Australia, Perth, Australia, 63–76.

ALLEN, J.R.L. 1965. Late Quaternary Niger Delta and adjacent areas: sedimentary environments and lithofacies. *American Association of Petroleum Geologists Bulletin*, **49**, 547–600.

ALLEN, J.R.L. 1983. Studies in fluviatile sedimentation: Bars, bar complexes and sandstone sheets (low sinuosity braided streams) in the Brownstones (L. Devonian), Welsh Borders. *Sedimentary Geology*, **33**, 237–293.

BACKHOUSE, J. & BALME, B.E. 2002. *Late Triassic Palynology of the Northern Carnarvon Basin*. Minerals and Energy Institute of Western Australia Report 226. Minerals & Energy Research Institute of Western Australia, Perth, Australia.

BACKHOUSE, J., BALME, B.E., HELBY, R., MARSHALL, N.G. & MORGAN, R. 2002. Palynological zonation and correlation of the latest Triassic, Northern Carnarvon Basin. *In*: KEEP, M. & MOSS, S.J. (eds) *The Sedimentary Basins of Western Australia 3. Proceedings of the Petroleum Exploration Society of Australia Symposium*. Petroleum Exploration Society of Australia, Perth, Australia, 179–201.

BHATTACHARYA, J.P. 2010. Deltas. *In*: JAMES, N.P. & DALRYMPLE, R.W. (eds) *Facies Models 4*. GEOtext, **6**. Geological Association of Canada, St John's, Newfoundland, Canada, 233–264.

BHATTACHARYA, J.P. 2011. Practical problems in the application of the sequence stratigraphic method and key surfaces: integrating observations from ancient fluvial-deltaic wedges with Quaternary and modelling studies. *Sedimentology*, **58**, 120–169.

BHATTACHARYA, J.P. & MACEACHERN, J.A. 2009. Hyperpycnal rivers and prodelatic shelves in the Cretaceous seaway of North America. *Journal of Sedimentary Research*, **79**, 184–209.

BLUM, M.D. 1990. Climatic and eustatic controls on Gulf coastal plain fluvial sedimentation: an example from the late Quaternary of the Colorado River, Texas. *In*: ARMENTROUT, J.M. & PERKINS, B.F. (eds) *Sequence Stratigraphy as an Exploration Tool, Concepts and Practices in the Gulf Coast. Gulf Coast Section of SEPM Eleventh Annual Research Conference Program with Abstracts*. Gulf Coast Section of the Society of Economic Paleontologists and Mineralogists Foundation, Houston, TX, 71–83.

BLUM, M.D. & TORNQVIST, T.E. 2000. Fluvial responses to climate and sea-level change: a review and look forward. *Sedimentology*, **47**, 2–48.

BLUM, M.D., MARTIN, J., MILLIKEN, K. & GARVIN, M. 2013. Paleovalley systems: insights from Quaternary analogs and experiments. *Earth-Science Reviews*, **116**, 128–169.

BOYD, R. 2010. Transgressive wave-dominated coasts. *In*: JAMES, N.P. & DALRYMPLE, R.W. (eds) *Facies Models 4*. GEOtext, **6**. Geological Association of Canada, St John's, Newfoundland, Canada, 265–294.

BRADSHAW, M.T., YEATES, A.N., BEYNON, R.M., BRAKEL, A.T., LANGFORD, R.P., TOTTERDELL, J.M. & YEUNG, M. 1988. Palaeogeographic evolution of the North West Shelf Region. *In*: PURCELL, P.G. & PURCELL, R.R. (eds) *The Sedimentary Basins of Western Australia. Proceedings of the Petroleum Exploration Society of Australia Symposium*. Petroleum Exploration Society of Australia, Perth, Australia, 29–54.

BRENNEMAN, R.J., MILLER, N.W.D., TEAKLE, R.P., SEEBURGER, D., VAN NISPEN, D.J. & HARRISON, C.J. 1998. Boolean modelling of the fluvial Mungaroo Formation, Gorgon Field, Northwest Shelf, Australia. *In*: *Proceedings of the SPE Asia and Pacific Oil and Gas Conference and Exhibition, Perth, WA*, 305–314, https://doi.org/10.2118/50084-MS

BRIDGE, J.S. 1984. Large scale facies sequences in alluvial overbank environments. *Journal of Sedimentary Petrology*, **54**, 583–588.

BRIDGE, J.S. 1993. The interaction between channel geometry, water flow, sediment transport and deposition in braided rivers. *In*: BEST, J.L. & BROSTOW, C.S. (eds) *Braided Rivers*. Geological Society, London, Special Publications, **75**, 13–71, https://doi.org/10.1144/GSL.SP.1993.075.01.02

BRIDGE, J.S. 2003. *Rivers and Floodplains*. Blackwell, Oxford.

BRIDGE, J.S. 2006. Fluvial facies models: recent developments. *In*: POSAMENTIER, H.W. & WALKER, R.G. (eds) *Facies Models Revisited*. SEPM (Society for Sedimentary Geology), Special Publications, **84**, 85–170.

BRIDGE, J.S. & TYE, R.S. 2000. Interpreting the dimensions of ancient fluvial channel bars, channels and channel belts from wireline-logs and cores. *American Association of Petroleum Geologists Bulletin*, **84**, 1205–1228.

BUATOIS, L. & MANGANO, M.G. 2011. *Ichnology: Organism–Substrate Interactions in Space and Time*. Cambridge University Press, Cambridge.

CATUNEANU, O. 2002. Sequence stratigraphy of clastic systems: concepts, merits, and pitfalls. *Journal of African Earth Sciences*, **35**, 1–43.

CECIL, C.B. 2003. The concept of autocyclic and allocyclic controls on sedimentation and stratigraphy, emphasizing the climatic variable. *In*: CECIL, C.B. & EDGAR, N.T. (eds) *Climate Controls on Stratigraphy*. SEPM (Society for Sedimentary Geology), Special Publications, **77**, 13–20.

CHATANANTAVET, P., LAMB, M.P. & NITTROUER, J.A. 2012. Backwater controls of avulsion locations on deltas. *Geophysical Research Letters*, **39**, 1–6.

DALRYMPLE, M. 1997. *Sedimentological evolution of the Statfjord Formation fluvial hydrocarbon reservoirs of the northern North Sea*. PhD thesis, University of Aberdeen, Aberdeen, UK.

Dalrymple, R.W., Zaitlin, B.A. & Boyd, R. 1992. Estuarine facies models: conceptual basins and stratigraphic implications. *Journal of Sedimentary Research*, **62**, 1130–1146.

Dalrymple, R.W., Boyd, R. & Zaitlin, B.A. (eds). 1994. History of research, types and internal organisation of incised-valley systems; Introduction to the Volume. *In*: *Incised-Valley Systems: Origin and Sedimentary Sequences*. SEPM (Society for Sedimentary Geology), Special Publications, **51**, 3–10.

Dixon, M., Morgan, R., Goodall, J. & van den Berg, M. 2012. Higher resolution palynostratigraphy of the Norian–Carnian (Triassic) Upper Mungaroo Formation, offshore Carnarvon Basin. *In*: Mackie, S. (ed.) *AAPEA Journal*, **52**, 1–6.

Elliott, T. 1974. Interdistributary bay sequences and their genesis. *Sedimentology*, **21**, 611–622.

Ellis, C., Boothby, P., Clague, A. & Rarey, P. 2010. *Clio-2, Carnarvon Basin*. Well Completion Report (Interpretive Data) WA-205-P. Unpublished well completion report, Chevron Australia Pty Ltd.

Ford, G.L. & Pyles, D.R. 2014. A hierarchical approach for evaluating fluvial systems: Architectural analysis and sequential evolution of the high net-sand content, middle Wasatch Formation, Uinta Basin, Utah. *American Association of Petroleum Geologists Bulletin*, **98**, 1273–1304.

Gerard, J. & Bromley, R. 2008. *Ichnofabrics in Clastic Sediments: Applications to Sedimentological Core Studies*. Jean R.F Gerard, Madrid.

Ghori, K.A.R. 2011. *Geology and Petroleum Prospectivity of State Acreage Release Areas T11-3/L11-6 and L11-7, Northern Carnarvon Basin, Western Australia*. Geological Survey of Western Australia, Perth.

Gibling, M.R. 2006. Width and thickness of fluvial channel bodies and valley fills in the geological record: A literature compilation and classification. *Journal of Sedimentary Research*, **76**, 731–770.

Gugliotta, M., Flint, S.S., Hodgson, D.M. & Veiga, G.D. 2015. Stratigraphic record of river dominated crevasse sub-deltas with tidal influence (Lajas Formation, Argentina). *Journal of Sedimentary Research*, **85**, 265–284.

Hampson, G.J. & Howell, J.A. 2005. Sedimentologic and geomorphic characterisation of ancient wave-dominated deltaic shorelines: Upper Cretaceous Blackhawk Formation, Book Cliffs, Utah, U.S.A. *In*: Giosan, L. & Bhattacharya, J.P. (eds) *River Deltas: Concepts, Models and Examples*. SEPM (Society for Sedimentary Geology), Special Publications, **83**, 133–154.

Hampson, G.J., Davies, S.J., Elliott, T., Flint, S.S. & Stollhofen, H. 1999. Incised valley fill sandstone bodies in Upper Carboniferous fluvio-deltaic strata: recognition and reservoir characterisation of Southern North Sea analogues. *In*: Fleet, A.J. & Boldy, S.A.R. (eds) *Petroleum Geology of Northwest Europe: Proceedings of the 5th Conference*. Geological Society, London, 771–788, https://doi.org/10.1144/0050771

Heldreich, G., Redfern, J., Gerdes, K., Legler, B., Taylor, S., Hodgetts, D. & Williams, B. 2013. Analysis of Geobody Geometries within the Fluvio-Deltaic Mungaroo Formation, NW Australia. *In*: Keep, M. & Moss, S.J. (eds) *The Sedimentary Basins of Western Australia IV. Proceedings of the Petroleum Exploration Society of Australia Symposium*. Petroleum Exploration Society of Australia, Perth, Australia, 1–16.

Hocking, R.M. 1988. Regional Geology of the Northern Carnarvon Basin. *In*: Purcell, P.G. & Purcell, R.R. (eds) *The North West Shelf of Australia. Proceedings of the Petroleum Exploration Society of Australia Symposium*. Petroleum Exploration Society of Australia, Perth, Australia, 97–114.

Hocking, R.M., Moors, H.T. & Van de Graaff, W.J.E. 1987. *Geology of the Carnarvon Basin, Western Australia*. Geological Society of Western Australia Bulletin, **133**.

Holbrook, J. & Wanas, H. 2014. A fulcrum approach to assessing source-to-sink mass balance using channel paleohydrologic parameters derivable from common fluvial data sets with an example from the Cretaceous of Egypt. *Journal of Sedimentary Research*, **84**, 349–372.

Holbrook, J.M. 2001. Origin, genetic interrelationships and stratigraphy over the continuum of fluvial channel-form bounding surfaces: an illustration from middle Cretaceous strata, south-eastern Colorado. *Sedimentary Geology*, **144**, 179–222.

Holbrook, J.M. & Bhattacharya, J.P. 2012. Reappraisal of the sequence boundary in time and space: case and considerations for an SU (subaerial unconformity) that is not sediment bypass surface, a time barrier or an unconformity. *Earth-Science Reviews*, **133**, 271–302.

Holbrook, J.M., Scott, R.W. & Oboh-Ikuenobe, F.E. 2006. Base level buffers and buttresses: a model for upstream v. downstream control on fluvial geometry and architecture within sequences. *Journal of Sedimentary Research*, **76**, 162–174.

Ichaso, A.A. & Dalrymple, R.W. 2009. Tide- and wave-generated fluid mud deposits on the Tilje Formation (Jurassic), offshore Norway. *Geology*, **37, 539–542**.

Jablonski, D. 1997. Recent advances in the sequence stratigraphy of the Triassic to Lower Cretaceous succession in the northern Carnarvon Basin, Australia. *AAPEA Journal*, **37**, 429–454.

Jerolmack, D.J. & Paola, C. 2010. Shredding of environmental signals by sediment transport. *Geophysical Research Letters*, **37**, 1–5.

Jerolmack, D.J. & Swenson, J.B. 2007. Scaling relationships and evolution of distributary networks on wave-influenced deltas. *Geophysical Research Letters*, **34**, 1–5.

Kosters, E.C. & Suter, J.R. 1993. Facies relationships and systems tracts in the late Holocene Mississippi delta plain. *Journal of Sedimentary Petrology*, **63**, 727–733.

Kulp, M., Fitzgerald, D. & Penland, S. 2005. Sand-rich lithosomes of the Holocene Mississippi River Delta Plain. *In*: Giosan, L. &. Bhattacharya, J.P. (eds) *River Deltas – Concepts, Models and Examples*. SEPM (Society for Sedimentary Geology), Special Publications, **83**, 279–293.

LeClair, S.F. & Bridge, J.S. 2001. Quantitative interpretation of sedimentary structures formed by river dunes. *Journal of Sedimentary Research*, **71**, 713–716.

Lewis, C.J. & Sircombe, K.N. 2013. Use of U–Pb geochronology to delineate provenance of North West Shelf sediments, Australia. *In*: Keep, M. & Moss, S.J. (eds) *The Sedimentary Basins of Western Australia IV. Proceedings of the Petroleum Exploration Society of Australia Symposium*. Petroleum Exploration Society of Australia, Perth, Australia, 1–27.

Longley, I.M., Buessenschuett, C. *et al.* 2002. The North West Shelf of Australia – a Woodside perspective. *In*: Keep, M. & Moss, S.J. (eds) *The Sedimentary Basins of Western Australia 3. Proceedings of the Petroleum Exploration Society of Australia Symposium*. Petroleum Exploration Society of Australia, Perth, Australia, 27–88.

Lorenz, J.C., Heinze, D.M., Clark, J.A. & Searls, C.A. 1985. Determination of widths of meander belt sandstone reservoirs from vertical downhole data, Mesaverde Group, Piceance Creek Basin, Colorado. *American Association of Petroleum Geologists Bulletin*, **69**, 710–721.

Marshall, N.G. & Lang, S.C. 2013. A new sequence stratigraphic framework for the North West Shelf, Australia. *In*: Keep, M. & Moss, S.J. (eds) *The Sedimentary Basins of Western Australia IV. Proceedings of the Petroleum Exploration Society of Australia Symposium*. Petroleum Exploration Society of Australia, Perth, Australia, 1–32.

McLaurin, B.T. & Steel, R.J. 2000. Fourth-order nonmarine to marine sequences, middle Castlegate Formation, Book Cliffs, Utah. *Geology*, **28**, 359–362.

Metcalfe, I. 1996. Gondwanaland dispersion, Asian accretion and evolution of eastern Tethys. *Australian Journal of Earth Sciences*, **43**, https://doi.org/10.1080/08120099608728282

Metcalfe, I. 1999. Gondwana dispersion and Asian accretion: An overview. *In*: Metcalfe, I. (ed.) *Gondwana Dispersion and Asian Accretion: IGCP 321 Final Results Volume*. A.A Balkema, Rotterdam, 9–28.

Miall, A. 1980. Cyclicity and the facies model concept in geology. *Bulletin of Canadian Petroleum Geology*, **28**, 59–80.

Miall, A. 1996. *The Geology of Fluvial Deposits*. Springer, Berlin.

Miall, A. 2010. Alluvial deposits. *In*: James, N.P. & Dalrymple, R.W. (eds) *Facies Models 4*. GEOtext, **6**. Geological Association of Canada, St John's, Newfoundland, Canada, 105–137.

Moiola, R.J., Welton, J.E., Wagner, J.B., Fearn, L.B., Farrell, M.E., Enrico, R.J. & Echols, R.J. 2004. Integrated analysis of the Upper Ferron Deltaic Complex, Southern Castle Valley, Utah. *In*: Chidsey, T.C., Jr., Adams, R.D. & Morris, T.H. (eds) *Regional to Wellbore Analog for Fluvial-Deltaic Reservoir Modeling: The Ferron Sandstone of Utah*. American Association of Petroleum Geologists, Studies in Geology, **50**, 79–92.

Muto, T. & Steel, R.J. 2004. Autogenic response of fluvial deltas to steady sea-level fall: Implications from flume-tank experiments. *Geology*, **32**, 401–404.

Muto, T. & Swenson, J.B. 2005. Large-scale fluvial grade as a non-equilibrium state in linked depositional systems: Theory and experiment. *Journal of Geophysical Research*, **110**, 1–15.

Muto, T., Steel, R.J. & Swenson, J.B. 2007. Autostratigraphy: A framework norm for genetic stratigraphy. *Journal of Sedimentary Research*, **77**, 2–22.

Nicoll, R.S. 2002. Conodont biostratigraphy and palaeogeography of the Triassic on the western, northwestern and northern margins of the Australian Plate. *In*: Keep, M. & Moss, S.J. (eds) *The Sedimentary Basins of Western Australia 3. Proceedings of the Petroleum Exploration Society of Australia Symposium*. Petroleum Exploration Society of Australia, Perth, Australia, 167–177.

Nittrouer, J.A., Mohrig, D. & Allison, M. 2011. Punctuated sand transport in the lowermost Mississippi River. *Journal of Geophysical Research*, **116**, 1–24.

Nittrouer, J.A., Shaw, J., Lamb, M.P. & Mohrig, D. 2012. Spatial and temporal trends for water flow velocity and bed-material transport in the lower Mississippi River. *Geological Society of America Bulletin*, **124**, 400–414.

Olariu, C., Steel, R.J. & Petter, A.J. 2010. Delta-front hyperpycnal bed geometry and implications for reservoir modelling: Cretaceous Panther Tongue delta, Book Cliffs, Utah. *American Association of Petroleum Geologists Bulletin*, **94**, 819–845.

Patterson, P.E., Jones, C.R. & Skelly, R.L. 2010. Sequence stratigraphy of Alluvial systems: Hierarchical description and sequence stratigraphy of Cretaceous Alluvial Strata, Hawkins Field, Texas. *In*: Abreu, V., Neal, J.E., Bohacs, K.M. & Kalbas, J.L. (eds) *Sequence Stratigraphy of Siliciclastic Systems – the ExxonMobil Methodology: Atlas of Exercises*. SEPM (Society of Sedimentary Geology) Concepts in Sedimentology and Paleontology, **9**, 131–143.

Payenberg, T.H.D. & Lang, S.C. 2003. Reservoir geometry of fluvial distributary channels – Implications for Northwest Shelf, Australia, deltaic successions. *APPEA Journal*, **43**, 325–337.

Payenberg, T.H.D., Howe, H. *et al.* 2013. An integrated regional Triassic stratigraphic framework for the Carnarvon Basin, NWS, Australia. *In*: Keep, M. & Moss, S.J. (eds) *The Sedimentary Basins of Western Australia IV. Proceedings of the Petroleum Exploration Society of Australia Symposium*. Petroleum Exploration Society of Australia, Perth, Australia, 1–24.

Payenberg, T.H.D., Willis, B. *et al.* 2014. Channel belt rugosity in reservoir characterisation. Search and discovery article 41420, AAPG Annual Conference & Exhibition, 6–9 April 2014, Houston, TX, USA.

Plint, A.G. 2010. Wave and storm-dominated shoreline and shallow marine systems. *In*: James, N.P. & Dalrymple, R.W. (eds) *Facies Models 4*. GEOtext, **6**. Geological Association of Canada, St John's, Newfoundland, Canada, 167–199.

Posamentier, H.W. 2001. Lowstand alluvial bypass systems: Incised vs unincised. *American Association of Petroleum Geologists Bulletin*, **85**, 1771–1793.

Posamentier, H.W. & Allen, G.P. 1999. *Siliciclastic Sequence Stratigraphy – Concepts and Applications*. SEPM (Society for Sedimentary Geology), Concepts in Sedimentology and Paleontology, **7**.

Posamentier, H.W. & Vail, P.R. 1988. Eustatic controls on clastic deposition II – sequence and systems tract models. *In*: Wilgus, C.K., Hastings, B.S., Kendall, C.G.St.C., Posamentier, H.W., Ross, C.A. & Van Wagoner, J.C. (eds) *Sea Level Changes: an Integrated Approach*. SEPM (Society for Sedimentary Geology), Special Publications, **42**, 125–154.

Pranter, M.J., Cole, R.D., Panjaitan, H. & Sommer, N.K. 2009. Sandstone-body dimensions in a lower coastal-plain depositional setting: Lower Williams Fork Formation, Coal Canyon, Piceance Basin, Colorado. *American Association of Petroleum Geologists Bulletin*, **93**, 1379–1401.

Ratcliffe, K.T., Wright, A.M. *et al.* 2010. Application of chemostratigraphy to the Mungaroo Formation, the Gorgon Field, offshore northwest Australia. *AAPEA Journal*, **50**, 371–388.

Reineck, H.-E. & Singh, I.B. 1980. *Depositional Sedimentary Environments; with Reference to Terrigenous Clastics*. Springer, Heidelberg.

Reynolds, A.D. 1999. Dimensions of paralic sandstone bodies. *American Association of Petroleum Geologists Bulletin*, **83**, 211–229.

Robinson, J.W. & McCabe, P.J. 1997. Sandstone-body and shale-body dimensions in a braided fluvial system: Salt Wash Sandstone Member (Morrison Formation), Garfield County, Utah. *American Association of Petroleum Geologists Bulletin*, **81**, 1267–1291.

Ryer, T.A. & Anderson, P.B. 2004. Facies of the Ferron sandstone, East-Central Utah. *In*: Chidsey, T.C., Jr., Adams, R.D. & Morris, T.H. (eds) *Regional to Wellbore Analog for Fluvial-Deltaic Reservoir Modeling: The Ferron Sandstone of Utah*. American Association of Petroleum Geologists, Studies in Geology. **50**, 59–78.

Schlager, W. 1993. Accommodation and supply – a dual control on stratigraphic sequences. *Sedimentary Geology*, **86**, 111–136.

Schumm, S.A. 1981. Evolution and response of the fluvial system, sedimentological implications. *In*: Ethridge, F.G. & Flores, R.M. (eds) *Recent and Ancient Nonmarine Depositional Environments*. SEPM (Society for Sedimentary Geology), Special Publications, **31**, 19–29.

Shanley, K.W. & McCabe, P.J. 1994. Perspectives on the sequence stratigraphy of continental strata. *American Association of Petroleum Geologists Bulletin*, **78**, 544–568.

Shanley, K.W., McCabe, P.J. & Hettinger, R.D. 1992. Significance of tidal influence in fluvial deposits for interpreting sequence stratigraphy. *Sedimentology*, **39**, 905–930.

Sixsmith, P. 2009. *Sedimentology of the Clio-2 Core Triassic Mungaroo Formation, Northern Carnarvon Basin, Offshore Australia: Core description and Interpretation*. Well Completion Report (Interpretive Data) WA-205-P. Unpublished well completion report, Chevron Australia Pty Ltd.

Tucker, M.E. 2001. *Sedimentary Petrology*. Blackwell Scientific, Oxford.

Van Wagoner, J.C., Mitchum, R.M., Campion, K.M. & Rahmanian, V.D. (eds) 1990. *Siliciclastic Sequence Stratigraphy in Well Logs, Core and Outcrops: Concepts for High Resolution Correlation of Time and Facies*. American Association of Petroleum Geologists, Methods in Exploration Series, **7**.

Wood, J.D. 2013. Sedimentological characterisation and regional palaeo-environmental implications of the Messak Fm, SW Libya. *Unpublished PhD thesis*, University of Manchester, Manchester, UK.

Wright, P.V. & Marriott, S.B. 1993. The sequence stratigraphy of fluvial depositional systems: the role of floodplain sediment storage. *Sedimentary Geology*, **86**, 203–210.

Yoshida, S. 2000. Sequence and facies architecture of the upper Blackhawk Formation and the Lower Castlegate sandstone (Upper Cretaceous), Book Cliffs, Utah, USA. *Sedimentary Geology*, **136**, 239–276.

Zaitlin, B.A., Dalrymple, R.W. & Boyd, R. 1994. The stratigraphic organisation of incised-valley systems associated with relative sea level change. *In*: Dalrymple, R.W., Boyd, R. & Zaitlin, B.A. (eds) *Incised-Valley Systems: Origin and Sedimentary Sequences*. SEPM (Society for Sedimentary Geology), Special Publications, **51**, 45–60.

Analysis of floodplain sedimentation, avulsion style and channelized fluvial sandbody distribution in an upper coastal plain reservoir: Middle Jurassic Ness Formation, Brent Field, UK North Sea

YVETTE S. FLOOD & GARY J. HAMPSON*

Department of Earth Science and Engineering, Imperial College London, South Kensington Campus, London SW7 2AZ, UK

**Correspondence: g.j.hampson@imperial.ac.uk*

Abstract: Numerical models and recent outcrop case studies of alluvial-to-coastal-plain strata suggest that autogenic avulsion can control the stacking density and architecture of channelized fluvial sandbodies. The application of these models to subsurface well data was tested by the analysis of upper coastal plain deposits of the late Bajocian Ness Formation in the Brent Field reservoir, UK North Sea. These coastal plain deposits accumulated during the progradation and retrogradation of the wave-dominated 'Brent Delta'. Sedimentological facies analysis and palaeosol characterization in cores were used to interpret the styles of palaeochannel avulsion. These results were then compared with the dimensions and distributions of channelized fluvial sandbodies that had been quantified by spatial statistical tools (lacunarity, Besag's *L* function) applied to interpretative correlation panels between closely spaced wells. The results indicate that the distributions of channelized sandbodies may plausibly have been generated by avulsions and that they influence sandbody connectivity and pressure depletion patterns. Intervals of upper coastal plain strata with relatively wide sandbodies that display some clustering in their stratigraphic architecture are associated with a high proportion of avulsions by incision and annexation in core samples. Such intervals display relatively good vertical pressure communication and relatively slow, uniform pressure depletion.

Numerical models that investigate the dimensions and connectivity of channelized fluvial sandstone bodies suggest that avulsion frequency and the sediment accumulation rate control the spatial distribution of such bodies in alluvial-to-coastal-plain strata (Allen 1978; Leeder 1978; Bridge & Leeder 1979; Mackey & Bridge 1995; Heller & Paola 1996; Törnqvist & Bridge 2002; Jerolmack & Paola 2007). However, alluvial-to-coastal-plain successions are commonly interpreted with reference to sequence stratigraphic models that relate the character and distribution of channelized fluvial sandbodies to allogenic (external) controls such as tectonic subsidence, base level, and changes in sediment and water supply (e.g. Wright & Marriott 1993; Shanley & McCabe 1994). In this context, autogenic (internal) behaviours are often presumed to represent relatively small-scale, high-frequency 'noise' that modulates the effects of larger scale allogenic controls (Slingerland & Smith 2004; Hajek & Wolinsky 2012). Recent physical and numerical modelling experiments and selected outcrop case studies of channel belt stacking patterns suggest that autogenic behaviours can generate large-scale self-organization while allogenic forcing remains relatively constant (Mackey & Bridge 1995; Blum & Törnqvist 2000; Jerolmack & Paola 2007; Bridge 2008; Straub *et al.* 2009; Hajek *et al.* 2010; Hajek & Wolinsky 2012; Flood & Hampson 2015). However, there have been few attempts to apply the concept of autogenic stratigraphic organization to alluvial-to-coastal-plain reservoirs (Hofmann *et al.* 2011).

Avulsion is an autogenic process that occurs during active alluvial sedimentation. It involves the relatively rapid diversion of flow out of an established channel belt, either by the reoccupation of a pre-existing channel or relocation to a new permanent position on the floodplain (Allen 1978; Mohrig *et al.* 2000; Slingerland & Smith 2004; Jones & Hajek 2007). Avulsion typically controls the long-term distribution of sediment and water on the alluvial plain (Mohrig *et al.* 2000) and thus plays an important part in controlling the fluvial stratigraphic architecture (Allen 1978; Leeder 1978; Bridge & Leeder 1979; Mackey & Bridge 1995; Heller & Paola 1996; Mohrig *et al.* 2000). Two patterns of channel belt stacking pattern can be generated by avulsion: clustering and compensational stacking (Straub *et al.* 2009; Hajek *et al.* 2010; Hofmann *et al.* 2011). Channel belt clustering results from locally confined accommodation space and/or sediment supply (Leeder 1978; Shanley & McCabe 1994; Hajek *et al.* 2010; Hofmann *et al.* 2011). In contrast, compensational stacking results from

From: Hampson, G. J., Reynolds, A. D., Kostic, B. & Wells, M. R. (eds) 2017. *Sedimentology of Paralic Reservoirs: Recent Advances*. Geological Society, London, Special Publications, **444**, 109–140.
First published online June 29, 2016, https://doi.org/10.1144/SP444.3

preferential channel relocation into topographically lower positions on the floodplain as a result of differential sedimentation rates (Jerolmack & Paola 2007; Straub *et al.* 2009).

Palaeosol analysis can be used to determine variations in the floodplain sediment accumulation rate (Kraus & Gwinn 1997). The spatial relationships between channelized fluvial sandbodies and the surrounding overbank deposits, including palaeosols, can help to determine avulsion style and variability in alluvial successions (Kraus & Aslan 1993; Kraus 1996). Three styles of avulsion have been documented in previous studies: avulsion by annexation (avulsion by reoccupation *sensu* Slingerland & Smith 2004); avulsion by erosion (avulsion by incision *sensu* Slingerland & Smith 2004), and avulsion by progradation (Kraus & Wells 1999; Mohrig *et al.* 2000; Slingerland & Smith 2004; Flood & Hampson 2014). The style of avulsion may be controlled by floodplain topography, sedimentation processes, base level, and the distribution and stacking density of channels on the alluvial plain (Kraus & Wells 1999; Mohrig *et al.* 2000; Jones & Hajek 2007). These models of avulsion style have not previously been applied to subsurface data.

In the work reported here, we used data from an alluvial-to-coastal plain reservoir (late Bajocian Ness Formation, Brent Field, UK North Sea) to: (1) analyse the detailed sedimentological character of overbank deposits and palaeosols in core samples to interpret the stratigraphic and palaeogeographical variation in avulsion style; (2) quantitatively analyse the previously interpreted (Livera 1989) dimensions and spatial distribution of channelized fluvial sandbodies; and (3) compare patterns of interpreted sandbody distribution with core-based sedimentological analysis to assess the degree of consistency between the two characterization approaches and their utility in reservoir characterization.

Geological context and stratigraphic framework

The Middle Jurassic (Aalenian–Bathonian) Brent Group was deposited across the East Shetland Platform, North Viking Graben and Horda Platform and forms a reservoir in over 65 fields in the northern North Sea (Fig. 1a; Husmo *et al.* 2003). The Brent Group was deposited over *c.* 12 Ma during a period of tectonic quiescence and passive thermal subsidence following Triassic rifting and prior to the main phase of late Jurassic rifting (Husmo *et al.* 2003). The Brent Field is situated in a gently dipping (8°), westerly rotated fault block on the eastern margin of the East Shetland Basin (Taylor *et al.* 2003; Fig. 1b). The field was initially discovered in 1971 (Taylor *et al.* 2003) and had produced just over 2000 MMSTB of oil by December 2014 (Department of Energy and Climate Change 2015). Planning of field abandonment is currently taking place (Shell UK 2015), but the Brent Group reservoir serves as a data-rich analogue for many less mature reservoirs in other Brent Province fields.

The Ness Formation constitutes the coastal plain deposits of the Brent Group, with the underlying Rannoch and Etive formations and overlying Tarbert Formation representing partly coeval shallow marine deposits (Fig. 2; Deegan & Scull 1977). In combination, the Rannoch, Etive, Ness and Tarbert formations record the overall regression and subsequent transgression of a wave-dominated delta (Fig. 2; Deegan & Scull 1977; Budding & Inglin 1981; Johnson & Stewart 1985). Various low- and high-resolution sequence stratigraphic schemes have been constructed for the Brent Group (e.g. Helland-Hansen *et al.* 1992; Mitchener *et al.* 1992; Fjellanger *et al.* 1996; Hampson *et al.* 2004). High-resolution sequence stratigraphic schemes tend to contain units that are comparable in scale to reservoir zones at a field scale (e.g. Flint *et al.* 1998; Morris *et al.* 2003). In the Brent Field reservoir, the Ness Formation is subdivided into lower (reservoir zones 3.1–3.3; Fig. 3), middle (reservoir zone 2.5; Fig. 4; 'Mid-Ness Shale'; Budding & Inglin 1981) and upper intervals (reservoir zones 2.1–2.4 and 1.2–1.4; Fig. 4) based on the distribution of field-wide coal zones and other facies associations (Livera 1989). The distribution of channelized fluvial sandbodies in the Ness Formation of the Brent Field is interpreted to reflect a combination of allogenic, high-frequency relative sea-level changes and associated variations in distance from the regional palaeoshoreline (e.g. Hampson *et al.* 2004) and localized, autogenic variations in sediment supply and basinal processes (e.g. Livera 1989). The Ness Formation was deposited under a relatively stable humid climate (e.g. Dreyer 1990).

Dataset and methodology

The dataset for this study consisted of core from four wells (211/29-2, 211/29-3, 211/29-BC06 and 211/29-BA16; Fig. 1) and the facies architectural interpretations of Livera (1989), which are based on wireline log and core data from 72 wells. In total, 570 m of core from four wells (211/29-2, 211/29-3, 211/29-BA16 and 211/29-BC06; Fig. 1b) were logged for facies analysis and palaeosol characterization. The apparent widths and thicknesses of channelized fluvial sandbodies were measured from two cross-sections aligned approximately perpendicular to the axes of channelized sandbodies, through the lower (Fig. 3), middle and upper Ness Formation (Fig. 4) (after figs 7 & 8 of Livera

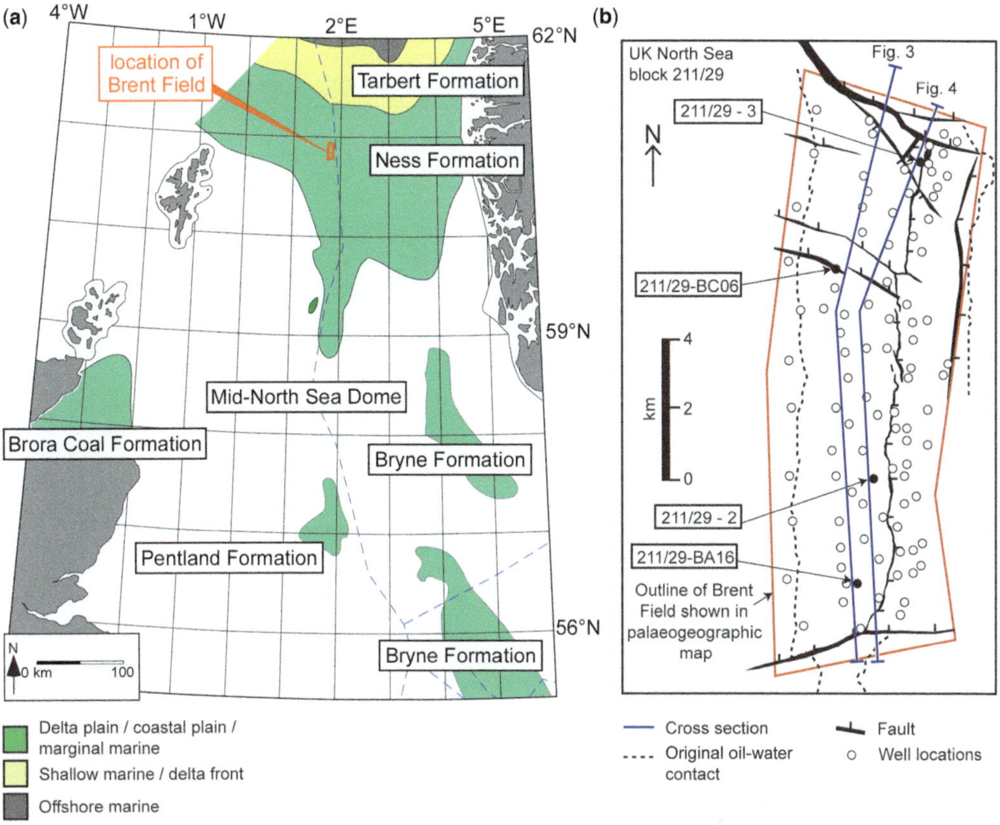

Fig. 1. (**a**) Palaeogeographical map for maximum regression of the Brent Group during the Late Bajocian (after Husmo *et al.* 2003 and references cited therein). Late Bajocian strata are absent over the Mid-North Sea Dome and adjacent areas (shown in white).The location of the Brent Field is indicated. (**b**) Map of the Brent Field locating the position of stratigraphic cross-sections (Figs 3 & 4), selected cored wells (211/29-2, 211/29-3, 211/29-BA16 and 211/29-BC06; Figures 8, 10, 11 and 12), original oil–water contact (after Taylor *et al.* 2003) and faults.

1989). These cross-sections were constructed using core and wireline log data from 51 wells projected into the lines of the cross-section (Livera 1989). Both cross-sections are interpretative given the limitations imposed by the well spacing (*c.* 200–1000 m, as projected into the lines of cross-section) and it has previously been inferred that the channelized sandbodies are oriented west–east, are nearly linear and are parallel to each other in plan view (Livera 1989) (Figs 3b & 4b). These inferences are consistent with subsequent well and production data and have been retained in several generations of reservoir models (e.g. Bryant & Livera 1991; Bryant *et al.* 1991; Abbotts & Van Kuijk 1997; James *et al.* 1999). The cross-sections of Livera (1989) are also consistent with a sequence stratigraphic interpretation of the studied strata based on core data from the four studied Brent Field wells (211/29-2, 211/29-3, 211/29-BA16 and 211/29-BC06; Fig. 1b) and additional regional well and biostratigraphic data (Hampson *et al.* 2004). However, alternative interpretations are clearly possible given that the sandbodies are narrow relative to the well spacing in the cross-sections (Figs 3a & 4a). It is not our intention to develop such interpretations in this paper, but instead to test and appraise the architectural interpretations of Livera (1989) and the sequence stratigraphic interpretations of Hampson *et al.* (2004). However, our results are discussed in the light of the potential variability in sandbody orientation and plan view geometry and in the sequence stratigraphic interpretation. The two cross-sections were subdivided into their constituent reservoir zones (Fig. 5) to measure the dimensions and spatial distributions of the channelized sandbodies in the upper coastal plain strata. The lower coastal plain strata contain too few channelized fluvial sandbodies for a meaningful analysis of their spatial distribution (Fig. 5). The dimensions and orientations of the panels are summarized in Table 1.

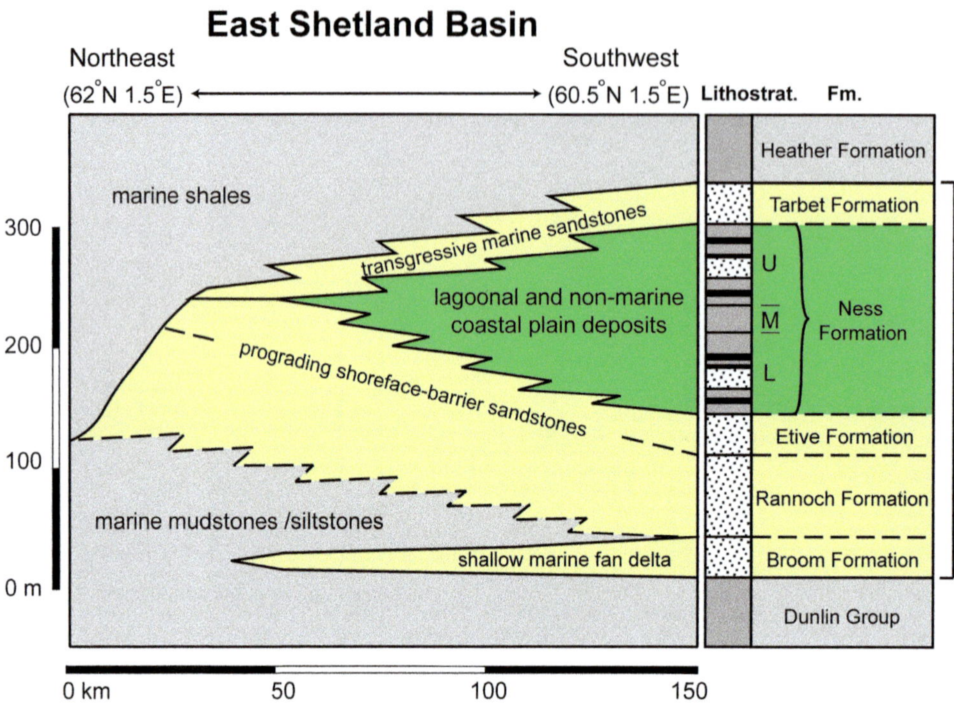

Fig. 2. Summary lithostratigraphic scheme for the Middle Jurassic Brent Group (after Deegan & Scull 1977).

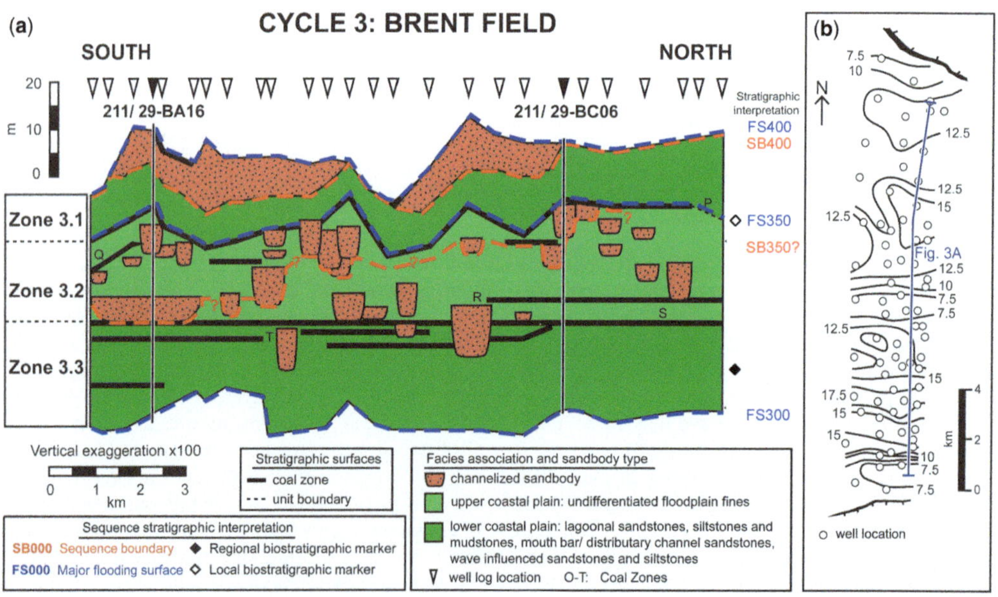

Fig. 3. (**a**) Stratigraphic cross-section of the lower Ness Formation in the Brent Field (Fig. 1) showing the distribution of channelized fluvial sandbodies and field-wide coal zones, which are used to define reservoir zones 3.1, 3.2 and 3.3 (after Livera 1989, fig. 3). Cored wells 211/29-BA16 and 211/29-BC06, shown in Figures 8 and 11, are located. A high-resolution sequence stratigraphic interpretation (Hampson *et al.* 2004) is also shown. The cross-section is vertically exaggerated by ×100. (**b**) Hand-contoured sandstone isopach map (in metres) of reservoir zone 3.1 (after Livera 1989, fig. 7). Large sandstone thicknesses (*c.* >12 m) are interpreted as a series of parallel, west–east-trending channelized fluvial sandbodies (Livera 1989).

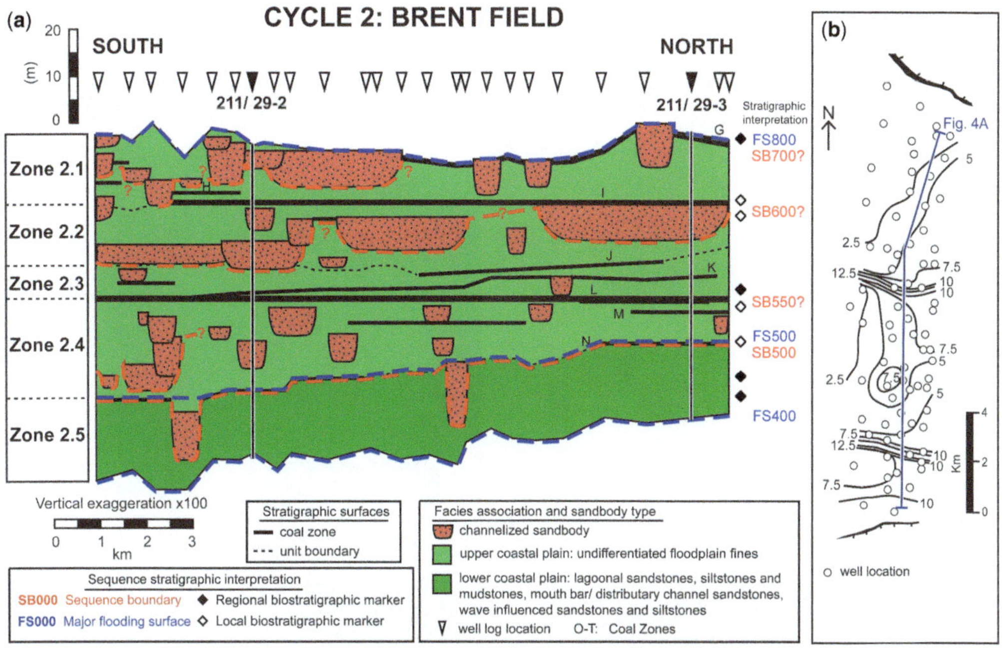

Fig. 4. (**a**) Stratigraphic cross-section of the middle and upper Ness Formation in the Brent Field (Fig. 1) showing the distribution of channelized fluvial sandbodies and field-wide coal zones, which are used to define reservoir zones 2.1–2.5 (after Livera 1989, fig. 4). Cored wells 211/29-2 and 211/29-3, shown in Figures 10 and 12, are located. A high-resolution sequence stratigraphic interpretation (Hampson *et al.* 2004) is also shown. The cross-section is vertically exaggerated by ×100. (**b**) Hand-contoured sandstone isopach map (in metres) of reservoir zone 2.5 (after Livera 1989, fig. 8). Large sandstone thicknesses (*c.* >12 m) are interpreted as a series of parallel, west–east-trending channelized fluvial sandbodies (Livera 1989).

Measurement of sandbody dimensions

The apparent width and thickness of each channelized fluvial sandbody was measured from the cross-sections for the upper coastal plain in reservoir zones 2.1–2.4 and 3.2 (Fig. 5). The errors associated with the panel images are small and reflect the image resolution and measurement repeatability, although there is a large uncertainty in sandbody correlation, extent and orientation between wells in the original work of Livera (1989). Sandstone isopach maps indicate that the channelized sandbodies are oriented approximately west–east, nearly perpendicular to the cross-sections (Figs 3b & 4b). However, the measured widths of sandbodies in the cross-sections are apparent values that may slightly over-estimate the true sandbody widths by up to 6% for the portions of the cross-sections that are oriented N019.

Measurement of sandbody distributions

Spatial statistical methods are applied to analyse the distributions of points and objects in various scientific disciplines. Several widely used methods are inappropriate for our analysis because they only compare results for areas of a similar size (e.g. the nearest neighbour distance method; Clark & Evans 1954) or require the robust identification of a chronologically ordered series of depositional horizons (e.g. the compensation index; Straub *et al.* 2009). For the purpose of this study, lacunarity and Besag's *L* function (Fig. 6) were used because these methods can be applied to geological datasets to generate clear and easily interpretable outputs (e.g. Plotnick 1999; Rankey 2002; Hajek *et al.* 2010; Roy *et al.* 2010; Zhao *et al.* 2011; Flood & Hampson 2015).

Lacunarity. Lacunarity is a scale-dependent measure of spatial dispersion (Plotnick *et al.* 1996). We used the gliding box algorithm method (Allain & Cloitre 1991) to calculate lacunarity because it is a relatively straightforward and computationally simple technique (Plotnick *et al.* 1996). A binary, black and white image of each panel was first generated (Fig. 6a). The gliding box algorithm then uses square boxes of different sizes to sample the binary image. A box of a given length is placed at the top left of the image and the number of pixels representing sandstone (black in Fig. 6a) within the box is

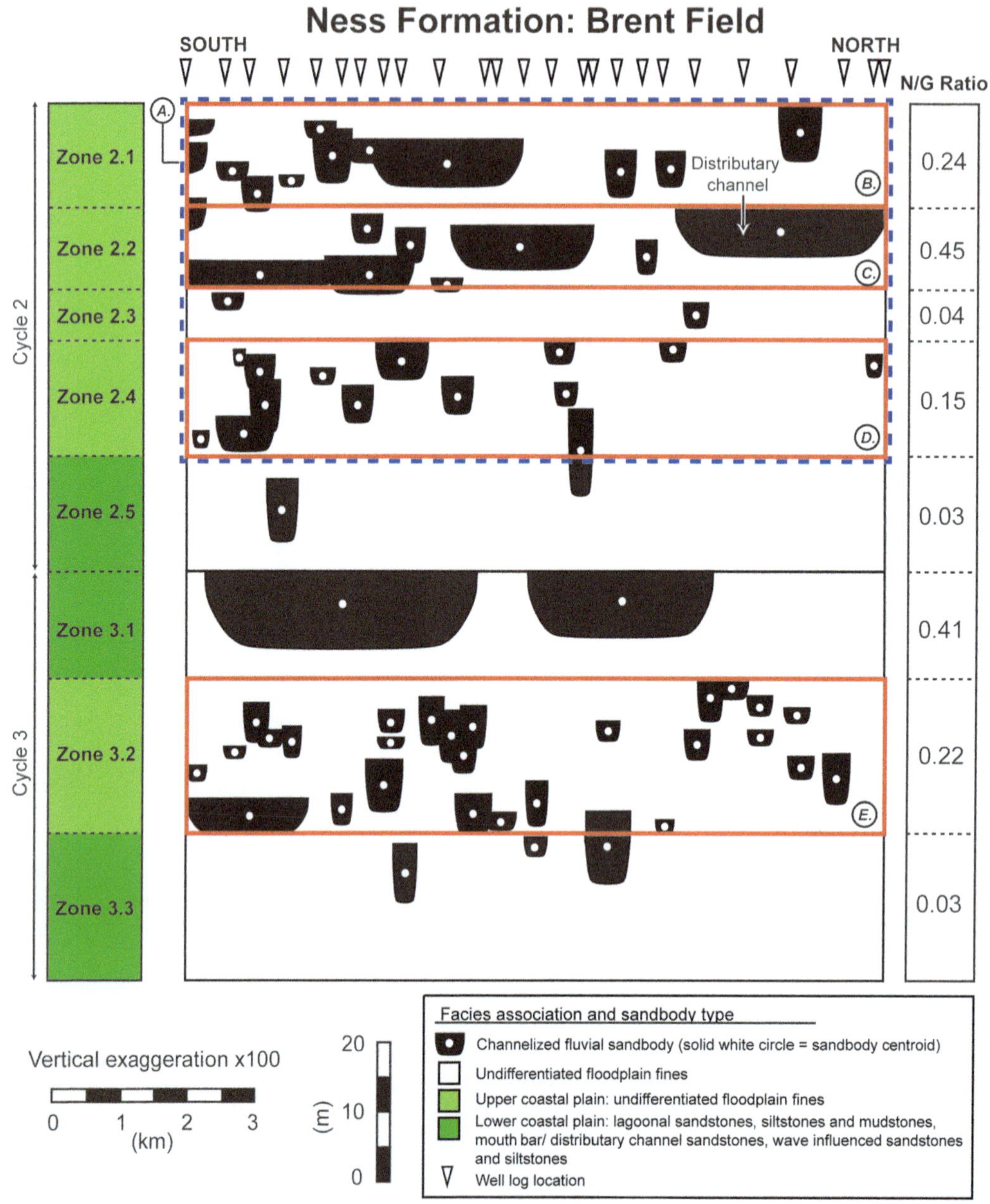

Fig. 5. Cross-section panel for stratigraphic subdivisions of the Ness Formation in the Brent Field. Panel A is 'cycle 2'; panel B is reservoir zone 2.1; panel C is reservoir zone 2.2; panel D is reservoir zone 2.4; and panel E is reservoir zone 3.2 (after Figs 3 & 4) (after Livera 1989). Net-to-gross (N/G) ratios within each reservoir zone in the panels are taken from Livera (1989). Each panel is converted into a binary image in which the 'foreground' channelized sandbodies (black) are distinguished from the 'background' floodplain and lagoonal deposits (white) to measure the lacunarity. A white point represents the centroid of each channelized fluvial sandbody and the distribution of centroids is used for our application of Besag's L function.

counted. The box is then moved one column along to the right and the process is repeated over all rows and columns until the entire area of the panel has been scanned and counted. Twelve box sizes were used to scan each panel with minimum and maximum box sizes of 2 and 45% of the panel

Table 1. *Dimensions and orientations of panels A–E in the Ness Formation (Fig. 5)*

Panel	Stratigraphic unit or cycle	Panel thickness (m)	Panel width (km)	Panel orientation	Number of sandbodies
A	Cycle 2	52	12	N 019/N 175	34
B	Unit 2.1	15	12	N 019/N 175	13
C	Unit 2.2	12	12	N 019/N 175	9
D	Unit 2.4	17	12	N 019/N 175	14
E	Unit 3.2	23	12	N 014/N 175	28

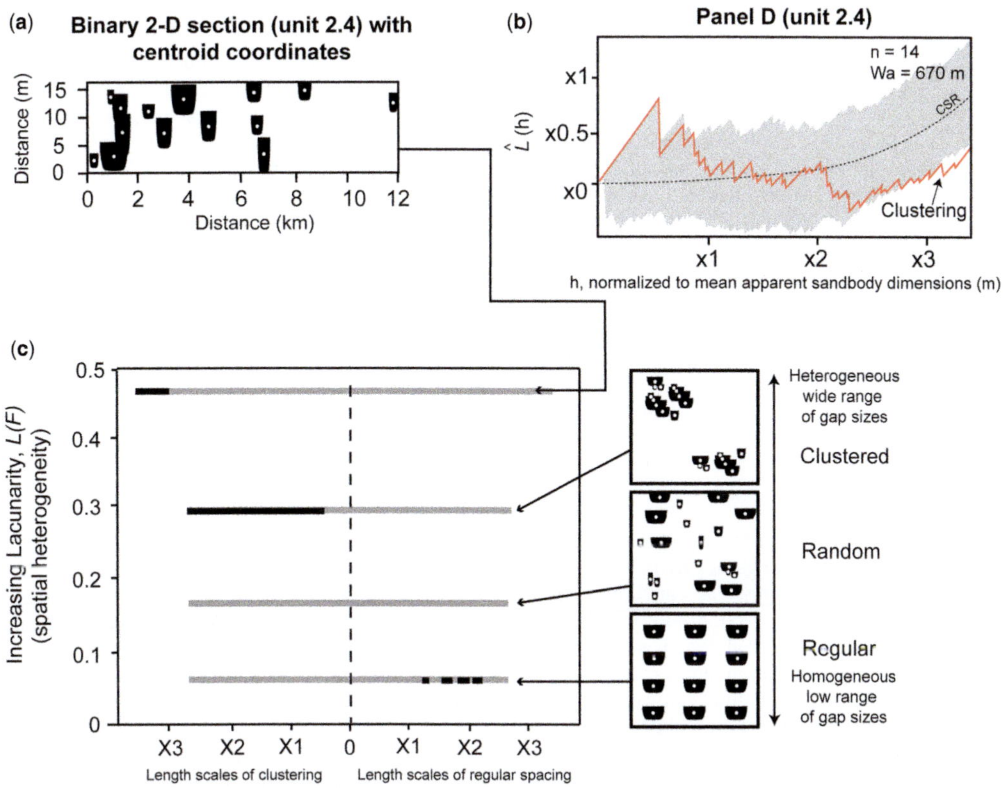

Fig. 6. Diagrams illustrating the application of lacunarity and the *L* function in this study. (**a**) Binary image of panel of reservoir zone 2.4 (Fig. 5, panel D) in which channelized fluvial sandbodies (black) are distinguished from floodplain deposits (white) to measure the lacunarity. The centroid of each channelized fluvial sandbody is illustrated as a white point and their distribution is used for our application of Besag's *L* function. The panel is vertically exaggerated by ×123 (i.e. the ratio of the mean apparent sandbody width of 990 m to the mean sandbody thickness of 8 m in reservoir zones 3.3–2.1). (**b**) Plot of *L* function for panel of reservoir zone 2.4. The horizontal and vertical axes show distances expressed as multiples of the mean apparent sandbody dimensions to minimize the effects of anisotropy in sandbody dimensions. Random distributions plot within the envelope (grey area) for complete spatial randomness (CSR) defined by 99 Monte Carlo simulations of centroid distributions. Clustered and regularly spaced centroid distributions plot below and above this envelope, respectively (after convention of Rosenberg & Anderson 2011). (**c**) Plot of lacunarity v. inhomogeneity in spatial positioning of sandbody centroids as identified using the *L* function. Data are shown for three cartoons that illustrate type examples of spatial patterns (right of plot) and for panel of reservoir zone 2.4. Grey bars represent the spatial extent of data for each image and the superimposed black bars show the length scales of sandbody centroid clustering or regular spacing. Length scales not represented by the black portions of the grey and black bars correspond to random spacing of sandbody centroids. Length scales are expressed as multiples of mean apparent sandbody dimensions. Lacunarity is dimensionless.

area, respectively. The maximum box size was chosen to be <50% of the panel area because larger sizes introduce statistical errors (Karperien 1999–2013). The gliding box algorithm is typically used to generate a frequency distribution of lacunarity against box size (e.g. Plotnick *et al.* 1996). As little variation in lacunarity exists for the different box lengths used in this study, a single mean value of lacunarity was calculated for each panel by averaging across the length scales of all grid box sizes and over all grid orientations, $L(F)$ (Karperien 1999–2013) (e.g. on the vertical axis of Fig. 6c) using the following equation:

$$L(F) = \frac{\sum[F\Lambda]}{\text{Grids}}. \quad (1)$$

where Λ is the mean value of lacunarity for all grid box sizes applied to a particular panel and F refers to the total number of pixels that are considered as sandstone (black in Fig. 6a) in the scanned part of the image per box count. A low value of lacunarity (minimum = 0) is suggestive of a homogeneous and translationally invariant pattern containing gaps of similar size (Fig. 6c). A high value of lacunarity (maximum = 1) indicates a heterogeneous pattern with a varied range of gap sizes (Fig. 6c) (Plotnick *et al.* 1996). Values of lacunarity depend only on the assignment of pixels to sandstone or shale, not on the interpretation of sandbody type or hierarchy.

Ripley's K *function and Besag's* L *function.* Analysis of second-order spatial point patterns commonly involves the use of Ripley's K function (Ripley 1977), which measures the extent of clustering and spatial dispersion at different length scales. Here, we used Ripley's K function to analyse the distribution of the centroids of channelized sandbodies in each panel of interpreted facies architecture (white dots in Figs 5 & 6a) (cf. Hajek *et al.* 2010).

Ripley's K function, $K(h)$, is obtained in a plane by comparing the predicted number of points (e.g. sandbody centroids) within a distance (h) of each point to the average rate of the point process (λ), as outlined in the following equation:

$$\hat{K}(h) = \lambda^{-1} E\,(N(h)) \quad \text{for} \quad h > 0 \quad (2)$$

where ë is the number of centroid points in the two-dimensional area of radius $h(N)$ divided by the area of the study region, and $E(N(h))$ is the expected number of points in the same region (e.g. Cressie 1993; Rosenberg & Anderson 2011). If the number of points found at a certain value of h is equal to the number of points expected, taking into account the intensity of the point process, then the resulting distribution pattern is random. If more points are found within a given value of h than the number expected, then this indicates clustering. If fewer points are found, then points are distributed regularly. We used Besag's L function (Besag 1977), $\hat{L}(h)$, a variance-stabilized version of Ripley's K function, so that the K function can be compared with its expected value and against a benchmark of zero (Besag & Diggle 1977; Rosenberg & Anderson 2011):

$$\hat{L}(h) = h - \sqrt{\frac{\hat{K}(h)}{\pi}} \quad (3)$$

If the expected value of $L(h)$ found at a certain distance is equal to the number of points estimated, taking into account the intensity of the point process, then the distribution pattern is close to zero and represents complete spatial randomness ($\hat{L}(h) = 0$; Besag 1977). We used 99 Monte Carlo simulations of a completely spatially random point process to establish a probability distribution for the number of points expected for the studied range of h at a 95% confidence level (Rosenberg & Anderson 2011). If the points are regularly dispersed, then $\hat{L}(h) > 0$ and the L function plots positively above the complete spatial randomness envelope (Fig. 6b; Besag 1977). In contrast, if the points are clustered, then $\hat{L}(h) < 0$ and the L function plots negatively below the complete spatial randomness envelope (Fig. 6b; Besag 1977).

In our analysis of the facies architectural panels of Livera (1989), each panel was vertically exaggerated by ×123 (i.e. the ratio of mean apparent sandbody width to mean maximum sandbody thickness in reservoir zones 3.3–2.1; Fig. 5) to minimize the effects of anisotropy in the sandbody dimensions on the results (cf. Flood & Hampson 2015). Consequently, the expected spacing of sandbody centroids displays a constant mean and constant variance in all directions. Length scale is expressed in multiples of mean apparent sandbody dimensions (labelled '×1', '×2', etc. on the horizontal axis of Fig. 6c) and the vertical and horizontal spacings of sandbody centroids are scaled according to the mean maximum sandbody thickness and the mean apparent sandbody width, respectively. To avoid distortion by edge effects, we used Ripley's weighted method (Ripley 1988) such that the maximum distance between the points considered in our application of the L function was 25% of the width or height of each panel (Rosenberg & Anderson 2011). Sandbody centroids that lie outside the limits of the panels were excluded from our analysis. The position of sandbody centroids within each panel is independent of the three-dimensional orientation of the sandbodies, but their horizontal spacing may vary according to the orientation of the panel.

The total number of centroids in each panel (Fig. 5) ranges between nine (panel C; Table 1) and 34 (panel A; Table 1). The identification of sandbody centroids is sensitive to the interpretation of sandbody type and hierarchy. The results will be most robust if only sandbodies of a particular hierarchical level (e.g. channel belts) are included in the analysis.

Sensitivity of spatial statistical tools. Villamizar *et al.* (2015) carried out several tests on cross-sections extracted from three-dimensional object-based reservoir models of channelized sandbodies to investigate the sensitivity of lacunarity and Ripley's *K* function to various parameters that may be poorly constrained in typical outcrop and subsurface datasets. Both lacunarity and Ripley's *K* function are relatively insensitive to the orientation of the cross-section in which they are measured, relative to the mean sandbody orientation, provided that the cross-section is oriented parallel or oblique to the depositional strike. Increasing the range of sandbody orientations and/or sandbody sinuosities results in a greater number of sandbody intersections and amalgamations and a wider range of gap sizes between sandbodies. The lacunarity is thus increased. Ripley's *K* function is not significantly affected because the sandbody centroid distributions show little variation as the range of sandbody orientations is increased.

Both lacunarity and Ripley's *K* function are strongly affected by stochastic variations in sandbody position (Villamizar *et al.* 2015), which depends on the number and distribution of well conditioning data. The relatively wide well spacing in the Brent Field dataset relative to interpreted sandbody dimensions (*c.* 200–1000 m; Fig. 1b) therefore implies that alternative sandbody distributions to those in the architectural interpretations of Livera (1989) (Figs 3–5) may give significantly different spatial statistical results.

Facies analysis

Facies analysis of the Ness Formation in the four studied cored wells has identified six facies associations, which are summarized in Table 2 (after Richards & Brown 1986; Livera 1989; Hampson *et al.* 2004). Three facies associations (FA 1–3; Table 2) characterize upper coastal plain strata and form the focus of this study. Facies associations that characterize lower coastal plain strata (FA 4–6; Table 2) are treated only briefly as they are not the focus of this study. Trace fossil assemblages and the intensity of bioturbation are described using the ichnofacies scheme of Pemberton *et al.* (1992) and the bioturbation index (BI) of Taylor & Goldring (1993), respectively, whereas the intensity of pedogenic modification is described using the palaeosol maturity index (MI) of Bown & Kraus (1987).

FA1: Channelized fluvial sandbodies

Description. Facies association 1 (FA1) consists of erosionally based, sharp-topped units that are 0.9–13 m thick in core (Table 2, Figs 7 & 8). Basal erosion surfaces are directly overlain by thin (<10 cm), pebble-grade mudstone intraclast lag deposits or carbonaceous debris, followed by a fining-upwards succession (Fig. 7a, b). Fining-upwards successions consist of trough and planar cross-bedded sandstone, horizontally laminated sandstone and siltstone, current ripple cross-laminated sandstone and siltstone (Fig. 7c), massive sandstone and siltstone, and root-penetrated beds. Soft sediment deformation structures – in the form of convolute lamination – and abundant plant debris are locally present in medium- and coarse-grained sandstones. Bioturbation is generally absent (BI 0), but monospecific *Planolites* and Scoyenia ichnofacies (*Taenidium*) (BI 1–2) occur locally in the upper portion of each fining-upwards succession, as do roots, wood fragments, and siderite and pyrite concretions (MI 0–3; Fig. 8b, c). Channelized fluvial sandbodies either consist of a single fining-upwards succession (i.e. single-storey sandbodies *sensu* Gibling 2006; e.g. Fig. 8a) or a series of stacked, fining-upwards successions, each with a major erosion surface at its base (i.e. multi-storey sandbodies *sensu* Gibling 2006; e.g. Fig. 8b, c). Single-storey channelized sandbodies are 0.9–11.1 m thick (e.g. Fig. 8a) and stacked multi-storey channelized sandbodies reach up to 16.3 m in thickness (e.g. Fig. 8b, c).

Interpretation. FA1 represents fluvial channel fill and barform deposition (Livera 1989). Basal erosion surfaces and the overlying mudclast and carbonaceous lags record initial channel scour (Allen 1978). Soft sediment deformation structures situated towards the base of upwards-fining successions, or immediately overlying internal erosional scour surfaces, provide evidence for bank collapse and/or the rapid loading of sediment in the channel following a period of non-deposition (e.g. Alexander & Gawthorpe 1993). During the early stages of channel filling, when channel depth and sediment input were high, cross-bedded and current ripple cross-laminated sandstones provide evidence for the migration of dunes and ripples in response to unidirectional currents (Harms *et al.* 1975; Bristow 1993). Towards the top of each upwards-fining succession, the abundance of horizontally laminated and current ripple cross-laminated sandstones and siltstones indicates that the later stages of channel filling occurred under decreasing flow velocities and

Table 2. *Summary of facies associations*

Facies association		Description	Thickness (m)	Bioturbation and pedology	Interpretation
Upper coastal plain	**FA-1**: channelized fluvial sandstone	Erosionally based, fine- to coarse-grained sandstone with pebble-grade intra-formational lags and carbonaceous debris. Trough and planar cross-bedding, horizontal lamination, current ripple cross-lamination and soft sediment deformation are common. Successions generally fine upwards. Roots and pedogenic modification in some multi-storey (stacked) successions. Pyrite and siderite concretions	0.9–13.0	Generally absent (BI 0), but locally sparse to low (BI 1–2; *Taenidium*). MI 0–2	Fluvial channel fill deposits contain internal cross-stratification produced by the migration of bars, dunes and ripples in response to unidirectional currents. Periods of non-deposition and subaerial exposure characterized by root traces
	FA-2: non-channelized fluvial sandstones and siltstones	Siltstone to medium-grained sandstone. Planar cross-bedding, current ripple cross-lamination, structureless bedding and carbonaceous debris are common. Modification by soft sediment deformation, bioturbation and pedogenesis. Pyrite and siderite concretions	0.6–6.0	Absent to moderate (BI 0–3; *Skolithos*, *Taenidium*). MI 0–3	Sandstones and siltstones record waxing and/or waning of unconfined unidirectional currents in crevasse splays. Stacked coarsening-upwards and fining-upwards successions of beds record progradation and retreat of main fluvial channel (e.g. Miall 1985)
	FA-3: non-channelized floodplain fines (including coals)	Mudstone, siltstone and coals. Pedogenic modification, bioturbation, soft sediment deformation, carbonaceous debris, and pyrite and siderite nodules are common. Rare desiccation cracks and synaeresis cracks	0.3–3.0	Generally absent (BI 0), but locally low (BI 2; *Arenicolites*). MI 0–4	Vertical aggradation on a vegetated floodplain via intermittent influx of fine-grained sediment from suspension during river flooding, followed by period of subaerial exposure characterized by rootlets, pedogenic horizons, bioturbation and coal seams (e.g. Miall 1996). Synaeresis cracks indicate variations in salinity during deposition

Lower coastal plain	**FA-4**: lagoonal mouth bar sandstones	Siltstone to coarse-grained sandstone; sandstones have sharp to irregular basal erosion surfaces lined by mudstone intraclasts. Cross-bedding, wave and current ripple cross-lamination, planar lamination and soft sediment deformation	0.9–10.0	Generally absent (BI 0), but locally sparse to moderate (BI 1–3; *Skolithos*, *Arenicolites*). MI 0–2	Sharp-based sandstone beds record deposition of deltaic mouth bars by migration of dunes and ripples in response to unidirectional currents, with later wave-reworking and modification
	FA-5: lagoonal wave-influenced sandstones and siltstones	Siltstone to coarse-grained sandstone. Wave and current ripple cross-lamination, micro-scale hummocky cross-stratification, planar cross-bedding and horizontal bedding. Additional features include pyrite and siderite concretions, carbonaceous debris, synaeresis cracks and rooting	0.6–21.0	Sparse to moderate (BI 1–3; *Skolithos*, *Diplocraterion*, *Arenicolites*, *Teichichnus*). MI 0–3	Wave-influenced lagoon or bay associated with episodic storms; migration of dunes and ripples in response to unidirectional currents. Low diversity ichnofauna suggests brackish conditions. Synaeresis cracks indicate variations in salinity during deposition
	FA-6: lagoonal sandstones, siltstones and mudstones	Parallel-laminated mudstone and siltstone with rare very fine- to medium-grained sandstone beds containing micro-scale hummocky cross-stratification and current ripple cross-lamination. Additional features include concretions, carbonaceous debris, synaeresis cracks and rooting	0.3–14.0	Sparse to high (BI 1–4; Skolithos ichnofacies: *Arenicolites*, *Teichichnus*). MI 0–4	Mudstones and siltstones deposited from suspension settling in lagoon or embayment. Sandstone beds record rare episodic sand influx by unidirectional currents and reworking by storm waves. Low diversity ichnofauna suggests brackish conditions

After Livera (1989). Trace fossil assemblages, intensity of bioturbation, and palaeosol maturity are described using the ichnofacies scheme of Pemberton *et al.* (1992), the bioturbation index (BI) of Taylor & Goldring (1993) and the palaeosol maturity index (MI) of Bown & Kraus (1987), respectively.

Fig. 7. Photographs illustrating key features of selected facies associations (Table 2) in well 211/29-BC06 (Fig. 8). Facies association 1 (FA1), consisting of channelized fluvial sandbodies: (**a**) basal erosion surface (at 11 952′), (**b**) mudstone and carbonaceous intraclast lag (at 11 599′) and (**c**) current ripple cross-laminated sandstones and siltstones (at 11 908′). Facies associations 2 and 3 (FA2, 3) consisting of non-channelized floodplain deposits: (**d**) root-penetrated sandstone (at 11 454′), (**e**) root-penetrated siltstone containing monospecific *Planolites montanus* (labelled *Pm*) (at 11 877′) and (**f**) coal (at 11 899′).

reduced water depths. The development of root traces indicates channel abandonment and/or the exposure of bar-tops (e.g. Fig. 8b; Miall 1977, 1985; Olsen 1988). The occurrence of abundant erosion surfaces, palaeosol horizons and lenses of aggradational floodplain fines (FA3) in successions containing channelized sandbodies (Fig. 8b, c) is suggestive of an environment that received variable sediment input and discharge and implies that there was sufficient time between the deposition of successive storeys for floodplain and fine-grained channel fill deposits to accumulate (e.g. Kraus & Davies-Vollum 2004). Occurrences of monospecific assemblages of simple trace fossils (*Planolites*) and the Scoyenia ichnofacies (*Taenidium)* indicate temporary burrowing by deposit feeders (Pemberton *et al.* 1992). *Taenidium* may be associated with abandoned or inactive fluvial channels and desiccated overbank settings (Buatois & Mangano 2011). Single-storey and multi-storey channelized sandbodies, containing one or multiple basal erosional scours respectively, are suggestive of a channel belt architecture (e.g. Miall 1996). From this analysis we can assume that the majority of channelized sandbodies illustrated in the well correlation panels of Livera (1989) through the Ness Formation are channel belts (Figs 3a & 4a). Individual storeys are difficult to distinguish in multi-storey channelized sandbodies because only the uppermost storey is usually fully preserved (Rubidge *et al.* 2000).

FA2: Non-channelized fluvial sandbodies

Description. Facies association 2 (FA2) consists of sandstones and siltstones with sharp upper and lower boundaries that occur as individual beds and vertically amalgamated beds that are 0.6–6 m thick in core (Table 2). Sandstone and siltstone beds are variously planar cross-bedded, current ripple cross-laminated (including climbing ripples), massive and bioturbated, and many beds are penetrated by roots or exhibit weak palaeosol development (MI 0–2) (Figs 7d & 8a). Twenty-eight percent of cored sandstone bedsets in the facies association exhibit a fining-upwards grain size trend, typically consisting of planar cross-bedding overlain by current ripple cross-lamination and capped by roots, whereas 28% exhibit a coarsening-upwards grain size trend, consisting of structureless sandstone overlain by current ripple cross-lamination and planar cross-bedding. The remaining 44% of cored beds consist of structureless sandstone or bioturbated sandstone and siltstone. Trace fossil assemblages are of low diversity and constitute monospecific *Planolites montanus* (BI 0–3), an impoverished Skolithos ichnofacies

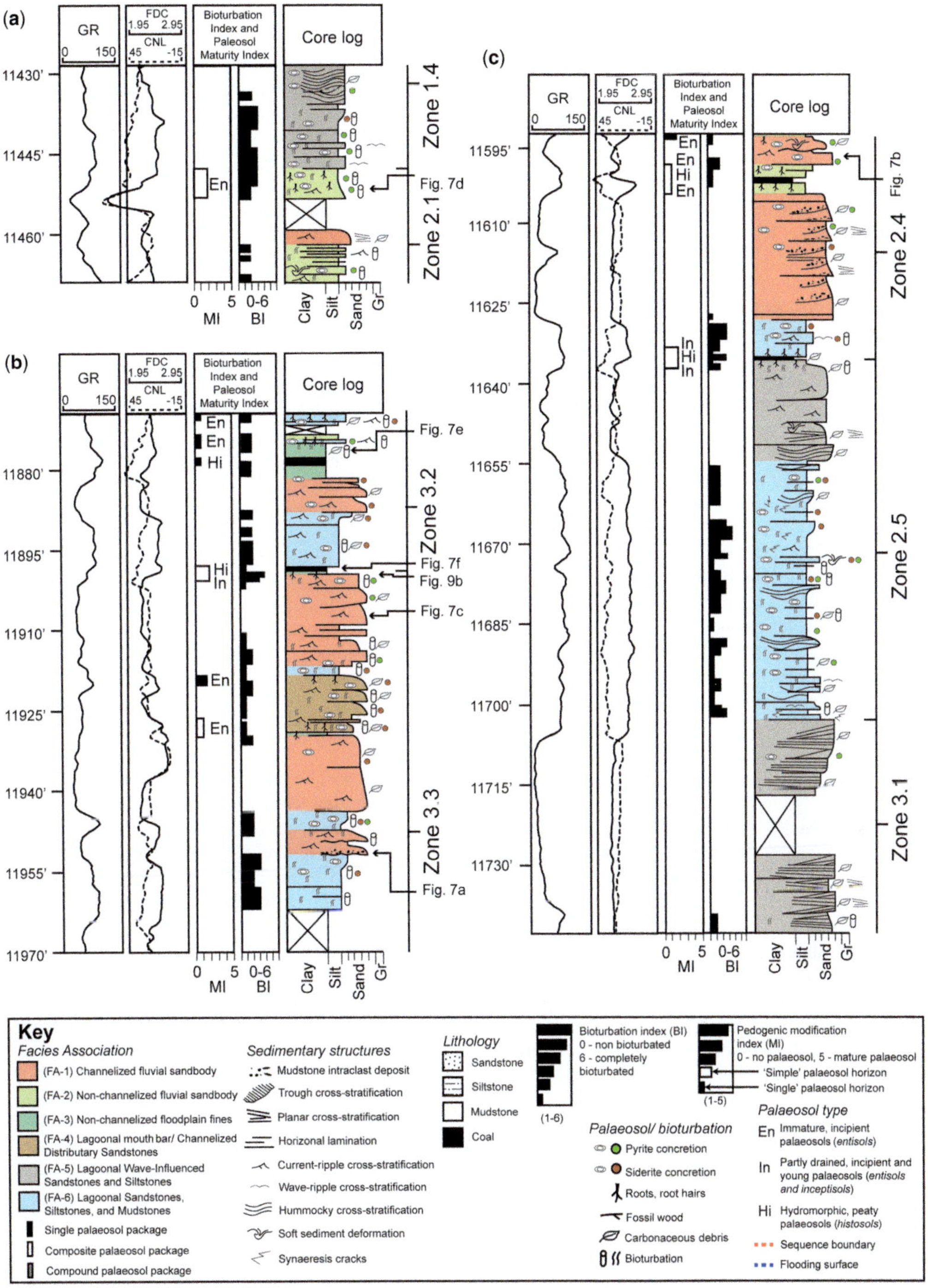

Fig. 8. Representative core descriptions from well 211/29-BC06 illustrating facies associations (Table 2), wireline log trends, bioturbation intensity and intensity of pedogenic modification. The locations of photographs in Figures 7 and 9b are shown. Well 211/29-BC06 is located in Figures 1 and 3.

(*Skolithos;* BI 0–2) or the Scoyenia ichnofacies (*Taenidium*; BI 0–2).

Interpretation. The small thickness of beds and bedsets, the predominance of structures indicating unidirectional currents and the close association with channelized fluvial sandbodies (FA1) implies that FA2 was deposited by crevasse splays and levees, which developed in response to the breaching of an active channel during overbank flooding (Livera 1989; cf. Fielding 1986). The fining-upwards grain size trend is suggestive of waning unidirectional flow conditions and the gradual abandonment of the crevasse splay (Bridge 1984). In contrast, coarsening-upwards grain size trends suggest crevasse splay or levee progradation into aggradational floodplain fines (Elliott 1974; Farrell 2001). The structureless nature of some beds implies rapid sediment-laden fallout from a turbulent suspension (Shultz 1984) or, more likely, modification by soft sediment deformation, bioturbation and/or pedogenic processes that destroyed the original sedimentary fabric. The development of roots, palaeosols and bioturbation towards the top of most crevasse splay and/or levee deposits reflects repeated intervals of non-deposition and rapid colonization during breaks in sedimentation (cf. Fielding 1986). *Planolites* and *Taenidium* are commonly associated with subaqueous and desiccated floodplain environments, respectively (Buatois & Mangano 2011).

FA3: Non-channelized floodplain fines

Description. Facies association 3 (FA3) consists of coal-bearing and root-penetrated, bioturbated and structureless siltstone and mudstone successions that are 0.3–3.0 m thick in core (Table 2, Figs 7e, f & 8b). Weakly developed palaeosols (MI 0–3) and a low diversity trace fossil assemblage of monospecific *Planolites montanus* or an impoverished Skolithos ichnofacies (*Planolites montanus, Arenicolites*; BI 0–2) occur locally. Additional features include soft sediment deformation in the form of convolute lamination and pyrite and siderite concretions.

Interpretation. FA3 was deposited under low flow regimes via suspension fallout during intermittent overbank flooding events (Livera 1989; cf. Miall 1977; Fielding 1986). The occurrence of roots provides evidence for repeated periods of subaerial exposure following periods of overbank flooding (cf. Fielding 1986; Melvin 1987). Bioturbation and pedogenesis lead to the development of structureless siltstones and mudstones. Coals are associated with the accumulation and preservation of carbonaceous plant material in water-saturated peat swamps, which requires a sustained period of reduced clastic input and a high water table (Haszeldine 1989; Bohacs & Suter 1997), which probably occurred in between avulsion events (e.g. Davies-Vollum & Smith 2008). The low diversity of trace fossil assemblages suggests a restricted, shallow water environment of deposition, such as a floodplain lake, whereas the sporadic occurrence of *Arenicolites* suggests that the water column was intermittently stirred and oxygenated by strong currents and/or waves (cf. Buatois & Mangano 2011).

FA 4–6: Lagoonal deposits

Description. Facies associations 4, 5 and 6 (FA4, FA5, FA6; Table 2) generally occur as upwards-coarsening successions of variably interbedded mudstones siltstones and sandstones. The thicknesses of successions for each facies association range between 0.9 and 10 m (FA4; Table 2), 0.6 and 21 m (FA5; Table 2) and 0.3 and 14 m (FA6; Table 2). Units are sharp-based and exhibit flat upper contacts. Common features include hummocky cross-stratification, wave ripples, bioturbation and/or rooting, pyrite and siderite concretions (<5 cm in diameter), synaeresis cracks and soft sediment deformation in the form of convolute laminations. Fining-upwards grain size trends also occur (e.g. Table 2, Figs 8c). Sparse to intense bioturbation (BI 1–4) by an impoverished Skolithos or mixed Skolithos–Cruziana ichnofacies is typical.

Interpretation. The prevalence of hummocky cross-stratification and symmetrical wave ripples in FA 5–6 (Table 2) suggests deposition in a shallow water body at or above storm wave base and within close proximity to the coeval shoreline, to allow for wave and storm reworking of the sediment into sandy shoals and deposition under combined and oscillatory flows (Livera 1989; Tye *et al.* 1999). In this context, the sandstones of FA4 (Table 2) contain predominantly cross-bedding and current ripple cross-lamination, which indicate deposition from unidirectional currents, and are interpreted as mouth bars (Budding & Inglin 1981; Livera 1989). Interbedded siltstone and mudstone intervals (e.g. Fig. 8b) indicate periods of reduced flow velocities during the temporary abandonment of mouth bars, distributary channels and floodplain lagoonal shoals. Palaeosol profiles and abundant root traces (e.g. Fig. 8b) record the episodic emergence of the mouth bars and lagoonal shoals (Livera 1989, Tye *et al.* 1999). Synaeresis cracks in FA 4–6 indicate fluctuations in salinity during deposition (Livera 1989). In combination with the impoverished character of trace fossil assemblages, which implies a physico-chemical stress (MacEachern & Bann 2008), synaeresis cracks are consistent with a brackish, lagoonal setting (Livera 1989) or a sheltered embayment or estuarine basin (cf. the 'estuarine basin fill/outer estuary' facies association of Løseth *et al.* 2009).

Palaeosol characterization

Eighty-eight palaeosols were identified in the studied core intervals (reservoir zones 3.2 and 2.4–2.1) and their characteristics are summarized in Table 3. The studied palaeosols were assigned to palaeosol maturity stages 1–5 (Figs 9 & 10) using the palaeosol maturity scheme of Bown & Kraus (1987). Stage 1 palaeosols are very weakly developed and contain >80% of the primary depositional fabric. Stage 2 palaeosols retain 60–70% of their primary depositional fabric and are weakly developed. Stage 3 palaeosols are weakly to moderately developed and contain 30–60% of their primary depositional fabric. Stage 4 palaeosols are moderately to strongly developed and retain <30% of their primary depositional fabric. Stage 5 palaeosols were not present in the studied cores.

Description

Three types of palaeosol were recognized in the studied cores. Palaeosols of the first type exhibit a maturity stage of 1–2, are 15–150 cm thick and constitute 37% of the palaeosols in the studied cored intervals. Stage 1 palaeosols of this type exhibit very little evidence of palaeosol development, are commonly bioturbated, retain their original primary depositional colour and fabric, and contain pyrite and siderite concretions and small (<5 cm long by <0.5 cm wide) carbonaceous root traces (e.g. Fig. 9a). Stage 2 palaeosols of this type are grey–white in colour, contain abundant root traces and are found in a single weakly developed horizon.

Palaeosols of the second type exhibit maturity stages 1–4, are 15–290 cm thick, contain pyrite and siderite concretions, and constitute 34% of the studied palaeosols. Stage 1 palaeosols of this type consist of a grey to white rooted horizon, with a greater degree of coloration than palaeosols of the first type. Stage 2 palaeosols of this type contain a red–brown horizon that grades upwards into a grey–white rooted horizon (e.g. Fig. 9b). Stage 3 palaeosols consist of a clayey or organic-rich lower horizon, which passes upwards into a root-penetrated, orange–brown or green–grey upper horizon that exhibits some degree of red and purple coloration (e.g. Fig. 9c). Stage 4 palaeosols exhibit a greater degree of red–purple coloration. The colour contacts in this type of palaeosol are diffuse.

Palaeosols of the third type exhibit maturity stages 1–4, are organic-rich, dark grey–black in colour and contain coal horizons of 15–150 cm thickness (e.g. Figs 7f & 9d); they constitute 29% of the studied palaeosols. Palaeosols of this type also contain root traces, yellow patches of sulphur staining (up to 5 cm in diameter) and carbonaceous lenses.

The palaeosols are stacked into single, compound or composite profiles. A single palaeosol profile is 15–150 cm thick and exhibits a maturity stage of 1–2 (21% of palaeosols in the studied cored intervals; represented by black bars in Figs 8 & 10–12). Compound profiles (*sensu* Kraus 1996) are 2.1–5.7 m thick (represented by grey bars in Figs 8 & 10–12), exhibit a maturity stage of 2–4, consist of multiple horizons that are each separated by sediment, and are bounded above and below by channelized fluvial sandbodies of FA1 (11% of palaeosols in the studied cored intervals). Composite palaeosol horizons occur as a series of vertically stacked successive profiles (*sensu* Morrison 1967; Kraus 1999), 0.9–4.2 m thick, and have a maturity stage 2–4 (68% of studied palaeosols; represented by white bars in Figs 8 & 10–12). Weakly to moderately developed compound palaeosols are commonly associated with non-channelized floodplain levees and crevasse splays (FA2; Table 2, e.g. Fig. 9a, c) and lagoonal sandstones, sandstones, siltstones and mudstones (FA6; Table 2, e.g. Fig. 10a). Compound profiles are also located at the boundaries between two facies associations such as channelized fluvial sandbodies and aggradational floodplain fines (FA1, FA3; Table 2; Fig. 10b).

Interpretation

The first type of palaeosol is immature entisols (Mack *et al.* 1993), which record intermittent and relatively short-lived plant colonization of floodplain sub-environments such as levees, lagoonal shoals, floodplain lakes and crevasse splays (Figs 9a & 10). The dark grey–black colour of stage 1 palaeosols of this type, and the abundance of carbonaceous material within them, are suggestive of soil formation under reducing conditions in poorly drained, permanently waterlogged areas of the floodplain (e.g. Besly & Fielding 1989; Retallack 2001). The grey–white colour of stage 2 palaeosols may record gleization, which implies development under generally waterlogged conditions – for example, as a result of a high water table (Besly & Fielding 1989; Retallack 2001).

The second type of palaeosol is inceptisols (Mack *et al.* 1993). Waterlogged and incipient inceptisols developed in an environment that exhibited variations in the water table and occupied reducing conditions, as indicated by their dark grey colour, absence of desiccation cracks and high abundance of carbonaceous material (Figs 9b, c & 10) (e.g. Besly & Fielding 1989). In contrast, the

Table 3. *Summary of palaeosol characteristics*

Palaeosol type	Maturity stage	Thickness (cm)	Description	Interpretation
Entisols	MI 1	15–150	>80% of primary depositional fabric preserved. Single pedogenic horizon of dark grey–black colour. Bioturbation and mottling, rare carbonaceous root traces, pyrite and siderite concretions	Short-lived, intermittent plant colonization under reducing conditions on poorly drained or waterlogged floodplain
	MI 2		60–70% of primary depositional fabric preserved. Single pedogenic horizon of grey–white colour. Abundant carbonaceous root traces	Weak soil development under reducing conditions on poorly drained or waterlogged floodplain, resulting in development of characteristic (gleyed) colour
Inceptisols	MI 1	15–290	>80% of primary depositional fabric preserved. Single pedogenic horizon of grey–white colour. Rare carbonaceous root traces, pyrite and siderite concretions	Very weak soil development under generally reducing conditions on poorly drained floodplain
	MI 2		60–70% of primary depositional fabric preserved. Two pedogenic horizons of red–brown (lower) and grey–white (upper) colour. Abundant carbonaceous root traces, pyrite and siderite concretions	Weak soil development under generally reducing conditions on poorly drained floodplain
	MI 3		30–60% of primary depositional fabric preserved. Lower pedogenic horizon is clayey or organic-rich; upper pedogenic horizon is orange–brown or green–grey in colour with weak red and purple mottling. Abundant carbonaceous root traces, pyrite and siderite concretions	Moderate soil development under episodic conditions of poor drainage, partial drainage and oxidation on the floodplain
	MI 4		<30% of primary depositional fabric preserved. Lower pedogenic horizon is clayey or organic-rich; upper pedogenic horizon is orange–brown or green–grey in colour with strong red and purple mottling. Carbonaceous root traces, pyrite and siderite concretions	Strong soil development under episodic conditions of poor drainage, partial drainage and oxidation on the floodplain
Histosols	MI 1–4	15–150	Variable preservation (>80% to <30%) of primary depositional fabric. Lower pedogenic horizon is organic-rich and dark grey–black in colour; upper horizon consists of black coal. Carbonaceous root traces, sulphur staining	Very weak to strong soil development under reducing conditions on permanently waterlogged floodplain. Peat accumulation (to form coal) records sustained high water table and clastic sediment starvation

Stages of palaeosol maturity are described using the palaeosol maturity index (MI) of Bown & Kraus (1987). Palaeosols of various types and maturities occur as isolated (single) palaeosols, as palaeosols densely stacked vertically to form overprinted (composite) palaeosols, or as palaeosols loosely stacked vertically to form partially overlapping (compound) palaeosols (e.g. Kraus 1999).

Fig. 9. Photographs of representative palaeosols (Table 3). (**a**) Entisol of stage 1 palaeosol maturity (*sensu* Bown & Kraus 1987) with no horizon development, minor root hairs and little alteration of the primary depositional fabric (at 9980′ in well 211/29-BA16). (**b**) Inceptisol of stage 2 palaeosol maturity in which more intense pedogenesis has led to horizon development (defined by diffuse grey and red coloration) and overprinted the primary sedimentary fabric (at 11 900′ in well 211/29-BC06). (**c**) Inceptisol of stage 3 palaeosol maturity that exhibits development of diffuse green, grey and purple horizons, root hairs and siderite nodules (at 8830′ in well 211/29-2). (**d**) Histosol of stage 2 palaeosol maturity with an upper horizon marked by a coal and a lower horizon consisting of a clay-rich layer containing root traces and a yellow to white coloured concretion (at 9094′ in well 211/29-3). Black arrows in the centre of the core in (**d**) have been drawn to indicate way up. The locations of these photos are shown in Figures 8 and 10.

purple–red coloration in the upper part of some palaeosol profiles implies development under prolonged partially drained and oxidized conditions (Besly & Fielding 1989; Retallack 2001), possibly during episodic lowering of the water table (cf. Bown & Kraus 1987). This second type of palaeosol therefore records a wider range of environmental conditions than the first type.

The third type of palaeosol consists of hydromorphic and peaty histosols (Mack *et al.* 1993), which developed in shallow, waterlogged areas of the floodplain with high concentrations of vegetation (e.g. Besly & Fielding 1989; Retallack 2001). The dark grey colour (Fig. 9d) is characteristic of the preservation of organic matter under poorly drained, reducing surface and subsurface conditions (Duchaufour 1982). The development of coals requires a sustained high water table and limited clastic input (Haszeldine 1989; Bohacs & Suter 1997), most likely during periods of sediment starvation between avulsion-related deposition (Davies-Vollum & Kraus 2001).

The general abundance of moderately developed palaeosols (maturity stages 2–4 of Bown & Kraus 1987) and the prevalence of stacked profiles consisting of entisols, inceptisols and histosols (68% of profiles are composite and 11% are compound) imply relatively sustained periods of non-deposition (Kraus & Bown 1993). Compound palaeosols are suggestive of rapid sedimentation and developed adjacent to channel margins where sedimentation was rapid and episodic and erosion was minor (e.g. Wright & Marriott 1993; Kraus 1999; Kraus & Aslan 1993). Compound palaeosols may be associated with the avulsion of a main channel (Kraus & Aslan 1993; Kraus 1996; Kraus & Gwinn 1997). Composite palaeosols developed where the rate of pedogenesis was higher than the rate of deposition, or where erosion and channel incision were pronounced (Kraus 1992, 1999; Wright 1992; Kraus & Bown 1993). As a main channel migrated and avulsed laterally over time, a series of vertically and partially overlapping palaeosol profiles developed due to relatively short-lived pauses in pedogenesis (e.g. Morrison 1967). Single palaeosols that consist of a single rooted horizon (21% of palaeosol profiles) suggest that palaeosol development was relatively short-lived and aggradation rates

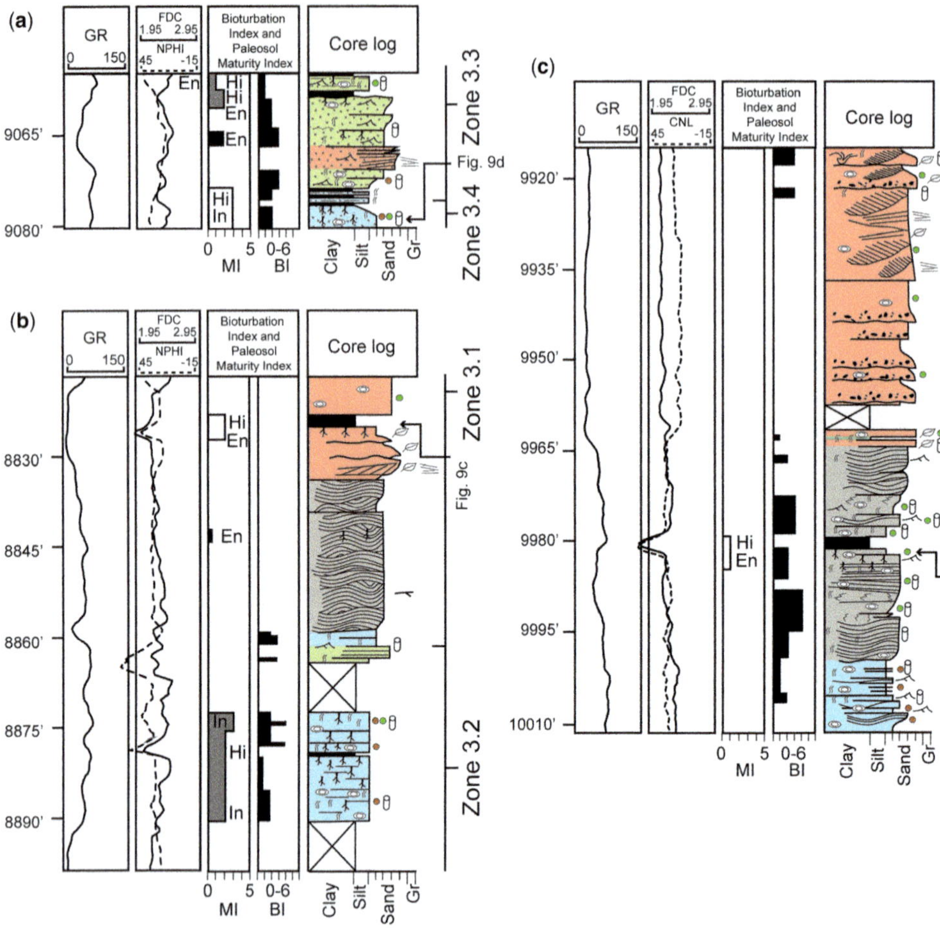

Fig. 10. Core descriptions of representative palaeosols in the Ness Formation. (**a**) One single palaeosol succession, one composite palaeosol profile and one compound palaeosol succession (well 211/29-3). (**b**) One single palaeosol succession, one compound palaeosol succession and one composite package consisting of inceptisols and histosols (well 211/29-2). (**c**) One composite palaeosol succession consisting of entisols and histosols (well 211/29-BA16). Refer to Figure 8 for key to core descriptions.

were sufficiently high to limit the development of mature palaeosols (e.g. Kraus & Bown 1993; Kraus 2002).

Avulsion style

Three styles of avulsion were interpreted from the vertical facies context and palaeosol types associated with 41 channelized fluvial sandbodies in the studied core dataset: (1) avulsion by annexation; (2) avulsion by progradation; and (3) avulsion by incision (cf. Flood & Hampson 2014, after Mohrig *et al.* 2000; Slingerland & Smith 2004). There is uncertainty in our interpretation of avulsion style for three reasons. First, only vertical facies relationships (e.g. Figs 11 & 12) can be assessed, rather than the lateral facies relationships between channelized sandbodies and neighbouring deposits that can also be observed at outcrop. Second, several of the channelized sandbodies were interpreted to overlie sequence boundaries (Figs 11 & 12), implying that there is no genetic linkage between the sandbodies and underlying deposits. The interpretation of these sequence boundaries is not unequivocal and several of the sequence boundaries cannot be readily traced across the field (SB350 in Fig. 3a; SB550, SB600 and SB700 in Fig. 4a). An avulsion interpretation is presented for these sandbodies in Figures 11 and 12, which assumes that the sandbody bases are not sequence boundaries; the implications of these alternative sequence stratigraphic and avulsion-based

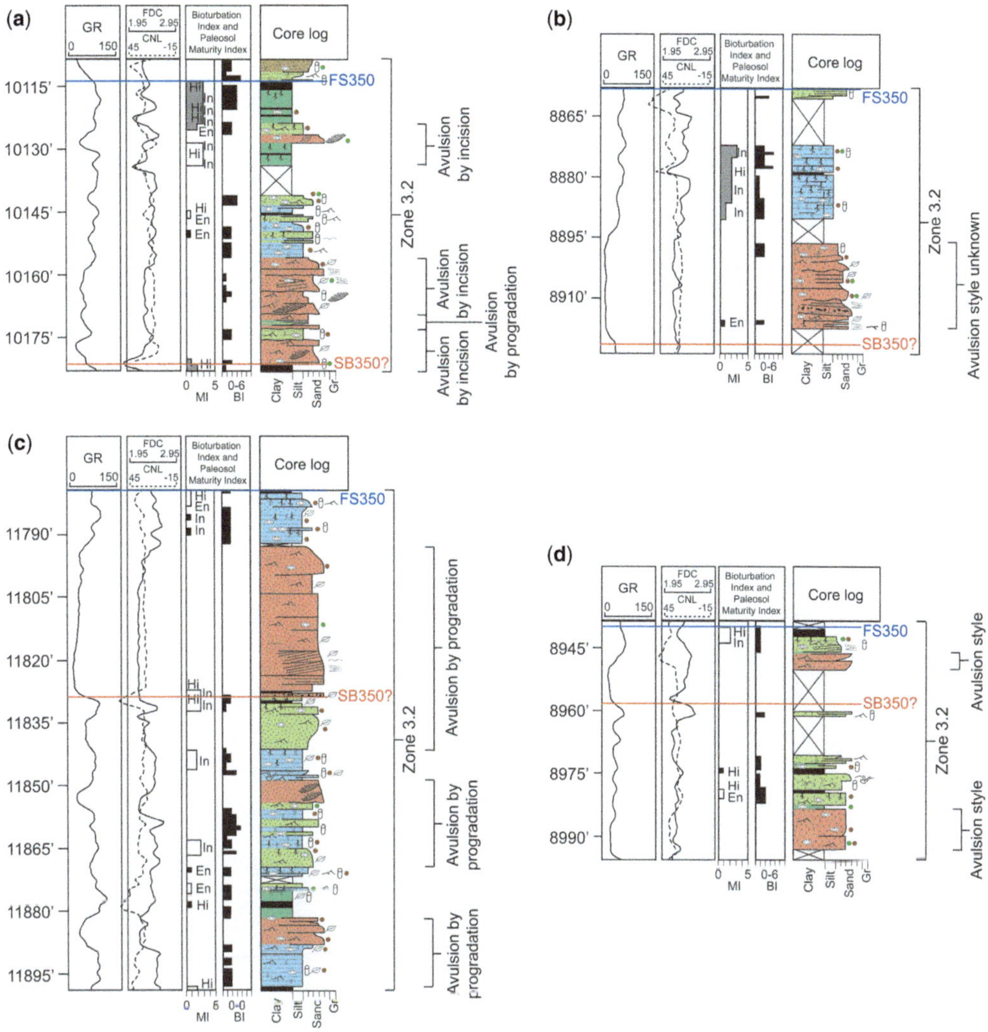

Fig. 11. Core descriptions of upper coastal plain successions in cycle 3 (Fig. 3) in wells (**a**) 211/29-BA16, (**b**) 211/29-2, (**c**) 211/29-BC06 and (d) 211/29-3 showing facies successions, interpreted avulsion style and channelized sandbody distributions (Fig. 16). Refer to Figure 8 for key to core descriptions.

interpretations are discussed later in this paper. The style of avulsion cannot be determined for *c.* 17% of the channelized sandbodies as a result of the absence of core over their bases.

(1) Twelve percent of the channelized fluvial sandbodies (FA1) contain stacked, vertically amalgamated stories (e.g. Fig. 12a, d). This vertical succession implies repeated reoccupation of the same site by an avulsing channel form, involving repeated phases of channel abandonment and subaerial exposure (i.e. avulsion by annexation; Slingerland & Smith 2004). A previously abandoned channel acts as a partially infilled conduit for redirection during a later avulsion, or contains a more easily eroded lithology into which a newly avulsed channel can scour (e.g. Aslan & Blum 1999; Mohrig *et al.* 2000).

(2) Forty-nine percent of the channelized fluvial sandbodies (FA1) cut into an upwards-coarsening (e.g. Figs 10a & 12a, c–d), upwards-fining (e.g. Fig. 12a) or other succession (e.g. Figs 11c & 12a) of crevasse splay and/or levee deposits (FA3), or lagoonal deposits (FA4–6) that contain weakly to moderately developed palaeosols. This vertical succession implies the gradual progradation of a

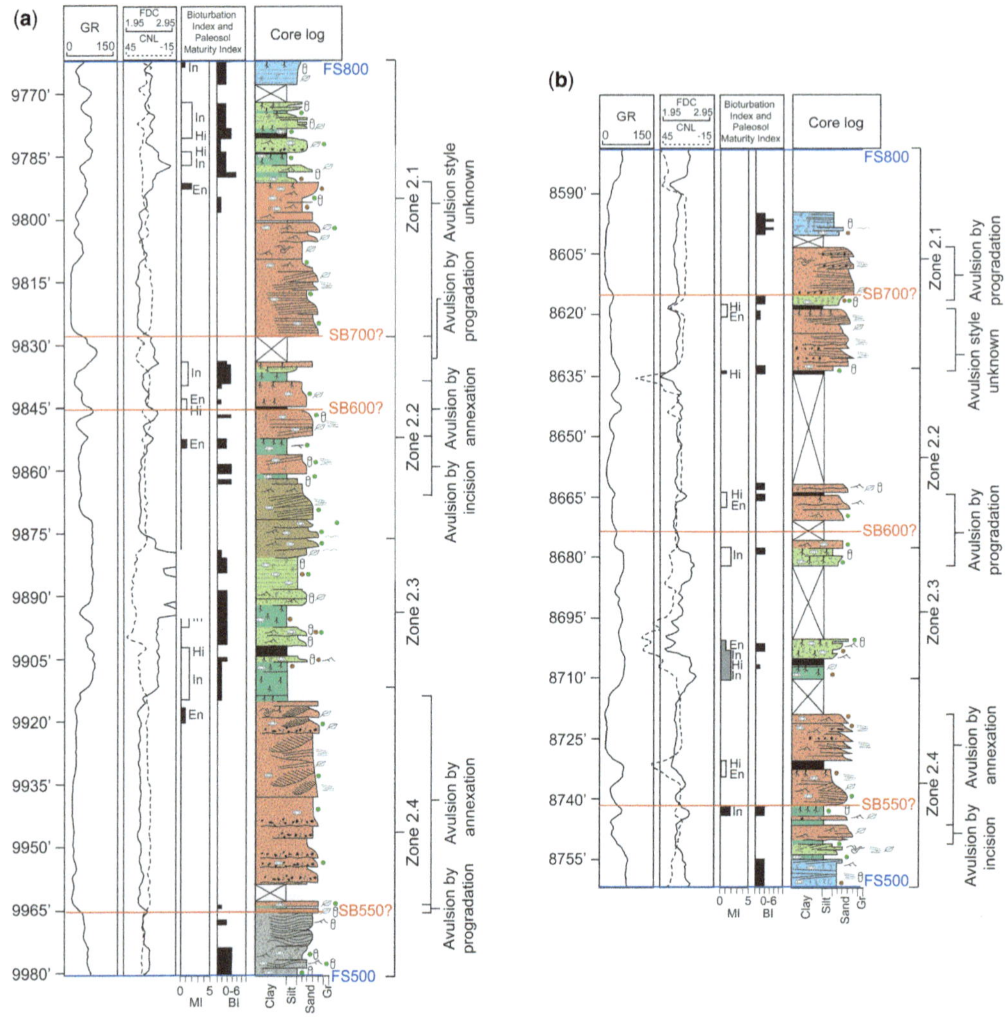

Fig. 12. Core descriptions of upper coastal plain successions in cycle 2 (Fig. 4) in wells (**a**) 211/29-BA16, (**b**) 211/29-2.

fluvial channel into a topographically low part of a floodplain, as recorded by precursor crevasse splays or levees, prior to channel avulsion (i.e. avulsion by progradation; Mohrig *et al.* 2000), equivalent to 'stratigraphically transitional avulsion' (*sensu* Jones & Hajek 2007). Avulsion by progradation occurs in locations proximal to the original parent channel, where precursor crevasse splays accumulate to form a downstream-thinning wedge of sediment (Slingerland & Smith 2004).

(3) Twenty-two percent of the channelized fluvial sandbodies (FA1) lie abruptly above aggradational floodplain fines (FA3) that contain palaeosols and coals, some of which are moderately to strongly developed (e.g. Figs 11d & 12a, d). This vertical succession implies erosion and non-deposition prior to channel avulsion (i.e. avulsion by incision; Slingerland & Smith 2004), equivalent to 'stratigraphically abrupt avulsion' (*sensu* Jones & Hajek 2007). Avulsion by incision may occur in locations distal to the original parent channel, where precursor crevasse splays are absent, or in floodplain settings with very slow or no aggradation (Slingerland & Smith 2004).

There are too few data to confidently identify and interpret apparent lateral trends within reservoir zones in the proportion of channelized fluvial

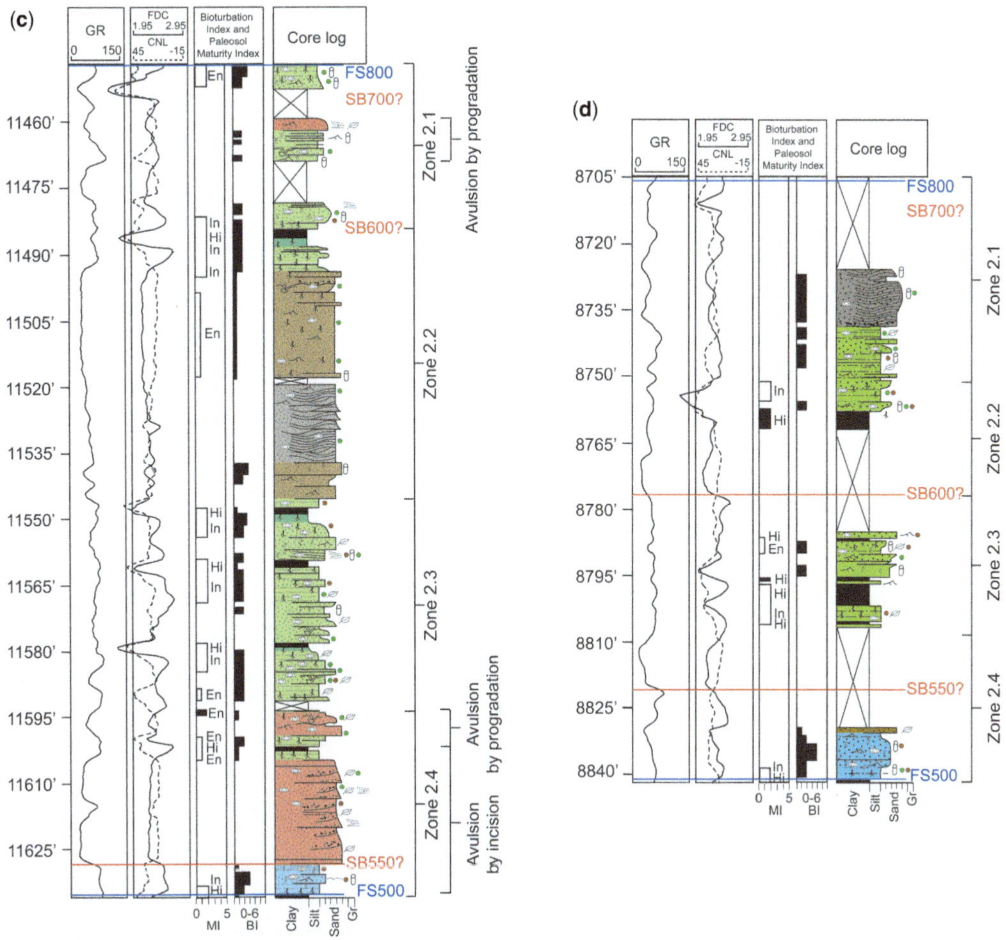

Fig. 12. (*Continued*) (**c**) 211/29-BC06 and (**d**) 211/29-3 showing facies successions, interpreted avulsion style and channelized sandbody distributions (Fig. 16). Refer to Figure 8 for key to core descriptions.

sandbodies (FA1) (Fig. 13a), palaeosol type (Fig. 13b), palaeosol stacking (Fig. 13c) and avulsion style (Fig. 13d) between cored wells. However, from base to top of the studied interval (reservoir zones 3.1–3.3, 2.1–2.5) there is an overall apparent decrease in the proportion of lower coastal plain facies associations (FA 4–6; Table 2) in the studied cores and an associated increase in the proportion of upper coastal plain facies associations (FA 1–3; Table 2) (Fig. 13a). There is little apparent stratigraphic variation in the type of palaeosols (Fig. 13b) and histosols are abundant throughout. From base to top of the studied interval, the overall proportion of stacked, composite palaeosol profiles increases upwards as the proportion of single palaeosols decreases (Fig. 13c). An apparent stratigraphic trend in avulsion style is observed, with an upwards increase in avulsion by incision and a corresponding decrease in avulsion by progradation (Fig. 13d), although data are sparse at some stratigraphic levels. The upwards increases in the proportion of upper coastal plain facies associations, composite palaeosols and avulsion by incision are consistent with increasing distance from the coeval shoreline, as documented in previous studies (Mitchener *et al.* 1992; Fjellanger *et al.* 1996; Husmo *et al.* 2003; Hampson *et al.* 2004). More speculatively, these trends may also reflect an upwards decrease in tectonic subsidence rate, which may have forced overall progradation of the 'Brent Delta'; however, tectonic subsidence rates are not resolved in sufficient temporal detail to support this speculation. The occurrence of histosols throughout the studied interval suggests that the water table remained

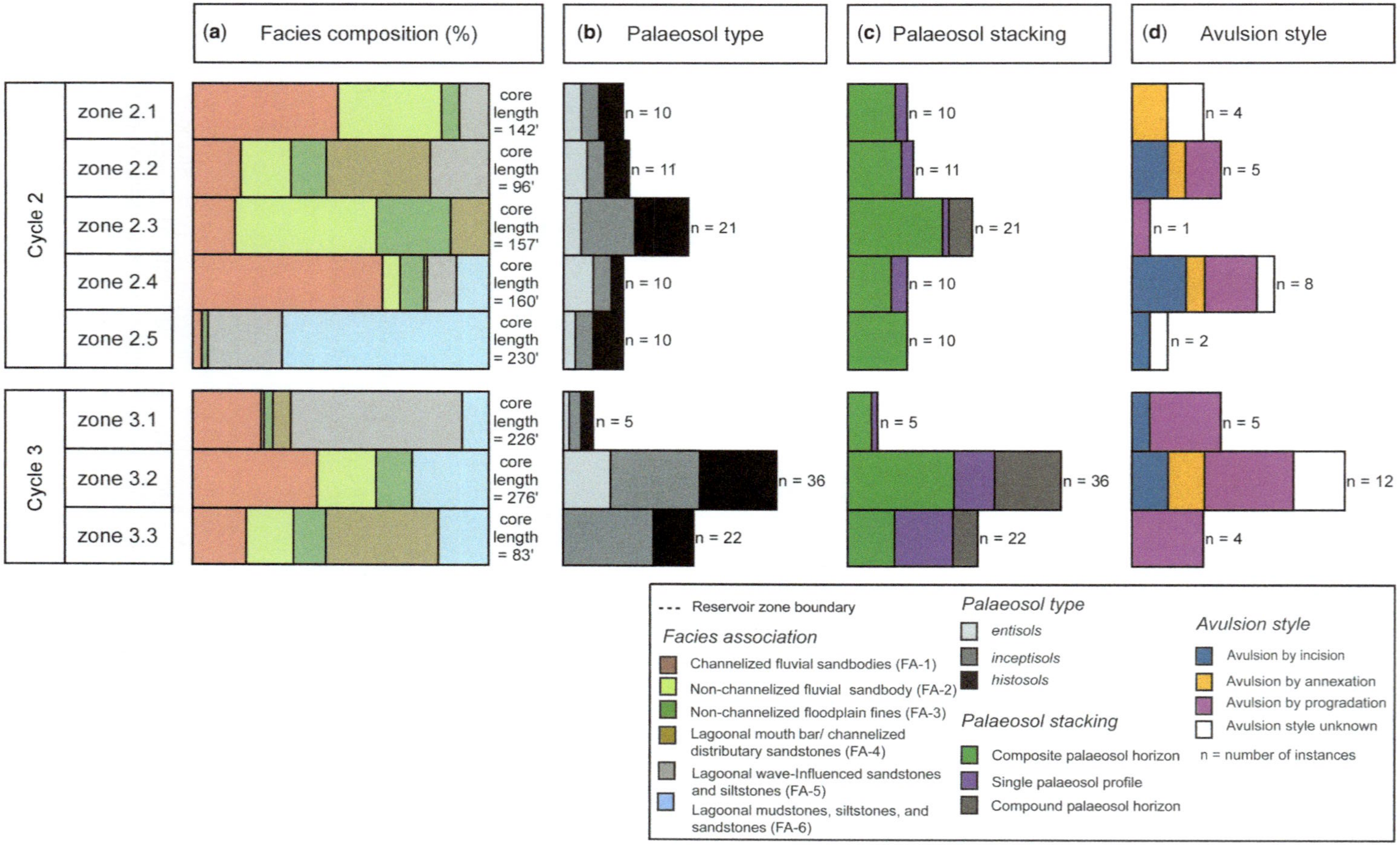

Fig. 13. Comparison of selected parameters in cored successions from wells 211/29-BA16, 211/29-2, 211/29-BC06 and 211/29-3 (e.g. Figs 11 & 12) for stratigraphic subdivisions of lower (reservoir zones 3.1, 3.3 and 2.5; Figs 3a & 4a) and upper coastal plain strata (reservoir zones 3.2 and 2.1–2.4; Figs 3a & 4a): (**a**) proportions of facies associations; (**b**) palaeosol type; (**c**) palaeosol stacking; and (**d**) interpreted avulsion style.

sufficiently high during deposition to enable the development of mires.

Sandbody dimensions

Description

Sixty-four channelized fluvial sandbodies (FA1) were interpreted in the upper coastal plain strata of cycles 2 and 3 in the studied cross-sections by Livera (1989) (Figs 3a, 4a, & 5; Table 1). The mean apparent width of channelized fluvial sandbodies in reservoir zones 3.2, 2.4, 2.2 and 2.1 is 740 m (standard deviation 630 m), although this value is highly interpretative given the well spacing of *c.* 200–1000 m in the lines of cross-section and the mean thickness of these sandbodies of 5 m (standard deviation 2 m). The mean apparent sandbody width and thickness are greater (990 and 8 m, respectively) in cycles 2 and 3 as a whole.

The channelized fluvial sandbodies in the upper coastal plain strata of cycle 3 (mean apparent widths of 580 m; Fig. 14j) are interpreted to be narrower than in cycle 2 (mean apparent widths of 780 m; Fig. 14f). Channelized sandbodies generally become wider in successively younger reservoir zones (mean apparent widths of 580, 670 and 830 m, respectively, in zones 3.2, 2.4 and 2.1; Fig. 14j, i & g), except for reservoir zone 2.2, which contains the greatest value of mean apparent sandbody width (1330 m; Fig. 14h). Reservoir zone 2.2 contains an unusually wide sandbody (2830 wide by 8 m thick; Fig. 5) in the northern part of the field, which distorts the statistical trends that are based only on a small number of data points. This sandbody has been interpreted previously as a trunk distributary channel (Livera 1989) or an incised valley fill (Hampson *et al.* 2004). The mean thickness values for upper coastal plain channelized sandbodies in cycles 2 and 3 and their associated stratigraphic subdivisions (reservoir zones 2.1–2.2 and 2.4) are similar (mean thicknesses of 5 m; Fig. 14a–e).

Interpretation

The interpreted general upwards increase in mean apparent sandbody width in upper coastal plain strata of the Ness Formation (Fig. 14g–j) may have resulted from a greater degree of lateral channel migration and/or widening of channel belts (cf. Helland-Hansen *et al.* 1992; Wright & Marriott 1993), an increase in the supply of coarse-grained sediment (cf. Törnqvist 1994) and/or an upwards decrease in subsidence rate and differential compaction (cf. Allen 1978; Livera 1989; Mitchener *et al.* 1992; Hampson *et al.* 2004).

Sandbody distributions and stratigraphic trends

Description

The value of lacunarity in cycle 2 (0.32) is lower than the value of lacunarity in cycle 3 (0.38), indicating a greater spatial heterogeneity in sandbody distribution in the latter (cf. Fig. 6c). Values of lacunarity for each stratigraphic interval (reservoir zones 3.2, 2.4, 2.2 and 2.1) show no apparent relationship with stratigraphic position (black open circles in Fig. 15). Similarly, there is no apparent trend between stratigraphic position and net-to-gross ratio (grey filled circles in Fig. 15a). However, there is an upwards-decreasing trend in the number of sandbodies per unit area (green filled circles in Fig. 15b) and an upwards-increasing trend in apparent width of channelized fluvial sandbodies (blue filled circles in Fig. 15c). There is no apparent trend in the proportion of palaeosol type (orange symbols in Fig. 15d) or avulsion style (purple symbols in Fig. 15f) with stratigraphic position, but the proportion of composite (stacked) palaeosol profiles generally increases upwards (red crosses in Fig. 15e).

Reservoir zones 2.1, 2.2 and 2.4 and cycle 2 (Fig. 5) display clustering of sandbody centroids over length scales that lie between *c.* 1.5 and 3.5 times the mean sandbody dimensions (Figs 16 & 17). Reservoir zone 3.2 (Fig. 5) displays a random distribution of sandbody centroids over length scales of up to *c.* 3.5 times the mean sandbody dimensions (Figs 16 & 17). Cycle 2 and reservoir zones 2.2 and 3.2 (Fig. 5) display spatial regularity in sandbody centroid positions over length scales of up to *c.* 0.8 times the mean sandbody dimensions (Figs 16 & 17). Thus randomly distributed and clustered patterns of sandbody centroids appear to be dominant and there is no strong variation with stratigraphic position.

Interpretation

Spatial patterns of channelized sandbody distribution are more apparent when large stratigraphic intervals (e.g. cycles 2 and 3 of Livera 1989) are broken down into stratigraphic subdivisions (e.g. reservoir zones 3.2, 2.4, 2.2 and 2.1) (Fig. 16). This in part reflects the smaller number of sandbodies (i.e. smaller sample size) within the stratigraphic subdivisions. However, a similar trend is noted in outcrop datasets of comparable length scale, in which the stratigraphic subdivisions are related to variations in allogenic controls such as tectonic subsidence rate and distance from the coeval shoreline (e.g. Flood & Hampson 2015). Variations in spatial patterns of sandbody distribution between stratigraphic subdivisions may thus be obscured by

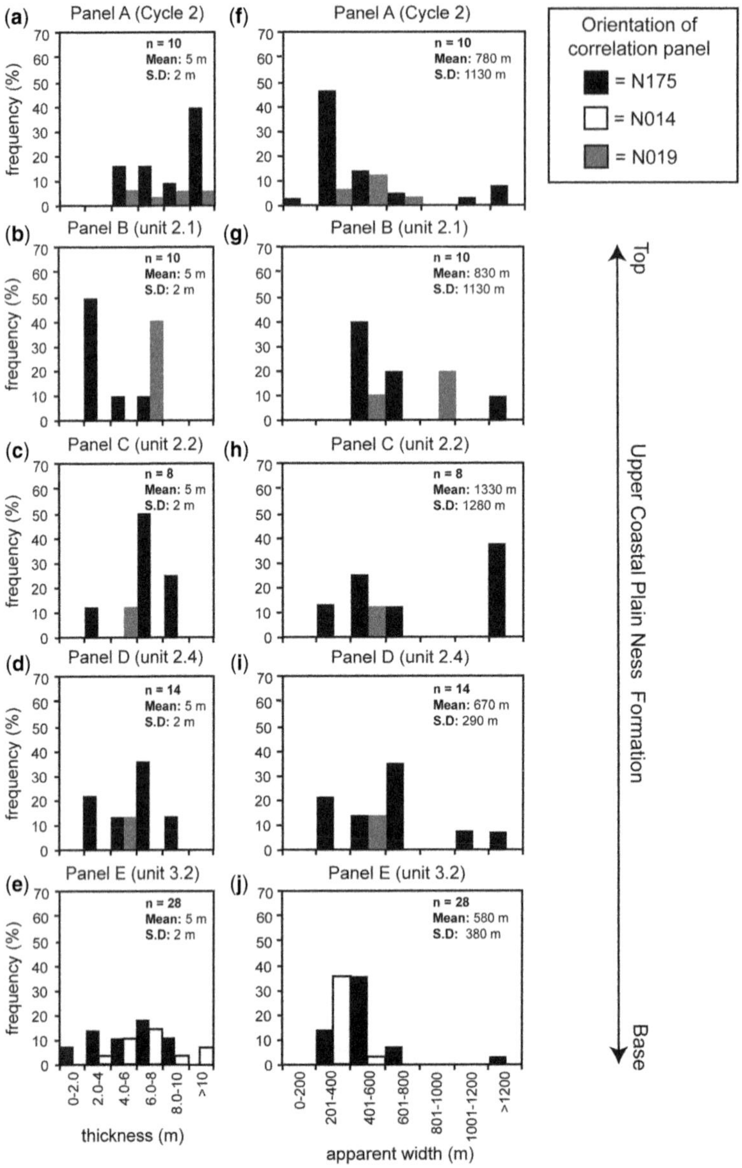

Fig. 14. Graphs illustrating the **(a–e)** thicknesses and **(f–j)** apparent widths of channelized fluvial sandbodies in stratigraphic subdivisions of the Ness Formation (panels A–E in Fig. 5; after the interpretations of Livera 1989): **(a, b)** cycle 2; **(c, d)** reservoir zone 2.1; **(e, f)** reservoir zone 2.2; **(g, h)** reservoir zone 2.4; and **(i, j)** reservoir zone 3.2. The number of channelized fluvial sandbodies (*n*), their mean apparent width and mean thickness, and values of standard deviation (S.D.) for apparent sandbody width and thickness are listed at the top right of each graph. Black, white and grey bars represent channelized fluvial sandbodies that intersect panels that are oriented at N 175, N 014 and N 019, respectively.

averaging over larger stratigraphic intervals. In these same outcrop datasets, increasing values of lacunarity are associated with trends of increasing number of sandbodies per unit area, decreasing net-to-gross ratio and decreasing apparent sandbody width (Flood & Hampson 2015), although no such relationship is apparent in the Brent Field reservoir (Fig. 15a–c). The general absence of trends in palaeosol type and avulsion style with stratigraphic position is also noted in outcrop datasets that show

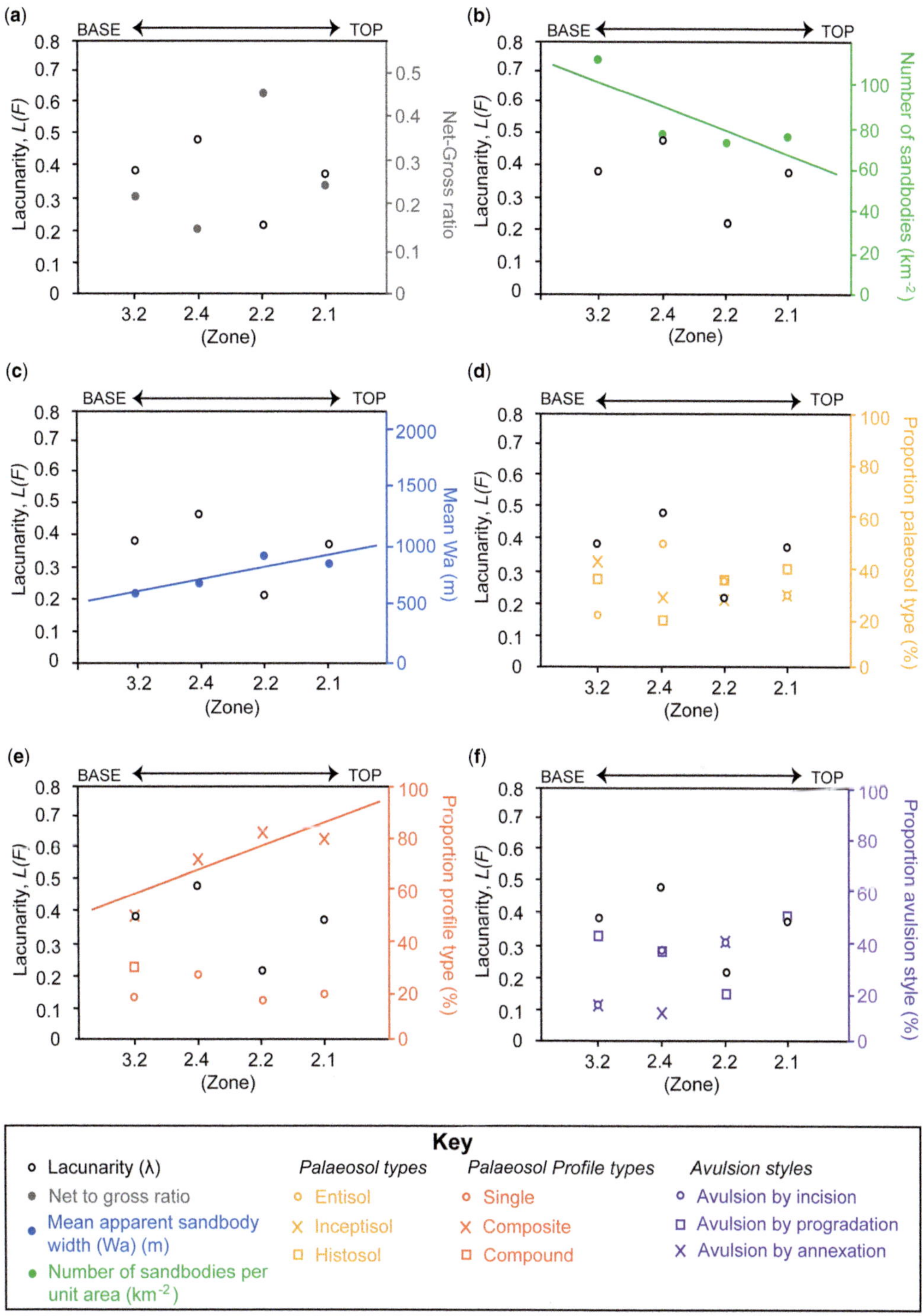

Fig. 15. Cross-plots of lacunarity and (**a**) net-to-gross ratio, (**b**) number of sandbodies per unit area, (**c**) mean apparent sandbody width, (**d**) proportion of palaeosol type, (**e**) palaeosol stacking and (**f**) avulsion style against stratigraphic subdivisions of upper coastal plain strata in the Ness Formation (reservoir zones 3.2, 2.1, 2.2 and 2.4 in Fig. 5). Interpreted trends are shown as coloured lines in (**b**) and (**c**) and for composite palaeosols in (**e**).

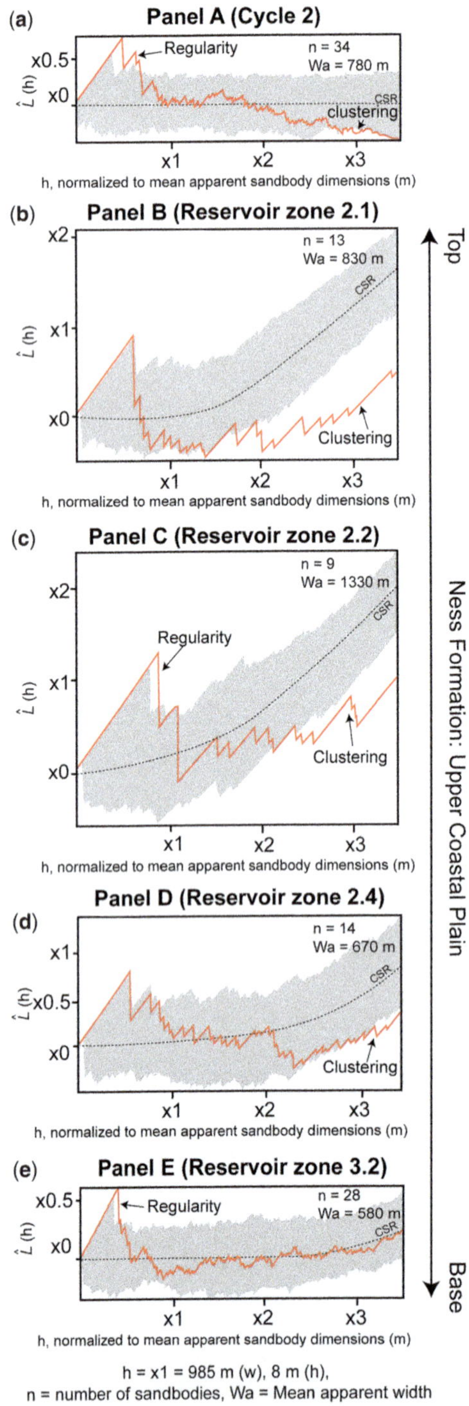

Fig. 16. Graphs of *L* function for sandbody centroids in upper coastal plain strata of the Ness Formation for (**a**) cycle 2 (Fig. 5) and (**b–e**) for each stratigraphic subdivision (reservoir zones 3.2, 2.1, 2.2 and 2.4 in Fig. 5) positioned to illustrate variability from base to top of the study area. Randomly distributed centroids plot in the envelope for complete spatial randomness (CSR) defined by 99 Monte Carlo simulations (grey). Clustered and regularly spaced sandbody centroids plot beneath and above the Monte Carlo envelope (grey), respectively.

consistent stratigraphic variations in patterns of channelized sandbody distribution related to distance from the coeval shoreline (Flood & Hampson 2014, 2015), which implies that localized controls on palaeosol character and avulsion style (e.g. in specific wells or groups of wells) dominate over field-wide stratigraphic variations in sandbody distributions (e.g. between reservoir zones).

Discussion

Uncertainty in interpretations of sandbody dimensions, geometries and distributions

Given the limitations imposed by well spacing (*c.* 200–1000 m, as projected into the lines of cross-section) and distribution, there is significant uncertainty in the interpretation of stratigraphic architecture in the study dataset, including the dimensions, geometries and distributions of channelized fluvial sandbodies (Figs 3 & 4). Here we used the architectural interpretations of Livera (1989) (Figs 3a & 4a), which have withstood the integration of additional well and production data to remain the stratigraphic template for reservoir modelling and management in the Brent Field (e.g. Bryant & Livera 1991; Bryant *et al.* 1991; Abbotts & van Kuijk 1997; James *et al.* 1999). The reservoir modelling experiments of Villamizar *et al.* (2015) suggest that variations in sandbody dimensions and positions would have a greater influence on the spatial statistical patterns measured here using lacunarity and Besag's *L* function than variations in sandbody sinuosity and orientation. The interpretation of Livera (1989) is not based on sequence stratigraphic concepts, but implies that avulsion was the dominant control on sandbody distribution. Sequence stratigraphic interpretations (e.g. Hampson *et al.* 2004) and avulsion-based interpretations such as those presented herein (cf. Jones & Hajek 2007) are both based principally on vertical patterns of sandbody stacking and facies architecture in wells. Both approaches to architectural interpretation are uncertain because the lateral architectural relationships that would constrain the degree of genetic linkage between channelized sandbodies and underlying deposits are absent. We therefore consider them to be interpretative to the same degree in the context of the study dataset (i.e. in the absence of high-resolution three-dimensional seismic, palynological and/or chemostratigraphic data).

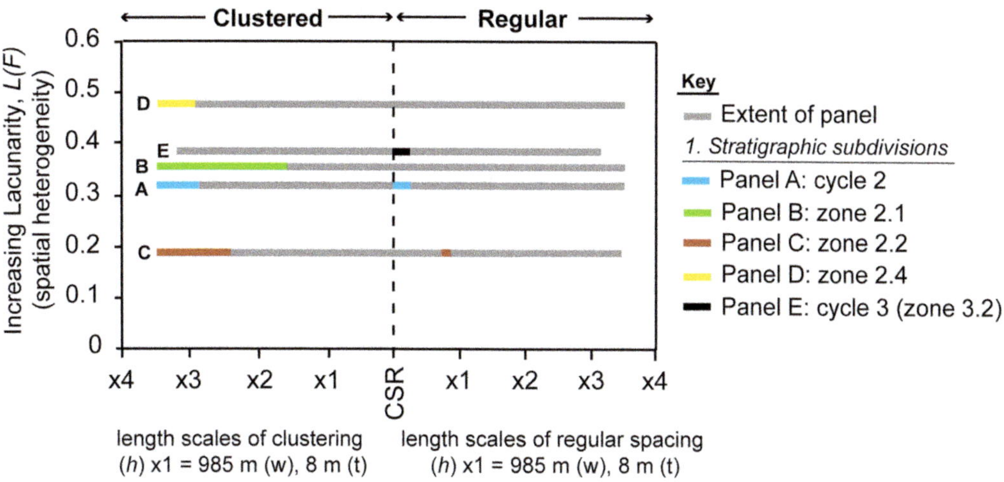

Fig. 17. Plot of lacunarity v. inhomogeneity in spatial positioning of sandbody centroids (cf. Fig. 6c) for stratigraphic subdivisions of upper coastal plain strata in the Ness Formation (cycles 2 and 3 and reservoir zones 3.2, 2.1, 2.2 and 2.4 in Fig. 5). Grey bars represent the spatial extent of data for panels of each stratigraphic interval and superimposed coloured bars show the length scales of sandbody centroid clustering or regular spacing. Length scales not represented by coloured portions of the grey bars correspond to random spacing of sandbody centroids. Length scales of *L* function results are expressed as multiples of mean apparent sandbody dimensions. Lacunarity is dimensionless.

Relationships between sandbody distributions, sandbody dimensions and avulsion style

In the studied upper coastal plain strata (reservoir zones 3.2 and 2.1–2.4; Figs 3a & 4a), clustered and random distributions of sandbody centroids in the interpreted architectural cross-sections (Figs 16 & 17) are similar to those generated by the avulsion of deltaic distributary channels in numerical models of delta plain strata (Mackey & Bridge 1995; Karssenberg & Bridge 2008). Furthermore, values of lacunarity and patterns of sandbody centroid distribution are directly comparable with those measured in large outcrops of coastal plain strata in which the avulsion of deltaic distributary channels has been interpreted (Flood & Hampson 2015, fig. 12b). In this context, the upwards-increasing trend in mean apparent sandbody width (Fig. 15c) can be interpreted to reflect a greater degree of lateral channel migration and/or widening of channel belts (cf. Wright & Marriott 1993) and the upwards-decreasing trend in the number of sandbodies per unit area (Fig. 15b) can be attributed to increased proximity to the upstream avulsion node(s) of a major trunk channel (cf. Mackey & Bridge 1995; Karssenberg & Bridge 2008; Flood & Hampson 2015). Sandbody distributions can alternatively be interpreted in terms of temporal variations in accommodation and/or sediment supply using sequence stratigraphic models, but, as noted earlier, this approach is as interpretive as an avulsion-based approach. For example, it is difficult to interpret the position of sequence boundaries in regions between major channelized sandbodies (Figs 3a & 4a), which would represent interfluves between incised valley fills in a sequence stratigraphic interpretation.

There is no apparent relationship between sandbody dimensions and distribution patterns in the upper coastal plain intervals of the interpreted architectural cross-sections (Figs 3a & 4a) and the proportions of palaeosol type, palaeosol stacking and avulsion style interpreted in cores (Fig. 15d–f). The absence of such relationships may reflect the predominance of localized controls on palaeosol development and avulsion (e.g. palaeotopography, vertical sedimentation rate) that operated at interwell scales, rather than field-wide stratigraphic controls. However, this absence may also arise from difficulties in defining avulsion style from vertical facies successions because the degree of genetic linkage between a channelized fluvial sandbody and the underlying deposits is uncertain in the absence of lateral architectural relationships. This uncertainty is most pronounced where a sequence boundary might plausibly be interpreted at the base of a sandbody (e.g. across 'SB350?', 'SB550?', 'SB600?' and 'SB700?' in Figs 11 & 12). However, it should be noted that avulsion by incision and avulsion by annexation are interpreted to be more common in upper coastal plain strata (reservoir zones 3.2 and 2.1–2.4) than in lower coastal plain strata (reservoir zones 3.1, 3.3 and 2.5) (Fig. 13d), which implies that core-based interpretations of avulsion style can be made that are consistent with interpretations of depositional environment from facies analysis.

Linking sandbody distribution patterns with palaeosol type, palaeosol stacking and avulsion style may require a denser spacing of cored wells than is available in the Brent Field dataset to: (1) constrain sandbody dimensions and positions more tightly; (2) characterize localized variations in palaeosol development and avulsion style with greater lateral resolution; and (3) generate larger datasets that contain more samples and allow more robust statistical analysis. The acquisition of such core datasets is unrealistic for most reservoirs. Instead, it may be more plausible to integrate sparse core and well data with three-dimensional seismic data of sufficient resolution and quality to constrain aspects of sandbody dimensions and distributions.

Implications for reservoir characterization

In cycles 2 and 3 of the Ness Formation in the Brent Field reservoir, pressure data indicate the presence of hydraulically isolated sandstones (flow units) that are separated by laterally extensive shales (Johnson & Stewart 1985; Livera 1989; Bryant *et al.* 1991; James *et al.* 1999). The vertical communication of channelized fluvial sandbodies in reservoir zones 3.3–3.1 (Fig. 3) is relatively poor (Taylor *et al.* 2003). These reservoir zones are associated with field-wide lagoonal shales in lower coastal plain strata (reservoir zones 3.1 and 3.3) (Livera 1989), a higher proportion of avulsions generated by progradation in lower coastal plain strata (Fig. 13d) and contain randomly distributed channelized sandbodies (Figs 16e & 17). Channelized sandbodies are better connected in parts of reservoir zones 2.1–2.4 (Fig. 4) (Bryant *et al.* 1991). In these zones, channelized sandbodies in the southern part of the Brent Field display better vertical communication than in the northern part of the field and also less rapid and more uniform pressure depletion (Taylor *et al.* 2003). Cycle 2 is associated with a higher proportion of avulsions generated by incision and annexation (Fig. 13d) and generally contains relatively large, randomly spaced and clustered sandbodies (Figs 16a–d & 17). Alternatively, the relatively high connectivity of channelized sandbodies in reservoir zones 2.1–2.4 may be attributed to the occurrence of three sequence boundaries ('SB550?', 'SB600?' and 'SB700?' in Figs 4a & 12), each overlain by a multi-storey incised valley fill that is in contact with smaller sandbodies below the sequence boundary (Hampson *et al.* 2004).

Based on facies analysis and palaeosol characterization of the four studied cored wells in the Brent Field, it is possible to interpret lower and upper coastal plain deposits that have different spatial distribution patterns and connectivities of channelized fluvial sandbodies. Furthermore, it is possible to link these patterns of sandbody distribution to differences in avulsion style interpreted in sparse core data. However, it is not possible to accurately predict the positions of individual channelized fluvial sandbodies or clusters of such sandbodies (cf. Villamizar *et al.* 2015). Thus uncertainty in the precise positions of sandbodies needs to be incorporated into reservoir modelling efforts, even though appropriate patterns of sandbody distribution in lower and upper coastal plain deposits may be interpreted from the core-based evaluation of depositional environment and avulsion style during the early stages of field development. This will probably require the simulation of stochastic variability in sandbody positions within the context of a 'template' of their spatial distribution for a given depositional environment or range of avulsion styles.

Conclusions

The late Bajocian Ness Formation constitutes an alluvial-to-coastal-plain succession situated in the Brent Field, UK North Sea. This study has characterized the distribution of channelized sandbodies in upper coastal plain strata of the reservoir (reservoir zones 3.2 and 2.1–2.4) based on previously interpreted facies architectural panels.

Facies analysis of cored wells indicate that the upper coastal plain deposits consist of three facies associations: FA1, channelized fluvial sandbodies; FA2, non-channelized fluvial sandbodies; and FA3, non-channelized floodplain fines. Palaeosols consist of poorly to strongly developed entisols, inceptisols and histosols associated with single, compound (overprinted) and composite (stacked) palaeosol profiles. Three styles of avulsion are identified on the basis of vertical facies relationships in cored wells. Avulsion by annexation is represented by vertically stacked (i.e. multi-storey) channelized fluvial sandbodies and records the reoccupation of previously abandoned palaeochannels. Avulsion by progradation is represented by a channelized fluvial sandbody that directly overlies non-channelized fluvial sandbodies deposited by crevasse splays and levees as a precursor to avulsion. Avulsion by incision is represented by a channelized fluvial sandbody that directly overlies non-channelized floodplain fines, indicating channel erosion into a relatively distal area of the floodplain. Spatial statistical measures (Besag's L function, lacunarity) applied to facies architectural interpretations of well correlation panels indicate that the dominant patterns of sandbody distribution are random and clustered, the latter over a range of length scales (1.5× to 3.5× the mean sandbody dimensions).

These sandbody distribution patterns can be attributed to the autogenic avulsion of deltaic distributary channels downstream of a major avulsion node(s).

These avulsion-generated patterns of sandbody distribution influence sandbody connectivity and pressure communication within the Brent Field reservoir. There is limited vertical communication between channelized fluvial sandbodies in lower coastal plain deposits (reservoir zones 3.1, 3.3 and 2.5) as a result of the presence of field-wide lagoonal shales and thick floodplain intervals, the latter consistent with avulsion by progradation. In comparison, upper coastal plain deposits (e.g. reservoir zones 2.1–2.4) display better vertical communication and slower, more uniform pressure depletion, consistent with increased sandbody connectivity as a result of a higher proportion of avulsions generated by channel incision and annexation. More detailed patterns of sandbody distribution and connectivity (e.g. within upper coastal plain deposits) cannot be readily linked to interpretations of palaeosol type, palaeosol stacking or avulsion style using the sparse core data. Additional uncertainty arises from the interpretation of the degree of genetic linkage between a channelized fluvial sandbody and the underlying deposits, which is uncertain from vertical facies successions in core; this interpreted degree of genetic linkage is fundamental to any interpretation of avulsion style or sequence stratigraphy.

We thank two anonymous reviewers and Tony Reynolds for their constructive and critical reviews and editorial comments. We also acknowledge the Department of Earth Science and Engineering, Imperial College London for support of YSF via a Janet Watson PhD Scholarship, Chevron Energy Technology Company for additional support and the British Geological Survey for access to cores. ImageJ and FracLac were used for lacunarity analysis and PASSaGE v2 to apply Besag's *L* function. We acknowledge Wayne Rasband for developing ImageJ (Research Services Branch, National Institute of Mental Health, USA), Audrey Karperien for creating the FracLac plug-in for ImageJ (Charles Sturt University, Australia), and Michael Rosenberg and Corey Anderson for developing PASSaGE v2 (Arizona State University, USA). We appreciate these researchers making the software readily available for public use.

References

Abbotts, F.V. & van Kuijk, A.D. 1997. Using 3D geological modelling and connectivity analysis to locate remaining oil targets in the Brent reservoir of the mature Brent Field. Paper SPE 38473, presented at the Society of Petroleum Engineers Offshore Europe conference, 9–12 September 1997, Aberdeen, UK.

Alexander, J. & Gawthorpe, R.L. 1993. The complex nature of a Jurassic multi-storey, alluvial sandstone body, Whitby, North Yorkshire. *In*: North, C.P. & Prosser, D.J. (eds) *Characterization of Fluvial and Aeolian Reservoirs*. Geological Society, London, Special Publications, **73**, 123–142, https://doi.org/10.1144/GSL.SP.1993.073.01.08

Allain, C. & Cloitre, M. 1991. Characterizing the lacunarity of random and deterministic fractal sets. *Physical Review A*, **44**, 3552–3558.

Allen, J.R.L. 1978. Studies in fluviatile sedimentation: an exploratory quantitative model for the architecture of avulsion controlled alluvial suites. *Sedimentary Geology*, **21**, 129–147.

Aslan, A. & Blum, M.D. 1999. Contrasting styles of Holocene avulsion, Texas Gulf Coastal Plain, USA. *In*: Smith, N.D. & Rogers, J. (eds) *Fluvial Sedimentology VI*. International Association of Sedimentologists, Special Publications, **28**, 193–209.

Besag, J.E. 1977. Comments on Ripley's paper. *Royal Statistical Society Journal*, **B39**, 193–195.

Besag, J.E. & Diggle, P.J. 1977. Simple Monte Carlo tests for spatial pattern. *Applied Statistics*, **26**, 327–333.

Besly, B.M. & Fielding, C.R. 1989. Palaeosols in Westphalian coal-bearing and red-bed sequences, central and northern England. *Palaeogeography, Palaeoclimatology, Palaeoecology*, **70**, 303–330.

Blum, M.D. & Törnqvist, T.E. 2000. Fluvial responses to climate and sea-level change: a review and look forward. *Sedimentology*, **47**, 2–48.

Bohacs, K.M. & Suter, J.R. 1997. Sequence stratigraphic distribution of coaly rocks: fundamental controls and paralic examples. *American Association of Petroleum Geologists Bulletin*, **81**, 1612–1639.

Bown, T.M. & Kraus, M.J. 1987. Integration of channel and floodplain suites, I. Developmental sequence and lateral relations of alluvial palaeosols. *Journal of Sedimentary Petrology*, **57**, 587–601.

Bridge, J.S. 1984. Large-scale facies sequences in alluvial overbank environments. *Journal of Sedimentary Petrology*, **54**, 583–588.

Bridge, J.S. 2008. Numerical modelling of alluvial deposits: recent developments. *In*: De Boer, P., Postma, G., van der Zwan, K., Burgess, P.M. & Kukla, P. (eds) *Analogue and Numeroical Modelling of Sedimentary Systems: From Understanding to Prediction*. International Association of Sedimentologists, Special Publications, **40**, 97–138.

Bridge, J.S. & Leeder, M.R. 1979. A simulation model of alluvial stratigraphy. *Sedimentology*, **26**, 599–623.

Bristow, C.S. 1993. Sedimentary structures exposed in bar tops in the Brahmaputra River, Bangladesh. *In*: Best, J.L. & Bristow, C.S. (eds) *Braided Rivers*. Geological Society, London, Special Publications, **75**, 277–289, https://doi.org/10.1144/GSL.SP.1993.075.01.17

Bryant, I.D. & Livera, S.E. 1991. Identification of unswept oil volumes in a mature field by using integrated data analysis: Ness Formation, Brent field, UK North Sea. *In*: Spencer, A.M. (ed.) *Generation, Accumulation and Production of Europe's Hydrocarbons*. European Association of Petroleum Geoscientists, Special Publications, **1**, 75–88.

Bryant, I.D., Paardekan, A.H.M., Davies, P. & Budding, M.C. 1991. Integrated reservoir characterization of Cycle III, Brent Group, Brent Field, UK North Sea for reservoir management. *In*: Schneider, R.M. (ed.) *The Integration of Geology, Geophysics, and Reservoir Engineering in Reservoir Delineation, Description and Management*. American Association of Petroleum Geologists Memoir, **26**, 405–422.

Buatois, L.A. & Mangano, M.G. 2011. *Ichnology: Organism–Substrate Interactions in Space and Time*. Cambridge University Press, Cambridge.

Budding, M.C. & Inglin, H.F. 1981. A reservoir geological model of the Brent sands in Southern Cormorant. *In*: Illing, L.V. & Hobson, G.D. (eds) *Petroleum Geology of the Continental Shelf of North-West Europe*. Heyden, London, 326–334.

Clark, P.J. & Evans, F.C. 1954. Distance to nearest neighbor as a measure of spatial relationships in populations. *Ecology*, **35**, 445–453.

Cressie, N. 1993. *Statistics for Spatial Data*. Revised edn. Wiley, New York.

Davies-Vollum, K.S. & Kraus, M.J. 2001. A relationship between alluvial backswamps and avulsion cycles: an example from the Willwood Formation of the Bighorn Basin, Wyoming. *Sedimentary Geology*, **140**, 235–249.

Davies-Vollum, K.S. & Smith, N.D. 2008. Factors affecting the accumulation of organic-rich deposits in a modern avulsive floodplain: examples from the Cumberland Marshes, Saskatchewan, Canada. *Journal of Sedimentary Research*, **78**, 683–692.

Deegan, C.E. & Scull, B.J. 1977. *A Standard Lithologic Nomenclature for the Central and Northern North Sea*. Institute of Geological Sciences Report **77/25**.

Department of Energy and Climate Change 2015. UK annual oil production sorted by field. https://www.gov.uk/guidance/oil-and-gas-uk-field-data#uk-production-data

Dreyer, T. 1990. Sandbody dimensions and infill sequences of stable humid-climate delta plain channels. *In*: Buller, A.T., Berg, E., Hjelmeland, O., Kleppe, J., Torsaeter, O. & Aasen, J.O. (eds) *North Sea Oil and Gas Reservoirs, II*. Graham & Trotman, London, 337–351.

Duchaufour, P. 1982. *Pedology: Pedogenesis and Classification*. Springer, Dordrecht.

Elliott, T. 1974. Interdistributary bay sequences and their genesis. *Sedimentology*, **21**, 611–622.

Farrell, K.M. 2001. Geomorphology, facies architecture, and high-resolution, non-marine sequence stratigraphy in avulsion deposits, Cumberland Marshes, Sasketchewan. *Sedimentary Geology*, **139**, 93–150.

Fielding, C.R. 1986. Fluvial channel and overbank deposits from the Westphalian of the Durham coalfield, NE England. *Sedimentology*, **33**, 119–140.

Fjellanger, E., Olsen, T.R. & Rubino, J.-L. 1996. Sequence stratigraphy and palaeogeography of the Middle Jurassic Brent and Vestland deltaic systems, northern North Sea. *Norsk Geologisk Tidsskrift*, **76**, 75–106.

Flint, S.S., Knight, S. & Tilbrook, A. 1998. Application of high-resolution sequence stratigraphy to the Northwest Hutton Field, northern North Sea: implications for management of a mature Brent Group field. *American Association of Petroleum Geologists Bulletin*, **82**, 1416–1436.

Flood, Y.S. & Hampson, G.J. 2014. Facies and architectural analysis to interpret avulsion style and variability: Upper Cretaceous Blackhawk Formation, Wasatch Plateau, central Utah, USA. *Journal of Sedimentary Research*, **84**, 743–762.

Flood, Y.S. & Hampson, G.J. 2015. Quantitative analysis of the dimensions and distribution of channelised fluvial sandbodies within a large outcrop dataset: Upper Cretaceous Blackhawk Formation, Wasatch Plateau, central Utah, USA. *Journal of Sedimentary Research*, **85**, 315–336.

Gibling, M.R. 2006. Width and thickness of fluvial channel bodies and valley fills in the geological record: a literature compilation and classification. *Journal of Sedimentary Research*, **76**, 731–770.

Hajek, E.A. & Wolinsky, M.A. 2012. Simplified process modelling of river avulsion and alluvial architecture: connecting models and field data. *Sedimentary Geology*, **257-260**, 1–30.

Hajek, E.A., Heller, P.L. & Sheets, B.A. 2010. Significance of channel-belt clustering in alluvial basins. *Geology*, **38**, 535–538.

Hampson, G.H., Sixsmith, P.J. & Johnson, H.D. 2004. A sedimentological approach to refining reservoir architecture in a mature hydrocarbon province: the Brent Province, UK North Sea. *Marine and Petroleum Geology*, **21**, 457–484.

Harms, J.C., Southard, J.B., Spearing, D.R. & Walker, R.G. 1975. *Depositional Environments as Interpreted from Primary Sedimentary Structures and Stratification Sequences*. Society for Sedimentary Geology Short Course Notes **2**.

Haszeldine, R.S. 1989. Coal reviewed: depositional controls, modern analogues and ancient climates. *In*: Whateley, M.K.G. & Pickering, K.T. (eds) *Deltas: Sites and Traps for Fossil Fuels*. Geological Society, London, Special Publications, **41**, 289–308, https://doi.org/10.1144/GSL.SP.1989.041.01.20

Helland-Hansen, W., Ashton, M., Lømo, L. & Steel, R. 1992. Advance and retreat of the Brent Delta: recent contributions to the depositional model. *In*: Morton, A.C., Haszeldine, R.S., Giles, M.R. & Brown, S. (eds) *Geology of the Brent Group*. Geological Society, London, Special Publications, **61**, 109–127, https://doi.org/10.1144/GSL.SP.1992.061.01.07

Heller, P.L. & Paola, C. 1996. Downstream changes in alluvial architecture: an exploration of controls on channel-stacking patterns. *Journal of Sedimentary Research*, **66**, 297–306.

Hofmann, M.H., Wroblewski, A. & Boyd, R. 2011. Mechanisms controlling the clustering of fluvial channels and the compensational stacking of cluster belts. *Journal of Sedimentary Research*, **81**, 670–685.

Husmo, T., Hamar, G.P., Høiland, O., Johannessen, E.P., Rømuld, A., Spencer, A.M. & Titterton, R. 2003. Lower and Middle Jurassic. *In*: Evans, D., Graham, C., Armour, A. & Bathurst, P. (eds) *The Millennium Atlas: Petroleum Geology of the Central and Northern North Sea*. Geological Society, London, 129–155.

James, D., Pronk, D., Abbotts, F., Ward, V., Van Dierendonck, A. & Stevens, D. 1999. The Brent Field: improving subsurface characterization for late field life management. *In*: Fleet, A.J. & Boldy, S.A.R. (eds) *Petroleum Geology of Northwest Europe:*

Proceedings of the 5th Conference. Geological Society, London, 1039–1049, https://doi.org/10.1144/0051039

JEROLMACK, D.J. & PAOLA, C. 2007. Complexity in cellular model of river avulsion. *Geomorphology*, **91**, 259–270.

JOHNSON, H.D. & STEWART, D.J. 1985. Role of clastic sedimentology in the exploration and production of oil and gas in the North Sea. *In*: BRENCHLEY, P.J. & WILLIAMS, B.P.J. (eds) *Sedimentology: Recent Developments and Applied Aspects*. Geological Society, London, Special Publications, **18**, 249–310, https://doi.org/10.1144/GSL.SP.1985.018.01.12

JONES, H.J. & HAJEK, E.A. 2007. Characterising avulsion stratigraphy in ancient alluvial deposits. *Sedimentology Geology*, **202**, 124–137.

KARPERIEN, A. 1999–2013. FracLac for ImageJ, http://rsb.info.nih.gov/ij/plugins/fraclac/FLHelp/Introduction.htm

KARSSENBERG, D. & BRIDGE, J.S. 2008. A three-dimensional numerical model of sediment transport, erosion and deposition within a network of channel belts, floodplain and hill slope: extrinsic and intrinsic controls on floodplain dynamics and alluvial architecture. *Sedimentology*, **55**, 1717–1745.

KRAUS, M.J. 1992. Mesozoic and Tertiary palaeosols. *In*: MARTINI, I.P. & CHESWORTH, W. (eds) *Weathering, Soils, Paleosols*. Developments in Earth Surface Processes, **2**. Elsevier, Amsterdam, 525–542.

KRAUS, M.J. 1996. Avulsion deposits in lower Eocene alluvial rocks, Bighorn Basin. *Journal of Sedimentary Research*, **66**, 354–366.

KRAUS, M.J. 1999. Palaeosols in clastic sedimentary rocks: their geologic applications. *Earth-Science Reviews*, **47**, 41–70.

KRAUS, M.J. 2002. Basin-scale changes in floodplain palaeosols: implications for interpreting alluvial architecture. *Journal of Sedimentary Research*, **72**, 500–509.

KRAUS, M.J. & ASLAN, A. 1993. Eocene hydromorphic palaeosols: significance for interpreting ancient floodplain processes. *Journal of Sedimentary Petrology*, **63**, 453–463.

KRAUS, M.J. & BOWN, T.M. 1993. Palaeosols and sandbody prediction in alluvial sequences. *In*: NORTH, C.P. & PROSSER, D.J. (eds) *Characterization of Fluvial and Aeolian Reservoirs*. Geological Society, London, Special Publications, **73**, 23–31, https://doi.org/10.1144/GSL.SP.1993.073.01.03

KRAUS, M.J. & DAVIES-VOLLUM, K.S. 2004. Mudrock-dominated fills formed in avulsion splay channels: examples from the Willwood Formation, Wyoming. *Sedimentology*, **51**, 1127–1144.

KRAUS, M.J. & GWINN, B. 1997. Facies and facies architecture of Paleogene floodplain deposits, Willwood Formation, Bighorn Basin, Wyoming, USA. *Sedimentary Geology*, **114**, 33–54.

KRAUS, M.J. & WELLS, T.M. 1999. Recognizing avulsion deposits in the ancient stratigraphical record. *In*: SMITH, N.D. & ROGERS, J. (eds) *Fluvial Sedimentology VI*. International Association of Sedimentologists, Special Publications, **28**, 251–268.

LEEDER, M.R. 1978. A quantitative stratigraphic model for alluvium, with special reference to channel deposit density and interconnectedness. *In*: MIALL, A.D. (ed.) *Fluvial Sedimentology*. Canadian Society of Petroleum Geologists Memoirs, **5**, 587–596.

LIVERA, S.E. 1989. Facies associations and sand-body geometries in the Ness Formation of the Brent Group, Brent Field. *In*: WHATELEY, M.K.G. & PICKERING, K.T. (eds) *Deltas: Sites and Traps for Fossil Fuels*. Geological Society, London, Special Publications, **41**, 269–286, https://doi.org/10.1144/GSL.SP.1989.041.01.19

LØSETH, T.M., RYSETH, A.E. & YOUNG, M. 2009. Sedimentology and sequence stratigraphy of the middle Jurassic Tarbert Formation, Oseberg South area (northern North Sea). *Basin Research*, **21**, 597–619.

MACEACHERN, J.A. & BANN, K.L. 2008. The role of ichnology in refining shallow marine facies models. *In*: HAMPSON, G.J., STEEL, R.J., BURGESS, P.M. & DALRYMPLE, R.W. (eds) *Recent Advances in Models of Siliciclastic Shallow-Marine Stratigraphy*. Society for Sedimentary Geology, Special Publications, **90**, 149–116.

MACK, G.H., JAMES, W.C. & MONGER, H.C. 1993. Classification of palaeosols. *Geological Society of America Bulletin*, **105**, 129–136.

MACKEY, S.D. & BRIDGE, J.S. 1995. Three-dimensional model of alluvial stratigraphy: theory and application. *Journal of Sedimentology Research*, **65**, 7–31.

MELVIN, J. 1987. Fluvio-paludal deposits in the lower Kekiktuk Formation (Mississippian), Endicott Field, northeast Alaska. *In*: ETHRIDGE, F.G., FLORES, R.M. & HARVEY, M.D. (eds) *Recent Developments in Fluvial Sedimentology: Contributions from the Third International Fluvial Sedimentology Conference*. Society for Sedimentary Geology, Special Publications, **39**, 343–352.

MIALL, A.D. 1977. A review of the braided-river depositional environment. *Earth-Science Reviews*, **13**, 1–62.

MIALL, A.D. 1985. Architectural-element analysis: a new method of facies analysis applied to fluvial deposits. *Earth-Science Reviews*, **22**, 261–308.

MIALL, A.D. 1996. *The Geology of Fluvial Deposits*. Springer, Berlin.

MITCHENER, B.C., LAWRENCE, D.A., PARTINGTON, M.A., BOWMAN, M.B.J. & GLUYAS, J. 1992. Brent Group: sequence stratigraphy and regional implications. *In*: MORTON, A.C., HASZELDINE, R.S., GILES, M.R. & BROWN, S. (eds) *Geology of the Brent Group*. Geological Society, London, Special Publications, **61**, 45–80, https://doi.org/10.1144/GSL.SP.1992.061.01.05

MOHRIG, D., HELLER, P.L., PAOLA, C. & LYONS, W.J. 2000. Interpreting avulsion process from ancient alluvial sequences: Guadalupe–Matarranya (Northern Spain) and Wasatch Formation (Western Colorado). *Geological Society of America Bulletin*, **112**, 1787–1803.

MORRIS, J.E., HAMPSON, G.J. & MAXWELL, G. 2003. Controls on facies architecture in the Brent Group, Strathspey Field, UK North Sea: implications for reservoir characterization. *Petroleum Geoscience*, **9**, 209–220, https://doi.org/10.1144/1354-079302-504

MORRISON, R.B. 1967. Principles of Quaternary soil stratigraphy. *In*: MORRISON, R.B. & WRIGHT, H.E. (eds) *Means of Correlation of Quaternary Successions*. International Association of Quaternary Research, Proceedings, **9**. University of Utah Press, Salt Lake City, 1–69.

OLSEN, H. 1988. The architecture of a sandy braided-meandering river system: an example from the Lower Triassic Solling Formation (M. *Buntsandstein) in W-Germany. Geologische Rundschau*, **77**, 797–814.

PEMBERTON, S.G., MACEACHERN, J.A. & FREY, R.W. 1992. Trace fossil facies models: environmental and allostratigraphic significance. *In*: WALKER, R.G. & JAMES, N. (eds) *Facies Models: Responses to Sea Level Change*. Geological Association of Canada, St John's, Newfoundland, Canada, 47–72.

PLOTNICK, R.E. 1999. Landscape ecology and quantitative stratigraphy: parallel perspectives on spatial heterogeneity. *In*: HARBAUGH, J.W., WATNEY, W.L., RANKEY, E.C., SLINGERLAND, R., GOLDSTEIN, R.H. & FRANSEEN, E.K. (eds) *Numerical Experiments in Stratigraphy: Recent Advances in Stratigraphic and Sedimentologic Computer Simulations*. Society for Sedimentary Geology, Special Publications, **62**, 271–278.

PLOTNICK, R.E., GARDENER, R.H., HARGROVE, W.W., PRESTEGAARD, K. & PERLMUTTER, M. 1996. Lacunarity analysis: a general technique for the analysis of spatial patterns. *Physical Review E*, **53**, 5461–5468.

RANKEY, E.C. 2002. Spatial patterns of sediment accumulation on a Holocene carbonate tidal flat, northwest Andros Island, Bahamas. *Journal of Sedimentary Research*, **72**, 591–601.

RETALLACK, G.J. 2001. *Soils of the Past: an Introduction to Paleopedology*. 2nd edn. Wiley-Blackwell, Hoboken, NJ.

RICHARDS, P.C. & BROWN, S. 1986. Shoreface storm deposits in the Rannoch Formation (Middle Jurassic), North West Hutton oilfield. *Scottish Journal of Geology*, **22**, 367–375, https://doi.org/10.1144/sjg22030367

RIPLEY, B.D. 1977. Modelling spatial patterns. *Royal Statistical Society Journal, Series B (Methodological)*, **39**, 172–212.

RIPLEY, B.D. 1988. *Statistical Inference for Spatial Processes*. Cambridge University Press, Cambridge.

ROSENBERG, M.S. & ANDERSON, C.D. 2011. PASSaGE: pattern analysis, spatial statistics and geographic exegesis, version 2. *Methods in Ecology and Evolution*, **2**, 229–232.

ROY, A., PERFECT, E., DUNNE, W.M., ODLING, N. & KIM, J-W. 2010. Lacunarity analysis of fracture networks: evidence for scale-dependent clustering. *Journal of Structural Geology*, **32**, 1444–1449.

RUBIDGE, B.S., HANCOX, P.J. & CATUNEANU, O. 2000. Sequence analysis of the Ecca-Beaufort contact in the southern Karoo of South Africa. *South African Journal of Geology*, **103**, 81–96.

SHANLEY, K.W. & MCCABE, P.J. 1994. Perspectives on the sequence stratigraphy of continental strata. *American Association of Petroleum Geologists Bulletin*, **78**, 554–568.

SHELL UK 2015. Brent Field decommissioning, http://www.shell.co.uk/sustainability/decommissioning/brent-field-decommissioning/brent-field-timeline.html

SHULTZ, A.W. 1984. Subaerial debris-flow deposition in the Upper Paleozoic Cutler Formation, Western Colorado. *Journal of Sedimentary Petrology*, **54**, 749–772.

SLINGERLAND, R.L. & SMITH, N.D. 2004. River avulsions and their deposits. *Annual Reviews of Earth and Planetary Science*, **32**, 257–285.

STRAUB, K.M., PAOLA, C., MOHRIG, D., WOLINSKY, M.A. & GEORGE, T. 2009. Compensational stacking of channelised sedimentary deposits. *Journal of Sedimentary Research*, **79**, 673–688.

TAYLOR, A.M. & GOLDRING, R. 1993. Description and analysis of bioturbation and ichnofabric. *Journal of the Geological Society, London*, **150**, 151–148, https://doi.org/10.1144/gsjgs.150.1.0141

TAYLOR, S.R., ALMOND, J., ARNOTT, S., KEMSHELL, D. & TAYLOR, D. 2003. The Brent Field, Block 211/29, UK North Sea. *In*: GLUYAS, J.G. & HICHENS, H.M. (eds) *United Kingdom Oil and Gas Fields, Commemorative Millennium Volume*. Geological Society, London, Memoirs, **20**, 233–250, https://doi.org/10.1144/GSL.MEM.2003.020.01.20

TÖRNQVIST, T.E. 1994. Middle and late Holocene avulsion history of the River Rhine (Rhine-Meuse delta, Netherlands). *Geology*, **22**, 711–714.

TÖRNQVIST, T.E. & BRIDGE, J.S. 2002. Spatial variation of overbank aggradation rate and its influence on avulsion frequency. *Sedimentology*, **49**, 891–905.

TYE, R.S., BHATTACHARYA, J.P., LORSONG, J.A., SINDELAR, S.T., KNOCK, D.G., PULS, D.D. & LEVINSON, R.A. 1999. Geology and stratigraphy of fluvio-deltaic deposits in the Ivishak formation: applications for development of Prudhoe Bay Field, Alaska. *American Association of Petroleum Geologists Bulletin*, **83**, 1588–1623.

VILLAMIZAR, C.A., HAMPSON, G.J., FLOOD, Y.S. & FITCH, P.J.R. 2015. Object-based modelling of avulsion-generated sandbody distributions and connectivity in a fluvial reservoir analogue of low to moderate net-to-gross ratio. *Petroleum Geoscience*, **21**, 249–270, https://doi.org/10.1144/petgeo2015-004

WRIGHT, V.P. 1992. Paleopedology: stratigraphic relationship and empirical models. *In*: MARTINI, I.P. & CHESWORTH, W. (eds) *Weathering, Soils, Paleosols*. Developments in Earth Surface Processes, **2**. Elsevier, Amsterdam, 475–499.

WRIGHT, V.P. & MARRIOTT, S.B. 1993. The sequence stratigraphy of fluvial depositional systems: the role of floodplain storage. *Sedimentary Geology*, **86**, 203–210.

ZHAO, J., CHEN, S., ZUO, R. & CARRANZA, E.J.M. 2011. Mapping complexity of spatial distribution of faults using fractal and multifractal models: vectoring towards exploration targets. *Computers & Geosciences*, **37**, 1958–1966.

Deflection of the progradational axis and asymmetry in tidal seaway and strait deltas: insights from two outcrop case studies

SERGIO G. LONGHITANO[1]* & RON J. STEEL[2]

[1]*Department of Sciences, University of Basilicata, Potenza, Italy*

[2]*Jackson School of Geosciences, University of Texas at Austin, Texas, USA*

**Correspondnce: sergio.longhitano@unibas.it*

Abstract: Deltas represent the major sediment source for tectonically confined, tide-dominated seaways or straits. Modern examples show how along-shore tidal currents are able to modify the impinging delta shape, generating asymmetrical coastal plains, deflected delta fronts and elongate sandbanks. Seaway or strait deltas can become become tide-influenced or tide-dominated, assuming physical attributes that may depart from classical models. Ancient deltas in seaways and straits can also reveal unexpected facies stacking and stratigraphies, which can be misinterpreted or attributed to different depositional settings. Two ancient analogues of deltas that prograded into elongate basins dominated by amplified tidal currents are presented here. A common element in these deltas is the progressive-upwards change in the dominant process of sediment dispersion recorded in the delta facies. Early stages of progradation are dominated by river- and wave-influenced lithofacies, whereas late deltaic advancements occur with a dominance of tidal current circulation on the delta fronts and the consequent morphologies are deflected/elongated in the direction of tidal flow. This study provides the basis for a preliminary stratigraphic framework for the depositional style of these types of delta. The studied deposits also suggest analogies with the spatial distribution of many hydrocarbon reservoirs investigated along the margins of confined, narrow, linear basins, the interpretation of which is still debated.

River deltas, the main vehicle of water and sediment transport from continents to ocean basins, are among the most investigated depositional systems in the world. Their historical importance stems from their great natural and economic potential for society because of their strategic importance in land use (Syvitski *et al.* 2005; Bhattacharya 2006). In modern times, ancient deltaic deposits have revealed important georesources (Tyler & Finley 1991) and their architectural and facies elements, from outcrop to experimental studies, have been highlighted in increasing detail in the last few decades as a result of innovative investigation techniques.

For these reasons, river delta systems have benefited from an impressive number of depositional models. These are mostly based on: (1) the dominant hydrodynamic processes governing sediment delivery along the deltaic coastlines (e.g. Galloway 1975; Orton & Reading 1993); (2) their diagnostic morphologies assumed at the shoreline (Fisher *et al.* 1969; Bhattacharya & Giosan 2003); or (3) the distinctive depositional geometries that deltas generate in a cross-sectional view when they enter a standing body of water (i.e. basin) (e.g. Wescott & Ethridge 1990). Despite this proliferation of models and associated classifications, the Galloway-based terminology has rightly received preferential use among sedimentologists because it has been improved through time (e.g. Dalrymple *et al.* 1992; Orton & Reading 1993; Ainsworth *et al.* 2011).

Deltas have been described in a wide range of tectonic settings and have been used as proxies to reconstruct basinal history around the world because they are considered formidably sensitive to high-frequency sea-level changes, process regime variations and structural movements of the basin margins (Muto & Steel 2002; Yoshida *et al.* 2007; Longhitano 2008).

River deltas can depart from a classical symmetrical coastal shape as a result of the influence of channel avulsion in the delta plain area or an oblique approach of waves at the coastline (Bhattacharya & Giosan 2003). In wave-dominated coastlines, strongly asymmetrical delta morphologies result from the influence of wind-induced along-shore currents, which tend to deflect the deltaic components at the river mouth and also favour the formation of spits and bar complexes (Wright 1977; Orton & Reading 1993; Reading & Collinson 1996). Asymmetrical river mouth entrances are also attributed to autocyclic distributary switching in meandering river channels on the delta plain (Longhitano & Colella 2007). In these settings, the river mouth is often bypassed when a river flood transports sediment onto a delta mouth bar and beyond.

From: Hampson, G. J., Reynolds, A. D., Kostic, B. & Wells, M. R. (eds) 2017. *Sedimentology of Paralic Reservoirs: Recent Advances*. Geological Society, London, Special Publications, **444**, 141–172.
First published online July 13, 2016, https://doi.org/10.1144/SP444.8

Sand is mostly stored in the mouth bar and slowly reworked and re-entrained in the littoral sand stream. Fine sand, silt and clay are carried as a buoyant half-jet and rained onto the shelf floor (Swift & Thorne 1991). These studies focused on the observation of subaerial features, but paid less attention to sediment routing and distribution at the delta front, which represent a volumetrically important deltaic component.

Fewer studies have documented river deltas impinging upon, and interacting with, straits or elongate seaways (Keller & Richards 1967; Galloway 2002; Willis 2005). These deltas often prograde into confined marine settings where the effects of waves, flood and/or ebb tides and river input are subordinate to the dominant influence of tidal current patterns running parallel to the coastline, increasing in their strength as a result of the mass convergence of water (Defant 1961; Pugh 1987; Pratt 1990). As a consequence, river deltas sourcing tidal straits or seaways can assume strongly skewed or strike-elongate morphologies, departing substantially from the expected river delta morphologies (e.g. Weise 1980; Korus & Fielding 2015; Ayranci & Dashtgard 2016).

Surprisingly, a general observation along the world's coastlines reveals that >20% of modern deltas prograde into tide-dominated straits and seaways, although many of these examples are not necessarily subject to significant deflections or asymmetries in their subaerial environments (www.eoearth.org). Based on this evaluation, a significant number of ancient deltas may have been accumulated and preserved in the rock record of tidal seaways and straits, especially where tectonics shaped parts of the sedimentary basins into elongate narrow sectors (e.g. rift, thrust-top or strike-slip basins). These types of deltas have possibly assumed important morphological asymmetries and deflections of their subaqueous environments, as elongate basins usually experience an oceanographic circulation with tidal or other currents flowing axially. This is one of the main issues that has inspired the present research, which investigated the basis for a new stratigraphic model for such deltas

These general considerations point towards types of deltas that differ from 'classical' wave-, river- or tide-dominated models. Therefore the possible importance of paying more attention to river deltas sourcing tidal elongate seaways or straits can be highlighted by the following questions.

(1) What are the distinctive morphological features and sediment zonations of tide-dominated seaway/strait deltas?
(2) How can we recognize them in the rock record?
(3) What kind of vertical facies successions and stratigraphic sequences can they assume?
(4) Are tide-dominated seaway/strait deltas the possible interpretative key in unravelling sand-rich hydrocarbon reservoirs entrapped in confined settings?

We critically review two outcrop case studies in which siliciclastic-dominated, sand-rich sedimentary successions have been interpreted as the record of deltaic sequences prograding into narrow and elongate settings. Their stratigraphic and sedimentary facies stacking patterns record delta front accumulation strongly influenced by tidal currents, causing progressive sandbody and distributary channel deflection, the likely key with which to detect other ancient deltas building into a similar setting. The two field study cases derive from: (1) the Campanian of the Western Interior Seaway of southern Wyoming, USA; and (2) the Tortonian of the Calabrian Arc in southern Italy.

Tidal seaways and straits

Tidal seaways are elongate marine passageways, thousands of kilometres wide, connecting two wider basins. The dominant hydrodynamic force in tidal seaways results from tidal currents flowing parallel to strike. Tidal straits are narrower than seaways and are governed by the convergence and amplification of tidal currents as a result of narrowing of the cross-sectional area of the water body (Pugh 1987). Recent studies on ancient tidal strait-fill successions in southern Italy have attempted to demonstrate that a critical cross-sectional area is the fundamental quantitative condition through which a strait becomes dominated by the amplification of a tidal flow (Longhitano *et al.* 2014). This condition, favourable to the onset of a tidal circulation, can be also matched during late transgressive stages, as has occurred recently along many of the world's coastal systems at the end of the post-Last Glacial Maximum relative sea-level rise (e.g. Longhitano *et al.* 2016).

Seaways and straits are mainly sourced by river deltas. Other minor sediment supply can be provided by: (1) lateral debris aprons or landslides shed from steep margins; (2) clastic sediments derived from the erosion of older lowstand deposits by strong tidal flows in the proximity of the tidal maxima at the narrowest strait zone; and (3) *in situ* carbonate factories, well adapted to high-energy conditions (Anastas *et al.* 2006; Longhitano 2013). As observable in many modern examples (e.g. the Juan de Fuca Strait, Malacca Strait, San Francisco Strait, Messina Strait, Dover Strait and Torres Strait; see Fig. 1), coaxial tidal currents strongly influence the sea bottom. They flow in reversal phases, often in opposition between the two interlinked basins, favouring the development of wide tidal bedform

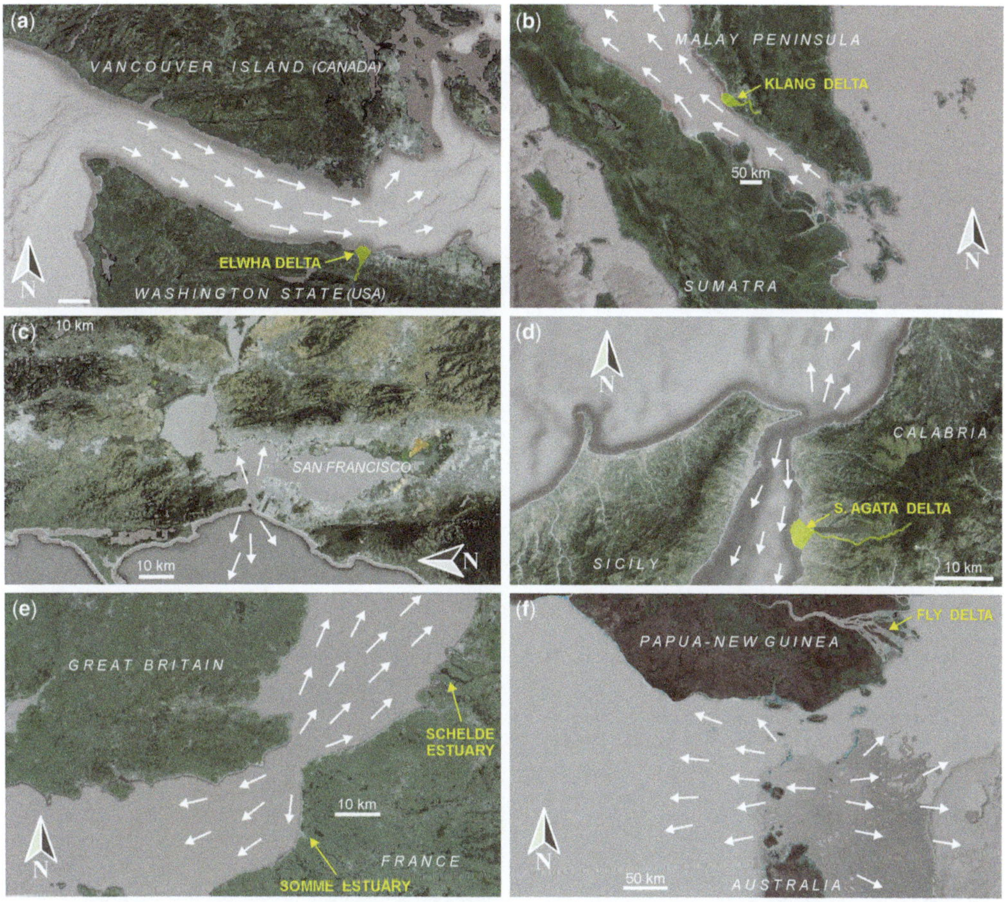

Fig. 1. Examples of modern tidal straits and seaways. (**a**) The Juan de Fuca Stait at the northwestern boundary of the USA, where the Elwha Delta is developing (see Fig. 19a). (**b**) The Malacca Seaway, Indonesia, with the Klang Delta prograding from the northeastern border (Fig. 19d). (**c**) The San Francisco (or Golden Gate) Strait, California. (**d**) The Messina Strait in the central Mediterranean Sea; the Sant'Agata River Delta is highlighted. (**e**) The Dover Strait, western Europe. (**f**) The Bass-Torres Strait, northern Australia. Arrows indicate the main directions of tidal currents.

fields, including two- and three-dimensional, simple and compound dunes, passing downcurrent and laterally to rippled, highly bioturbated fine-grained sheets (e.g. Dalrymple 1984, 2010; Ashley 1990; Anastas *et al.* 1997; Bastos *et al.* 2003). Tidal currents are able to produce deposition and bedform migration outwards of the seaway or strait, rather than inwards. This is mainly due to the energy condition of the opposing flow during each phase of tidal reversal, which is usually in a weaker, but increasing, stage (in the strait side from which it flows in) and is thus unable to move significant sand volumes. However, many stratigraphic sections observed in ancient tidal strait-fill successions occasionally show the occurrence of thinner cross-strata migrating in an opposing direction to the dominant palaeocurrent trend as a result of the momentary interruption of the dominant tidal flow and the consequent increasing influence of the opposite, subordinate current.

Along the margins, tidal dunes and ripples merge with shoreface and deltaic sediments (e.g. Longhitano 2012, 2013; Steel *et al.* 2012; Longhitano *et al.* 2014). The Coriolis effect can be an important additional process regulating the tidal circulation in wide seaways (e.g. the modern Taiwan Strait), whereas tidal straits, which are considered to be narrower sea branches between islands or promontories, usually generate a tidal amplification as a result of coastal narrowing and the consequent tidal current convergence (Reynaud & Dalrymple 2012).

River deltas sourcing tide-dominated passageways can be strongly influenced during their evolution by the effect of tidal currents flowing landwards and seawards through the delta channel systems and tidal currents flowing parallel to the strait coastline (Uroza 2008; Longhitano 2015; Longhitano & Steel 2015). This type of delta can be strongly skewed in its coastal morphology and asymmetrically deflected towards the dominant tidal current direction (e.g. the Klang Delta in the Malacca Strait; the Elwha Delta in the Juan de Fuca Strait). Importantly, deltas that undergo such tidal influence in their subaqueous reaches can be intensely modified on their delta front (e.g. the Fraser River Delta; Ayranci & Dashtgard 2016). Sediments that are significantly reworked under the effect of the dominant tidal current form elongate sandbodies, pseudo-spits or sand ribbons. In the case of consistent sediment supply and persistent strong tidal circulation, sand can be transported far downcurrent, forming 'detached' shoals and macroforms.

Geological settings of study areas

Two outcrop study areas were considered in this work with the aim of reviewing tide-influenced, deltaic successions developed across the margin of a wide seaway and in a narrower tidal strait. The two field study areas expose strata deposited in the Cretaceous Western Interior Seaway of the USA and the Tortonian Calabrian Arc. The Cretaceous Western Interior Seaway is an extensive marine passageway that connected the Gulf of Mexico to the south with the Arctic Sea to the north, where there was a tidal circulation governed by the overall oceanographic setting and by the narrowing and widening of this seaway as the relative sea-level repeatedly rose and fell. The Tortonian Calabrian Arc is a small orogen that hosted a number of tectonically controlled corridors. These corridors produced significant tidal current amplification and the consequent exchange of sediment dispersal between the interlinked Tyrrhenian and Ionian basins.

Cretaceous Western Interior Seaway of North America

The Cretaceous Western Interior Seaway of North America was a 500–1000 km wide and *c.* 4500 km long tidal seaway that occupied a retro-arc foreland basin during the easterly migration of the Sevier fold–thrust belt towards the Colorado Plateau (Kauffman 1977; Kauffman & Caldwell 1993) (Fig. 2a). The Western Interior Seaway was characterized by relatively shallow water depths along its length, probably rarely exceeding 100 m, and with low-gradient margins (e.g. Kauffman 1977), except for localized areas in the proximity of the forebulge zone or isolated intra-basinal highs (Cole & Young 1991; Plint *et al.* 1993). The best documented aspect of the Western Interior Seaway is the repeated regressive–transgressive transits of deltas (usually on a 100–300 ka timescale) that built transversely out into the seaway from the western margin (Asquith 1970; Plint *et al.* 1986; Plint 1990, 1991; Roehler 1990; Hettinger & Kirschbaum 2002; Plint & Kreitner 2007; Aschoff & Steel 2011*a*, *b*). A notable feature of these deltas, particularly in the Campanian, was that the deltas developed during highstand periods (i.e. those sited furthest west when the seaway was wider) were generally wave-dominated on relatively linear coasts (Martinsen 2001, 2003; Hampson & Howell 2005), whereas the lowstand deltas (developed up to 150 km further east when the seaway was narrower) developed on invaginated coastlines and were more strongly tide-influenced (Mellere & Steel 1995*b*; Hampson *et al.* 2008; Aschoff & Steel 2011*a*, *b*). The large-scale effect of tidal waves entering the seaway from the Gulf of Mexico produced an overall current-dominated oceanography and significant anticlockwise tidal flow patterns that particularly affected the distal, subaqueous segments of the lowstand deltas (Klein & Ryer 1978; Dalrymple 2010) as they built into the seaway from its western margin. The width of the Western Interior Seaway changed dramatically over any relative sea-level cycle. At relative highstands of sea-level, the seaway was 500–1000 km wide, whereas it narrowed by several hundred kilometres during relative lowstands, favouring the progradation of falling-stage deltas (Fig. 2b) (Steel *et al.* 2012). This pattern of a changing shoreline process regime with seaway width was especially notable across Utah, Colorado and Wyoming during the Campanian period, the interval of the present case study (Fig. 2c).

The tendency for basinwards regime change from decreasing wave influence to increasing tide influence as the deltas migrated from highstand inner to lowstand outer reaches of the pre-existing platform of the seaway is, interestingly, the opposite of the typical trend on passive continental margins, where the strongest wave climate is commonly observed along the outermost shelf reaches (Porębski & Steel 2006; Yoshida *et al.* 2007).

A Campanian cross-section from Rock Springs to Denver (see Fig. 2c for location) illustrates this basinwards change in shoreline/delta type from west to east into the Western Interior Seaway and illustrates the series of tide-dominated lowstand deltas (e.g. the Airport, Blair, Meeker, Morapos, Kremmling, Muddy Buttes, Hygiene and Gunsight Pass sandstone units) that developed far basinwards of the updip wave-dominated deltas (Fig. 3).

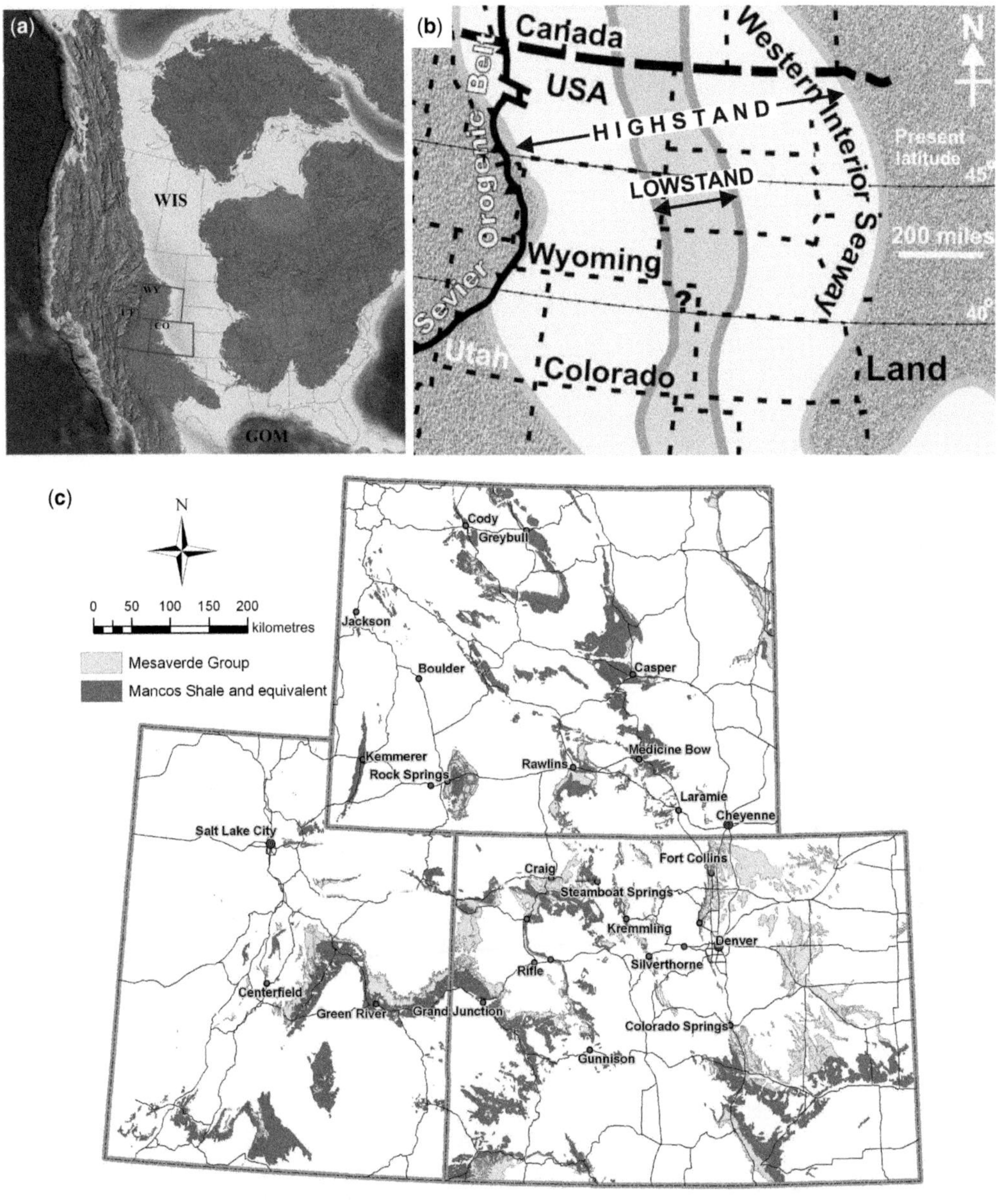

Fig. 2. (**a**) Reconstruction of the Western Interior Seaway (WIS) (from Blakey & Ranney 2008). The study areas of Utah, Colorado and Wyoming are highlighted. This setting represents the widest dimensions of the seaway during the sea-level highstand at 75 Ma. GOM, Gulf of Mexico. (**b**) Palaeogeographical detail of the study area (modified from Yoshida *et al.* (2007), after Kauffman (1977)). The shallow seaway changed its width significantly over each third- and fourth-order relative sea-level cycle. (**c**) Overview of the Mesaverde Group outcrops in Utah, Wyoming and Colorado, providing evidence for the eastwards and southeastwards progradation of the Campanian delta systems into the seaway over >600 km (see panel b) over 10 Ma. There was a later landwards retreat of the same extent (Fig. 3). The most distal segment of any delta system was the lowstand part and the example highlighted here is the O'Brien Spring Sandstone, documented from the Rawlins area in Wyoming.

A particular feature of many of the lowstand, tide-dominated deltaic units of the Campanian is that their palaeocurrents indicate that they prograded not eastwards, like the highstand shorelines, but southwards, i.e. they became skewed southwards to run sub-parallel to the highstand shorelines. We

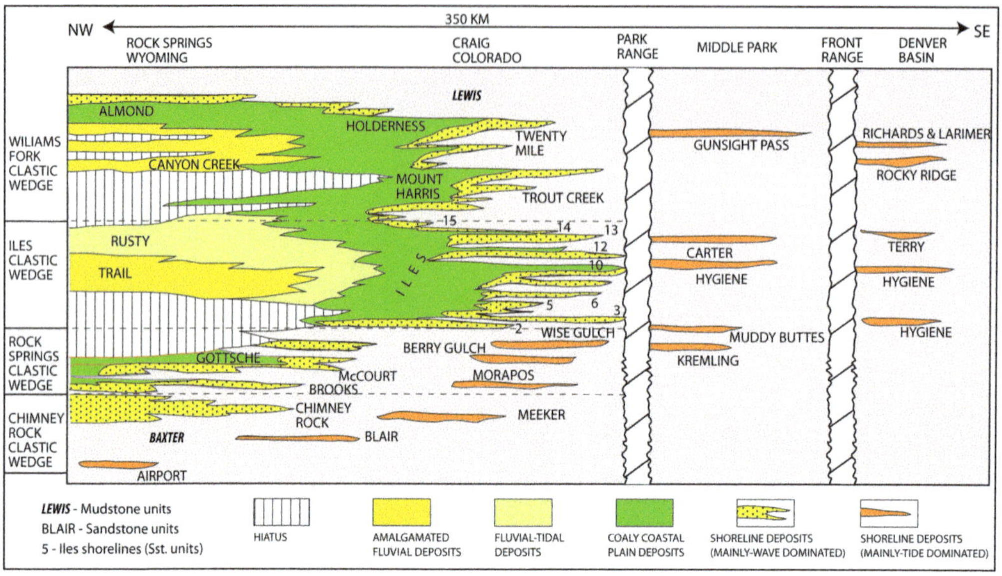

Fig. 3. Overview of the Campanian stratigraphy in a schematic cross-section from SW Wyoming through to the Denver Basin, Colorado. Note the systematic progradation, at all levels, of lowstand tide-dominated deltas that entered the narrowed Western Interior Seaway. These lowstand deltas swing from an eastwards to a southwards pathway as they meet the southwards-flowing tidal current gyre dominating the seaway at these times (from Steel *et al.* 2012). The lowstand O'Brien Spring Sandstone discussed here occurs northwards along-strike from this cross-section, broadly coeval with the Morapos or Berry Gulch sandstones.

have chosen a single example, the O'Brien Springs Sandstone (Fig. 4) in southern Wyoming (broadly coeval with the Brooks and Morapos sandstone units in Fig. 3) to illustrate how this happened, as the delta built out into the tide-dominated lowstand seaway.

O'Brien Springs Sandstone of the Haystack Mountains Formation, Wyoming

The O'Brien Springs Sandstone is a member of the Campanian Haystack Mountains Formation (Martinsen & Tillman 1989; Mellere & Steel 1995*a*, *b*) cropping out in the Rawlins area of southern Wyoming. It was described and interpreted in detail by Uroza (2008). The sandstone unit is about 55 m thick and coarsens upwards (in two parasequences) from thin, current-rippled sandstone sets alternating with extensive mudstone layers to thicker cross-stratified sandstone sets that also alternate with thin mudstones, although the mudstones becomes thinner upwards (Fig. 5a). The dune cross-strata that dominate the sandstone succession tend to change upwards from planar to trough cross-strata, as shown in Figure 5b. The cross-strata imply that the O'Brien Springs delta front was covered by southwards-migrating subaqueous dunes, probably at times of strong reworking by the north to south tidal currents of the seaway (see Fig. 8). The mud layers, on the other hand, are likely to have been generated at times of muddy river effluent flooding down onto the delta front. The key evidence that the sandstone beds were deposited mainly by tidal currents is the presence of abundant slack water mud drapes, the tidal bundling of strata and bidirectional cross-strata (Fig. 6) (Mellere & Steel 1995*a*; Steel *et al.* 2012)

The hypothesis of mud–sand interbedding caused by the interaction of influences from muddy river flooding and from the reworking action of the sandy tidal seaway is supported by: (1) the marked bimodality of grain sizes in the measured succession, with alternating mudstones and upper fine- to medium-grained sandstones (Fig. 7a); (2) the gentle lensing and channelling of the sandstone beds (Fig. 5a), which indicates how the tidal current reworking was taking place; (3) the occasional abundance of *Ophiomorpha* traces in the clean sandstones (Fig. 7b), suggesting open marine conditions; and (4) the marked change in dominant palaeocurrent direction from easterly (in the western outcrops) to southerly (in the eastern outcrops) (Uroza 2008) (Fig. 8).

The schematic reconstruction of how the deltas prograding eastwards into the narrowed lowstand seaway were markedly deflected southwards by the Western Interior Seaway tidal current circulation is given in Figure 8. Because the lowstand tidal

Fig. 4. The O'Brien Springs Sandstone (only the upper 25 m is exposed) cropping out near the North Platte River, NE of Rawlins.

sandstone units of the Haystack Mountains Formation are, in places, associated with storm wave-generated sandbodies (Mellere & Steel 1995*a*, *b*; Uroza 2008), it is possible that the deflected deltas were protected on their eastern side by wave-formed spits, as shown in Figure 8.

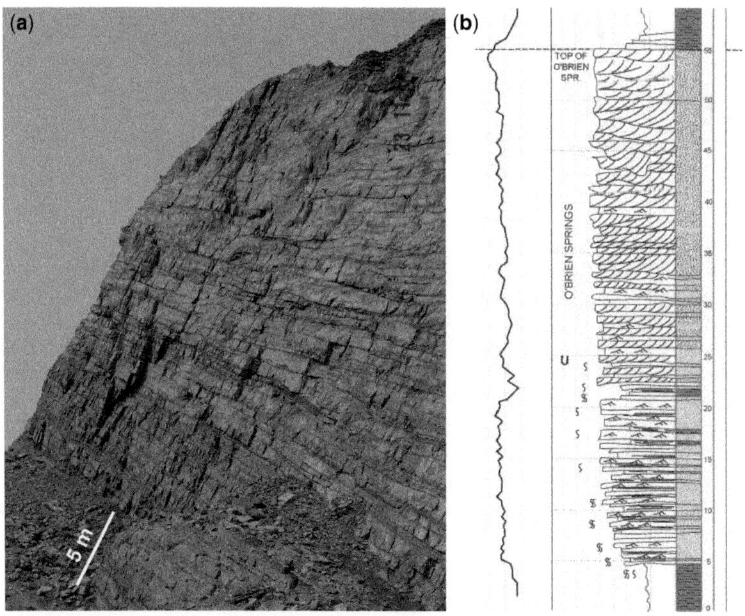

Fig. 5. (**a**) Outcrop of upper 35 m of O'Brien Springs Sandstone on Sinclair Road near North Platte River. The thin-bedded lower part is current rippled, whereas the thicker bedded upper part is mainly cross-stratified. Note that the sandstone beds are lensing in places and the very thin mud layers are pervasive through much of the succession. (**b**) Measured section of O'Brien Springs Sandstone, including a hand-held gamma ray profile that accentuates the presence of two parasequences.

Fig. 6. Clear bidirectional (herringbone) cross-strata in O'Brien Springs sandstones at the westernmost location on Haystack Mountains. A total of 117 readings on cross-stratal azimuths showed clear eastwards and westwards palaeocurrents (Fig. 8).

Miocene to Quaternary of the Calabrian Arc, central Mediterranean, southern Italy

The Calabrian Arc is a restricted orogen, presently located in southern Italy, which originated after crustal subduction beneath the Ionian lithosphere during the middle Miocene and the opening of the Tyrrhenian back-arc basin (Malinverno & Ryan 1986; Dewey *et al.* 1989; Gueguen *et al.* 1997; Bonardi *et al.* 2001). The Calabrian Arc consists mainly of Hercynian metamorphic and intrusive rocks, tectonically superposed on ophiolite-bearing units of Tethyan affinity, in turn overlying Mesozoic carbonate platform limestones and Cenozoic clastic cover (Fig. 9) (Ogniben 1969; Amodio Morelli *et al.* 1976; Tortorici 1982).

During its tectonic translation towards the Ionian Sea due to the westwards subduction of the oceanic Ionian crust caused by convergence of Africa and Europe, the Calabrian Arc was affected by the onset of intense thrusting, as a result of the activation of regional SE- and ESE-trending strike-slip fault zones and differently oriented normal fault sets (Ghisetti 1979; Ghisetti & Vezzani 1981; Rehault *et al.* 1987; Turco *et al.* 1990; Knott & Turco 1991; Monaco & Tansi 1992; Catalano *et al.* 1993). Strike-slip and extensional tectonics, possibly related to block rotation, induced a structural fragmentation of the Calabrian Arc from the Middle Miocene onwards (Turco *et al.* 1990; Martini *et al.* 2001). This tectonic phase favoured the formation of structural highs, which separated graben and half-graben filled by alluvial, shallow water and deeper marine sediments (Monaco & Tansi 1992; Catalano *et al.* 1993). Many of these basins originated across some of the major strike-slip zones of the Calabrian Arc, forming narrow, linear, tectonically controlled marine straits (Fig. 9). They connected the Tyrrhenian and Ionian seas, generating important marine water exchanges and consequent current convergence, with an oceanographic setting very similar to that occurring along the present day Messina Strait separating Sicily from the Italian peninsula (Colella & D'Alessandro 1988; Colella 1995; Longhitano & Nemec 2005).

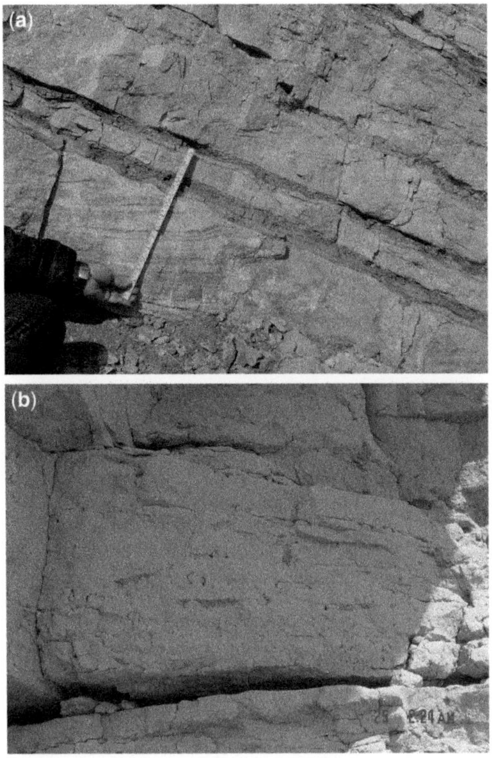

Fig. 7. (**a**) Marked bimodality of grain size between mudstone beds and up to medium-grained cross-stratified sandstone sets. (**b**) Some of the cross-stratified sandstone beds are thoroughly reworked by *Ophiomorpha* burrows.

These extensional basins are the Miocene Amantea Basin (Colella 1995; Longhitano & Nemec 2005) and a series of Plio-Pleistocene graben, including the Catanzaro Basin (Longhitano *et al.* 2014), the Crotone Basin (Zecchin *et al.* 2004; Zecchin 2005), the Siderno Basin (Colella & D'Alessandro 1988) and the proto-Messina Strait (Mercier *et al.* 1987; Longhitano *et al.* 2012*b*) (Fig. 9). Large sediment volumes, mostly derived from river deltas and tidal current abrasion of the sea bottom, filled these straits under the influence of amplified tidal currents (Longhitano 2012, 2013). Today, these successions commonly exhibit sandstones with recurrent large-scale (>3 m thick) cross-stratified sets (Longhitano *et al.* 2012*a*, *b*) as well as a suite of tidal sedimentary structures, such as herringbone cross-strata, tidal bundles, reactivation surfaces and heterolithic lamina sets (Longhitano 2011). Along the tectonically controlled strait margins, a number of transversely impinging deltaic successions can be detected in many of these Calabrian tidal straits, although they were not interpreted as such in the past and were more generically considered as the result of deposition along the strait margin zones (Longhitano 2012, 2013).

Belmonte Calabro section in the Amantea Basin

Many intervals of these strait-fill successions are today well exposed in Calabria, including transgressive deltaic complexes. Transgressions occurred due to the high subsidence rates imparted by marginal faults that were active during sedimentation, producing coastal back-stepping, with strata lapping against block-faulted margins (Longhitano & Nemec 2005; Longhitano *et al.* 2012*b*; Longhitano 2015). For this reason, the lowermost stratigraphic intervals of many of these deltas are tectonically deformed, passing upwards to less inclined stratal sequences through a series of progressive internal unconformities.

A well-exposed example is the Belmonte Calabro section in the Miocene Amantea Basin (Fig. 10a). This is a triangular-shaped, peri-Tyrrhenian Miocene tectonic depression, presently exposed on the western coast of Calabria (Fig. 10a). The Amantea Basin formed in the Serravalian, after the extensional faulting of this part of the Calabrian Arc and consequent marine inundation (Argentieri *et al.* 1998). During the middle Tortonian to early Messinian, this basin was part of a wider tidal strait, the oceanography of which was possibly dominated by strong tidal currents flowing towards the SW (dominant) and NE (subordinate) (Colella 1995; Longhitano & Nemec 2005). The sedimentary record of this setting is preserved in a basin-scale transgressive succession (Fig. 10b) up to 260 m thick, consisting of basal, polymict conglomerates and sandstones of deltaic and shallow marine origin. A basin-wide transgressive lag separates these deposits from the overlying tidal cross-stratified mixed arenites (Longhitano *et al.* 2012*b*), which are of middle to late Tortonian age and up to 120 m thick. They are overlain by latest Tortonian to early Messinian mudstones, which are nearly 100 m thick and interbedded with calc-arenites and relics of evaporite deposits (Fig. 10b) (Di Nocera *et al.* 1974; Ortolani *et al.* 1979; Tansi 1991; Colella 1995; Argentieri *et al.* 1998; Colella & Longhitano 1998; Longhitano & Colella 1998; Mattei *et al.* 1999; Speranza *et al.* 2000; Martini *et al.* 2001; Mattei *et al.* 2002; Muto & Perri 2002).

Although the average palaeocurrent direction detectable from the foresets of the tidal cross-strata indicates a dominant direction towards the SW (e.g. Colella 1995), the Amantea Basin may have had a different orientation during the Tortonian. In fact, the tidal succession preserved today along this coastal sector of Calabria might represent only a

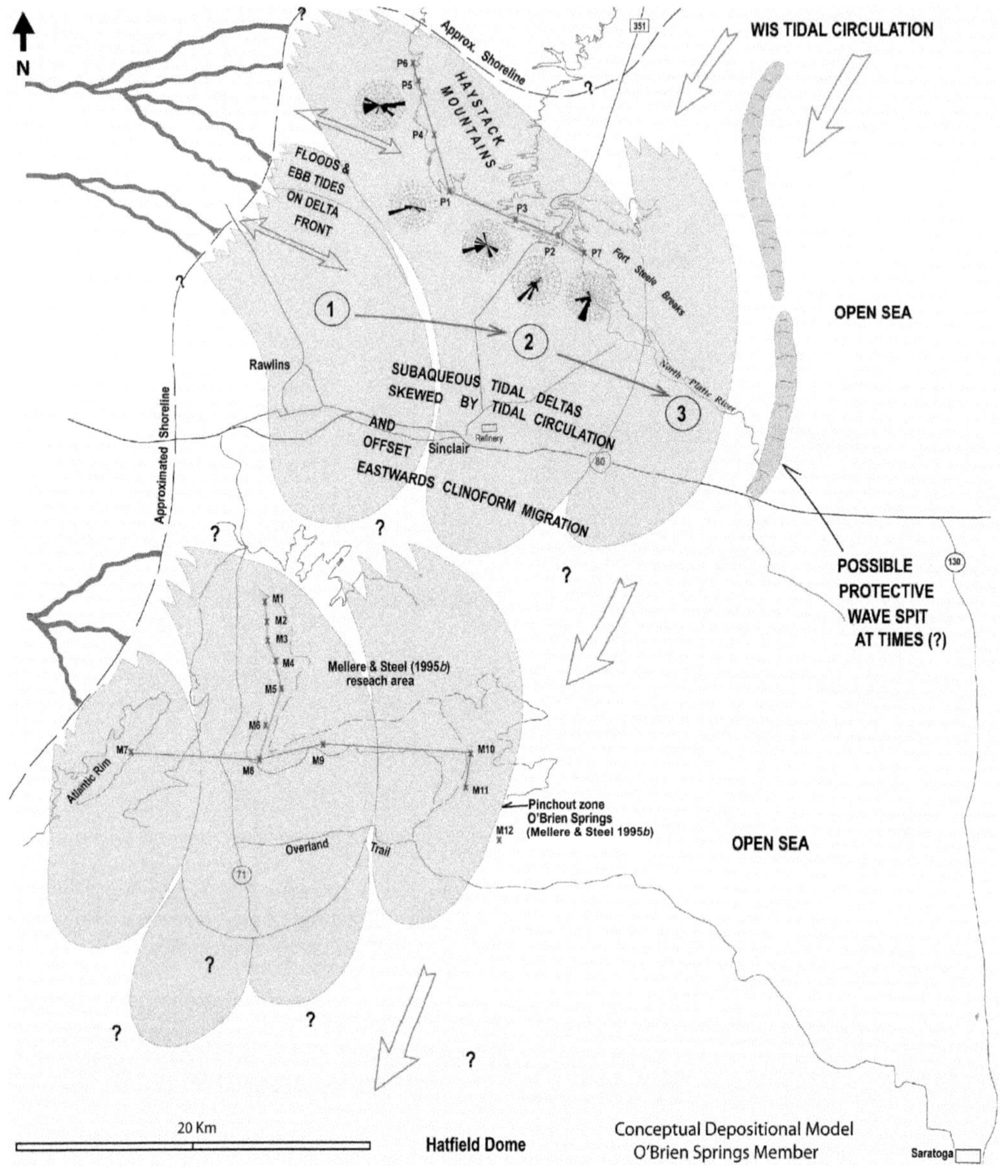

Fig. 8. Palaeogeographical reconstruction of the O'Brien Springs Member, Haystack Mountains Formation, SW Wyoming during the Cretaceous. The palaeocurrents measured at five sites along the front of the Haystack Mountains clearly show a change from bimodal west–east directions in the west to southerly directed currents further east, indicating that the deltas impinging into the tidal seaway were preferentially deflected to the south (modified from Uroza 2008).

marginal portion of a wider tidal strait system, the main strait centre deposits of which would presently be tectonically downthrown to the west after Plio-Pleistocene faulting of the Tyrrhenian Sea margin (Catalano *et al.* 1993). As a consequence, the original orientation and shape of the Amantea tidal strait still remains uncertain.

The Belmonte Calabro section is a 500 m long and 200 m thick, north–east-oriented cliff adjacent to the northwestern margin of the Amantea Basin (Fig. 11a). It exposes a stratified succession exhibiting evident synsedimentary tectonic deformation attributed to the active faulting of the margin during the middle Tortonian (Fig. 11b) (Turco *et al.* 1990;

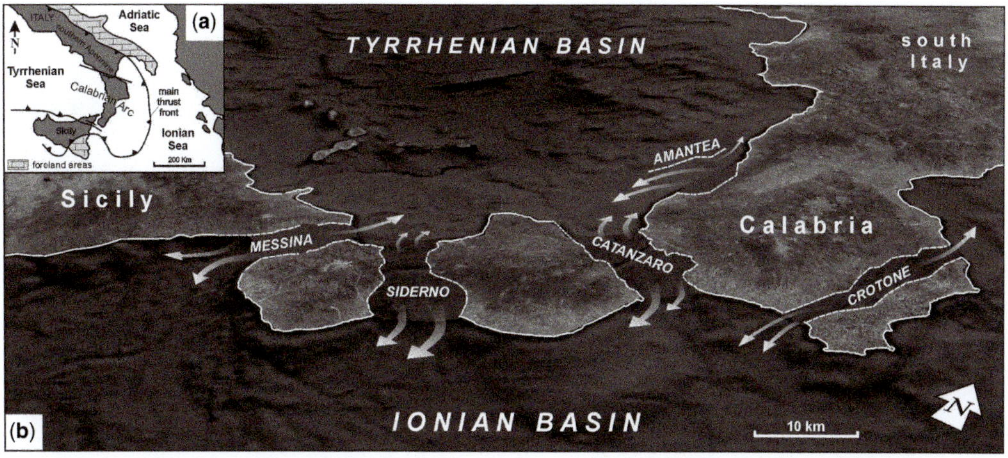

Fig. 9. Reconstruction of the palaeogeographical setting of the Calabrian Arc in southern Italy during the Neogene–Quaternary. This small orogen experienced a strike-slip tectonic fragmentation, which favoured the opening of a number of block-faulted narrow basins. These basins acted as tidal straits linking the Tyrrhenian and the Ionian seas and were filled by thick, cross-stratified deposits (modified from Longhitano 2013).

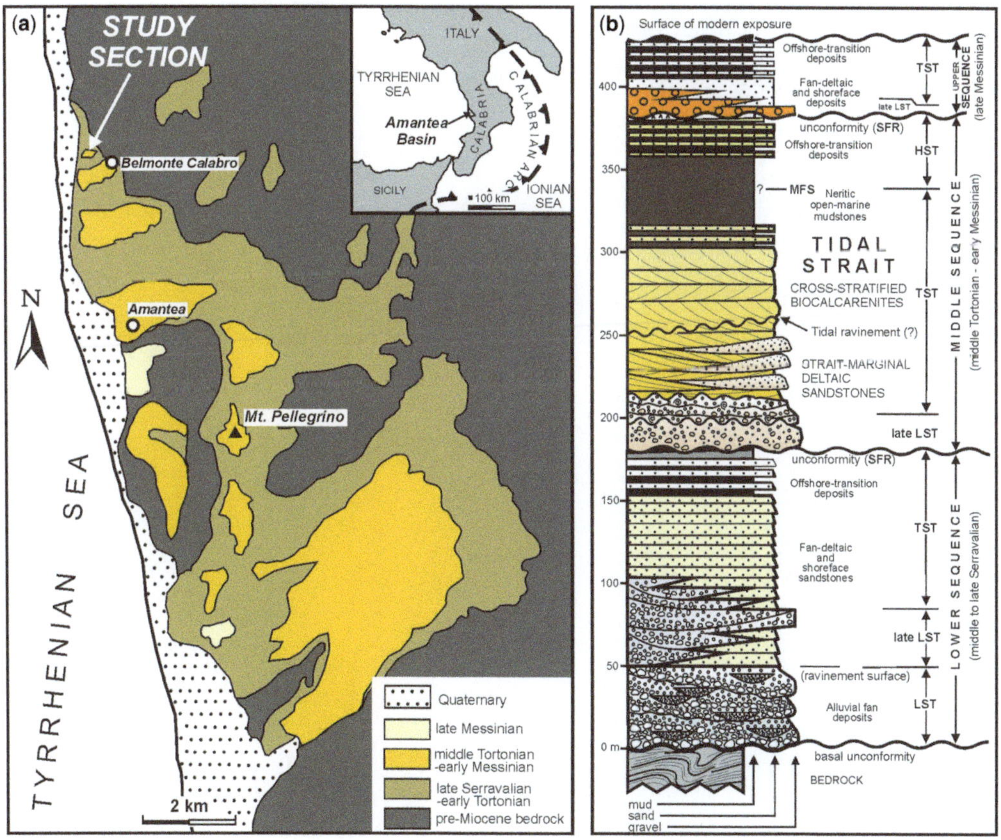

Fig. 10. (**a**) Simplified geological map of the Amantea Basin. (**b**) Stratigraphy of the Amantea basin-fill succession. The studied interval corresponds to part of the intermediate upper Tortonian–lower Messinian depositional sequence (modified from Longhitano & Nemec 2005).

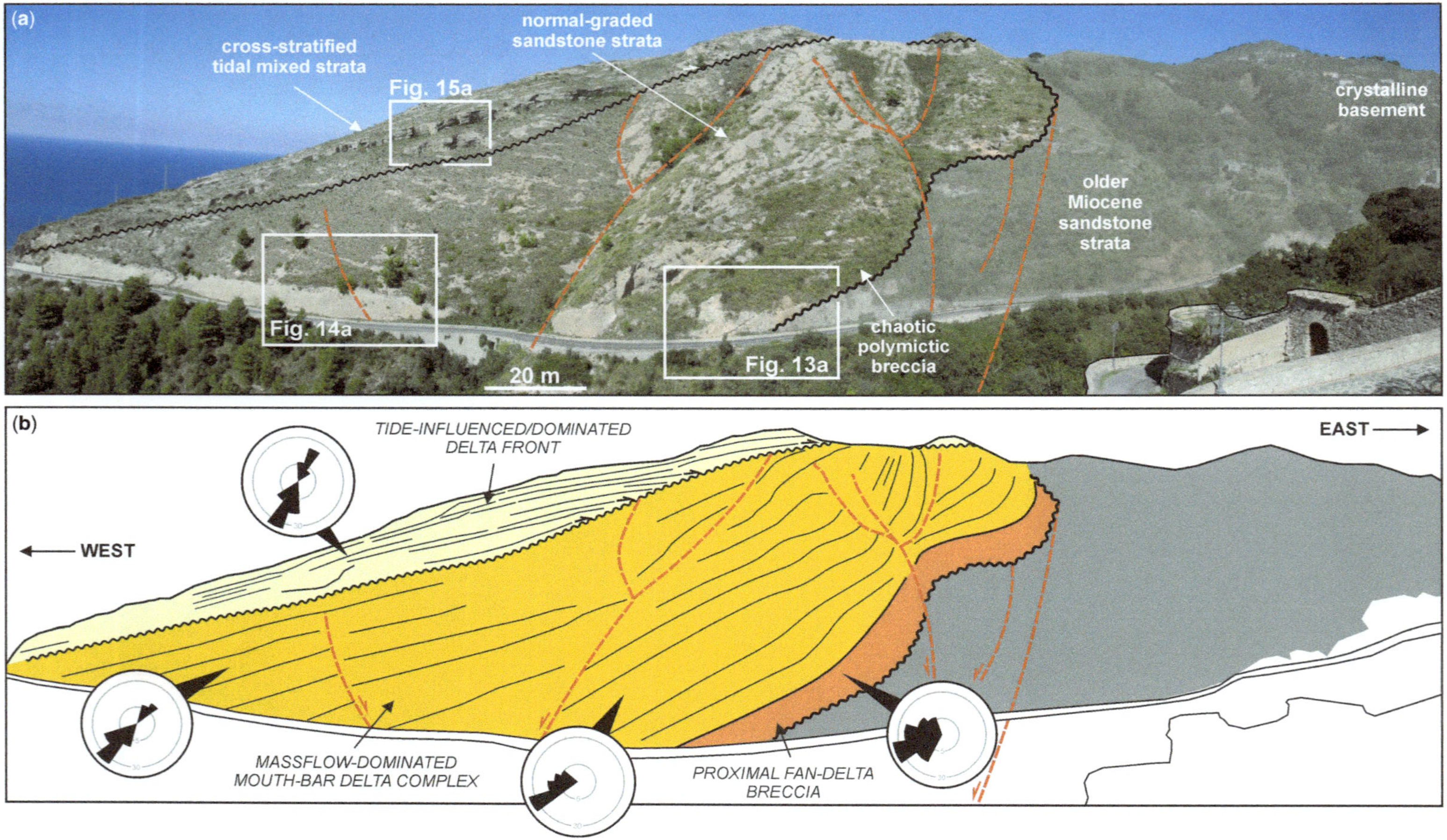

Fig. 11. (**a**) Photomosaic and (**b**) interpreted line drawing of the Belmonte Calabro section, north of the Amantea Basin. The succession, interpreted as a deltaic transgressive unit, shows the basal coarse-grained sediments strongly deformed as a result of synsedimentary normal faults and the occurrence of the overlying tide-dominated cross-stratified sandbodies that lie erosionally on the older deposits. The overall progradation of the deltaic body is to the west. Rectangles indicate details shown in the following figures.

Martini *et al.* 2001). The succession included in this section is bounded by a basin-scale basal angular unconformity, incised into an older marine sequence (Colella 1995). In general, from the bottom to the top, sediments consist of basal fossiliferous breccias, rapidly passing upwards to normally graded, indistinctly laminated sandstones and a series of vertically stacked sets of tidal cross-stratified mixed arenites with different stratal thicknesses and internal foreset architectures (Longhitano & Nemec 2005; Longhitano *et al.* 2012*b*). These deposits can be grouped into three main facies associations, which are vertically and laterally stacked (Fig. 12).

Coarse-grained breccia and conglomerate (proximal fan delta). The lowermost stratigraphic interval exposed at the Belmonte Calabro section is *c.* 25 m thick and lies on the basal unconformity, which separates this middle Tortonian sequence from the underlying Serravallian sequence (Fig. 13a) (Di Nocera *et al.* 1974; Ortolani *et al.* 1979; Colella 1995; Colella & Longhitano 1998; Longhitano & Colella 1998; Muto & Perri 2002; Longhitano & Nemec 2005). This deposit is organized into a series of vertically stacked, 4–5 m thick, normal-graded strata, the inclination of which is a combination of an original depositional dip enhanced by a syn- to post-depositional tectonic tilt. Sediments consist of monomict and polymict breccias, including dominant granitic and metamorphic angular clasts, associated with subordinate well-rounded sandstone and limestone clasts, in many places interspersed in a sandy matrix containing fragments of bivalves, bryozoans, echinoids and serpulids (Fig. 13b). Palaeocurrents detected from imbricated clasts indicate an average eastwards sediment transport direction. Upwards, this coarse-grained facies passes abruptly to a 40–50 cm thick interval of mixed siliciclastic and/or bioclastic coarse-grained sandstone containing low-angle cross-lamination with N 260° E palaeocurrents, capped by remnants of a highly bioturbated mudstone interval (Fig. 13c).

This coarse-grained basal interval shows the classical sedimentary assemblage of a base of slope deposit, produced by sediment accumulation

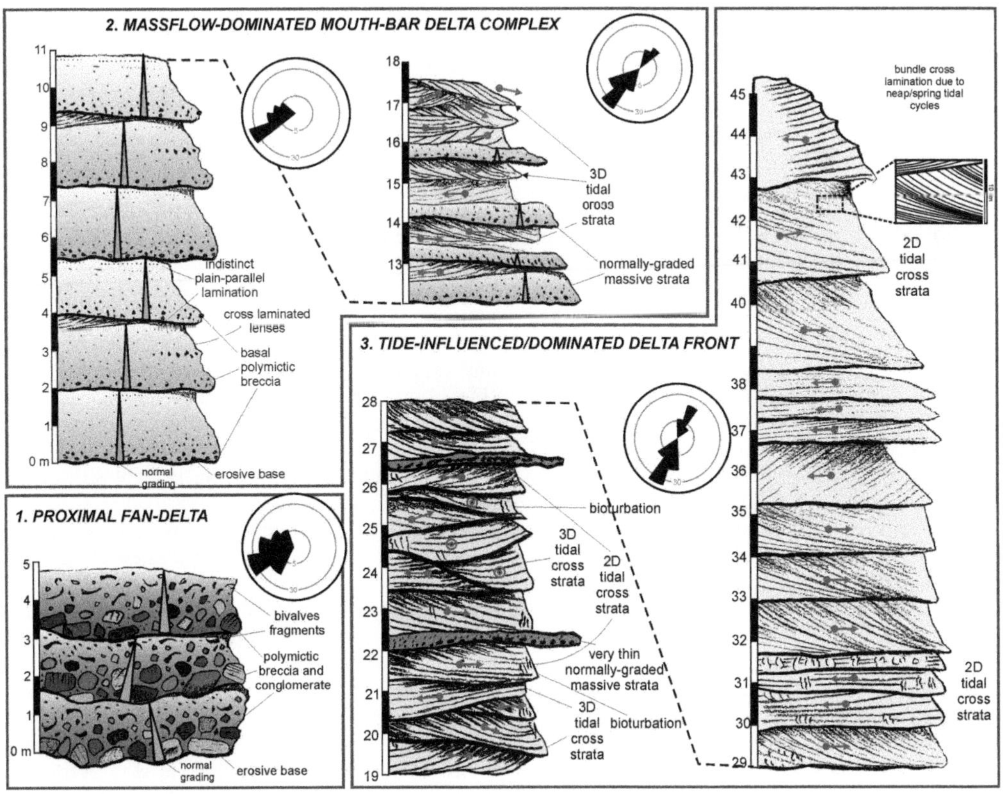

Fig. 12. Schematic framework of the main facies associations recognized in the studied succession. Numbers 1–3 indicate the progressive occurrence of these deposits, from older (1) to younger (2). Note the progressive-upwards rotation of palaeocurrent patterns.

Fig. 13. (**a**) Outcrop showing the deltaic basal interval of the studied succession lying unconformably on older open marine sediments. (**b**) Details of the fossiliferous breccia. (**c**) Fine-grained, laminated intervals and muddy horizons (arrow) are often preserved on top of these strata.

at the toe of the subaqueous slope, against a tectonically active, faulted basin margin (see Fig. 16a). Angular monomict clasts showing encrusting submarine organisms suggest that these deposits had a brief period under the water table before being incorporated into seacliff collapses. However, there are also well-rounded polymict clasts, which possibly indicate derivation from a marginal alluvial fan. Sediments included in this interval thus indicate mass transport by rockfall and debris flow processes, recording the subaqueous toe of a cliff-attached fan delta system dominated, at that time, by repeated gravitational mass flow events (see Fig. 16a).

Normally graded sandstones (mouth bar). The overlying, conformable stratigraphic interval is *c.* 100 m in thickness. Sediments are mainly siliciclastic and form a monotonous succession of normally graded, 6–7 m thick units, the erosional bases of which appear planar to slightly undulating in an along-dip cross-sectional view (Fig. 14a). These units contain three main intervals. The first of these is a basal microbreccia and conglomeratic interval up to 1 m thick (Fig. 14b), passing upwards into a *c.* 4 m thick structureless coarse-grained sandstone with scattered granules and pebbles and scarce, undifferentiated fossil fragments. In the uppermost part of these intervals, indistinct cross-lamination can be seen, associated with bioturbation and rare fluid escape soft sediment deformation (Fig. 14c). The third (topmost) interval of the units can be occupied by 1–2 m thick sets of tidal cross-strata consisting of coarse- to medium-grained arenites and exhibiting evident siliciclastic/bioclastic tidal bundles and reactivation surfaces (Fig. 14d). Palaeocurrent directions show N 260° E to N 240° E trends. Upwards, the thickness of the second massive interval decreases, being progressively replaced by the tidal cross-stratified deposits with a counter-clockwise progressive change of the palaeocurrent directions.

This unit is interpreted as a deltaic mouth bar complex (see Fig. 16b). A basal erosion surface suggests an initial river jet, followed by a rapidly accumulated, siliciclastic sandy interval with normally graded beds and fluid escape structures. During the waning of energy in this initial growth stage of the mouth bars, the sand accumulated near the river mouth was tidally reworked. At the end of

Fig. 14. (**a**) Vertically stacked, normally graded sandstone strata, interpreted as a deltaic mouth bar complex. These strata are intensely dissected by a system of normal faults related to the development of this margin of the Amantea Basin after deposition. (**b**) Detail of the basal micro-conglomerate lying above an erosional surface. (**c**) Soft sediment deformation structures also occur in this facies. (**d**) Along-strike view of these deposits showing alternating normally graded and cross-stratified intervals with different palaeocurrent directions.

deposition, the top of the mouth bars became occupied by small- to medium-sized tidal dunes, as this part of the basin had become tidally influenced (see Fig. 16b). The mouth bars thus developed after a first important step of tectonically induced marine transgression, which changed the main sedimentary processes into river-dominated, but also tidally influenced (see Fig. 16b).

Large cross-stratified mixed arenites (delta front). At the very top of the Belmonte Calabro section, a *c.* 40 m thick tabular succession erosionally overlies the rest of the previously described deposits. The erosional surface can be traced laterally throughout the entire Amantea Basin for several kilometres and is interpreted as a ravinement surface as it recorded the regional-scale onset of a dominant tidal circulation during the middle Tortonian (Longhitano & Nemec 2005) (Figs 10b & 11a). Mixed siliciclastic/bioclastic arenites overlie the erosional surface and are composed of >65% polycrystalline quartz fragments, mica and glauconitic crystals, with fragments of bivalves, corals, red algae and echinoids comprising the remaining percentage. The bioclastic content increases upwards to as much as 70%. This succession consists of five to six vertically stacked, retrogradational wedge-shaped units, separated by discontinuity surfaces marked by shell and pebble lags. The wedges, which include thinning-upwards strata sets (Fig. 15a), pinch-out towards the west–SW (basinwards) and lap against the underlying strata (landwards) (Fig. 11b). They mostly include two main facies. In the first facies, faintly laminated sandstone strata form 2–3 m thick and 30–40 m wide lenticular beds, including siliciclastic-dominated deposits, which commonly show normal grading, soft sediment deformation and cross-stratification, but lack other small-scale tidal signatures. In places, these lenses are encased within, or alternate with, the second facies of tidally cross-stratified, 3–4 m thick tabular bedsets, in which the content of the bioclastic fraction increases dramatically compared with the other lithofacies. Cross-sets exhibit two- and three-dimensional geometries (Fig. 15b), dominantly unidirectional

Fig. 15. (**a**) Topmost stratigraphic interval observable in the Belmonte Calabro section showing vertically stacked, thinning-upwards cross-stratified units, interpreted as delta front wedges prograding towards the west (left in the photo). (**b**) Detail of tidal trough cross-strata observable in an along-strike view of the section. (**c**) Tabular tidal cross-strata, forming simple and compound foresets.

foresets dipping towards N 220° E, and numerous tidal structures such as herringbone cross-strata, neap–spring tidal bundles, heterolithic lamina sets and reactivation surfaces (Fig. 15c).

This third stratigraphic interval caps a transgressive sedimentary succession, which records the development of a steep margin fan delta that prograded onto the eastern margin of the Amantea Strait. Deposits belonging to this uppermost unit developed above the deltaic mouth bar complex after a significant change in the coastal setting of this area, possibly related to the evolution of the block-faulted basin margin. Tidal cross-stratified wedges accumulated on this surface record a more mature stage of a fan delta system and a backstepping against the basin margin. After this further dramatic flooding of the coastal area, the delta had become tidally dominated (Fig. 16c). Tidal wedges are interpreted as part of the delta front that prograded into a strait margin zone where tidal currents flowing axially to the strait were the dominant hydrodynamic force (Fig. 16c). They are stacked in a back-stepped fashion, possibly recording steps of normal regression that punctuated the overall relative sea-level rise. Internally, the wedges indicate tidal dominance, but also an intermittent river influence. The indistinctly laminated sandstone intercalations indicate brief river flood influxes and the deposition of siliciclastic-rich plumes entering the delta front too rapidly to be completely reworked by tidal currents. The cross-stratal sets, the volumetrically dominant element in these sandbodies, represent two- and three-dimensional tidal dunes in the delta front environment (Fig. 16c). Palaeocurrents suggest that the main progradation of this fan delta was rapidly 'adjusted' to the dominant tidal current trend. Thus the thinning-upwards observed in the delta front wedges might reflect a progressive deflection of this part of the delta and a wider sand distribution into elongated sandbodies towards the strait axis (Fig. 16c).

Discussion

The field examples documented in the previous sections indicate that deltas prograding into tidal seaways or straits diverge slightly from classical deltaic models because they leave different diagnostic signatures in the rock record. The recognition of such deltas in ancient stratigraphic successions may have significant potential in palaeogeographical reconstructions of past scenarios and can be of crucial importance in clastic reservoir characterizations of subsurface sandstone-rich deposits accumulated in tectonically confined settings or elongate marine basins.

Deltas sourcing tidal straits and elongate seaways: sedimentary signature and diagnostic features

River deltas impinging tidal elongate seaways or straits are marginal depositional systems, the dimensions, features and processes of which are strictly dependent on the coastal gradient of the seaway borders (Fig. 17). Tectonically active, steep margin straits are commonly characterized by isolated, rare fan–deltaic deposits, often associated with debris aprons, as a result of the high gradient of their coastal areas, which prevents significant sediment accumulation on the shelf and favours the direct transfer of the river discharge towards the deeper zones of the straits (e.g. the modern Messina Strait). Conversely, river deltas can extend much further across gently sloping seaway margins because tectonically stable borders favour clinoform progradation and long-lived deltaic sequences (e.g. the modern Klang Delta in the Malacca Strait) (Fig. 17).

The gradient of the margins may interplay with the tidal oceanography in shaping deltas impinging seaways and straits. It is well known that the morpho-tectonic setting of the basin borders governs the type of the delta nucleated from that margin, especially with respect to its internal architecture, facies distribution and sediment calibre (e.g. Orton & Reading 1993). A possible reconstruction of how deflected deltas differ between steep and gently sloping seaway/strait margins is summarized in Figure 18.

One recurrent aspect of structurally controlled steep margins on a seaway is that deltas can be coarse-grained and of small dimensions (with 1–5 km wide and 2–3 km long delta plains) (Fig. 18a). The geological record indicates that such deltas are short-lived and/or deposit multiple vertically stacked sequences because of the continuous change in the morphological configuration of the coastline (e.g. Colella 1988). These deltas often originate in single units separated by internal unconformities recording tectonic pulses (e.g. Longhitano 2008). Consequently, tidal currents oriented along-shore may be able to rework only the distal environments of these deltas, where fine-grained deposits are delivered (Fig. 18a). As the result, tidal seaways or straits that have tectonically active margins may potentially generate deltas that are sensitive to the local tidal circulation only in their sandstone-dominated delta front areas or during episodes of transgression, as the delta documented in the Amantea Basin demonstrates.

Tidal seaways characterized by more tectonically stable margins often exhibit gently sloping shelves (Fig. 18b). In such settings, deltas form more extensive systems (with 10–50 km wide and 50–100 km long delta plains), but, importantly, they

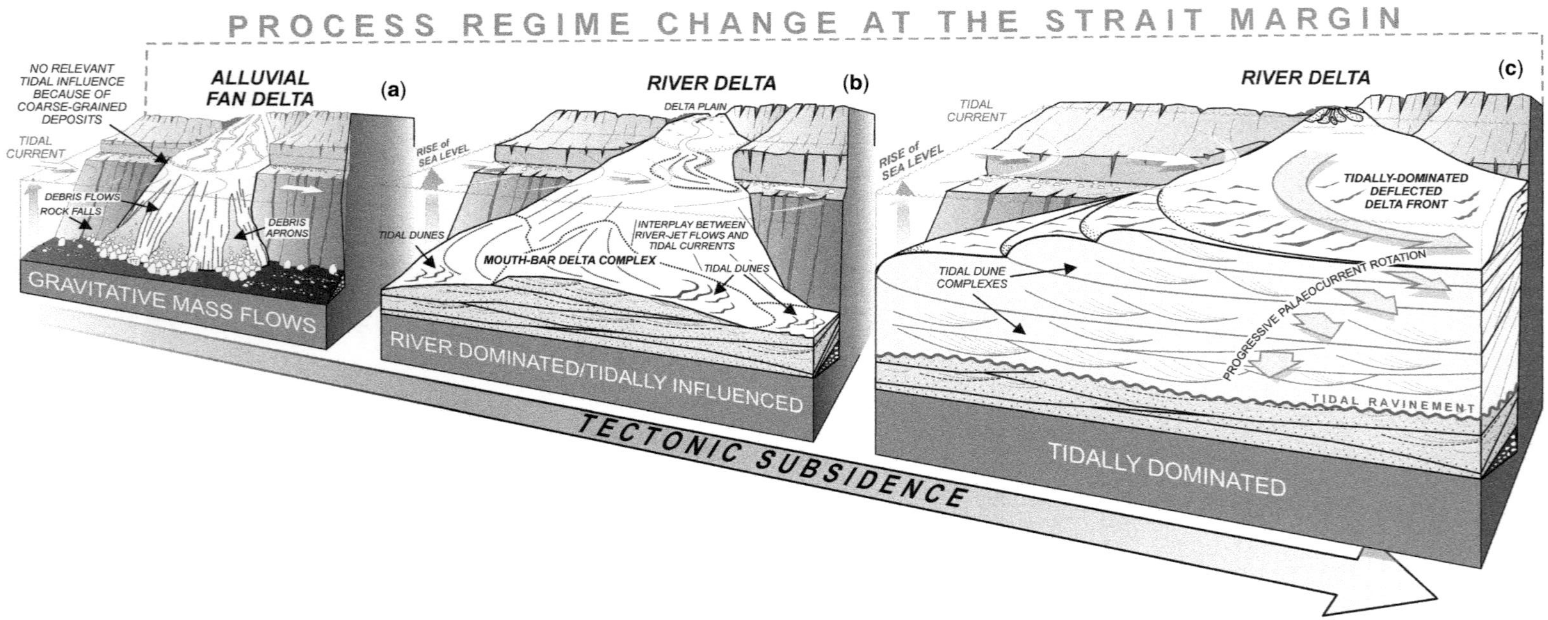

Fig. 16. Block diagram showing the inferred evolution of the Belmonte Calabro fan delta. (**a**) An early stage of delta progradation occurred through the emplacement of gravity-dominated mass flows, accumulating coarse-grained deposits. (**b**) A tectonically induced transgression generated back-stepping of the delta and a more mature deltaic stage, through the development of a tide-influenced mouth bar delta complex. (**c**) The late stage of progradation produced tide-dominated sandbanks in the delta front, with deflected morphologies and tidal dunes.

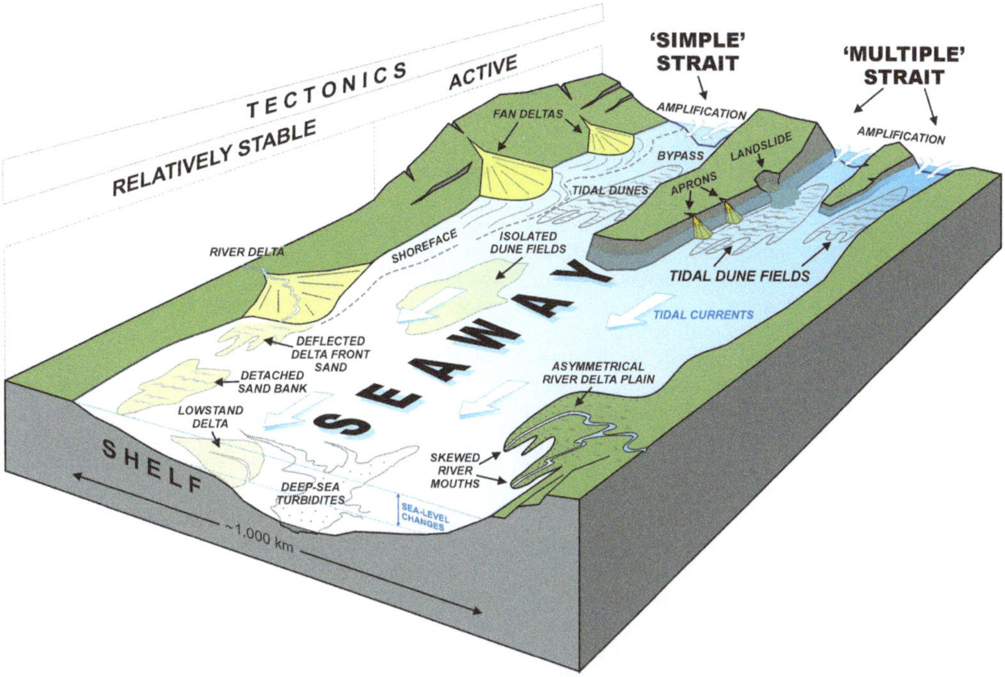

Fig. 17. Schematic representation of a tidal seaway including single and multiple narrower straits. Tidal currents are the major hydrodynamic controlling factor on the processes of sediment distribution. Deltas prograding into the seaway can be related to steep, tectonically active margins or gently sloping shelves. In both cases, tidal currents are able to rework and deflect delta front deposits, forming elongate subaqueous sandbodies and detached sandbanks, including large tidal dunes. In some cases, the delta plain area is also skewed in the sense of the dominant tidal flow.

form long-lived systems that are recorded in extensive, laterally continuous clinoform sets, including abundant volumes of sandstone-dominated delta front deposits (Uroza 2008). Under these conditions, tidal currents may interact to different degrees with the prograding deltas, interplaying with waves and generating facies belts with different attributes (Steel *et al.* 2012) (Fig. 18b). The outcrop example documented in the Western Interior Seaway indicates that these deltas are very sensitive to tidal

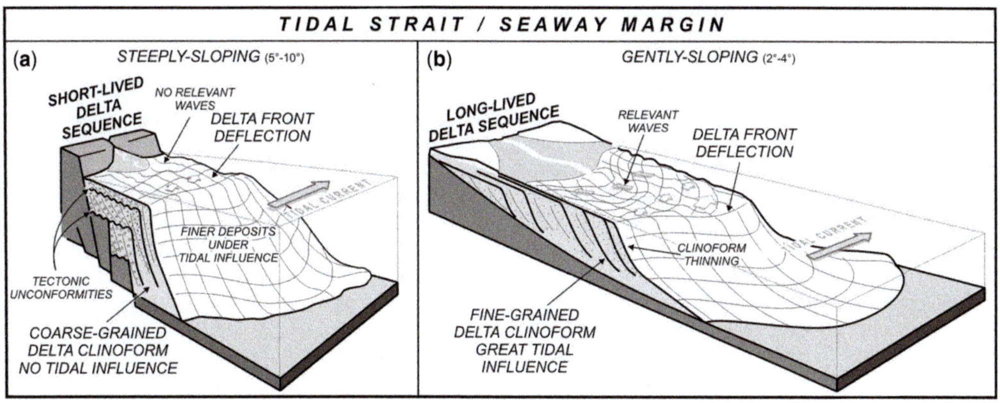

Fig. 18. Morpho-sedimentary models reconstructing deltaic deposition over (**a**) a steeply and (**b**) a gently sloping margin of a seaway/strait governed by tidal currents.

deflection and are also able to preserve the signature of the increasing tidal flow energy as the seaway is progressively narrowed by their progradation.

Deflection in the progradational axis and asymmetry of the delta front sand distribution

Although reconstruction of the plan view shape of deltas at the shoreline is often a rather speculative aspect of ancient stratigraphic analysis, the primary coastal morphology of tidal strait deltas is one of their most distinctive elements.

Modern examples of deltas prograding into seaways or straits in which tidal currents are the dominant hydrodynamic process show that the tides act differently from those in 'classical' tidal-dominated deltas – for example, those in macro- to mega-tidal oceanographic settings (Table 1) (e.g. Longhitano *et al.* 2012*a*). In these latter well-documented systems, the repeated influence of flood- and ebb-directed tidal currents acting at the coastline usually reshapes the delta plain and delta front ('delta platform') environments, forming sandbodies striking perpendicular to the shoreline (Table 1) (e.g. Swenson *et al.* 2005; Willis 2005; Wright & Friedrichs 2006; Fagherazzi & Overeem 2007; Slingerland *et al.* 2008; Goodbred & Saito 2012; Plink-Björklund 2012). In very general terms, tide-dominated deltas are usually subject to a landwards-directed incoming ocean tide, which first interacts with the clinoformal delta front. Modern examples, such as the Changjiang River Delta or the Ganges–Brahmaputra River Delta, show shallow water settings, from 20–90 m at the bottomsets to 5–30 m at the topset/foreset transition (Goodbred *et al.* 2003). Tidal currents thus accelerate across this zone, producing an important morphodynamic influence on the compound clinoform typical of most tide-dominated delta systems (Goodbred & Saito 2012). After crossing the delta front, the tidal prism becomes compressed through the channelized distributary mouths and can be subject to upstream amplification, generating a second phase of increase in the strength of the tidal current and consequently an enhanced bed shear stress. This phenomenon, known as a 'hyper-synchronous' condition (Dalrymple & Choi 2007), produces a general upstream increase in the tidal range and current velocities, although also a subsequent decline as the tidal energy becomes gradually attenuated by bottom frictional forces (Dalrymple *et al.* 1992).

Recent decades have provided a growing body of documentation on tide-dominated and tide-influenced deltas in the ancient record (e.g. Mutti *et al.* 1985; Maguregui & Tyler 1991; Mellere & Steel 1995*a*, *b*; Willis *et al.* 1999; Bhattacharya & Willis 2001; Martinius *et al.* 2001; Willis & Gabel 2001, 2003; Willis 2005; Pontén & Plink-Björklund 2007, 2009; Tänavsuu-Milkeviciene & Plink-Björklund 2009; Plink-Björklund 2012). In contrast with these systems, river deltas prograding into seaways or straits are usually subject to a strong tidal influence as a result of the narrowing of marine passageways and a significant tidal current convergence and amplification (Longhitano 2013). Importantly, these deltas may prograde into microtidal oceanographic settings. The dominance of the tidal hydrodynamics on the deltaic systems thus occurs with very different modalities compared with the systems described previously (Table 1). Tidal currents usually flow with a general direction parallel to the coastline, very similar to along-shore current circulation (Reading & Collinson 1996). Although characterized by high velocities and strengths, tidal currents passing through straits or seaways can have less influence on the coastal and shoreline morphology of these basins, favouring the development of symmetrical delta plain geometries and near-linear shorelines. Rather, they seem to exert a more dominant effect across the subaqueous areas of the seaway margins (e.g. Chiocci *et al.* 2008; Walsh & Nittrouer 2009; Ayranci & Dashtgard 2016).

Modern analogues of deltas entering tidal seaways and straits

The Elwha River Delta, along the southern margin of the Juan de Fuca Strait, separating the state of Washington from the Vancouver Island (USA), represents one of the most noticeable modern examples of a delta impinging a tidal strait (Foreman *et al.* 2004; Gelfenbaum *et al.* 2015). The delta coastline exhibits a quasi-symmetrical protuberance (Fig. 19a), despite the influence of eastwards-directed tidal currents that flow parallel to the strait margin. This tide-induced littoral drift is also responsible for the building of the Ediz Hook, which is a *c.* 5 km long, coast-attached sandy spit developed a short distance from the subaerial Elwha Delta. However, the most impressive geometric response to the dominant tidal current pattern on this delta system is the geometry assumed by the submerged delta, the most volumetrically important part of the entire system, which is strongly and asymmetrically deflected in a direction parallel to the coastline (Fig. 19a). At the easternmost periphery of the Elwha delta front, sediment accumulations assume a digitate shape, forming multiple subaqueous sandbodies attached to the main deltaic body, but elongate for several hundreds of metres (Fig. 19a). This seafloor morphology possibly reflects a long-term along-shore accretion of the subaqueous delta after various episodes of Pleistocene glaciation and, in particular, the latest Wisconsinan phase, known locally as the Fraser Stade.

Another modern example is represented by the Fraser River Delta, prograding into the Strait of Georgia in southwestern British Columbia, Canada (Fig. 19b). The Fraser River Delta has been investigated by Ayranci & Dashtgard (2016) who, based on a wide core and grab dataset of sediment samples, detected a distinct asymmetry in the sedimentological and ichnological characteristics of the updrift (south) and downdrift (north) sides of the main distributary channel in water depths below storm wave base. These researchers attributed the asymmetry in the subaqueous delta environments to the strong influence of the net north-directed tidal flow acting along the coastlines (Ayranci & Dashtgard 2016).

An interesting modern analogue of extremely deflected and elongate subaqueous deltas is the Po Delta, which enters the northern Adriatic Sea on the Italian coastline (Fig. 19c). Although the Adriatic Sea does not represent a tidal strait, because it is closed in its northern border the tides exert a relevant influence along the northern coastlines. A semi-diurnal tidal prism is, in fact, considerably amplified in the northern part of the basin, promoting enhanced tidal ranges and secondary thermohaline currents (Bondesan *et al.* 1995). In the southern Adriatic Sea, 350 km south of the Po River delta, high-resolution seismic data have revealed the occurrence of a mud-prone clinoform system, which is retained as the expression of downcurrent along-shore transport of the sediment debouched by the Po River and other additional source points (Cattaneo *et al.* 2003). Sediment drift is governed by shore-parallel bottom currents, the transport capacity of which is possibly enhanced by the narrowing of the Adriatic Sea and by the tidal influx amplified by the northern engulfed basin (Fig. 19c). This recent example demonstrates an extreme, current-driven deflection in delta front or prodelta deposits that may influence sediment accumulation over large distances from the main sediment point source.

A more mature stage of delta front deflection due to amplified tidal currents may be seen in the subaqueous Klang Delta in the Malacca Strait, a segment of the Sunda shelf sea lying between Sumatra and the Malay Peninsula (Fig. 19d). Keller & Richards (1967), Emery (1971) and, subsequently, Galloway (2002) identified this as a tide-dominated delta front platform, which actively supplies sediment to the strait, occurs in a strike-elongate fashion downcurrent of the delta coastline, and is remarkably re-aligned parallel to the northern border (Coleman *et al.* 1970). This area consists of an axial sandy shoal complex, about 150 × 100 km in extent, located where the narrow southern strait opens into the relatively broad central strait, creating a 50–80 m deep shelf. Sediments of the shoal complex are reworked into large tidal bars separated by deep oval to elongate scour troughs and hollows (Fig. 19d). Tidal bars have amplitudes of 15–25 m and form elongate ridges 0.3–1 km wide and 15–70 km long. As the Malacca Strait widens northwestwards, another wide sand bank occurs detached from the main Klang deltaic body (Fig. 19d). This additional sandbank can be interpreted to result from the persistent bed shear stress imparted by the dominant tidal circulation and the consequent long-term reworking of the delta front deposits in an offset location with respect to the main delta.

Diagnostic elements in ancient deflected tide-dominated strait/seaway deltas

Deflection of the progadational axis and asymmetry in the delta front sand distribution (Fig. 20) is a feature that can be detected in ancient tidal seaway deltas based on the following diagnostic elements.

(i) The systematic measurement of palaeocurrent directions across the stratigraphic succession that records the delta front progradation. In tide-dominated deltas this character is usually detectable in large-scale cross-stratified complexes, which represent the result of the migration of tidal dunes across the delta front area. Smaller structures, including medium and small tidal cross-strata, and tidal bars filling distributary channels and their axes are also suitable for palaeocurrent measurements (Rahmani 1988; Nio & Yang 1991; Dalrymple 1992; Shanley *et al.* 1992; Fenies *et al.* 1999; Willis 2005; Dalrymple 2010; Steel *et al.* 2012). In deltas subject to a strong tidal deflection during progradation, the palaeocurrent patterns record a progressive upwards-rotatory trend, changing from an initial shore-perpendicular direction to a shore-parallel direction, according to the dominant tidal current. This change can be detectable within a single deltaic succession, such as the O'Brien Spring Sandstone and the Belmonte Calabro delta facies described here.

(ii) The general absence of volumetrically significant prodelta mudstone deposits in the distal deltaic environments. This feature is more common in deltas prograding into tidal straits rather than into seaways. This is because straits generally reproduce conditions of stronger tidal current amplification as a result of the restriction of the cross-sectional area in the narrowest strait centre and consequent water mass convergence (Pugh 1987; Longhitano 2013). Accordingly, deltas entering straits dominated by tidal currents encounter a high-energy oceanographic setting and thus the fine-grained suspended load is transported away for several

Table 1. *Comparison between the main morpho-sedimentological features among 'classic' tide-dominated deltas and deflected deltas prograding in seaways and/or straits developed under the influence of along-shore tidal currents*

Delta features	Tide-dominated deltas		Tidal seaway/strait-deflected deltas	
In plan		Modern examples		Modern examples
Delta plain width/length	3:4	Changjiang Delta (Goodbred *et al.* 2003)	4:3 to 2:1	• **Elwha Delta**, Juan de Fuca Strait (Foreman *et al.* 2004; Gelfenbaum *et al.* 2015)
Delta front platform width/length	5:2	Ganges–Brahmaputra Delta (Goodbred & Kuehl 1999; Kuehl *et al.* 2005)	2:1 to 3:1	• **Fraser Delta**, Strait of Georgia (Ayranci & Dashtgard 2016)
Prodelta width/length	4:1 to 6:1	Fly Delta (Wolanski *et al.* 1995; Harris *et al.* 1996) Colorado Delta (Thompson 1968; Carriquiry & Sanchez 1999)	Up to 7:2	• **Klang Delta**, Malacca Seaway (Keller & Richards 1967; Emery 1971; Galloway 2002) • **Sant'Andrea Delta**, Messina Strait (Chiocci *et al.* 2008)
Delta front sandbodies	Dip elongated, seawards diverging (parallel to ebb/flood tidal currents)	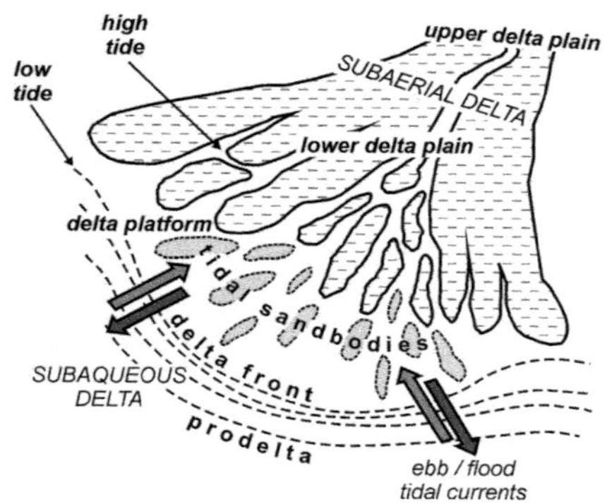 Redrawn, after Goodbred & Saito (2012)	From oblique to the coastline to strike-elongated	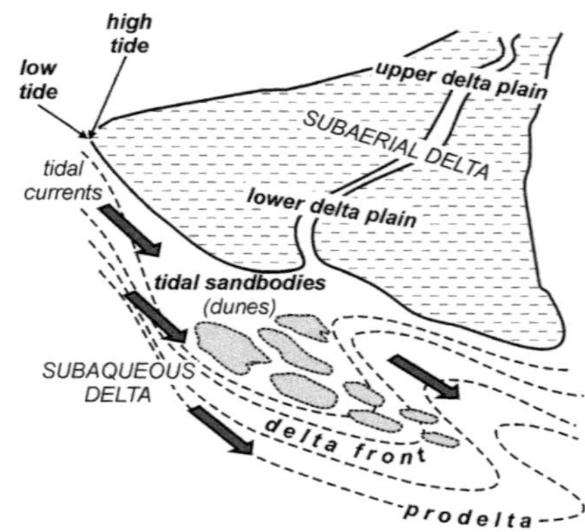

Cross-section	*Ancient examples*		*Ancient examples*	
Palaeocurrent trend	Perpendicular to the coastline or seawards diverging	• Maracaibo Basin, Venezuela Middle Eocene Lagunullas Formation (Maguregui & Tyler 1991) • Central Wyoming, USA Mid-Cretaceous Frontier Formation (Willis *et al.* 1999) • Halten Terrace, offshore Norway, Early Jurassic,Tilje Formation (Martinius *et al.* 2001) • Baltic Basin, Estonia, Latvia, Lithuania, Middle Devonian Kernave–Aruküla Formation (Plink-Björklund 2012)	Rotating from oblique to the coastline to parallel to the coastline (or to the seaway/strait axis) Wave influence at the delta coastline	• **Amantea Basin**, southern Italy, Tortonian Belmonte Calabro section (Longhitano & Nemec 2005; Longhitano *et al.* 2012*a*, *b*; this work) • **Messina Strait**, southern Italy, Lower Pleistocene Calanna section (Barrier 1987; Mercier *et al.* 1987; Longhitano *et al.* 2012*a*, *b*) • **Southern Wyoming**, USA Campanian Haystack Mountains (Martinsen & Tillman 1989; Mellere & Steel 1995*a*, *b*; this work)
Dominant process regime	Ebb/flood tidal currents	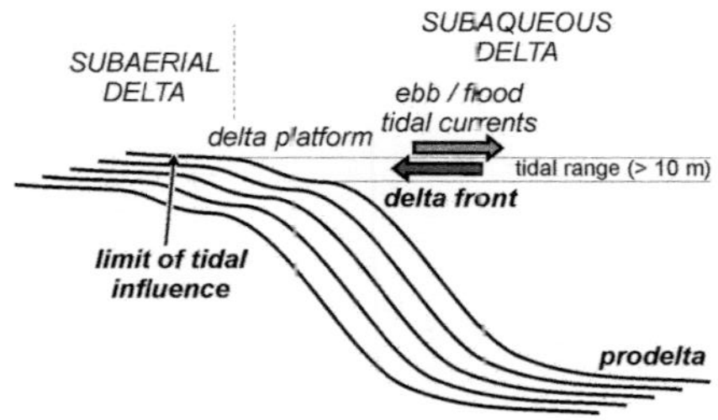	Along-shore tidal current dominance at the delta front/prodelta	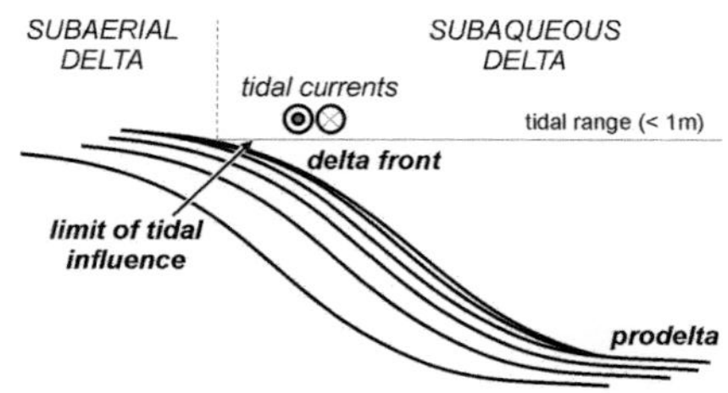
Clinoform thickness/geometry	Thick topsets because of tidal reworking on the delta platform	Redrawn, after Goodbred & Saito (2012)	Thinning-upwards, in case of strong deflection	

Examples are based on modern and ancient case studies.

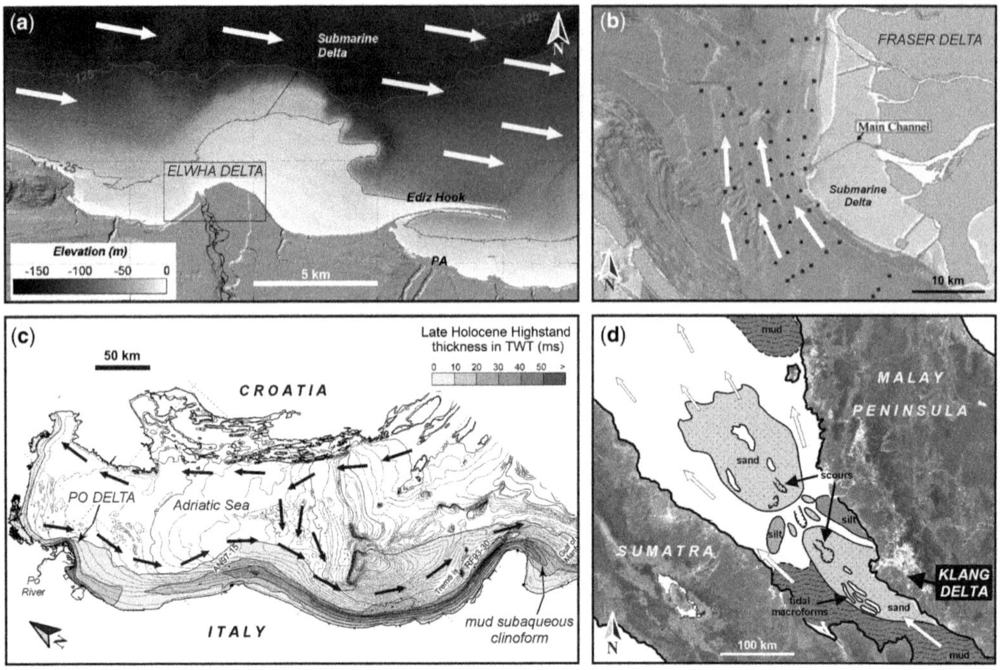

Fig. 19. (**a**) The Elwha Delta in the state of Washington, at the northwestern border of the USA, represents a modern example of deflected river delta prograding into the Juan de Fuca tidal strait (Fig. 1a). Although its coastal plain has no evident skewed profile, the subaqueous delta shows multiple elongate sand ridges as a result of the influence of along-shore transport due to the dominant east-directed tidal current (modified from Gelfenbaum *et al.* 2015). (**b**) The Fraser Delta prograded over the eastern margin of the same tidal strait as the previous example (i.e. the Strait of Georgia) and assumes an asymmetrical shape due to the north-directed tidal prism (modified from Ayranci & Dashtgard 2016). (**c**) The Po Delta in the northern Adriatic Sea undergoes an impressive deflection and detachment for over 350 km of prodelta muds, which feed a subaqueous clinoform in the southern Adriatic Sea (modified from Cattaneo *et al.* 2003). (**d**) Sediment pattern distribution in the subaqueous Klang Delta of the Malacca Strait (modified from Galloway 2002). Delta front sands are strongly deflected according to the dominant tidal current. Sediments are detached downcurrent from the main deltaic body to form an isolated sandbank that contains scours and tidal macroforms.

kilometres before accumulating in protected or sheltered adjacent coastal areas. Modern examples documenting an analogous process are observable in the Sant'Agata River Delta, which represents a small system prograding along the southwestern margin of the Messina Strait (Fig. 1d) in southern Italy. In the rock record, strait deltas can thus exhibit tidal cross-stratification motifs that generally lack fine-grained sediment, being deprived of 'mud-bearing' tidal structures such as double mud drapes, flaser bedding or fluid mud drapes (Longhitano 2011; Longhitano *et al.* 2012*a*).

(iii) An upwards change in the dominant process regime acting on the deltas. This feature is common in many ancient deltaic successions accumulated in basins that changed their oceanographic setting – for example, due to phases of relative sea-level rise (e.g. Porębski & Steel 2003, 2006; Yoshida *et al.* 2007; Plink-Björklund 2008). Transgressive coastlines commonly tend to be more tide-influenced than regressive coastlines (e.g. Dalrymple 1992; Dalrymple *et al.* 1992; Shanley *et al.* 1992; Allen & Posamentier 1993; Dalrymple & Zaitlin 1994; Shanley & McCabe 1994; Shennan & Andrews 2000; Yoshida 2000) and the effects of tidal processes in delta front deposits is more easily detectable in successions accumulated in incised valleys or embayed coastal areas flooded after relative sea-level rise. Ancient tidal straits documented from very different geological settings often experienced tectonically controlled transgressions (e.g. Allen & Bass 1993; Colella 1995; Mellere & Steel 1996; Anastas *et al.* 1997;

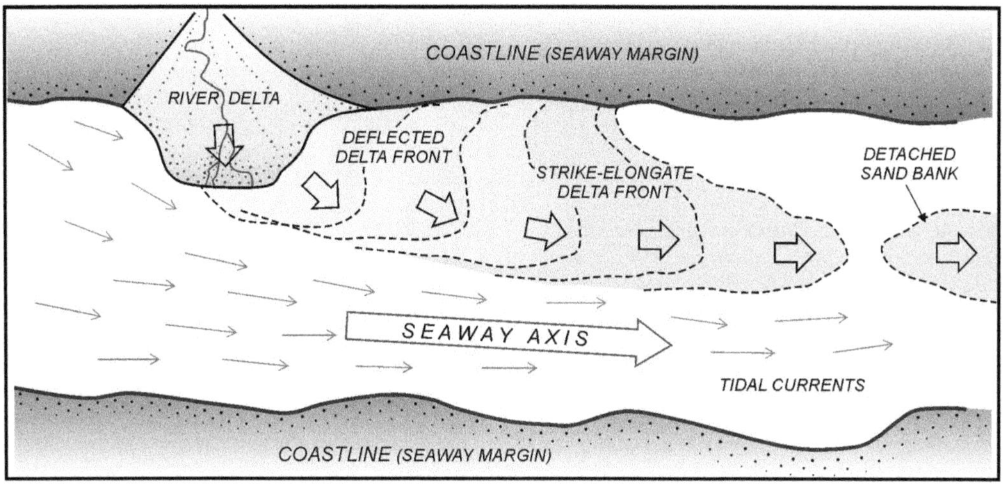

Fig. 20. Theoretical model explaining the process and the possible evolution through time of a river delta impinging a tidal strait or seaway. The deposits can assume deflected morphologies during an initial stage of progradation, but they may evolve into strike-elongate bodies and, finally, into detached sandbanks.

Longhitano & Nemec 2005; Olariu *et al.* 2012; Longhitano *et al.* 2012*b*, 2014; Longhitano 2013; Reynaud *et al.* 2013). In case of deltas prograding into these settings, their early transgressive stage was possibly characterized by the dominance of gravity-driven mass wasting processes or river flood jets superimposed onto other local marine influences. As the strait enlarged and tidal circulation was more pronounced at the basinal scale, deltas became even more tidally influenced/dominated in their vertical successions (Fig. 12) (Longhitano & Nemec 2005).

(iv) As documented in the Belmonte Calabro delta in the Tortonian Amantea Strait (Fig. 11), strait/seaway deltas often exhibit vertically stacked, thinning-upwards delta front units. This architectural arrangement was also described in current-deflected deltaic wedges of the Brazos, Danube, Fraser, Ganges–Brahmaputra, Kura, Po and many other deltas (see review in Korus & Fielding 2015). As observed in other case studies of coastal systems documented during the last 20 000 years (e.g. Longhitano *et al.* 2016), tidal influence has been enhanced during the late post-Last Glacial Maximum transgression in response to the deceleration of the rate of relative sea-level rise, which results in decreased accommodation space in the receiving basin (e.g. Cattaneo & Steel 2003) and the amplification of tidal processes at coastlines. However, in the observed sections, as well as in many documented deflected modern deltas, the progressive-upwards reduction in the thickness of cross-stratified wedges can be interpreted as the dip-oriented cross-sectional expression of the deflection of the delta front units and the consequent vertical juxtaposition of downcurrent locations in successive wedges. This effect and the related geometrical configurations have been documented for a number of Holocene deltas by Korus & Fielding (2015), who quantified delta front and prodelta deflection as consequence of discharge partitioning, wave reworking, plume deflection and variable subsidence (e.g. the Brazos, Danube, Fraser and Ganges–Brahmaputra deltas) or due to changing progradation rates and irregular palaeobathymetry (e.g. the Ba Lat, Kura, Llobregat, Mekong and Po deltas).

However, although recent studies have quantified delta deflection due to oceanic interference with the deltaic depositional processes, information is still lacking in our comprehension of river deltas that are tidally deflected because of their seaway and strait settings. For example, it is not fully clear whether delta deflection occurs simultaneously with sediment distribution (i.e. delta accretion) or is a secondary effect of tidal reworking of the frontal sectors of such deltas during periods of sediment starvation. Moreover, ichnofacies studies of the deltaic component of these systems are today very scarce. A systematic characterization of the distribution of the lifestyles of the organisms populating tidally deflected deltaic environments would reveal important insights into the dynamics of seaway/strait deltas and also into the interplay between subordinate factors such as turbidity, waves and occasional high-energy storm events.

Conclusions

This study defines the main diagnostic sedimentological and stratigraphic elements of river deltas impinging tide-dominated seaways and straits. As these deltas prograde into confined marine settings, the effects of waves, flood/ebb tides and river input are usually less important than the dominant influence of tidal current patterns running parallel to the coastline. These currents represent the major hydrodynamics in tidal seaways and straits and originate from tidal reversals between two interconnected wider basins. As a consequence of the restriction, the tidal prism increases noticeably in strength due to water mass convergence, exerting a strong influence along the seaway/strait margin and related depositional systems.

Many modern examples of marine passageways governed by tidal currents were described in this work as present day analogues, including the Juan de Fuca, Malacca, San Francisco, Messina, Dover and Torres straits, because they often show that river deltas prograding from their margins can assume strongly skewed or strike-elongate morphologies, departing substantially from the expected river delta settings. Outcrop examples from two ancient basins were identified as comparable ancient models for tidal seaway/strait deltas: the Cretaceous Western Interior Seaway of the USA and the Tortonian Calabrian Arc of southern Italy. These two areas represent, respectively, an ancient extensive marine passageway connecting the Gulf of Mexico to the south with the Arctic Sea to the north in which there was a tidal circulation governed by the overall oceanographic setting, and a small orogen that hosted a number of tectonically controlled corridors, which produced significant tidal current amplification and consequent sediment dispersal exchange between the interlinked Tyrrhenian and Ionian basins.

The stratigraphic sections described from these two areas demonstrated that river deltas can generate different stratal architectures depending on the gradient of the margin they prograded from. In the case of gently sloping borders, deltas may be characterized by long-lived sequences, with clinoforms prograding for large distances and assuming a progressive change in their dip direction according to the dominant tidal current pattern. Conversely, tidal straits with steep margins tend to generate single, short-lived deltas, which are usually involved in active tectonics and form unconformity-bounded sequences with a single, uniform direction of progradation.

Both these examples indicate that the delta front is the sector of the delta system most sensitive to the short-term influence of tidal currents. Here, tidal currents are able to rework sand to create wide tidal dune fields, preserved as thick cross-stratified complexes in the deltaic stratigraphic record. Although there is no clear field evidence in either of our two ancient case studies, modern tidally deflected deltas prograding into straits, such as the Elwha, Fraser and Klang deltas, show that prodelta fines can form elongate depositional features until they are detached from the main deltaic body and may assume the connotations of different systems.

Therefore we suggest some key diagnostic elements with which to detect ancient deflected tide-dominated seaway/strait deltas in the stratigraphic record.

(1) The palaeocurrent patterns record gradual upwards-rotatory trends, changing direction from initially shore-perpendicular to progressively shore-parallel, following the dominant tidal current direction.

(2) Prodeltaic fines are scarcely preserved in tide-dominated straits because the deltas advance rapidly into a high-energy setting where mud is perennially transported in suspension. In seaway deltas, fine-grained prodelta deposits tend to be strike-elongated, oriented parallel to the dominant tidal current and can be transported for large distances across the shelf, as in the case of the modern Po River delta in the Adriatic Sea.

(3) Seaway/strait deltas can experience an upwards change in the dominant process regime, transiting from river- or wave-dominated during early progradation, progressively to tide-influenced and, during their late stages of progradation, tide-dominated. This variance results in an asymmetrically deflected delta front and (where present) prodelta deposits and tidal sandbodies elongated in the direction of the dominant tidal current. In cross-section, this process can be reflected by the accretion of thinning-upwards clinoform units, reflecting a progressive downcurrent shift of the delta front sandbodies.

These considerations establish a basis to promote further studies and reinterpretations of the elongate, coastline-attached sandbodies detected in a number of narrow confined settings in many petroliferous basins of economic significance around the world (e.g. around the margins of many tectonic highs recognized offshore the North Atlantic Ocean). The results of these analyses are crucial in predicting and modelling sandbody distributions and in reviewing many environmental interpretations of ancient settings that may be related to tide-dominated seaways or straits.

This study resulted from the convergence of two initially separate research programmes, which were contemporaneously pointing towards common observations and conclusions. The authors express their gratitude to all the collaborators and colleagues who assisted them in the field and with whom they exchanged fruitful discussions that have allowed the writing of this paper. The authors are also grateful to Allard Martinius, an anonymous reviewer and Gary Hampson for their constructive criticisms of the content of this work.

References

Ainsworth, R.B., Vakarelov, B.K. & Nanson, R.A. 2011. Dynamic spatial and temporal prediction of changes in depositional processes on clastic shorelines: toward improved subsurface uncertainty reduction and management. *American Association of Petroleum Geologists Bulletin*, **95**, 267–297.

Allen, G.P. & Posamentier, H.W. 1993. Sequence stratigraphy and facies models of an incised valley fill: the Gironde estuary, France. *Journal of Sedimentary Petrology*, **63**, 378–391.

Allen, P.A. & Bass, J. 1993. Sedimentology of the upper marine molasse of the Rhone-Alp region, eastern France: implications for basin evolution. *Eclogae Geologicae Helvetiae*, **86**, 121–172.

Amodio Morelli, L., Bonardi, G. et al. 1976. L'arco Calabro – Peloritano nell'orogene Appenninico – Maghrebide. *Memorie Società Geologica Italiana*, **17**, 1–60.

Anastas, A.S., Dalrymple, R.W., James, N.P. & Nelson, C.S. 1997. Cross-stratified calcarenites from New Zealand: subaqueous dunes in a cool-water, Oligo-Miocene seaway. *Sedimentology*, **44**, 869–891.

Anastas, A.S., Dalrymple, R.W., James, N.P. & Campbell, S.N. 2006. Lithofacies and dynamics of a cool-water carbonate seaway: mid-Tertiary, Te Kuiti Group, New Zealand. *In*: Pedley, H.M. & Carannante, G. (eds) *Cool-Water Carbonates: Depositional Systems and Palaeoenvironmental Controls*. Geological Society, London, Special Publications, **255**, 245–268, https://doi.org/10.1144/GSL.SP.2006.255.01.15

Argentieri, A., Mattei, M., Rossetti, F., Argnani, A., Salvini, F. & Funiciello, R. 1998. Tectonic evolution of the Amantea basin (Calabria, southern Italy): comparing inland and offshore data. *Annales Tectonicae*, **12**, 79–96.

Aschoff, J. & Steel, R.J. 2011*a*. Anatomy and development of a low-accommodation clastic wedge, Upper Cretaceous, Cordilleran Foreland Basin, USA. *Sedimentary Geology*, **236**, 1–24.

Aschoff, J. & Steel, R.J. 2011*b*. Anomalous clastic wedge development during the Sevier-Laramide transition, North American Cordilleran Foreland basin, USA. *Geological Society of America Bulletin*, **123**, 1822–1835.

Ashley, G.M. 1990. Classification of large-scale subaqueous bedforms: a new look at an old problem. *Journal of Sedimentary Petrology*, **60**, 160–172.

Asquith, D.O. 1970. Depositional topography and major marine environments, Late Cretaceous, Wyoming. *American Association of Petroleum Geologists Bulletin*, **54**, 1184–1224.

Ayranci, K. & Dashtgard, S.E. 2016. Asymmetrical deltas below wave base: insights from the Fraser River Delta, Canada. *Sedimentology*, **63**, 761–779, https://doi.org/10.1111/sed.12237

Barrier, P. 1987. Stratigraphie des dépôts pliocènes et quaternaires du Détroit de Messine. *Documents et Travaux de l'Institut Géologique Albert Lapparent, Paris*, **11**, 59–81.

Bastos, A.C., Collins, M. & Kenyon, N.H. 2003. Morphology and internal structure of sand shoals and sandbanks off the Dorset coast, English Channel. *Sedimentology*, **50**, 1105–1122.

Bhattacharya, J.P. 2006. Deltas. *In*: Posamentier, H.W. & Walker, R.G. (eds) *Facies Models Revisited*. SEPM Special Publications, **84**, 237–292, https://doi.org/10.2110/pec.06.84.0237

Bhattacharya, J.P. & Giosan, L. 2003. Wave-influenced deltas: geomorphological implications for facies reconstruction. *Sedimentology*, **50**, 187–210.

Bhattacharya, J.P. & Willis, B.J. 2001. Lowstand deltas in the Frontier Formation, Powder River Basin, Wyoming: implications for sequence stratigraphic models, USA. *American Association of Petroleum Geologists Bulletin*, **80**, 139–150.

Blakey, R.C. & Ranney, W. 2008. *Ancient Landscapes of the Colorado Plateau*. Grand Canyon Association, Arizona.

Bonardi, G., Cavazza, W., Perrone, V. & Rossi, S. 2001. Calabria-Peloritani terrane and northern Ionian Sea. *In*: Vai, G.B. & Martini, I.P. (eds) *Anatomy of an Orogen: the Apennines and Adjacent Mediterranean Basin*. Kluwer Academic, Dordrecht, 287–306.

Bondesan, M., Castiglioni, G.B., Elmi, C., Gabbianelli, G., Marocco, R., Pirazzoli, P.A. & Tomasin, A. 1995. Coastal areas at risk from storm surges and sea-level rise in northeastern Italy. *Journal of Coastal Research*, **11**, 1354–1379.

Carriquiry, J.D. & Sanchez, A. 1999. Sedimentation in the Colorado River delta and upper Gulf of California after nearly a century of discharge loss. *Marine Geology*, **158**, 125–145.

Catalano, S., Monaco, C., Tortorici, L. & Tansi, C. 1993. Pleistocene strike–slip tectonics in the Lucanian Apennine (Southern Italy). *Tectonics*, **12**, 656–665.

Cattaneo, A. & Steel, R.J. 2003. Transgressive deposits: a review of their variability. *Earth-Science Reviews*, **62**, 187–228.

Cattaneo, A., Correggiari, A., Langone, L. & Trincardi, F. 2003. The late-Holocene Gargano subaqueous delta, Adriatic shelf: sediment pathways and supply fluctuations. *Marine Geology*, **193**, 61–91.

Chiocci, F.L., Catalano, R. et al. 2008. *Note illustrative aree sommerse della carta geologica d'italia alla scala 1:50 000, Foglio 601. 'Messina – Reggio di Calabria'*. Agenzia per la protezione dell'ambiente e i servizi tecnici, Rome.

Cole, R.D. & Young, R.G. 1991. Facies characterization and architecture of a muddy shelf-sandstone complex – Mancos B interval of the Upper Cretaceous Mancos Shale, northwest Colorado–northeast Utah, *In*: Miall, A.D. & Tyler, N. (eds) *The Three-Dimensional Facies Architecture of Terrigenous Clastic Sediments and its Implications for Hydrocarbon Discovery and Recovery*. SEPM Concepts in Sedimentology and Paleontology, **3**, 277–286.

Colella, A. 1988. Fault-controlled marine Gilbert-type fan deltas. *Geology*, **16**, 1031–1034.

Colella, A. 1995. Sedimentation, deformational events and eustacy in the perityrrhenian Amantea Basin: preliminary synthesis. *Giornale di Geologia*, **57**, 179–193.

Colella, A. & D'Alessandro, A. 1988. Sand waves, Echinocardium traces and their bathyal depositional setting (Monte Torre Palaeostrait, Plio-Pleistocene, southern Italy). *Sedimentology*, **35**, 219–237.

Colella, A. & Longhitano, S.G. 1998. Evidenze di tettonica transpressiva tortoniana nel Bacino di Amantea (Calabria). *Atti del 79 Congresso della SGI (Pt. A)*, 6–9 October 1998, Bologna, Italy, 322–324.

Coleman, J.M., Gagliano, S.M. & Smith, W.G. 1970. Sedimentation in a Malaysian high tide tropical delta. *In*: Morgan, J.P. (ed.) *Deltaic Sedimentation: Modern and Ancient*. SEPM Special Publications, **15**, 185–197.

Dalrymple, R.W. 1984. Morphology and internal structure of sandwaves in the Bay of Fundy. *Sedimentology*, **31**, 365–382.

Dalrymple, R.W. 1992. Tidal depositional systems. *In*: Walker, R.G. & James, N.P. (eds) *Facies Models: Response to Sea Level Change*. Geological Association of Canada, St John's, 195–218.

Dalrymple, R.W. 2010. Tidal depositional systems. *In*: James, N.P. & Dalrymple, R.W. (eds) *Facies Models 4*. Geological Association of Canada, St John's, 199–208.

Dalrymple, R.W. & Choi, K.S. 2007. Morphologic and facies trends through the fluvial–marine transition in tide-dominated depositional systems: a schematic framework for environmental and sequence-stratigraphic interpretation. *Earth-Science Reviews*, **81**, 135–174.

Dalrymple, R.W. & Zaitlin, B.A. 1994. High-resolution sequence stratigraphy of a complex, incised valley succession, Cobequid Bay-Salmon River estuary, Bay of Fundy, Canada. *Sedimentology*, **41**, 1069–1091.

Dalrymple, R.W., Zaitlin, B.A. & Boyd, R.A. 1992. A conceptual model of estuarine sedimentation. *Journal of Sedimentary Petrology*, **62**, 1130–1146.

Defant, A. 1961. *Physical Oceanography*. Pergamon Press, Oxford.

Dewey, J.F., Helman, M.L., Turco, E., Hutton, D.H.W. & Knott, S.D. 1989. Kinematics of the western Mediterranean. *In*: Coward, M.P., Dietrich, D. & Park, R.G. (eds) *Alpine Tectonics*. Geological Society, London, Special Publications, **45**, 265–283, https://doi.org/10.1144/GSL.SP.1989.045.01.15

Di Nocera, S., Ortolani, F., Russo, M. & Torre, M. 1974. Successioni sedimentarie messiniane e limite Miocene–Pliocene nella Calabria settentrionale. *Bollettino della Societa Geologica Italiana*, **93**, 575–607.

Emery, K.O. 1971. Bottom sediment map of Malacca Strait. *ECAFE, Technical Bulletin*, **4**, 149–152.

Fagherazzi, S. & Overeem, I. 2007. Models of deltaic and inner continental shelf evolution. *Annual Review of Earth and Planetary Sciences*, **35**, 685–715.

Fenies, H., de Ressequier, A. & Tastet, J.P. 1999. Intertidal clay-drape couplets (Gironde Estuary, France). *Sedimentology*, **46**, 1–15.

Fisher, W.L., Brown, L.F., Scott, A.J. & McGowen, J.H. 1969. *Delta Systems in the Exploration for Oil and Gas, a Research Colloquium*. Bureau of Economic Geology, Austin, Texas.

Foreman, M.G.G., Sutherland, G. & Cummins, P.F. 2004. Tidal dissipation around Vancouver island: an inverse approach. *Continental Shelf Research*, **24**, 2167–2185.

Galloway, W.E. 1975. Process framework for describing the morphologic and stratigraphic evolution of deltaic depositional systems. *In*: Broussard, M.L. (ed.) *Deltas, Models for Exploration*. Houston Geological Society, Houston, TX, 87–98.

Galloway, W.E. 2002. Paleogeographic setting and depositional architecture of a sand dominated shelf depositional system, Miocene Utsira Formation, North Sea Basin. *Journal of Sedimentary Research*, **72**, 476–490.

Gelfenbaum, G., Stevens, A.W., Miller, I., Warrick, J.A., Ogston, A.S. & Eidam, E. 2015. Large-scale dam removal on the Elwha River, Washington, USA: coastal geomorphic change. *Geomorphology*, **246**, 729–750.

Ghisetti, F. 1979. Evoluzione neotettonica dei principali sistemi di faglie della Calabria centrale. *Bollettino della Societa Geologica Italiana*, **98**, 387–430.

Ghisetti, F. & Vezzani, L. 1981. Contribution of structural analysis to understanding the geodynamic evolution of the Calabrian Arc (southern Italy). *Journal of Structural Geology*, **3**, 371–381.

Goodbred, S.L., Jr. & Kuehl, S.A. 1999. Holocene and modern sediment budgets for the Ganges–Brahmaputra River: evidence for highstand dispersal to floodplain, shelf, and deep-sea depocenters. *Geology*, **27**, 559–562.

Goodbred, S.L., Jr. & Saito, Y. 2012. Tide-dominated deltas. *In*: Davis, A., Jr. & Dalrymple, R.W. (eds) *Principles of Tidal Sedimentology*. Springer, Dordrecht, 129–149.

Goodbred, S.L., Jr, Kuehl, S.A., Steckler, M.S. & Sarker, M.H. 2003. Controls on facies distribution and stratigraphic preservation in the Ganges–Brahmaputra delta sequence. *Sedimentary Geology*, **155**, 301–316.

Gueguen, E., Doglioni, C. & Fernandez, M. 1997. Lithospheric boudinage in the Western Mediterranean back-arc basin. *Terra Nova*, **9**, 184–187.

Hampson, G.J. & Howell, J.A. 2005. Sedimentologic and geomorphic characterization of ancient wave-dominated shorelines: examples from the Late Cretaceous Blackhawk Formation, Book Cliffs, Utah, *In*: Bhattacharya, J.P. & Giosan, L. (eds) *Deltas Old and New*. SEPM Special Publications, **83**, 133–154.

Hampson, G.J., Procter, E.J. & Kelly, C. 2008. Controls on shallow marine sandstone deposition and shelf construction: late Cretaceous, Western Interior Seaway,

N. Utah and Colorado, USA. *In*: HAMPSON, G.J., STEEL, R.J., BURGESS, P.M. & DALRYMPLE, R.W. (eds) *Recent Advances in Models of Siliciclastic Shallow Marine Stratigraphy*. SEPM Special Publications, **90**, 355–389.

HARRIS, P.T., PATTIARATCHI, C.B. *ET AL*. 1996. Late Quaternary deltaic and carbonate sedimentation in the Gulf of Papua foreland basin: response to sea-level change. *Journal of Sedimentary Research*, **66**, 801–819.

HETTINGER, R.D. & KIRSCHBAUM, M.A. 2002. *Stratigraphy of the Upper Cretaceous Mancos Shale (Upper Part) and Mesaverde Group in the Southern Part of the Uinta and Piceance Basins*, US Geological Survey, Utah and Colorado, Geologic Investigation Series, I-2764.

KAUFFMAN, E.G. 1977. Geological and biological overview, Western Interior Cretaceous basin. *The Mountain Geologist* (Rocky Mountain Association of Geologists), **14**, 75–99.

KAUFFMAN, E.G. & CALDWELL, W.G.E. 1993. The Western Interior Basin in space and time. *In*: CALDWELL, W.G.E. & KAUFFMAN, E.G. (eds) *Evolution of the Western Interior Basin*. Geological Association of Canada Special Paper, **39**, 1–30.

KELLER, G.H. & RICHARDS, A.F. 1967. Sediments of the Malacca Strait, Southeast Asia. *Journal of Sedimentary Petrology*, **37**, 102–127.

KLEIN, G.D. & RYER, T.A. 1978. Tidal circulation patterns in Precambrian, Paleozoic and Cretaceous epeiric and mioclinal shelf seas. *Geological Society of America Bulletin*, **89**, 1050–1058.

KNOTT, S.D. & TURCO, E. 1991. Late Cenozoic kinematics of the Calabrian Arc. *Tectonics*, **10**, 1164–1172.

KORUS, J.T. & FIELDING, C.R. 2015. Asymmetry in Holocene river deltas: patterns, controls, and stratigraphic effects. *Earth-Science Reviews*, **150**, 219–242.

KUEHL, S.A., ALLISON, M.A., GOODBRED, S.L. & KUDRASS, H.R. 2005. The Ganges-Brahmaputa Delta. *In*: GIOSAN, L. & BHATTACHARYA, J.P. (eds) *River Deltas – Concepts, Models, and Examples*. SEPM Special Publications, **83**, 87–129.

LONGHITANO, S.G. 2008. Sedimentary facies and sequence stratigraphy of coarse-grained Gilbert-type deltas within the Pliocene thrust-top Potenza Basin (Southern Apennines, Italy). *Sedimentary Geology*, **210**, 87–110.

LONGHITANO, S.G. 2011. The record of tidal cycles in mixed silici–bioclastic deposits: examples from small Plio-Pleistocene peripheral basins of the microtidal Central Mediterranean Sea. *Sedimentology*, **58**, 691–719.

LONGHITANO, S.G. 2012. Microtidal straits: outcrop analogues from Calabria, south Italy. *Rendiconti Online della Società Geologica Italiana*, **21**, 937–939.

LONGHITANO, S.G. 2013. A facies-based depositional model for ancient and modern, tectonically-confined tidal straits. *Terra Nova*, **25**, 446–452.

LONGHITANO, S.G. 2015. Deflected steep-marginal deltas in confined tidal straits an outcrop analogue from the upper Miocene Amantea Basin, southern Italy. *In*: *Proceedings of the 'Sedimentology of Paralic Reservoirs: Recent Advances and their Applications' conference*, 18–19 May, Burlington House, Piccadilly, London, 94–95.

LONGHITANO, S.G. & COLELLA, A. 1998. Tettonica transpressiva tortoniana nel Bacino di Amantea (Calabria) e record sedimentario. *Atti della Riunione Annuale del GIS, Bologna. Giornale di Geologia*, **60**, 45–47.

LONGHITANO, S.G. & COLELLA, A. 2007. Geomorphology, sedimentology and recent evolution of the anthropogenically modified Simeto river delta system (Eastern Sicily, Italy). *Sedimentary Geology*, **194**, 195–221.

LONGHITANO, S.G. & NEMEC, W. 2005. Statistical analysis of bed-thickness variation in a Tortonian succession of biocalcarenitic tidal dunes, Amantea Basin, Calabria, southern Italy. *Sedimentary Geology*, **179**, 195–224.

LONGHITANO, S.G. & STEEL, R.J. 2015. Deltas sourcing tidal straits: observations from some field case studies. Paper presented at the 31st Meeting of the International Association of Sedimentologists, 22–25 June, Krakow, Poland.

LONGHITANO, S.G., MELLERE, D., STEEL, R.J. & AINSWORTH, R.B. 2012*a*. Tidal depositional systems in the rock record: a review and new insights. *Sedimentary Geology*, **279**, 2–22.

LONGHITANO, S.G., CHIARELLA, D., DI STEFANO, A., MESSINA, C., SABATO, L. & TROPEANO, M. 2012*b*. Tidal signatures in Neogene to Quaternary mixed deposits of southern Italy straits and bays. *Sedimentary Geology*, **279**, 74–96.

LONGHITANO, S.G., CHIARELLA, D. & MUTO, F. 2014. Three-dimensional to two-dimensional cross-strata transition in the lower Pleistocene Catanzaro tidal strait transgressive succession (southern Italy). *Sedimentology*, **61**, 2136–2171.

LONGHITANO, S.G., DELLA LUNA, R., MILONE, A.L., CILUMBRIELLO, C., CAFFAU, M. & SPILOTRO, G. 2016. The 20,000-years-long sedimentary record of the Lesina coastal system (southern Italy): from alluvial, to tidal, to wave process regime change. *The Holocene*, **26**, 678–698, https://doi.org/10.1177/0959683615618256

MAGUREGUI, J. & TYLER, N. 1991. Evolution of Middle Eocene tide-dominated deltaic sandstones, Lagunullas Field, Maracaibo Basin, Western Venezuela. *In*: MIALL, A.D. & TYLER, N. (eds) *Concepts in Sedimentology and Paleontology*. SEPM, Tulsa, OK, **3**, 233–244.

MALINVERNO, A. & RYAN, W.B.F. 1986. Extension in Tyrrhenian Sea and shortening in the Apennines as a result of arc migration driven by sinking of the lithosphere. *Tectonics*, **5**, 227–254.

MARTINI, I.P., SAGRI, M. & COLELLA, A. 2001. Neogene–Quaternary basins of the inner Apennines and Calabrian arc. *In*: VAI, G.B. & MARTINI, I.P. (eds) *Anatomy of an Orogen. The Apennines and Adjacent Mediterranean Basins*. Kluwer Academic, Dordrecht, 375–400.

MARTINIUS, A.W., KAAS, I., NAESS, A., HELGESEN, G., KJAEREFJORD, J.M. & KEITH, D.A. 2001. Sedimentology of heterolithic and tide-dominated Tilje Formation (Early Jurassic, Halten Terrace, offshore mid-Norway). *In*: MARTINSEN, O.J. & DREYER, T. (eds) *Sedimentary Environments Offshore Norway – Paleozoic to Recent*. Norwegian Petroleum Foundation, Special Publications, **10**, 103–144.

MARTINSEN, R.S. 2001. Wave dominated v. current dominated shorelines in the Cretaceous Western Interior

Seaway. *Abstract volume, AAPG/SEPM Annual Meeting*, 3–6 June 2001, Denver, Colorado.

Martinsen, R.S. 2003. Depositional remnants; part 2, examples from the Western Interior Cretaceous basin of North America. *American Association of Petroleum Geologists Bulletin*, **87**, 1883–1909.

Martinsen, R.S. & Tillman, R.W. 1989. *Haystack Mountain Formation*. 40th Annual Field Conference. Wyoming Geological Association Guidebook, Casper.

Mattei, M., Argentieri, A., Rossetti, F., Speranza, F., Sagnotti, L. & Funiciello, R. 1999. Extensional tectonics in the Amantea Basin: a comparison between AMS and structural data. *Tectonophysics*, **307**, 33–49.

Mattei, M., Cipollari, P., Cosentino, D., Argentieri, A., Rossetti, F., Speranza, F. & Di Bella, L. 2002. The Miocene tectono-sedimentary evolution of the southern Tyrrhenian Sea: stratigraphy, structural and paleomagnetic data from the on-shore Amantea Basin (Calabrian Arc, Italy). *Basin Research*, **14**, 147–168.

Mellere, D. & Steel, R.J. 1995*a*. Facies architecture and sequentiality of nearshore and shelf sandbodies; Haystack Mountains Formation, Wyoming, USA. *Sedimentology*, **42**, 551–574.

Mellere, D. & Steel, R.J. 1995*b*. Variability of lowstand wedges and their distinction from forced-regressive wedges in the Mesaverde Group, southeast Wyoming. *Geology*, **23**, 803–806.

Mellere, D. & Steel, R.J. 1996. Tidal sedimentation in Inner Hebrides half graben, Scotland: the Mid-Jurassic Bearreraig Sandstone Formation. *In*: Batist, M.D. & Jacobs, P. (eds) *Geology of Siliciclastic Shelf Seas*. Geological Society, London, Special Publications, **117**, 49–79, https://doi.org/10.1144/GSL.SP.1996.117.01.04

Mercier, D., Barrier, P., Beaudoin, B., Didier, S., Montenat, J.L. & Salinas Zuniga, E. 1987. Les facteurs hydrodynamiques dans la sédimentazion plio-quaternaire du Détroit de Messine. *Documents et Travaux de l'Institut Géologique Albert Lapparent, Paris*, **11**, 171–183.

Monaco, C. & Tansi, C. 1992. Strutture transpressive lungo la zona trascorrente sinistra nel versante orientale del Pollino (Appennino calabro-lucano). *Bollettino della Societa Geologica Italiana*, **111**, 291–301.

Muto, F. & Perri, E. 2002. Evoluzione tettono-sedimentaria del bacino di Amantea, Calabria occidentale. *Bollettino della Societa Geologica Italiana*, **121**, 391–409.

Muto, T. & Steel, R.J. 2002. In defense of shelf-edge delta development during falling and lowstand of relative sea level. *The Journal of Geology*, **110**, 421–436.

Mutti, E., Rosell, J., Allen, G.P., Fonnesu, F. & Sgavetti, M. 1985. The Eocene Baronia tide dominated delta-shelf system in the Ager Basin. *In*: Mila, M.D. & Rosell, J. (eds) *Excursion Guidebook, 6th European Regional Meeting of the International Association of Sedimentologists*, Lleida, Spain, 579–600.

Nio, S.D. & Yang, C.S. 1991. Diagnostic attributes of clastic tidal deposits: a review. *In*: Smith, D.G., Reinson, G.E. et al. (eds) *Clastic Tidal Sedimentology*. Canadian Society of Petroleum Geologists Memoirs, **16**, 3–28.

Ogniben, L. 1969. Schema introduttivo alla geologia del confine calabro-lucano. *Memorie Società Geologica Italiana*, **8**, 453–763.

Olariu, C., Steel, R.J., Dalrymple, R.W. & Gingras, M.K. 2012. Tidal dunes v. tidal bars: the sedimentological and architectural characteristics of compound dunes in a tidal seaway, the lower Baronia Sandstone (Lower Eocene), Ager Basin, Spain. *Sedimentary Geology*, **279**, 134–155.

Ortolani, F., Torre, M. & Di Nocera, S. 1979. I depositi altomiocenici del Bacino di Amantea (Catena Costiera Calabra). *Bollettino della Societa Geologica Italiana*, **98**, 559–587.

Orton, G. & Reading, H.G. 1993. Variability of deltaic processes in terms of sediment supply, with particular emphasis on grain size. *Sedimentology*, **40**, 475–512.

Plink-Björklund, P. 2008. Wave-to-tide facies change in a Campanian shoreline complex, Chimney Rock Tongue, Wyoming–Utah, USA. *In*: Hampson, G., Steel, R., Burgess, P. & Dalrymple, R.W. (eds) *Recent Advances in Models of Siliciclastic Shallow-Marine Stratigraphy*. Society for Sedimentary Geology Special Publications, **90**, 265–291.

Plink-Björklund, P. 2012. Effects of tides on deltaic deposition: causes and responses. *Sedimentary Geology*, **279**, 107–133.

Plint, A.G. 1990. An allostratigraphic correlation of the Muskiki and Marshybank Formations (Coniacian-Santonian) in the foothills and subsurface of the Alberta Basin. *Bulletin of Canadian Petroleum Geology*, **38**, 288–306.

Plint, A.G. 1991. High-frequency relative sea-level oscillations in Upper Cretaceous shelf clastics of the Alberta foreland basin: possible evidence for a glacio-eustatic control? *In*: MacDonald, D.I.M. (ed.) *Sedimentation, Tectonics and Eustasy: Sea-Level Changes at Active Margins*. International Association of Sedimentologists, Special Publications, **12**, 409–428.

Plint, A.G. & Kreitner, M.A. 2007. Extensive, thin sequence spanning Cretaceous foredeep suggest high-frequency eustatic control: Late Cenomanian, Western Canada foreland basin. *Geology*, **35**, 735–738.

Plint, A.G., Walker, R.G. & Bergman, K.M. 1986. Cardium Formation 6: stratigraphic framework of the Cardium in subsurface. *Bulletin of Canadian Petroleum Geology*, **34**, 213–225.

Plint, A.G., Hart, B.S. & Donaldson, W.S. 1993. Lithospheric flexure as a control on stratal geometry and facies distribution in Upper Cretaceous rocks of the Alberta foreland basin. *Basin Research*, **5**, 69–77.

Pontén, A. & Plink-Björklund, P. 2007. Depositional environments in an extensive tide influenced delta plain, Middle Devonian Gauja Formation, Devonian Baltic Basin. *Sedimentology*, **54**, 969–1006.

Pontén, A. & Plink-Björklund, P. 2009. Regressive to transgressive transits reflected in tidal bars, Middle Devonian Baltic Basin. *Sedimentary Geology*, **218**, 48–60.

Porębski, S.J. & Steel, R.J. 2003. Shelf-margin deltas: their stratigraphic significance and relation to deep-water sands. *Earth-Science Reviews*, **62**, 283–326.

Porębski, S.J. & Steel, R.J. 2006. Deltas and sea-level change. *Journal of Sedimentary Research*, **76**, 1–14.

Pratt, L. 1990. *The Physical Oceanography of Sea Straits*. Kluwer, Dordrecht.

Pugh, D.T. 1987. *Tides, Surges and Mean Sea-Level*. Wiley, Chichester.

Rahmani, R.A. 1988. Estuarine tidal channel and nearshore sedimentation of a Late Cretaceous epicontinental sea, Drumheller, Alberta, Canada. *In*: de Boer, P.L., van Gelder, A. & Nio, S.D. (eds) *Tide-Influenced Sedimentary Environments and Facies*. Reidel, Dordrecht, 433–481.

Reading, H.D. & Collinson, J.D. 1996. Clastic coasts. *In*: Reading, H.G. (ed.) *Sedimentary Environments. Processes, Facies and Stratigraphy*. 3rd edn. Wiley-Blackwell, Oxford, 154–231.

Rehault, J.P., Moussat, E. & Fabbri, A. 1987. Structural evolution of the Tyrrhenian back-arc basin. *Marine Geology*, **74**, 123–150.

Reynaud, J.-Y. & Dalrymple, R.W. 2012. Shallow-marine tidal deposits. *In*: Davis, R.A., Jr. & Dalrymple, R.W. (eds) *Principles of Tidal Sedimentology*. Springer, New York, 335–370.

Reynaud, J.-Y., Ferrandini, M. et al. 2013. From non-tidal shelf to tide-dominated strait: the Miocene Bonifacio Basin, Southern Corsica. *Sedimentology*, **60**, 599–623.

Roehler, H.W. 1990. *Stratigraphy of the Mesaverde Group in the Central and Eastern Greater Green River Basin, Wyoming, Colorado, and Utah, United States*. US Geological Survey Professional Paper, **1508**.

Shanley, K.W. & McCabe, P.J. 1994. Perspectives on the sequence stratigraphy of continental strata. *American Association of Petroleum Geologists Bulletin*, **78**, 544–568.

Shanley, K.W., McCabe, P.J. & Hettinger, R.D. 1992. Tidal influence in Cretaceous fluvial strata from Utah, USA: a key to sequence stratigraphic interpretation. *Sedimentology*, **39**, 905–930.

Shennan, I. & Andrews, J. (eds) 2000. *Holocene Land–Ocean Interaction and Environmental Change around the North Sea*. Geological Society, London, Special Publications, **166**, http://sp.lyellcollection.org/content/166/1

Slingerland, R., Driscoll, N.W., Milliman, J.D., Miller, S.R. & Johnstone, E.A. 2008. Anatomy and growth of a Holocene clinothem in the Gulf of Papua. *Journal of Geophysical Research*, **113**, F01S13.

Speranza, F., Mattei, M., Sagnotti, L. & Grasso, F. 2000. Rotational differences between the northern and southern Tyrrhenian domains: palaeomagnetic constraints from the Amantea basin (Calabria, Italy). *Journal of the Geological Society, London*, **157**, 327–334, https://doi.org/10.1144/jgs.157.2.327

Steel, R.J., Plink-Bjorklund, P. & Aschoff, J. 2012. Tidal deposits of the Campanian Western Interior Seaway, Wyoming, Utah and Colorado, USA. *In*: Davis, A., Jr. & Dalrymple, R.W. (eds) *Principles of Tidal Sedimentology*. Springer, Berlin, 437–472.

Swenson,. J.B., Paola, C., Pratson, L., Voller, V.R. & Murray, A.B. 2005. Fluvial and marine controls on combined subaerial and subaqueous delta progradation: morphodynamic modeling of compound-clinoform development. *Journal of Geophysical Research*, **110**, 1–16.

Swift, D.J.P. & Thorne, J.A. 1991. Sedimentation on continental margins, I: a general model for shelf sedimentation. *In*: Swift, P., Oertel, G.F., Tillman, R.W. & Thorne, J.A. (eds) *Shelf Sand and Sandstone Bodies: Geometry, Facies and Sequence Stratigraphy*. International Association of Sedimentologists, Special Publications, **14**, 3–31.

Syvitski, J.P.M., Harvey, N., Wollanski, E., Burnett, W.C., Perillo, G.M.E. & Gornitz, V. 2005. Dynamics of the Coastal Zone. *In*: Crossland, C.J., Kremer, H.H., Lindebloom, H.J., Marshall-Crossland, J.I. & Le Tissier, M.D.A. (eds) *Global Fluxes in the Anthropocene*. Springer, Berlin, 39–94.

Tänavsuu-Milkeviciene, K. & Plink-Björklund, P. 2009. Recognizing tide-dominated v. tide-influenced deltas: middle Devonian strata of the Baltic Basin. *Journal of Sedimentary Research*, **79**, 887–905.

Tansi, C. 1991. La tettonica recente dell'area di Amantea (Catena Costiera, Calabria). *Rendiconti della Società Geologica Italiana*, **2**, 33–35.

Thompson, R.W. 1968. *Tidal Flat Sedimentation on the Colorado River Delta, North-western Gulf of California*. Geological Society of America Memoir, **107**.

Tortorici, L. 1982. Lineamenti geologico-strutturali dell'Arco Calabro Peloritano. *Società Italiana di Mineralogia e Petrografia*, **38**, 927–940.

Turco, E., Maresca, R. & Cappadona, P. 1990. La tettonica plio-pleistocenica del confine calabro–lucano: modello cinematico. *Memorie Società Geologica Italiana*, **45**, 519–529.

Tyler, N. & Finley, R.J. 1991. Architectural controls on the recovery of hydrocarbons from sandstone reservoirs. *In*: Miall, A.D. & Tyler, N. (eds) *The Three-dimensional Facies Architecture of Terrigenous Clastic Sediments, and Its Implications for Hydrocarbon Discovery and Recovery*. SEPM, Concepts and Models in Sedimentology and Paleontology, **3**, 1–5.

Uroza, C.A. 2008. *Processes and architectures of deltas in shelf-break and ramp platforms: examples from the Eocene of West Spitsbergen (Norway), the Pliocene paleo-Orinoco delta (SE Trinidad), and the Cretaceous Western Interior Seaway (S Wyoming and NE Utah)*. PhD thesis, University of Texas at Austin.

Walsh, J.P. & Nittrouer, C.A. 2009. Understanding fine-grained river-sediment dispersal on continental margins. *Marine Geology*, **263**, 34–45.

Weise, B.R. 1980. *Wave-dominated Delta Systems of the Upper Cretaceous San Miguel Formation, Maverick Basin, South Texas*. Bureau of Economic Geology, Report of Investigations **107**.

Wescott, W.A. & Ethridge, F.G. 1990. Fan deltas – alluvial fans in coastal settings. *In*: Rachocki, A.H. & Church, M. (eds) *Alluvial Fans: a Field Approach*. Wiley, Chichester, 195–211.

Willis, B.J. 2005. Deposits of tide-influenced river deltas. *In*: Giosan, L. & Bhattacharya, J.P. (eds) *River Deltas – Concepts, Models and Examples*. Society for Sedimentary Geologists, Special Publications, **83**, 87–129.

Willis, B.J. & Gabel, S.L. 2001. Sharp-based, tide-dominated deltas of the Sego sandstone, Book Cliffs, Utah, USA. *Sedimentology*, **46**, 479–506.

Willis, B.J. & Gabel, S.L. 2003. Formation of deep incisions into tide-dominated river deltas: implications for the stratigraphy of the Sego Sandstone, Book Cliffs, Utah, USA. *Journal of Sedimentary Research*, **73**, 246–263.

Willis, B.J., Bhattacharya, J.P., Gabel, S.L. & White, C.D. 1999. Architecture of a tide-influenced river delta in the Frontier Formation of central Wyoming, USA. *Sedimentology*, **46**, 667–688.

Wolanski, E., King, B. & Galloway, D. 1995. Dynamics of the turbidity maximum in the Fly River estuary, Papua New Guinea. *Estuarine, Coastal and Shelf Science*, **40**, 321–337.

Wright, L.D. 1977. Sediment transport and deposition at river mouths: a synthesis. *Geological Society of America Bulletin*, **88**, 857–868.

Wright, L.D. & Friedrichs, C.T. 2006. Gravity-driven sediment transport on continental shelves: a status report. *Continental Shelf Research*, **26**, 2092–2107.

Yoshida, S. 2000. Sequence and facies architecture of the upper Blackhawk Formation and the Lower Castlegate Sandstone (Upper Cretaceous), Book Cliffs, Utah, USA. *Sedimentary Geology*, **136**, 239–276.

Yoshida, S., Steel, R.J. & Dalrymple, R.W. 2007. Changes in depositional processes – an ingredient of a new generation of stratigraphic models. *Journal of Sedimentary Research*, **77**, 447–460.

Zecchin, M. 2005. Relationships between fault-controlled subsidence and preservation of shallow-marine small-scale cycles: example from the lower Pliocene of the Crotone Basin (southern Italy). *Journal of Sedimentary Research*, **75**, 300–312.

Zecchin, M., Massari, F., Mellere, D. & Prosser, G. 2004. Anatomy and evolution of a Mediterranean-type fault bounded basin: the Lower Pliocene of the northern Crotone Basin (Southern Italy). *Basin Research*, **16**, 117–143.

Tidally influenced shoal water delta and estuary in the Middle Jurassic of the Søgne Basin, Norwegian North Sea: sedimentary response to rift initiation and salt tectonics

DONATELLA MELLERE[1]*, ARUNA MANNIE[1], SERGIO LONGHITANO[2], MIKE MAZUR[1], HYELNI KULAUSA[1], SAMME BROUGH[1] & JAMES COTTON[1]

[1]*Premier-Oil Exploration plc, 23 Lower Belgrave Street, London, UK*

[2]*Universita' della Basilicata, Department of Sciences, Potenza, Italy*

**Correspondence: dmellere@premier-oil.com*

Abstract: Recent studies in the Middle Jurassic Bryne and Sandnes formations, primary reservoirs in several fields across the Norwegian and Danish North Sea, show the widespread occurrence of tidal-influenced and tide-dominated deposits. Aalenian–Bajocian Bryne cores reflect deposition by a shoal water, tidally influenced delta onto a low wave energy tidal platform (both supratidal and intertidal) that probably occupied the majority of the Søgne Basin, a narrow rift system connected to the Central and Danish graben and transgressed from an open-marine basin, possibly located to the south. At the Bathonian–Callovian boundary, a new phase of rifting and progressive salt movements led to the deposition of the upper Bryne and Sandnes formations within an 80–100 km long composite estuarine valley. Basin tilting to the south and continued transgression resulted in tidal deltas that offlap the northern margin of the basin. The basin was fully transgressed by the end of the Callovian. From the Late Bathonian onwards, differential tectonic movements along the broadly interconnected Middle Jurassic rift basins led to a change in the transgression direction from south to north, with an open-marine basin located in the Central and Viking graben.

Both ancient and modern examples of deltas influenced by tides and waves and associated with heterolithic, muddy/sandy deposits have been widely described (e.g. Galloway 1976; Ta *et al.* 2002; Dalrymple *et al.* 2003; Roberts & Sydow 2003; Willis & Gabel 2003; Plink-Björklund 2005, 2012; Amir Hassan *et al.* 2013; Ichaso & Dalrymple 2014). These models document shoal water deltas, which represent systems developing on low gradient coastal areas, where the main sediment deposition often occurs in shallow-marine settings, protected from powerful alongshore circulation and typified by a delta 'platform'. The delta platform is a shallow and relatively flat area extending from the subaerial lower delta plain up to the rollover point of the delta front (Roberts & Sydow 2003; Swenson *et al.* 2005) and thus it includes tidally reworked mouth bars, subtidal channels and shorefaces (Helland-Hansen & Hampson 2009; Rossi & Steel 2016). In a cross-sectional view, the delta platform forms the topmost segment of a delta clinoform, between the subaerial and delta front portions (Kuehl *et al.* 1997; Pirmez *et al.* 1998; Roberts & Sydow 2003; Swenson *et al.* 2005; Xue *et al.* 2010). In shoal water deltas with minimal clinoform geometries (i.e. lacking significant delta slopes because of the limited accommodation basin into which they encroach), the platform may represent the volumetrically most important segment of the delta.

This paper addresses the sedimentology of the Middle Jurassic deposits of the coal-bearing Bryne and the tidally influenced shallow-marine Sandnes formations in the Norwegian sector of the Søgne Basin, North Sea (Fig. 1). Both formations, which bear important hydrocarbon reservoirs, are widespread across the North Sea from Norway (where the Bryne–Sandnes pair are well known as the Sleipner and Hugin formations in the area of the Sleipner Field; Vollset & Doré 1984; Folkestad & Satur 2008; Kieft *et al.* 2010), Denmark (the Lulu Formation in the Lulu Field; Frandsen 1986; Damtoft *et al.* 1992; Andsbjerg 2003; Andsbjerg & Dybkjær 2003; Michelsen *et al.* 2003) and the UK. In the UK, the Bryne and Sandnes formations have their equivalents in the Ness and Tarbert formations of the Brent Group, about 150 km further to the north into the Viking Graben (Graue *et al.* 1987; Fält *et al.* 1989; Cockings *et al.* 1992; Mitchener *et al.* 1992; Milner & Olsen 1998; Hampson *et al.* 2004, Fig. 2). In the Norwegian Sea, a petroleum province offshore mid-Norway, the Bryne and Sandnes formations have their equivalents in the younger Aalenian Åre Formation sandstones and in the tide-dominated Tilje Formation (Martinius *et al.* 2001, 2005; Ichaso & Dalrymple 2014). Understanding the sedimentary processes, channel dimensions and networks, the depositional environment and the partitioning of the sediment load in the

From: Hampson, G. J., Reynolds, A. D., Kostic, B. & Wells, M. R. (eds) 2017. *Sedimentology of Paralic Reservoirs: Recent Advances*. Geological Society, London, Special Publications, **444**, 173–213.
First published online September 21, 2016, https://doi.org/10.1144/SP444.11

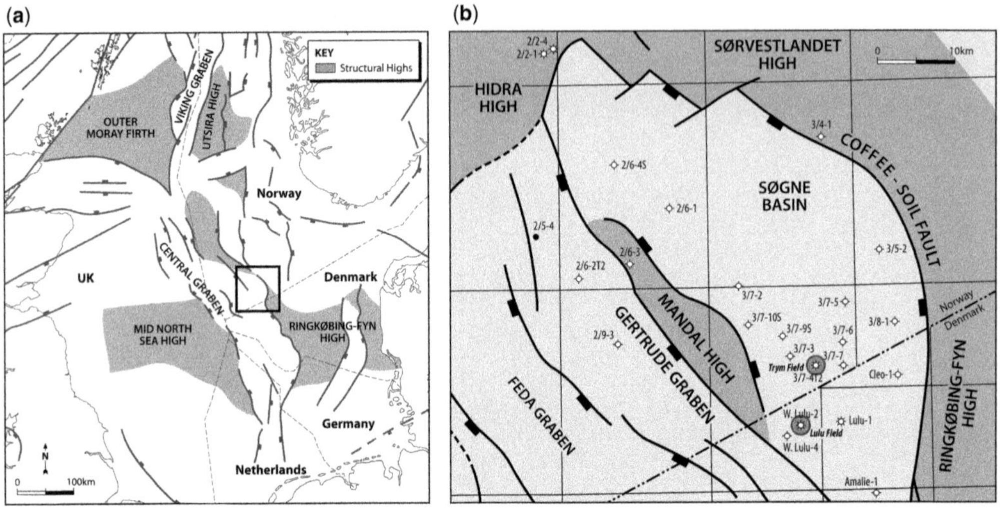

Fig. 1. (**a**) Structural element map of the North Sea and (**b**) location of the Søgne Basin with location of wells used in this study. Wells 3/7-6, 3/7-7 and 3/8-1 are shown for reference. The Søgne Basin represents an embayment of the Central Graben constrained by the Mandal High to the east and the Sørvestlandet–Rynkøbing–Fyn High to the west.

Bryne and Sandnes formations are, therefore, important from an exploration and development perspective to predict reservoir occurrence within, ahead and beyond the river delta mouths and, most importantly, to assess the regional palaeogeography.

With the exception of Andsbjerg (2003), who first recognized the presence of marine floodings within the Bryne Formation in the Danish North Sea, the general interpretation is that of a paralic fluvial-dominated system, with a shoreline possibly coinciding with the Brent delta shoreline (Vollset & Doré 1984), and a system that became shallow marine during the deposition of the Sandnes Formation. We recognize widespread tidal influences in the deposits throughout the overall succession, from the Aalenian to the Callovian. An important break occurs near the top of the Bryne Formation, below the base of the Callovian Sandnes Formation. The break is marked by a regional unconformity filled with stacked channel sandstones, interpreted by Andsbjerg (2003) as the fill of nested estuarine valleys and capped by a coal horizon and tidal deltas of the late Sandnes Formation.

This makes the Bryne–Sandnes succession an excellent example of an overall regressive–transgressive, mixed-energy delta–estuarine system, similar to those described by Tänavsuu-Milkeviciene & Plink-Björklund (2009) in the Baltic Sea and in the Oligo-Miocene deposits of the Gabus-Aran/Udang successions observed offshore Indonesia and Vietnam (see also Shultz & Ways 2004). Process variability can be observed at different scales within the Bryne and Sandnes units, from bed to sequence to megasequence scales. Instead of applying the commonly used term 'coastal plain' to describe the widespread occurrence of coaly shales and paralic deposits, we interpret the facies belt to be much closer to a delta distributary system. We interpret the Bryne succession to be deposited on a supratidal and subtidal platform within a low gradient, tidally influenced shoal delta subjected to several transgressive events, with a nearby seaway that was physically constrained by the early rift systems. The delta was then transformed into a large tide-dominated estuarine system (*sensu* Dalrymple *et al.* 1992) before becoming fully marine. The direction of the transgression is discussed in the following sections.

The aims of this paper are three-fold: (1) to reconstruct the palaeogeography of the Bryne (and Sandnes) deposits within the Søgne Basin using detailed core descriptions, wireline log correlations and state-of-the-art three-dimensional seismic interpretation and attribute analysis; (2) to identify the role of salt tectonics in sedimentation and the moment of inception of the Late Jurassic rifting; and (3) to understand the reasons for the widespread deposition of a low net-to-gross succession across vast regions of the North Sea during the Middle Jurassic.

Our study extends from the framework of Johannessen & Andsbjerg (1993), Andsbjerg (2003) and Johannessen *et al.* (2010), who extensively analysed the Bryne–Sandnes system in the Danish sector of the Søgne Basin. We mapped the system from Denmark into the northernmost tip of the Søgne Basin in

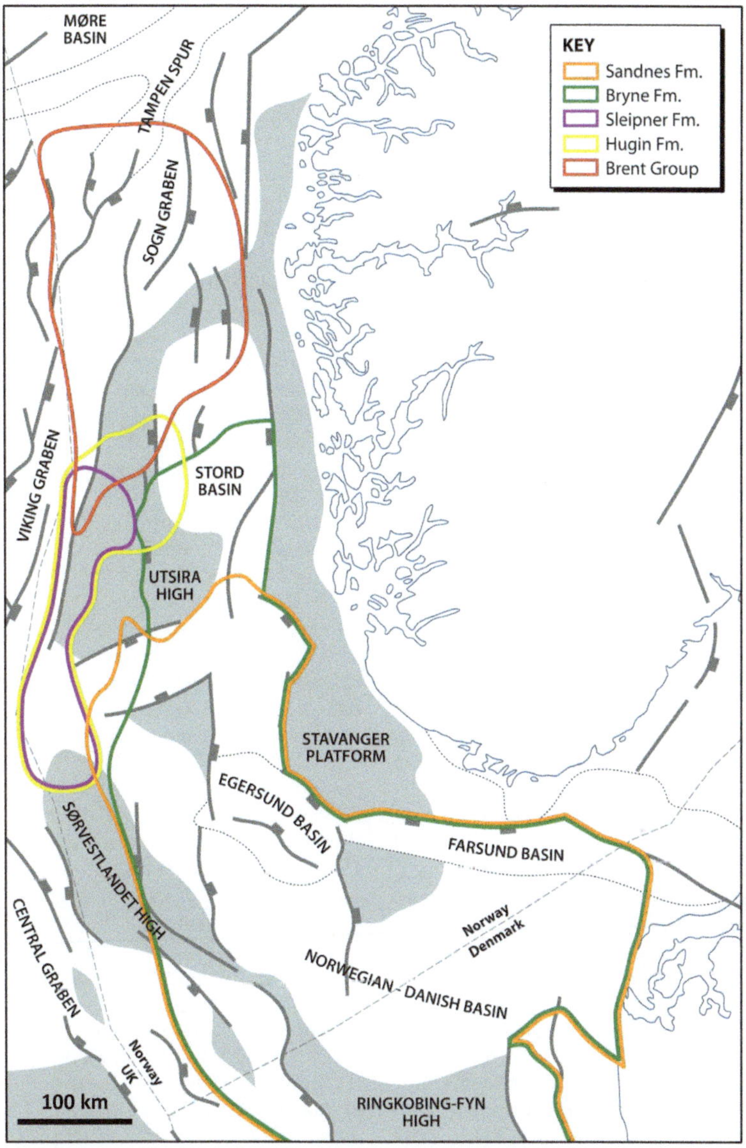

Fig. 2. Spatial distribution of the Bryne–Sandnes formations and coeval systems (Hugin and Sleipner formations and Brent Group) (modified from Blystad *et al.* 1995; Halland *et al.* 2014). The mid-Jurassic succession is widespread across the North Sea. Sandnes and Bryne are the lithostratigraphic units defined in the southern part of the basin.

offshore Norway, extending the palaeogeography across the national borders and tentatively across the North Sea.

Background geology

The Søgne Basin extends from south to north across the Norwegian–Danish boundary in the Southern North Sea (Fig. 1). The basin is bounded to the east and the north by the Coffee Soil Fault, which delineates the eastern sides of the Sørvestlandet High and its southern prolongation, the Ringkøbing/Fyn High. The basin is bounded to the west by the metamorphic basement of the Mandal High and, to the south, the Søgne Basin merges into the Tail End Graben.

The Mandal High and part of the Sørvestlandet–Ringkøbing/Fyn horst formed during Permian rifting and were reactivated during Middle and Late

Jurassic rifting (Gowers & Sæbøe 1985; Møller 1986; Cartwright 1991; Michelsen *et al.* 1992; Korstgaard *et al.* 1993; Stemmerik *et al.* 2000; Røssland 2011). These highs greatly influenced sedimentation in the basin, constraining the basin physiography and restricting longshore currents and the advection of sediment plumes.

The Permian rift occurred during the late sedimentation of the Rotliegendes Formation and was accompanied by footwall erosion and the deposition of the Zechstein evaporites. The basin underwent a renewed rifting phase during the early Jurassic, with tilting and uplifting of the Mandal High and Sørvestlandet High, enhanced salt movement and a change in climate (Røssland 2011). It is in this context that the Aalenian to Bathonian coal-bearing Bryne and the Callovian shallow-marine Sandnes strata were deposited (Fig. 3). On seismic data, these strata thin onto salt diapirs and onto the bounding Mandal and Sørvestlandet highs, where they unconformably overlie Triassic and Permian arid peneplains (Fig. 4b). To the south, the stratal package thickens dramatically on the flank of the salt diapir below well 3/7-5 (Fig. 4c). It is either conformably overlain by shale-prone marine deposits of the Upper Jurassic (in the basin) or unconformably by the Cretaceous Chalk of the Shetland Group on structural highs (Figs 3 & 4). The eventual presence of Middle Jurassic deposits on top of the Mandal High is highly questionable. The deposits are not present in the wells drilled on the high, at least not in its southern part. According to Røssland (2011), the first subaerial exposure of the Mandal horst had already occurred during the Triassic and lasted until Late Jurassic flooding. The Late Jurassic rifting event was probably initiated during Callovian time during the deposition of the Sandnes Formation (see also Møller & Rasmussen 2003; Andsbjerg 2013). This tectonic phase records the progressive deepening of the basin and the change in sedimentation from tidal to open marine (Oxfordian) and, finally, during Volgian time, to deep marine. Across most of the basin, the Base Upper Jurassic reflector corresponds to a widespread flooding surface. The Upper Cretaceous Chalk deposits lie unconformably on the Upper Jurassic, with the base of the Chalk merging with the so-called Base Cretaceous Unconformity.

Bryne and Sandnes formations

The Bryne and Sandnes formations are defined on their lithostratigraphic characteristics. In the wells of the Søgne Basin, the Bryne Formation varies between 130 and 300 m in thickness (see also Andsbjerg 2003) and it forms a heterolithic package of coaly shales, coal beds, palaeosols, thin prograding heterolithic units and channelized sandstone bodies. In the Søgne Basin and in the Danish Central Graben, the formation has been interpreted as representing deposition in alluvial and coastal plain (Koch 1983; Vollset & Doré 1984; Damtoft *et al.* 1992; Andsbjerg 2003) or deltaic interdistributary bay settings (in Lulu-1 well; Frandsen 1986). Further to the north, the Bryne Formation has its equivalent in the coal-bearing Ness Formation, interpreted to represent the paralic system of the Brent Group (Livera 1989).

The Sandnes Formation is a sand-prone interval formed by a series of tidally influenced, blocky to coarsening-upwards cross-stratified units interpreted, in the Danish Søgne Basin, to record the outbuilding of flood and ebb tidal deltas over a beach barrier system during the Callovian to Oxfordian transgression of the Viking Graben (Vollset & Doré 1984; Andsbjerg 2003). The Sandnes Formation has its equivalent in the Lulu Formation of the Danish Basin (Andsbjerg 2003; Michelsen *et al.* 2003) and in the Hugin Formation of the Sleipner area (see Folkestad & Satur 2008; Kieft *et al.* 2010; Fig. 2). According to previously published data, the Hugin Formation represents the southern extension of the Brent Group, (Graue *et al.* 1987; Fält *et al.* 1989; Cockings *et al.* 1992; Mitchener *et al.* 1992; Sneider *et al.* 1995; Milner & Olsen 1998; Folkestad & Satur 2008). It developed during the drowning of the Brent delta because of the rapid deepening of the Viking Graben–Central Graben rift systems (Cockings *et al.* 1992; Hodgson *et al.* 1992). The Hugin Formation forms a transgressive sequence set of stacked transgressive–regressive units in the Sleipner area (Folkestad & Satur 2008; Kieft *et al.* 2010). It has been interpreted as wave- and storm-dominated shorefaces attached to a beach barrier lagoonal and coastal plain system (Cockings *et al.* 1992; Milner & Olsen 1998). New investigations indicate an important tidal influence (Folkestad & Satur 2008; Kieft *et al.* 2010), similar to that observed in the Sandnes deposits of the Søgne Basin (see also Andsbjerg 2003).

As traditionally defined, the boundary between the two formations is highly debatable and it is taken as the base of a blocky clean sandstone unit on top of the first coal bed, or its correlative coaly mudstone in the upper part of the Middle Jurassic succession (Vollset & Doré 1984; Michelsen *et al.* 2003). Bryne and Sandnes deposits have not been distinguished on regional seismic data because of the limited thickness of the Callovian with respect to the thicker Aalenian to Bathonian succession, although local aspects of the internal architecture have been unravelled by seismic attributes.

The zonation scheme used in this study follows the Partington *et al.* (1993) sequence stratigraphic scheme for the North Sea based on

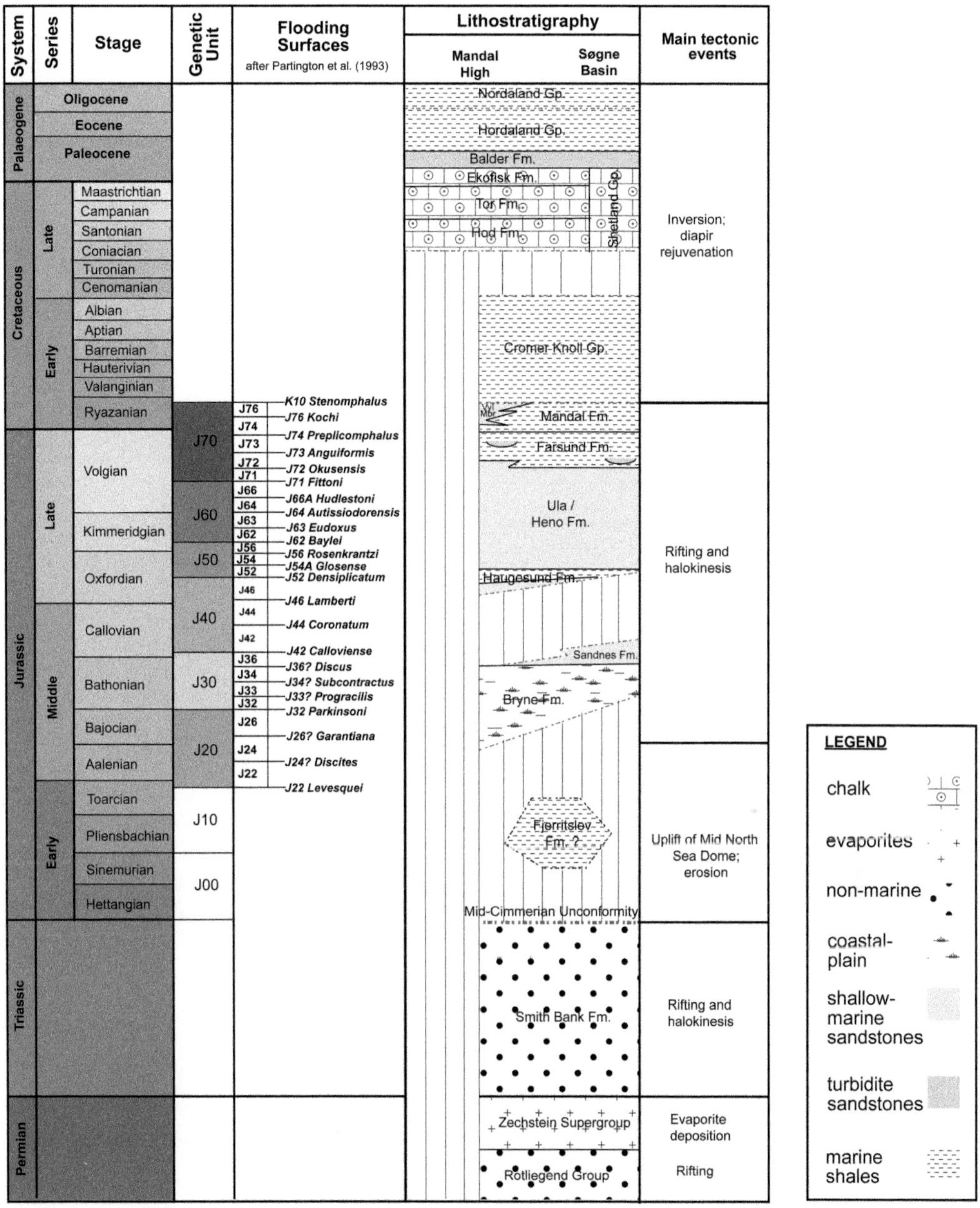

Fig. 3. Stratigraphic summary. The zonation scheme used in this study is based on the Partington *et al.* (1993) sequence stratigraphic (J-sequence) scheme for the North Sea and uses sequence boundary and maximum flooding surface picks. For the Jurassic section, a new chronostratigraphy based on palynology enabled the differentiation of four chronostratigraphic zones within the cored section of well 3/7-4: (1) Bathonian (J32–J34); (2) Early Callovian (J36); (3) Middle to Late Callovian (J42–J44); and (4) Middle Oxfordian (J46).

sequence boundary and maximum flooding surface picks (J-sequence) coupled with a more recent biostratigraphic evaluation based on palynology. The principal markers are shown in Figure 3. We identified four chronostratigraphic zones in well 3/7-4: Bathonian (J32–J34); Early Callovian (J36); Middle to Late Callovian (J42–J44); and Middle Oxfordian (J46).

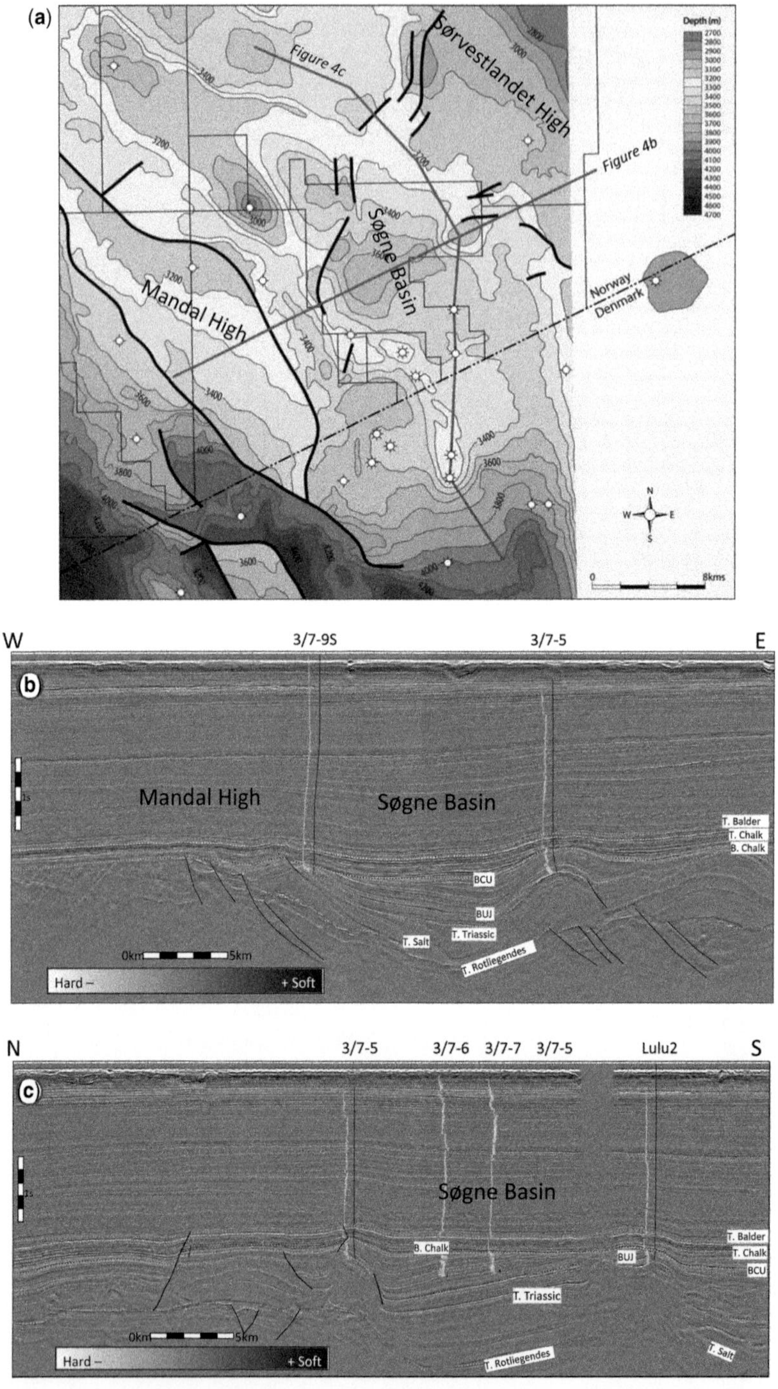

Fig. 4. (**a**) Structural map at Base Cretaceous Unconformity (BCU) with outline of crossline and inline shown in (**b**) and (**c**), respectively. (**b**) East–west crossline across the Søgne Basin from the Mandal High to the Sørvestlandet High. The study interval lies between the Base Upper Jurassic (BUJ) horizon and the top-Triassic (T. Triassic). Note the thick salt diapir below the 3/7-5 well and a salt wall below the 3/7-9S well. The succession thins both eastwards and westwards. (**c**) North–south inline along the Søgne Basin. The Middle Jurassic interval thins dramatically north of the 3/7-5 well due to salt growth and tectonic tilting.

Database and methodology

We examined 20 wells spread across the Søgne Basin, the Mandal High and the Gertrude Graben (Fig. 1b). These wells have a complete suite of geophysical logs and detailed final well reports. The cores were scattered because most of the target intervals were Upper Jurassic units. Well 3/7-4 in the Trym Field had the most comprehensive core suite of the Bryne Formation; we chose the cores from this well to illustrate the Bryne and Sandnes facies trends. In the Danish sector, extensive coring was performed in the Lulita and Lulu fields, as described by Andsbjerg (2003).

Core facies were linked to petrophysical logs and correlated across the basin, from the Danish sector to the north. Chronostratigraphy, a major issue within the Bryne Formation, especially in its lower interval, was assessed through internal reports, publications (Johannessen & Andsbjerg 1993; Andsbjerg 2003; Johannessen *et al.* 2010) and public online data from the Norwegian Petroleum Directorate.

The PGS geostreamer multi-client MC3D-CGR2013RM TWT seismic reflectivity dataset was widely used. Shot in 2010, 2011 and 2012, the entire broadband survey covers an area of 1100 km^2, which extends over the Mandal High, Søgne Basin and Gertrude Graben in the Norwegian North Sea, as well as south into the Danish sector. The quality of data was generally excellent, despite problems with Base Chalk and/or Base Cretaceous Unconformity peg-leg multiples that prevented the accurate imaging of deeper reflectors. After assessing the quality of the data and applying noise removal and spectral enhancement, the wells were tied in and a three-dimensional geomodel was built using Paleoscan software. Horizon attributes were then extracted and examined on a large number of iso-age surfaces, the time interval of which was less than a seismic loop. This allowed the identification of structural and sedimentary features that are difficult to detect with traditional interpretation techniques. Frequency extractions at key modelled horizons were then blended and used semi-quantitatively to attempt a spatial classification of the facies based on calibration with well data. Structural and isopach maps were used to assess the distribution of the Bryne and Sandnes formations.

Facies associations and depositional environments

The deposits of the Bryne and Sandnes formations show tidal influence processes throughout the succession. Based on cores from wells 3/7-4, 3/7-2, 3/7-3 and 2/6-1 and previous work in the Danish sector by Andsbjerg (2003), we recognized a set of facies associations from broadly 'brackish–continental' to tidal–marine that could be grouped into four major depositional systems. From base to top, these facies associations are:

(1) a regressive, river-dominated, tide-influenced delta platform – this is the dominant depositional environment of the Bryne Formation;
(2) a transgressive, tide-dominated estuary, which shows continuous interactions between tidal, riverine and subordinate wave and storm processes – this depositional system characterizes the Sandnes Formation and specific intervals of the Bryne successions;
(3) regressive, tide-dominated and wave-influenced deltas that occur in the upper part of the Sandnes Formation;
(4) offshore marine siltstones of the uppermost Oxfordian section.

A summary of the overall facies succession for well 3/7-4 is presented in Figure 5, including a sequence stratigraphic interpretation presented in a later section of the paper. Within each environment, facies associations are described in more detail and illustrated in Figures 6–11. Individual facies are summarized in Tables 1–4.

FA1: Regressive, river-dominated and tidal-influenced delta platform

This depositional system includes facies associations of the proximal, distal and subaqueous tidal delta platform (Table 1).

FA1.1: Proximal, supratidal delta platform. Facies association 1 consists of palaeosols and fluvial channels normally ascribed to the upper delta plain

Facies association 1.1a: palaeosols
Description. Facies association 1.1a (FA1.1a) consists of brecciated, massive and mottled shales and siltstones up to 1 m thick. The association is encountered in the most proximal reaches of the delta platform, where it generally overlies the muddy, vegetated coaly shales and channels of the intertidal delta platform. Colour mottling varies from dark grey to green and occasionally to red, often containing iron-cemented nodules. Red mottling is usually overlain by vertical green/grey mottling and coal. Bioturbation is essentially related to rooting.

Interpretation. The association is interpreted to represent dry and wet vertisols developed on the margin of the supratidal platform and strongly subjected to changes in the water table and to churning of the sediments by roots (see also Folkestad & Satur 2008). These deposits are normally described as

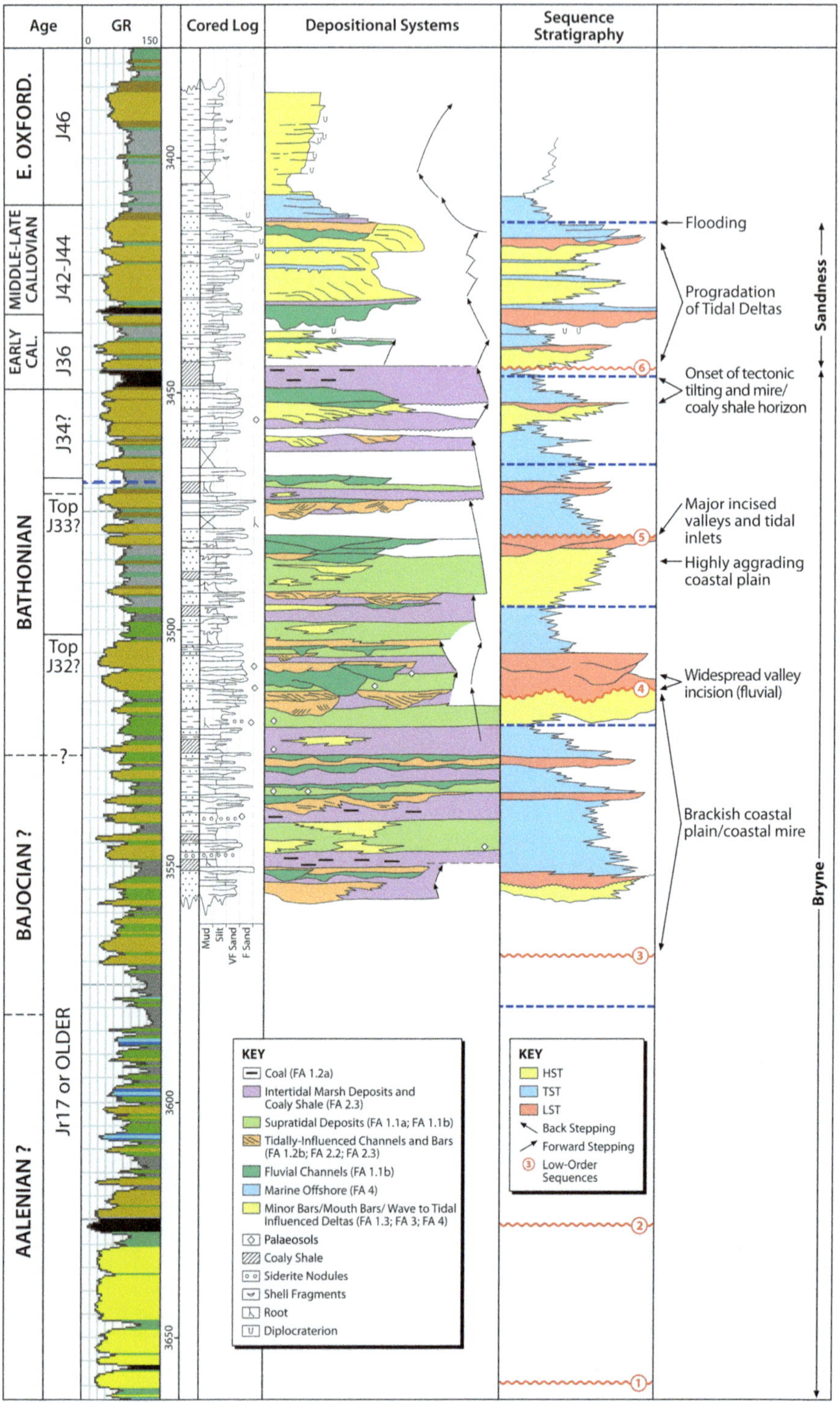

Fig. 5. Core to well log correlation and reconstructed sequence stratigraphic framework for the 3/7-4 cored interval. Numbers indicate the megasequences discussed in this work.

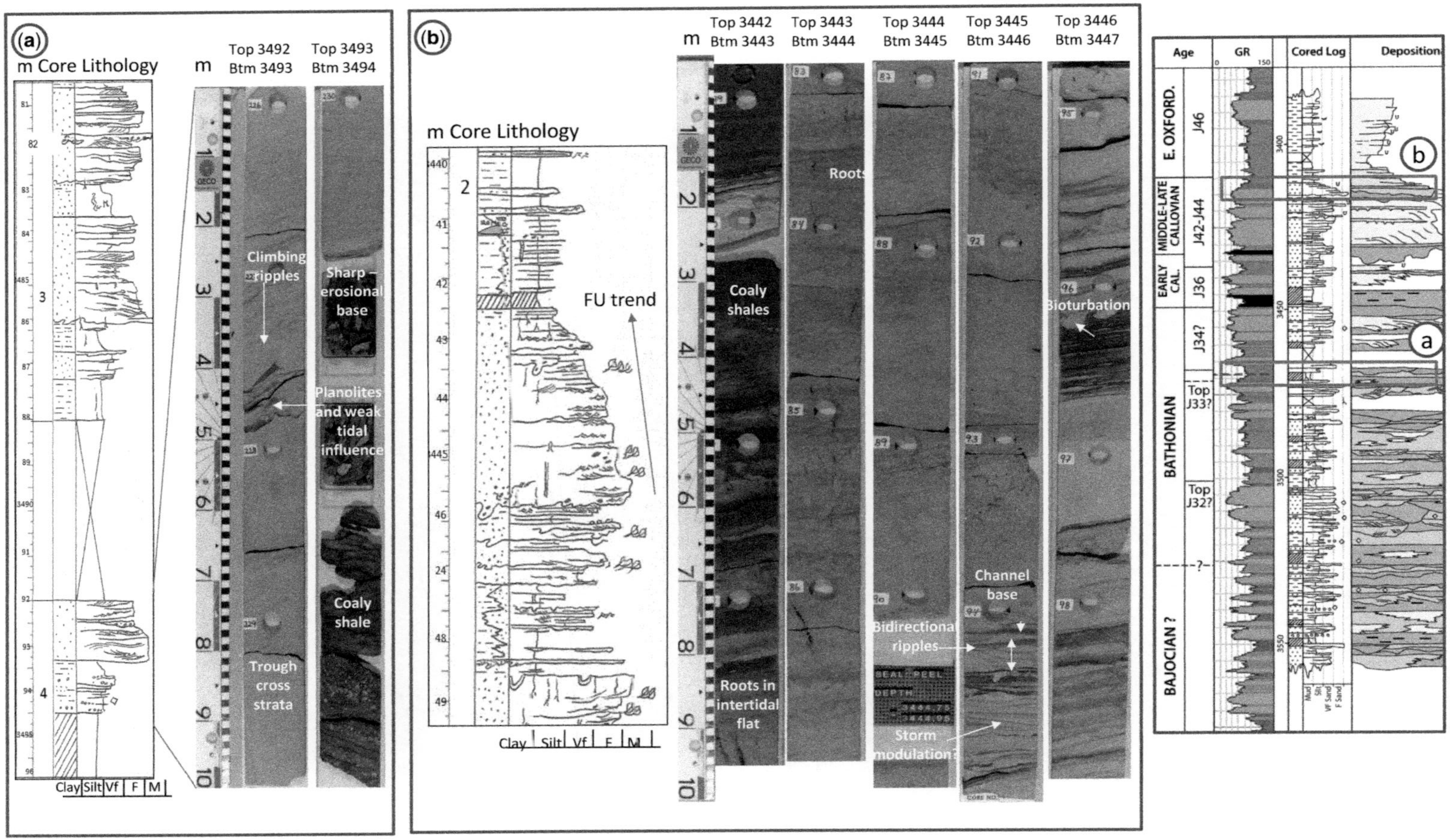

Fig. 6. Bryne delta, 3/7-4 well. (**a, b**) Examples of fluvial channels of facies association 1.1b of the supratidal delta platform (location on the log to the right).The channels are sharply based and typically fine upwards. They display cross-stratification and current ripple lamination. Roots, coal and plant fragments are ubiquitous. The channel in (b) overlies transgressive tidal flat deposits and attests a relative basinwards shift in facies.

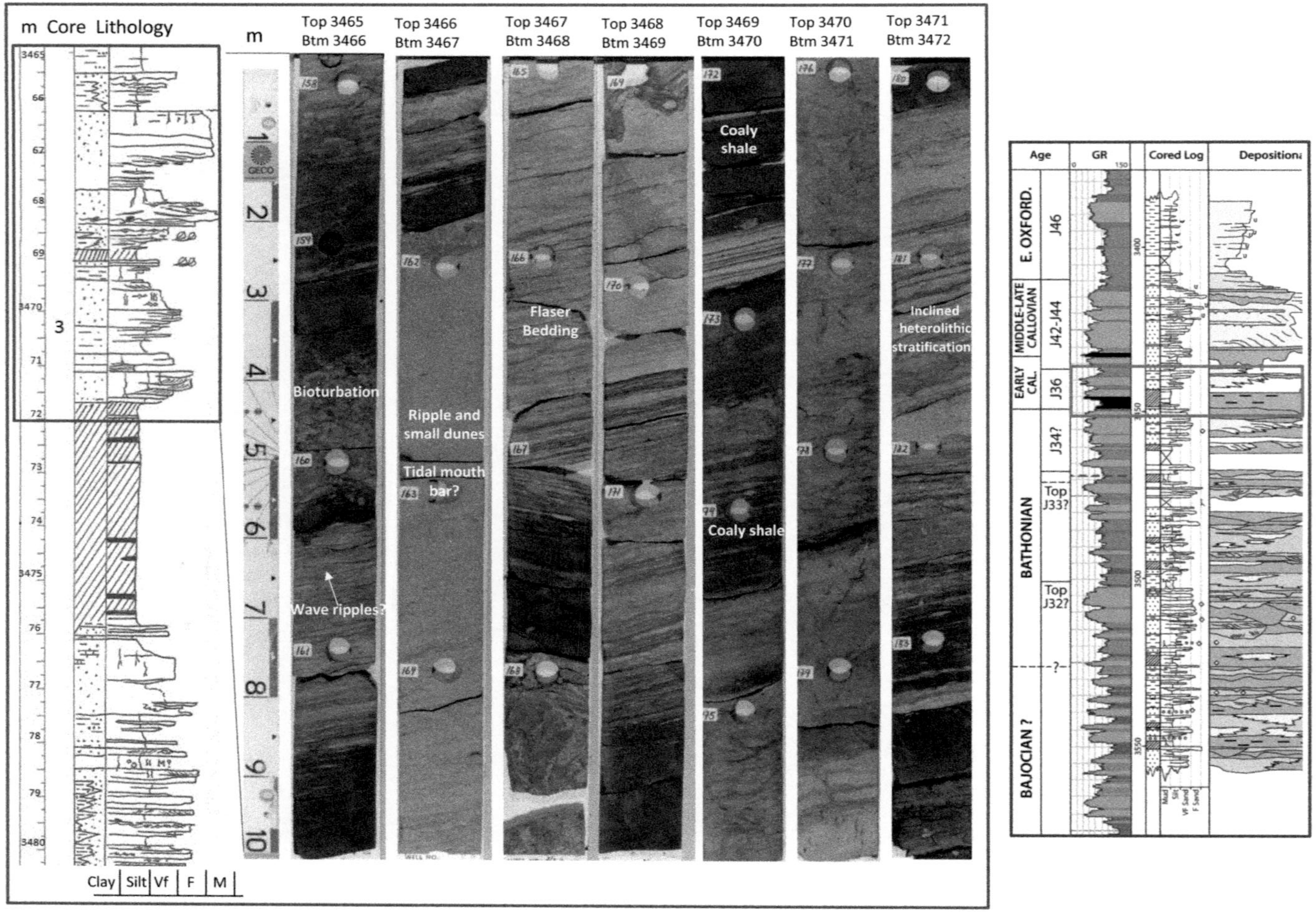

Fig. 7. Bryne delta, 3/7-4 well. Example of intertidal delta platform with the inclined heterolithic strata of facies association 1.2c. The deposits are encased in regressive coaly shales. The *Diplocraterion* traces observed in the upper part of the core at 3465.5 m represent a flooding event.

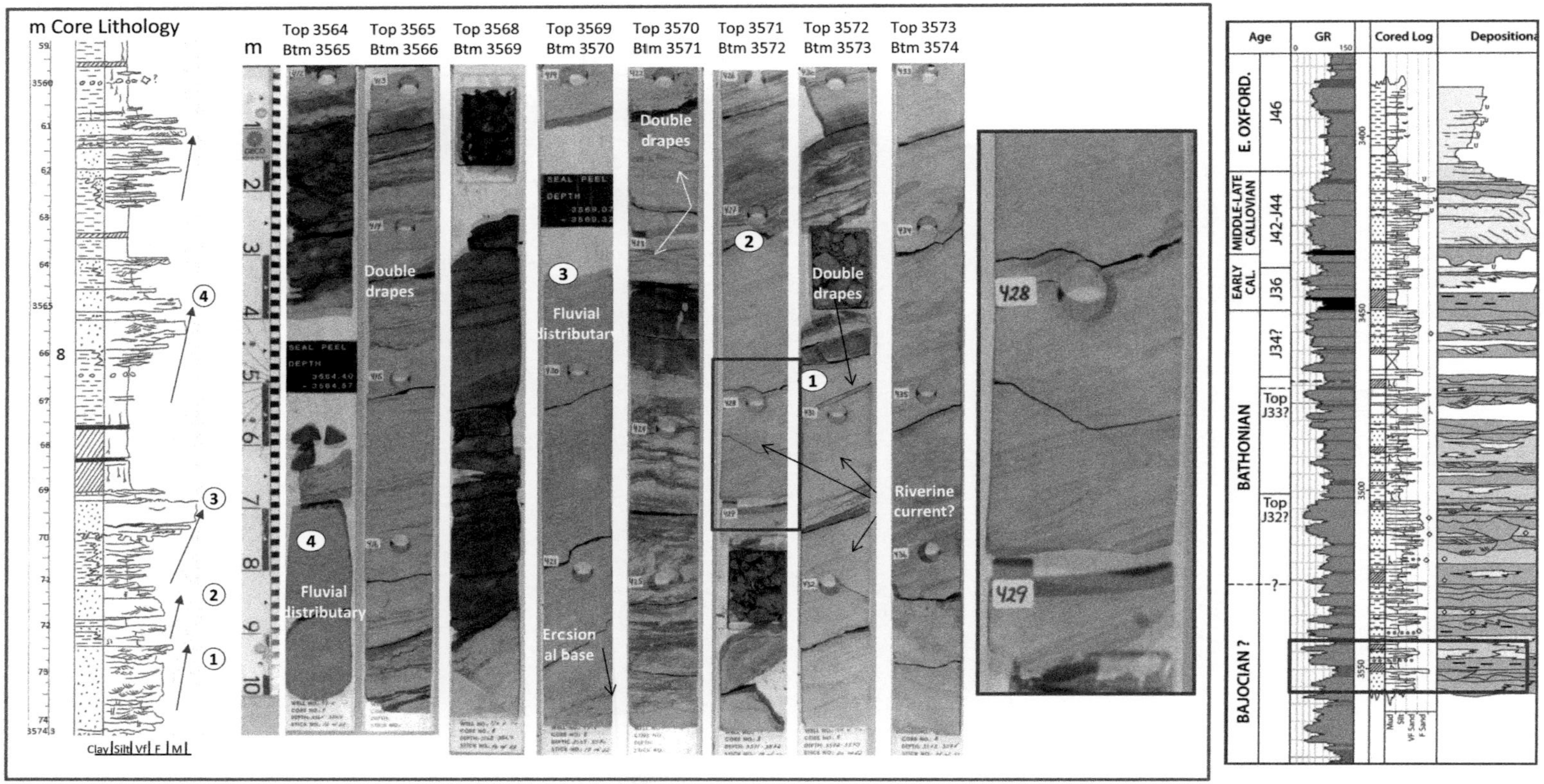

Fig. 8. Bryne delta, 3/7-4 well. Example of the subtidal delta platform deposits with stacked coarsening-upwards sandstone units of facies association 1.3, interpreted to represent small mouth bars in interdistributary bays. The deposits are low-angle to ripple-laminated, with abundant double drapes along ripple sets. Low-angle erosional surfaces may suggest storm modulation.

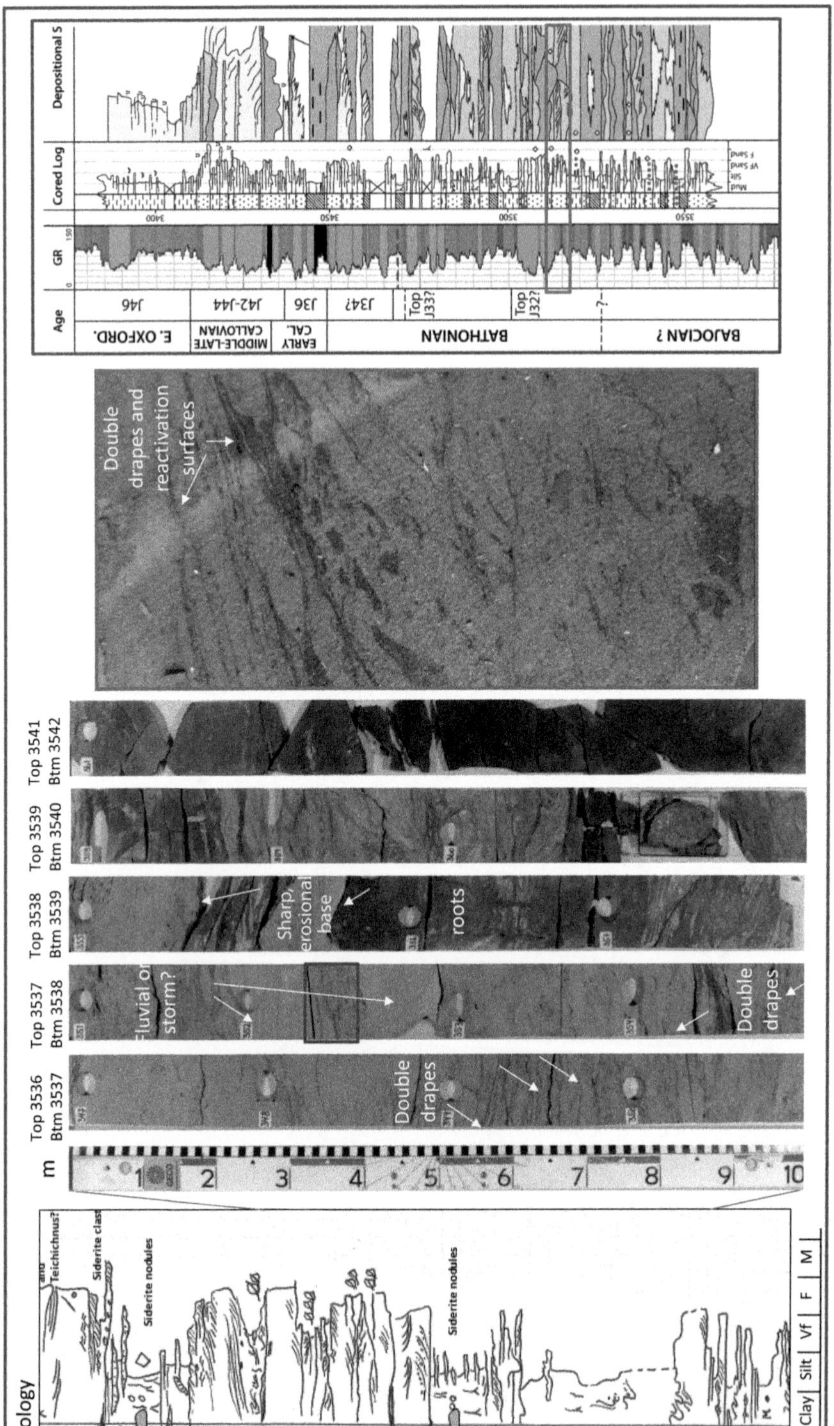

Fig. 9. Bryne delta, 3/7-4 well. Example of compound dunes in tidally influenced channels of facies association 2.1. The deposits sharply overlie regressive coal and coaly shales of FA1.2a and FA1.2d and are characterized by tabular, trough and sigmoidal cross-strata. Double mud drapes line reactivation surfaces and toesets. Bioturbation is common.

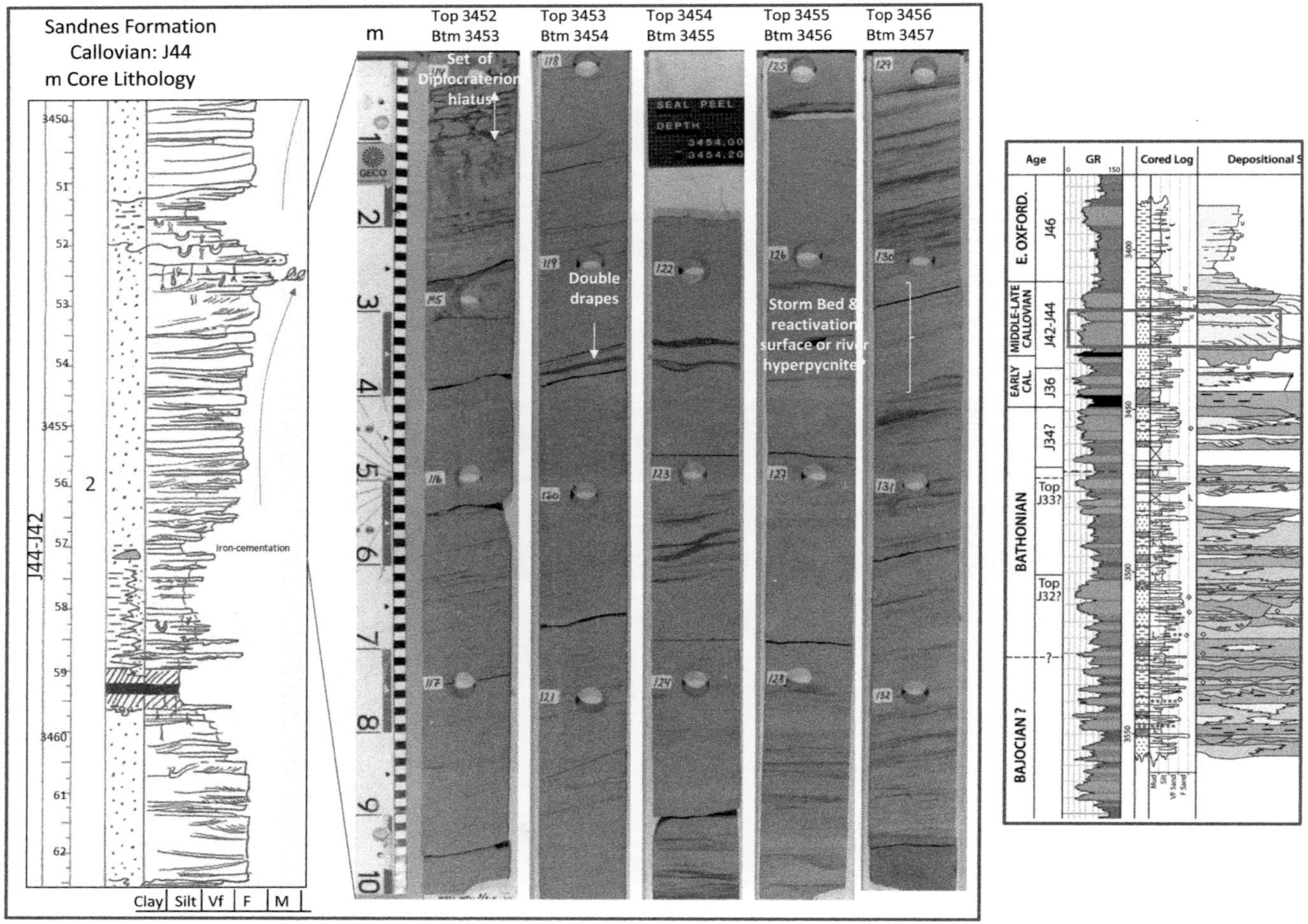

Fig. 10. Sandnes Formation, 3/7-4 well. Example of the coarsening-upwards, clean, well-sorted sandstone units representing tidally dominated, wave-influenced deltas of facies association 3. The coarsening-upwards motif is bounded at its top by *Diplocraterion* horizons, which represent a small hiatus associated with transgression.

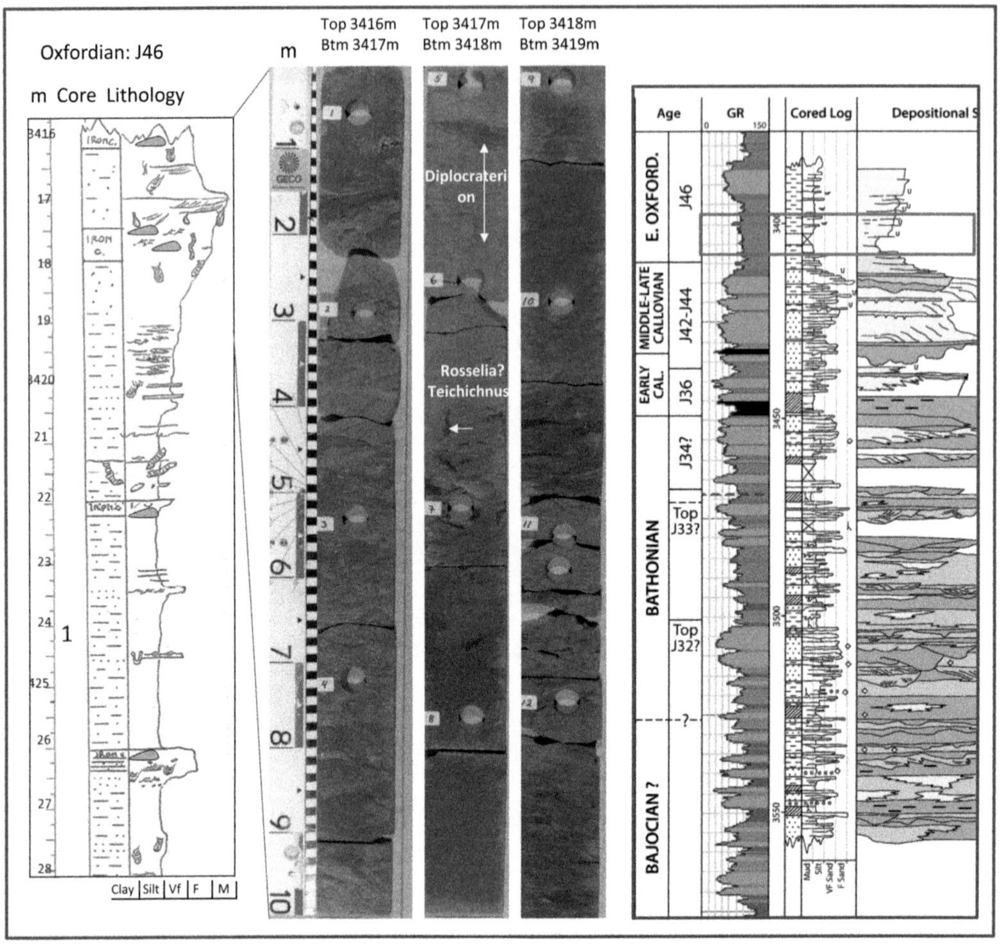

Fig. 11. Oxfordian outer shelf (distal tidal delta lobe?) deposits of facies association FA4. The deposits are heavily bioturbated with a diverse marine ichnofabric. Sedimentary structures are almost obliterated. Where present, they are in the form of cross-beds, 5–8 cm thick, probably representing small dunes.

part of the upper coastal plain (Swenson *et al.* 2005; Folkestad & Satur 2008; Kieft *et al.* 2010; Mannie *et al.* 2014; Rossi & Steel 2016), but we choose to ascribe them to the supratidal delta platform to emphasize the overall tectonically constrained, low gradient depositional system.

Facies association 1.1b: fluvial channel-fills
Description. Facies association 1.1b (FA1.1b) is not common in the cored sections and the representative examples are scattered (Fig. 6). It consists of sharply based, fining-upwards sandstone units up to 4 m thick. Some of the sandstone units are structureless with dispersed shale rip-up clasts, plant material and medium sand grains concentrated at the base. Sedimentary structures vary from planar to trough cross-stratification, climbing and ripple drift lamination. Bioturbation is absent. These sandstone units are associated with the palaeosols of FA1.1a and are often overlain by the coal and coaly shales of FA1.2.

Interpretation. The erosionally based, fining-upwards, unbioturbated sandstone units associated with palaeosols are interpreted as fluvial channel-fills in the landward reaches of the delta platform. Sedimentary structures are consistent with river processes under supercritical (plane beds) and subcritical (trough, planar and ripple lamination, Gibling 2006; Gibling *et al.* 2011; Table 1) flow regime conditions. The ripple-laminated sandstones are interpreted to mark channel abandonment. The structureless intervals are interpreted as the product of flash floods and high river discharge.

Similar deposits have been described by Alexander & Gawthorpe (1993; their facies S4) and by

Table 1. *Characteristics of the regressive Bryne stratigraphy: FA1 fluvial to tidally influenced delta depositional environment/system*

Stratigraphy	Depositional environment/system		Facies association	Description	Facies	Thickness	Biogenic/mechanical structures	Sedimentary process/facies interpretation
Regressive Bryne	FA1 Fluvial to tidally influenced delta	FA 1.1: supratidal delta platform	1.1a: palaeosols	Brecciated, massive and mottled shales and siltstones, up to 1 m thick. Mottling colour varies from dark grey to green and occasionally to red, often containing iron-cemented nodules	Mottled shales and siltstones	20 cm–1 m	Roots, syneresis cracks	Variation in water table
					Brecciated siltstones, massive and cemented	20–50 cm		Pedogenesis related to prolonged exposure
			1.1b: fluvial channel-fills	Sharply based, fining-upwards sandstone units up to 4 m thick. Sandstones are planar to cross-stratified or structureless with dispersed shale, rip-up clasts, plant material and medium sandstone grains concentrated at the base	Structureless sandstones with dispersed shale rip-up clasts	10–100 cm	Plant debris	Sandy debris flow
					Planar cross-stratified sandstones	20–30 cm		Traction, 2D dunes
					Trough cross-stratified sandstones	Up to 50 cm		Traction, 3D dunes
					Climbing ripple sandstones	Up to 40 cm		Traction and waning flow
					Ripple drift laminated sandstones	2–8 cm		Waning flow
		FA1.2: distal, intertidal delta platform	1.2a: coal and green mudstones	Splintery coal and silty coals with a large amount of plant debris and *in situ* roots and grey to green mudstones. Mudstones are grey–green, structureless or faintly planar laminated	Coal	3–20 cm	Roots	Plant accumulation in anoxic ponds in lower delta plain
					Structureless mudstones	Up to 3 m	Absent, roots	Suspension fallout of fines from channel spillover flow
					Planar laminated mudstones	Up to 1 m	Absent	River overflow in coastal plain
					Structureless to ripple-laminated sandstones	5–100 cm	Bioturbation low, with *Planolites*	Interdistributary bays and distal crevasse splay in lower delta plain
			1.2b: fluvial-dominated, tidally influenced distributary channel-fills	Erosional based, moderately to poorly sorted, fine- to very fine-grained sandstone, fining-upwards units up to 4 m thick. Trough cross-stratification, ripple and climbing ripple lamination and locally low-angle and planar lamination. Cross-bedded sets are separated by broad scours, shale rip-up clasts. Locally, the basal boundary displays contorted bedding, soft sediment deformation and chaotic texture	Trough cross-stratified sandstones	50 cm–2 m, if stacked	Bioturbation low, very impoverished association with *Planolites* and roots at top	Traction, 3D dunes in channels
					Current ripple and climbing ripple lamination, unidirectional	5 cm–1.5 m, if stacked		Sustained to waning current flow with high density low. Rapid sedimentation within channel
					Soft sediment deformation and contorted beds	Up to 40 cm		Channel bank collapse and possible dewatering structures
			1.2c: high sinuousity tidal-dominated channel-fills	Sharply to gradually based, fining-upwards sandstone/mudstone units with inclined heterolithic strata and sigmoidal bedding. Sandstone bands are massive, ripple-laminated or parallel to low-angle laminated	Inclined heterolithic stratification with sigmoidal bedding	0.5–1.5 m	*Planolites* and small *Skolithos*; plant and wood fragments	Lateral accretion in tidally influenced point bars
					Mud drapes, double mud drapes	1–3 cm		Neap–spring cycles by tidal periodicity
					Water escape and convolute lamination	5–15 cm		Soft sediment deformation following rapid deposition
			1.2d: regressive coaly shales	Fissile to laminated coaly shales, forming laterally extensive horizons that are used as markers. Overall similar to facies association 2.3	Massive to laminated	Up to 6 m	Absent	Anoxic swamp, regions of turbidity maximum
		FA1.3: subaqueous delta platform	1.3: fluvial, tidally influenced mouth bar of interdistributary bay	Coarsening- to fining-upwards sandstones, moderately to well-sorted, very fine- to lower fine-grained, micaceous with wave ripples and small-scale low-angle cross-stratification. Coal and shale drapes, single and double, mark ripple sets	Coarsening to fining-upwards sandstones	0.5–2.5 m	*Arenicolites*, *Planolites*, possible *Skolithos*; plants and coal common	Progradation and retrogradation of mouth bar
					Current ripple	3 cm up to 1 m		Unidirectional flow on bars
					Wave ripple	1–3 cm		Wave modulation
					Low-angle cross-stratification	Up to 1 m		Storm modulation on bar front

Table 2. *Characteristics of the transgressive Bryne and Sandnes stratigraphy: FA2 transgressive estuary depositional environment/system*

Stratigraphy	Depositional environment/system	Facies association	Description	Facies	Thickness	Biogenic/mechanical structures	Sedimentary process/facies interpretation
Transgressive Bryne & Sandnes	FA 2 Transgressive Estuary	FA2.1: tidal bars	Stacked fine- to medium-grained cross-bedded sandstone sets organized into fining to coarsening-upwards compound cross-strata and sigmoidal beds separated by scoured surfaces draped by thick shale laminae and climbing ripples. Thin double mud and coaly drapes are common along co-sets and set boundaries	Tabular cross-strata with double drapes	0.2–1 m	Moderate to heavily bioturbated with *Skolithos*; roots, plant and wood fragments	Bi-dimensional dunes from tidally influenced currents
				Trough cross-strata with double drapes	0.2–1 m		3D dunes, traction from tidally influenced currents
				Sigmoidal bedding	Up to 1 m		Tidal currents
		FA2.2: tidal-influenced channel-fills	Erosional based, blocky to fining-upwards sandstone units with trough cross-strata, sigmoidal, low-angle cross-stratification, ripple and climbing ripple lamination. Ripple laminae show double mud drapes and, in places, apparent opposite flow direction	Trough cross-strata with double drapes along set		*Planolites*, small *Skolithos* and U-shaped burrows; siderite mineralization	Traction, 3D dunes, primarily fluvial current-related, influenced by tidal currents
				Sigmoidal bedding	Up to 1 m		Tidal currents
				Low-angle cross-stratification on scour surfaces paved by shale and with double laminae	20–50 cm		Dunes migrating onto tidal reactivation surfaces
				Climbing ripple	Up to 1 m		Rapid deposition
				Ripple lamination with double drapes	1–8 cm		Waning flow at slack water period between tides
		FA2.3: transgressive coaly shales and lenticular beds	Fissile to laminated coaly shales containing sandstone stringers locally with starved ripples forming small coarsening- to fining-upwards units. Sandstones are very fine-grained with unidirectional and possibly bidirectional ripples. Mud drapes, single and double can be dominant	Laminated coaly shales	Up to 6 m	*Planolites* and U-shaped burrows; siderite mineralization	Sedimentation in anoxic swamps and mires
				Starved ripples and sandstone stringers in coarsening-upwards units, lenticular bedding	0.3–1.5 m		Crevassing and overbank from tidal channels, with tidal current reworking
				Current ripples, locally bidirectional			Tidal currents
				Double mud drapes	1–2 cm	Escape burrows	Deposition by slack water during tidal cycle

Table 3. *Characteristics of regressive Sandnes stratigraphy: FA3 tidal-dominated, wave-influenced delta depositional environment/system*

Stratigraphy	Depositional Environment/ system	Facies association	Description	Facies	Thickness	Biogenic/ mechanical structures	Sedimentary process/facies interpretation
Regressive Sandnes	FA 3 Tidal-dominated, wave-influenced delta	FA3: tidally dominated, wave-influenced mouth bars	Coarsening- and thickening-upwards sandstone units up to 7 m thick, of very clean, very fine- to upper fine-grained, rippled to trough and low-angle cross-stratified sandstones. Base is either sharp onto coal and coaly shales or gradational. Top is planar and marked by roots and *Diplocraterion* traces	Low-angle cross-stratification onto scour surfaces with double drapes	30 cm–1.5 m	Bioturbation low to moderate with *Diplocraterion* and roots	Dune migrating onto tidal reactivation surfaces by tidal modulated currents.
				Cuspate, wave ripple sandstones	1–3 cm		Wave modification on mouth bar
				Trough to swaley cross-strata	Up to 3 m		Bar migration, possible modulation by storm

Table 4. *Characteristics of transgressive/regressive Oxfordian units: FA4 distal delta depositional environment/system*

Stratigraphy	Depositional Environment/ system	Facies association	Description	Facies	Thickness	Biogenic/ mechanical structures	Sedimentary process/facies interpretation
Transgressive/ Regressive Oxfordian units	FA 4 Distal delta	FA4: marine, offshore siltstone and silty sandstones	Poorly sorted, silty to very fine-grained, grey sandstones, heavily bioturbated. Sedimentary structures are mostly obliterated, where preserved they are small-scale, 5–8 cm thick trough cross-strata and ripple-laminated beds	Poorly sorted, bioturbated silty sandstones	4 m	*Teichichnus, Terebellina, Rossellia, Cylindrichnus, Thalassinoides, Chondrites*	Quiet setting, offshore shelf
				Trough cross-strata	5–8 cm		Current-generated (tidal?), 3D dunes
				Current ripple-laminated beds	3–5 cm		Unidirectional currents, possibly tidal

Guion *et al.* (1995) as part of their minor channel facies in fluvial settings. These deposits may occur tens of kilometres upstream from the delta front (Dashtgard *et al.* 2012). They lack any prominent marine or brackish influence, whereas terrestrial and riverine influences dominate.

FA1.2 Distal, intertidal delta platform

Facies association 1.2a: coal and green mudstones

Description. Facies association 1.2a (FA1.2a) is widespread across the cored sections and can be up to 4 m thick. It overlies the regressive deposits of FA1 and FA3 and, locally, the tidal bars of FA2. It consists of splintery coal and silty coals with a large amount of plant debris, *in situ* roots and grey to green mudstones. The mudstones are structureless or faintly planar laminated. Locally, they contain thin structureless to ripple-laminated sandstones. The deposits are often eroded by overlying fluvial and tidally influenced channels and reworked as shale rip-up clasts into the overlying deposits (Figs 7 & 8).

Interpretation. The coal deposits originated from extensive accumulations of organic-rich plant material in anoxic swamps and mires, with limited clastic input and rapid burial that prevented decomposition, as indicated by the dark colour of the deposits and the presence of roots and pyrite. Coal and silty coals are interpreted as having been deposited in the wet, vegetated distal part of the delta platform, as suggested by *in situ* roots below the coal horizons (see also Kieft *et al.* 2010; Mannie *et al.* 2014). The green mudstones and associated heterolithic sandy units suggest deposition in a low-energy floodplain environment, sporadically reached by fluvial current overflows that produced breaching of the overbank, generating the rapid spread-out of sediments into the inter-channel areas (Collinson 1996; Miall 1996; Bristow *et al.* 1999; Bridge 2003, 2006; North & Davidson 2012).

Facies association 1.2b: fluvial-dominated, tidally influenced distributary channel-fills

Description. Facies association 1.2b (FA1.2b) consists of erosional based, moderately to poorly sorted, fine- to very fine-grained sandstones organized in fining-upwards units up to 4 m thick. Trough cross-stratification, ripple and climbing ripple lamination and locally low-angle and planar lamination are the dominant structures (Table 1). Cross-bedded sets are separated by broad scours. Compound cross-beds or inclined heterolithic beds are not present. Shale rip-up clasts pave the erosional bases of the units (i.e. at 3482 m; Fig. 6a in the upper part of the measured log). Locally, the basal boundary displays contorted bedding, soft sediment deformation and a chaotic texture defined by abundant plant debris and intraformational mudstone clasts. The upper part of the deposits shows evidence of bioturbation, with an impoverished trace fossil assemblage dominated by *Planolites* and often capped by roots. Ripples often display double mud drapes in these upper parts.

Interpretation. Fining-upwards, erosionally based sandstone units with multiple scours are interpreted as fluvial-dominated channels subjected to repeated episodes of incision and infill. Like the channels of FA1.2b, sedimentary structures are consistent with river processes (Table 1). The limited marine bioturbation and abundance of coal and wood debris indicate a strong terrestrial influence and the apparent lack of inclined heterolithic strata, which are widely observed in transgressive estuarine deposits, may suggest that these channels had a relatively low sinuosity (Bridge *et al.* 2000). The presence of double mud drapes on ripple sets and bioturbation especially at their tops, followed by rooted coaly shales, suggest that these channels were originally fluvial, but influenced and reworked by tidal processes in their upper part (Visser 1980). Based on these observations, the facies association is interpreted as distributary channels in which the marine influence gradually increases upwards, possibly as result of lower river discharge periods (Dalrymple & Choi 2007; Rossi & Steel 2016).

Facies association 1.2c: high-sinuosity tidal-dominated channel-fills

Description. Association 1.2c (FA1.2c) forms a small part of the intertidal platform and it consists of inclined heterolithic strata in very fine-grained sandstones (Fig. 7). The units are 0.5–1.5 m thick. The base is locally sharp and the units commonly overlie the coaly shale deposits of facies association 2. Beds consist of inclined and slightly sigmoidal 1–3 cm thick sandstone–mudstone couplets. Sandstone layers are massive, ripple-laminated or parallel to low-angle laminated. They can be erosionally based (centimetre-scale in core) and seem to be organized in sets 5–15 cm thick with smaller sets of ripple cross-lamination migrating up and down the bedding plane and separated by single and double mud-draped erosional surfaces. Shales are typically coaly and form thin stringers and layers. Water escape structures and convolute lamination are present. Plants, wood fragments and roots are common in the upper part of the facies units as coal layers and root horizons mark their tops. Bioturbation is present with *Planolites* and small *Skolithos*.

Interpretation. The inclined heterolithic strata are interpreted as lateral accretion deposits of tidal point bars in tidal creeks and tidally influenced channels, as suggested by the compound cross-stratification, the well-defined mud drapes along the inclined strata, the ubiquitous single and double mud drapes along set and bedset boundaries, and the brackish ichnofabric (Thomas *et al.* 1987; Choi *et al.* 2004). The fine interlamination of sandstones and mudstones is interpreted as reflecting deposition during single tidal cycles (Bridge & Leeder 1976; Visser 1980). The widespread presence of mud drapes may reflect deposition within the maximum turbidity zone, with coarse and fine sediments deposited by abrupt changes in shear velocity as a result of salinity mixing of fluvial freshwater and marine waters (Nichols & Biggs 1985; Dalrymple *et al.* 1990, 1992; Allen 1991). Rooted and coal horizons mark channel abandonment.

Facies association 1.2d: regressive coaly shales
Description. Facies association 1.2d (FA1.2d) is one of the most prominent facies observed within the core succession, especially within the Bryne interval, where it represents the majority of the finest-grained deposits. The facies forms intervals up to 6 m thick with laterally continuous horizons, which are excellent correlation markers. FA1.2d consists of fissile to laminated coaly shales that overlie FA1.1 and often enclose the deposits of FA1.2c (Figs 7 & 8).

Interpretation. The coaly shale deposits of the Bryne–Sandnes pair and analogous systems have been variously interpreted as regressive, formed in the intertidal delta platform above and in between distributary channels and minor delta lobes (Kieft *et al.* 2010; Fitzgerald *et al.* 2012; Hughes 2012; Mannie *et al.* 2014), or as transgressive (Petersen & Andsbjerg 1996; Folkestad & Satur 2008), or both (Andsbjerg 2003). We tend to agree with Andsbjerg (2003) because the coaly shales are present in both highstand and transgressive systems tracts, as explained later. The coaly shales were probably deposited in the turbidity maximum zone at the freshwater to brackish water transition, probably reflecting mud flocculation as a result of the maximum turbidity (La Croix & Dashtgard 2014). Recent work in the Fraser River in British Colombia showed that the mudstone intervals reach their maximum thickness at the freshwater to brackish-water transition, decreasing in both seawards and landwards directions (La Croix & Dashtgard 2014).

FA1.3: Subaqueous delta platform
Facies association 1.3: fluvial, tidally influenced mouth bar of interdistributary bay
Description. Facies association 1.3 (FA1.3) consists of coarsening- to fining-upwards 0.5–2.5 m thick sandstones and sandy siltstones that are generally seen to prograde onto the coaly shale deposits of the distal delta platform (Fig. 8). They can be capped by small channels of FA1.2b. Sandstones are moderately to well-sorted, very fine- to lower fine-grained and micaceous. This facies displays current and possibly wave ripples and small-scale low-angle cross-stratification and even small hummocks, seen in the thickest intervals in the lower part of the succession (e.g. at 3574 m, Fig. 8). Single and double coal and shale drapes mark ripple sets; double mud drapes generally occur in the upper part of facies units. Plant and coal debris are common. Bioturbation is weak, with *Arenicolites*, *Planolites*, possible *Skolithos* and escape structures.

Interpretation. These deposits are interpreted to represent mouth bars prograding onto a shallow intertidal to subtidal delta platform, into which channels of FA1.2b incise. The deposits show evidence of interacting processes, from riverine (the large amount of coal and plant debris), tidal (double mud drapes) and subordinate wave (wavy lenticular beds) and storm, the latter especially observed in the thickest coarsening-upwards units. Storm modulation was probably responsible for the presence of low-angle stratification, similar to small hummocky cross-strata. Gentle scours and associated thin laminations are attributed to lobe switching at the mouth of fluvial distributaries. These deposits resemble those observed in the lower part of the Middle Jurassic Lajas Formation of the Neuquén Basin by Rossi & Steel (2016), interpreted as crevasse splays in interdistributary bays. These deposits are the most distal observed within the Bryne delta platform.

FA2: transgressive, estuary

This association includes tidal bars and channels, coal and coaly shales, and sinuous tidal-dominated channels (Table 2).

Facies association 2.1: tidal bars
Description. Units of tabular and trough cross-strata are present within and outside channels, which are marked by erosional basal surfaces. These cross-stratified units are up to 4 m thick and erosionally or gradationally overlie the prograding interdistributary bay units of FA1.3 or the deposits of FA1.2c (Fig. 9). They consist of stacked 0.2–1 m thick, fine- to medium-grained cross-bedded sandstone sets organized into fining- to coarsening-upwards compound cross-strata separated by scoured surfaces, which are often draped by thick shale laminae and climbing ripples (Fig. 9). Thin double mud and coaly drapes are common along co-sets and set boundaries. Locally, cross-bedded sets are sigmoidal, with angles varying between 5° (along topsets and toesets) to 15–30° (along

foresets). Cyclical repetition of lower to higher inclined laminae sets is also present. Low-angle scours overlain by low-angle draping laminae are locally present. In the sandbody shown in Figure 9, an increase in thickness of the cross-bedded sets is observed, with the thickest (1 m) at the top. In this unit, the laminae at the base of the cross-beds merge into bioturbated and deformed fine-grained sandstones. The 1 m thick cross-laminae set is reworked at its top by current ripples and possible wave ripples. The deposits contain abundant plant and woody fragments and are moderately to heavily bioturbated with *Skolithos* traces and roots, especially in their upper part. These sandbodies are overlain by bioturbated sandstones and eventually by palaeosols (Fig. 9).

Interpretation. Facies association 2.1 (FA2.1) is characterized by deposits with a compound architecture typical of tidal bedforms (Allen & Homewood 1984; Dalrymple 1984). The change in angularity of successive cross-bedded sets mimics sigmoidal cross-stratification; the low-angle reactivation surfaces are probably related to reversals of flow direction in a subtidal environment (Visser 1980; Boersma & Terwindt 1981; Allen & Homewood 1984; Dalrymple 1984; Shanley *et al.* 1992). The local presence of gentle scours draped by low-angle cross-stratification can be related to swaley or small hummocky cross-strata, possibly representing storm events. FA2.1 is interpreted to represent subtidal compound bars within or outside major tidal channels, developed during transgression by tidal reworking of the mouth bars in the subaqueous delta platform. The sharply based compound bedforms that overlie FA1.2c in Figure 9 are interpreted as having been formed within a fluvial channel during the bank-full stage and combined tidal flood in the upstream reaches of the estuary embayment.

Facies association 2.2: tidal-influenced channel-fills

Description. Facies association 2.2 (FA2.2) consists of erosionally based, blocky to fining-upwards sandstone units identified in the Callovian upper part of the cored succession (Fig. 10, base of the measured log) and locally in the lower part of the succession, where they overlie the mouth bar complexes of the subaqueous delta platform (FA1.3, Fig. 7). Trough cross-strata, sigmoidal, low-angle cross-stratification, and ripple and climbing ripple lamination are the dominant sedimentary structures. Ripple laminae show double mud drapes and, in places, opposite apparent flow directions. Erosional surfaces, locally lined by shale drapes and climbing ripples, separate the cross-bedded sets. Bioturbation is present, with small *Planolites* traces.

Interpretation. The erosional base and the fining-upwards trend suggest that the sandbodies represent fluvial channels scoured into the delta platform or the mouth bar systems (Ichaso & Dalrymple 2014). Tidal influences are suggested by the abundance of cross-strata of different kinds, the presence of double mud drapes, apparent reverse currents, and erosional surfaces interpreted as reactivation surfaces lined by shales and double mud drapes. The mud drapes indicate deposition during slack water periods and are interpreted as tidal bundles, accumulated and modified in response to neap–spring–neap tide fluctuations (Boersma & Terwindt 1981; Dalrymple 1984; Allen 1991).

Facies association 2.3: transgressive coaly shale and lenticular beds

Description. Facies association 2.3 (FA2.3) is similar to FA1.2d, forming laterally continuous horizons. In transgressive strata, the coaly shale deposits tend to contain sandstone stringers, locally with starved ripples forming small coarsening- to fining-upwards units 0.3–1.5 m thick (Fig. 6b, upper part & Fig. 7). Sandstones are very fine-grained and display very limited bioturbation by *Planolites* and U-shaped traces, which are locally concentrated at the top of the fining-upwards intervals. Ripples are unidirectional, but locally display opposite apparent directions in adjacent sets. Mud drapes, single and double, can be dominant. Small-scale deformation from escape burrows and rooting are common. The deposits are locally associated with sinuous tidal channels of FA1.2c and the inter-distributary mouth bars of FA1.3. Siderite mineralization occurs locally.

Interpretation. The transgressive coaly shales occur above coals and rooted horizons and underlie prograding mouth bar deposits and the tidally dominated and wave-influenced deltas of facies association 3. They are interpreted as having been deposited within the estuarine system, in the central part of the estuarine basin (*sensu* Dalrymple 1992) or in transgressive lagoons (Andsbjerg 2003).

The heterolithic sandy units indicate fluctuations in the energy regime within a very restricted tidally influenced environment, as suggested by the presence of tidal indicators (double drapes and apparent bidirectional ripple beds) and very restricted ichnofabric. Thin coarsening-upwards sandstones overlying lenticular bedding testify to crevassing and overbank deposition from adjacent subtidal channels and sand re-distribution by tides. Similar deposits have been ascribed to tidally influenced inner estuary settings (Kuecher *et al.* 1990) or to estuary-flanking channels (Folkestad & Satur 2008).

Facies association 3: tidally dominated, wave-influenced mouth bars

Description. Facies association 3 (FA 3) forms coarsening- and thickening-upwards sandstone units up to 7 m thick, best represented in the Callovian part of the succession (3450–3460 m, Fig. 10). They overlie the deposits of FA2.3 and, locally, the coals of FA1.2a. They consist of very clean, very fine- to upper fine-grained, rippled to trough and low-angle cross-stratified sandstones. Their base is either sharp onto coal and coaly shales or gradational. Their top is planar and marked by roots and *Diplocraterion* traces. Current ripple sets, 3–10 cm thick, fill gentle scours and broad troughs, lined by shale laminae. Double shale drapes are common. Cuspate, wavy ripples are locally present. Ripples are mostly unidirectional, but locally opposite apparent directions are observed (e.g. at 3456.2 m, Fig. 10). Gentler troughs and swale-like cross-strata overlie the ripple sets. Reactivation surfaces are common. Bioturbation is low to moderate, present especially at the base and in the upper part of the facies units.

Interpretation. FA3 is interpreted as a tide-dominated delta system that progressively filled the underlying estuary embayment. The tidal signature is evidenced by the presence of double mud drapes, reactivation surfaces, compound bedforms and bidirectional currents. The tidal action was, however, modulated by wave dynamics, as suggested by the presence of cuspate wave ripples. Similar deposits have been interpreted by Andsbjerg (2003) as ebb tidal delta systems. Rooted horizons on top of some of these coarsening-upwards units indicate that the mouth bars built upwards to sea-level with temporary colonization by vegetation (Fig. 10). Eventually, the mouth bar complexes would have joined laterally to form broad shorelines, as recognized in Denmark (Andsbjerg 2003) and in the coeval Hugin Formation in the Viking Graben and in the Egersund Basin (Folkestad & Satur 2008; Kieft *et al.* 2010; Mannie *et al.* 2014, 2016).

Facies association 4: marine, offshore siltstones and silty sandstones

Description. This association displays a broad aggradational to coarsening-upwards trend. It consists of poorly sorted, silty to very fine-grained grey sandstones developed in the uppermost part of the cored succession within the back-stepping sequence set that characterizes the Oxfordian strata (Fig. 11). The deposits are heavily bioturbated with a diverse ichnofabric characterized by *Teichichnus*, *Terebellina*, *Rosselia*, *Cylindrichnus*, *Thalassinoides* and *Chondrites*. Sedimentary structures are mostly obliterated; however, where preserved, they are represented by small-scale, 5–8 cm thick trough cross-strata and ripple-laminated beds. Shales are common and mark some of the cross-bedded sets and infill burrows.

Interpretation. The laminated to bioturbated coarsening-upwards siltstones and silty sandstones are interpreted to represent either (1) the distal reaches of a tidal(?) delta mouth bar system or (2) current-generated bedforms (tidal?) in a marine shelf. The deposits were emplaced by weak traction currents in a relatively quiet setting of the offshore shelf and subsequently completely colonized by organisms. This interpretation is based on the lack of any storm event beds and by the presence, where preserved, of cross-bedded sets and shale laminae. The highly diverse ichnofabric suggests marine conditions. Coarsening-upwards systems attached to shorelines have been described by Andsbjerg (2003) in the same stratigraphic section in the Danish Basin. However, we do not have this evidence in the Norwegian side of the Søgne Basin. The deposits recall the distal cross-bedded units described by Mellere & Steel (1995) in the Haystack Mountain Formation of Wyoming, interpreted as the distal reaches of tide-dominated delta lobes deposited during the early stages of transgression. Stratigraphically, highly bioturbated siltstones can be both transgressive and regressive.

Depositional models

The Bryne Formation is dominated by the prograding delta system of FA1 that culminates, in the study area, with the outbuilding of small mouth bar systems in interdistributary bays by distributary channels (Fig. 12). The observed depositional processes vary from purely riverine to tidally influenced. Tidal influence is almost absent in the uppermost reaches of the delta platform, but it becomes more prominent within the distributary channels and the heterolithic fine-grained sediments of the interdistributary bays. Storm and wave modulation is expressed by small hummocky cross-strata and the local presence of cuspate wave ripples. However, the paucity of these sedimentary features indicates that storm and wave action was subordinate with respect to riverine and tidal action. Based on the described facies associations, the Bryne delta is an excellent example of a shale-rich, mixed-energy, tide-influenced delta system, with facies models similar to those described by Willis (2005) and Rossi & Steel (2016).

The delta prograded onto a very low gradient, tectonically confined delta platform (*sensu* Roberts & Sydow 2003; Swenson *et al.* 2005) that extended from its subaerial part, through a vast intertidal section, eventually (and locally) reaching the subtidal environment with subtidal bars and channels. The thin character of the mouth bar deposits in the interdistributary bays suggests that delta clinoforms were negligible, probably due to the continuous equilibrium between accommodation and sediment supply. The platform was protected from powerful alongshore circulation and recorded little wave influence. The envisaged Bryne depositional environment is similar to the modern day coastal plain of the Mahakam, Mekong and Brahmaputra deltas, dominated by fine-grained deposition, where a large part of the sediment yield is stored in the lower delta plain and progradation occurs on a very shallow tidally influenced platform before reaching the delta front (Goodbred & Kuehl 1999; Bachtiar *et al.* 2010; Xue *et al.* 2010; Goodbred & Saito 2012; Amir Hassan *et al.* 2013). In the Mahakam delta, the platform is 2–10 m thick (Bachtiar *et al.* 2010), similar to the thicknesses of the mouth bar successions observed in the Bryne system. In the modern delta examples, the majority of sediments bypass the tidal platform to be deposited further offshore in the delta front, where accommodation increases. On the platform, active sand transportation and deposition occur on the slopes of point bars and side bars. In the Bryne delta system, heavily influenced by rift and salt tectonics, sediments were instead dispersed into the broad intertidal platform, with possibly little sediment reaching the delta front.

The deposits show similarities to the deltaic coarsening-upwards units of the Hugin Formation (Folkestad & Satur 2008; Kieft *et al.* 2010; Mannie *et al.* 2014). However, it appears that in the Bryne–Sandnes pair the tidal influence is much stronger and not just limited to the upper part of the succession. Progradation was punctuated by transgressions, during which the delta system was transformed into a broad, shallow estuary. During transgression, the distal reaches of the tidal estuarine system were dominated by tidal bars, which dissected the estuarine basin. Landwards, tidal action would have reworked the fluvial channels.

A complete tide-dominated estuary system is observed at the time of deposition of the Sandnes Formation. Estuarine channels and bars are overlain by coaly shales and eventually the estuarine valley is filled by the tidally dominated and wave-influenced mouth bars of FA3 and the coal deposits of FA1.1 (Fig. 12b). It is difficult to establish whether the estuarine system was tide- or wave-dominated (Dalrymple 1992; Dalrymple *et al.* 1992). Probably both systems were present, with more wave dominance in the Danish sector where a system of lagoons, beach barriers and tidal deltas has been observed (Andsbjerg 2003) and tidal dominance in the large, most restricted embayment in the north of the basin.

Correlations and sequence stratigraphy

The chronostratigraphic and sequence stratigraphic framework of the Bryne Formation is shown in Figure 5 and in the correlation diagrams of Figures 13 and 14. There are only few markers across the succession. In the northernmost part of the basin, the wells present a poorly defined J32 top Bajocian marker and a good J33 Bathonian marker. The J34 Lower Callovian zone is represented by a coaly shale horizon, considered as the principal datum for the correlation panels in the Danish sector (Andsbjerg 2003).

In this work, the datum for the correlation is the Base Oxfordian flooding (top J44) and the blocky sandstone units at the top of the succession (wells 2/2-4, 3/5-2, 3/6-1, 3/7-2 and 3/7-9S) are considered to be part of the latest Callovian sequence.

Based on biostratigraphy, log correlation and the observed alternation of facies associations, the succession has been subdivided into seven low-order sequences (*sensu* Van Wagoner *et al.* 1988, 1990) or megasequences, formed by higher-order sequences and cycles (Figs 13 & 14).

Stratigraphic definitions

The low-order sequences are 20–80 m thick and can be traced across the Søgne Basin. Following the definitions of Van Wagoner *et al.* (1990), each sequence consists of: (1) lowstand deposits, just above a major erosion surface (sequence boundary); (2) transgressive deposits with landwards-stepping tide-dominated and tidally influenced estuarine systems; and (3) highstand deposits with aggradational to seawards-stepping, tidal deltas and mouth bar systems. The highstand deposits are best documented in the Callovian uppermost part of the succession. An idealized regressive–transgressive cycle is shown in Figure 12. Major surfaces and systems tracts are discussed in the following sections.

Sequence boundaries. The basal surface of each sequence erodes into the tidally influenced mouth bars of FA1.3 and FA3 and the meandering channel deposits of FA1.2c. The local erosional relief is 2–3 m (at the core scale), but the erosion can be up to 5–15 m (e.g. the base of sequence 5 in the Amalie and West Lulu 2 wells, Figs 13 & 14). Laterally, the surfaces merge into the rooted horizons and palaeosols of FA1.1 (e.g. sequence 5 at wells 2/2–1, 2/2–4 and 3/7-2, Figs 13 & 14). These surfaces are interpreted as sequence boundaries. They can be traced across the basin for more than 80 km; they are associated with erosion, or extensive rooting and pedogenesis in the interfluves, and they are overlain by deposits that attest to a seawards shift of sediment supply.

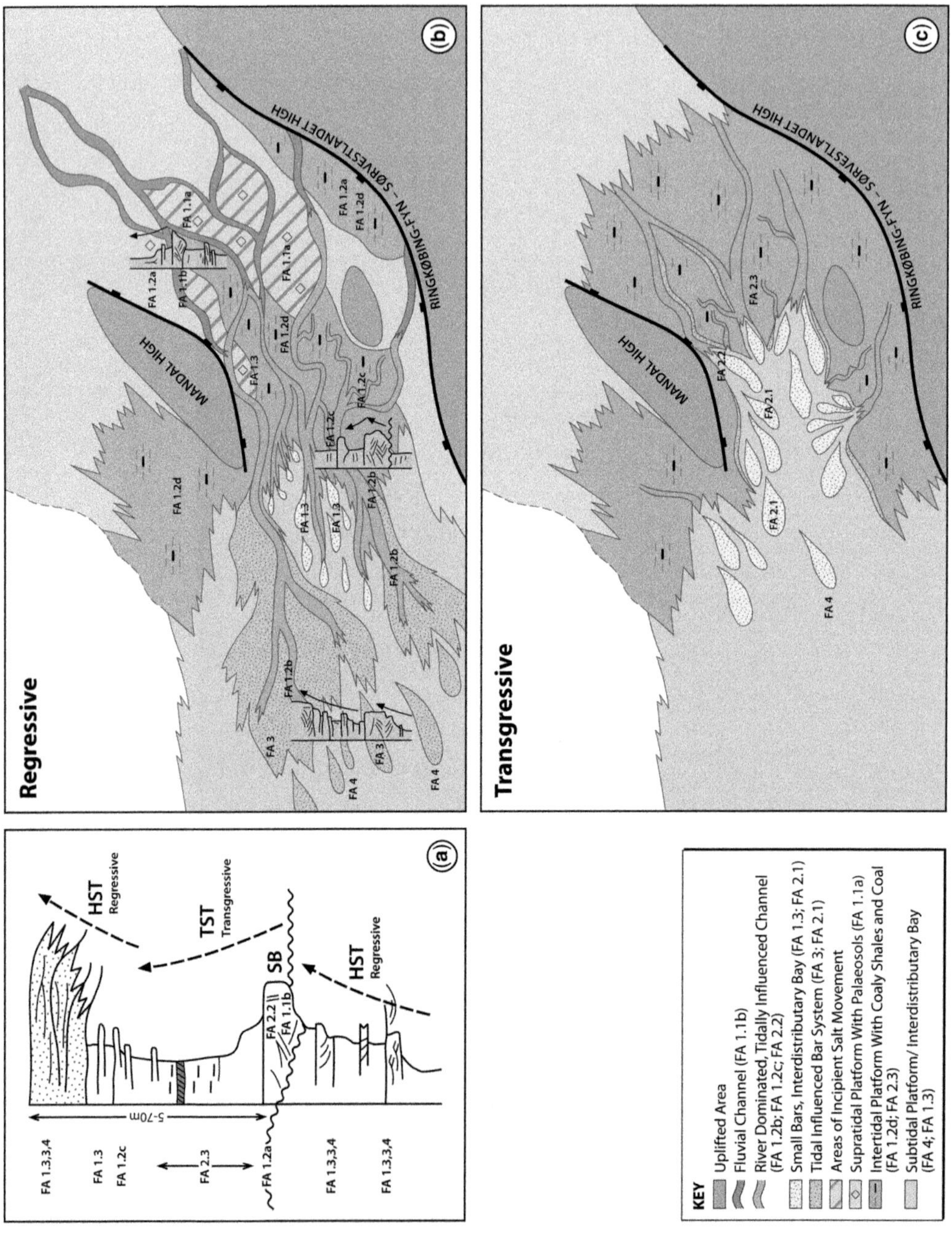
(a)
HST
Regressive
TST
Transgressive
SB
HST
Regressive
FA 1.3,3,4
FA 1.3
FA 1.2c
FA 2.3
FA 1.2a
FA 1.3,3,4
FA 1.3,3,4
5-70m
FA 2.2
FA 1.1b
Regressive
(b)
MANDAL HIGH
RINGKØBING-FYN – SØRVESTLANDET HIGH
Transgressive
(c)
MANDAL HIGH
RINGKØBING-FYN – SØRVESTLANDET HIGH
FA 2.1
FA 2.2
FA 2.3
FA 4
KEY
Uplifted Area
Fluvial Channel (FA 1.1b)
River Dominated, Tidally Influenced Channel (FA 1.2b; FA 1.2c; FA 2.2)
Small Bars, Interdistributary Bay (FA 1.3; FA 2.1)
Tidal Influenced Bar System (FA 3; FA 2.1)
Areas of Incipient Salt Movement
Supratidal Platform With Palaeosols (FA 1.1a)
Intertidal Platform With Coaly Shales and Coal (FA 1.2d; FA 2.3)
Subtidal Platform/ Interdistributary Bay (FA 4; FA 1.3)

Lowstand deposits. Most of the sediments immediately overlying the sequence boundaries consist of fluvial channel belts (FA1.2b), which are generally reworked by tidal processes in their upper part. They are interpreted as lowstand deposits. The local upwards transition from fluvial to tidal influence suggests that some of these channels were deposited during the initial relative sea-level rise, with the basal channel-fill representing the early lowstand and the upper channel deposits the late lowstand. Associated rooted horizons and pedogenically reworked intervals indicate that the areas between the active channel belts underwent prolonged subaerial exposure.

Transgressive deposits. Transgressive deposits are associated with facies association 2. A tidal ravinement surface is detected locally, at places coinciding with the sequence boundary (e.g. at 3538.5 m, Fig. 9). The overlying landwards shift of tidal influence is marked by coaly shales (FA2.3). The transgressive deposits volumetrically dominate the depositional sequences. In contrast with the lowstand deposits, the transgressive deposits are mud-prone.

Maximum flooding surfaces. Maximum flooding surfaces are defined at the turnaround from landwards- to seawards-stepping in the seawards reaches of the tidal platform. Landwards, they have been traced within the coaly shales or, where these are absent, on top of the coal layers (e.g. sequence 6).

Highstand deposits. Highstand deposits are formed of the prograding mouth bars of FA1.3 and FA4. Landwards, the highstand deposits are difficult to discriminate because they are similar to those of the underlying transgressive tract. The distinction is based on evidence of a seawards shift in facies, often marked by an increase in sediment supply onto stagnant coaly shale intervals, such as thin coarsening-upwards heterolithics or meandering tidal creeks (FA1.2c).

Megasequences and sediment partitioning

The seven identified megasequences form a clastic wedge that thins northwards (Fig. 15). For simplicity, and as a result of a lack of biostratigraphic markers, we define megasequences 1 and 2 as the oldest sedimentary succession below the J32 sequence boundary, a time framework that, according to Andsbjerg (2003), embraces the Aalenian (megasequence 1) and Bajocian (megasequence 2). No core is available in this interval.

Megasequence 1 is present in the Amalie-1 well and possibly, based on log correlation, in the basal part of the 3/7-4, 3/7-5 and 3/5-2 wells (Figs 13 & 14). It is about 40 m thick. It consists of a 15 m thick blocky to fining-upwards sandstone package at the base, interpreted to represent stacked fluvial deposits. It onlaps Triassic deposits and is probably not present in the northernmost part of the basin.

Megasequence 2 is more widespread and well developed across the western part of the basin (Fig. 13). It is up to 70 m thick in the Amalie-1 well. It onlaps the shale-prone Triassic Smith Bank Formation towards the north. It is characterized by a 30 m thick blocky to fining-upwards sandstone package at the base, interpreted to represent stacked fluvial deposits. The sequence is potentially present in the northernmost reaches of the Søgne Basin in a 10 m thick unit between the J32 sequence boundary and top-Triassic deposits (Figs 13 & 14). In the 3/7-4 and Amalie wells, megasequence 2 is developed as a 35 m thick back-stepping transgressive tract.

Megasequence 3 is Bajocian to Bathonian in age. It is up to 70 m thick in the 3/7-5 and Amalie-1 wells, thinning to the north to about 15 m at well 2/2-1. It is marked at the base by a thick sandstone unit in the Amalie-1 well, interpreted to represent stacked fluvial channels. This basal unit can be traced laterally into the West Lulu 4 and 2 wells and into the Trym area (3/7-4 well), suggesting thick and widespread fluvial channel belts across the southern part of the basin (Figs 13 & 14). To the north, the basal sandstone package is only a few metres thick. The core section in well 3/7-4 shows a thick development of tidally influenced bars and coaly shales interpreted to represent the transgressive part of the sequence (Fig. 5). Core to well log correlation suggests that bar forms were also probably extensively present in the 3/5-2 and, possibly, Cleo-1 wells (Fig. 14).

Megasequence 4 has a Bathonian age (its top corresponds to the top-J33 marker) and it is well

Fig. 12. (**a**) Simplified facies arrangements in sequences and megasequences and (**b, c**) depositional models for the river and tidally influenced mixed-energy Bryne–Sandnes delta system during (b) regression and (c) transgression. The delta occupied a shallow, low gradient platform, tectonically constrained by the Mandal and the Sørvestlandet highs and open to the south (Fig. 1b). During regression (e.g. highstand systems tract (HST) in (a)), a mixed fluvial-dominated and tidally influenced delta system existed, spread across the supratidal to subtidal platform. No delta front was observed within the Bryne cores. Salt growth and related subsidence had a major influence on sediment dispersal, with palaeosols and overbanks developed on salt highs and fluvial channels in intra-salt lows. In the Sandnes section, wave influence is present at the delta front. During transgression (e.g. transgressive systems tract (TST) in (a)), the delta is transformed into a tide-dominated, wave-influenced estuary.

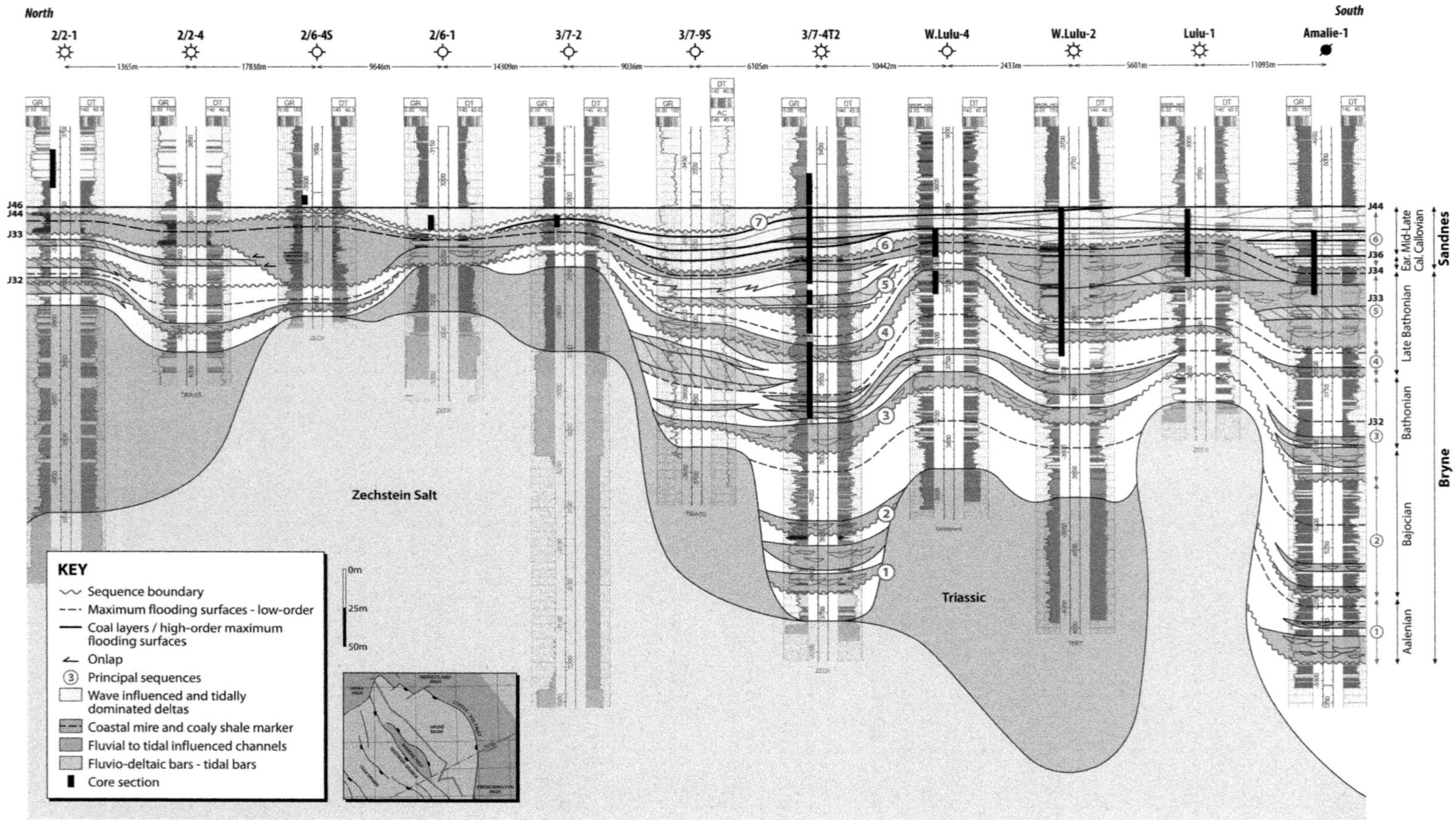

Fig. 13. Well correlation along the westernmost part of the Søgne Basin, where most of the wells were drilled on top of salt highs. The principal sequence boundaries can be traced with the help of the biostratigraphic markers, but other aspects of the correlation are uncertain. Correlation across the Danish wells has been made according to Andsbjerg (2003). Numbers refer to megasequences.

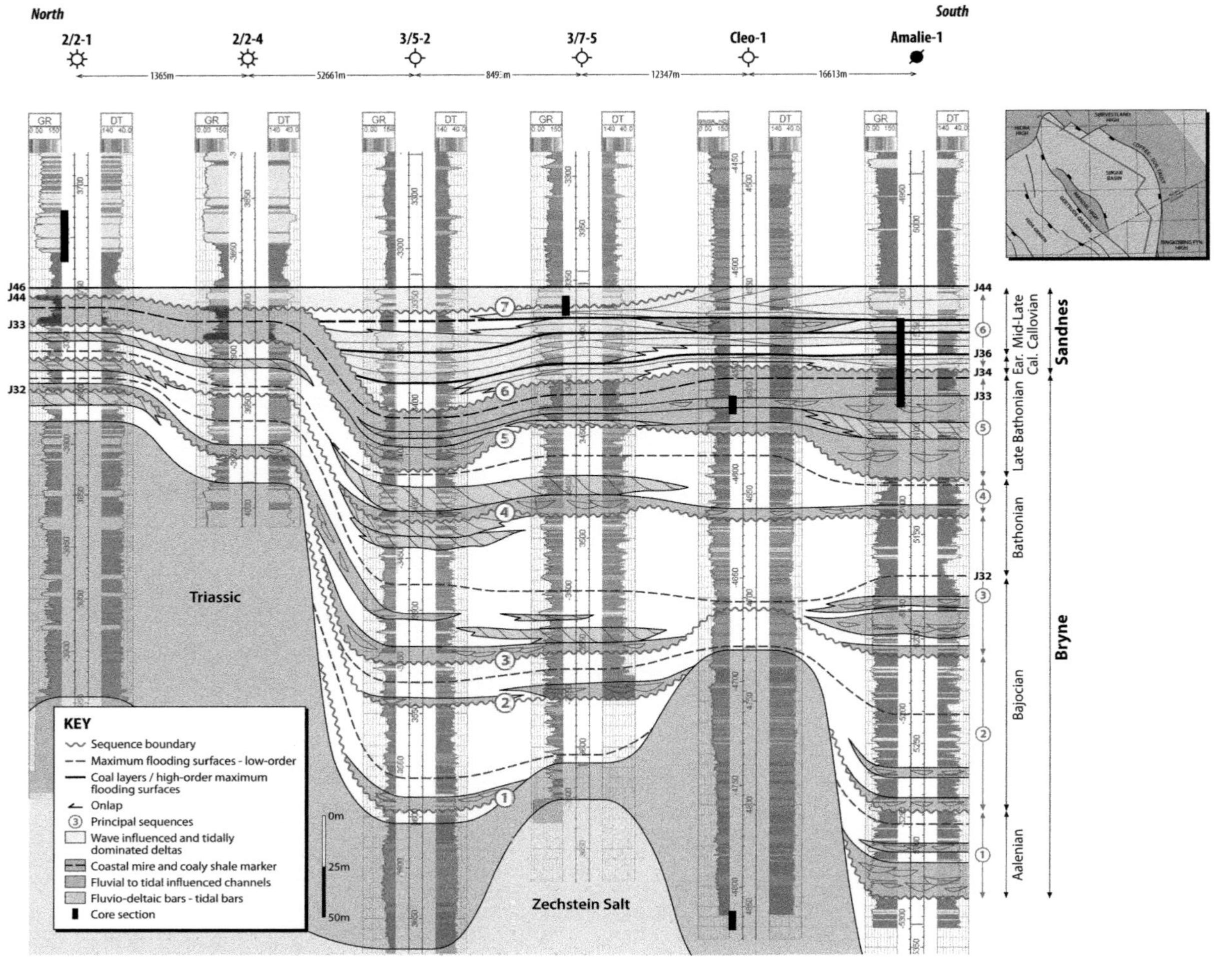

Fig. 14. Well correlation of the Bryne succession in the eastern part of the Søgne Basin away from salt diapirs. Sequence boundaries and flooding surfaces are easier to trace between wells than in Figure 13. Correlation between Cleo-1 and Amalie-1 wells is model-driven.

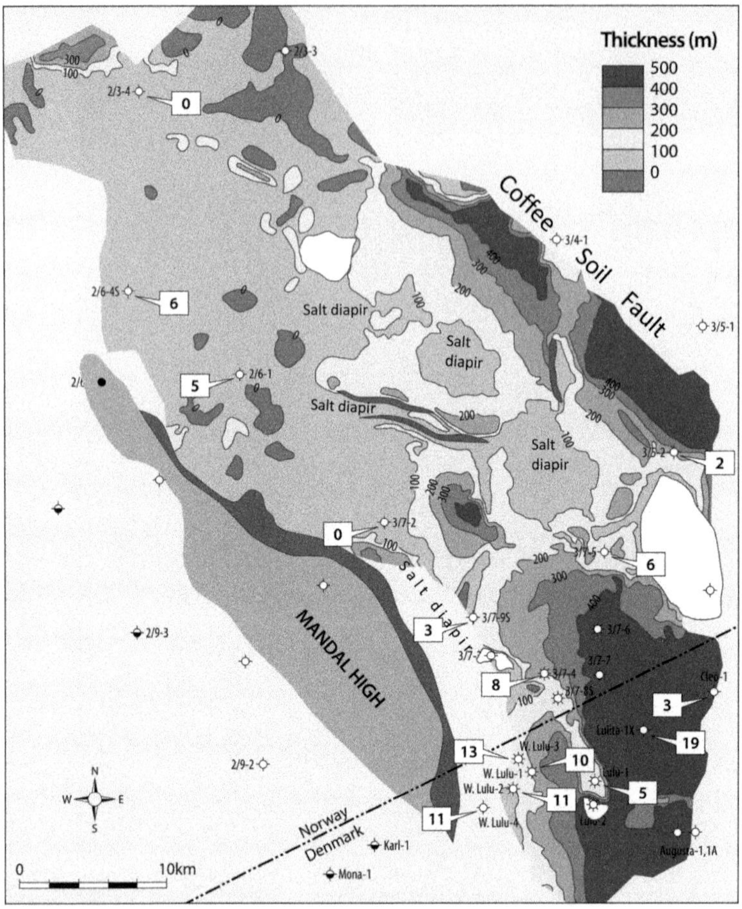

Fig. 15. Seismically derived isopach map of the Bryne Formation and thickness of coal layers penetrated by wells. The map shows the thinning of the succession to the north and the main depocentres to the south and east in the hanging wall of the Coffee Soil Fault. These depocentres favoured coaly shale deposition and probably the formation of a backwater prism.

illustrated in the core section at Trym, where fluvial deposits overlie tidally influenced bars. Compared with megasequence 3, and based on the Trym cores, fluvial and lower coastal plain deposits are widely developed, suggesting a strong migration of the fluvial facies belt across the underlying tidal platform of megasequence 3. Megasequence 4 is about 30 m thick to the south and 15 m thick to the north. Locally, it is completely eroded by the basal sequence boundary of megasequence 5 (2/6-4S, 2/6-1 and 3/7-2 wells; Fig. 13).

Megasequence 5 is Late Bathonian to early Callovian as it contains the J34 marker. Its thickness is highly variable, from 10 m (northernmost wells) to 50 m (Amalie-1 well). The basal erosional surface represents a sequence boundary beneath a widespread incised valley filled by fluvial and estuarine deposits (wells 3/7-4, West Lulu 1, 2 and Amalie wells, Figs 13 & 14). Andsbjerg (2003), in the core sections of the Lulu 1 and Amalie-1 wells, interpreted these as tidal inlets. The valley-fill is draped by a 3–20 m thick coaly shale marker, continuous across the basin. In the correlation panel of Figure 14, the marker is almost constantly 10 m thick; to the west, the thickness is more variable due to erosion at the base-megasequence 6 sequence boundary. The coaly shale interval is interpreted to represent an increment of tectonically induced accommodation that led to widespread flooding of the Bryne delta platform. The top of the coaly shales is assessed as top-J34.

Megasequence 6 encompasses the J36 to J44 Callovian sequences and it is marked at its base by a combination of sequence boundary and transgressive surface with little incision into the underlying coaly shale marker (in the upper part of megasequence 5). We postulate that the basal sequence boundary of megasequence 6 is a tilted surface onlapped by the Callovian shallow-marine prism. Progradation of the Callovian tidal delta system was not continuous, but was punctuated by high-frequency cycles of delta progradation and back-stepping estuarine units. At least three high-frequency cycles can be recognized, separated by coal and coaly shale markers.

Megasequence 7 has a J46 age and it is present in the northernmost part of the basin (wells 3/7-9S to 2/2-4). It represents the final stage of the Callovian bay-fill, with tidal and wave-influenced delta sandstones onlapping the northern tip of the basin. The basal boundary is again a combination of sequence boundary and transgressive surface that might be traced further to the south to a horizon of *Glossifungites* ichnofabric at the top of the core section in well 3/7-4.

Discussion

Palaeogeographical implications

Aalenian–Lower Bathonian: megasequences 1–3. No delta front was encountered in the Aalenian–Bathonian. We do not have cores in the Norwegian part of the Søgne Basin for the deepest Aalenian section. The 3/7-4 core shows paralic conditions in its lowermost part, dated at early Late Bajocian/Early Bathonian. The thick blocky to fining-upwards channels at the base of megasequence 3 in Amalie, Trym, West Lulu may represent a fluvial to estuarine valley-fill, attesting to the transgression of the basin (Fig. 16a).

Bathonian: megasequences 3 and 4. There is a progressive decrease in the thickness of brackish deposits and transgressive coaly shales from megasequence 3 to megasequence 4, accompanied by a progressive increase in the thickness of aggradational mudstones of FA1.2a (Fig. 5). Seismic attribute analysis of the deepest Bryne horizons did not reveal much as a result of the the large burial depth of the succession. Better imaging has been obtained in Late Bathonian to Early Callovian strata in the deposits of megasequence 5.

Late Bathonian to Early Callovian: megasequences 5 and 6. Renewal of sediment input and strong tidal reworking is recorded at this time. There is an overall switch from a mixed-energy fluvial and tidally influenced delta to clear estuarine and marine conditions. Megasequences 5 and 6 can be linked together, with megasequence 5 representing a fluvial to estuarine valley-fill and megasequence 6 the latest stage of infill of the large Søgne embayment. All the diagnostic elements of estuarine valley-fills are present (Dalrymple *et al.* 1990, 1992; Dalrymple 1992): (1) the erosional surface that defines the valley shape (base of megasequence 5); (2) the initial fluvial infill; (3) a widespread central estuary coaly shale and mire horizon; (4) the transgressive tidal ravinement (base of megasequence 6); and (5) the prograding tidal and wave-influenced delta systems. Mouth bar deposits are thicker, 6–8 m thick, suggesting that the supply of river sediment was high in relation to the ability of tidal currents to redistribute sediments. A marine ichnofabric that contains the *Glossifungites* ichnofacies at hiatal ravinement surfaces indicates the seawards movement of the depositional system and the onset of more open-marine conditions.

The seismic attributes in Figure 17 show the network of Bryne channels during approximately late J33 times and probably represent a snapshot of the fluvial valley architecture at lowstand. Channels are 1–3 km wide and have moderate to low sinuosity. They narrow northwards and seem to become wider to the south. Figures 18 and 19 show attribute snapshots for the Callovian J36 and J38 transgressions. Most of the central part of the basin is open marine, but embayments with possible tidal bars can be envisaged along the margins of the Mandal High, the Coffee Soil Fault and to the north (see also Fig. 16b). Despite the continued transgression, coal and coaly shales were largely preserved below the associated transgressive surfaces of marine erosion (i.e. megasequence 6, Figs 13 & 14).

Late Callovian: megasequence 7. The rapid transition from paralic sediments to offshore mudstones and siltstones indicates the onset of a very rapid transgression at Late Callovian times (Fig. 16c). The sediment supply was progressively reduced as the basin entered into the Oxfordian and was no longer sufficient to keep pace with the increased rate of subsidence. Shelfal mudstones took over. The shoreline observed by Andsbjerg (2003) in the Danish sector probably bordered the basin.

Tectonic controls on sedimentation

Sedimentation and salt tectonics. The initiation of halokinesis in the Central North Sea occurred during the Early Triassic in response to differential loading of the salt by prograding clastic fluvial wedges from the north and by thin-skinned extension that caused reactive diapirism (Stewart & Clark 1999). In the Søgne Basin, salt tectonics was also acting almost continuously during the Triassic and across the Middle and Late Jurassic, as shown by multiple unconformities and the thinning and thickening of the succession at and away from salt diapirs (Figs 4b & 4c). Halokinesis is a function of multiple variables, including the rate of vertical salt rise, the

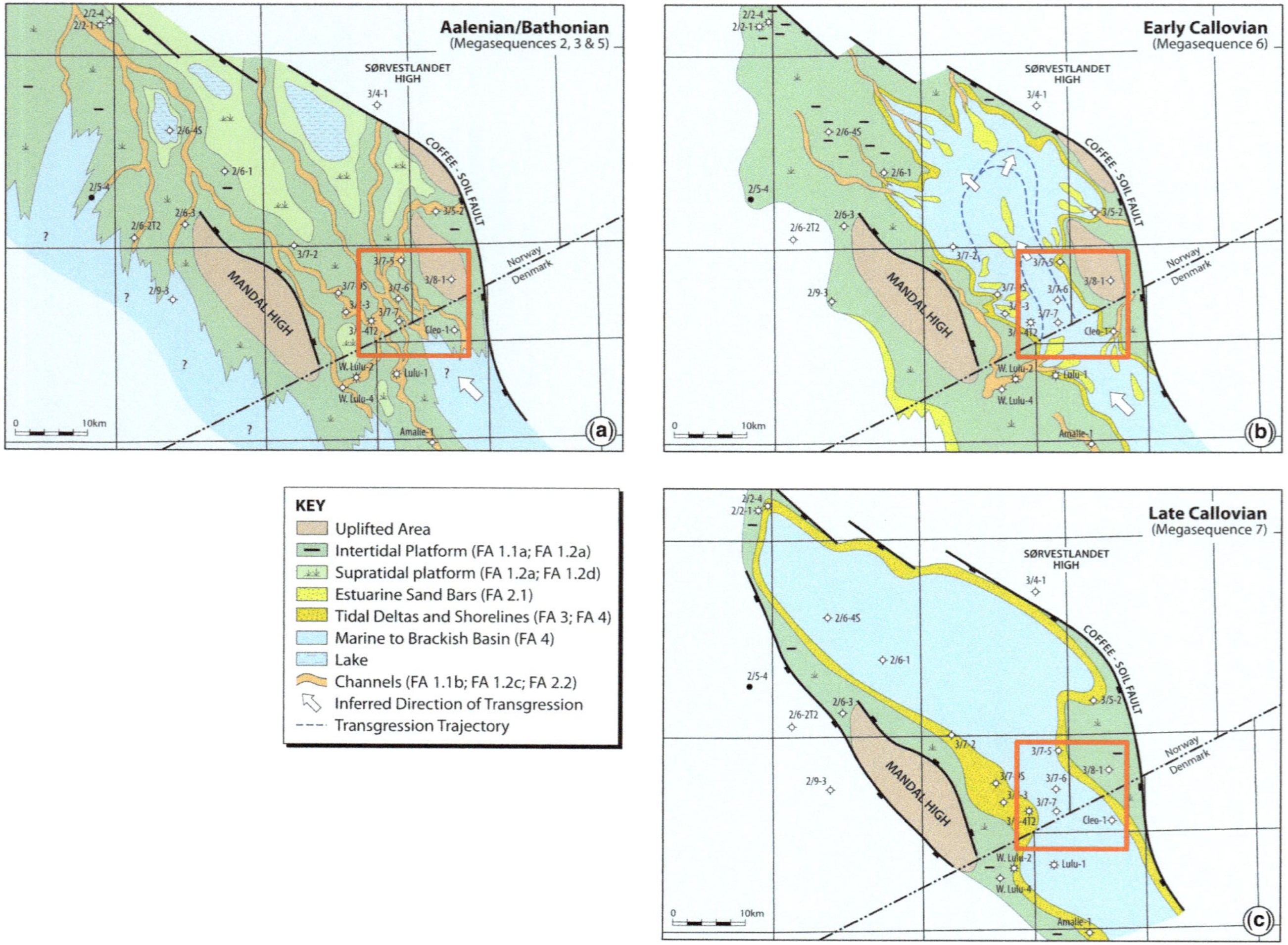

Fig. 16. Reconstructed transgressive palaeogeographies across the Søgne Basin during: (**a**) the Aalenian/Bathonian (megasequences 2, 3 & 5); (**b**) the Early Callovian (Megasequence 6); and (**c**) the Late Callovian (Megasequence 7). Red inset shows the region of seismic attributes shown in Figures 17, 18 and 19.

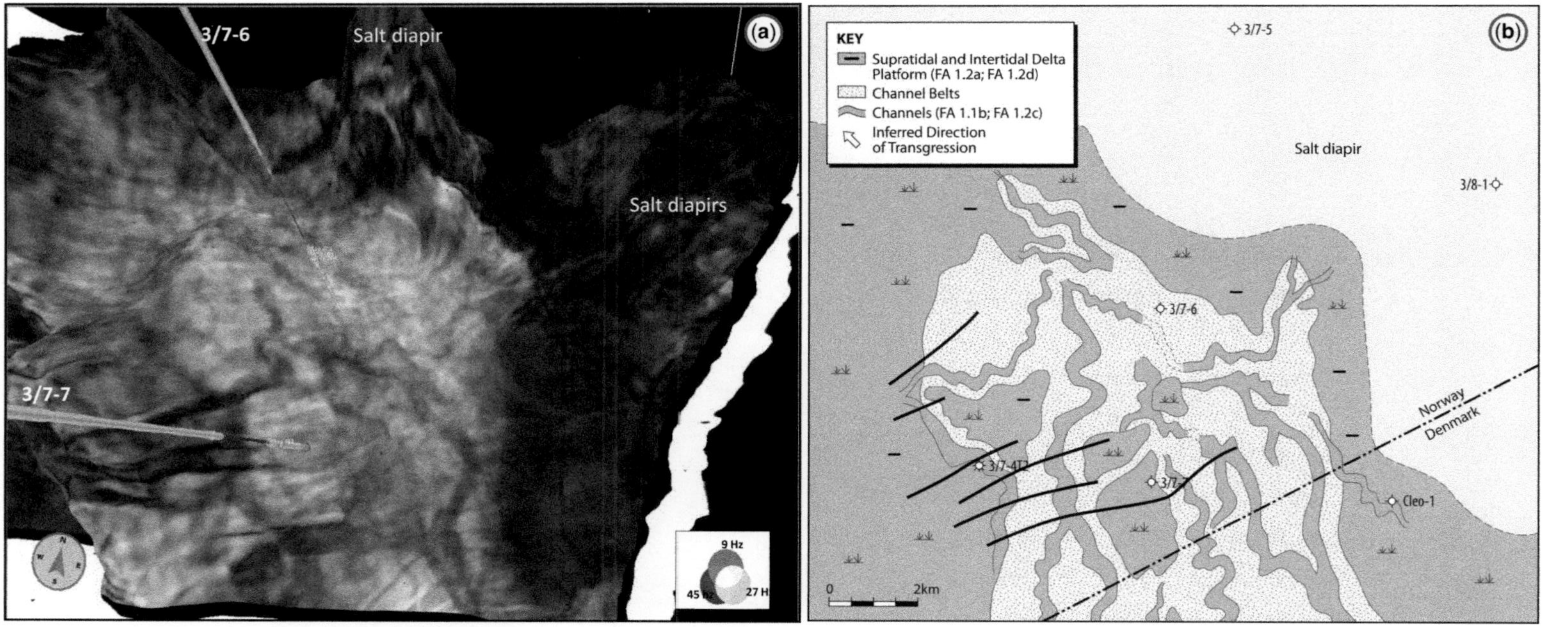

Fig. 17. (**a**) Frequency decomposition at J34 sequence boundary (megasequence 5) and (**b**) palaeogeographical reconstruction in the southern part of the Søgne Basin with wells 3/7-7 and 3/7-6 for reference (see Fig. 1 for location). The picture represents a snapshot of the fluvial channel network at lowstand within the intra-salt diapir minibasin in the central part of the Søgne Basin.

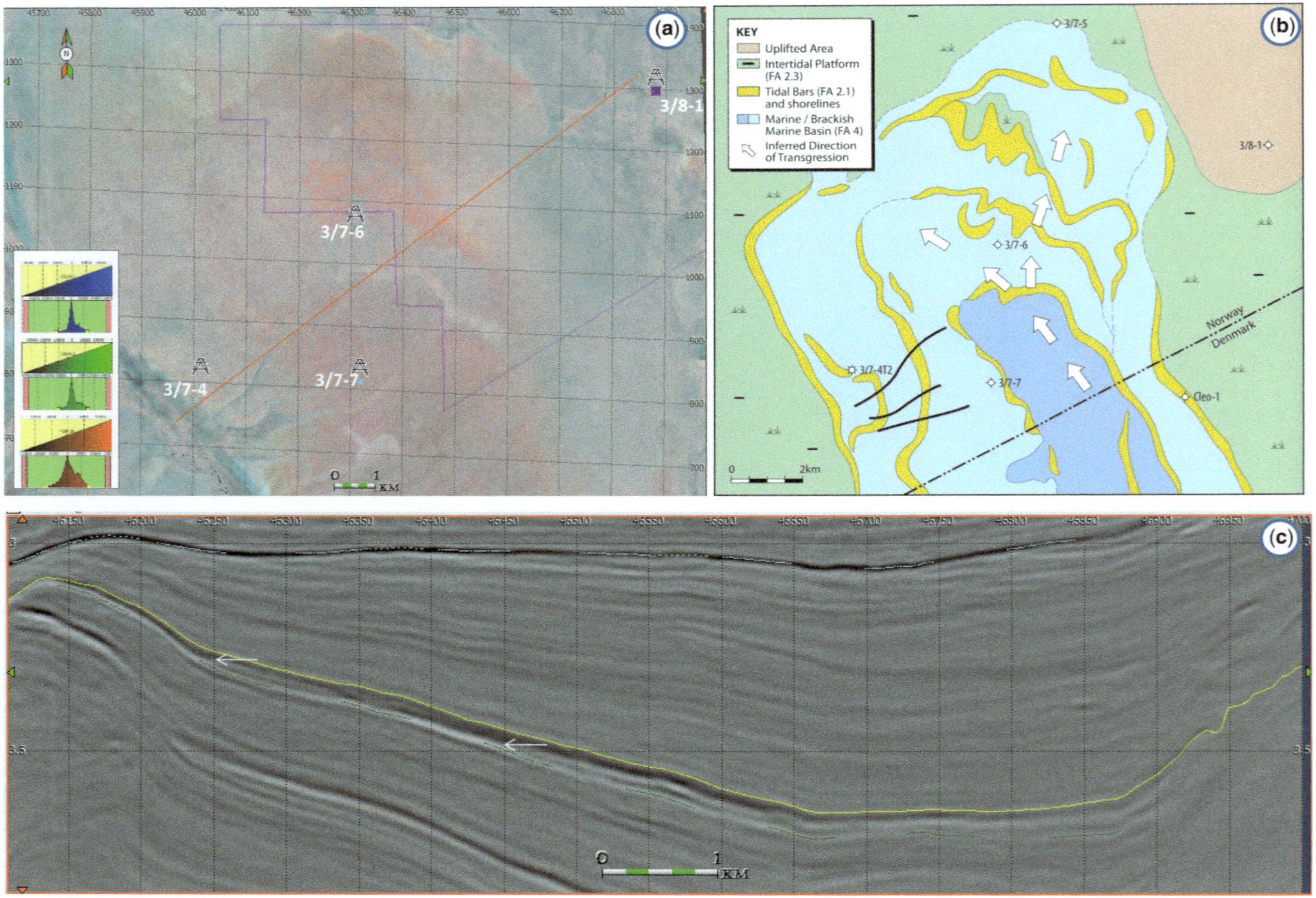

Fig. 18. (**a**) Horizon attribute map at approximately the J36 transgression (Megasequence 6) with position of the seismic line shown in (c). Histograms show colour frequency used. (**b**) Palaeogeographical reconstruction, in the southern part of the Søgne Basin. (**c**) Reference seismic line. Attributes were made on the yellow horizon. Arrows highlight onlap relationships of the basal part of Megasequence 6 on to the underlying J34 coaly shale marker.

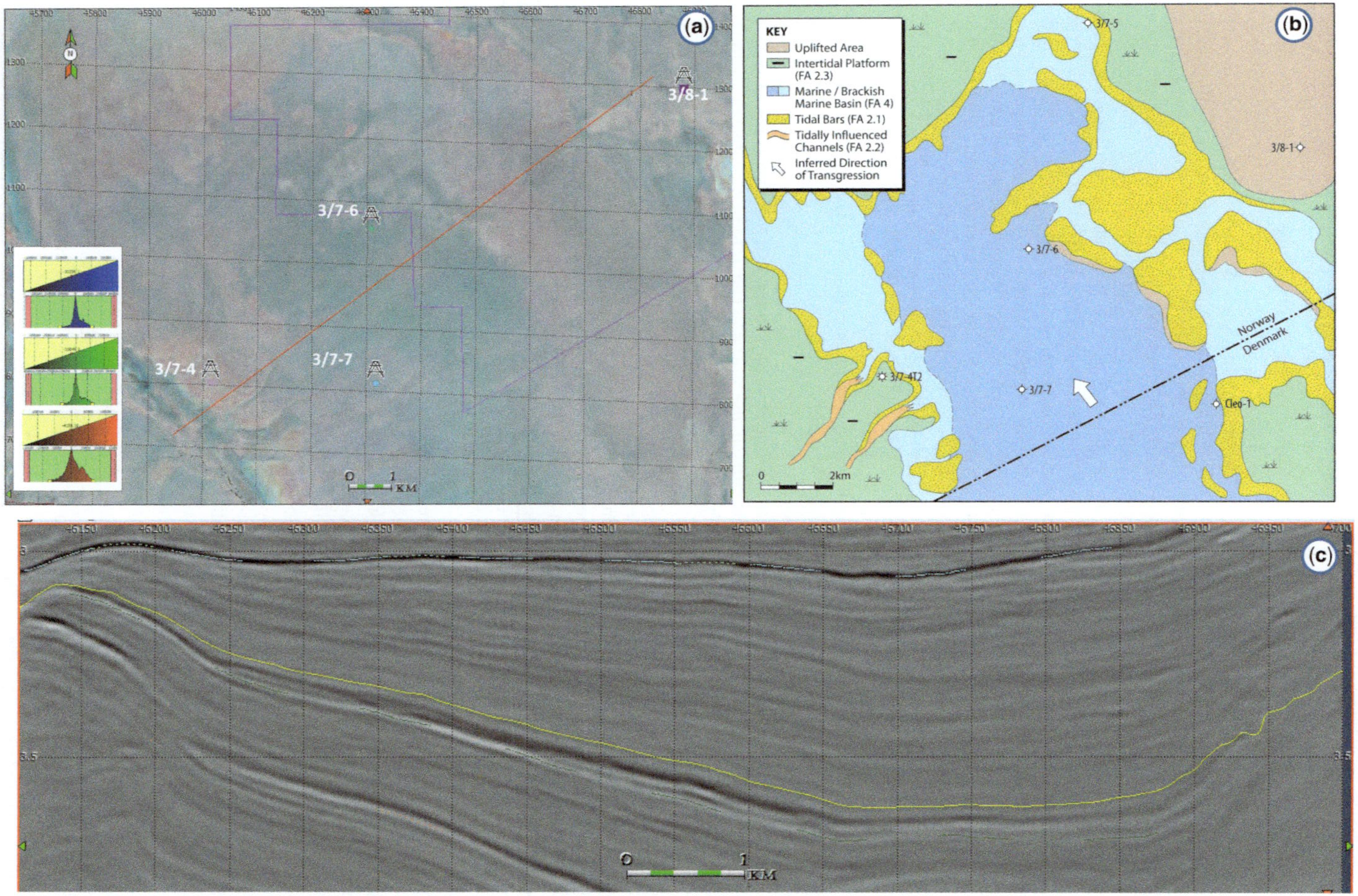

Fig. 19. (**a**) Horizon attribute map and (**b**) palaeogeographical reconstruction at the J38 (?) Callovian transgression in the southern part of the Søgne Basin. (**c**) Reference seismic line (location in (a)). Attributes were made on the yellow horizon.

rate of sediment accumulation, diapiric dissolution and subsidence (Weislogel & Lawton 2000). In the case of the Bryne Formation, the rate of net subsidence is the sum of subsidence due to salt withdrawal plus subsidence resulting from extensional movements along the Coffee Soil Fault.

At a basin scale, the succession thins to the west, east and north, as evidenced in the seismic lines of Figure 4 and in the isopach map in Figure 15. The thickest section is present to the south, at the Amalie-1 well, and to the east on the footwall of the Coffee Soil Fault. The Søgne Basin margins are flanked by a thick salt wall (to the west) and a ring of salt diapirs to the east. During Aalenian to Bajocian times, accommodation space was probably created by salt withdrawal at the top-Triassic level (Figs 13 & 14 at the Amalie-1, 3/7-4 and 3/5-2 wells). This space was infilled by the lowstand deposits of megasequences 1 and 2. By Bathonian time, salt movement in the southern part of the basin was responsible for gentle onlap of the sedimentary succession onto salt diapirs and thick channelized fluvial deposits on their flanks. The impact of these processes on sediment dispersal was that the majority of the sediment load was sequestered on the delta platform, in the space between major salt highs and dispersed through a network of channels and shallow mouth bars in the intra-salt interdistributary bays, as shown in Figure 17. On the salt highs, the lack of accommodation led to the deposition of thin channels, overbanks and palaeosols. The Bathonian fluvial channels identified in the 3/7-9S, 3/7-4 and West Lulu wells flank the salt wall adjacent to the Mandal High. Their drainage became preferentially aligned parallel to the rising and moving salt, indicating that sufficient topography was generated to capture drainage pathways.

The high accommodation space recorded in the hanging wall of the Coffee Soil Fault and to the south (Fig. 16) was most probably created by fault movements, coupled with subsurface salt movement. To the north, the Bryne succession developed a series of progressive angular unconformities where the tops of megasequences 2, 3, 4 and 5 converged (Fig. 14). We interpret the northwards thinning of the succession as the consequence of salt growth and basin tilting to the south.

Upper Jurassic rift inception. The 80–100 km long valley-fill deposits of the Late Bathonian megasequence 5, the widespread distribution of coaly shales within the valley-fill and the overlying southwards offlap of the Sandnes tidal delta systems are, in combination, interpreted as the sedimentary response to the inception of the Late Jurassic rift.

The southwards tilting of the basin was possibly controlled by the footwall rotation of the Coffee Soil Fault. A tectonic pulse is recorded in the Danish Central Graben during Callovian time (Mogensen *et al.* 1992; Møller & Rasmussen 2003; Andsbjerg 2013), characterized by subsidence along north–south-trending faults and with the most pronounced fault-controlled subsidence to the east around the Ringkøbing–Fyn High and Sørvestlandet High. According to Møller & Rasmussen (2003), the activity along the north–south-trending faults ceased during the Oxfordian and was renewed during Early Volgian and Ryazanian times. Based on our dating, we observed the initiation of extensional tectonics earlier, during the deposition of megasequence 5 in the Late Bathonian. The timing of the rift event is in accordance to that described in the Viking Graben (Steel 1993; Thomas & Coward 1996) and in the Sleipner area, where thickness variations of the Bryne-equivalent Sleipner Formation reflect the initiation of extensive rifting (Richards 1991; Sneider *et al.* 1995; Folkestad & Satur 2008). The rifting was accompanied by relative sea-level rise and the deposition of net-transgressive successions (Hampson *et al.* 2004; Mannie *et al.* 2014).

Direction of transgression: from the south or from the north?. The interplay between tectonics and sedimentation exerted a fundamental control on the palaeogeomorphology and sediment transport routes. We consider that the drainage systems changed through time.

During the Aalenian and Bajocian (megasequences 1–3), drainage was oriented southwards (Fig. 20a). This was a time period that followed the ‘Mid-Cimmerian’ tectonic phase that led to the uplift of the Central North Sea dome (Rattey & Hayward 1993; Underhill & Partington 1993). According to Herngreen *et al.* (2003), the deformation front shifted gradually southwards to the northern Netherlands, where open-marine sedimentation ceased during the Bajocian (Van Adrichem Boogaert & Kouwe 1993). The widespread brackish conditions observed in megasequence 3 suggest that the restricted shallow-marine incursions were proceeding from the southernmost Tethyan Sea, while the central part of the Central Graben recorded thermal subsidence and deposition of the Brent Group. The closure of the southern marine corridor is recorded by the widespread supratidal deposits of megasequence 4 around Bathonian J32 time.

The renewal of brackish to open-marine conditions during megasequences 5–6 indicates the re-opening of a new marine corridor, following the initiation of the Late Jurassic rifting. Transgression was rapid and from the north, along the Viking Graben and Central Graben (Fig. 20b). The marine incursion bordered both flanks of the Utsira High

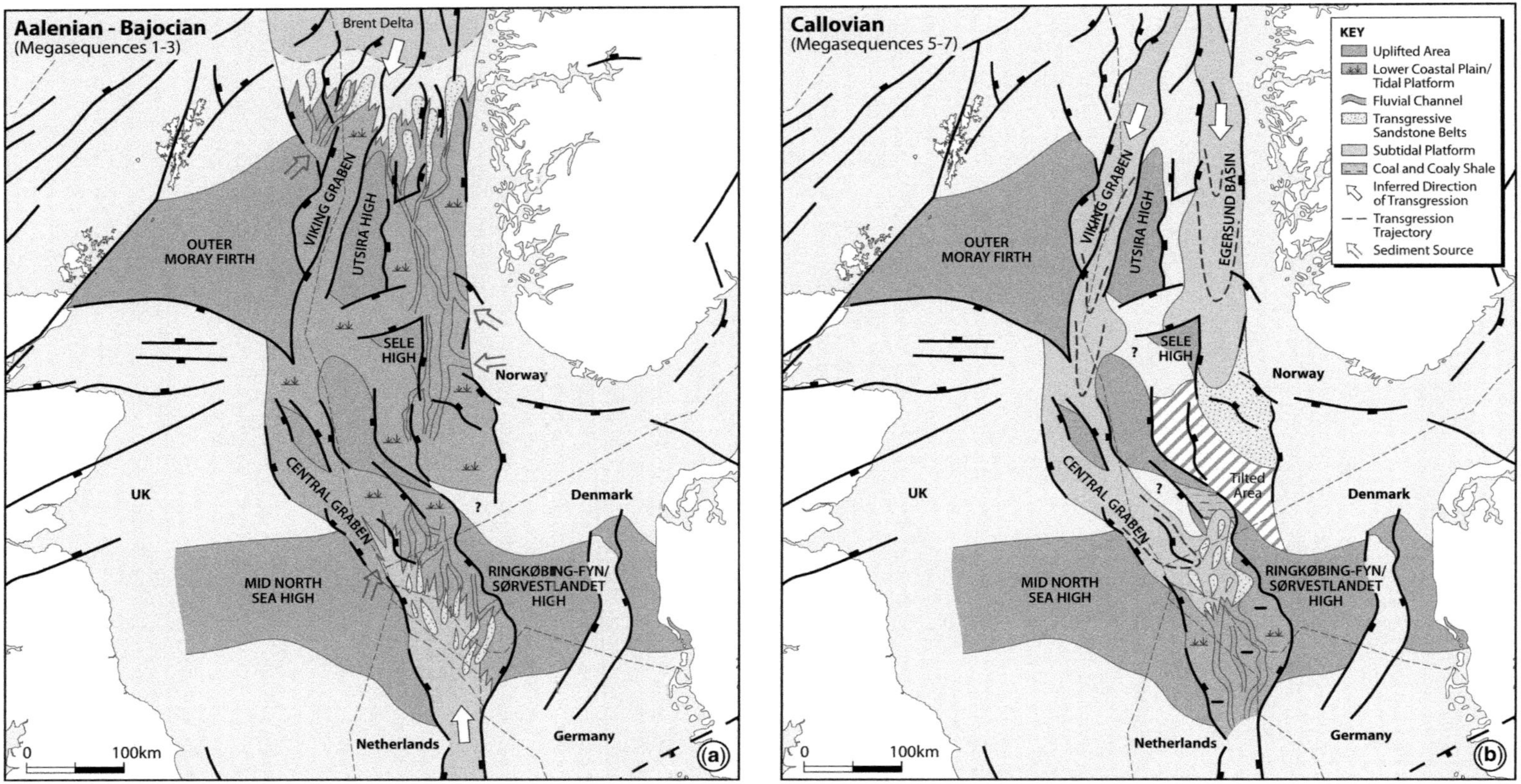

Fig. 20. Reconstructed regional palaeogeography during (**a**) Aalenian to Bajocian and (**b**) Callovian times. (a) A lower coastal plain and tidal platform occupies most of the Søgne Basin with a north–south drainage and inferred open-marine conditions in the southern part of the Danish–Netherlands basins. An open-marine basin is also present to the north in the Viking Graben, into which the early Brent delta prograded. By the early Bathonian (megasequence 4), the southern corridor was closed and a large coastal plain probably occupied most of the Central Graben. (b) Drainage changes during the Callovian in response to the inception of the Late Jurassic rifting. Marine transgression from the north proceeded southwards along the Central Graben into the Søgne Basin. Note the marine corridor between the Utsira–Sele High and the Norwegian mainland. The uplift of the Sørvestlandet–Ringkøbing–Fyn High blocked the transgression to penetrate directly (at least during Callovian) into the Søgne Basin. During the Oxfordian, this high was completely transgressed.

and penetrated southwards through the Egersund Basin, which was located between the Sele High and the Norwegian mainland. Andsbjerg (2003) suggested that this marine corridor linked the Søgne Basin to the north. However, this study and recent investigations by Mannie *et al.* (2016) indicate that the Sørvestlandet–Fyn High was uplifted during Callovian, with engulfed east–west shorelines identified to the north. To the south, deposition is recorded by a condensed section of coaly shales and the tidal delta units of megasequence 7. Only during Oxfordian times was the Sørvestlandet High completely transgressed.

The onset of diffuse low-energy paralic depositional systems across an extended narrow segment of the Norwegian continental shelf during the middle Jurassic indicates general conditions of low subsidence and the consequent absence of coarse-grained sediment due to the low rates of tectonic exhumation of the basin margins (Doré *et al.* 2002). This condition, which is common in low subsidence areas (e.g. Einsele 2013), favoured the development of tide-influenced shoal water deltas and estuaries. The tidal control on these systems was possibly enhanced by a number of concurrent processes, including: (1) the morphotectonic sheltering of many coastal systems, which were protected from fetch-driven waves and alongshore circulation, and generated embayments, coastal invaginations or gulfs, all settings favourable for the development of tidal-influenced and tide-dominated conditions (e.g. Longhitano *et al.* 2012; Plink-Björklund 2012); (2) shallow depths of the continental shelf, which were possibly responsible for the damping of tidal prisms across the coastal areas (e.g. Steel *et al.* 2012); and (3) a regional-scale transgression, which typically serves as a forcing element in enhancing the tidal influence on the coastal sediment dispersion (e.g. Dalrymple *et al.* 2012). Consequently, this setting produced specific conditions for tidal resonance and amplification across the shelf and the entrapment of large volumes of mud-prone deposits across the platform of shoal water deltas, as in the example documented in the present study.

Conclusions

Sedimentological and stratigraphic analyses based on core descriptions provided unexpected insights into the depositional systems of the Middle Jurassic succession of the Søgne Basin, especially within the Bryne Formation.

At well 3/7-4, the Aalenian to Bathonian Bryne cores reflect deposition by a mixed-energy, river-dominated, tidally influenced delta system onto a shallow tidal platform that extended from supratidal reaches with palaeosols and fluvial channels to an intertidal mud-dominated section and subtidal interdistributary mouth bars. The platform, which was tectonically constrained by the Mandal High to the west and the Sørvestlandet–Ringkøbing–Fyn High to the east, extended across the Søgne Basin, and probably across most of the Central Graben. The structural configuration limited wind fetch and prevented the development of longshore currents. Sediment was mostly dispersed within the platform, where salt growth controlled sediment dispersal. Delta progradation was punctuated by transgressive episodes, during which the delta platform was transformed into a broad brackish central estuarine basin and, eventually, during the deposition of the Sandnes Formation, into a tide-dominated, wave-influenced estuary.

Seven genetically related megasequences were recognized. During Aalenian to Bajocian times (megasequences 1–3), transgression proceeded from the south, from a marine basin assumed to be located in the northern part of the Netherlands. This marine corridor, responsible for the thick brackish deposits recognized, was closed at the time of deposition of megasequence 4 (early–middle? Bathonian), which was dominated by thick fluvial and supratidal deposits. An important physiographic change, marked by extensive palaeovalleys and renewed brackish and estuarine conditions, occurred in the upper part of the Bryne Formation and throughout the Sandnes Formation during Late Bathonian and Callovian times (megasequences 5–7). This change was induced by the inception of the Late Jurassic rift event that led to the tilting of the basin to the south, with subsequent offlap of the Callovian strata. The tectonic tilting was accompanied by generalized transgression that eventually led to a fully marine system. During the Callovian, transgression was from the north, across the Viking Graben and Central Graben. The northern tip of the Søgne Basin was uplifted and is believed not to have been connected to the marine corridor present in the Egersund Basin. Only during the Oxfordian did transgression progress across the drowned high.

The presence of diffuse, low-energy paralic depositional systems across the UK and Norwegian continental shelf during the Middle Jurassic indicates that subsidence was accompanied by low rates of tectonic exhumation of the basin margins. The latter favoured the development of elongated, shallow and low gradient platforms that responded to relative changes in base level with the deposition of regressive tide-influenced shoal water deltas and transgressive estuaries.

This study was initiated as part of the work commitment of Premier-Oil Norge AS for Licenses PL725 and 726, and continued independently by Premier-Oil plc. We thank Premier-Oil and the partners of the PL725 (Ithaca, now MOL) and 726 (Fortis Petroleum and Petoro) Licences for allowing us to publish the results of our study in the Søgne Basin. We are deeply indebted to Fredrik Bockelie of Ithaca Petroleum and Andrew Taylor for the discussion at a core workshop organized during License work and to our referees Jan Andsbjerg and Steve Hubbard, who greatly improved the manuscript with insightful comments. Finally, we thank the Editor, Gary Hampson, for his encouragement and his knowledge of the coeval systems of the North Sea that helped us to broaden our local perspective.

References

ALEXANDER, J. & GAWTHORPE, R.L. 1993. The complex nature of a Jurassic multi-storey alluvial sandstone body, Whitby, North Yorkshire. *In*: NORTH, C.P. & PROSSER, D.J. (eds) *Characterization of Fluvial and Aeolian Reservoirs*. Geological Society, London, Special Publications, **73**, 123–142, https://doi.org/10.1144/GSL.SP.1993.073.01.08

ALLEN, J.R.L. 1991. Sedimentary processes and facies in the Gironde estuary: a recent model for macrotidal estuarine systems. *In*: SMITH, D.G., REINSON, G.E., ZAITLIN, B.A. & RAHMANI, R.A. (eds) *Clastic Tidal Sedimentology*. Canadian Society of Petroleum Geology, Memoirs, **16**, 29–39.

ALLEN, P.A. & HOMEWOOD, P. 1984. Evolution and mechanics of a Miocene tidal sand wave. *Sedimentology*, **31**, 63–81.

AMIR HASSAN, M.H., JOHNSON, H.D., ALLISON, P.A. & ABDULLAH, W.H. 2013. Sedimentology and stratigraphic development of the upper Nyalau Formation (Early Miocene), Sarawak, Malaysia: a mixed wave- and tide-influenced coastal system. *Journal of Asian Earth Sciences*, **76**, 301–311.

ANDSBJERG, I. 2003. Sedimentology and sequence stratigraphy of the Bryne and Lulu Formations, Middle Jurassic, northern Danish Central Graben. *In*: INESON, J.R. & SURLYK, F. (eds) *The Jurassic of Denmark and Greenland*. Geological Survey of Denmark and Greenland Bulletin, **1**, 301–347.

ANDSBJERG, I. 2013. Sedimentological and architectural characteristics of a rift initiation package: Middle–Upper Jurassic, North Sea. AAPG Search and Discovery Article 30279, presented at the AAPG Annual Convention and Exhibition, 19–22 May 2013, Pittsburgh, PA.

ANDSBJERG, J. & DYBKJÆR, K. 2003. Sequence stratigraphy of the Jurassic of the Danish Central Graben. *In*: INESON, J.R. & SURLYK, F. (eds) *The Jurassic of Denmark and Greenland*. Geological Survey of Denmark and Greenland Bulletin, **1**, 265–300.

BACHTIAR, A., WIOUNO, J., SYAFUL, M., PURNAMA, Y.S., ROZALLI, M., KRISYUIANTO, A. & PURNAMA, A.S. 2010. The dynamics of Mahakam Delta – Indonesia, based on spatial and temporal variations of grab samples, cores and salinity. AAPG Search and Discovery Article 50363, presented at the AAPG International Conference and Exhibition, 12–15 September 2010.

BLYSTAD, P., BREKKE, H., FÆRSETH, R.B., LARSEN, B.T., SKOGSEID, J. & TØRUDBAKKEN, B. 1995. Structural elements of the Norwegian continental shelf. Part II: The Norwegian Sea Region. *Norwegian Petroleum Directorate (NPD) Bulletin*, **8**, 1–44.

BOERSMA, J.R. & TERWINDT, J.H.J. 1981. Neap–spring tide sequences of intertidal shoal deposits in a mesotidal estuary. *Sedimentology*, **28**, 151–170.

BRIDGE, J.S. 2003. *Rivers and Floodplains: Forms, Processes, and Sedimentary Record*. Blackwell Publishing, Oxford.

BRIDGE, J.S. 2006. Fluvial facies models: recent developments. *In*: POSAMENTIER, H.W. & WALKER, R.G. (eds) *Facies Models Revisited*. SEPM Special Publications, **84**, 85–170.

BRIDGE, J.S., JALFIN, G.A. & GEORGIEFF, S.M. 2000. Geometry, lithofacies and spatial distribution of Cretaceous fluvial sandstone bodies, San Jorge basin, Argentina: outcrop analog for the hydrocarbon-bearing Chubut Group. *Journal of Sedimentary Research*, **70**, 341–359.

BRIDGE, P.H. & LEEDER, M.R. 1976. Sedimentary model for intertidal mudflat channels, with examples from the Solway Firth, Scotland. *Sedimentology*, **23**, 533–552.

BRISTOW, C.S., SKELLY, R.L. & ETHRIDGE, F.G. 1999. Crevasse splays: from the rapidly aggrading, sand-bed, braided Niobrara River, Nebraska: effect of base lever rise. *Sedimentology*, **46**, 1029–1047.

CARTWRIGHT, J. 1991. The kinematic evolution of the Coffee Soil Fault. *In*: ROBERTS, A.M., YIELDING, G. & FREEMAN, B. (eds) *The Geometry of Normal Faults*. Geological Society, London, Special Publications, **56**, 29–40, https://doi.org/10.1144/GSL.SP.1991.056.01.03

CHOI, K.S., DALRYMPLE, R.W., CHUN, S.S. & KIM, S.-P. 2004. Sedimentology of modern, inclined heterolithic stratification (HIS) in the macrotidal Han River delta, Korea. *Journal of Sedimentary Research*, **74**, 677–689.

COCKINGS, J.H., GIFFORD KESSLER, L., II, MAZZA, T.A. & RILEY, L.A. 1992. Bathonian to mid-Oxfordian sequence stratigraphy of the South Viking Graben, North Sea. *In*: HARDMAN, R.F.P. (ed.) *Exploration Britain: Geological Insights for the Next Decade*. Geological Society, London, Special Publications, **67**, 65–105, https://doi.org/10.1144/GSL.SP.1992.067.01.04

COLLINSON, J.D. 1996. Alluvial sediments. *In*: READING, H.G. (ed.) *Sedimentary Environments: Processes, Facies and Stratigraphy*. Blackwell Science, Oxford.

DALRYMPLE, R.W. 1984. Morphology and internal structure of sandwaves in the Bay of Fundy. *Sedimentology*, **31**, 365–382.

DALRYMPLE, R.W. 1992. Tidal depositional systems. *In*: WALKER, R.G. & JAMES, N.P. (eds) *Facies Models: Response to Sea Level Changes*. Geological Association of Canada, St John's, 195–218.

DALRYMPLE, R.W. & CHOI, K. 2007. Morphologic and facies trends through the fluvial-marine transition in tide-dominated depositional systems: a schematic framework for environmental and sequence-stratigraphic interpretation. *Earth-Science Reviews*, **81**, 135–174.

DALRYMPLE, R.W., KNIGHT, R.J., ZAITLIN, B.A. & MIDDLETON, G.V. 1990. Dynamics and facies model of a macrotidal sandbar complex, Cobequid Bay – Salmon River Estuary (Bay of Fundy). *Sedimentology*, **37**, 577–612.

DALRYMPLE, R.W., ZAITLIN, B.A. & BOYD, R. 1992. Perspective: estuarine facies models: conceptual basis and stratigraphic implications. *Journal of Sedimentary Research*, **62**, 1130–1146.

Dalrymple, R.W., Baker, E.K. & Harris, P.T. 2003. Sedimentology and stratigraphy of a tide-dominated, foreland-basin delta (Fly River, Papua New Guinea). *In*: Sidi, F.H., Nummedal, D., Imbert, P., Darman, H. & Posamentier, H.W. (eds) *Tropical Deltas of Southeast Asia. Sedimentology, Stratigraphy, and Petroleum Geology*. SEPM Special Publications, **76**, 147–173.

Dalrymple, R.W., Mackay, D.A., Ichaso, A.A. & Choi, K.S. 2012. Processes, morphodynamics, and facies of tide-dominated estuaries. *In*: Davis, R.A., Jr. & Dalrymple, R.W. (eds) *Principles of Tidal Sedimentology*. Springer, New York, 79–108.

Damtoft, K., Nielsen, L.H., Johannessen, P.N., Thomsen, E. & Andersen, P.R. 1992. Hydrocarbon plays of the Danish Central Trough. *In*: Spencer, A.M. (ed.) *Generation, Accumulation and Production of Europe's Hydrocarbons II*. European Association of Petroleum Geoscientists, Special Publications, **2**, 35–58.

Dashtgard, S., Venditti, J., Hill, P., Sisulak, C., Johnson, S. & La Croix, A. 2012. Sedimentation across the tidal fluvial transition in the lower Fraser River, Canada. *The Sedimentary Record*, **10**, 4–9.

Doré, A.G., Cartwright, J.A., Stoker, M.S., Turner, J.P. & White, N. (eds) 2002. *Exhumation of the North Atlantic Margin: Timing, Mechanisms and Implications for Petroleum Exploration*. Geological Society, London, Special Publications, **196**, http://sp.lyellcollection.org/content/196/1

Einsele, G. 2013. *Sedimentary Basins: Evolution, Facies, and Sediment Budget*. Springer, Berlin.

Fält, L.M., Helland, R., Jacobsen, V.W. & Renshaw, D. 1989. Correlation of transgressive–regressive depositional sequences in the Middle Jurassic Brent/Vestland Group megacycle, Viking Graben, Norwegian North Sea. *In*: Collinson, J.D. (ed.) *Correlation in Hydrocarbon Exploration*. Graham & Trotman, London, 191–200.

Fitzgerald, D., Buynevich, I. & Hein, C. 2012. Morphodynamics and facies architecture of tidal inlets and tidal deltas. *In*: Davis, R.A., Jr. & Dalrymple, R.W. (eds) *Principles of Tidal Sedimentology*. Springer, New York, 301–334.

Folkestad, A. & Satur, N. 2008. Regressive and transgressive cycles in a rift-basin: depositional model and sedimentary partitioning of the Middle Jurassic Hugin Formation, southern Viking Graben, North Sea. *Sedimentary Geology*, **207**, 1–21.

Frandsen, N. 1986. *Middle Jurassic Deltaic and Coastal Deposits in the Lulu-1 Well of the Danish Central Trough*. Danmarks Geologiske Undersøgelse Serie A, **9**.

Galloway, W.E. 1976. Sediments and stratigraphic framework of the Copper River fan-delta, Alaska. *Journal of Sedimentary Research*, **46**, 726–737.

Gibling, M.R. 2006. Width and thickness of fluvial channel bodies and valley fills in the geological record: a literature compilation and classification. *Journal of Sedimentary Research*, **76**, 731–770.

Gibling, M.R., Fielding, C.R. & Sinha, R. 2011. Alluvial valleys and alluvial sequences: towards a geomorphic assessment. *In*: Davidson, S.K., Leleu, S. & North, C.P. (eds) *From River to Rock Record: the Preservation of Fluvial Sediments and their Subsequent Interpretation*. Society for Sedimentary Geology, Special Publications, **97**, 423–447.

Goodbred, S.L. & Kuehl, S.A. 1999. Holocene and modern sediment budgets for the Ganges–Brahmaputra river system: evidence for highstand dispersal to floodplain, shelf, and deepsea depocenters. *Geology*, **27**, 559–562.

Goodbred, S.L., Jr. & Saito, Y. 2012. Tide-dominated deltas. *In*: *Principles of Tidal Sedimentology*. Springer, New York, 129–149.

Gowers, M.B. & Sæbøe, A. 1985. On the structural evolution of the Central Trough in the Norwegian and Danish sectors of the North Sea. *Marine and Petroleum Geology*, **2**, 298–318.

Graue, E., Helland-Hansen, W. *et al.* 1987. Advance and retreat of the Brent Delta System, Norwegian North Sea. *In*: Brooks, J. & Glennie, K. (eds) *Petroleum Geology of North West Europe*. Graham & Trotman, London, 915–937.

Guion, P.D., Fulton, I.M. & Jones, N.S. 1995. Sedimentary facies of the coal-bearing Westphalian A and B of the Wales–Brabant High. *In*: Whateley, M.K.G. & Spears, D.A. (eds) *European Coal Geology*. Geological Society, London, Special Publications, **82**, 45–78, https://doi.org/10.1144/GSL.SP.1995.082.01.03

Halland, K.E., Gjeldvik, T.I. *et al.* 2014. *The CO_2 Storage Atlas of the Norwegian Continental Shelf*. Norwegian Petroleum Directorate Publications, 30–73.

Hampson, G.J., Sixsmith, P.J. & Johnson, H.D. 2004. A sedimentological approach to refining reservoir architecture in a mature hydrocarbon province: the Brent Province, UK North Sea. *Marine and Petroleum Geology*, **21**, 457–484.

Helland-Hansen, W. & Hampson, G.J. 2009. Trajectory analysis: concepts and applications. *Basin Research*, **21**, 454–483.

Herngreen, G.F.W., Kouwe, W.F.P. & Wong, T.E. 2003. The Jurassic of the Netherlands. *Geological Survey of Denmark and Greenland Bulletin*, **1**, 217–229.

Hodgson, N.A., Farnsworth, J. & Fraser, A.J. 1992. Salt-related tectonics, sedimentation and hydrocarbon plays in the Central Graben, North Sea, UKCS. *In*: Hardman, R.F.P. (ed.) *Exploration in Britain: Geological Insights for the Next Decade*. Geological Society, London, Special Publications, **67**, 31–63, https://doi.org/10.1144/GSL.SP.1992.067.01.03

Hughes, Z.J. 2012. Tidal channels on tidal flats and marshes. *In*: Davis, R.A., Jr. & Dalrymple, R.W. (eds) *Principles of Tidal Sedimentology*. Springer, New York, 269–300.

Ichaso, A.A. & Dalrymple, R.W. 2014. Eustatic, tectonic and climatic controls on an early synrift mixed-energy delta, Tilje Formation (early Jurassic, Smørbukk Field, offshore mid-Norway). *In*: Martinius, A.W., Ravnaas, R., Howell, J.A., Steel, R.J. & Wonham, J.P. (eds) *Depositional Systems to Sedimentary Successions on the Norwegian Continental Shelf*. International Association of Sedimentology, Special Publications, **46**, 339–388.

Johannessen, P.N. & Andsbjerg, J. 1993. Middle to Late Jurassic basin evolution and sandstone reservoir distribution in the Danish Central Trough. *In*: Parker,

J.R. (ed.) *Petroleum Geology of Northwest Europe: Proceedings of the 4th Conference*. Geological Society, London, 271–283, https://doi.org/10.1144/0040271

Johannessen, P.N., Dybkjær, K., Andersen, C., Kristensen, L., Hovikoski, J. & Vosgerau, H. 2010. Upper Jurassic reservoir sandstones in the Danish Central Graben: new insights on distribution and depositional environments. *In*: Vining, B.A. & Pickering, S.C. (eds) *Petroleum Geology: From Mature Basins to New Frontiers–Proceedings of the 7th Petroleum Geology Conference*. Geological Society, London, 127–143, https://doi.org/10.1144/0070127

Kieft, R.L., Jackson, C.A.-L., Hampson, G.J. & Larsen, E. 2010. Sedimentology and sequence stratigraphy of the Hugin Formation, Quadrant 15, Norwegian sector, South Viking Graben. *In*: Vining, B.A. & Pickering, S.C. (eds) *Petroleum Geology: From Mature Basins to New Frontiers–Proceedings of the 7th Petroleum Geology Conference*. Geological Society, London, 157–176, https://doi.org/10.1144/0070157

Koch, J.O. 1983. Sedimentology of Middle and Upper Jurassic sandstone reservoirs of Denmark. *In*: Kaasschieter, J.P.H. & Reijers, T.J.A. (eds) *Petroleum Geology of the Southeastern North Sea and the Adjacent Onshore Areas. Geologie en Mijnbouw*, **62**, 115–129.

Korstgaard, J.A., Lerche, I., Mogensen, T.E. & Thomsen, R.O. 1993. Salt and fault interactions in the northeastern Danish Central Graben: observations and inferences. *Bulletin of the Geological Society of Denmark*, **40**, 197–255.

Kuecher, G.J., Woodland, B.G. & Broadhurst, F.M. 1990. Evidence of deposition from individual tides and of tidal cycles from the Francis Creek Shale, Westphalian D (Pennsylvanian), north-eastern Illinois. *Sedimentary Geology*, **68**, 211–221.

Kuehl, S.A., Levy, B.M., Moore, W.S. & Allison, M.A. 1997. Subaqueous delta of the Ganges–Brahmaputra river system. *Marine Geology*, **144**, 81–96.

La Croix, A.D. & Dashtgard, S.E. 2014. Of sand and mud: sedimentological criteria for identifying the turbidity maximum zone in a tidally influenced river. *Sedimentology*, **61**, 1961–1981.

Livera, S.E. 1989. Facies association and sand-body geometries in the Ness Formation of the Brent Group, Brent Field. *In*: Whateley, M.K.G. & Pickering, K.T. (eds) *Deltas: Sites and Traps for Fossil Fuels*. Geological Society, London, Special Publications, **41**, 269–286, https://doi.org/10.1144/GSL.SP.1989.041.01.19

Longhitano, S.G., Mellere, D., Steel, R.J. & Ainsworth, R.B. 2012. Tidal depositional systems in the rock record: a review and new insights. *In*: Longhitano, S.G., Mellere, D. & Ainsworth, R.B. (eds) *Modern and Ancient Depositional Systems: Perspectives, Models and Signatures. Sedimentary Geology*, **279**, 2–22.

Mannie, A.S., Jackson, C.A.-L. & Hampson, G.J. 2014. Structural controls on the stratigraphic architecture of net-transgressive shallow-marine strata in a salt-influenced rift basin: Middle-to-Upper Jurassic Egersund Basin, Norwegian North Sea. *Basin Research*, **26**, 675–700.

Mannie, A.S., Jackson, C.L., Hampson, G.J. & Fraser, A.J. 2016. Tectonic controls on the spatial distribution and stratigraphic architecture of a net-transgressive shallow-marine syn-rift succession in a salt-influenced rift basin: middle to Upper Jurassic, Norwegian Central North Sea. *Journal of the Geological Society, London*, first published online May 13, 2016, https://doi.org/10.1144/jgs2016-033

Martinius, A.W., Kaas, I., Næss, A., Helgesen, G., Kjærefjord, J.M. & Leith, D.A. 2001. Sedimentology of the heterolithic and tide-dominated Tilje Formation (Early Jurassic, Halten Terrace, offshore Mid-Norway). *In*: Martinsen, O.J. & Dreyer, T. (eds) *Sedimentary Environments Offshore Norway – Palaeozoic to Recent*. Norwegian Petroleum Society, Special Publications, **10**, 103–144.

Martinius, A.W., Ringrose, P.S., Brostrøm, C., Elfenbein, C., Næss, A. & Ringås, J.E. 2005. Reservoir challenges of heterolithic tidal sandstone reservoirs in the Halten Terrace, mid-Norway. *Petroleum Geoscience*, **11**, 3–16, https://doi.org/10.1144/1354-079304-629

Mellere, D. & Steel, R.J. 1995. Facies architecture and sequentiality of nearshore/shelf sandbodies (Haystack Mountains Formation – Wyoming, USA). *Sedimentology*, **42**, 551–574.

Miall, A.D. 1996. *The Geology of Fluvial Deposits*. Springer, Berlin.

Michelsen, O., Mogensen, T.E. & Korstgaard, J.A. 1992. Pre Cretaceous structural development of the Danish Central Trough and its implications for the distribution of Jurassic sands. *In*: Larsen, R.M. (ed.) *Structural and Tectonic Modelling and its Application to Petroleum Geology*. Norwegian Petroleum Society, Special Publications, **1**, 495–506.

Michelsen, O., Nielsen, L.H., Johannessen, P.N., Andsbjerg, J. & Surlyk, F. 2003. Jurassic lithostratigraphy and stratigraphic development onshore and offshore Denmark. *In*: Ineson, J.R. & Surlyk, F. (eds) *The Jurassic of Denmark and Greenland*. Geological Survey of Denmark and Greenland Bulletin, **1**, 147–216.

Milner, P.S. & Olsen, T. 1998. Predicted distribution of the Hugin Formation reservoir interval in the Sleipner Øst field, South Viking Graben; the testing of a three dimensional sequence stratigraphic model. *In*: Gradstein, F.M., Sandvik, K.O. & Milton, N.J. (eds) *Sequence Stratigraphy: Concepts and Applications*. Norwegian Petroleum Society, Special Publications, **8**, 337–354.

Mitchener, B.C., Lawrence, D.A., Partington, M.A., Bowman, M.B.J. & Gluyas, J. 1992. Brent Group; sequence stratigraphy and regional implications. *In*: Morton, A.C., Haszeldine, R.S. & Giles, M.R. & Brown, S. (eds) *Geology of the Brent Group*. Geological Society, London, Special Publications, **61**, 45–80, https://doi.org/10.1144/GSL.SP.1992.061.01.05

Mogensen, T.E., Korstgaard, J.A. & Geil, K. 1992. Salt tectonics and faulting in the NE Danish Central Graben. *In*: Spencer, A.M. (ed.) *Generation, Accumulation and Production of Europe's Hydrocarbons II*. European Association of Petroleum Geoscientists, Special Publications, **2**, 163–173.

Møller, J.J. 1986. *Seismic Structural Mapping of the Middle and Upper Jurassic in the Danish*

Central Trough. Danmarks Geologiske Undersøgelse Serie A, **13**.

Møller, J.J. & Rasmussen, E.S. 2003. Middle Jurassic–Early Cretaceous rifting of the Danish Central Graben. *In*: Ineson, J.R. & Surlyk, F. (eds) *The Jurassic of Denmark and Greenland*. Geological Survey of Denmark and Greenland Bulletin, **1**, 247–264.

Nichols, M.M. & Biggs, R.B. 1985. Estuaries. *In*: Davis, R.A. (ed.) *Coastal Sedimentary Environments*. Springer, New York, 77–186.

North, C.P. & Davidson, S.K. 2012. Unconfined alluvial flow processes: recognition and interpretation of their deposits and the significance for paleogeographic reconstruction. *Earth-Science Reviews*, **111**, 199–223.

Partington, M.A., Copestake, P., Mitchener, B.C. & Underhill, J.R. 1993. Biostratigraphic calibration of genetic stratigraphic sequences in the Jurassic–lowermost Cretaceous (Hettangian to Ryazanian) of the North Sea and adjacent areas. *In*: Parker, J.R. (ed.) *Petroleum Geology of Northwest Europe: Proceedings of the 4th Conference*. Geological Society, London, 371–386, https://doi.org/10.1144/0040371

Petersen, H.I. & Andsbjerg, J. 1996. Organic facies development within Middle Jurassic coal seams, Danish Central Graben, and evidence for relative sea-level control on peat accumulation in a coastal plain environment. *Sedimentary Geology*, **106**, 259–277.

Plink-Björklund, P. 2005. Stacked fluvial and tide-dominated estuarine deposits in high frequency (fourth-order) sequences of the Eocene Central Basin, Spitsbergen. *Sedimentology*, **52**, 391–428.

Plink-Björklund, P. 2012. Effects of tides on deltaic deposition: causes and responses. *In*: Longhitano, S.G., Mellere, D. & Ainsworth, R.B. (eds) *Modern and Ancient Depositional Systems: Perspectives, Models and Signatures. Sedimentary Geology*, **279**, 107–133.

Pirmez, C., Pratson, L.F. & Steckler, M.S. 1998. Clinoform development by advection-diffusion of suspended sediment: modelling and comparison to natural systems. *Journal of Geophysical Research*, **103**, 24141–24157.

Rattey, R.P. & Hayward, A.B. 1993. Sequence stratigraphy of a failed rift system: the Middle Jurassic to Early Cretaceous basin evolution of the central and northern North Sea. *In*: Parker, J.R. (ed.) *Petroleum Geology of Northwest Europe: Proceedings of the 4th Conference*. Geological Society, London, 215–249, https://doi.org/10.1144/0040215

Richards, P.C. 1991. An estuarine facies model for the Middle Jurassic Sleipner Formation, Beryl Embayment, North Sea. *Journal of the Geological Society, London*, **148**, 459–471, https://doi.org/10.1144/gsjgs.148.3.0459

Roberts, H.H. & Sydow, J. 2003. Late Quaternary stratigraphy and sedimentology of the offshore Mahakam delta, East Kalimantan (Indonesia). *In*: Sidi, F.H., Nummedal, F.H., Imbert, P., Darman, H. & Posamentier, H.W. (eds) *Tropical Deltas of Southeast Asia. Sedimentology, Stratigraphy, and Petroleum Geology*. SEPM Special Publications, **76**, 125–145.

Rossi, V.M. & Steel, R.J. 2016. The role of tidal, wave and river currents in the evolution of mixed-energy deltas: example from the Lajas Formation (Argentina). *Sedimentology*, **63**, 824–864.

Røssland, A. 2011. *The tectono-stratigraphic evolution of the greater Mandal High area, southern North Sea, Norway*. Master's thesis, University of Stavanger.

Shanley, K.W., McCabe, P.J. & Hettinger, R.D. 1992. Significance of tidal influence in fluvial deposits for interpreting sequence stratigraphy. *Sedimentology*, **39**, 905–930.

Shultz, A.W. & Ways, L.C. 2004. Reservoir architecture of Gabus and Udang formations, West Natuna Basin, offshore Indonesia. AAPG Search & Discovery article 90034 presented at the AAPG Herberg Conference: Sandstone Deposition in Lacustrine Environment, 18–21 May, Baku, Azerbaijan.

Sneider, J.S., de Clarens, P. & Vail, P.R. 1995. Sequence stratigraphy of the Middle and Upper Jurassic, Viking Graben, North Sea. *In*: Steel, R.J., Felt, V.L., Johannessen, E.P. & Mathieu, C. (eds) *Sequence Stratigraphy on the Northwest European Margin*. Norwegian Petroleum Society, Special Publications, **5**, 167–198.

Steel, R.J. 1993. Triassic–Jurassic megasequence stratigraphy in the northern North Sea: rift to post rift evolution. *In*: Parker, J.R. (ed.) *Petroleum Geology of Northwest Europe: Proceedings of the 4th Conference*. Geological Society, London, 299–315, https://doi.org/10.1144/0040299

Steel, R.J., Plink-Bjorklund, P. & Aschoff, J. 2012. Tidal deposits of the Campanian Western Interior Seaway, Wyoming, Utah and Colorado, USA. *In*: DavisA., Jr. & Dalrymple, R.W. (eds) *Principles of Tidal Sedimentology*. Springer, New York, 437–472.

Stemmerik, L., Ineson, J.R. & Mitchell, J.G. 2000. Stratigraphy of the Rotliegend Group in the Danish part of the northern Permian Basin, North Sea. *Journal of the Geological Society, London*, **157**, 1127–1136, https://doi.org/10.1144/jgs.157.6.1127

Stewart, S.A. & Clark, J.A. 1999. Impact of salt on the structure of the Central North Sea hydrocarbon fairways. *In*: Fleet, A.J. & Boldy, S.A.R. (eds) *Petroleum Geology of Northwest Europe: Proceedings of the 5th Conference*. Geological Society, London, 179–200, https://doi.org/10.1144/0050179

Swenson, J.B., Paola, C., Pratson, L., Voller, V.R. & Murray, A.B. 2005. Fluvial and marine controls on combined subaerial and subaqueous delta progradation: morphodynamic modelling of compound clinoform development. *Journal of Geophysical Research*, **110**, F02013.

Ta, T.K.O., Nguyen, V.L., Tateishi, M., Kobayashi, I., Saito, Y. & Nakamura, T. 2002. Sediment facies and Late Holocene progradation of the Mekong River Delta in Bentre Province, southern Vietnam: an example of evolution from a tide-dominated to a tide- and wave-dominated delta. *Sedimentary Geology*, **152**, 313–325.

Tänavsuu-Milkeviciene, K. & Plink-Björklund, P. 2009. Recognizing tide-dominated v. tide-influenced deltas: middle Devonian strata of the Baltic Basin. *Journal of Sedimentary Research*, **79**, 887–905.

Thomas, D.W. & Coward, M.P. 1996. Mesozoic regional tectonics and South Viking Graben formation: evidence for localised thin skinned detachments during rift development and inversion. *Marine and Petroleum Geology*, **13**, 149–177.

THOMAS, R.G., SMITH, D.G., WOOD, D.G., VISSER, J., CALVERLY-RANGE, E.A. & KOSTER, E.H. 1987. Inclined heterolithic stratification – terminology, description, interpretation and significance. *Sedimentary Geology*, **53**, 123–179.

UNDERHILL, J.R. & PARTINGTON, M.A. 1993. Jurassic thermal doming and deflation in the North Sea: implication of the sequence stratigraphic evidence. *In*: PARKER, J.R. (ed.) *Petroleum Geology of Northwest Europe: Proceedings of the 4th Conference*. Geological Society, London, 337–345, https://doi.org/10.1144/0040337

VAN ADRICHEM BOOGAERT, H.A. & KOUWE, W.F.P. (eds) 1993. Lower and Middle Jurassic (Altena Group). *In*: VAN ADRICHEM BOOGAERT, H.A. & KOUWE, W.F.P. (eds) *Stratigraphic Nomenclature of the Netherlands, Revision and Update by Rijks Geologische Dienst and Netherlands Oil and Gas Exploration and Production Association*. Mededelingen Rijks Geologische Dienst, **50**, 20.

VAN WAGONER, J.C., MITCHUM, R.M., CAMPION, K.M. & RAHMANIAN, V.D. 1990. *Siliciclastic Sequence Stratigraphy in Well Logs, Cores and Outcrops: Concepts for High-resolution Correlation of Time and Facies*. American Association of Petroleum Geologists, Methods in Exploration Series, **7**.

VAN WAGONER, J.C., POSAMENTIER, H.W., MITCHUM, R.M., VAIL, P.R., SARG, J.F., LOUTIT, T.S. & HARDENBOL, J. 1988. An overview of the fundamentals of sequence stratigraphy and key definitions. *In*: WILGUS, C.K., Hastings, B.S. ET AL. (eds) *Sea-Level Changes – an Integrated Approach*. SEPM Special Publications, **42**, 39–45.

VISSER, M.J. 1980. Neap–spring cycles reflected in Holocene subtidal large-scale bedform deposits: a preliminary note. *Geology*, **8**, 543–546.

VOLLSET, J. & DORÉ, A.G. 1984. A revised Triassic and Jurassic lithostratigraphic nomenclature for the Norwegian North Sea. *Norwegian Petroleum Directorate Bulletin*, **3**, 1–53.

WEISLOGEL, A.L. & LAWTON, T.F. 2000. Diapir-flank facies assemblages and stratal geometries within the tidally influenced deltaic system of the Maastrichtian Muerto Formation, La Popa Basin, Northeastern Mexico: an analogue for the Gulf of Mexico subsurface. *Gulf Coast Association of Geological Societies Transactions*, **50**, 669–678.

WILLIS, B.J. 2005. Deposits of tide-influenced river-deltas. *In*: GIOSAN, L. & BHATTACHARYA, J.P. (eds) *River Deltas – Concepts, Models, and Examples*. Society for Sedimentary Geology, Special Publications, **83**, 87–129.

WILLIS, B.J. & GABEL, S.L. 2003. Formation of deep incisions into tide-dominated river deltas – implications for the stratigraphy of the Sego Sandstone, Book Cliffs, Utah, U.S.A. *Journal of Sedimentary Research*, **73**, 246–263.

XUE, Z., LIU, J.P., DEMASTER, D., VAN NGUYEN, L. & TA, T.K.O. 2010. Late Holocene evolution of the Mekong subaqueous delta, southern Vietnam. *Marine Geology*, **269**, 46–60.

Sedimentology and stratigraphic architecture of a Miocene retrogradational, tide-dominated delta system: Balingian Province, offshore Sarawak, Malaysia

MEOR H. AMIR HASSAN[1]*, HOWARD D. JOHNSON[2], PETER A. ALLISON[2] & WAN H. ABDULLAH[1]

[1]*Geology Department, University of Malaya, 50603 Kuala Lumpur, Malaysia*

[2]*Department of Earth Science and Engineering, Imperial College London, London SW7 2AZ, UK*

**Correspondence: meorhakif@um.edu.my*

Abstract: The Balingian Province of NW Borneo is a major hydrocarbon-bearing region with production dominantly from Early Miocene (cycle II) coastal to lower coastal plain deposits. A regional sedimentological synthesis is outlined from the main productive interval of three offshore oilfields, which are representative of the depositional environments and stratigraphic architecture in cycle II. Four types of vertical facies successions are recognized: (1) fluvio-tidal channel; (2) tide-dominated delta front; (3) wave-dominated shoreface; and (4) barrier lagoon. The facies successions demonstrate that cycle II consists of a tide-dominated delta system that is partly analogous to the modern Rajang Delta and Lupar Embayment of southern Sarawak. Fluvio-tidal channel and tide-dominated delta successions represent periods of river-fed progradation. Wave-dominated shoreface and barrier lagoon facies successions represent depositional systems that developed during periods of transgression and/or delta lobe abandonment. Cycle II consists of several high-order sequences stacked together into two lower-order, *c.* 100–300 m thick fining-upwards, unconformity-bounded megasequences. This stratigraphic architecture reflects a combination of hinterland denudation and long-term eustatic sea-level rise. These trends were interrupted by frequent, higher-order relative sea-level falls, which were a result of movement along the West Balingian Line fault zone. Variations in the thickness of megasequences between the three oilfields reflect fluctuations in tectonically driven subsidence rates. This exerted a fundamental control on the reservoir architecture, zonation and correlation and provides a basis for comparing field behaviour.

The initial focus in tidal depositional systems and their reservoir characteristics was the establishment of reliable criteria for their recognition (Raaf & Boersma 1971; Visser 1980; Middleton 1991; Nio & Yang 1991; Klein 1998; Martinius & Van den Berg 2011; Longhitano *et al.* 2012). The early emphasis on high-latitude/temperate climate tidal flats established the tidal flat depositional model, which remains widely applicable to many of the world's present day transgressive shorelines (Straaten 1954; Evans 1965; Oertel 1972; Reineck 1972; Hayes 1975, 1979; Reineck & Singh 1975). Sequence stratigraphic concepts highlighted the high preservation potential of tidal deposits within estuaries, especially as part of the transgressive infill of drowned incised valleys, and their recognition in this setting rapidly expanded (Dalrymple *et al.* 1990, 1992; Allen 1991; Allen & Posamentier 1993, 1994; Johnson & Levell 1995; Wonham & Elliott 1996; Cattaneo & Steel 2003; Boyd *et al.* 2006; Dalrymple 2006; Longhitano *et al.* 2012). The tide-dominated estuary facies model was established by Dalrymple *et al.* (1992) and has been widely applied by many workers in a diverse range of geological settings around the world.

For a long period of time, the interpretation of ancient deltaic successions was strongly skewed towards the better known fluvial-dominated variety, in part reflecting the early pre-eminence of the modern Mississippi Delta (Fisk *et al.* 1954). Wave-dominated delta models based on systems such as the Rhône (Kruit 1955) closely followed. The establishment of a tide-dominated delta model has lagged behind the fluvial- and wave-dominated equivalents. Depositional models of tide-dominated to tide-influenced deltas based on modern examples have been developed mainly in low-latitude settings, often in mud-rich systems (Dalrymple & Choi 2007; Goodbred & Saito 2012), most notably the Colorado (Meckel 1975), Mahakam (Gastaldo *et al.* 1995; Allen & Chambers 1998), Mekong (Tanabe *et al.* 2003), Orinoco (Warne *et al.* 2002; Aslan *et al.* 2003), Fly (Dalrymple *et al.* 2003; Roberts & Sydow 2003; Harris *et al.* 2004), Ganges–Brahmaputra (Kuehl *et al.* 1997, 2005; Allison *et al.* 2003) and parts of the mixed-energy Niger

From: Hampson, G. J., Reynolds, A. D., Kostic, B. & Wells, M. R. (eds) 2017. *Sedimentology of Paralic Reservoirs: Recent Advances*. Geological Society, London, Special Publications, **444**, 215–250.
First published online August 30, 2016, https://doi.org/10.1144/SP444.12

(Allen 1965; Oomkens 1974) delta systems. Somewhat belatedly, there is now a rapidly growing body of literature on ancient tide-dominated deltas, which highlights their variability and provides criteria for their recognition (Willis *et al.* 1999; Mellere & Steel 2000; Willis & Gabel 2001; Willis 2005; Pontén & Plink-Björklund 2007; Tänavsuu-Milkeviciene & Plink-Björklund 2009; Willis & Fitris 2012; Legler *et al.* 2013, 2014; Chen *et al.* 2014). Many ancient examples are sand-dominated, for which there are negligible modern analogues (Goodbred & Saito 2012). In some cases, ancient deltaic successions that were originally considered to be fluvial-dominated have now been re-assessed in the light of an improved understanding of tide-influenced systems and a set of recognition criteria has been developed for tide-influenced systems 20 years after those for fluvial- and wave-dominated systems. We present here a case study that fits into this category.

This paper is part of a regional-scale facies analysis of the Early Miocene cycle II stratigraphic interval of the Balingian Province, offshore Sarawak, NW Borneo (Malaysia). This paper covers aspects derived mainly from subsurface and regional modelling considerations and integrates this with an earlier facies analysis from adjacent outcrops (Amir Hassan *et al.* 2013). This paper has two main objectives: (1) to present a subsurface sedimentary facies analysis of a clastic coastal–deltaic depositional system (the Early Miocene cycle II succession); and (2) to reconstruct the stratigraphic architecture and palaeogeography of the West Balingian region during the Early Miocene, assessing the relative interplay between autogeneic and allogeneic processes.

To set the context, a fluvial-dominated delta depositional model, analogous to the modern Mississippi lower delta plain, was initially proposed for the cycle II succession in the D18 field (Almond *et al.* 1990). However, this paper aims to show that both the onshore (Nyalau Formation) and the subsurface (cycle II) display a plethora of features indicative of a more variable depositional system, including significant tide and wave/storm influences. The paper also considers the stratigraphic architecture of cycle II, including the genetic significance of the large-scale (100s of metres thick), fining- upwards megasequences and the regional controls on this long-lived and extensive paralic depositional system.

Geological setting

The Balingian Province is a distinctive tectonostratigraphic region of NW Borneo; it covers an extensive onshore region and continues into the adjacent offshore area (Fig. 1). The latter area is a major proved hydrocarbon-bearing province with around 30 oil and gas accumulations, which produce largely from Early Miocene (cycle II) coastal to lower coastal plain sandstones (Madon & Abolins 1999). The fields have been the subject of numerous studies aimed at reservoir characterization and production optimization (Finol *et al.* 2005), including reservoir sedimentology (Almond *et al.* 1990), heterogeneity characterization and three-dimensional reservoir geological modelling (Budding *et al.* 1992; Dronamraju *et al.* 2005), three-dimensional seismic analysis of channel sand body distributions (Lasman 1998) and seismic inversion analysis (Rajput *et al.* 2016). Ismail & Hamza (2013) conducted a range of geophysical, geological and reservoir engineering studies designed to better assess the impact of complex channel architecture and faulting on compartmentalization. Zainuddin (1998) also described the application of quantitative three-dimensional seismic interpretation methods to guide infill drilling in one of the fields.

The Balingian Province (Leichti *et al.* 1960; Wolfenden 1960; Madon & Abolins 1999; Hutchison 2005) is part of a relatively stable structural and tectonic block located mainly offshore Sarawak, Malaysia (Madon 1999*a*; Tate 2001; Cullen 2010). The province covers a triangular-shaped area of *c.* 27 000 km^2, of which 70% is offshore (Hutchison 2005). The basin-fill consists of up to 6 km of Oligocene–Recent siliciclastic sedimentary rocks, which unconformably overlie strongly folded, deep-water deposits of the Late Cretaceous to Late Eocene Rajang Group (Hamilton 1974, 1979; Hall & Nichols 2002) (Fig. 2). The province is bounded by two major NW–SE-trending tectonic lineaments (Fig. 1a): (1) the West Baram Line, which separates the Balingian Province from the thick Neogene coastal–shallow-marine deposits of the West Baram Delta Province (Fig. 1a); and (2) the West Balingian Line fault zone, which separates the Balingian and Tatau provinces, the latter including the complexly faulted Penian High (Fig. 1b). Both lineaments were active throughout the deposition of most of the Neogene successions in NW Borneo (Hutchison 2005; Cullen 2010, 2014). The southern boundary is marked by the Tatau-Bukit Mersing Line (Fig. 1a; Mohd Idrus & Redzuan 1999). The Balingian Province grades northwards into the Central Luconia Province, which is characterized by the development of extensive Miocene carbonates (Mohammad Yamin & Abolins 1999). Both the Central Luconia and Balingian provinces are underlain by attenuated continental crust of Sundaland (Hutchison 2005). The tectonic origin of the Balingian Province is complex and uncertainties remain. It has been interpreted as part of a foreland basin that developed in response to the collision of

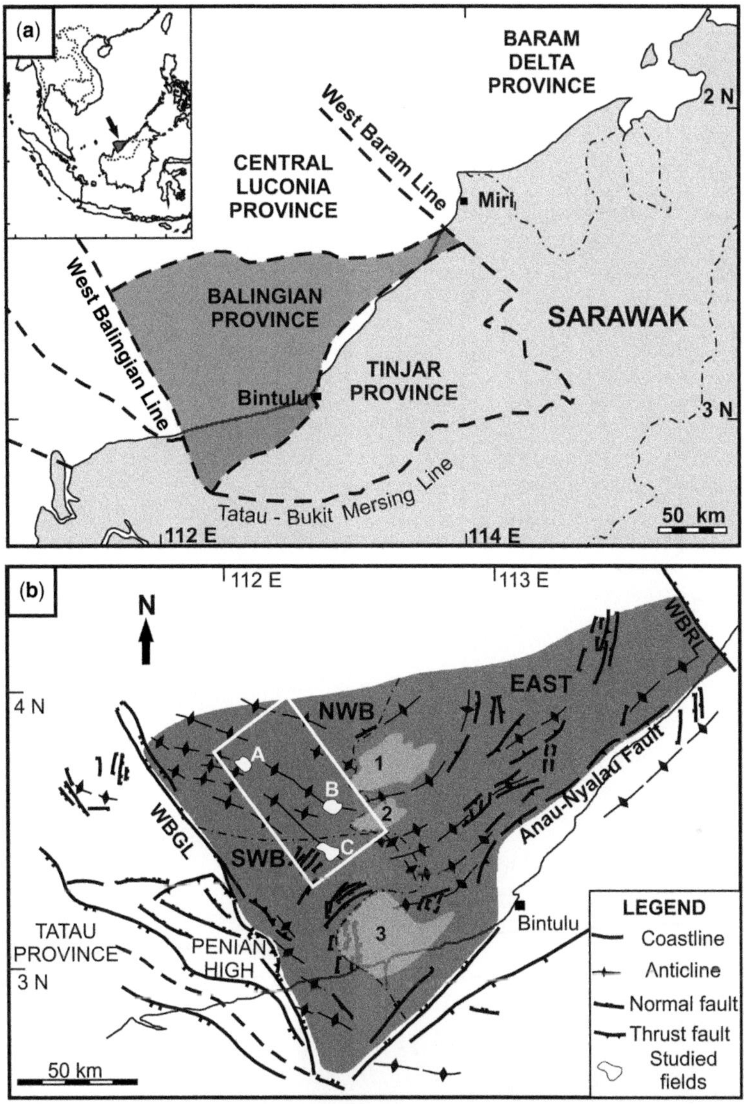

Fig. 1. Maps showing the location of the Balingian Province in the Sarawak Basin of SE Asia. (**a**) Offshore and onshore extent of the Balingian Province (dark shading) and its position in relation to other basin provinces of NW Borneo. Modified from Mohd Idrus & Redzuan (1999). (**b**) Structural elements of the Balingian Province. The locations of the major depocentres (1, 2, 3) and the three studied fields (A, B, C) are marked. Modified from Madon (1999*a*, *b*) and Madon & Abolins (1999). NWB, NW Balingian sub-province; SWB, SW Balingian sub-province; WBL, West Balingian Line; WBRL, West Baram Line.

the Luconia–Balingian Province with the West Borneo Basement during the Eocene to Early Miocene. (e.g. James 1984; Koopman 1996; Hall 2002).

The stratigraphic framework of the Balingian Province, and the other provinces of offshore Sarawak, was originally reported to consist of a series of regionally extensive cycles (Fig. 2a) that defined thick (hundreds to thousands of metres) regressive sediment packages bounded by regional flooding surfaces (Ho 1978). These cycles were defined by microfauna and gross facies characteristics and predated the modern concepts of sequence stratigraphy, but were closely aligned with the genetic stratigraphic sequence concept of Galloway (1989). Eight cycles were recognized and referred to as cycles I–VIII (Oligocene to Pliocene), based mainly on their occurrence within the Baram Delta Province towards the NE of Balingian Province

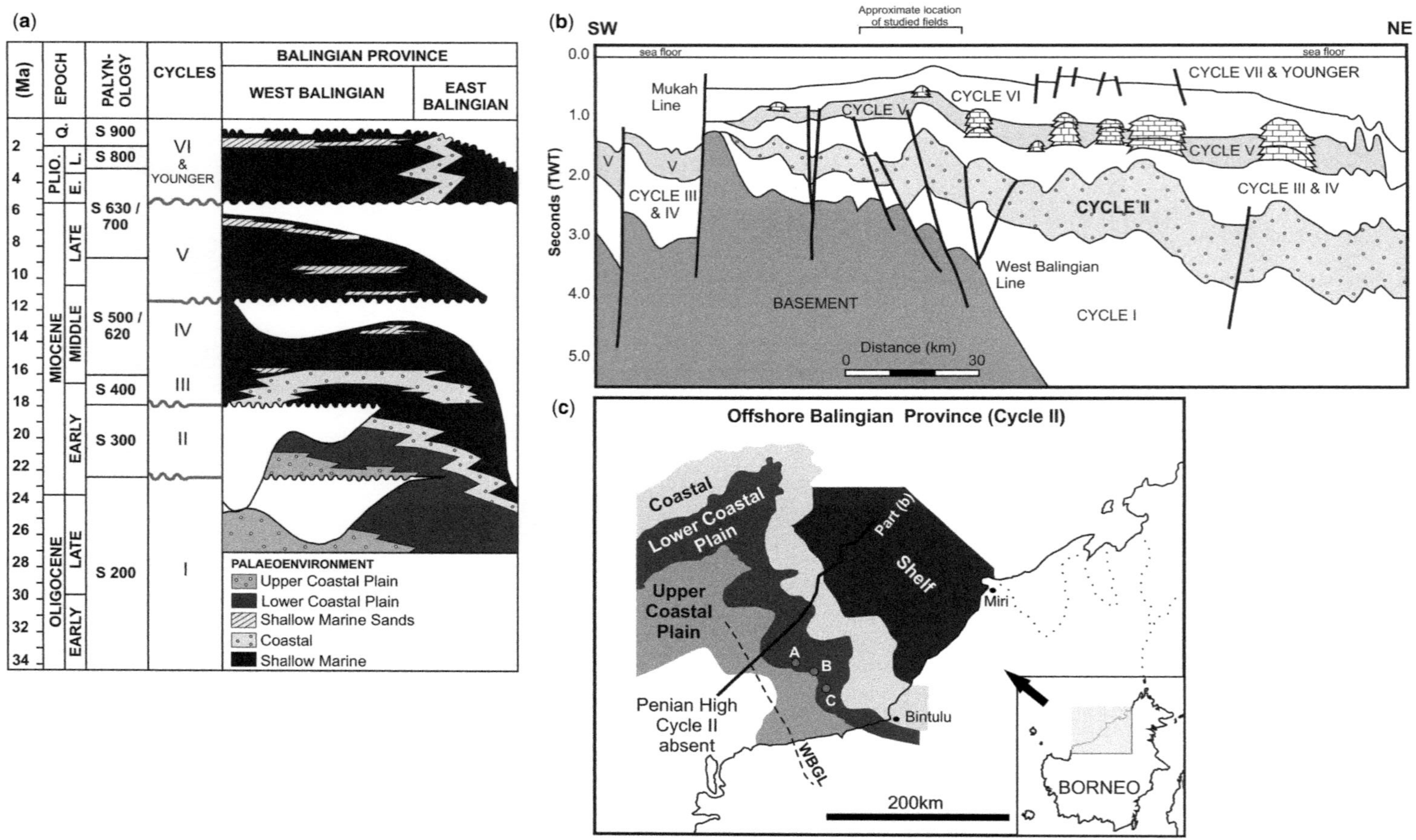

Fig. 2. Stratigraphic framework of the Balingian Province. (**a**) Generalized stratigraphy of the Balingian Province. From Madon & Abolins (1999). (**b**) Geoseismic section oriented SW–NE across the Balingian Province. Note the thickness variations in cycle II associated with syndepositional faulting. Note also the thinning of cycle II towards the West Balingian Line fault zone. Redrawn from Ismail & Tucker (1999). (**c**) Composite Early Miocene palaeogeography of the Sarawak Basin, modified from Amir Hassan *et al.* (2013). Location of studied fields shown as red dots. Location of drawn seismic section in (b) marked as bold, solid line. WBGL, West Balingian Line.

(Figs 1 & 2). However, the cycle boundaries were more difficult to identify in the Balingian Province as a result of (1) the difficulty of identifying flooding surfaces in successions dominated by non-marine/lower coastal plain sediments (Ismail & Tucker 1999) and (2) increased tectonic activity identified in the form of regional (seismic-scale) unconformities (Ismail & Swarbrick 1997). Nonetheless, the biostratigraphic calibration of cycle boundaries defined by flooding surfaces and seismically defined unconformities provides a widely applied stratigraphic framework (e.g. Hageman 1987) that has been adopted in this paper.

This study focused on rocks of Early Miocene cycle II in the West Balingian sub-province, which is laterally equivalent to the onshore Nyalau Formation (Leichti *et al.* 1960; Ho 1978; Hageman 1987; Madon & Abolins 1999; Wan Hasiah 1999; Amir Hassan *et al.* 2013). The cycle II–Nyalau Formation interval is characterized by a thick succession (200–2000 m) of coastal and coastal plain deposits, often with *in situ* coals and minor interbedded shallow-marine shales and limestones (ASCOPE 1981). Thickness variations in cycle II reflect a combination of (1) the irregular topography of the basal unconformity, (2) the top unconformity and (3) syndepositional faulting (Fig. 2b). The base and top of cycle II are marked by angular unconformities in the West Balingian sub-province, which merge into correlative conformities eastwards (basinwards) (Hageman 1987; Madon & Abolins 1999).

During the Early Miocene (cycle II), the study area was located along the western margin of a large embayment. (Amir Hassan *et al.* 2013) (Fig. 2c). The palaeocoastline trended roughly NW–SE, with the basin deepening towards the north and NE (Hageman 1987; Ismail & Swarbrick 1997; Madon 1999*b*; Mansor *et al.* 1999). The palaeoshoreline trend changed to broadly east–west further south in the onshore Balingian/Tinjar Province area (Amir Hassan *et al.* 2013). The sediment source area consisted of the uplifted and fault-bounded Penian High in the SW (Fig. 2c), from where rivers flowed to the NE into the embayment (Almond *et al.* 1990; Madon & Abolins 1999). A transect perpendicular to the palaeocoastline (SW–NE) displays a gradual basinwards change from coastal plain, through coastal and into marine depositional environments (Fig. 2c).

Data and methodology

The project covered an area of 60 × 25 km, which included three producing oilfields in the West Balingian sub-province (referred to here as fields A, B and C; see Fig. 1b for location).

Cores from 12 exploration and appraisal wells were studied, which contained a total of 1100 m of cycle II strata. The cores were described and logged at a 1:30 vertical scale, with an emphasis on lithology, grain size and primary sedimentary structures, trace fossil types and degree of bioturbation. The latter is measured semi-quantitatively in the form of the ichnofabric index (abbreviated as II in this paper; Droser & Bottjer 1986), with II = 1 meaning no bioturbation and the maximum II = 5 meaning completely or thoroughly bioturbated. The rocks have been classified into their distinct descriptive facies (Anderton 1985) and grouped together into facies associations (Figs 3–6). The facies associations are closely tied to different depositional environments. The calibration of facies and facies associations with wireline logs allows the identification of larger scale vertical patterns (lower order facies successions). The core–log calibration results led to the conclusion that the gamma ray log is a reliable guide to predicting the lithologies and facies associations in uncored wells.

Sedimentary facies characteristics

Facies associations

Cycle II in the study area consists of 11 facies associations (Fig. 7), which are summarized in the following sections.

FA1: offshore facies association

Sedimentology. FA1 forms 0.4–12 m thick intervals, mainly consisting of silty bioturbated mudstones interbedded with thin siltstone and sandstone beds, and occasional thin limestone layers (Fig. 8a). The laminated mudstones display interbedded sandstone and siltstone streaks, commonly with normal grading. Thicker siltstone layers (1–4 cm thick) also display erosional bases and normal grading. Thin beds (2–15 cm thick) of micro-hummocky cross-stratified sandstones and laminated sandstones are also commonly interbedded with mudstones. Millimetre-thick lenses of very fine to fine-grained sandstones form symmetrical ripples in lenticular-bedded mudstones, displaying bundled up-building of cross-laminae, swollen lens-like sets and offshoot cross-laminae drapes. Bioturbated sandstones (centimetre-thick) displaying relict hummocky cross-stratification are less commonly intercalated between the mudstones. Limestone occurs as rare, centimetre-thick fine-grained layers. Occasional fossil bivalve shell impressions are preserved in the mudstones.

Ichnology. FA1 is mainly moderately bioturbated (II 2–3), but varies between strong (II 4–5) and weakly to non-bioturbated. The mudstones display

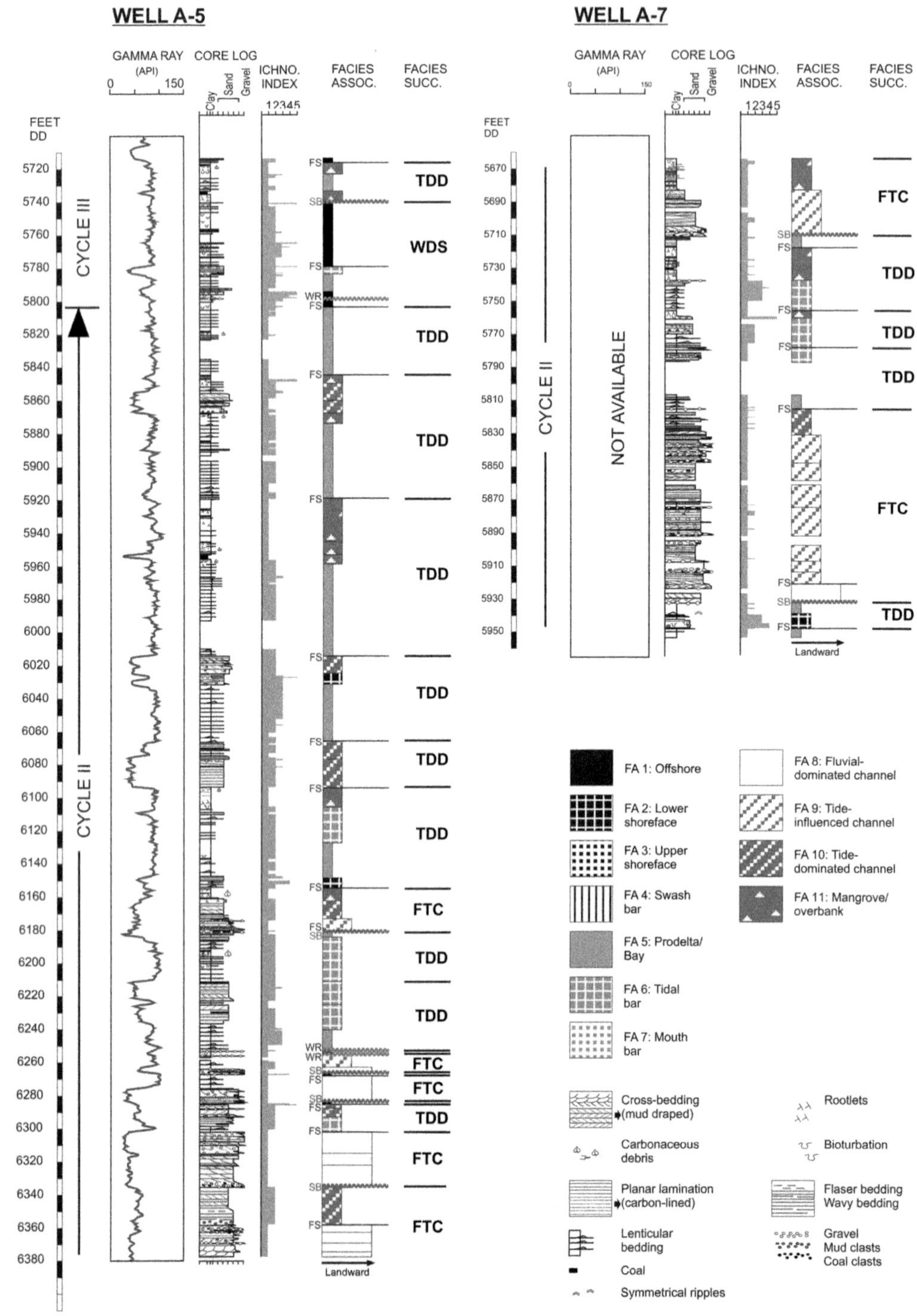

Fig. 3. Sedimentary logs of cores from field A, from Early Miocene cycle II of the Balingian Province, with calibrated gamma ray log, facies associations, facies successions and interpreted sequence stratigraphic surfaces. DD, driller's depth; FTC, fluvio-tidal channel facies succession; TDD, tide-dominated delta front facies succession; WDS, wave-dominated shoreface facies succession; BR, barrier lagoon facies succession; FS, flooding surface; SB, sequence boundary; WR, wave ravinement surface.

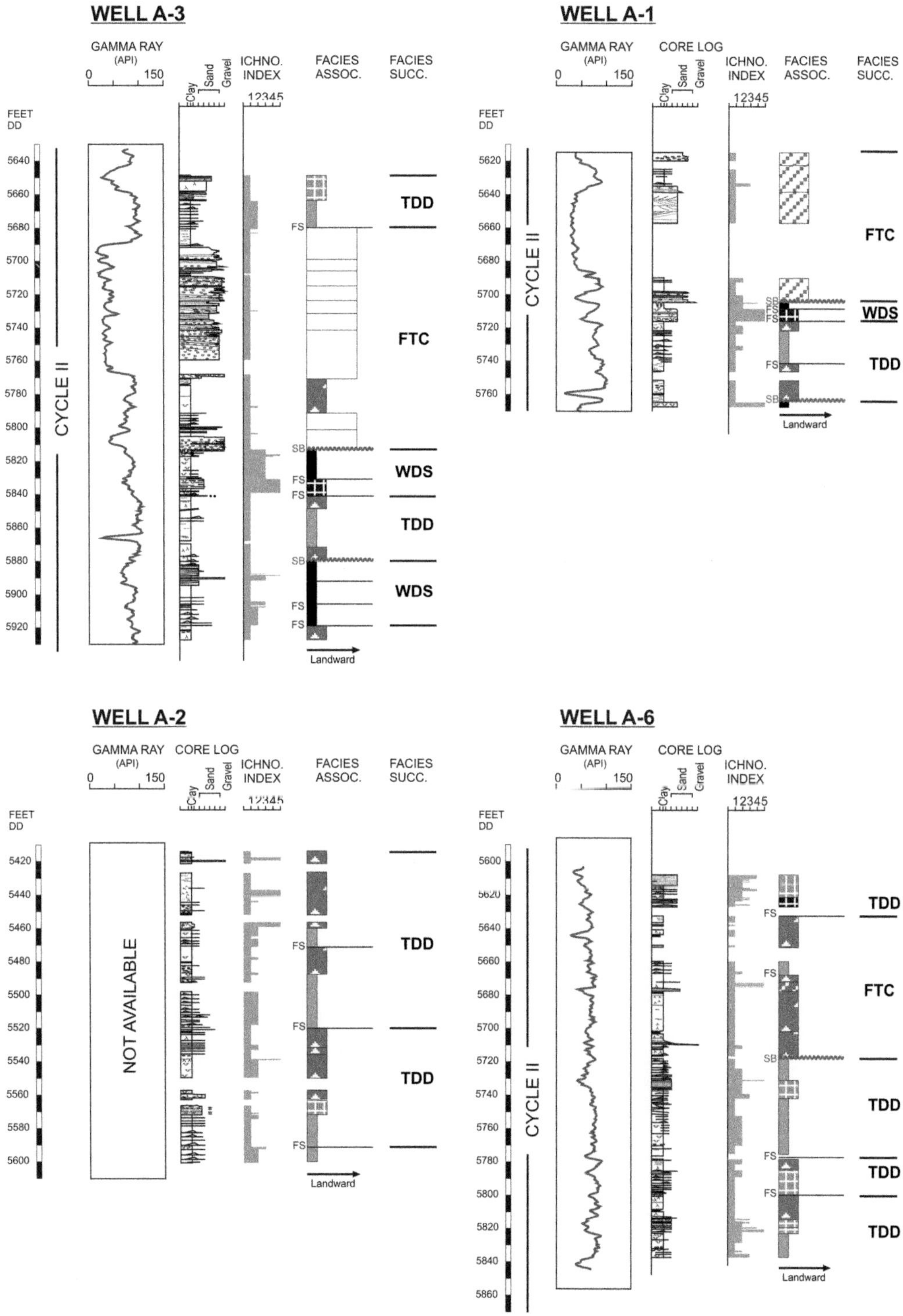

Fig. 4. Sedimentary logs of cores from field A, from Early Miocene cycle II of the Balingian Province, with calibrated gamma ray log, facies associations, facies successions and interpreted sequence stratigraphic surfaces. Refer to Figure 3 for explanation of the abbreviations and key.

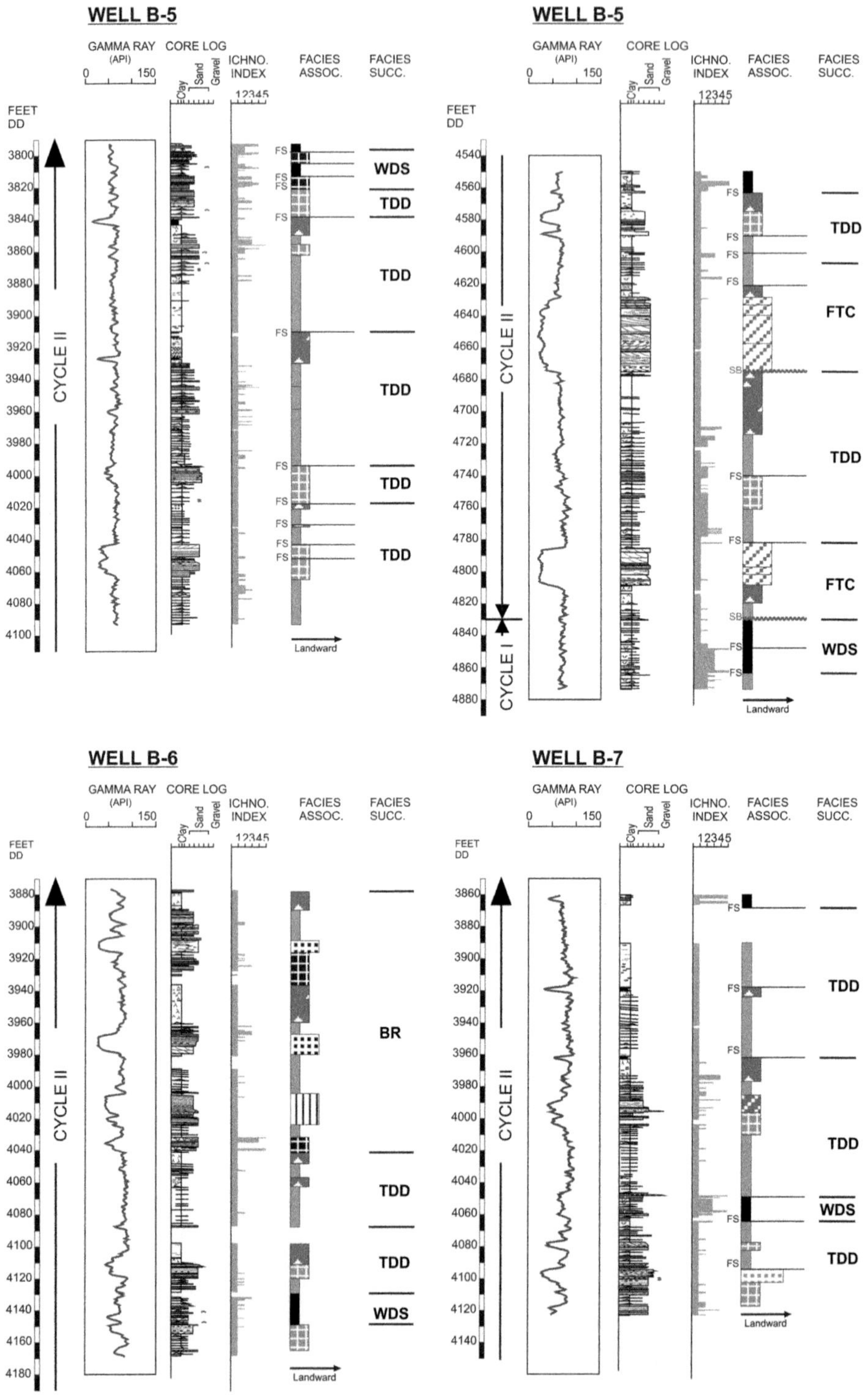

Fig. 5. Sedimentary logs of cores from field B, from Early Miocene cycle II of the Balingian Province, with calibrated gamma ray log, facies associations, facies successions and interpreted sequence stratigraphic surfaces. Refer to Figure 3 for explanation of the abbreviations and key.

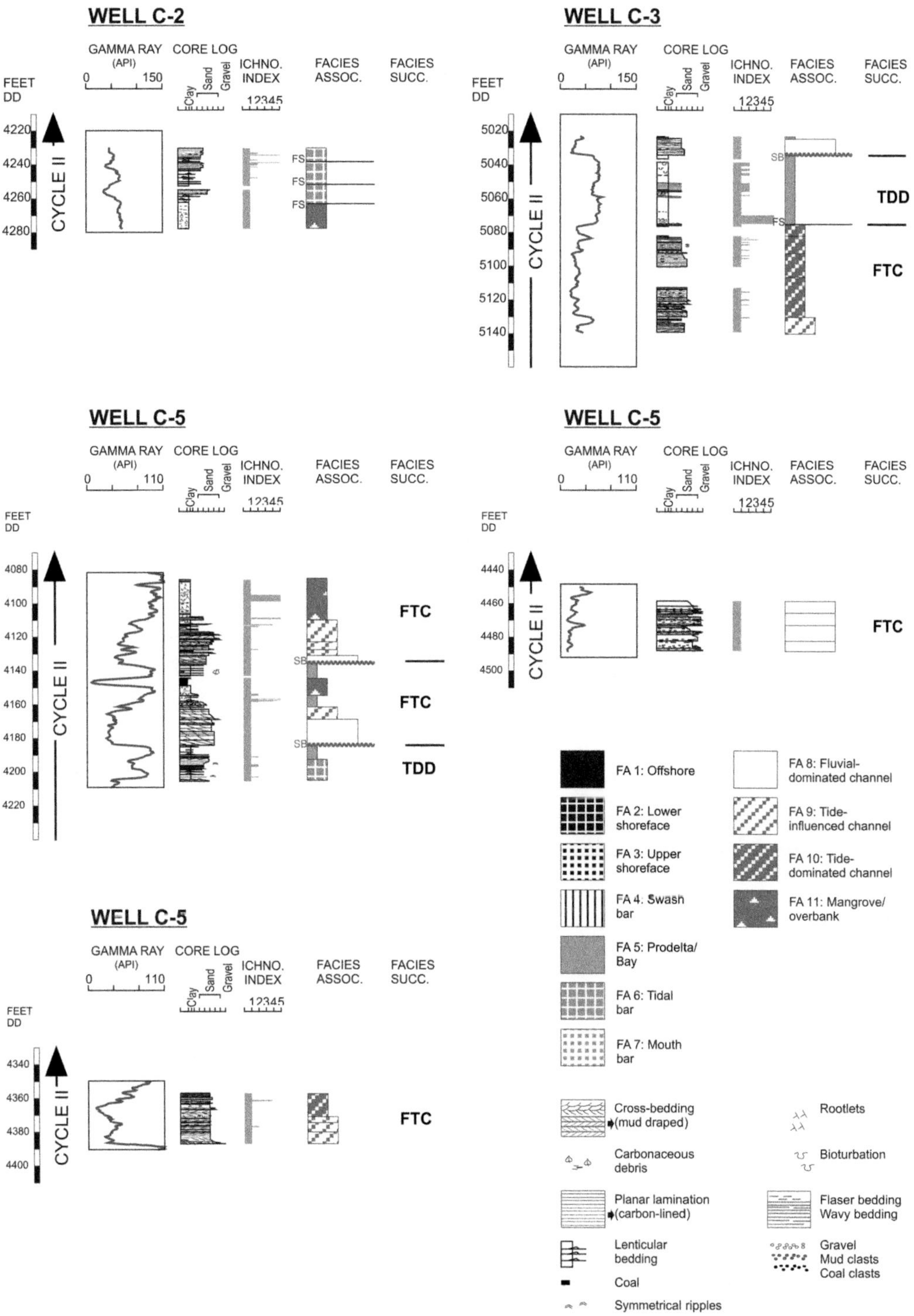

Fig. 6. Sedimentary logs of cores from field C, from Early Miocene cycle II of the Balingian Province, with calibrated gamma ray log, facies associations, facies successions and interpreted sequence stratigraphic surfaces. Refer to Figure 3 for explanation of the abbreviations.

Facies Association		Characteristics	Bioturbation	Interpretation
FA 1: Offshore	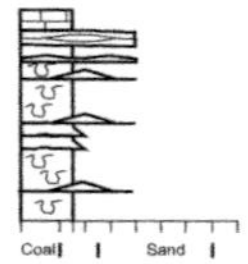	0.3-12 m thick intervals of silty bioturbated mudstone, interbedded with thin, graded siltstone, micro-hummocky cross-stratification, wave ripples, bioturbated sandstone and minor limestone. Sand content <50%	Moderate to intense (II 2-5). Local intervals of none to sparse bioturbation Diverse trace fossil assemblage	Fairweather suspension deposits and thin storm deposits laid down below fairweather wave base in a wave-dominated environment
FA 2: Lower Shoreface	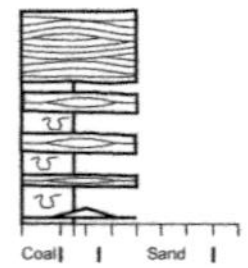	0.3-3 m thick intervals of cm to m-thick, fine-grained HCS sandstone interbedded with thinner bioturbated mudstone and wave rippled layers. Interval sometimes coarsens upward. Moderate bioturbation. Sand content 50-75%.	Sparse to intense (II 2-5). Moderate-low diversity trace fossil assemblage	Alternating fairweather suspension and oscillatory flow deposits, and more proximal storm deposits laid down above fairweather wave base in a wave-dominated environment
FA 3: Upper Shoreface	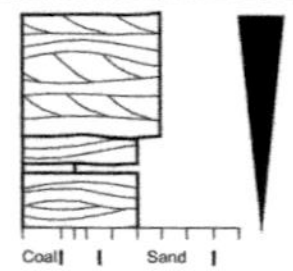	0.3-3 m thick, coarsening and thickening upward intervals of amalgamated, cm- to m-thick, fine-medium-grained HCS sandstone interbedded with cross-bedded sandstone and thin mudstone. Moderate bioturbation. Sand content >75%.	Sparse (II 1-2). Local intervals of intense bioturbation Diverse trace fossil assemblage	Fairweather oscillatory flow deposits, proximal storm deposits, and longshore current deposits laid down in shallower waters above fairweather wave base in a wave-dominated environment
FA 4: Swash bar	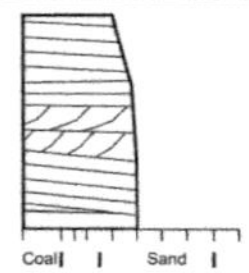	1.5-4 m thick intervals of parallel laminated sandstone interbedded with low angle cross-bedded sandstone and thin heteroliths. Absent-sparse bioturbation. Sand content >75%.	Absent to sparse (II 1-2). Low diversity trace fossil assemblage	Breaking wave swash deposits and possible longshore drift and/or minor tidal current deposit
FA 5: Prodelta	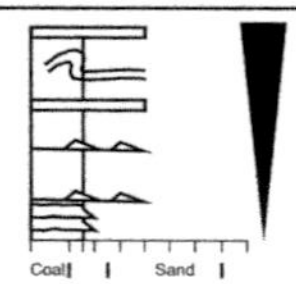	1-17 m thick, coarsening upward intervals of muddy heterolithic bedding, structureless mudstone and graded siltstone. Sand content <50%	Absent to sparse (II 1-2). Low diversity trace fossil assemblage	Prograding sedimentary body composed of low energy tidal current and deposits, fluid mud deposits and occasional storm deposits

FA 6: Tidal Bar	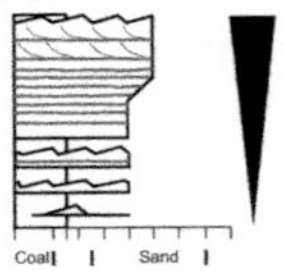	0.5-6 m thick coarsening upward or blocky successions of heteroliths and cross-bedded sandstone. Mud drapes. Sand content 20-80%	Sparse to intense (II 1-5). Low diversity trace fossil assemblage	Prograding barform composed of heterolithic tidal deposits
FA 7: Mouth Bar	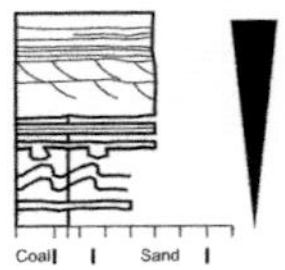	1-5 m thick coarsening upward successions of massive sandstone, cross-bedded sandstone, heteroliths and rippled sandstone. Fluid escape structures and soft sediment deformation. Sand content 20-80%	Absent (II 1).	Prograding barform composed of fluvial deposits mixed with marine suspension, wave and current deposits
FA 8: Fluvial-Dominated Channel	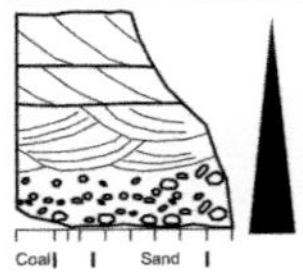	1-10 m thick erosive based, blocky or fining upward intervals of conglomerates and/or trough cross-bedded sandstone with intercalated parallel laminated sandstone and heteroliths. Sand content >75%	Absent (II 1)	Fluvial-dominated channel fill bodies, freshwater rivers
FA 9: Tide-Influenced Channel	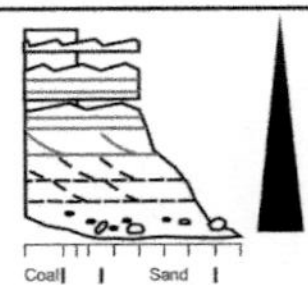	3-6 m thick, erosive based, fining upward intervals of cross-bedded sandstone and heteroliths, sometimes with conglomeratic or pebbly basal lag. Irregularly spaced, thin mud drapes. Abundant carbonaceous debris. Ichnofossils sparse-absent. Sand content 20-80%	Sparse to absent (II 1-2). Low diversity trace fossil assemblage	Fluvial-dominated channel fill bodies, but with significant tidal currents. Dominantly freshwater, but with periodic brackish marine incursions
FA 10: Tide-Dominated Channel	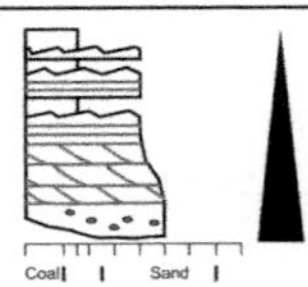	3-6 m thick, erosive based, fining upward successions of fine-medium-grained, cross-bedded sandstone and heteroliths. Regularly spaced mud drapes. Sparse-moderate, impoverished marine ichnofauna. Sand content 20-80%	Sparse to moderate (II 2-3) Low diversity trace fossil assemblage	Tide-dominated channel fill bodies. Brackish water conditions
FA 11: Mangrove / Overbank	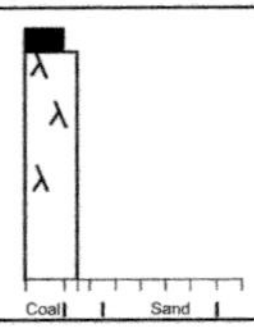	1.5-8 m thick successions of rooted carbonaceous mudstone capped by thin coal seams, and light grey mudstone forming seat earth below coaly layers. Sand content <50%	Absent to intense (II 1-5)	Tide-dominated mangrove flats and swamp or supratidal, freshwater overbank fines and swamp

Fig. 7. Summary of the 11 facies associations identified in studied cores of the Early Miocene cycle II, Balingian Province, offshore Sarawak. HCS, hummocky cross-stratified.

Fig. 8. Photographs of cores from the Early Miocene cycle II interval, Balingian Province, offshore Sarawak. (**a**) Bioturbated mudstone, offshore facies association (FA1). (**b**) Hummocky cross-stratified sandstone, lower shoreface facies association (FA2). (**c**) Amalgamated parallel-laminated and cross-bedded sandstone, upper shoreface facies association (FA3). (**d**) Parallel-laminated sandstone of the swash bar facies association (FA4). (**e**) Lenticular bedding, prodelta/bay facies association (FA5). (**f**) Graded siltstone beds in prodelta/bay facies association (FA5). (**g**) Lenticular bedding associated with the tidal bar facies association (FA6). (**h**) Wavy and flaser bedding associated with the tidal bar facies association (FA6). (**i**) Structureless sandstone of the mouth bar facies association (FA7). (**j**) Parallel-laminated sandstone with abundant organic debris laminae, mouth bar facies association (FA7). Scale bar spacing 2 cm. *Dip*, *Diplocraterion*; *Oph*, *Ophiomorpha*; *Pa*, *Palaeophycus*; *Pla*, *Planolites*; *Si*, *Siphonichnus*; *Te*, *Teichichnus*.

a diverse range of horizontal and vertical trace fossils, including *Arenicolites*, *Siphonichnus*, *Teichichnus*, *Thalassinoides*, *Zoophycos*, *Gyrolithes*, *Phycosiphon* and mantle and swirl structures (Fig. 8a). The sandstone facies are sparsely to moderately bioturbated, mainly consisting of *Conichnus*, *Skolithos*, *Thalassinoides*, *Diplocraterion* and escape trace fossils. Intense bioturbation in sandstones is characterized by general mottling.

Interpretation. FA1 was deposited in a relatively low-energy setting, dominated by fine-grained deposition interspersed with storm events, as evidenced by the interbedded micro-hummocky cross-stratified sandstone beds (Harms *et al.* 1975; Dott & Bourgeois 1982; Duke 1985). The millimetre-thick sandstone and siltstone streaks displaying undulating parallel lamination, and the thicker siltstone beds displaying normal grading, may also represent storm deposition (Aigner 1982), with some events transforming into weak gravity flows travelling below the storm wave base (Walker 1985; Davis *et al.* 1989). The wave ripples are interpreted as the result of reworking by waves active during the waning storm period (Raaf *et al.* 1977).

The variable bioturbation indicates a low-energy, shallow-marine setting. The trace fossil assemblage represents the *Cruziana* ichnofacies, composed of a mixture of suspension and deposit feeding traces, and indicates a subtidal, low- to moderate-energy environment (Pemberton *et al.* 1992*a*). The association of vertical trace fossils of the *Skolithos* ichnofacies within the laminated and hummocky cross-stratified sandstones is consistent with high-energy storm deposits (Pemberton *et al.* 1992*b*). The dominance of mudstone facies and thin storm beds indicates an offshore depositional environment below the fair weather wave base (e.g. Hampson & Storms 2003).

FA2: lower shoreface facies association

Sedimentology. FA2 forms 0.3–3 m thick, coarsening-upwards intervals composed of bioturbated, laminated and sand-streaked mudstones, interbedded with fine- to medium-grained, hummocky cross-stratified, laminated, rippled or bioturbated sandstones. FA2 differs from FA1 in having a higher sand content (around 50–75%). Hummocky cross-stratified sandstone beds are sharp-based and 10–30 cm in thickness. Hummocky cross-stratified and parallel-laminated sandstone beds occur either as individual beds encased in mudstone, or as amalgamated beds reaching up to 1 m in thickness (Fig. 8b). Rippled sandstone beds form units up to 20 cm thick, which are intercalated between muddy facies and hummocky cross-stratified beds. The ripples are symmetrical, display swollen outlines and are associated with sandstone layers displaying undulating internal lamination. The sandy facies are mainly interbedded between laminated, streaked and normal graded silty mudstone beds, bioturbated mudstones and structureless mudstones.

Ichnology. The muddy facies ranges from sparsely to strongly bioturbated (II 2–5). The hummocky cross-stratified beds are generally sparsely to moderately bioturbated by an assemblage composed of *Ophiomorpha*, *Diplocraterion*, *Palaeophycus*, *Skolithos* and escape burrows (Fig. 8b). Some sandstone beds in FA2 are intensely bioturbated, preserving only relict planar lamination. These bioturbated beds can form successions up to 3 m thick and are characterized by indistinct mottling and bored siderite, with occasional identifiable trace fossils, such as *Planolites*, *Thalassinoides* and large *Diplocraterion*.

Interpretation. The facies composition of FA2 generally resembles FA1, with the mudstone facies representing suspension fallout and sandstone beds representing storm deposits. The trace fossil assemblage is also similar, with a *Cruziana* ichnofacies in the mudstone facies and a *Skolithos* ichnofacies in the sandstone facies. However, the relatively higher sand content of FA2, in the form of thicker hummocky cross-stratified sandstones with thin, intercalated muddy facies, indicates a more proximal, lower shoreface setting (e.g. Hampson 2010).

FA3: upper shoreface facies association

Sedimentology. FA3 forms coarsening- and thickening-upwards intervals up to 3 m thick. It is dominated by fine- to medium-grained sandstones (>75% sand) with decimetre- to metre-thick cross-bed sets frequently interbedded with amalgamated beds of parallel lamination and hummocky cross-stratification (Fig. 8c). The cross-bed foresets are occasionally draped by silt and layers of organic debris. Vertical fluid escape structures occasionally cross-cut the primary bedding.

Ichnology. Bioturbation is sparse (II 1–2) and dominated by large *Ophiomorpha* (up to 3 cm in diameter) trace fossils.

Interpretation. The predominance of cross-bedded sandstones intercalated with amalgamated, parallel-laminated and hummocky cross-stratified sandstones, a high sand content and associated marine trace fossils supports a high-energy, shallow-marine setting. Hummocky cross-stratified beds indicate storm deposition (Harms *et al.* 1975; Dott & Bourgeois 1982; Duke 1985; Dumas & Arnott 2006). Parallel-laminated sandstones associated with the hummocky cross-stratified beds result from wave- and storm-generated processes (Duke 1990; Dumas *et al.* 2005). The common occurrence of cross-bedding indicates subaqueous dune bedforms, possibly in longshore troughs and rip-current channels associated with elongate, shoreline parallel bars. The occurrence of rare mud drapes could suggest either a tidal or fluvial influence following storm events. FA3 represents the shallower subtidal shoreface region of a prograding, wave- or storm-influenced shoreline. These features, together with *Ophiomorpha* trace fossils and the *Skolithos* ichnofacies, suggest an upper shoreface depositional environment (Reinson 1984; Hampson 2010).

FA4: swash bar facies association

Sedimentology. FA4 is observed only in cores from well B6 (Fig. 8d). It is characterized by 1.5–4 m thick units of fine- to medium-grained, well-sorted, parallel-laminated sandstones and low-angle cross-bedded sandstones, with minor intercalations of mudstones and heterolithic sediments (>75% sand).

Mud drapes and organic debris laminae occasionally line cross-bed foresets. Mud clasts and coal fragments occur (up to 35 cm in diameter), as do thin intervals of wavy bedding. The intercalated mudstones are commonly sand-streaked with gradational tops.

Ichnology. Bioturbation is absent to sparse (II 1–2) and includes small trace fossils (*Palaeophycus*).

Interpretation. The well-sorted, parallel-laminated sandstones are interpreted as the deposits of swash processes produced by breaking waves. Low-angle cross-bedding may reflect the seawards dip of the swash zone, with the truncation of cross-beds reflecting changes in the beachface slope angles as a result of varying wave conditions. Cross-beds may also represent traction current deposition in ridge and runnel systems. Common mud drapes, mud clasts and thin, wavy-bedded intervals indicate a minor tidal influence. This suggests a protected, back-barrier position, rather than an open marine facing foreshore. Sand-streaked mudstone beds in between thick successions of parallel-laminated sandstone may indicate the presence of locally protected areas (backshore lagoon, small bays). Similar facies have been reported from modern swash bars associated with tidal inlets (Chang *et al.* 2006).

FA5: prodelta/bay facies association

Sedimentology. FA5 is characterized by 1–17 m thick successions of mud-dominated, heterolithic bedding. Lenticular bedding (Reineck & Wunderlich 1968) is dominant, consisting of millimetre-thick, very fine- to fine-grained, cross-laminated, rippled sandstone lenses embedded within thicker mudstone layers (1–3 cm thick) (Fig. 8e). Asymmetrical ripples are dominant, but symmetrical forms are also present. Silt- and sand-streaked and laminated mudstones contain undulating, parallel-laminated sandstone layers (up to 2 cm thick), which fine upwards into mudstones. Thicker beds of graded siltstone (1–3 cm thick) have sharp bases, with either normal or reverse grading (Fig. 8f) and symmetrical ripples with swollen outlines. Thin layers of coaly debris are intercalated between thick intervals of muddy facies. Soft sediment deformation, including convolute lamination, small-scale load structures, load-casted ripples, flame structures and synaeresis cracks, is abundant (Fig. 8f). Siderite bands and nodules are occasionally present.

Ichnology. Bioturbation in FA5 is generally sparse to absent (II 1–2), typically consisting of a mixture of simple vertical and horizontal trace fossils, mainly *Planolites* and *Teichichnus* (Fig. 8e) or *Planolites* associated with *Chondrites* and rare *Zoophycos*. Mantle and swirl structures are common within the thick mudstone layers.

Interpretation. The dominance of mudstones and frequent 'starved' current ripples reflects a combination of limited sand availability, low current and/or wave velocities and alternating periods of suspension deposition of mud and the bedload transport of sand (Reineck & Wunderlich 1968; Raaf *et al.* 1977). In some settings, lenticular bedding can reflect alternating ebb and flood tidal currents, separated by slack water periods, in a tidal setting. However, in this example the mudstone thickness, internal homogeneity and limited bioturbation (mantle and swirl structures) are more consistent with the deposition of fluid mud. The millimetre- to centimetre-thick sandstone and siltstone streaks and the thicker, normally graded sandstone beds all resemble waning flow events, such as either gravity flows (hyperpycnites) or distal storm-generated deposits (Aigner 1982; Davis *et al.* 1989; Walker & Plint 1992). The normal and inverse graded beds indicate sustained flow, which is consistent with a hyperpycnite interpretation (Mulder *et al.* 2003; Bhattacharya & MacEachern 2009). Swollen, symmetrical ripples and sandstones with undulating laminae suggest a significant wave influence in some examples. Coal debris is indicative of reworked terrestrial material and may reflect the settling of organic debris associated with the hyperpycnal flows. The abundance of soft sediment deformation reflects rapid sand emplacement onto uncompacted muds. Synaeresis cracks can be induced by variations in salinity, sediment compaction and/or temperature (Plummer & Gostin 1981). Early (pre-compaction) siderite nodules and bands suggest significant mixing of freshwater and seawater.

The trace fossil suite is an impoverished, distal expression of the *Cruziana* ichnofacies (MacEachern *et al.* 2005; MacEachern & Bann 2008). The small size, low diversity, low abundance and sporadic occurrence of trace fossils indicates a stressed environment, probably due to fluctuating salinities and/or high sedimentation rates. Mantle and swirl structures in the thick mudstones indicate soupy substrates (fluid mud) (Lobza & Schieber 1999). Overall, the trace fossil assemblage is consistent with ichnological facies models for river-dominated deltas and tide-influenced to tide-dominated delta successions (MacEachern *et al.* 2005; MacEachern & Bann 2008).

Together, these observations support a prodelta depositional environment that was influenced by periodic storm and/or river flood events.

FA6: tidal bar facies association

Sedimentology. FA6 mainly consists of 0.5–6 m thick, coarsening- and thickening-upwards successions (Fig. 8g, h), with the following vertical trends (bottom to top): (1) lenticular-bedded, muddy heteroliths (1–5 mm thick mud layers); (2) wavy and flaser bedded heteroliths; and (3) medium-grained, cross-bedded sandstone (e.g. Fig. 3, Well A-5, between 6240 and 6210 ft). The lenticular bedding contains fine- to medium-grained sandstones within asymmetrical ripple lenses, sometimes with small-scale load structures and load-cast ripples. The sandy heteroliths consist of wavy and flaser bedding (Fig. 8h) with 2–5 cm thick sandstone layers with rippled tops and asymmetrical cross-lamination. The mud drapes are thin (1–10 mm). Thicker layers of graded siltstones (5 mm to 2 cm thick) are sometimes interbedded between the sandstone beds. Mud clasts occur at the base of some of the thicker sandstone layers. Thin intervals of parallel-laminated sandstones and hummocky cross-stratified sandstones are also intercalated between some wavy beds. Cross-bedded sandstone beds up to 10 cm thick are occasionally intercalated between heterolithic beds and become more common upwards. Cross-bed foresets are mud-draped or lined with mud clasts.

Ichnology. Bioturbation is generally low (II 1–2) and is mainly represented by small horizontal and vertical trace fossils. The muddy heteroliths are sparsely bioturbated by sporadic *Planolites* with rare, small *Siphonichnus*/*Skolithos*. Mantle and swirl structures are common in the thicker mudstone layers and drapes. Occasional *Zoophycos* are also present. Some muddy intervals display a denser (II 3–5) and more diverse assemblage, which includes *Planolites*, *Skolithos* and other ichnotaxa, such as *Asterosoma*, *Teichichnus*, *Thalassinoides* and *Gyrolithes*. The sandy heteroliths and cross-bedded sandstones are dominated by *Arenicolites*, *Diplocraterion* and *Palaeophycus*, with *Planolites* and occasional *Teichichnus* in the mud drapes. Some of the laminated sandstone intervals contain organic debris laminae with organic debris-lined *Palaeophycus*.

Interpretation. The upwards coarsening and thickening of beds reflects either progradation or the lateral accretion of a sand body composed of current ripples and larger subaqueous dune bedforms. The latter become more common upwards in response to gradual shoaling and an increase in current strength. Tidal processes are inferred from the abundant and regularly interlayered mud drapes in wavy and lenticular bedding, and in the cross-bed foresets and rhythmites (Reineck & Wunderlich 1968; Nio & Yang 1991). The thicker mudstone layers in the lenticular-bedded facies are interpreted as fluid muds, based on their internal homogeneity and negligible bioturbation, except for mantle and swirl structures (Lobza & Schieber 1999; Schieber 2003). The normally graded siltstone beds represent waning flow events, such as either gravity flows (hyperpycnites) or distal storm-generated deposits (Aigner 1982; Davis *et al.* 1989; Walker & Plint 1992; Mulder *et al.* 2003). The intercalated hummocky cross-stratified beds suggest occasional wave and storm influences.

A brackish-water setting is supported by the small size, low abundance, low diversity and variable intensity of bioturbation (MacEachern *et al.* 2005; MacEachern & Bann 2008). The trace fossil assemblage is also consistent with an estuarine to tide-influenced delta depositional setting (MacEachern *et al.* 2005), but most closely matches the tide-influenced delta front ichnological model of MacEachern *et al.* (2005) and MacEachern & Bann (2008).

The overall facies characteristics of FA6 most closely resemble tide-influenced, prograding or laterally accreting sandbodies, such as those formed in estuary or delta distributary mouth settings (Dalrymple & Choi 2007). Modern analogues for FA6 include estuarine tidal bars in the Gironde estuary (Fenies & Tastet 1998), tidal bars in the distributary channels in the Fly river delta (Dalrymple *et al.* 2003) and the bayhead deltas in Brunei Bay (Damit 2001; Lambiase *et al.* 2003).

FA7: mouth bar facies association

Sedimentology. FA7 consists of 1–5 m thick, coarsening-upwards successions, mainly composed of structureless sandstones interbedded with cross-bedded sandstones and wavy-bedded or current ripple cross-laminated sandstones. Structureless sandstones form poorly stratified intervals up to 2.5 m thick (Fig. 8i). Pebbles 'float' sporadically in the sandstone matrix and there are intercalations of decimetre-thick, fining-upwards units of medium- to coarse-grained, cross-bedded sandstones. The cross-bed foresets are occasionally lined by mud clasts, mud drapes or organic debris. Parallel-laminated sandstones are also present and contain laminae of organic debris (Fig. 8j). Thinly interbedded sandstone and mudstone layers commonly display soft sediment deformation, including folded and faulted beds and load structures (Fig. 8j). FA7 typically forms the top of coarsening-upwards facies successions, commonly in gradational contact with prodelta (FA5) and/or tidal bar (FA6) deposits.

Ichnology. Bioturbation is generally absent (II1).

Interpretation. The thick, structureless sandstone beds with abundant fluid escape structures and 'floating' pebbles are interpreted as the result of fluidization and liquefaction due, in part, to rapid deposition. Laminated sandstone beds are interpreted as plane bed, upper flow regime deposits. The abundance of laminae rich in organic debris is consistent with a nearby terrigenous sediment source. The presence of cross-bedded sandstones indicates sufficient current flow to generate subaqueous dunes (Baas *et al.* 2015). Thin wave ripples intercalated with the other facies indicate occasional reworking by waves. The absence of any marine bioturbation supports a highly stressed, possibly freshwater, environment (Pemberton *et al.* 1992*a*). These features suggest that FA7 was deposited in a river-dominated mouth bar setting (e.g. Elliott 1974; Wellner *et al.* 2005; Olariu & Bhattacharya 2006; Olariu *et al.* 2010).

FA8: fluvial-dominated channel facies association

Sedimentology. FA8 consists of 2–3 m thick, fining-upwards successions with erosional bases, which are overlain by *c.* 0.3–1 m thick conglomerates or pebbly sandstone lags (Fig. 9a). The conglomerates are composed of sub-angular to rounded pebbles and granules in a coarse- to very coarse-grained sandstone matrix. Clasts are composed of quartz, sedimentary rock fragments, mud clasts and coal fragments. The conglomerates are cross-stratified and usually fine upwards, with co-sets up to 1 m thick and individual sets *c.* 2–15 cm thick. Cross-bedded sandstones are the dominant facies, consisting of decimetre- to metre-thick beds of well-sorted, coarse- to very coarse-grained sandstones displaying trough cross-bedding (Fig. 9b). Thin beds of parallel-laminated sandstones (up to 5 cm thick) are intercalated between the cross-beds. Mud drapes, mud layers (centimetre-thick), laminae rich in organic debris, and thin coal layers are common. Some sandstone intervals fine upwards into lenticular and wavy-bedded sand streaks or lenses and rhythmically laminated mudstones. Thin conglomerate layers (0.5–4 cm thick) occasionally occur between mudstone intervals.

Ichnology. Bioturbation is generally absent in FA8 (II 1), apart from rare, small, deformed sandstone layers in the muddy heteroliths, which may represent mantle and swirl structures.

Interpretation. A fluvial-dominated channel-fill interpretation is supported by the erosional base, fining-upwards trend, a combination of bedforms generated by lower and upper flow regimes, and the general absence of bioturbation. Sandstone and conglomerate packages with sharp, scoured, erosional bases are interpreted as representing active channel-fills. Erosive-based, fining-upwards successions associated with basal scours have been interpreted to be the result of the lateral migration of channels (Allen 1964). Thick cross-stratified granule to pebble conglomerates and pebbly sandstones may represent the lateral or downstream accretion of channel bars. The graded foresets are probably a result of stage fluctuations over the crest of the bar. Trough cross-bedded sandstones indicate bedload transport in the form of the migration of three-dimensional subaqueous dunes (Ashley 1990; Baas *et al.* 2015). The irregular occurrence of mud-draped foresets suggests that they were deposited during slack water periods, but of uncertain origin (e.g. possibly post-flood events). The rare mud-draped foresets probably indicate fluctuations in river stage. Occasional thin beds of parallel-laminated sandstones are interpreted as plane bed, upper flow regime deposits, probably laid down during floods (Stear 1985). The fine-grained facies capping FA8 are interpreted as mud plugs of abandoned channels (Jordan & Pryor 1992). The general absence of marine bioturbation is consistent with a freshwater, fluvial channel environment (Pemberton *et al.* 1992*a*).

FA9: tide-influenced channel facies association

Sedimentology. FA9 forms 3–6 m thick, sharp-based, fining-upwards successions (e.g. Fig. 3, Well A-7 between 5680 and 5710 ft). These successions fine upwards from a basal conglomerate lag, through very coarse-grained sandstones and into heterolithic facies (wavy to lenticular bedding). Medium- to very coarse-grained, trough cross-bedded sandstones are the dominant primary sedimentary structure. Cross-beds display abundant and regularly spaced mud drapes. Organic debris occurs as foreset drapes or dispersed throughout the sandstone. Laminated sandstone beds (up to 1 m thick) are intercalated between cross-bedded sandstones and heterolithic intervals. Heterolithic wavy bedding increases upwards, forming packages up to 2 m thick. The sandstone layers are mainly medium- to very coarse-grained. Mud drapes associated with the wavy beds are mainly of millimetre-scale thickness, but can reach 2–4 cm thick. Some of the mud drapes form paired associations (Fig. 9c, d). Synaeresis cracks penetrate the mud drapes from the overlying sandstone layers (Fig. 9d). Organic debris also lines the foresets of current ripple cross-laminations. Some heterolithic beds are folded and overlie basal breccias.

Ichnology. Bioturbation is sparse to absent (II 1–2). *Planolites* occurs sporadically in the mud drapes and organic debris layers display a churned texture, with circular, carbon-lined trace fossils (*Palaeophycus*).

Fig. 9. Photographs of cores from the Early Miocene cycle II interval, Balingian Province, offshore Sarawak. (**a**) Conglomerate and cross-bedded sandstone of the fluvial-dominated channel facies association (FA8). (**b**) Cross-bedded sandstone with coaly layer, fluvial-dominated channel facies association (FA8). (**c**, **d**) Tidal indicators in the form of paired mud drapes, but with no bioturbation, tide-influenced channel facies association (FA9). (**e**) Mud drapes and moderate bioturbation in beds of tide-dominated channel facies association (FA10). Note opposing orientations of cross-bed foresets (white arrows). (**f**) Rhythmic lenticular bedding with sparse bioturbation, tide-dominated channel facies association (FA10). (**g**) Rhythmic, sandy heterolithic bedding with thin paired mud drapes, tide-dominated channel facies association (FA10). (**h**) Mudstone with rootlets, mangrove/overbank facies association (FA11). Scale bar spacing 2 cm. *Pa*, *Palaeophycus*; *Pla*, *Planolites*; *Tha*, *Thalassinoides*.

Interpretation. The fining-upwards, erosively based successions, which display both tidal- and fluvial-generated structures, are interpreted as tide-influenced channel-fills. The fining-upwards successions resemble laterally migrating channels. The abundant and regularly spaced mud drapes along cross-bed foresets are suggestive of a tidal influence. Beds displaying paired mud drapes and rhythmic sand–mud couplets resemble tidal cyclicity (Visser 1980; Nio & Yang 1991). The irregular spacing of some of the mud drapes could indicate a combination of both tidal and fluvial processes. The wavy and lenticular bedding is also interpreted as representing tidal deposition (Reineck & Wunderlich 1968). Deformed beds are probably the product of channel bank collapse, as seen in the Nyalau Formation outcrops (Amir Hassan *et al.* 2013).

The paucity of bioturbation, combined with the impoverished marine ichnofauna, indicates a stressed environment, probably with fluctuations in salinity (Gingras *et al.* 2011). This is consistent with a marginal marine depositional setting with both fluvial and tidal influences. The extra-formational nature of the conglomerate lags supports a direct connection to a fluvial sediment source.

FA10: tide-dominated channel facies association

Sedimentology. FA10 forms 3–6 m thick, erosively based, fining and thinning upwards facies succession (e.g. Fig. 3, Well A-5, between 6330 and 6360 ft). The sharp, irregular base is commonly overlain by a mud clast conglomerate. Complete successions fine upwards from fine-, medium- or coarse-grained, cross-bedded sandstones into fine-grained wavy to lenticular-bedded heteroliths. Cross-beds are commonly several centimetres to decimetres thick (mainly 8–15 cm; up to 1 m). Cross-bed foresets are lined by thin (up to 5 mm thick) mud drapes, mud pebbles and organic debris drapes (Fig. 9e). Flaser, wavy and lenticular bedding becomes more dominant up-section, as the mud content and thickness of the mud layers increase (Fig. 9f, g). Wavy and lenticular beds commonly display successive ripple layers with opposing current ripple cross-lamination foreset orientations.

Ichnology. Bioturbation is generally sparse to moderate (II 2–3). *Planolites* and *Teichichnus* are commonly associated with the lenticular-bedded mudstones and mud drapes (Fig. 9e, f). *Palaeophycus, Skolithos, Conichnus* and *Rosselia* occur in the sandier facies (Fig. 9e). *Thalassinoides* with *Siphonichnus* occurs in some heterolithic intervals.

Interpretation. The erosively based, fining-upwards facies successions with abundant tide-generated sedimentary structures and an impoverished marine trace fossil assemblage are interpreted as laterally migrating tidal channels (Allen 1964, 1970; Oomkens 1974). The regularly spaced sand–mud couplets in both the cross-bed foresets and heterolithic beds reflect tidal cyclicity (Reineck & Wunderlich 1968; Visser 1980; Nio & Yang 1991). The impoverished marine ichnofauna, displaying low diversity and simple, diminutive ichnotaxa, is typical of a brackish water setting (MacEachern *et al.* 2005). However, the relatively higher diversity of trace fossils in FA10 compared with FA9 suggests a stronger marine influence, possibly in a more downstream (seawards) location. The thinner mud drapes are also consistent with a location seawards of the zone of elevated suspended sediment concentrations, where mud deposition is inhibited by more regular wave and tidal reworking (Allen & Posamentier 1993; Dalrymple & Choi 2007).

FA11: mangrove/overbank facies association

Sedimentology. FA11 forms 1.5–8 m thick mud-dominated successions and is mainly composed of rooted, coal-bearing dark grey to light grey carbonaceous mudstones (Fig. 9h). Finely dispersed carbon debris, leaf fossils and wood fragments are abundant. Roots and rootlets originate from overlying *in situ* coal layers (*c.* 0.15–1.00 m thick) (Fig. 9h). Slickensided fractures are also common, which, together with the extensive rootlets and organic debris, give the mudstone a crumbly texture in core. Light grey carbonaceous mudstones directly underlie coal seams. Remnants of rhythmic lamination or lenticular bedding and mottled muddy sandstone or silty sandstone beds are present in the lower parts of the successions. However, most of the original stratification has been destroyed by a combination of bioturbation and rooting. There are also occasional thin layers of brecciated mudstones.

Ichnology. Bioturbation is highly variable (II 1–5), but is usually in the form of general mottling and root disturbance of the beds. There are also occasional large vertical trace fossils (*Skolithos*).

Interpretation. The *in situ* soil profiles, pedogenic reworking and extensive bioturbation are consistent with an intertidal mangrove swamp depositional environment. The mud-dominated succession indicates mainly low-energy conditions, with the deposition of mud from suspension. The relict preservation of sand streaks, laminae and lenticular bedding indicates periodic higher energy tidal current flows within an otherwise low-energy tidal environment. There is abundant evidence for subaerial exposure, including palaeosol horizons with slickensides and rootlets. The high organic content of the dark mudstone and the presence of fossil wood, rootlets and thin coals indicates a densely vegetated environment. Organic petrological studies on the coals indicate that they are mangrove-derived (Wan Hasiah & Abolins 2006). A mangrove interpretation is consistent with the close association of FA11 with the underlying tide-dominated and tide-influenced facies associations. However, FA11 also gradually overlies fluvial-dominated channel-fill deposits, which suggests that some intervals represent supratidal vegetated floodplains.

Large-scale facies successions

Large-scale (up to 30 m thick) stacking patterns of successive facies associations, which have been calibrated by distinctive gamma ray log patterns (Figs 10–12), define the following four core-based facies successions: (1) fluvio-tidal channel facies successions; (2) tide-dominated delta front facies successions; (3) wave-dominated shoreface facies successions; and (4) barrier lagoon facies successions. The four facies successions are interpreted within the framework of a tide-dominated deltaic depositional system, with two distinct facies trends: (1) proximal to distal (fluvial → mixed fluvio-tidal → tide-dominated); and (2) axial to lateral (tide-dominated → tide-influenced → wave-dominated) (Fig. 10).

Fluvio-tidal channel facies succession

This facies succession consists of 10–30 m thick, erosively based, channel-fill deposits, which fine upwards from fluvial-dominated into tide-influenced and tide-dominated channel-fill facies associations (FA8, FA9 or FA10). Mangrove facies associations (FA 11) are occasionally preserved between the channel-fills, although many of the observed fluvio-tidal channel facies successions are incompletely preserved. The gamma ray profiles typically display a blocky or generally bell-shaped pattern (e.g. Fig. 3, Well A-7, between 5810 and 5930 ft). Fluvio-tidal channel facies successions are commonly overlain by tide-dominated delta front facies successions.

Complete fining-upwards, fluvio-tidal channel facies successions can be interpreted as the result of two possible mechanisms within a tide-influenced lower delta plain setting: (1) infill during rising base level (e.g. by allocyclic/transgressive drowning); or (2) the lateral migration of channel belts (e.g. by autocyclic channel shifting/channel abandonment). In the former mechanism, channels evolve from a braided river system into a gradually more tidally influenced river system, located in a more seawards position relative to the fluvial-dominated channel deposits (e.g. Allen & Posamentier 1994). In the latter mechanism, channel switching in a lower delta plain setting results in a progressive increase in the degree of tidal reworking (e.g. Allen & Chambers 1998). The tide-influenced channel-fills probably represent the deposits of bank-attached point bars, lateral bars in straight distributaries and mid-channel, laterally accreting tidal bars. The gradual upwards increase in bioturbation from FA8 to FA10 is consistent with either of these depositional models. The fluvio-tidal channel facies succession could represent a combination of the following allocyclic and autocyclic processes: (1) decreasing discharge; (2) peneplanation of the hinterland; (3) avulsion; or (4) a rise in relative sea-level.

Tide-dominated delta front facies succession

Tide-dominated delta front facies successions form 5–20 m thick, coarsening-upwards packages, which pass upwards from the prodelta (FA5) into the tidal bar facies association (FA6). Some successions have a thin offshore (FA1) and/or lower shoreface facies association (FA2) underlying the prodelta deposits, possibly indicating a more lateral position within the delta front. The uppermost part of these successions may be capped by a combination of tidal bar (FA6), mouth bar (FA7) and tide-dominated channel (FA10) deposits, indicting a more axial facies succession. A muddier variant of the tide-dominated delta front facies succession grades upwards from prodelta/bay deposits, directly into mangrove/overbank deposits. This reflects areas of the lower delta plain that were lateral to the main sand fairways, such as drowned embayments and/or abandoned delta lobes.

The gamma ray profiles display a general funnel-shaped pattern and an upwards reduction in gamma ray values (e.g. Fig. 3, Well A-5, between 6210 and 6260 ft). The facies succession is interpreted as representing the deposits of prograding tide-dominated deltas, based on the coarsening-upwards trend, the presence of deltaic facies and the abundance of tidal indicators. The degree of tidal reworking (e.g. tide-dominated v. tide-influenced) in the delta front mainly reflects the location relative to the fluvial input point (axial v. lateral), the relative degree of fluvial and tidal reworking (proximal to distal) and water depth. Thick successions with basal offshore deposits indicate distal deltaic progradation directly onto the marine shelf. Successions consisting of prodelta deposits at their base and stacked tidal bars towards the top reflect more axial to proximal delta front locations. The stacked tidal bars were probably deposited in shallow water on the subaqueous delta platform and in tide-dominated distributary channels, similar to those in the modern day Fly (Dalrymple *et al.* 2003) and Ganges–Brahmaputra deltas (Kuehl *et al.* 1997). Tide-dominated channel-fills associated with the elongate tidal bars probably represent inter-bar channels and/or the distal extension of distributary channels (Fig. 10).

Intercalated fluvial- and wave-dominated deposits within the tide-dominated delta front facies successions reflect axial to lateral variability on the delta front. The upwards transition of some tidal bars into proximal mouth bar deposits indicates proximity to fluvial-dominated distributary mouths. Thin hummocky cross-stratified sandstone units (1–2 m thick) in the tide-dominated delta facies successions represent periodic storm reworking on the delta front. The mainly sparsely bioturbated hummocky cross-stratified beds in these examples differ from their equivalents within wave-dominated shoreface systems, which are typically more strongly bioturbated (Pemberton *et al.* 1992*b*; MacEachern *et al.* 2005). This may reflect the brackish waters and more abundant sediment supply within a deltaic depositional setting (MacEachern *et al.* 2005). Prodelta/bay deposits (FA5) directly capped by mangroves (FA11) represent embayed muddy shorelines located in the interdistributary

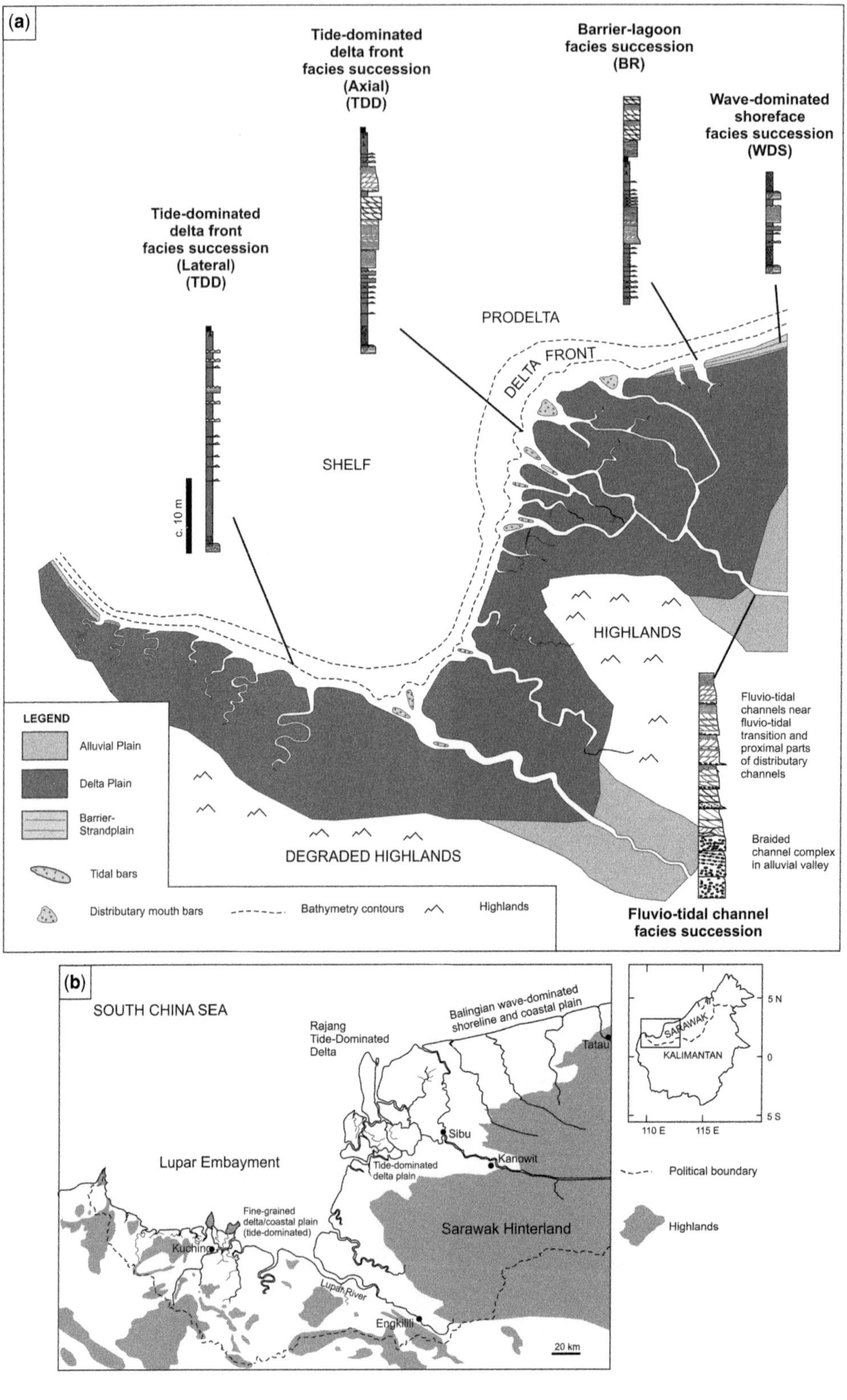
(a)
Tide-dominated delta front facies succession (Axial) (TDD)
Barrier-lagoon facies succession (BR)
Wave-dominated shoreface facies succession (WDS)
Tide-dominated delta front facies succession (Lateral) (TDD)
c. 10 m
PRODELTA
DELTA FRONT
SHELF
HIGHLANDS
DEGRADED HIGHLANDS
Fluvio-tidal channels near fluvio-tidal transition and proximal parts of distributary channels
Braided channel complex in alluvial valley
Fluvio-tidal channel facies succession
LEGEND
Alluvial Plain
Delta Plain
Barrier-Strandplain
Tidal bars
Distributary mouth bars
Bathymetry contours
Highlands
(b)
SOUTH CHINA SEA
Balingian wave-dominated shoreline and coastal plain
Rajang Tide-Dominated Delta
Tatau
Sibu
Kanowit
Lupar Embayment
Tide-dominated delta plain
Fine-grained delta/coastal plain (tide-dominated)
Kuching
Sarawak Hinterland
Lupar River
Engkilili
20 km
SARAWAK
KALIMANTAN
5 N
0
5 S
110 E
115 E
Political boundary
Highlands

bays lateral from the main fluvial input points. Similarly, thinner tidal bars overlying muddy prodelta/bay deposits may also represent distal/lateral locations on a subaqueous delta platform. The tidal bars are inferred to have had elongate geometries, parallel to the dominant tidal currents, which transported sediments at high angles to the shoreline.

Wave-dominated shoreface facies succession

The wave-dominated shoreface facies successions are 3–20 m thick and typically overly a tide-dominated delta front facies succession. A typical wave-dominated shoreface facies succession displays a lower part that fines upwards from lower shoreface deposits (FA2) into offshore deposits (FA1), followed by a coarsening-upwards part, which grades from the offshore deposits into overlying lower shoreface deposits. Isolated tidal bar deposits may also be present in the lower part of the succession. The boundary between the basal lower shoreface deposits and the overlying offshore mudstones is commonly marked by an intensely bioturbated surface. Lower shoreface deposits are occasionally absent and the offshore mudstone is directly overlain by a fluvio-tidal channel or tide-dominated delta front facies succession. The gamma ray profile of a wave-dominated shoreface facies succession generally has a jagged to slightly bell-shaped pattern, with gamma ray values decreasing up-section (e.g. Fig. 4, Well A-3, between 5810 and 5840 ft).

The wave-dominated shoreface facies succession is interpreted as shoreface deposits based on the dominance of wave- and storm-generated facies. The thin basal lower shoreface sandstone represents a transgressive deposit formed by wave ravinement. Thin tidal deposits associated with the transgressive deposit may represent erosional remnants of tidal inlet or back-barrier deposits below the transgressive surface. Maximum transgression is marked by the strongly bioturbated surface at the top of the transgressive sand. This is followed by a prograding shoreline succession in which thick offshore mudstones coarsen upwards into lower shoreface deposits. More proximal facies (upper shoreface, foreshore) are rarely preserved, possibly due to relative sea-level fall and erosion by overlying deposits.

Barrier lagoon facies succession

There is only a single example of this facies succession preserved in core from well B-6 (Fig. 5, between 3880 and 4050 ft). The succession is *c.* 50 m thick and is composed of alternating packages of wave- and tide-dominated facies associations. The wave-influenced facies associations consist of deposits of lower shoreface, lower to upper shoreface (coarsening-upwards) and swash bar (fining-upwards) environments. The tide-dominated facies is represented by muddy prodelta/bay deposits. The gamma ray profile displays a generally serrated pattern, with high gamma ray intervals interspersed between thin, blocky, bell- and funnel-shaped low gamma ray value curves.

The presence of swash bar deposits, the absence of complete coarsening-upwards shoreface successions, the interbedding of the shoreface deposits with tidal prodelta/bay mudstones and the sharp contact between the wave and tidal facies association strongly suggests that the shoreface deposits represent remnants of transgressive shorelines.

This facies succession is interpreted as representing a transgressive barrier lagoon depositional environment with a wave-dominated barrier shoreface protecting a back-barrier lagoon represented by the bay mudstone deposits of FA5. The swash bar deposits may have been associated with tidal inlet channels, as is observed in modern day examples (Chang *et al.* 2006), but may also represent a swash platform associated with the barrier shoreface.

Depositional model

A mixed-energy depositional model is proposed that consists of two main depositional settings (Fig. 10): (1) a tide-dominated delta system; and (2) a wave-dominated to mixed wave- and tide-influenced coastal system. The dominant environment was the tide-dominated delta system. The fluvio-tidal channel facies successions are interpreted as having been genetically connected updip to river systems and downdip to distributaries and a tide-dominated delta front system. The stacked fluvial-dominated channels in the lower part of the fluvio-tidal channel facies successions are interpreted as bedload-dominated, braided river deposits, possibly filling

Fig. 10. Depositional model for the Early Miocene cycle II interval, Balingian Province, offshore Sarawak. (**a**) The four facies successions observed in cycle II were part of a mixed-energy depositional system divided into a dominant, tide-dominated delta system and a subordinate, wave-dominated to mixed wave- and tide-influenced coastal system. Sandier expressions of the tide-dominated delta front facies succession represent axial positions on the delta system near to major active distributaries, while muddier expressions of the succession represent more lateral positions, away from the main fluvial input. (**b**) Sketch map of the Lupar Embayment and Rajang Delta system of SW Sarawak, which is used as a modern analogue for the cycle II depositional model.

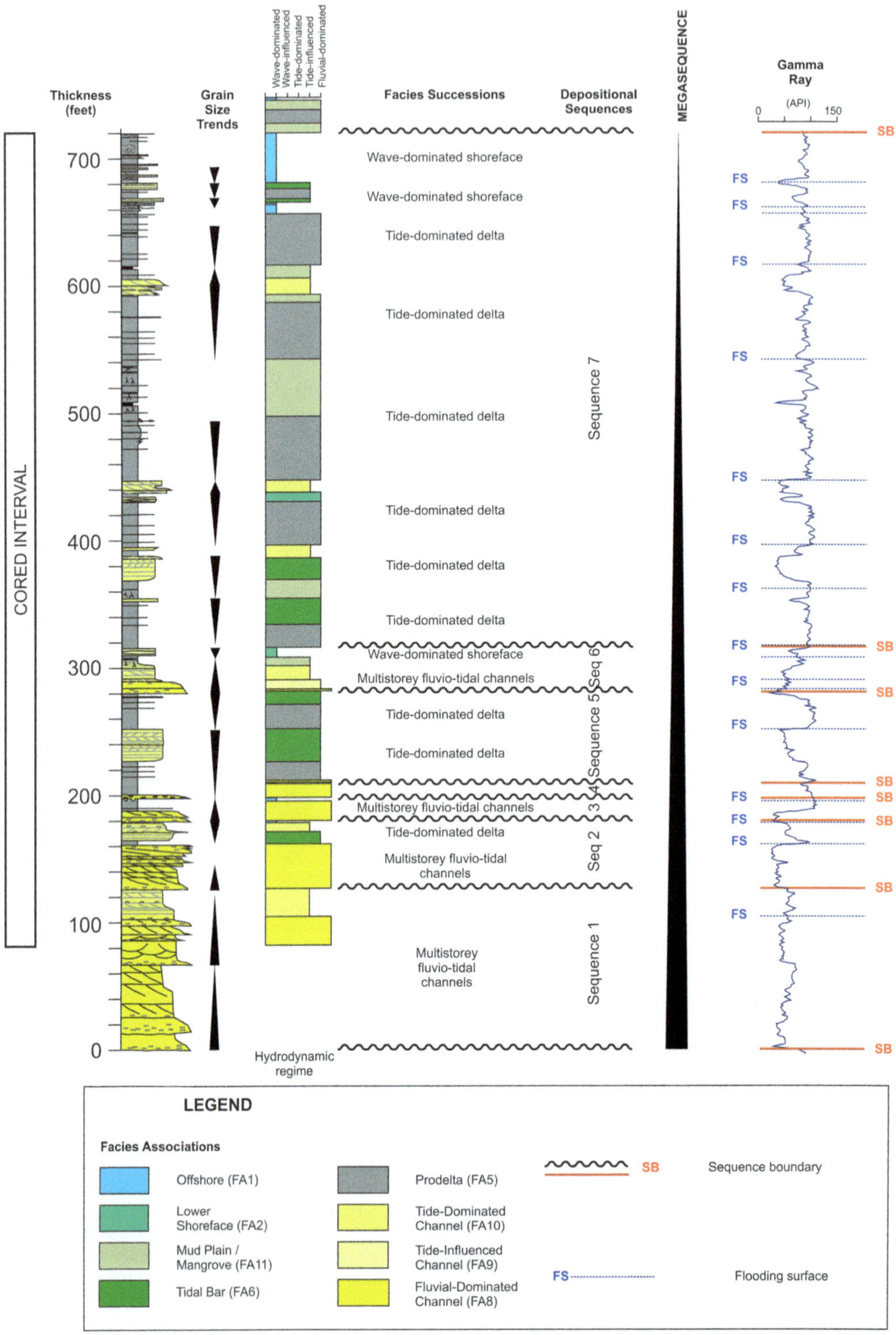

Fig. 11. Summary of the facies composition and stratigraphic architecture of a single, low-order megasequence in the Early Miocene cycle II interval of the Balingian Province, offshore Sarawak. Based on core from well A-5.

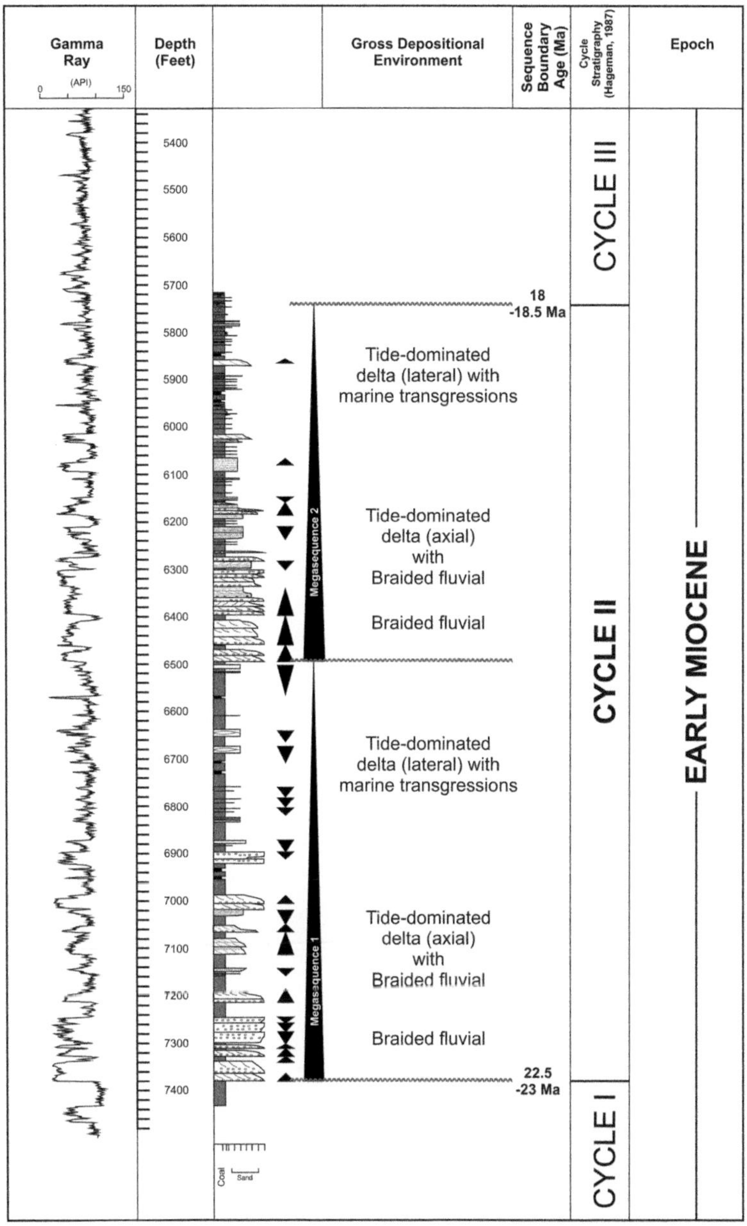

Fig. 12. Generalized log for the two low-order megasequences identified in the Early Miocene cycle II interval of the Balingian Province, offshore Sarawak, based on core descriptions and interpreted gamma ray logs.

broad, shallow alluvial valleys (Fig. 10). The overlying tide-influenced and tide-dominated channels are interpreted as the seawards extension of these channel complexes. A similar transition from gravelly alluvial plain to finer grained, tide-influenced and tide-dominated deltaic deposits is observed in the modern Rajang River delta of southern Sarawak (Staub & Gastaldo 2003).

The tide-dominated delta front facies succession resembles the coarsening-upwards profiles that typify the prograding clinothem of the subaqueous delta platform of modern tide-dominated and tide-influenced delta systems, such as the Ganges–Brahmaputra (Allison *et al.* 2003; Goodbred 2003), Mekong (Ta *et al.* 2002; Tanabe *et al.* 2003), Changjiang (Hori *et al.* 2001), Fly (Harris *et al.* 1993;

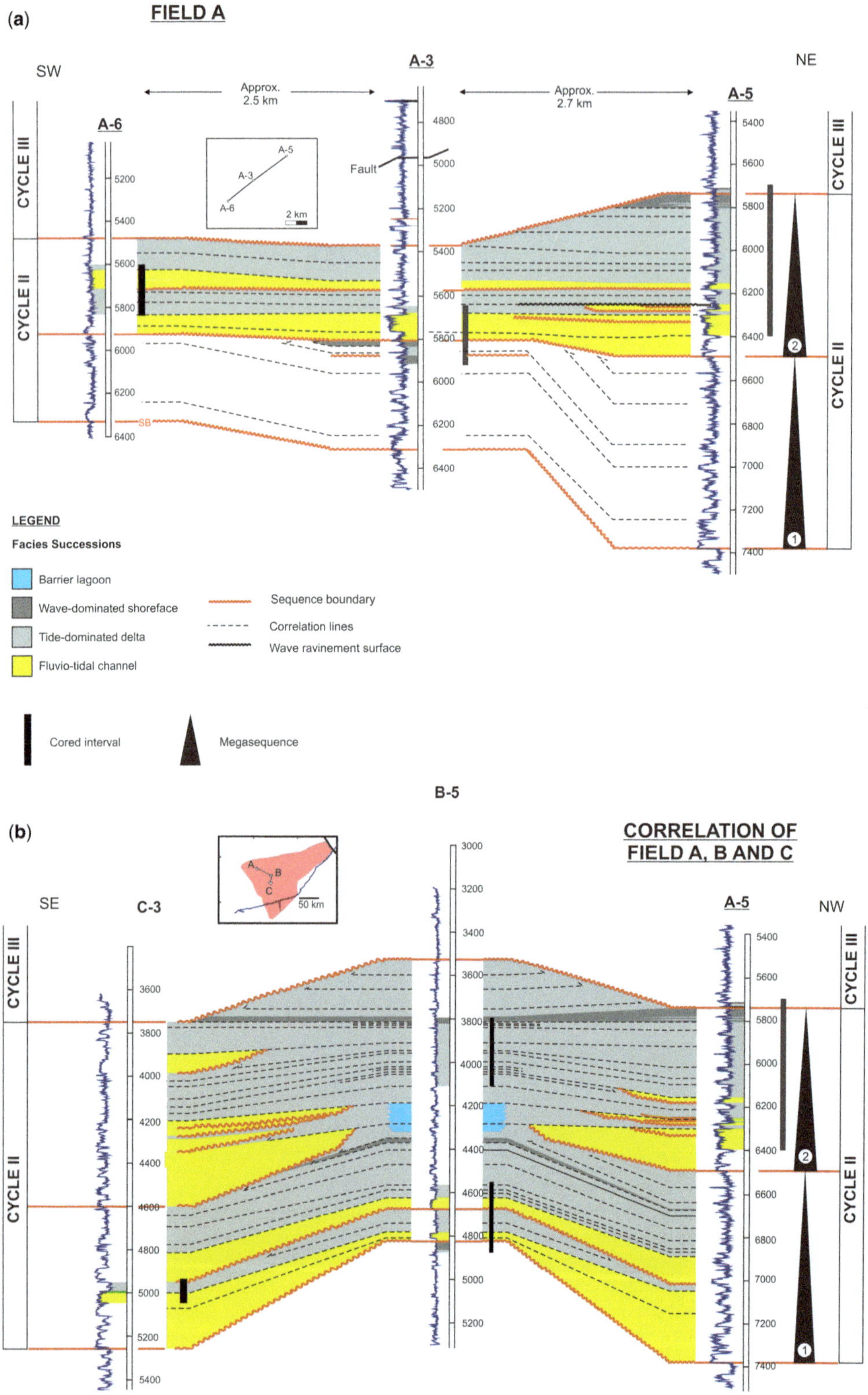
(a)
FIELD A
SW
NE
A-6
A-3
A-5
Approx. 2.5 km
Approx. 2.7 km
Fault
2 km
CYCLE III
CYCLE II
SB
LEGEND
Facies Successions
Barrier lagoon
Wave-dominated shoreface
Tide-dominated delta
Fluvio-tidal channel
Sequence boundary
Correlation lines
Wave ravinement surface
Cored interval
Megasequence
(b)
CORRELATION OF FIELD A, B AND C
SE
NW
C-3
B-5
A-5
50 km
CYCLE III
CYCLE II

Dalrymple *et al.* 2003), Rajang (Staub & Gastaldo 2003) and Mahakam (Allen & Chambers 1998; Roberts & Sydow 2003; Storms *et al.* 2005) deltas. Continued progradation results in the subaqueous delta platform being sharply overlain, and incised into, by the lower delta plain, which is dominated by laterally migrating, tide-influenced distributary channels. In ancient successions, the 'proximal' part of tide-dominated deltas typically consists of a complex assemblage of coarsening- and fining-upwards facies successions composed of variably preserved tidal delta front (subaqueous platform) and tide-influenced distributary channel deposits (e.g. Oomkens 1974; Goodbred 2003; Storms *et al.* 2005; Willis 2005).

Tidal bars (or tidal sand ridges), as interpreted herein, form narrow, elongate sand bodies, which are oriented at high angles to the shoreline and parallel to the direction of the dominant tidal currents (e.g. Willis 2005). They coalesce laterally to form the laterally extensive subaqueous delta platform, which is one of the most areally extensive of all subtidal depositional settings (e.g. Goodbred & Saito 2012). The tidal bars identified in this study consist of coarsening-upwards heterolithic bodies. Meckel (1975) described similar prograding, elongate tidal bars near the mouth of the Colorado River. Smaller scale coarsening-upwards, prograding, elongate tidal bars have also been described from bayhead deltas in the Gironde estuary (Fenies & Tastet 1998) and in Brunei Bay (Lambiase *et al.* 2003).

The close association of the tide-dominated delta front and wave-dominated shoreface facies successions suggests that these coastal depositional systems were geographically closely linked. Hence it is proposed that wave-dominated shorelines were located along-strike from the tide-dominated river mouths that occupied the axial part of the tide-dominated delta (Fig. 10). In such settings, some tidal bars may also have migrated across the adjacent marine shelf.

Stratigraphic architecture

The larger scale facies successions are arranged into a series of erosively bounded, fining-upwards megasequences (*c.* 120–300 m thick) (Fig. 11). The lower part is dominated by fluvio-tidal channel facies successions (*c.* 10–60 m thick and sand-dominated; *c.* 70–90% sand) and the upper part consists of several tide-dominated delta front facies successions (*c.* 60–90 m thick and mud-dominated; *c.* 10–25% sand). The basal contact of each megasequence is erosional, across which there is an abrupt basinwards shift in facies and environment; fluvial channels typically overlie offshore or prodelta deposits of the tide-dominated delta front or wave-dominated shoreface facies successions.

Some of the megasequence boundaries are calibrated with erosional truncations and angular unconformities seen on seismic sections, which define the bases of cycles II, III, V and VI (Fig. 13; Almond *et al.* 1990; Ismail 1997; Jong *et al.* 2014). These are widely interpreted as tectonically generated unconformities, which extended throughout the Balingian Province (Madon & Abolins 1999; Jong *et al.* 2014).

The fining-upwards megasequences display the following upwards changes: (1) progressively thicker prodelta mudstone deposits; (2) thinner sand-dominated tidal bar deposits; (3) an upwards reduction in the proportion of channelized deposits; and (3) a progressive increase in the degree of bioturbation up-section. Although each megasequence displays an overall more distal depositional character upwards, the continued occurrence of mangrove-related coal horizons suggests that the palaeo-water depth was not dramatically different. In contrast, the most striking difference was in the overall reduction in both gross sand content and average sand grain size. Hence the fining-upwards megasequences are interpreted as retrogradational successions.

The fining-upwards megasequences display a distinctive wireline log pattern (e.g. Fig. 12). The basal part starts with a low gamma ray interval, with either blocky or bell-shaped profiles, which represents the fluvio-tidal channel facies succession. This is gradually succeeded by stacked funnel-shaped profiles with interspersed jagged, bell-shaped and/or serrated profiles, which marks the tide-dominated delta front successions. These pass gradually into thin, jagged to slightly bell-shaped gamma ray profiles in the topmost part of the sequence, representing the wave-dominated shoreface and barrier lagoon facies successions of the transgressive systems tract.

In the case of incomplete depositional sequences, there is commonly no abrupt change from a high to low gamma ray profile across the sequence boundary. *In situ* coal layers occur sporadically throughout the fining-upwards megasequences.

Fig. 13. (**a**) Correlation of three wells in field A, Early Miocene cycle II, Balingian Province, offshore Sarawak. There are two large-scale fining-upwards megasequences in cycle II, each with a basal unconformity overlain by braided channel deposits. (**b**) Correlation of cycle II between the three studied fields (fields A, B and C). Each field is represented by single wells with the thickest cycle II sections and with good core data. Intervals with core data are coloured on the wireline logs.

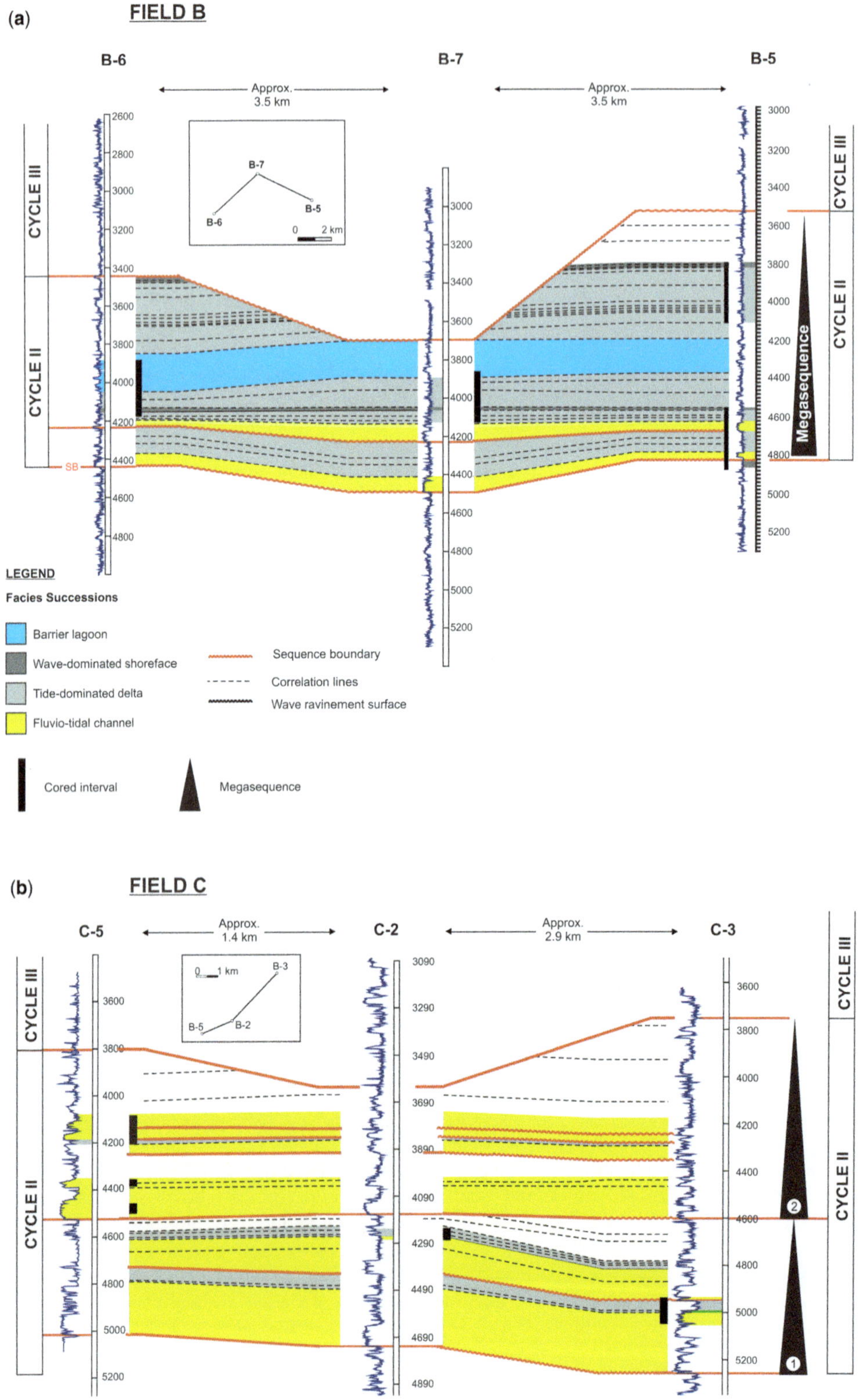
(a)
FIELD B
B-6
B-7
B-5
Approx. 3.5 km
Approx. 3.5 km
CYCLE III
CYCLE II
SB
Megasequence
LEGEND
Facies Successions
Barrier lagoon
Wave-dominated shoreface
Tide-dominated delta
Fluvio-tidal channel
Sequence boundary
Correlation lines
Wave ravinement surface
Cored interval
Megasequence
(b)
FIELD C
C-5
C-2
C-3
Approx. 1.4 km
Approx. 2.9 km
CYCLE III
CYCLE II

The base of the fining-upwards megasequences represent abrupt, erosively bounded basinwards facies dislocations, which are interpreted as subaerial unconformities or sequence boundaries (*sensu* Vail *et al.* 1977; Van Wagoner *et al.* 1988). This is supported by the following observations: (1) they formed by fluvial incision; (2) they result from a relative sea-level fall; and (3) they are of regional extent. Higher order erosion surfaces (e.g. intra-cycle II, separating megasequences 1 and 2; Figs 12 & 13) are likely to have a similar origin, albeit of a more limited lateral extent.

The sharp boundaries separating the basal fluvio-tidal channel facies successions from the overlying tide-dominated delta front facies successions are interpreted as flooding surfaces. Similarly, the sharp boundaries between the tide-dominated delta front facies successions and overlying wave-dominated shoreface/barrier lagoon facies successions are also interpreted as flooding surfaces, probably associated with wave ravinement. The bioturbated sandstones overlying the flooding surfaces probably represent transgressive lag deposits.

The fluvial deposits at the base of each megasequence (part of the fluvio-tidal channel facies succession) represent the lowstand systems tract, whereas the retrogradational package of tide-influenced to tide-dominated channel-fill deposits, stacked tide-dominated delta front successions, barrier lagoon and wave-dominated shoreface successions collectively represent the transgressive systems tract. The wave-dominated deposits in the topmost part of the sequence probably represent the maximum transgression, resulting in drowning of embayed tidal environments, thereby enabling greater access to open marine processes. The highstand systems tract is absent in these sequences. This can be explained either by: (1) high erosional relief of the unconformities due to large falls in relative sea-level, resulting in the complete removal of highstand deposits; or (2) very short regressive highstand periods, resulting in thin highstand systems tracts with low preservation potential. This is consistent with the very high rates of subsidence, as manifested in the extremely thick transgressive systems tracts and the unusually thick, retrogradational stacking of tide-dominated delta front successions. The absence, or poor development, of aggrading highstand deposits may be a further consequence of this tectonostratigraphic setting.

During the deposition of cycle II there were rapid fluctuations in sea-level operating on a timescale of hundreds of thousands of years, based on the high number of depositional sequences (up to six) recognized in vertical successions of cycle II (*c.* 4.5 myr duration; Ismail & Tucker 1999; Madon & Abolins 1999).

Megasequence correlation and lateral variability

Almond *et al.* (1990) recognized an unconformity-bounded, hundreds of metres thick, transgressive succession in cycle II, represented by a sharp-based, fining-upwards gamma ray log profile. These megasequences are *c.* 120–300 m thick and erosional truncations are visible on seismic sections (Ismail & Tucker 1999). One complete megasequence is almost completely cored in well A-5 (220 m) (Well A-5, Fig. 3 and Fig. 11). The lowermost sequence contains a thick fluvio-tidal channel succession, which overlies the basal unconformity. Lowstand fluvio-tidal channel successions gradually become thinner in successively overlying sequences and are absent in the topmost sequences. Tide-dominated delta front facies successions are finer grained and more mud-rich in the topmost depositional sequences. Wave-dominated shoreface and tide-dominated delta front facies successions become more common in the topmost sequences, marking maximum transgression in the megasequences.

In the A and C fields, the cycle II interval in the West Balingian Province can be stratigraphically divided into two megasequences (Figs 12–14). The base of the lower megasequence corresponds with the base of cycle II. Dip-oriented correlations indicate that the megasequences thin out landwards (present day westwards) (Fig. 13). This indicates that two large-scale retrogradational periods occurred during the deposition of cycle II deposits in these fields.

The cycle II interval in field B appears to be represented by a single megasequence, which is also finer grained and less developed than the other fields, with only a thin fluvio-tidal channel succession at the base. Correlation between fields A, B and C indicates that the basal megasequence 2 unconformity is probably present in the B field in the form of a subtle facies change from mudstones

Fig. 14. (**a**) Correlation of three wells in field B, Early Miocene cycle II, Balingian Province, offshore Sarawak. The boundary between the two megasequences in field B is not clear without comparison with logs from other fields, thus it gives the appearance of only a single megasequence in field B. (**b**) Correlation of three wells in field C, Early Miocene cycle II, Balingian Province, offshore Sarawak. Again, two large-scale fining-upwards megasequences are clearly observed in cycle II, each with a basal unconformity overlain by braided channel deposits.

into thin sandstones (at a depth of 4320 ft on the B-5 well log) (Figs 13b & 14a). Therefore cycle II in field B represents a relatively condensed section and the appearance of a single megasequence is due to the subtle unconformity.

The two extensive unconformities/sequence boundaries bounding the cycle II interval in the western Balingian sub-provinces (base of cycle II and base of cycle III) do not coincide with any fall in relative sea-level on the global eustatic curve of Haq *et al.* (1987) (Ismail & Tucker 1999). Hence these unconformities were probably caused by tectonic events, which affected the relative sea-level in the western Balingian (NW and SW Balingian sub-provinces) area (Ismail & Tucker 1999). Syndepositional tectonics were also influenced by the nearby West Balingian Line (Figs 1 & 2), which controlled the location and elevation of the Penian High to the SW of the study area, which was the main fluvial sediment source area during cycles I and II (Fig. 2a, b). The large-scale, retrogradational nature of the megasequences coincided with a large-scale eustatic sea-level rise during the Early Miocene (Haq *et al.* 1987), superimposed on high rates of tectonic subsidence. The unconformities bounding higher order sequences are difficult to correlate, even between wells that are only a few kilometres apart (e.g. Fig. 13), suggesting smaller scale, local tectonic events.

The variation in the thickness of the megasequences, and the relief of bounding unconformities between closely spaced fields, indicates that the western Balingian area displayed a varied topography during the Early Miocene. The thick, conglomeratic, braided channel deposits of fields C and A show the presence of a high-relief sediment source west of these fields (the Penian High; Fig. 2). In contrast, the intervening field B is relatively starved of coarse sediments. The thicker successions in the A and C fields indicate areas of high accommodation space, which were exploited by pebbly braided rivers. The main source of the conglomerates was probably the highlands comprising the Penian High (Figs 1 & 2), which was strongly uplifted during the deposition of cycle II (Mohd Idrus & Redzuan 1999). The thinner cycle II succession in field B indicates an area of low accommodation, which was avoided by the major river systems (Fig. 15).

Modern analogue

The tide-dominated Rajang Delta and the adjacent Lupar Embayment of present day southern Sarawak provide a modern analogue for the depositional systems and their evolution within the cycle II megasequences. The temporal change from active delta progradation to delta abandonment in cycle II megasequences can be visualized spatially by the change in environment from east to west along the Lupar Embayment of Sarawak (Fig. 10b).

During Early Miocene cycle II times, uplifted fault blocks supplied sediment that accumulated in alluvial valleys. The steep topography and high fluvial gradients resulted in the initial development of coarse-grained braided rivers (Fig. 15a, e). Gradual degradation of the highlands decreased the rate of sediment supply to the lower delta plain and river mouths, which is marked by a gradual increase in the tidal influence in the river channel deposits. The alluvial valley-fill gradually became more heterolithic, changing from a fluvial river system into a fluvio-tidal transition and, finally, into a tide-dominated delta plain distributary channel system (Fig. 15b, f). The Pleistocene–Recent Rajang Delta and alluvial valley-fill succession closely resembles the succession observed in the lower part of cycle II megasequences. The Rajang Delta alluvial valley-fill succession exceeds 45 m in thickness and fines upwards from fluvial gravels to sands, heteroliths and muds, with an upwards increase in tidal character (Staub & Gastaldo 2003).

It is inferred that a tide-dominated delta system developed within a large tide-dominated embayment, similar to the present day Lupar Embayment (Fig. 10b). Tidal mouth bars prograded and migrated laterally on top of a muddy prodelta. Coalescing tidal bars and mouth bars formed a sandy heterolithic delta front and a tide-dominated, mangrove-vegetated delta plain developed, criss-crossed by tide-influenced to tide-dominated distributary channels. A similar prograding succession is observed from the front of the Rajang Delta, which is up to 25 m thick (Staub & Gastaldo 2003). Periods of abrupt relative sea-level rise during cycle II deposition resulted in flooding of the delta system and progradation of successive delta lobes. The gradual depletion of sediment supply from the gradually denuded highlands resulted in the retrogradation of successive delta parasequences during cycle II deposition. Successive deltas became muddier and increasingly tide-dominated with time (Fig. 15c, g).

The Rajang Delta, protruding along the east coast of the Lupar Embayment, is supplied by sediment from the Sarawak hinterland, where the elevation averages around 2000 m (Fig. 10b). The Rajang Delta can be used to represent the cycle II delta system during the initial phase of fault block uplift and high sediment supply (Fig. 10b). The main source of sediment for cycle II was the Penian High fault block (Fig. 1b), which was located *c.* 40 km SW of the palaeoshoreline (Fig. 2c). Initially, steep gradient rivers developed, but the stream gradients gradually reduced as a result of degradation of the hinterland. This produced a lower gradient and

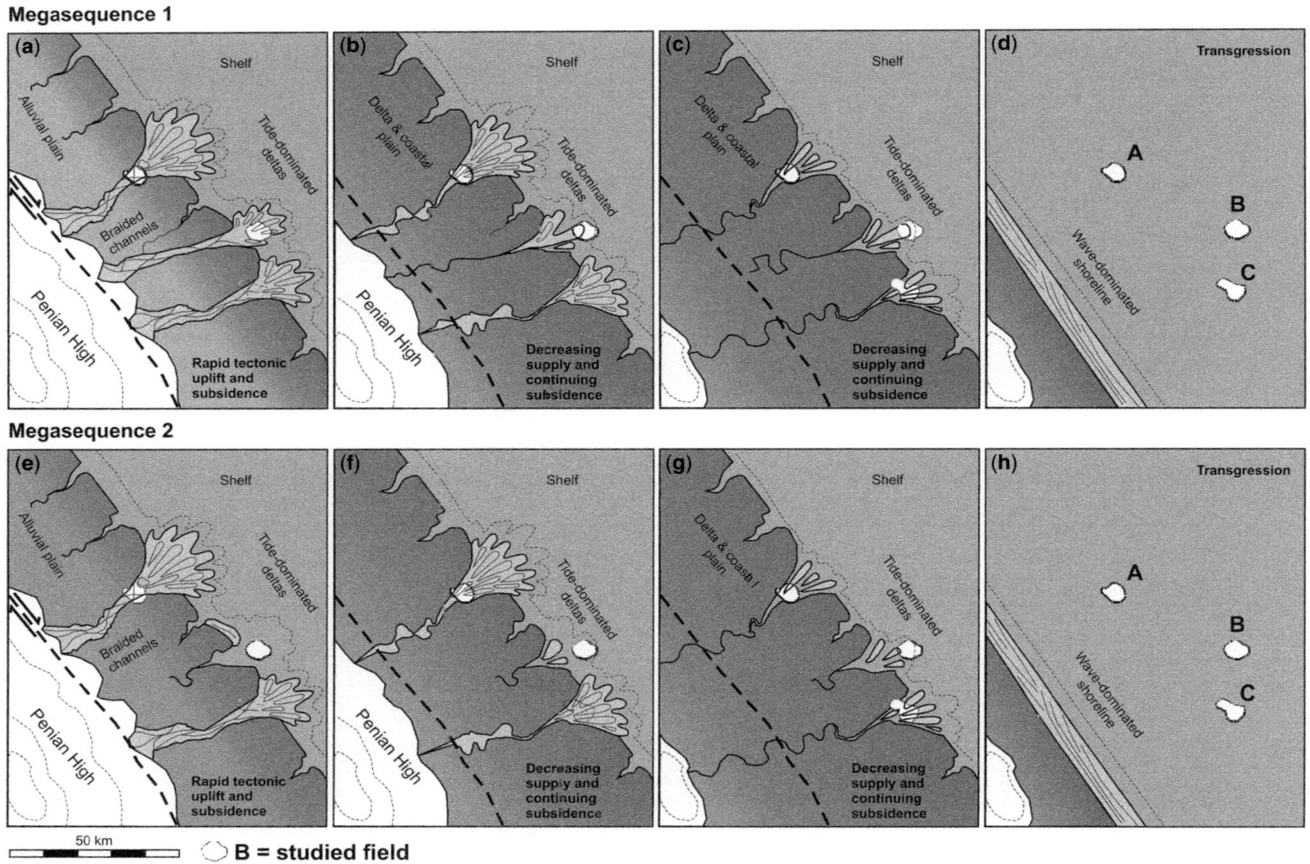

Fig. 15. Depositional evolution of the Early Miocene cycle II interval megasequences representing retrogradation of tide-dominated deltas. (**a**, **e**) Periods of rapid tectonic uplift and subsidence resulted in the aggradation of fluvial tidal deposits. (**b**, **c**, **f**, **g**) Denudation of the highlands resulted in a gradual decrease in the sediment supply. This, coupled with continued subsidence, resulted in the retrogradational succession overlying the basal fluvial deposits. Tide-dominated deltas developed along the shoreline. (**d**, **h**) A wave-dominated shoreline developed during periods of maximum transgression, with very minimal fluvial supply. Large pebbly rivers avoided the area of field B during cycle II deposition, which was an area of low accommodation.

finer-grained delta/coastal plain, with close similarities to the western coastline of the present day Lupar Embayment (Fig. 10b). The latter consists of several northwards-flowing rivers that drain a low elevation mountain range. These rivers discharge into the Lupar Embayment, where they feed multiple tide-dominated river mouths (Fig. 10b). The distance between the sediment source and river mouths is 25–70 km, which is similar to the cycle II rivers that drained the Penian High (Fig. 2c). The Lupar Embayment river mouths crisscross wide, flat, muddy, mangrove-vegetated tidal flats. This depositional setting compares favourably with the delta plain succession in the upper part of cycle II megasequences (Fig. 15).

Discussion

The cycle II tide-dominated delta system is associated with a regional embayment at the SW end of the South China Sea. This basin morphology was a key to amplifying tidal currents during the Early Miocene when the basin was open to the Pacific Ocean (to the NE), but closed in the SW (Collins *et al.* 2015) in the vicinity of the Penian High/Balingian Basin and its lateral continuation into the adjacent, tide-influenced Cuu Long and Nam Con Son basins, offshore south Vietnam (Lee *et al.* 2001). A similar basin morphology, albeit smaller, characterizes the sandstone reservoirs in the Miocene Sihapas Group in the Minas Field (Sumatra), which represents a tide-dominated delta system in a subsiding, low gradient, tide-dominated shelf (Willis & Fitris 2012). The Sihapas deltaic succession is also similar to cycle II in displaying a large-scale (hundreds of metres thick) retrogradational fluvio-deltaic succession with superimposed, smaller scale depositional sequences reflecting higher frequency relative sea-level fluctuations (Willis & Fitris 2012). The development and preservation of retrogradational deltaic systems is explained by subsidence associated with extensional tectonics slightly exceeding the rate of sediment supply. Doust & Sumner (2007) note that similar retrogradational deltaic successions are a common element of the late synrift phase on the Sundaland shelf. Preservation is enhanced by geological settings in which the high rates of creation of accommodation space occur simultaneously with high rates of fluvial sediment supply to the shorelines.

It is not surprising to find that many other ancient tide-dominated to tide-influenced deltaic deposits are part of large-scale, regressive sedimentary successions. This includes the Eocene Lower Dir Abu Lifa Member (Legler *et al.* 2013), the Devonian Kernave, Arukula and Gauja formations in the Baltic Basin (Milkeviciene & Plink-Björklund 2009; Pontén & Plink-Björklund 2007) and the Eocene Misoa Formation reservoirs in the Lagunilas Field in the Misoa Formation, Maracaibo Basin (Venezuela) (Maguregui & Tyler 1991). The Frewens and Sego sandstones (from the USA) both form tens of metres thick regressive, tide-dominated clastic wedges sandwiched between transgressive marine shales.

Other examples of tide-dominated to tide-influenced delta systems developed within compressional tectonic settings, such as the Eocene Dir Abu Lifa tide-dominated delta system (Western Desert, Egypt), which developed within a structurally controlled embayment (Legler *et al.* 2013). The Frewens Sandstone (Western Interior Basin, Wyoming) comprises a tide-influenced delta that was confined to a narrow embayment, associated with a foreland basin (Willis *et al.* 1999). The Cretaceous Sego Member tide-dominated delta system (Utah) is also associated with a foreland basin and formed where the regional shoreline developed an embayment (the 'Utah Bight') (Willis & Gabel 2001). All these examples support the concept of tidal amplification caused by constriction of the tidal wave within a 'blind gulf' type of basin setting.

The Sego Sandstone and Sihapas Group (Minas Field) deltas both display evidence of rapid relative sea-level fluctuations, with sharp-based and/or erosionally bounded sandstones overlying muddy shelf and prodelta deposits. These have been interpreted as forced regressive, tide-dominated delta successions, which overlie regressive surfaces of erosion (Willis & Gabel 2001; Berné *et al.* 2002; Willis & Fitris 2012).

The Balingian Province cycle II deltaic system preserves relatively thick, fine-grained delta plain strata directly above delta front and distributary channel deposits (up to 8 m thick). However, in other examples only the prodelta–delta front successions are preserved, with the delta plain strata absent (e.g. in the Minas Field and the Frewens Sandstone). The absence of any delta plain strata is often explained by their complete erosion during transgression, usually by wave ravinement. However, where the delta front deposits were located in basinwards positions, it is possible that the delta front was completely submerged. For example, the delta front in modern tide-dominated to tide-influenced settings (e.g. the Ganges–Brahmaputra Delta) is typically subaqueous and can extend tens of kilometres basinwards of the subaerial delta plain (e.g. Kuehl *et al.* 2005).

The fine-grained and heterolithic nature of the cycle II delta front deposits is a function of sediment type and availability, the humid–tropical climate and depositional setting. The deep weathering of the hinterlands, with their thick, deep marine mudstone-dominated successions (e.g. the Rajang

Group), under a humid–tropical climate was conducive to liberating large volumes of mud-grade sediments (Hall & Nichols 2002). The abundance of mud reaching the lower delta plain and delta front/prodelta areas would have been enhanced by the multiple river mouths and by the tendency of tidal systems to trap suspended sediment in nearshore areas, partly by mud flocculation (Dalrymple & Choi 2007). The widespread development and preservation of mangrove swamps would have further enhanced the trapping efficiency of mud throughout the lower delta plain. Hence the cycle II deltaic cycles, and many other Neogene humid–tropical, tide-dominated deltas, are associated with both thick prodelta mudstones and mud-rich delta plain successions.

Conclusions

(1) A facies analysis of the Early Miocene cycle II interval in the Balingian Province of NW Borneo indicates deposition within a tide-dominated deltaic system, which was flanked by a more wave-influenced barrier lagoon system.

(2) Cycle II consists of large-scale (hundreds of metres thick), unconformity-bounded, fining-upwards megasquences, which pass upwards from mainly fluvio-tidal channel facies at the base into more tide-dominated delta front facies successions towards the top.

(3) These megasequences define retrogradational coastal–deltaic stacking patterns resulting from the progressive denudation of the hinterland, which was initially triggered by tectonic uplift, within a humid–tropical depositional setting.

(4) The cycle II interval in the A and C fields consists of two megasequences developed in an area of creation of high accommodation. This contrasts with the cycle II interval in the B field, which is relatively condensed and consists of a single megasequence.

(5) The numerous localized, sequence-bounding unconformities and the more laterally extensive, megasequence-bounding unconformities were probably developed due to strike-slip related uplift associated with the West Balingian Line fault zone.

(6) The Miocene cycle II depositional system is compared with the modern Rajang Delta and adjacent Lupar Embayment in southern Sarawak. The latter stages of cycle II deposition are closely comparable with the western coastline of the Lupar Embayment, where multiple rivers feed sediment through funnel-shaped, tide-dominated river mouths to the delta front.

The majority of the work presented here forms part of Meor Hakif's PhD thesis at Imperial College London. We appreciate the sponsorship provided by the Academic Training Scheme (SLAB) from the Ministry of Higher Education, Government of Malaysia. Additional work was completed under a research project at the University of Malaya, supported by the University of Malaya Grant RU011-2013. We are also grateful to PETRONAS, in particular Mohd Hamka Ibrahim, Zakiah Zainol, Azhar Yusoff and Hamdan Mohamad, for enabling access to the subsurface data and for providing support and encouragement for this publication. We thank Herman Darman and an anonymous reviewer for providing constructive feedback, which has been incorporated into the final paper.

References

Aigner, T. 1982. Calcareous tempestites: storm-dominated stratification in Upper Muschelkalk limestones (Middle Trias, SW-Germany). *In*: Einsel, G. & Seilacher, A. (eds) *Cyclic and Event Stratification*. Springer, Berlin, 180–198.

Allen, G.P. 1991. Sedimentary processes and facies in the Gironde estuary: a Recent model for macrotidal estuarine systems. *In*: Smith, D.G., Reinson, G.E., Zaitlin, B.A. & Rahmani, R.A. (eds) *Clastic Tidal Sedimentology*. Canadian Society of Petroleum Geologists, Memoirs, **16**, 29–40.

Allen, G.P. & Chambers, J.L.C. 1998. *Sedimentation in the Modern and Miocene Mahakam Delta*. Indonesian Petroleum Association, Jakarta.

Allen, G.P. & Posamentier, H.W. 1993. Sequence stratigraphy and facies model of an incised valley fill: the Gironde Estuary, France. *Journal of Sedimentary Petrology*, **63**, 378–391.

Allen, G.P. & Posamentier, H.W. 1994. Transgressive facies and sequence architecture in mixed tide- and wave-dominated incised valleys: example from the Gironde Estuary, France. *In*: Dalrymple, R.W., Boyd, R. & Zaitlin, B.A. (eds) *Incised Valley Systems: Origin and Sedimentary Sequences*. SEPM Special Publications, **51**, 225–240.

Allen, J.R.L. 1964. Studies in fluviatile sedimentation: six cyclothems from the Lower Old Red Sandstone, Anglo-Welsh Basin. *Sedimentology*, **3**, 163–198.

Allen, J.R.L. 1965. Late Quaternary Niger Delta and adjacent areas: sedimentary environments and lithofacies. *American Association of Petroleum Geologists Bulletin*, **49**, 547–600.

Allen, J.R.L. 1970. Studies in fluviatile sedimentation: a comparison of fining-upwards cyclothems, with special reference to coarse-member composition and interpretation. *Journal of Sedimentary Petrology*, **40**, 298–323.

Allison, M.A., Khan, S.R., Goodbred, S.L. & Kuehl, S.A. 2003. Stratigraphic evolution of the late Holocene Ganges–Brahmaputra lower delta plain. *Sedimentary Geology*, **155**, 317–342.

Almond, J., Vincent, P. & Williams, L.R. 1990. The application of detailed reservoir geological studies in the D18 field, Balingian Province, offshore Sarawak. *Geological Society of Malaysia Bulletin*, **27**, 137–160.

Amir Hassan, M.H., Johnson, H.D., Allison, P.A. & Abdullah, W.H. 2013. Sedimentology and stratigraphic development of the upper Nyalau Formation (Early Miocene), Sarawak, Malaysia: a mixed wave- and tide-influenced coastal system. *Journal of Asian Earth Sciences*, **76**, 301–311.

Anderton, R. 1985. Clastic facies models and facies analysis. *In*: Brenchley, P.J. & Williams, B.P.J. (eds) *Sedimentology Recent Developments and Applied Aspects*. Geological Society, London, Special Publications, **18**, 31–47, https://doi.org/10.1144/GSL.SP.1985.018.01.03

Ashley, G.M. 1990. Classification of large-scale subaqueous bedforms: a new look at an old problem. *Journal of Sedimentary Petrology*, **60**, 160–172.

Aslan, A., White, W.A., Warne, A.G. & Guevara, E.H. 2003. Holocene evolution of the western Orinoco Delta, Venezuela. *Geological Society of America Bulletin*, **115**, 479–498.

ASCOPE (ASEAN Council on Petroleum) 1981. *Tertiary Sedimentary Basins of the Gulf of Thailand and South China Sea: Stratigraphy, Structure and Hydrocarbon Occurrences*. ASEAN Council on Petroleum, Jakarta.

Baas, J.H., Best, J.L. & Peakall, J. 2015. Predicting bedforms and primary current stratification in cohesive mixtures of mud and sand. *Journal of the Geological Society, London*, **173**, 12–45, https://doi.org/10.1144/jgs2015-024

Berné, S., Vagner, P. *et al.* 2002. Pleistocene forced regressions and tidal sand ridges in the East China Sea. *Marine Geology*, **188**, 293–315.

Bhattacharya, J.P. & MacEachern, J.A. 2009. Hyperpycnal rivers and prodeltaic shelves in the Cretaceous Seaway of North America. *Journal of Sedimentary Research*, **79**, 184–209.

Boyd, R., Dalrymple, R. & Zaitlin, B.A. 2006. Estuarine and incised-valley facies models. *In*: Posamentier, H.W. & Walker, R.G. (eds) *Facies Models Revisited*. SEPM Special Publications, **84**, 171–235.

Budding, M.C., Paardekam, A.H.M. & van Rossem, S.J. 1992. 3D connectivity and architecture in sandstone reservoirs. Paper SPE 22342, presented at the International Meeting on Petroleum Engineering, 24–27 March, Beijing, China.

Cattaneo, A. & Steel, R.J. 2003. Transgressive deposits, a review of their variability. *Earth Science Reviews*, **62**, 187–223.

Chang, T.S., Flemming, B.W., Tilch, E., Bartholomä, A. & Wöstmann, R. 2006. Late Holocene stratigraphic evolution of a back-barrier tidal basin in the East Frisian Wadden Sea, southern North Sea: transgressive deposition and its preservation potential. *Facies*, **52**, 329–340.

Chen, S., Steel, R.J., Dixon, J.F. & Osman, A. 2014. Facies and architecture of a tide-dominated segment of the Late Pliocene Orinoco Delta (Morne L'Enfer Formation) SW Trinidad. *Marine and Petroleum Geology*, **57**, 208–232.

Collins, D.S., Allison, P.A., Johnson, H.D., Avdis, A., Hill, J. & Piggott, M.D. 2015. Integrated palaeogeographic mapping and tidal modeling during the Oligo-Miocene, South China Sea. AAPG Search and Discovery Article No. 90236.

Cullen, A.B. 2010. Transverse segmentation of the Baram-Balabac Basin, NW Borneo: refining the model of Borneo's tectonic evolution. *Petroleum Geoscience*, **16**, 3–29, https://doi.org/10.1144/1354-079309-828

Cullen, A. 2014. Nature and significance of the West Baram and Tinjat Lines, NW Borneo. *Marine and Petroleum Geology*, **51**, 197–209.

Dalrymple, R.W. 2006. Incised valleys in time and space: introduction to the volume and an examination of the controls on valley formation and filling. *In*: Dalrymple, R.W., Leckie, D.A. & Tillman, R. (eds) *Incised Valleys in Time and Space*. SEPM Special Publications, **85**, 5–12.

Dalrymple, R.W. & Choi, K.S. 2007. Morphologic and facies trends through the fluvial-marine transition in tide-dominated depositional systems: a schematic framework for environmental and sequence-stratigraphic interpretation. *Earth Science Reviews*, **81**, 135–174.

Dalrymple, R.W., Knight, R.J., Zaitlin, B.A. & Middleton, G.V. 1990. Dynamics of facies model of a macrotidal sand-bar complex, Cobequid Bay – Salmon River Estuary (Bay of Fundy). *Sedimentology*, **37**, 577–612.

Dalrymple, R.W., Zaitlin, B.A. & Boyd, R. 1992. Estuarine facies models: conceptual basis and stratigraphic implications. *Journal of Sedimentary Petrology*, **62**, 1130–1146.

Dalrymple, R.W., Baker, E.K., Harris, P.T. & Hughes, M.G. 2003. Sedimentology and stratigraphy of a tide-dominated foreland-basin delta (Fly River, Papua New Guinea). *In*: Sidi, F.H., Nummedal, D., Imbert, P., Darman, H. & Posamentier, H.W. (eds) *Tropical Deltas of Southeast Asia – Sedimentology, Stratigraphy and Petroleum Geology*. SEPM Special Publications, **76**, 147–174.

Damit, A.R. 2001. *Brunei Bay, Northwest Borneo: depositional system*. PhD thesis, University of Aberdeen.

Davis, H.R., Byers, C.W. & Pratt, L.M. 1989. Depositional mechanisms and organic matter in Mowry Shale (Cretaceous), Wyoming. *American Association of Petroleum Geologists Bulletin*, **73**, 1103–1116.

Dott, R.H. & Bourgeois, J. 1982. Hummocky stratification; significance of its variable bedding sequences. *Geological Society of America Bulletin*, **93**, 663–680.

Dronamraju, S.V.C., Finol, J., Koraini, A.M. & Zakaria, A.A. 2005. Constraining geological heterogeneity in complex reservoirs: implications for stochastic modeling and reservoir management. Paper SPE 93853, presented at the SPE Asia Pacific Oil and Gas Conference and Exhibition, 5–7 April, Jakarta, Indonesia.

Doust, H. & Sumner, H.S. 2007. Petroleum systems in rift basins – a collective approach in Southeast Asian basins. *Petroleum Geoscience*, **13**, 127–144, https://doi.org/10.1144/1354-079307-746

Droser, M.L. & Bottjer, D.J. 1986. A semiquantitative field classification of ichnofabric. *Journal of Sedimentary Research*, **56**, 558–559.

Duke, W.L. 1985. Hummocky cross-stratification, tropical hurricanes, and intense winter storms. *Sedimentology*, **32**, 167–194.

Duke, W.L. 1990. Geostrophic circulation or shallow marine turbidity currents? The dilemma of paleoflow patterns in storm-influenced prograding shoreline systems. *Journal of Sedimentary Petrology*, **60**, 870–883.

Dumas, S. & Arnott, R.W.C. 2006. Origin of hummocky and swaley cross-stratification – the controlling influence of unidirectional current strength and aggradation rate. *Geology*, **34**, 1073–1076.

Dumas, S., Arnott, R.W.C. & Southard, J.B. 2005. Experiments on oscillatory-flow and combined-flow bed forms: implications for interpreting parts of the shallow-marine sedimentary record. *Journal of Sedimentary Research*, **75**, 501–513.

Evans, G. 1965. Intertidal flat sediments and their environments of deposition in the Wash. *Quarterly Journal of the Geological Society of London*, **121**, 209–245, https://doi.org/10.1144/gsjgs.121.1.0209

Elliott, T. 1974. Interdistributary bay sequences and their genesis. *Sedimentology*, **21**, 611–622.

Fenies, H. & Tastet, J.P. 1998. Facies and architecture of an estuarine tidal bar (the Trompeloup bar, Gironde Estuary, SW France). *Marine Geology*, **150**, 149–169.

Finol, J., Dronamraju, S., Zakaria, A.A. & Koraini, A.M. 2005. Combining geostatistics with dynamic modeling to improve reservoir management strategies: a case study from the Balingian Province. Paper SPE 93851, presented at the SPE Asia Pacific Oil and Gas Conference and Exhibition, 5–7 April, Jakarta, Indonesia.

Fisk, H.N., Kolb, C.R., McFarlan, E. & Wilbert, L.J. 1954. Sedimentary framework of the modern Mississippi delta (Louisiana). *Journal of Sedimentary Research*, **24**, 76–99.

Galloway, W.E. 1989. Genetic stratigraphic sequences in basin analysis, I. Architecture and genesis of flooding-surface bounded depositional units. *American Association of Petroleum Geologists Bulletin*, **73**, 125–142.

Gastaldo, R.A., Allen, G.P. & Huc, A.Y. 1995. The tidal character of fluvial sediments of the modern Mahakam River delta, Kalimantan, Indonesia. *In*: Flemming, B.W. & Bartholoma, A. (eds) *Tidal Signatures in Modern and Ancient Sediments*. International Association of Sedimentologists, Special Publications, **24**, 171–182.

Gingras, M.K., MacEachern, J.A. & Dashtgard, S.E. 2011. Process ichnology and the elucidation of physio-chemical stress. *Sedimentary Geology*, **237**, 115–134.

Goodbred, S.L., Jr. 2003. Response of the Ganges dispersal system to climate change: a source-to-sink view since the last interstade. *Sedimentary Geology*, **162**, 83–104.

Goodbred, S.L., Jr. & Saito, Y. 2012. Tide-dominated deltas. *In*: Davis, R.A., Jr. & Dalrymple, R.W. (eds) *Principles of Tidal Sedimentology*. Springer Science, Netherlands, 129–149.

Hageman, H. 1987. Palaeobathymetrical changes in NW Sarawak during Oligocene to Pliocene. *Geological Society of Malaysia Bulletin*, **21**, 91–102.

Hall, R. 2002. Cenozoic geological and plate tectonic evolution of SE Asia and the SW Pacific: computer-based reconstructions, model and animations. *Journal of Asian Earth Sciences*, **20**, 353–431.

Hall, R. & Nichols, G. 2002. Cenozoic sedimentation and tectonics in Borneo: climatic influences on orogenesis. *In*: Jones, S.J. & Frostick, L. (eds) *Sediment Flux to Basins: Causes, Controls and Consequences*. Geological Society, London, Special Publications, **191**, 5–22, https://doi.org/10.1144/GSL.SP.2002.191.01.02

Hamilton, W. 1974. *Map of Sedimentary Basins of the Indonesian Region*. United States Geological Survey, Map I-875-B. US Geological Survey, Reston, VA.

Hamilton, W. 1979. *Tectonics of the Indonesian Region*. US Geological Survey Professional Paper, **1078**.

Hampson, G.J. 2010. Sediment dispersal and quantitative stratigraphic architecture across an ancient shelf. *Sedimentology*, **57**, 96–141.

Hampson, G.J. & Storms, J.E. 2003. Geomorphological and sequence stratigraphic variability in wave-dominated, shoreface-shelf parasequences. *Sedimentology*, **50**, 667–701.

Haq, B.U., Hardenbol, J. & Vail, P.R. 1987. Chronology of fluctuating sea levels since the Triassic. *Science*, **235**, 1156–1167.

Harms, J.C., Southard, J.B., Spearing, D.R. & Walker, R.G. 1975. *Depositional Environments as Interpreted from Primary Sedimentary Structures and Stratification Sequences*. SEPM Short Course, **2**.

Harris, P.T., Baker, E.K., Cole, A.R. & Short, S.A. 1993. A preliminary study of sedimentation in the tidally dominated Fly River Delta, Gulf of Papua. *Continental Shelf Research*, **13**, 441–472.

Harris, P.T., Hughes, M.G., Baker, E.K., Dalrymple, R.W. & Keene, J.B. 2004. Sediment transport in distributary channels and its export to the pro-deltaic environment in a tidally dominated delta: Fly River, Papua New Guinea. *Continental Shelf Research*, **24**, 2431–2454.

Hayes, M.O. 1975. Morphology of sand accumulation in estuaries: an introduction to the symposium. *In*: Cronin, L.E. (ed.) *Estuarine Research 2. Geology and Engineeering*. Academic Press, New York, 3–22.

Hayes, M.O. 1979. Barrier island morphology as a function of tidal and wave regime. *In*: Leatherman, S.P. (ed.) *Barrier Islands – from the Gulf of St Lawrence to the Gulf of Mexico*. Academic Press, New York, 1–27.

Ho, K.F. 1978. Stratigraphic framework for oil exploration in Sarawak. *Geological Society of Malaysia Bulletin*, **10**, 1–13.

Hori, K., Saito, Y., Zhao, Q., Cheng, X., Wang, P., Sato, Y. & Li, C. 2001. Sedimentary facies and Holocene progradation rates of the Changjiang (Yangtze) delta, China. *Geomorphology*, **41**, 233–248.

Hutchison, C.S. 2005. *Geology of North-West Borneo: Sarawak, Brunei and Sabah*. Elsevier, Amsterdam.

Ismail, C.M.Z. 1997. Tectonics evolution and sedimentation history of the Sarawak Basin. *Geological Society of Malaysia Bulletin*, **41**, 41–52.

Ismail, C.M.Z. & Swarbrick, R.E. 1997. The tectonic evolution and associated sedimentation history of Sarawak Basin, eastern Malaysia: a guide for future hydrocarbon exploration. *In*: Fraser, A.J., Matthews, S.J.

& Murphy, R.W. (eds) *Petroleum Geology of Southeast Asia*. Geological Society, London, Special Publications, **126**, 237–245, https://doi.org/10.1144/GSL.SP.1997.126.01.15

Ismail, C.M.Z. & Tucker, M.E. 1999. An alternative stratigraphic scheme for the Sarawak Basin. *Journal of Asian Earth Sciences*, **17**, 215–232.

Ismail, M.S. & Hamza, N. 2013. Application of Low Pressure System at Balingian, Sarawak Brown Field; Candidate Selection and BY Field Case Study. Abstract IPTC 16569, presented at the International Petroleum Technology Conference, 26–28 March, Beijing, China.

James, D.M.D. 1984. Regional geological setting. *In*: James, D.M.D. (ed.) *The Geology and Hydrocarbon Resources of Negara Brunei Darussalam*. Muzium Brunei, Special Publications.

Johnson, H.D. & Levell, B.K. 1995. Sedimentology of a transgressive, estuarine sand complex: the Lower Cretaceous Woburn Sands (Lower Greensand), southern England. *In*: Plint, A.G. (ed.) *Sedimentary Facies Analysis: a Tribute to the Research and Teaching of Harold G. Reading*. International Association of Sedimentologists, Special Publications, **22**, 17–46.

Jong, J., Barker, S., Kessler, F.L. & Tan, T.T. 2014. *The Sarawak Bunguran Fold Belt: Structural Development in the Context of the South China Sea Tectonics*. Paper IPTC-18197-MS, presented at the International Petroleum Technology Conference, 10–12 December, Kuala Lumpur, Malaysia.

Jordan, D.W. & Pryor, W.A. 1992. Hierarchical levels of heterogeneity in a Mississippi River meander belt and application to reservoir systems. *American Association of Petroleum Geologists Bulletin*, **77**, 118–123.

Klein, G.D. 1998. Clastic tidalites – a partial retrospective view. *In*: Alexander, C.R., Davis, R.A., Jr. & Henry, V.J. (eds) *Tidalites: Processes & Products*. SEPM Special Publications, **61**, 5–14.

Koopman, A. 1996. Regional geological setting. *In*: Sandal, S.T. (ed.) *The Geology and Hydrocarbon Resources of Negara Brunei Darussalam*. Brunei Shell Petroleum, Seria, 49–63.

Kruit, C. 1955. *Sediments of the Rhone Delta*. Kononklijk Nederlandsch Geologisch Mijnbouwkundig Genootschap, The Hague.

Kuehl, S.A., Levy, B.M., Moore, W.S. & Allison, M.A. 1997. Subaqueous delta of the Ganges–Brahmaputra river system. *Marine Geology*, **144**, 81–96.

Kuehl, S.A., Allison, M.A., Goodbred, S.L. & Kudrass, H.R. 2005. The Ganges-Brahmaputa Delta. *In*: Giosan, L. & Bhattacharya, J.P. (eds) *River Deltas – Concepts, Models, and Examples*. SEPM Special Publications, **83**, 87–129.

Lambiase, J.J., Damit, A.R., Simmons, M.D., Abdoerrias, R. & Hussin, A. 2003. A depositional model and stratigraphic development of modern and ancient tide-dominated deltas in NW Borneo. *In*: Sidi, F.H., Nummedal, D., Imbert, P., Darman, H. & Posamentier, H.W. (eds) *Tropical Deltas of Southeast Asia – Sedimentology, Stratigraphy and Petroleum Geology*. SEPM Special Publications, **76**, 109–124.

Lasman, M.R. 1998. Channel chasing in the D35 field offshore Sarawak. *Geological Society of Malaysia Bulletin*, **42**, 39–45.

Lee, G.H., Lee, K. & Watkins, J.S. 2001. Geologic evolution of the Cuu Long and Nam Con Son Basins, offshore southern Vietnam, South China Sea. *American Association of Petroleum Geologists Bulletin*, **85**, 1055–1082.

Legler, B., Johnson, H.D., Hampson, G.J., Massart, B.Y.G., Jackson, C.A.-L. & Jackson, M.D. 2013. Facies model of a fine-grained, tide-dominated delta: Lower Dir Abu Lifa Member (Eocene), Western Desert, Egypt. *Sedimentology*, **60**, 1313–1356.

Legler, B., Hampson, G.J. *et al.* 2014. Facies relationships and stratigraphic architecture of distal, mixed tide- and wave-influenced deltaic deposits: Lower Sego Sandstone, western Colorado, U.S.A. *Journal of Sedimentary Research*, **84**, 605–625.

Leichti, P., Roe, F.W. & Haile, N.S. 1960. *The Geology of Sarawak, Brunei and the Western Part of North Borneo*. Geological Survey Department, British Territories in Borneo Bulletin, **3**. Government Printing Office, Kuching, Sarawak.

Lobza, V. & Schieber, J. 1999. Biogenic sedimentary structures produced by worms in soupy, soft muds: observations from the Chattanooga Shale (Upper Devonian) and experiments. *Journal of Sedimentary Research*, **69**, 1041–1049.

Longhitano, S.G., Mellere, D., Steel, R.J. & Ainsworth, R.B. 2012. Tidal depositional systems in the rock record: a review and new insights. *Sedimentary Geology*, **279**, 2–22.

MacEachern, J.A. & Bann, K.L. 2008. The role of ichnology in refining shallow marine facies models. *In*: Hampson, G.J., Steel, R.J., Burgess, P.M. & Dalrymple, R.W. (eds) *Recent Advances in Models of Siliciclastic Shallow-Marine Stratigraphy*. SEPM Special Publications, **90**, 73–116.

MacEachern, J.A., Bann, K.L., Bhattacharya, J.P. & Howell, C.D. 2005. Ichnology of deltas: organism responses to the dynamic interplay of rivers, waves, storms and tides. *In*: Giosan, L. & Bhattacharya, J.P. (eds) *River Deltas – Concepts, Models and Examples*. SEPM Special Publications, **83**, 49–85.

Madon, M. 1999*a*. Basin types, tectonostratigraphic provinces, and structural styles. *In*: Leong, K.M. (ed.) *The Petroleum Geology and Resources of Malaysia*. PETRONAS, Kuala Lumpur, 77–112.

Madon, M. 1999*b*. Geological setting of Sarawak. *In*: Leong, K.M. (ed.) *The Petroleum Geology and Resources of Malaysia*. PETRONAS, Kuala Lumpur, 275–290.

Madon, M. & Abolins, P. 1999. Balingian Province. *In*: Leong, K.M. (ed.) *The Petroleum Geology and Resources of Malaysia*. PETRONAS, Kuala Lumpur, 343–368.

Maguregui, J. & Tyler, N. 1991. Evolution of middle Eocene tide dominated deltaic sandstones, Launillas Field, Maracaibo Bain, western Venezuela. *In*: Miall, A.D. & Tyler, N. (eds) *The Three Dimensional Facies Architecture of Terrigenous Clastic Sediments and Its Implications for Hydrocarbon Discovery and Recovery*. SEPM Concepts in Sedimentology and Paleontology, **3**, 233–244.

Mansor, M.Y., Snedden, J.W., Sarg, J.F., Smith, B.S., Kolich, T. & Carter, M. 1999. Pre-drill predictions v. post-drill results: use of sequence stratigraphic methods in reduction of exploration risk, Sarawak Deep-water Blocks, Malaysia. *Journal of Asian Earth Sciences*, **17**, 247–254.

MARTINIUS, A.W. & VAN DEN BERG, J.H. 2011. *Atlas of Sedimentary Structures in Estuarine and Tidally-influenced River Deposits of the Rhine-Meuse-Scheldt System: their Application to the Interpretation of Analogous Outcrop and Subsurface Depositional Systems*. EAGE Publications, Houten.

MECKEL, L.D. 1975. Holocene sand bodies in the Colorado delta area, northern Gulf of California. *In*: BROUSSARD, M.L. (ed.) *Deltas, Models for Exploration*. Houston Geological Society, Houston, 239–265.

MELLERE, D. & STEEL, R.J. 2000. Style contrast between forced regressive and lowstand/transgressive wedges in the Campanian of south-central Wyoming (Hatfield Member of the Haystack Mountains Formation). *In*: HUNT, D. & GAWTHORPE, R.L. (eds) *Sedimentary Responses to Forced Regressions*. Geological Society, London, Special Publications, **172**, 141–162, https://doi.org/10.1144/GSL.SP.2000.172.01.07

MIDDLETON, G.V. 1991. A short historical review of clastic tidal sedimentology. *In*: SMITH, D.G., REINSON, G.E., ZAITLIN, B.A. & RAHMANI, R.A. (eds) *Clastic Tidal Sedimentology*. Canadian Society of Petroleum Geology, Memoirs, **16**, x–xv.

MILKEVICIENE, K.T. & PLINK-BJÖRKLUND, P. 2009. Recognizing tide-dominated versus tide-influenced deltas: Middle Devonian strata of the Baltic Basin. *Journal of Sedimentary Research*, **79**, 887–905.

MOHAMMAD YAMIN, B.A. & ABOLINS, P. 1999. Central Luconia Province. *In*: LEONG, K.M. (ed.) *The Petroleum Geology and Resources of Malaysia*. PETRONAS, Kuala Lumpur, 369–392.

MOHD IDRUS, H. & REDZUAN, A.H. 1999. Tinjar Province. *In*: LEONG, K.M. (ed.) *The Petroleum Geology and Resources of Malaysia*. PETRONAS, Kuala Lumpur, 393–410.

MULDER, T., SYVITSKI, J.P.M., MIGEON, S., FAUGERES, J.-C. & SAVOYE, B. 2003. Marine hyperpycnal flows: initiation, behaviour and related deposits. A review. *Marine and Petroleum Geology*, **20**, 861–882.

NIO, S.D. & YANG, C.S. 1991. Diagnostic attributes of clastic tidal deposits. *In*: SMITH, D.G., REINSON, G.E., ZAITLIN, B.A. & RAHMANI, R.A. (eds) *Clastic Tidal Sedimentology*. Canadian Society of Petroleum Geologists, Memoirs, **16**, 3–28.

OERTEL, G.F. 1972. *Patterns of Sedimentary Transport at Nearshore Zones Influenced by Wave and Tidal Currents: a Study Using Fluorescent Tracers*. Georgia Marine Science Center Technical Report 72/7.

OLARIU, C. & BHATTACHARYA, J.P. 2006. Terminal distributary channels and delta front architecture of river-dominated delta systems. *Journal of Sedimentary Research*, **76**, 212–233.

OLARIU, C., STEEL, R.J. & PETTER, A.L. 2010. Delta-front hyperpycnal bed geometry and implications for reservoir modeling: Cretaceous Panther Tongue delta, Book Cliffs, Utah. *American Association of Petroleum Geologists Bulletin*, **94**, 819–845.

OOMKENS, E. 1974. Lithofacies relations in the Late Quaternary Niger Delta complex. *Sedimentology*, **21**, 195–222.

PEMBERTON, S.G., MACEACHERN, J.A. & FREY, R.W. 1992*a*. Trace fossil facies models: environmental and allostratigraphic significance. *In*: WALKER, R.G. & JAMES, N.P. (eds) *Facies Models: Response to Sea Level Change*. Geological Association of Canada, St John's, 47–72.

PEMBERTON, S.G., VAN WAGONER, J.C. & WACH, G.D. 1992*b*. Ichnofacies of a wave-dominated shoreline. *In*: PEMBERTON, S.G. (ed.) *Application of Ichnology to Petroleum Exploration, a Core Workshop*. SEPM Core Workshop Notes, 17, 339–382.

PLUMMER, P.S. & GOSTIN, V.A. 1981. Shrinkage cracks: dessication or synaeresis? *Journal of Sedimentary Research*, **51**, 1147–1156.

PONTÉN, A. & PLINK-BJÖRKLUND, P. 2007. Depositional environments in an extensive tide-influenced delta plain, Middle Devonian Gauja Formation, Devonian Baltic Basin. *Sedimentology*, **54**, 969–1006.

RAAF, J.F.M. DE & BOERSMA, J.R. 1971. Tidal deposits and their sedimentary structures. *Geologie en Mijnbouw*, **50**, 479–503.

RAAF, J.F.M. DE, BOERSMA, J.R. & VAN GELDER, A. 1977. Wave-generated structures and sequences from a shallow marine succession, Lower Carboniferous, County Cork, Ireland. *Sedimentology*, **24**, 451–483.

RAJPUT, S., KHOR, Y.Y. *ET AL*. 2016. *Seismic Inversion Driven Improved Reservoir Characterisation and Production Optimisation in a Mature Field*. Paper OTC-26797-MS, presented at the Offshore Technology Conference Asia, 22–25 March, Kuala Lumpur, Malaysia.

REINECK, H.E. 1972. Tidal flats. *In*: RIGBY, J.K. & HAMBLIN, W.K. (eds) *Recognition of Ancient Sedimentary Environments*. SEPM Special Publications, **16**, 146–159.

REINECK, H.E. & SINGH, I.B. 1975. *Depositional Sedimentary Environments: With Reference to Terrigenous Clastics*. Springer, Berlin.

REINECK, H.E. & WUNDERLICH, F. 1968. Classification and origin of flaser and lenticular bedding. *Sedimentology*, **11**, 99–104.

REINSON, G.E. 1984. Barrier-island and associated strand-plain systems. *In*: WALKER, R.G. (ed.) *Facies Models*. Geological Association of Canada, Geosciences Canada Reprint Series, 1, 119–140.

ROBERTS, H.H. & SYDOW, J. 2003. Late Quaternary stratigraphy and sedimentology of the Offshore Mahakam Delta, East Kalimantan (Indonesia). *In*: SIDI, F.H., NUMMEDAL, D., IMBERT, P., DARMAN, H. & POSAMENTIER, H.W. (eds) *Tropical Deltas of Southeast Asia – Sedimentology, Stratigraphy and Petroleum Geology*. SEPM Special Publications, **76**, 125–146.

SCHIEBER, J. 2003. Simple gifts and buried treasures – implications of finding bioturbation and erosion surfaces in black shales. *The Sedimentary Record*, **12**, 4–8.

STAUB, J.R. & GASTALDO, R.A. 2003. Late Quaternary incised-valley fill and deltaic sediments in the Rajang River delta. *In*: SIDI, F.H., NUMMEDAL, D., IMBERT, P., DARMAN, H. & POSAMENTIER, H.W. (eds) *Tropical Deltas of Southeast Asia – Sedimentology, Stratigraphy and Petroleum Geology*. SEPM Special Publications, **76**, 71–88.

Stear, W.M. 1985. Comparison of the bedform distribution and dynamics of modern and ancient sandy ephemeral flood deposits in the southwestern Karoo region, South Africa. *Sedimentary Geology*, **45**, 209–230.

Storms, J.E.A., Hoogendoorn, R.M., Dam, R.A.C., Hoitink, A.J.F. & Kroonenberg, S.B. 2005. Late-Holocene evolution of the Mahakam delta, east Kalimantan, Indonesia. *Sedimentary Geology*, **180**, 149–166.

Straaten, L.M.J.U. van. 1954. Sedimentology of Recent tidal flat deposits and the Psammites du Condroz (Devonian). *Geologie en Mijnbouw*, **16**, 25–47.

Ta, T.K.O., Nguyen, V.L., Tateishi, M., Kobayashi, I., Saito, Y. & Nakamura, T. 2002. Sediment facies and late Holocene progradation of the Mekong River Delta in Bentre Province, southern Vietnam: an example of evolution from a tide-dominated to a tide- and wave-dominated delta. *Sedimentary Geology*, **152**, 313–325.

Tanabe, S., Ta, T.K.O., Nguyen, V.L., Tateishi, M., Kobayashi, I. & Saito, Y. 2003. Delta evolution model inferred from the Holocene Mekong Delta, southern Vietnam. *In*: Sidi, F.H., Nummedal, D., Imbert, P., Darman, H. & Posamentier, H.W. (eds) *Tropical Deltas of Southeast Asia – Sedimentology, Stratigraphy and Petroleum Geology*. SEPM Special Publications, **76**, 175–188.

Tänavsuu-Milkeviciene, K. & Plink-Björklund, P. 2009. Recognizing tide-dominated v. tide-influenced deltas: Middle Devonian strata of the Baltic Basin. *Journal of Sedimentary Research*, **79**, 887–905.

Tate, R.B. 2001. *The Geology of Borneo Island*. Geological Society of Malaysia, Kuala Lumpa, CD-ROM format.

Vail, P., Mitchum, R., Jr. & Thompson, S., III 1977. Seismic stratigraphy and global changes of sea level: part 4. Global cycles of relative changes of sea level. Section 2. Application of seismic reflection configuration to stratigraphic interpretation. *In*: Payton, C.E. (ed.) *Seismic Stratigraphy – Applications to Hydrocarbon Exploration*. American Association of Petroleum Geologists, Memoirs, **26**, 83–97.

Van Wagoner, J.C., Posamentier, H.W., Mitchum, R.M., Jr., Vail, P.R., Sarg, J.F., Loutit, T.S. & Hardenbol, J. 1988. An overview of the fundamentals of sequence stratigraphy and key definitions. *In*: Wilgus, C.K., Hastings, B.S., Kendall, C.G. St. C., Posamentier, H.W., Ross, C.A. & Van Wagoner, J.C. (eds) *Sea Level Changes – An Integrated Approach*. Society of Economic Paleontologists and Mineralogists (SEPM) Special Publication, **42**, 39–45.

Visser, M.J. 1980. Neap-spring cycles reflected in Holocene subtidal large-scale bedform deposits: a preliminary note. *Geology*, **8**, 543–546.

Walker, R.G. 1985. Geological evidence for storm transportation and deposition on ancient shelves. *In*: Tillman, R.W., Swift, D.J.P. & Walker, R.G. (eds) *Shelf Sands and Sandstone Reservoirs*. SEPM, Short Course Notes, 13, 243–302.

Walker, R.G. & Plint, A.G. 1992. Wave- and storm-dominated shallow marine systems. *In*: Walker, R.G. & James, N.P. (eds) *Facies Models: Response to Sea Level Change*. Geological Association of Canada, St. John's, 219–238.

Wan Hasiah, A. 1999. Oil-generating potential of Tertiary coals and other organic-rich sediments of the Nyalau Formation, onshore Sarawak. *Journal of Asian Earth Sciences*, **17**, 255–267.

Wan Hasiah, A. & Abolins, P. 2006. Oil-prone paralic coals: a case study from the Balingian Province of Sarawak, Malaysia. AAPG Search and Discovery article #90061 presented at the American Association of Petroleum Geologists International Conference and Exhibition, 5–8 November 2006, Perth, Australia, Technical Program.

Warne, A.G., Meade, R.H. *et al.* 2002. Regional controls on geomorphology, hydrology, and ecosystem integrity in the Orinoco Delta, Venezuela. *Geomorphology*, **44**, 273–307.

Wellner, R., Beaubouef, R., Van Wagoner, J., Roberts, H. & Sun, T. 2005. Jet-plume depositional bodies – the primary building blocks of Wax Lake delta. *Transactions of the Gulf Coast Association of Geological Societies*, **55**, 867–909.

Willis, B.J. 2005. Deposits of tide-influenced river deltas. *In*: Giosan, L. & Bhattacharya, J.P. (eds) *River Deltas – Concepts, Models and Examples*. SEPM Special Publications, **83**, 87–129.

Willis, B.J. & Fitris, F. 2012. Sequence stratigraphy of Miocene tide-influenced sandstone in the Minas Field, Sumatra, Indonesia. *Journal of Sedimentary Research*, **82**, 400–421.

Willis, B.J. & Gabel, S.L. 2001. Formation of deep incisions into tide-dominated river deltas: implications for the stratigraphy of the Sego Sandstone, Book Cliffs, Utah, U.S.A. *Journal of Sedimentary Research*, **73**, 246–263.

Willis, B.J., Bhattacharya, J.P., Gabel, S.L. & White, C.D. 1999. Architecture of a tide-influenced delta in the Frontier Formation of central Wyoming. *Sedimentology*, **46**, 667–688.

Wolfenden, E.B. 1960. *The Geology and Mineral Resources of the Lower Rajang Valley and Adjoining Areas, Sarawak*. Geological Survey Department British Territories in Borneo, Memoirs, **11**.

Wonham, J.P. & Elliott, T. 1996. High resolution sequence stratigraphy of a mid-Cretaceous estuarine complex: the Woburn sands of the Leighton Buzzard area, southern England. *In*: Hesselbo, S.P. & Parkinson, D.N. (eds) *Sequence Stratigraphy in British Geology*. Geological Society, London, Special Publications, **103**, 41–62, https://doi.org/10.1144/GSL.SP.1996.103.01.04

Zainuddin, M.J. 1998. Drilling using AvO to reduce uncertainties in D35 infill. *Geological Society of Malaysia Bulletin*, **42**, 47–53.

Stratigraphic evolution of an estuarine fill succession and the reservoir characterization of inclined heterolithic strata, Cretaceous of southern Utah, USA

CARI L. JOHNSON[1]*, L. STRIGHT[2], R. PURCELL[1] & P. DURKIN[3]

[1]*Department of Geology and Geophysics, University of Utah, Salt Lake City, UT, USA*

[2]*Warner College of Natural Resources, Colorado State University, Fort Collins, CO, USA*

[3]*Department of Geoscience, University of Calgary, Calgary, Alberta, Canada*

**Correspondence: cari.johnson@utah.edu*

Abstract: This study documented the architecture of a paralic succession using outcrops of the Upper Cretaceous John Henry Member of the Straight Cliffs Formation, southern Utah (USA). The 65 m thick interval of interest includes a basal succession of elongate tidal bars, which sits unconformably on the Calico Bed of the Smoky Hollow Member. These barforms are overlain by carbonaceous estuarine bay fill with tidal deposits, a bayhead delta and, ultimately, a coastal plain succession. The estuarine succession demonstrates a decreasing tidal influence through time and this is interpreted to reflect evolution from a mixed-energy to a wave-dominated estuary as a result of the changing morphology of the estuary and/or the migration of barrier islands across its mouth. The regional correlations are also discussed, with implications for sequence stratigraphic interpretations. A detailed interpretation of the ancient bayhead delta highlights the internal architecture and provides data for statistical comparison between this feature, which is dominated by inclined heterolithic strata, and a previously published study of fluvial point bar inclined heterolithic strata. The results emphasize distinct grain size trends from these two examples, as well as differences in the continuity and spatial distribution of mudstone drapes, which act as both baffles and barriers to fluid flow.

Paralic depositional environments represent a spectrum of fluvial, tidal, wave and storm influences, with lateral and vertical variability at all scales (Reynolds 1999). Despite the complexity of processes that define paralic environments, facies models tend to deal mainly with end-member scenarios and are commonly qualitative and descriptive rather than predictive (Coleman & Wright 1975; Galloway 1975; Boyd *et al.* 1992, 2006; Bhattacharya 2006; Plink-Björklund 2008). Studies of modern and ancient paralic systems (Bhattacharya & Giosan 2003; Willis 2005; Yang *et al.* 2005; Dalrymple & Choi 2007; Yoshida *et al.* 2007; Ainsworth *et al.* 2008) have begun to recognize and incorporate mixed process factors into their models. Ainsworth *et al.* (2011) developed a classification scheme for clastic shorelines based on the relative dominance of wave, tidal and fluvial processes and incorporated this into a set of matrices that characterized the relative effects of different coastal processes through time and space, thus providing a guide to their interpretation.

Estuaries are of particular interest in the Ainsworth *et al.* (2011) classification scheme as a result of the additional complexity that embayed coastal morphologies contribute to depositional architectures. Estuaries contain both transgressive and regressive fill (Dalrymple *et al.* 1992; Dalrymple 2006) and display a tripartite geomorphology, reflecting varying degrees of relative wave, tidal and fluvial energy (Dalrymple *et al.* 2012). Wave and tidal processes control the basic morphology of estuaries and thus represent two main morphological categories (wave- and tide-dominated). Fluvial processes mainly control the sediment flux entering the upstream portion of the estuary and do not have a significant impact on the fundamental morphology of the estuary itself (Boyd *et al.* 2006). Outcrop studies recording process regime changes in these paralic environments are thus critical to improving estuarine depositional models (Martinius *et al.* 2005).

In addition to their importance in understanding coastal depositional processes, paralic deposits host a significant portion of petroleum reserves (Terzuoli & Walker 1997; Martinius *et al.* 2001; Dreyer *et al.* 2005; Cummings *et al.* 2006; Tänavsuu-Milkeviciene *et al.* 2009; Feldman *et al.* 2013, 2014). Reservoir quality is a challenge as a result of the lithological heterogeneity common in these strata, particularly tidal deposits, which typically have fine-grained, low permeability layers

From: Hampson, G. J., Reynolds, A. D., Kostic, B. & Wells, M. R. (eds) 2017. *Sedimentology of Paralic Reservoirs: Recent Advances*. Geological Society, London, Special Publications, **444**, 251–285.
First published online June 29, 2016, https://doi.org/10.1144/SP444.1

deposited between porous and permeable sands, reflecting the cyclical variability in current velocity and direction (Visser 1980; Dalrymple & Choi 2007; Burton & Wood 2011). Such heterogeneity ranges from grain and bed to reservoir and field scales (Hassanpour *et al.* 2013), thus emphasizing the need for predictive, reservoir-scale, outcrop-derived facies models (White *et al.* 2004).

Inclined heterolithic strata (IHS) exemplify problematic reservoir-scale heterogeneity in paralic deposits. The term IHS refers to an arrangement of interbedded, fine- and coarser-grained facies with an inclined bedset architecture (Thomas *et al.* 1987). Despite the non-genetic terminology, there is a tendency to infer IHS as a tidal indicator (Choi *et al.* 2004). This is not surprising considering that tide-influenced environments of deposition account for the majority of published accounts of IHS (Shanley *et al.* 1992; Gingras *et al.* 1999, 2002; Martinius *et al.* 2001; Dalrymple *et al.* 2003; Hubbard *et al.* 2011; Labrecque *et al.* 2011; Musial *et al.* 2012). However, other settings for IHS are known, including deltas (Stanley & Surdam 1978; Martinius *et al.* 2001; Steel *et al.* 2012), meandering and mud-rich fluvial channels (Jackson 1981; Thomas *et al.* 1987; Smith *et al.* 2009) and submarine fans (Miall 1985*b*). Central to IHS interpretation is the lack of basic conceptual models for the full range of variability of these deposits.

As a result of their areal extent and preservation potential, bayhead deltas can form significant economic hydrocarbon reservoirs (Broger *et al.* 1997; Terzuoli & Walker 1997; Peijs-van Hilten *et al.* 1998; Hubbard *et al.* 2002). However, research focusing on bayhead deltas is scarce and these units are commonly lumped in with descriptions of IHS point bars (Thomas *et al.* 1987; Kirschbaum & Hettinger 2004), which they can closely resemble (Steel *et al.* 2012). Some of the most noteworthy examples of IHS deposits are the tide-influenced point bars of the Lower Cretaceous McMurray Formation, the main reservoir interval for the Alberta (Canada) Oil Sands, collectively one of the largest sources of *in situ* bitumen in the world (Strobl *et al.* 1997; Labrecque *et al.* 2011). Permeability is one of the most important physical properties influencing hydrocarbon recovery from these reservoirs and permeability heterogeneity caused by fine-grained beds may result in barriers and baffles that hamper heat and fluid flow at multiple scales (Willis & White 2000; Burton & Wood 2011; Labrecque *et al.* 2011; Pyrcz & Deutsch 2014). To investigate the variability in IHS deposits, this study highlights a comparison between IHS formed by a bayhead delta v. a fluvial point bar.

Given the importance of paralic depositional environments in understanding the dynamics of modern and ancient coastal processes in time and space, as well as the significant reservoir volumes that they provide, additional outcrop studies are needed to expand existing knowledge of clastic shoreline deposits. This study leverages excellent exposures in the Cretaceous Straight Cliffs Formation of southern Utah (USA) to understand process changes in an ancient estuarine system. Part 1 of this paper is an account of facies architecture and evolution of the estuarine system from a mixed-energy to a wave- and, ultimately, a fluvial-dominated estuary, with implications for stratigraphic correlation. Part 2 focuses on a quantitative analysis of an IHS bayhead delta within this overall succession and is presented to determine the characteristic differences between IHS bayhead deltas and IHS point bars. Grain size and other trends derived from outcrop measurements yield crucial insights into the various expressions of IHS deposits.

Geological background and previous work

This study focused on the John Henry Member of the Upper Cretaceous Straight Cliffs Formation (Turonian–early Campanian) of the Kaiparowits Plateau, which was located along the western margin of the Cretaceous Western Interior Seaway (Fig. 1). The Late Cretaceous Sevier fold–thrust belt formed in response to west–east crustal shortening driven by subduction of the Farallon plate beneath the North America plate (Armstrong 1968; Coney 1972; DeCelles & Coogan 2006). Crustal load-driven flexural subsidence (Jordan 1981; Currie 2002) as well as dynamic subsidence (Liu *et al.* 2014) led to foreland basin development east of the thrust belt. The global greenhouse climate and elevated rates of seafloor spreading favoured eustatic highstand conditions (Miller *et al.* 2005) and the consequent flooding of the foreland basin formed the Western Interior Seaway (Kauffman 1977; Hancock & Kauffman 1979).

The Straight Cliffs Formation is composed of siliciclastic sediments derived from distributive fluvial systems draining the Sevier fold–thrust belt, Mogollon Highlands and the Cordilleran Magmatic Arc (Eaton 1991; Lawton *et al.* 2014) (Fig. 2). A NE-flowing fluvial system, sub-parallel to the Sevier fold–thrust belt in southern Utah, transported sediment to the adjacent Kaiparowits Basin (Szwarc *et al.* 2014). Peterson (1969*a*, *b*) divided the Straight Cliffs Formation into four members: the Tibbet Canyon, Smoky Hollow, John Henry and Drip Tank members. Outcrops of the Straight Cliffs Formation are accessible around the plateau and record a broad spectrum of depositional environments, from fluvial to marginal marine. The John Henry Member is the thickest unit in the Straight Cliffs Formation (*c.* 200–500 m) and has

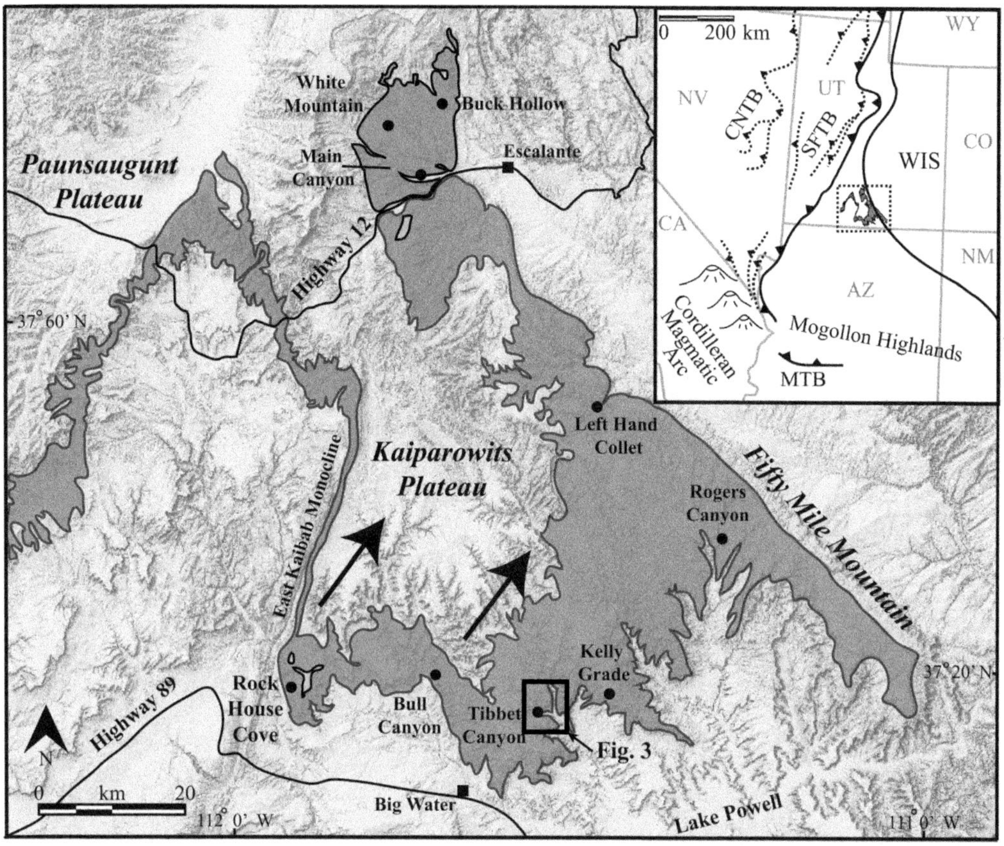

Fig. 1. Regional map of the Kaiparowits Plateau of southern Utah. The Straight Cliffs Formation is shown in grey. Black dots mark previous studies of the John Henry Member. Arrows indicate the general proximal to distal facies relationships in the John Henry Member, ranging from fluvial on the western margin to marine on the eastern margin, with tidal and paralic facies in between. The black box around Tibbet Canyon marks the primary field location of this study (Fig. 3). CNTB, Central Nevada Thrust Belt; MTB, Maria Thrust Belt; SFTB, Sevier fold–thrust belt; WIS, Western Interior Seaway. Modified from Chentnik *et al.* (2015).

the greatest degree of lateral variability (Mulhern & Johnson 2016). Generally, the southern and western plateau preserve fluvial and coastal plain strata and the northern and eastern plateau preserve shoreface, lagoonal and estuarine deposits (Peterson 1969*a*; Shanley & McCabe 1991, 1994). Peterson (1969*b*) categorized the marine shoreface units (A-G), which were used by subsequent studies to document the stratigraphic architecture in the eastern plateau (Allen & Johnson 2010*a*, *b*, 2011; Chentnik *et al.* 2015) (Fig. 2) and to explain fluvial channel stacking patterns in the western plateau (Gooley *et al.* 2016).

Part 1: stratigraphic analysis

Five sections were measured to assess the facies variability within the John Henry Member in Tibbet Canyon (Figs 3 & 4). One general measured section (RPTC-2) captures the full John Henry Member from the top of the Calico Bed to the Drip Tank Member. Several other sections focus on the spatial variability in the lower John Henry Member strata. These sections (RPTC-1, -3, -4 and -5) (Fig. 3a) are measured from the top of the Calico Bed to the last appearance of IHS or to the base of the first fluvial channel deposits lacking significant evidence of a tidal influence, as described in the following section. Thirteen additional sections (Fig. 3b) measured near the location of RPTC-2 detail the vertical and horizontal changes within a specific IHS package (see Part 2 of this paper). Palaeocurrent indicators ($N = 375$) were measured from planar, trough and ripple cross-stratified sandstone beds as well as barform accretion sets (Fig. 4).

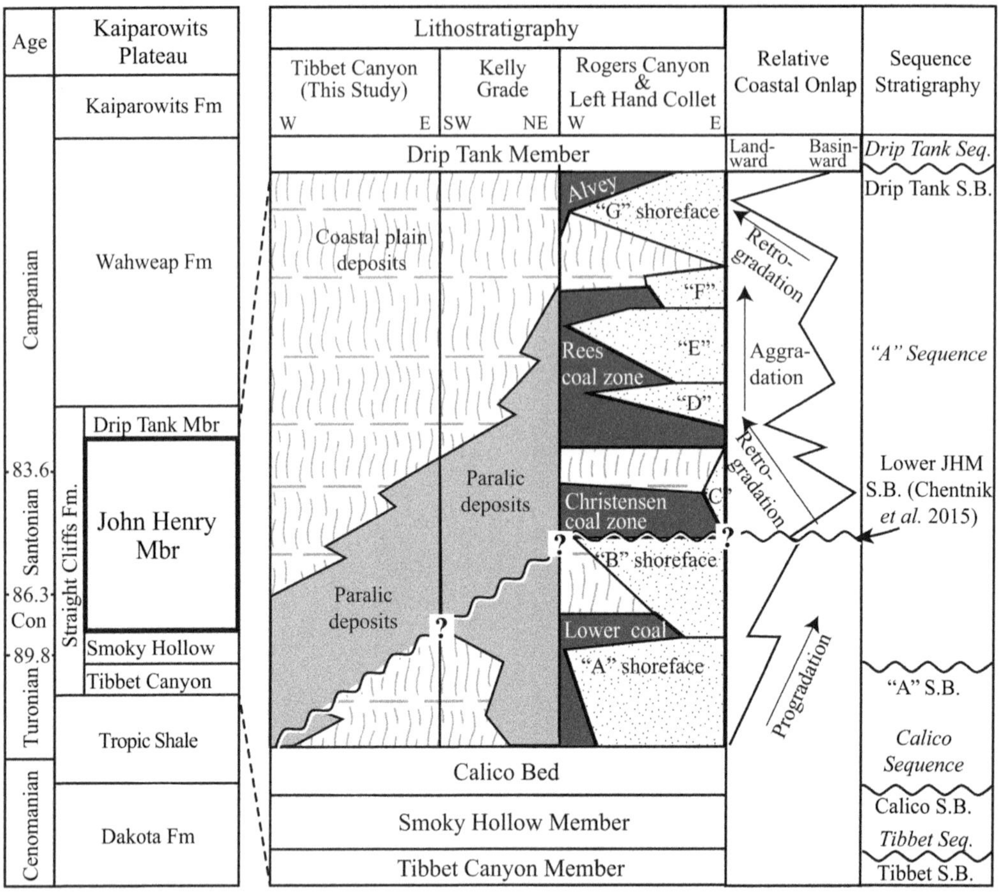

Fig. 2. Regional stratigraphy of the Kaiparowits region (left-hand panel), including detailed stratigraphic summary chart of the Turonian–Campanian Straight Cliffs Formation (right-hand panel). Timescale is from Gradstein *et al.* (2012). Lithostratigraphy of Kelly Grade (Gallin *et al.* 2010) documents relative tide- to fluvial-dominated paralic deposition. Marine sandstone packages A–G were originally defined by Peterson (1969*b*) and these pinch out landwards into coal zones and paralic and coastal plain facies. Relative coastal onlap is based on shoreface pinch-outs and marginal marine facies distributions at Rogers Canyon (Allen & Johnson 2011), Left Hand Collet (Dooling 2013) and Buck Hollow (Mulhern & Johnson 2016). Sequence stratigraphy is based on Shanley & McCabe (1991), except for the Lower John Henry Member Sequence Boundary, which is from Chentnik *et al.* (2015). Extension of the Lower John Henry Member Sequence Boundary to the west is inferential, as discussed in the text. JHM, John Henry Member; S.B., Sequence Boundary.

Eight lithofacies assemblages (LAs) were identified based on lithology, primary sedimentary structures, bedding geometries, palaeocurrent indicator measurements, trace and body fossils, and vertical and lateral relationships with other facies. Table 1 gives detailed descriptions of each lithofacies assemblage. Lithofacies assemblages are essentially architectural elements (*sensu* Miall 1985*a*) and consist of a group of facies that commonly occur together with recognizable spatial and geometric relationships. Lithofacies assemblages are grouped into two facies associations (FAs) according to their dominant depositional processes as described and interpreted in the following sections (Table 1).

Facies association 1

Description. Facies association 1 (FA-1) occurs mainly in the lower John Henry Member and has an average thickness of 37 m (ranging from 19 to 63 m). Deposits of LA 1.1 occur at the base of the John Henry Member in the main Tibbet Canyon section (RPTC-2) (Fig. 4), locally scouring down into the underlying Calico Bed, and consist of dipping

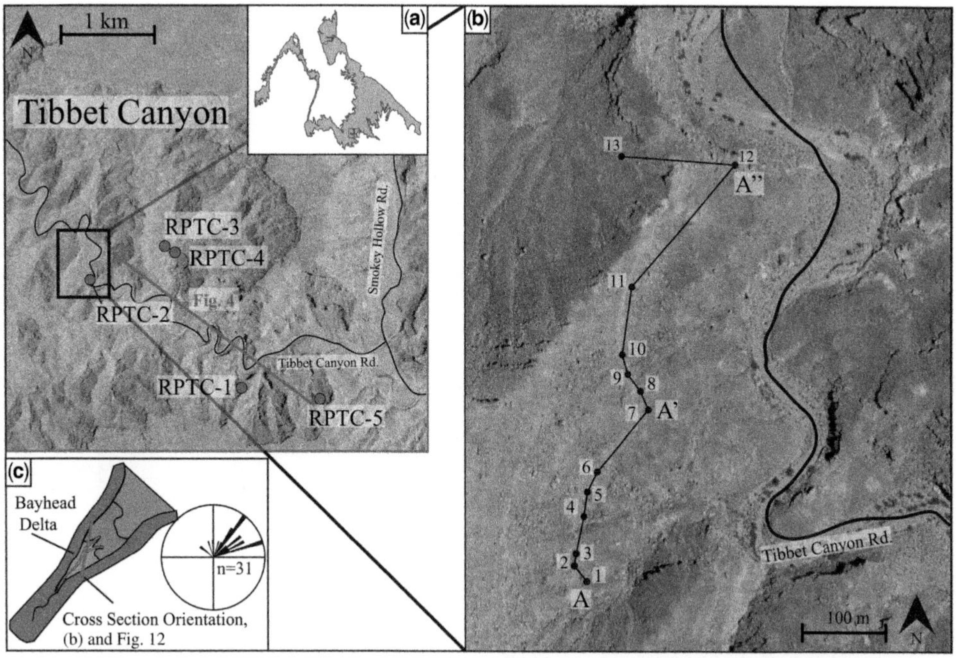

Fig. 3. Location map of Tibbet Canyon study area (see Fig. 1 and inset for location). (**a**) Aerial photograph of the southern Kaiparowits Plateau showing the Tibbet Canyon field locations. Closed circles show sections measured in this study (RPTC) projected on to the general cross-section of Figure 4. The black box around the location of RPTC-2 outlines the bayhead delta area inset shown in (b). (**b**) Aerial photograph showing the locations of measured sections TCIHS-1 to TCIHS-13 used to study the continuous bayhead delta outcrop. (**c**) Schematic diagram showing the interpreted relative orientation of the outcrop section in (b) to the proposed depositional direction of the bayhead delta (cf. Figs 8 & 12). The outcrop is exposed obliquely to the NE depositional trend of the bayhead delta, so moving north along the outcrop allows analysis of the bayhead delta downdip, from topset to toeset and towards the delta depositional axis. Rose diagram shows palaeocurrent measurements from beds in the bayhead delta deposits, indicating a dominant NE flow direction, supporting this interpretation.

heterolithic sandstosne and mudstosne (Figs 5 & 6). The sandstone units are very fine- to fine-grained, have concave-upward erosional bases and display trough cross-stratification, planar cross-laminations and ripple laminations. Trough cross-beds range from 0.20 to 0.60 m thick and typically transition into ripple laminations. Rippled beds, commonly draped, dominate LA 1.1 and range from 0.10 to 2 m thick. Double mud drapes and flaser bedding are also common (Fig. 5a, b). Siltstones, mudstones and carbonaceous shale interbeds are 0.05–0.10 m thick (Fig. 5c–e). Trace fossils include *Lockeia* and *Planolites* (Fig. 5a, f). Beds of LA 1.1 display sigmoidal bedding (Fig. 6a) and persistent mudstone drapes. Beds amalgamate to form lenticular bar complexes up to 8 m thick that extend more than 100 m laterally. Palaeocurrent indicators are weakly bidirectional with dominant east- to SE-directed flow (Fig. 4) and minor west-directed flow.

Deposits of LA 1.2 and 1.3 interfinger with LA 1.1. At its thickest, LA 1.2 is 40 m thick (Fig. 4). Deposits of LA 1.2 consist of mainly horizontal, tabular beds of mudstones, siltstones, carbonaceous shales and sulphur-rich coal (Fig. 6b). Tabular sandstone beds are also a minor constituent and these range from 0.10 to 0.70 m thick and contain ripple laminations (some climbing ripples), horizontal laminations, planar cross-laminations, convolute bedding, wood fragments and mudstone drapes. Fine-grained strata (siltstones and mudstones) dominate LA 1.2, and it contains abundant trace fossils (e.g. *Thalassinoides*, *Planolites* and *Lockeia*) (Fig. 5f, g) and bivalves.

Associated with the generally fine-grained, organic-rich LA 1.2 deposits are coarser grained facies of LA 1.3. These include isolated fine- to medium-grained sandstone bodies with erosional, concave-up bases, up to *c.* 7 m thick (most <2 m thick) and typically <20 m wide (Fig. 6b–f). These units fine upwards to mudstones and siltstones at the top and preserve trough cross-stratification, planar cross-stratification, ripple laminations, flaser

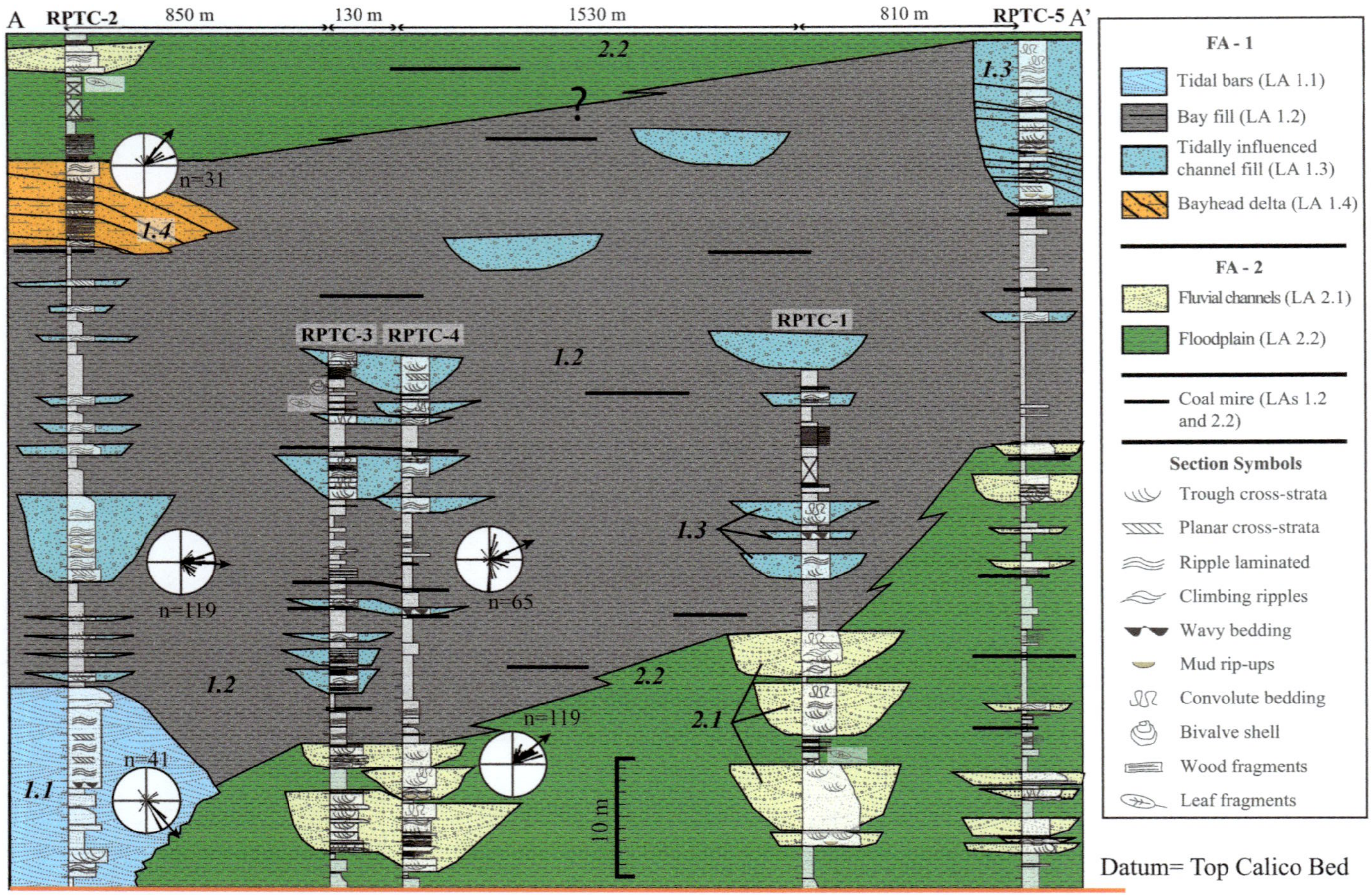

Fig. 4. Correlation of measured sections from NW to SE across Tibbet Canyon (location of cross-section shown in Fig. 3a). The figure highlights stratigraphic expressions of key lithofacies assemblages (LA) for each section and shows the general stacking patterns and relationships of the main facies associations (FA) summarized in the text and Table 1. The lateral relationships (particularly the apparent width and extent of isolated sandstone bodies) are shown schematically and the correlation of the upper facies boundary between FA-2 and FA-1 is inferential between control points (e.g. between sections RPTS-2 and RPTC-5).

laminations, clay rip-up clasts, convolute bedding and wood fragments. Palaeocurrent measurements from trough cross-beds in LA 1.3 show weak bidirectionality with dominant flow to the east (Fig. 4).

Some LA 1.3 units are notably heterolithic and contain IHS expressed as rhythmically interbedded siltstone, mudstone and sandstone units (Fig. 6a–c, g). Erosional surfaces within the IHS often dip more steeply than adjacent strata, creating complex internal facies relationships. IHS dip with angles between 5 and 15° and are continuous over 10s to 100s of metres laterally. Individual sandstone beds range from 0.05 to 1 m, are tabular and thin laterally over 10s to 100s of metres into surrounding mudstones. Mudstone and siltstone beds range from 0.02 to 0.50 m thick. Internal erosion surfaces are common within the sandstone beds, causing 0.10–2 m of local scour into the underlying mudstone layer (Fig. 6d–f). In addition to trough cross-bedding, these units contain flaser, wavy and lenticular bedding with pervasive mudstone drapes. Trace fossils include *Teredolites*, *Planolites*, *Lockeia*, *Thalassinoides* and *Psilonichnus* (Fig. 5).

LA 1.4 is composed of very fine- to medium-grained sandstone beds that coarsen and thicken upwards with interbedded mudstones and siltstones, creating a *c.* 8.5 m thick by 550 m wide complex (Fig. 6g). Mudstone and siltstone drapes are persistent and organic-rich, containing leaf and plant fragments. Sandstone beds dominantly contain ripple laminations, with flaser to wavy bedding, double mud drapes and climbing ripples common. LA 1.4 trace fossils include *Thalassinoides*, *Teredolites*, *Planolites* and *Lockeia* (Fig. 5f–i). The IHS are tabular to lenticular and inclined between 5 and 12° (Fig. 6g). Tabular sandstones range from 0.02 to 0.70 m thick. Sharp-based concave-up erosional surfaces with up to 2 m of relief are infilled with fine- to medium-grained sandstones that are characterized by trough and planar cross-stratification and incise into the underlying IHS beds. Palaeocurrent indicators from ripples and trough cross-beds are dominantly unidirectional flowing to the NE (Fig. 4).

Interpretation: estuary fill and tidal strata. FA-1 includes deposits of basal estuarine tidal bars (LA 1.1), central estuary bay fill (LA 1.2) with tidal channel and splay deposits (LA 1.3), and a prograding bayhead delta (LA 1.4) marking the transition from a mixed-energy estuary to a coastal plain. The concave-upwards geometry, bidirectional palaeocurrent indicators, sigmoidal bedding, lateral accretion surfaces and persistent mudstone drapes of LA 1.1 are indicative of deposition in tidal bars. Tidal bars are most commonly deposited as elongate bars in the outer estuary zone of tide-dominated estuaries or as elongate to lobate bars associated with bayhead deltas in mixed-energy estuaries (Steel *et al.* 2012).

LA 1.2 represents central estuary bay fill based on the brackish/marine trace and body fossils, the heterolithic nature of the deposits and the stratigraphic association with LA 1.1 and 1.3. Central bay fill is typically composed of organic-rich fine-grained sediment and occurs in the area of net bedload convergence as a result of interacting fluvial and marine energies (Elliott 1974; Boyd *et al.* 2006). In wave-dominated and mixed-energy estuaries, the central estuary bay fill is dominated by heterolithic mudstones, siltstones and sandstones (Mack *et al.* 2003). The prodelta facies of both flood-tidal deltas and bayhead deltas is equivalent to the central basin fill. Prodelta central basin fill may display fine-grained, organic-rich muds that are typically heavily bioturbated (Biggs 1967; Donaldson *et al.* 1970). The increased sulphur content observed in central basin coal beds represents the influence of brackish water (Banerjee *et al.* 1996). Coals of this assemblage formed in tidal flats and ponds as part of the central bay.

In addition to organic-rich, fine-grained deposits, the bay fill section contains, and is laterally associated with, sandstones in laterally confined, fining-upwards units with erosional, concave-up bases (LA 1.3), as well as tabular, thin sandstones and siltstones (part of LA 1.2) (Fig. 4). These are interpreted as small tidal channels, tide-influenced fluvial channels and associated crevasse splay deposits. Tide-influence is indicated by the rhythmical interbedding of sandstones and mudstones of LA 1.3, which reflects fluctuating current energies and variations in fluvial and tidal influence (Thomas *et al.* 1987; Shanley *et al.* 1992). Mudstone beds probably represent deposition during slack water periods between diurnal tidal cycles (Bridge & Leeder 1976) or seasonal fluctuations in discharge with changes in spring and neap tides (de Mowbray 1983; Dalrymple & Choi 2007). Such deposits have been identified in fluvial environments with no marine influence (Jackson 1981; Page *et al.* 2003). However, as a result of the presence of marine/brackish trace fossils (*Lockeia*, *Thalassinoides* and *Teredolites*) and double mud drapes, LA 1.3 is interpreted as mainly tide-influenced channel deposits (Fenies *et al.* 1999). Weakly bidirectional palaeocurrent indicators also support a tidal association, in this case with a dominant basinwards-directed (NE to east) ebb-tidal orientation and subordinate flood-tidal oriented flow (SW to west) (Fig. 4).

LA 1.4 consists of bayhead delta deposits based on the upwards-coarsening, inclined heterolithic nature, ubiquitous mudstone drapes, moderate thickness and the relationship with the organic-rich mudstones of LA 1.2 below and FA-2 above

Table 1. *Lithofacies descriptions and environmental interpretations*

Lithofacies assemblages	Description and architecture	Trace/body fossils	Interpretation
Facies Association 1: estuary fill and tidal strata			
1.1: Sigmoidal, channelized sandstones	Sigmoidal, lenticular, sandstone beds with concave-upwards, erosive bases, separated by persistent mudstone interbeds Very fine to fine-grained sandstones with trough cross-stratification, planar cross-stratification and ripple laminations Trough cross-beds ranging from 0.20 to 0.60 m thick and typically transitioning into ripple laminae Rippled beds dominate and range from 0.10 to 2 m thick. Ripples are commonly draped with mudstones or carbonaceous material Double mud drapes and flaser bedding common Scattered weakly bidirectional palaeocurrents (dominantly SE-directed) Beds amalgamate to form lenticular bar complexes up to 8 m thick	*Lockeia*, *Planolites*	Tidal bars
1.2: Carbonaceous shale and coal with minor sandstones	Predominantly carbonaceous shale and coal. Organic-rich throughout with common wood fragments Minor, tabular sandstone beds range from 0.10 to 0.70 m thick Sandstone beds have ripple laminations (some climbing), horizontal stratification, planar cross-stratification, convolute bedding, wood fragments and common mudstone drapes	*Thalassinoides*, *Planolites*, *Lockeia*. Rare bivalves	Bay fill including crevasse deposits
1.3: Channelized, fining-upwards sandstones and inclined heterolithic strata	Most commonly tabular/lenticular sandstone with erosive bases, lateral accretion sets and planar tops Fine- to coarse-grained, typically fining-upwards sandstone capped by mudstones or siltstones, up to 7 m thick Mud rip-up clasts, ripples, planar cross-stratification, trough cross-stratification and lateral accretion surfaces. Convolute and flaser to wavy bedding also present Bidirectional palaeoflow is dominantly towards the east (variable from NE to SE) with minor subordinate flow west Some channel form deposits include inclined heteolithoc strata: rhythmically interbedded mudstones, siltstones and sandstones inclined 5–15° and laterally continuous over 10s–100s of metres Mudstone drapes are pervasive and commonly organic-rich with plant material	*Teredolites, Planolites, Lockeia, Thalassinoides Psilonichnus*	Tide-influenced channel fill

1.4: Coarsening-upwards inclined heterolithic strata	Very fine- to medium-grained rippled to trough cross-bedded sandstones, ranging from 0.02 to 0.70 m thick Ripple-dominated, with flaser to wavy bedding, double mud drapes and climbing ripples common. Dominately north to NE palaeocurrent indicators Coarsening- and thickening-upwards inclined, tabular to lenticular, heterolithic beds Inclined heterolithic beds 5–12° cut laterally by erosive channel sandstones with up to 2 m relief Interbedded and draped organic-rich mudstone and siltstone drapes persistent throughout Wood fragments common and leaf fossils locally abundant within beds that cap the succession	*Thalassinoides, Teredolites, Planolites Lockeia*	Bayhead delta
Facies Association 2: coastal plain			
2.1: Channelized, upwards-fining, cross-stratified sandstones	Tabular to lenticular sandstones with erosive bases, lateral accretion sets and planar tops Fine- to coarse-grained sandstones typically fining-upwards and capped by mudstones or siltstones Sandstone is trough cross-stratified, planar cross-stratified, ripple and flaser laminated, with mud rip-up clasts, convolute bedding and wood fragments Coarse-grained pebble gravel lags and trough cross-stratified sandstones with pebble foresets Channels range from 1 to 6 m thick and are laterally continuous for 100s of metres	*Planolites*	Fluvial channels
2.2: Horizontal, tabular beds of interbedded mudstones, siltstones with coal and carbonaceous shales	Horizontal, tabular beds of interbedded mudstones and siltstones, predominantly massive or laminated with minor mottling and/or nodular concretions Organic-rich mud and carbonaceous shale common and these commonly grade vertically into coal Coal beds range from 0.05 to 0.50 m thick (amalgamating to 1–2 m thick). Deposits are typically sulphur-rich and contain abundant plant material and leaf fragments Very fine- to fine-grained sandstone beds range from 0.02 to 1 m thick with massive bedding, ripple laminations and horizontal stratification Root traces commonly penetrate tops of sandstone beds and plant fragments are common	*Planolites* present, but uncommon	Interdistributary deposits

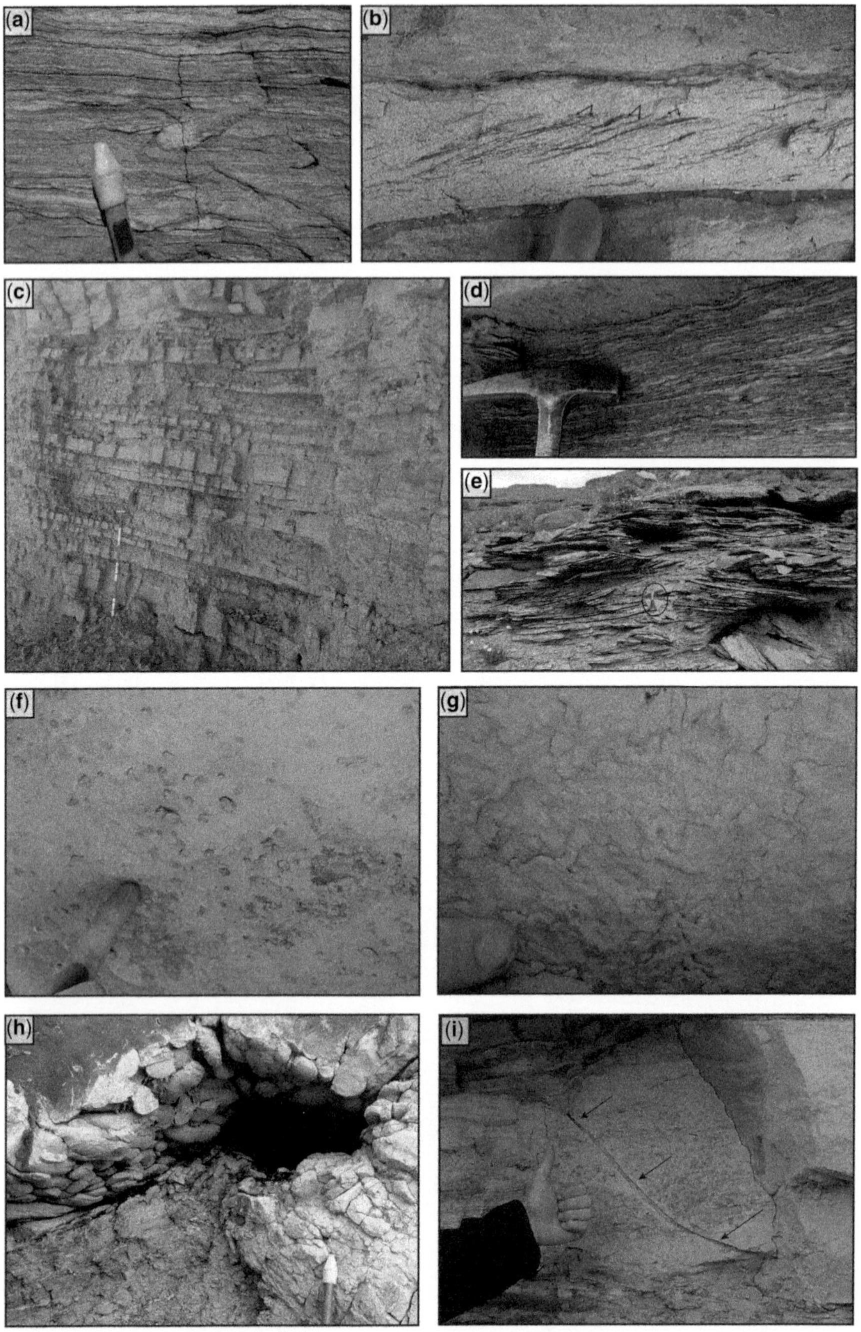

Fig. 5. Photographs of detailed sedimentology and ichnology within facies association 1 strata. (**a**) Flaser bedding and ripple laminations. Sub-horizontal *Planolites* trace featured prominently in the centre of the image. (**b**) Double mud drapes within rippled sandstone. Arrows mark possible couplet sets. (**c**) Heterolithic, tabular mudstone and very fine sandstone beds within LA 1.2 (bay fill). Staff is 1.5 m tall for scale. (**d**) Flaser to wavy bedding. (**e**) Multi- and bidirectional cross-bedding (outlined). Hammer for scale. (**f**) Almond-shaped *Lockeia* traces preserved on the base of a fine-grained sandstone bed. (**g**) Small, branching *Thalassanoides* burrows. (**h**) *Teredolites* trace fossils replacing a large log. (**i**) *Psilonichnus* trace fossil preserved in fine-grained sandstone bed of inclined heterolithic strata point bar.

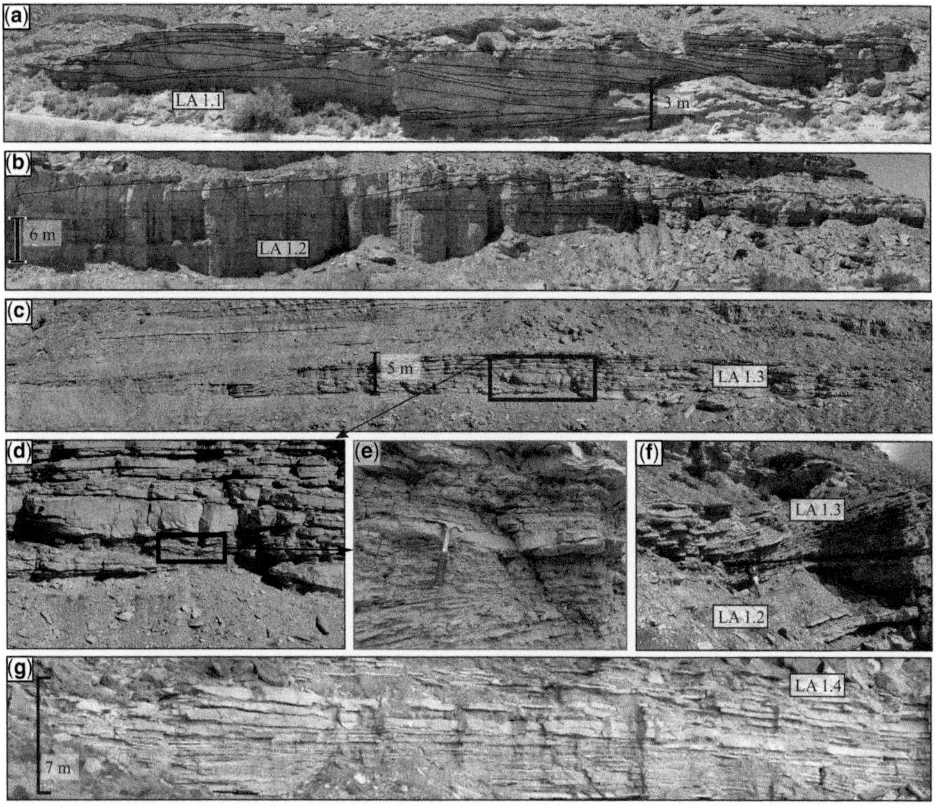

Fig. 6. Photographs of meso-scale architecture within estuary fill and tidal facies (FA-1). (**a**) Sigmoidal, heterolithic tidal bars found in LA 1.1 at the base of RPTC-2, directly overlying the Calico Bed. (**b**) Heterolithic bay fill and crevasse channel of LA 1.2. (**c–e**) Laterally amalgamated tide-influenced channel fill succession, with increasing levels of detail highlighting the varying scales of heterogeneity and local scour present in these deposits. (**f**) Inclined heterolithic strata of a tide-influenced point bar deposit, person for scale. (**g**) Inclined heterolithic strata of the bayhead delta succession.

(McEwen 1969; Dalrymple *et al.* 1992; Plink-Björklund 2008). This unit caps the bay fill succession and is surrounded by fine-grained units rather than channel forms, therefore it is probably not simply recording the lateral avulsion of a point bar from an adjacent channel. The presence of brackish trace and body fossils and the dominantly basinwards-directed palaeocurrent indicators suggest that the sediment was derived from a terrestrial source and was deposited by fluvial channels within a tide-influenced environment (Joeckel & Korus 2012). The bayhead delta deposits are addressed in more detail in Part 2 of this paper.

Facies association 2

Description. Facies association 2 (FA-2) (Fig. 7) dominates the upper John Henry Member of the main Tibbet Canyon section and is also present above the Calico Bed in some areas (Fig. 4). Measured section RPTC-2 is the only section that extends to the Drip Tank Member (thus capturing the full thickness of the John Henry Member). Although not central to the evolution of paralic strata, FA-2 is included here as it relates to regional correlations and the overall evolution of the system.

FA-2 contains lithofacies assemblages of cross-stratified sandstones (LA 2.1), interbedded mudstones, siltstones and sandstones, and coal and carbonaceous shales (LA 2.2) (Fig. 7a, b). LA 2.1 sandstone beds are capped by mudstones or siltstones, ranging from fine- to coarse-grained and typically fining-upwards. Coarse-grained pebble and gravel lags, as well as coarse-grained trough cross-beds with pebble lags along the foresets, are also present (Fig. 7c, h). Facies include trough cross-stratification, planar cross-stratification, ripple laminations, flaser laminations, clay rip-up clasts,

Fig. 7. Photographs of coastal plain facies association (FA-2). (**a**) Fluvial channel belt eroding into floodplain fines. (**b**) Fluvial channel belt displaying internal scour. Channel belt is eroding into underlying coal mire facies. (**c**) Coarse-grained, trough cross-stratified fluvial channels with gravel concentrated along cross-strata foresets. (**d**) Trough cross-stratified, fine-grained sandstone with ripples. (**e**) Convolute bedding in fine-grained sandstone displaying soft sediment deformation flame structures. (**f**) Carbonaceous shale. (**g**) Ripple-laminated, fine-grained sandstone. (**h**) Coarse-grained sandstone to granule bed at the base of a channel. (**i**) Mottled grey mudstone with possible oxidized rootlets.

convolute bedding and wood fragments (Fig. 7d, e). The sandstone bodies of LA 2.1 are tabular or lenticular and have concave, erosional bases with lateral accretion sets and planar tops. Individual lenticular units range from 1 to 6 m thick and are laterally continuous for 100s of metres. Palaeocurrent indicators show dominantly eastwards flow, but with a wide range from NE- to SE-directed flow (Fig. 4).

Tabular interbedded mudstone and siltstone beds dominate LA 2.2, with a few isolated sandstone beds (Fig. 7f). Mudstone and siltstone beds are predominantly massive or laminated. Some units display mottling or nodular concretions. Organic-rich mudstones and carbonaceous shales are also common in this assemblage and these facies commonly grade vertically into coal. Very fine- to fine-grained sandstones range from 0.02 to 1 m thick with massive bedding, ripple laminations and horizontal laminations. Root traces commonly penetrate the tops of sandstone beds (Fig. 7i) and plant fragments are present throughout this association. *Planolites* is present, but uncommon.

LA 2.2 also includes discontinuous, horizontal, tabular, organic-rich deposits of coal and carbonaceous shales (Fig. 7b). Deposits are typically sulphur-rich and contain abundant plant material and leaf fragments. Individual coals range from <0.05 to 0.50 m thick in places, but deposits of LA 2.2, including both coal and carbonaceous shale, may be a few metres thick.

Interpretation: coastal plain deposits. FA-2 preserves fluvial channel belts (LA 2.1) and associated interdistributary deposits of the coastal plain (LA 2.2). There are a few instances of tidal influence, mainly mud draping and rare *Teredolites* near the base of the section, but these are uncommon and ambiguous compared with FA-1. *Asthenopodichnium*, a freshwater trace fossil similar to *Teredolites* (Moran *et al.* 2010), was also identified in sections near Tibbet Canyon. Channel bases form sharp, erosional contacts with underlying floodplain facies assemblages or other channel forms. Channels may amalgamate laterally and/or vertically to form single- and multi-storey channel belts, respectively. Internal scour surfaces, lateral barform accretion surfaces and unidirectional trough cross-strata palaeocurrent indicators are indicative of fluvial channels (Miall 1985*a*). Floodplain and coal mire assemblages (LA 2.2) interfinger laterally and cap channel form deposits.

LA 2.2 represents interdistributary floodplain deposits as a result of their lateral relationship with channel fill deposits, the lack of marine indicators and the presence of root traces and abundant plant material. The architectural elements that comprise this assemblage include fine-grained sandstone beds, which can be levee or crevasse splay deposits (depending on their relationship with the channel deposits), floodplain fines and/or soil horizons (Bridge 2003; Miall 2006). Mottling and nodular concretions present in the tabular siltstone and mudstone beds are indicative of the early stages of diagenesis and potential soil development (Driese *et al.* 2010; Suarez *et al.* 2010). The heterolithic deposits of mudstones, siltstones and sandstones of LA 2.2 also form local abandoned channel deposits, which display the concave-upwards, erosional bases typical of fluvial channels. Sediment filling the abandoned channel is the result of suspended and bedload deposition during flood events and subsequent ponding in the topographic low created by the channel (Hooke 2004; Kraus & Davies-Vollum 2004). The coal and carbonaceous shales of LA 2.2 are coal mires, which can develop as part of floodplain, coastal plain, delta plain and back-barrier environments (Horne *et al.* 1978; Thornton 1979; Thomas 2012). Coals of this assemblage represent swampy interdistributary areas adjacent to fluvial channels.

Tibbet Canyon estuarine evolution

Strata cropping out within Tibbet Canyon at RPTC-2 show the characteristics of a mixed-energy estuary in the basal deposits, becoming more wave-dominated up-section. This 65 m thick fill pattern captures the development of transgressive tidal bars at the base of the section (T_1, LA 1.1), transgressive–regressive central basin fill (T_2, LA 1.2–1.3) and regressive bayhead delta overlain by regressive coastal plain deposits (T_3, LA 1.4) (Fig. 8). The change in facies architecture thus records a shift from mixed-energy to wave-dominated estuary settings.

The identification of inner estuary tidal bars deposited during T_1 is a key factor in interpreting the Tibbet Canyon area estuary as a mixed-energy system. The basal section of the John Henry Member measured at RPTC-2 preserves tidal bars of LA 1.1 overlying the Calico Bed (Fig. 4). The top of the Calico Bed itself is burrowed by traces of the *Glossifungites* ichnofacies (J. Primm pers. comm. 2015), which is representative of subaerial exposure followed by subsequent marine incursion and colonization during transgression (MacEachern *et al.* 1992). Therefore the basal tidal bars are probably transgressive deposits. These types of bars form by tidal currents that modify the fluvial-derived sediments of a bayhead delta, thereby giving it a tide-dominated delta morphology (Dalrymple & Choi 2007). Inner estuary tidal bars are not to be confused with the 'elongate tidal bars' that form at the mouth of tide-dominated estuaries (Dalrymple *et al.* 2012) and have relatively good sorting and coarse sandstone grain sizes as well as a general

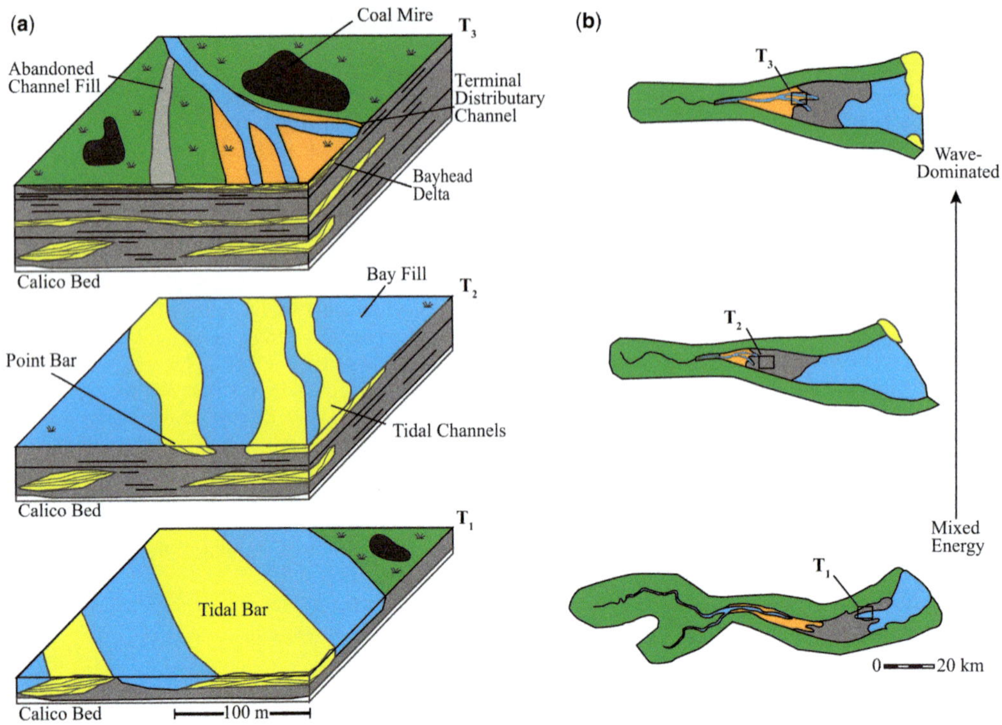

Fig. 8. Conceptual model summarizing evolution of an estuarine succession in the lower John Henry Member of the southern Kaiparowits Plateau. (**a**) Schematic depositional block model of Tibbet Canyon near the location of RPTC-2 (Fig. 3). T_1 (oldest deposits) shows deposition of mixed-energy estuarine tidal bars of LA 1.1. T_2 shows crevasse channel deposits amongst bay fill deposits of LA 1.2. T_3 represents the final stages of progradation of a bayhead delta into the wave-dominated estuary. (**b**) Plan view schematic diagram of estuary evolution illustrating the change from mixed-energy (T_1) to wave-dominated (T_3) estuary. Sequence shows widening of estuary funnel, decreasing coastal embayment and barrier island migration across the estuary mouth. Each of these changes may explain the decrease in relative tidal influence the estuary experienced during the time of deposition.

lack of mud drapes (Plink-Björklund 2008). Tidal bars are encased over 10s of metres laterally by heterolithic fine-grained deposits interpreted as bay fill. The same basal interval *c.* 1 km to the east and continuing several kilometres to the mouth of Tibbet Canyon is fluvial and thus 'outside' the estuary boundary (Figs 3 & 4).

Stratigraphically above the tidal bars are tabular, fine-grained strata including mudstones, siltstones, sandstones and coals, with brackish water indicators. No abrupt bounding surfaces exist, but a decrease in grain size and a lack of sigmoidal tidal bars marks the transition, with the central bay (LA 1.2, 1.3) forming as a result of a decrease in depositional energy. The bay fill overlies the tidal bars, which are depositionally downdip of the bayhead delta deposits. This probably represents the continued transgression and deposition of more basinwards bay fill deposits stratigraphically above the tidal bars (T_2, Fig. 8). The juxtaposition of the bayhead delta deposits of LA 1.4 above the bay fill indicates regressive fill into the estuary. The maximum transgressive surface probably exists within the bay fill and marks the transgressive to regressive turnaround (Allen & Posamentier 1994).

A bayhead delta is deposited above the bay fill (T_3, Fig. 8) showing fluvial-dominated characteristics common to wave-dominated estuaries (Dalrymple *et al.* 1992; Plink-Björklund 2008; Joeckel & Korus 2012; Steel *et al.* 2012). Heterolithic bay fill deposits exist lateral to the bayhead delta and coastal plain fluvial deposits overlie the delta itself. At about the same stratigraphic level as the bayhead delta, IHS fluvial point bar deposits have been interpreted *c.* 3 km away at the mouth of Tibbet Canyon (Shanley *et al.* 1992). The interpretation of IHS in bayhead deltas and IHS in point bars is further complicated because they can exist within the same stratigraphic interval.

Bayhead delta deposits (LA 1.4) are progradational, fluvial-dominated features that mark the transition from transgressive to regressive deposition

(Joeckel & Korus 2012). In wave-dominated estuaries, the relative fluvial energy is greater than the tidal energy, creating fluvial-dominated delta characteristics. However, tidal and brackish indicators are still present (e.g. double mud drapes, flaser bedding, brackish trace fossils). In mixed-energy estuaries, increased tidal energy in the inner estuary causes the winnowing and remobilization of fluvial sediments supplied to the bayhead delta. Tidal action leads to the formation of tidal bars within the estuary instead of a bayhead delta (Fenies & Tastet 1998; Billy *et al.* 2012; Chaumillon *et al.* 2013). The presence of mixed-energy estuary tidal bars (LA 1.1) at the base of the succession, and wave-dominated bayhead delta deposits (LA 1.4) at its top, suggest a relative decrease in tidal processes over the 65 m thick estuary succession (Fig. 8).

One probable scenario explaining the process regime change and decreasing tidal energy from mixed-energy to wave-dominated estuary could be changes in the morphology of the estuary, thereby interfering with tidal resonance and decreasing tidal wave amplification. During the initial transgression and deposition of the basal tidal bars, the estuary was probably more funnel-shaped, based on the geometries predicted by estuary facies models (Dalrymple *et al.* 1992; Martinsen & Helland-Hansen 1993; Dalrymple 2006). With continued transgression, tidal ravinement processes could have eroded the valley walls, thereby widening the estuary (Willis 1997; Willis & Gabel 2003; Li & Bhattacharya 2013; Chentnik *et al.* 2015). Decreased estuary constriction would have caused the tidal range to decrease and the estuary would be more wave-dominated (Plink-Björklund 2008). Continued deposition in the estuary would reflect decreased tidal energy as a result of a reduction in funnel constriction and the dampening of tidal amplification.

Another hypothesis explaining the changing tidal energy in the estuary is the formation of barrier islands across the mouth of the estuary (Dalrymple *et al.* 1992; Roy 1994; Plink-Björklund 2008). During the first stages of the transgression, the estuary mouth may have been more open, allowing a higher degree of tidal energy to penetrate into the estuary (Allen 1991; Allen & Posamentier 1993; Cattaneo & Steel 2003). However, longshore drift accounts for a significant component of sediment transport interpreted in the shoreface strata of the Straight Cliffs Formation along the eastern edge of the Kaiparowits Plateau (Allen & Johnson 2010*a*; Szwarc *et al.* 2014). In addition, barrier island deposits occur in Buck Hollow (Mulhern & Johnson 2016) and Left Hand Collet (Dooling 2013) (Fig. 1). Therefore it is reasonable to suggest that longshore drift could have supplied sufficient sediment across the estuary mouth, creating a barrier island or spit. Barrier islands decrease tidal energy within estuaries by restricting the estuary mouth and dissipating tidal power (Plink-Björklund 2008).

Stratigraphic correlations

This study is the first to document an estuary in the lower John Henry Member of the southern Kaiparowits Plateau, although its presence was predicted by Gallin *et al.* (2010) based on studies in nearby Kelly Grade (Figs 1 & 2). Tidal influence in fluvial deposits of the lower John Henry Member is noted at the mouth of Tibbet Canyon (RPTC-5) (Shanley *et al.* 1992) just a few kilometres east of the main study area (RPTC-2, Figs 3 & 4), where the full estuarine succession is preserved. A valley edge was not explicitly observed, so it is unclear whether the fluvial deposits at the eastern mouth of Tibbet Canyon predate the basal tidal bars (i.e. they were incised into during valley formation), or if they are lateral fluvial equivalents to the tidal bars (although this seems less likely in the context of an estuary model; Boyd *et al.* 2006): the former interpretation is favoured in Figure 4. This study shows that marine influence extends further west in the plateau, up depositional dip from the coeval shoreline, than previously recognized.

Plateau-wide correlations within the John Henry Member are difficult as a result of the complex stratigraphy and the lack of ash beds and other clear time markers, therefore the Tibbet Canyon estuarine succession has important sequence stratigraphic implications. The local removal of the Calico Bed, with tidal bars and estuary bay fill sitting directly on the coastal plain strata of the Smoky Hollow Member, suggests a major transgressive ravinement. Previous interpretations associate such an unconformity with the A sequence boundary (Shanley & McCabe 1991). However, it is now documented that a major, *c.* 14–17 km basinwards shift in the shoreline occurred between the B and C shoreface packages, more than twice the apparent basinwards shift recorded from the A to B shoreface packages in parts of the plateau (Allen & Johnson 2010*b*). This suggests a forced regression culminating in the Lower John Henry Member Sequence Boundary in the C interval (Chentnik *et al.* 2015; Mulhern & Johnson 2016) (Fig. 2). The C and D intervals are transgressive, marked by flooding of the northern plateau and the widespread deposition of major coal zones (Hettinger 1995; Dooling 2013; Mulhern & Johnson 2016).

Results from Tibbet Canyon suggest that the Lower John Henry Member Sequence Boundary of Chentnik *et al.* (2015) might extend to the southern plateau. Initial John Henry Member deposition in much of Tibbet Canyon was dominated by fluvial deposition (Fig. 4), which probably occurred during the net progradation of the A–B–C shoreface intervals. We suggest that erosion of the Calico Bed and subsequent estuarine deposition (e.g. RPTC-2; Fig. 4) might be the local expression of a combined subaerial unconformity and transgressive ravinement surface in the lower John Henry Member. Transgression may have been specifically associated with the retrogradational shoreface stacking patterns in the C and D intervals, but the unconformity also appears to be amalgamated with the Calico Sequence Boundary in places, as implied in Figure 2.

This interpretation is consistent with up depositional dip correlations into the tide-influenced coastal plain successions at Rock House Cove and Bull Canyon, 25 and 15 km west NW of Tibbet Canyon, respectively (Fig. 1) (Gooley *et al.* 2016). Similarly, down depositional dip correlations in the Kelly Grade area, *c.* 10 km east of Tibbet Canyon (Figs 1 & 2) indicate prominent marine 'signals' in the middle John Henry Member (40–115 m above the Calico Bed), with bayhead deltas, carbonaceous estuarine and lagoonal bay fill mudstones, isolated distributary channels, laterally restricted channels and laterally restricted channel belts (Gallin *et al.* 2010). These paralic deposits provide a logical downdip correlation with the Tibbet Canyon estuarine succession, again suggesting a major flooding succession, perhaps associated with transgressive C and D shoreface dynamics. In virtually all the locations of the southern Kaiparowits Plateau, the upper John Henry Member shows a decreasing tidal influence and the increasing amalgamation of channel belts transitionally into the overlying Drip Tank Member (Lawton *et al.* 2014; Gooley *et al.* 2016). This evolution of the coastal plain succession is consistent with the progradation of a distributive fluvial system and correlation with generally aggradational to progradational shoreface successions (Szwarc *et al.* 2014).

Part 2: bayhead delta analysis

Because a particular focus of this study was the meso-scale facies architecture and evolution of a bayhead delta, the bayhead delta lithofacies association (LA 1.4 of FA-1) was further divided into its constituent components, which are described in detail following the hierarchical scheme of Nardin *et al.* (2013) (Fig. 9). Thirteen detailed sections

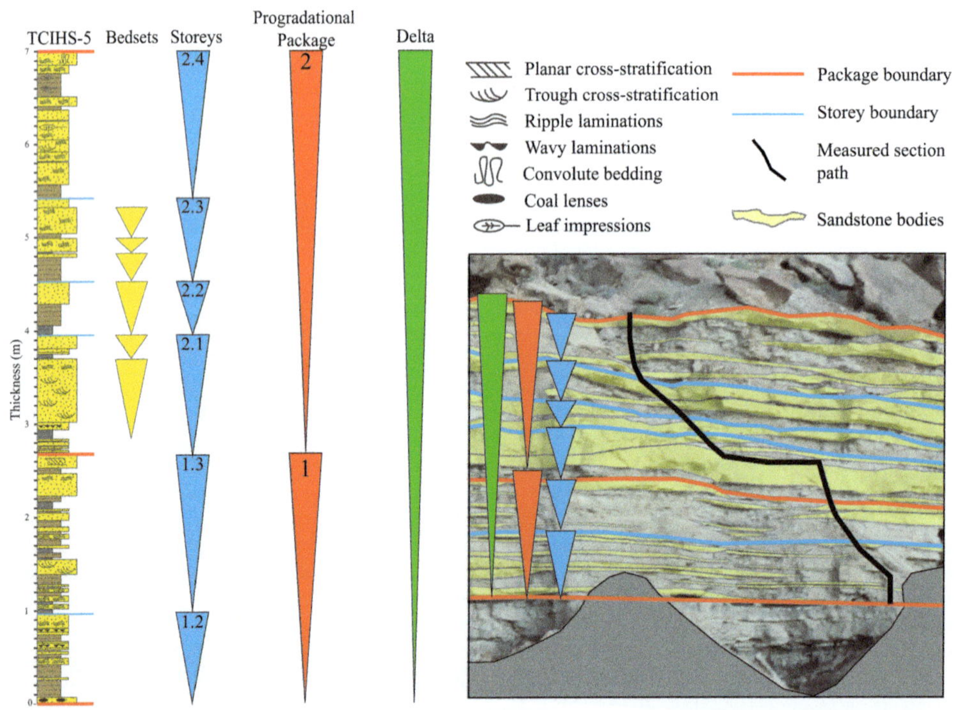

Fig. 9. Bayhead delta bedding hierarchy scheme detailing arrangement of beds, bedsets, storeys, progradational packages and deltas. Application of the hierarchy is shown on the outcrop image to the right, following the measured section path of TCIHS-5 with the beds of this section illustrated on the left. Measured section reaches a height of 7 m (scale of the 'Delta' triangle on the annotated photo).

(TCIHS-1 to TCIHS-13) (Fig. 3b) were measured from the top of the coal layer below the IHS to the top of the IHS package. The outcrop is exposed obliquely to the NE depositional trend of the bayhead delta (Fig. 3c). Thus moving north along the outcrop allows an analysis of the bayhead delta down depositional dip, from topset to toeset and towards the delta depositional axis (Figs 10 & 11). This approach focuses on stratal stacking patterns and discontinuities. The stratal units include, in order of increasing scale, beds, bedsets, storeys, progradational packages and delta. The beds and bedsets are defined as time-stratigraphic units (Cambell 1967; Nardin *et al.* 2013). However, in this study, the lateral accretion sets and bars of Nardin *et al.* (2013) become progradational packages and delta architectural elements to better reflect delta terminology (Fig. 9).

Multiple inclined surfaces, combined with measured sections, serve as the basis for the interpretation and model of the IHS strata. A grain size pseudo well-log was generated for each measured section according to the lithology profile and description. Each measured section was then analysed to determine the proportion of sandstone – as a proxy for the net-to-gross (NTG) ratio – and the average bed thickness according to the interpreted inclined surfaces. These metrics were then compared along the length of the outcrop, from the topset to the toeset of the inclined surfaces, to understand the spatial distribution of reservoir character. Bayhead delta deposits of FA-1 served as outcrop analogues to subsurface reservoirs and the statistical analysis of outcrop analogues provided a framework for interpretation support and feedstock for reservoir models (Doligez *et al.* 1999; Pringle *et al.* 2004; Falivene *et al.* 2006; Enge *et al.* 2007). Key outcomes of this study are used here as a foundation for potential modelling applications of heterolithic bayhead delta reservoirs.

Bayhead delta architectural elements

The basis for the statistical analysis was the lower John Henry Member bayhead delta interval at Tibbet Canyon, which was measured, characterized and interpreted at bed-scale detail. Specifically, the bayhead delta facies of LA 1.4 (described in Part 1) are here further divided into three architectural element categories.

The first architectural element within LA 1.4 is characterized by thinly interbedded tabular mudstone, siltstone and very fine-grained sandstone units, reaching a total amalgamated maximum thickness of 4 m in the study area. Ripple laminations (including climbing ripples) dominate the sandstone beds, which range in thickness from 0.10 to 0.27 m. The siltstones and mudstones are laminated to massive, often with abundant carbonaceous material, and range from 0.10 to 0.31 m thick. Brackish to tidal indicators include double mud drapes and trace fossils such as *Thalassinoides*, *Teredolites*, *Planolites* and *Lockeia* (Fig. 6).

These units were deposited in the proximal prodelta. This is the area of a delta where fine mud and silt are deposited through suspension settling or by hyperpycnal flows (Bhattacharya 2003). The preservation of silty and sandy laminations is thought to indicate the influence of fluctuations in fluvial discharge and sediment supply (Coleman & Prior 1982). Thin sandstone beds and thicker siltstones may represent frontal splays and slurry deposits fed from upstream channels and bars (McEwen 1969; Ahmed *et al.* 2014).

The second architectural element within LA 1.4 consists of very fine- to fine-grained sandstone beds with erosive bases, ranging from 0.02 to 1.90 m thick. Sedimentary structures include trough cross-stratification, ripple laminations, planar laminations, horizontal laminations, soft sediment deformation and mudstone rip-up clasts. Rippled beds display some mud draping as well as wavy to flaser bedding. Siltstone and mudstone units range from 0.01 to 0.52 m thick and are laminated to massive with common carbonaceous material.

These units of the second architectural element are interpreted to be delta front deposits, mainly terminal distributary channels represented by the very fine- to fine-grained sandstone beds. Terminal distributary channels are characterized by trough cross-stratification and mud rip-up clasts as well as erosive bases and a variable low topography (Olariu & Bhattacharya 2006). The delta front deposits of river-dominated systems commonly consist of a complex association of terminal distributary channels and mouth bars (van Heerden & Roberts 1988; Bhattacharya 2006). The apparent lack of mouth bars in this bayhead delta outcrop may be a result of erosion caused by the terminal distributary channels prograding basinwards. In low accommodation settings, where channel flow depths and water depth are on the same scale, channels can more easily cannibalize the underlying deposits (Holbrook 1996; Bhattacharya 2006).

The third architectural element within LA 1.4 features one erosive-based, flat-topped unit, *c.* 300 m in apparent width (Fig. 11). The lower contact scours up to 3 m into underlying delta front strata and the unit contains fining-upwards heterolithic beds of mudstone, siltstone and sandstone. Sandstone units are very fine- to fine-grained and have beds ranging from 0.02 to 0.38 m in thickness that display ripple and planar laminations. Some ripples are wavy and mud-draped, suggesting that there was some tidal influence. There are leaf impressions and organic fragments towards the top of the interval.

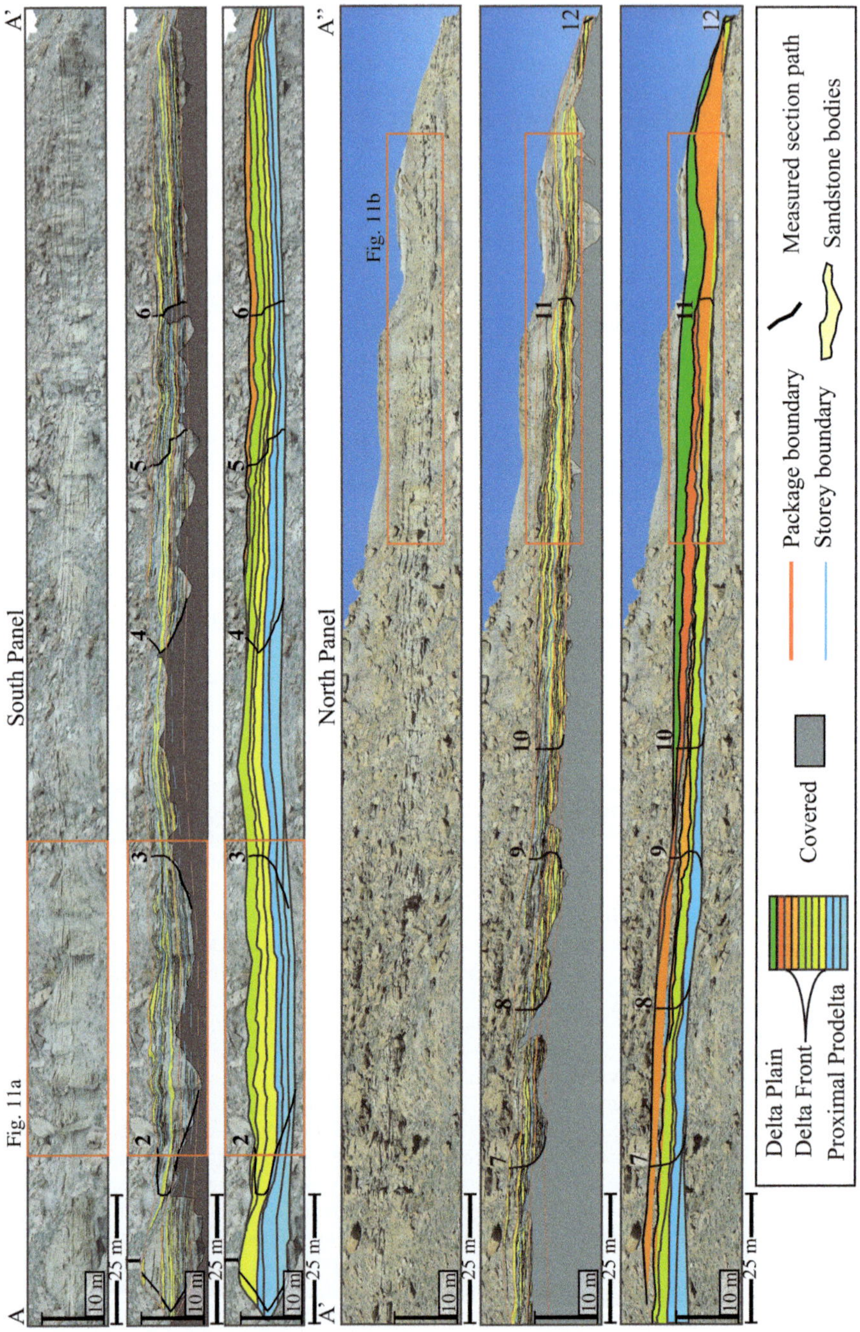

Fig. 10. Outcrop diagram of Tibbet Canyon bayhead delta used for stratigraphic analysis of inclined heterolithic strata. The top three images, labelled A to A′, show the southern outcrop face exposure and the bottom three, A′ to A″show the northern extent of the exposure resulting in *c.* 600 m of continuous outcrop in this image. Each panel consists of an uninterpreted outcrop image (top), a sand distribution overlay (middle) and a hierarchical storey interpretation (bottom). Black lines show the measured section paths of TCIHS 1–12 (numbered on each image) and measurements from these sections use grain size, architecture and spatial inputs for the bayhead delta model. Red boxes correspond to perspectives of the outcrop examining detailed bayhead delta facies in Figure 11.

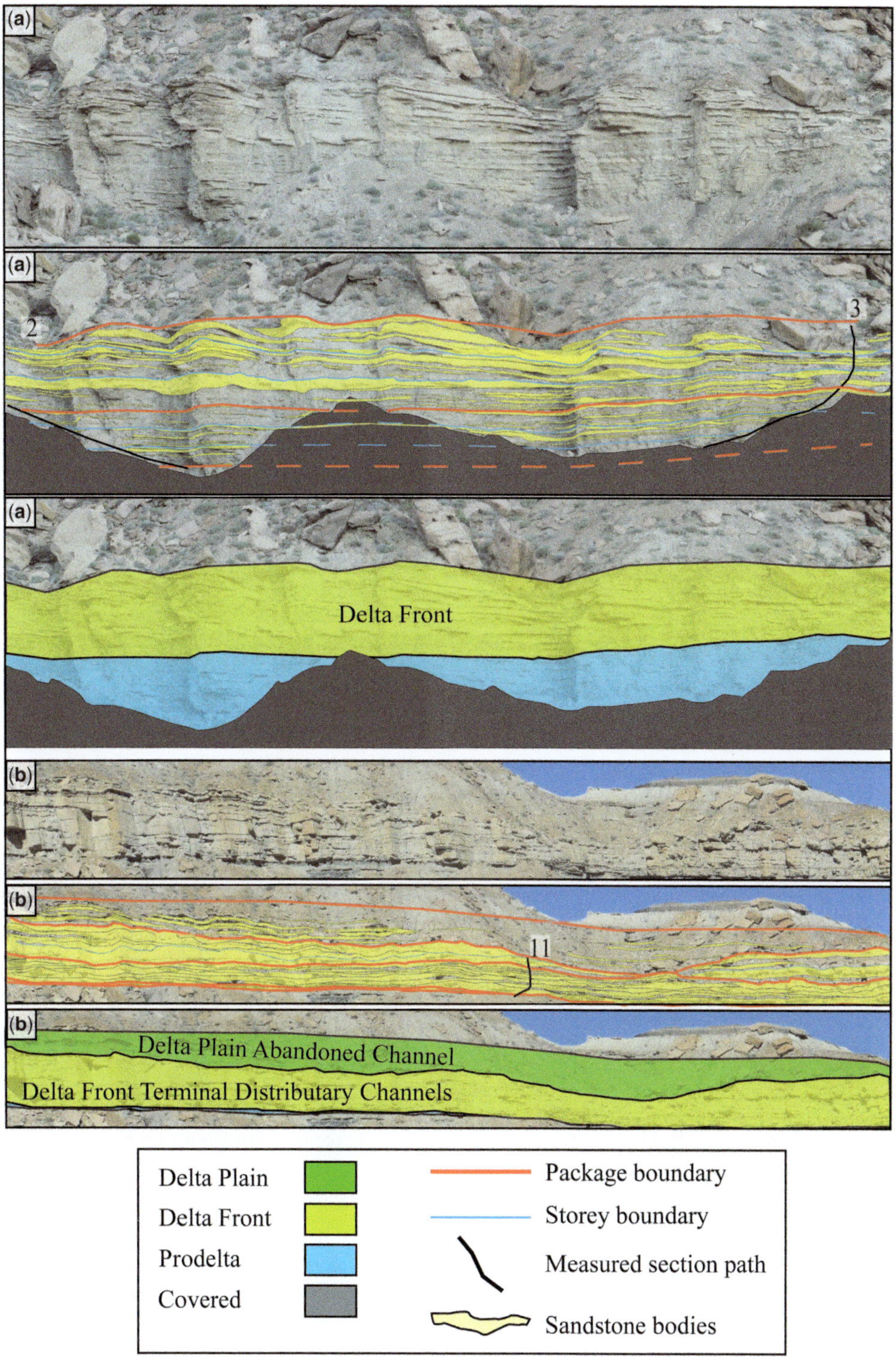

Fig. 11. Detailed interpretations of bayhead delta facies. Outcrop diagram of inset views from Figure 10 (**a** and **b** refer to different insets on the previous figure). Each image shows an uninterpreted (top) outcrop image, a sand distribution overlay (middle) and a depositional subenvironment interpretation (bottom).

This feature is interpreted to be an abandoned channel fill deposit associated with the delta plain architectural element, formed after channel switching and lobe abandonment (Bhattacharya 2006). Beds deposited within the abandoned channel are tabular, horizontal and onlap discordantly with the scour surface and the underlying delta front deposits. Grain size trends indicate a fining-upwards succession, consistent with typical channel fill facies models (e.g. Allen 1964).

Analysis of spatial data from the bayhead delta reservoir analogue

Five major surfaces define four progradational packages (P1, P2, P3 and P4) of the bayhead delta interval of interest. Eleven storeys exist within the four progradational packages shown in Figures 10 and 11. Each of the packages downlaps onto a continuous coal surface that extends across the length of the outcrop exposure. The top surface is a gradational boundary marking the transition from delta processes to coastal plain. Some measured sections were unable to capture the top surface as a result of poor outcrop exposure; in these situations, photomosaics allowed for the interpolation of the top surface.

A model was created using the five major surfaces defining the four progradational packages, intermediate stratigraphic layering from storeys (model zones), intra-storey layering relationships from the internal arrangement of beds and bedsets within storeys, and measured sections to define grain size changes within layers (Fig. 12). The objective was to visualize and quantify the spatial and temporal grain size distributions within the bayhead delta interval using vertical and horizontal facies proportion curves (cf. Ravenne *et al.* 2002) and summary statistical analyses. To track grain size variations as a function of horizontal or vertical position within each storey, measured sections were placed within the stratigraphic model framework that differentiated individual storeys and, within each storey, the base-conformable layer and position from topset to toeset (Fig. 12). The measured section bed resolution was sub-centimetre scale, but intra-storey-level layers were defined at a 0.1 m grid cell thickness. The resulting facies proportion curves provide sub-0.50 m accuracy to feed into subsurface modelling workflows and to be used for comparison with wireline logs, which have a similar resolution.

Statistical analyses include grain size analysis laterally within a package and vertically from package to package, average bedding thicknesses and the NTG ratio for each individual package, as well as the bayhead delta as a whole. The NTG ratio is defined as the thickness of sandstone beds with grain sizes greater than silt (fine- and medium-grained sandstones) divided by the gross thickness of the interval of interest from the top of the basal coal surface up to the top bayhead delta surface. The NTG ratio can also be described at the proportion of grain sizes greater than siltstone, or the total proportion of very fine-, fine- and medium-grained sandstone. Analysis was performed along the packages from the proximal to distal regions and then from package to package, in stratigraphic order, to describe the spatio-temporal evolution of the system. Data from individual storeys are shown in Figure 12 and are described in the following section; however, identifying trends between individual storeys is difficult because of the small amount of data sampling at the storey level. Therefore the bulk of the discussion of the statistical analyses focuses on package-level relationships.

Storey-level analysis. The storey-level data statistics revealed the data robustness and storey-level patterns. Although there is some variation in the thickness of each storey as a result of the erosion of the overlying packages, each storey is *c.* 2 m thick. The measured section data were not sampled evenly within each storey and the ratio of the amount of measured section to the total volume of each storey revealed that the packages in the middle of the section have the most robust data for analysis. This data robustness measure is skewed in storeys 2.3 and 3.2 as a result of the thicker nature of these sections and storey 4.1 is data-poor due to limited outcrop exposure. Therefore, in the following analysis, the delta plain (P4 defined by storey 4.1) was not analysed. Sections 12 and 13 were removed from further analysis because they are off-axis from the direction of deposition (Fig. 3), showing a different style of deposition and a different region of statistical stationarity. Within each package (e.g. P1), the average sandstone bed thickness increases from the base of the package to the top of the package. There is no clear proportion trend in grain size, with the exception of the proximal prodelta, where the percentage of mudstone increases, the percentage of siltstone decreases and the percentage of fine-grained sandstone increases from the base of the package to the top. The largest percentage of mudstone is in storey 3.1 and the smallest is in storeys 1.1, 2.4 and 4.1. The largest proportion of fine-grained sandstone is also in storey 3.1.

Vertical outcrop trends. Using the geometrical model derived from the outcrop interpretation, vertical proportion curves were computed from measured section data for the grain size within each layer of the model, grouped by storeys. The resulting curves provide the framework for a comparison from the top to the base of each storey and from storey to storey within each package to analyse

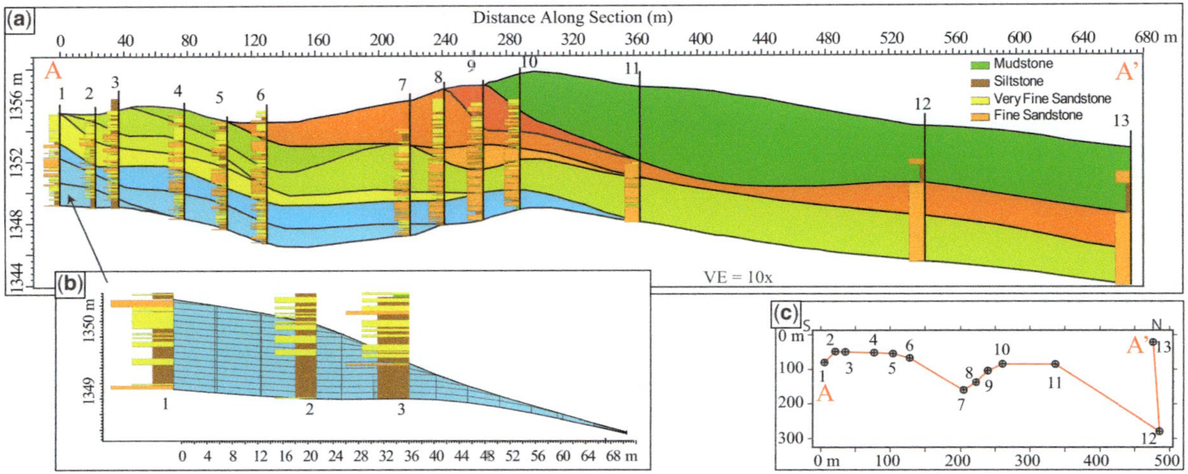

(d)

Zone	Package	Storey	Max Vertical Thickness	Number of Layers	Metres of Measured Section	Section to Zone Vol Ratio	Number of Sand Beds	Average Sandstone Thickness	Proportions: Mudstone	Proportions: Siltstone	Proportions: Very Fine Sandstone	Proportions: Fine Sandstone
Delta Plain	4	4.1	5.94	60	7.24	2%	4	0.16	0.1%	91.0%	7.9%	1.0%
Delta Front	3	3.3	2.91	30	4.12	12	11	0.17	6.7	47.3	29.4	16.5
		3.2	4.11	42	10.95	6	39	0.15	2.5	43.6	33.1	20.8
		3.1	1.61	17	3.23	13	16	0.14	14.1	15.9	6.6	63.4
	2	2.4	1.83	19	4.14	10	19	0.17	1.4	21.3	44.4	32.8
		2.3	3.2	32	15.35	5	60	0.15	5.1	17.5	24.4	53.0
		2.2	2.06	21	3.69	21	17	0.16	3.5	21.1	53.4	22.0
		2.1	2.03	21	9.21	12	59	0.12	6.5	17.0	51.7	24.8
Proximal Prodelta	1	1.3	1.77	18	11.78	11	68	0.13	6.8	17.5	24.2	51.5
		1.2	2.16	22	8.38	9	72	0.08	4.8	29.8	49.7	15.8
		1.1	2.27	23	2.58	7	15	0.07	0.0	59.9	34.5	5.6
			21.04									

Fig. 12. Measured section correlation of bayhead delta sections TCIHS 1–13. (**a**) South to north cross-section, from A to A′, of bayhead delta outcrop showing the distribution of packages and their orientation in space. Proximal prodelta storeys show progradational stacking with the top of each storey downlapping on to the bay floor. Delta front deposits are more lenticular, with individual storeys pinching out or eroded by overlying storeys (e.g. storeys 2.2, 3.1 and 3.3). A channel erodes into the delta front up to 3 m, removing delta front sandstones and replacing them with fine-grained sediments of the abandoned channel fill (storey 4.1). Package surfaces created with this model served as the guide for model layering used in generating the grain size proportion curves. (**b**) Example of package layers shown with storey 1.1 of package 1. (**c**) Locations of sections (cf. Fig. 3b). (**d**) Summary statistics of each storey.

how the grain sizes are distributed within each package and how these distributions change from storey to storey and package to package.

Each of the prodelta storeys (1.1–1.3; P1) coarsen upwards and there is an overall coarsening-upwards of the entire package (Fig. 13). The

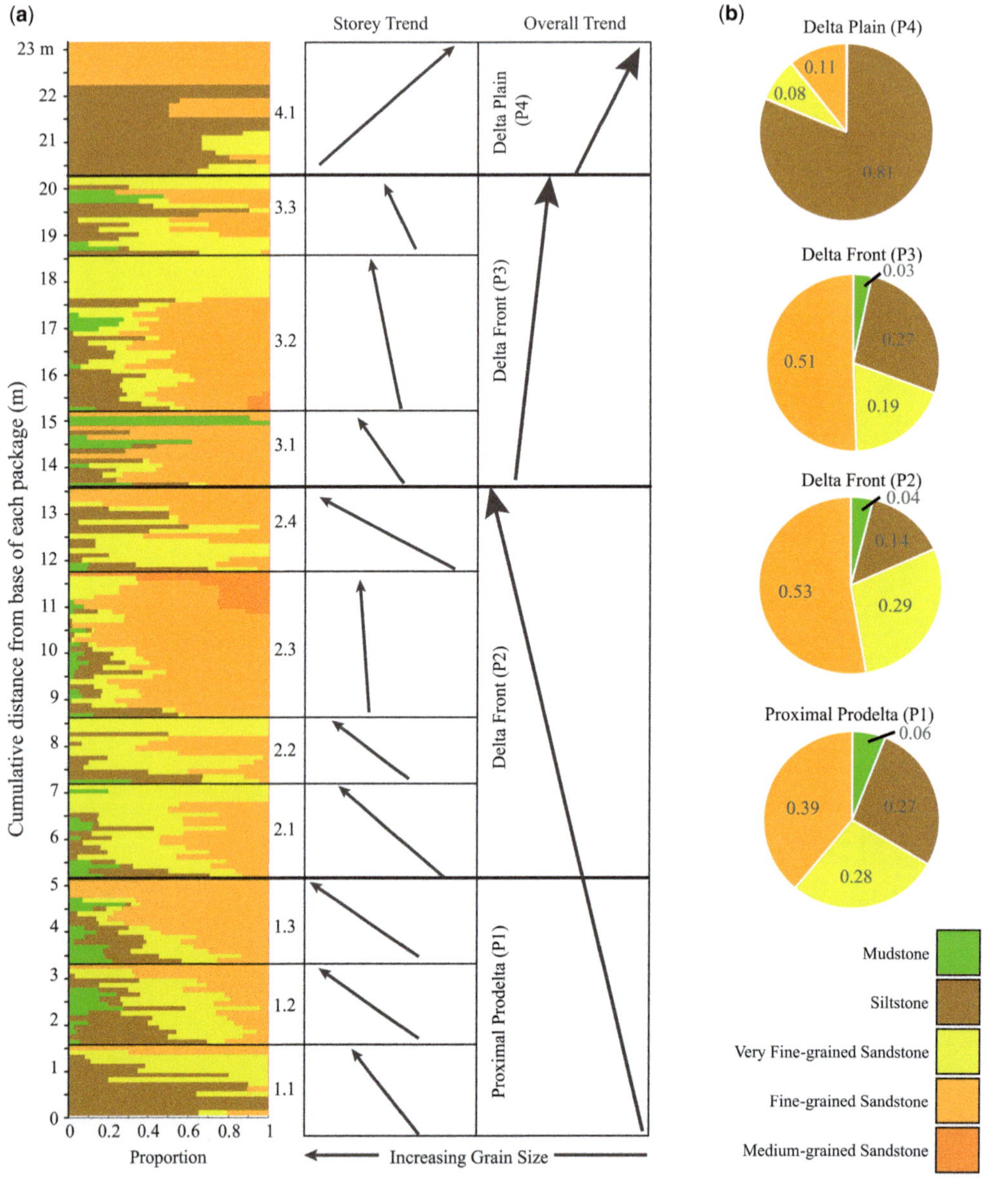

Fig. 13. Vertical grain size trends from measured sections (Fig. 10). (**a**) Grain size proportion curve of bayhead delta storeys. In general, grain size increases up-section until the top of storey 3.3 and then abruptly decreases in the abandoned channel fill of storey 4.1. Overall grain size trends indicate that P1 and P2 are coarsening-upwards, whereas P3 and P4 show overall fining-upwards trends. As a whole, the bayhead delta coarsens upwards to the top of P3 and then abruptly fines in the abandoned channel fill of P4. (**b**) Pie charts of package lithologies showing the percentage of each grain size that comprises the packages. P1, the proximal prodelta, is dominated by siltstones and very fine sandstones. P2 and P3 show very similar grain size percentages, both being delta front facies. These packages show the highest net-to-gross ratio of the packages, with high percentages of very fine and fine sandstones. Siltstones with minor amounts of sandstones dominates P4 (abandoned channel fill).

bottom portion of the delta front (2.1 and 2.2 of P2) is similar to the prodelta, with consistent coarsening-upwards storeys, but the trends become more variable towards the top (2.3 and 2.4 of P2) with a higher proportion of medium-grained sandstone in storey 2.3 and multiple cycles of very fine-grained sandstone in package 2.4. The upper delta front package (P3) shows a general fining-upwards trend and is finer grained overall than the lower delta front package (P2). The basal delta front storeys (3.2 and 3.3) individually coarsen upwards, but storey 3.3 fines upwards and each storey is finer grained than the underlying storey. Package four (P4), the delta plain abandoned channel fill, has a fining-upwards grain size proportion curve and is the finest grained of all the packages. The overall grain size trend of the bayhead delta generally coarsens upwards from the basal P1 prodelta to the top of the P2 delta front and is slightly finer grained in the P3 delta front, with abrupt fining occurring in the P4 abandoned channel (Fig. 14).

Lateral outcrop trends (down depositional dip). Figure 14 explores the spatial and stratigraphic relationships of the outcrop package statistics. Each *x*-axis shows the distance along the outcrop from the origin at measured section TCIHS-1. Because the outcrop exposure is oblique to the NE depositional trend of the bayhead delta, moving along the outcrop from the point of origin allows an analysis of the packages downdip, from topset to toeset and towards the delta depositional axis (Fig. 3). The number of beds described in each measured section and used in the calculation of the following metrics are given in Figure 14.

Average sandstone bed thickness. The distribution of sandstone bed thicknesses by package (Fig. 14a) and the spatial variation in the average sandstone bed thickness within a given package were calculated from the measured sections (Fig. 14b). Sandstone bed thicknesses were calculated for all beds with grain sizes greater than silt (i.e. all fine- and medium-grained sandstone beds). The minimum net sandstone bed thickness measured in each section was 0.01 m, which is the limit of resolution of this dataset. The average bed thickness (upside-down triangle in Fig. 14a) generally increases from the proximal prodelta (P1) to the delta front (P3) for all beds measured from the top to the base of the package and along the extent of the outcrop (Fig. 12). The proximal prodelta (P1) has a maximum bed thickness of 0.82 m, with an average bed thickness for all measured sections of 0.10 m. The thickest bed in the delta front (P2) reaches 0.70 m, with an average thickness of 0.16 m. The average bedding thickness of the upper delta front (P3) is 0.15 m, with a maximum

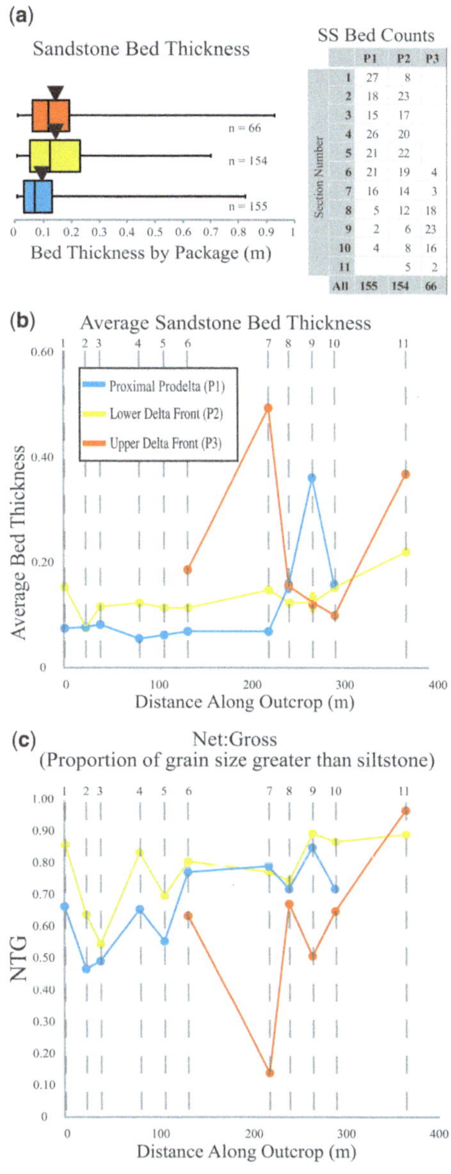

Section Number	P1	P2	P3
1	27	8	
2	18	23	
3	15	17	
4	26	20	
5	21	22	
6	21	19	4
7	16	14	3
8	5	12	18
9	2	6	23
10	4	8	16
11		5	2
All	155	154	66

Fig. 14. Graphs of outcrop statistical trends calculated for each measured section from the point of origin of the bayhead delta outcrop exposure (*x*-axis). Statistics are broken out by package: proximal prodelta (P1) and two packages of delta front (P2 and P3). The delta plain (P4) statistics were not included as a result of the sparse number of data points from the delta plain interval of TCIHS-12 and -13. The distribution of sandstone bed thicknesses are grouped by (**a**) package, (**b**) lateral trends in average bed thickness and (**c**) net-to-gross (NTG) ratio. See inset table for the number of sandstone bed measurements used in the calculations of these values for each measured section and package. Lower bed count values highlight the potential for outliers in these trend plots.

bedding thickness of 0.93 m, which is the thickest of the bayhead delta sandstone deposits (Fig. 14a). As the system progresses from the stratigraphically older proximal prodelta to the younger delta front (from P1 to P2), the average bedding thickness increases and a higher proportion of beds thicker than the median bed size are present. In each of these packages, the bed thickness statistics are derived from 155 and 154 bed thickness measurements; the average thickness calculations spatially within each package (Fig. 14b) are statistically reliable where the number of beds is larger. The more robust parts of the dataset (up to 200 m along the outcrop) show that sandstone beds in the proximal prodelta (P1) are generally about half the thickness of the sandstone beds in the lower delta front (P2). The relationship from the lower delta front (P2) to the upper delta front (P3) is not as clear and this is probably a result of the limited data from this portion of the depositional system. Limited available data from the delta plain (P4) makes statistical analysis difficult.

Net-to-gross ratio. The NTG ratio increases slightly from topset to toeset along the outcrop (increasing distance along the outcrop in Fig. 14c). The abandoned channel of P4 formed as the last phase of delta occupation and represents dominantly passive filling after the sediments had already been diverted away and the delta lobe was abandoned. The NTG ratios of the prodelta deposits (P1) are lower than the NTG ratios of the delta front deposits of P2, but higher than those of P3 delta front deposits. The prodelta deposits represent initial delta deposition and are comparatively distal to the sediment input. Therefore they are characteristically fine grained. However, proximal prodelta and delta front deposits show similar maximum NTG ratios, reaching maximum values of 0.85 and 0.89, respectively. In both cases, the minimum NTG ratio is close to 50%, at 0.47 and 0.54, respectively. Delta front deposits typically have higher NTG ratios than the prodelta because they are more proximal to sediment input and are deposited under higher energy conditions than the prodelta (Milton & Bertram 1995). The NTG ratio becomes more variable in the delta plain as the system moves towards the depositional axis.

Discussion

Outcropping paralic facies assemblages in the John Henry Member of Tibbet Canyon offer a case study of an estuarine fill succession. In this discussion, the statistical facies characterization of the bayhead delta in Tibbet Canyon is reviewed with a focus on reservoir implications. The bayhead delta IHS section is then compared with a study of similar lithofacies from tide-influenced point bar deposits (Durkin *et al.* 2015) (Fig. 15). This statistical comparison of IHS from two different depositional settings provides an excellent opportunity to investigate implications for interpreting IHS-dominated hydrocarbon reservoirs and a first-order discussion on the impact on reservoir quality and fluid flow potential.

Bayhead delta evolution. The overall grain size trends of the Tibbet Canyon bayhead delta packages (P1–P4) indicate that the proportion of fine sandstone increases within P1, is constant within P2, and decreases in P3 and P4 (Fig. 12). P1 and P2 have a coarsening-upwards grain size trend (Fig. 13). P3 and P4 show overall fining-upwards trends. P1, P2 and P3 have individual storeys that show coarsening-upwards trends. P2 and P3 are both representative of delta front facies. However, P3 has an overall finer grain size than P2. This could be due to a decrease in fluvial sediment discharge or, for example, a local avulsion. P4 represents abandoned channel fill, with a fining-upwards grain size trend. This package marks the final stage of delta occupation in Tibbet Canyon.

Individual storey trends show relative changes in grain size (Fig. 13). The trajectory of the trends within each storey may indicate relative accommodation v. sediment supply for this part of the delta. The majority of storeys show systematic increases in grain size and average sandstone bed thickness, which probably reflect normal progradational delta deposition (Bhattacharya 2006; Ahmed *et al.* 2014). Storeys 2.3, 3.2 and 3.3 have a grain size trajectory with a steeper angle and therefore appear more aggradational than progradational (Fig. 15). These aggradational events could reflect a combination of increased fluvial discharge accompanied by an increase in grain transport. Increased discharge may occur seasonally or might be caused by a single storm event (Nichol *et al.* 1997). Bayhead deltas are deposited in shallow water depths (<2 m) (Simms & Rodriguez 2014) and have low gradients downdip (Rodriguez *et al.* 2010), making them sensitive to small changes in local accommodation. Thus seasonal changes in discharge and central bay water depth could have a noticeable impact on grain size trends.

Reservoir implications. The reservoir quality for prograding deltas is expected to increase upwards and towards the sandstone body axis (White *et al.* 2004). A prevalence of amalgamated sandstone beds are observed in this area (>500 m from the origin, i.e. sections 12 and 13 of Fig. 14), meaning that mudstone drapes are less likely to be present that would inhibit vertical flow in the reservoir. The NTG ratio is also very high in these sections,

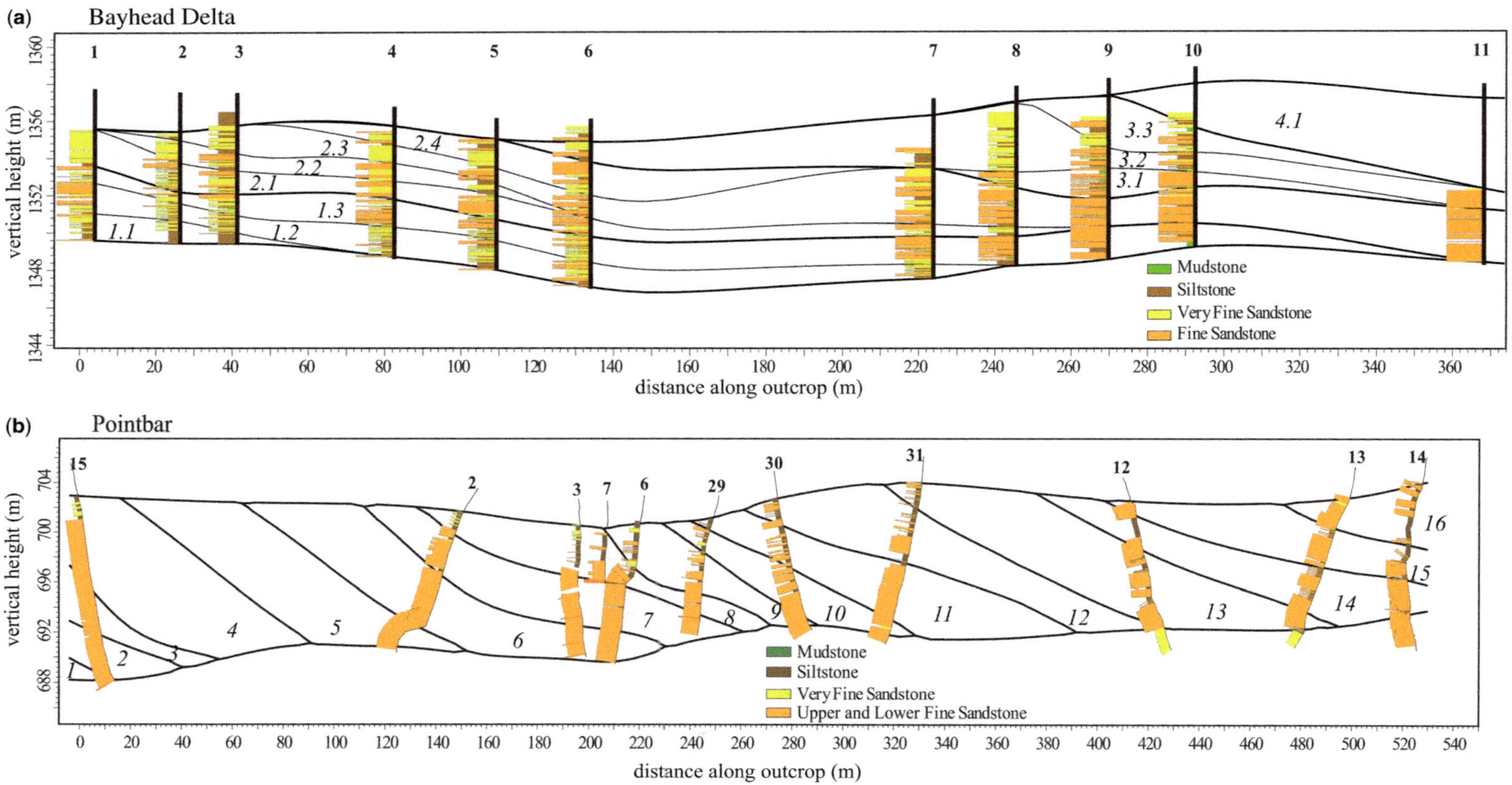

Fig. 15. Measured sections and surfaces that define storey-level correlations throughout inclined heterolithic strata intervals from (**a**) bayhead delta (cf. Fig. 10; this study) and (**b**) point bar outcrop examples (Durkin *et al.* 2015). Grain size measurements are shown overlain on the sections. Measured section numbers are listed in bold along the top of each section and the accretion set/storey numbers are in italics.

with very fine and fine-grained sandstone beds predominant.

Positions <500 m from the origin of the outcrop, the focus of the statistical analysis, represent a reservoir interval with a lower NTG ratio and average bed thickness and more vertically isolated sandstone beds. They are depositionally updip and off-axis from the main axis of sandstone deposition. Sandstone beds in this section are probably in communication with beds along the main channel axis and probably contain a significant portion of the connected reservoir sandstone. High proportions of continuous mudstone beds draping storey surfaces would inhibit the communication of these zones with the main axis of sandstone accumulation. Vertical facies proportion curves (Fig. 13) can be interrogated to ascertain the potential for continuous mudstone barriers or siltstone baffles. The coarsening-upwards trend places mostly siltstones at the base of each storey. From these curves, it is anticipated that anywhere between 60 and 80% of the base of each storey package is covered by siltstone and 0–30% covered by mudstone. Previous modelling studies have shown that these percentages of mudstone drapes are not enough to inhibit flow across a surface (e.g. Stright 2006), although the impact of the proportion of mudstone and continuity are highly dependent on the oil gravity.

Comparison of bayhead delta IHS with tide-influenced fluvial point bar IHS

Two models are used in this study to compare sedimentological trends in IHS-dominated deposits: one of the bayhead delta of interest in Tibbet Canyon and a similar model of a tide-influenced point bar (Durkin *et al.* 2015) (Fig. 15). The tide-influenced point bars of Durkin *et al.* (2015) are from south-central Alberta, in the Upper Cretaceous (Campanian–Maastrichtian) meander belt deposits of the Horseshoe Canyon Formation. The outcrops of the Horseshoe Canyon Formation have multiple strike and dip panels, allowing for reliable three-dimensional reconstruction, and offer insights into meander bend evolution that can be applied to subsurface reservoirs, such as the depositionally analogous McMurray Formation (Musial *et al.* 2012). The channel deposits of the Horseshoe Canyon Formation are 12–16 m thick and extend for over 600 m laterally, compared with the 8–10 m thick and 500 m laterally extensive bayhead delta deposits in Tibbet Canyon, Utah (Fig. 15). As a result of the scale differences, the intra-accretion package layers were defined at a 0.3 m grid cell thickness for the point bar model, compared with the 0.1 m intra-storey-level layers in the bayhead delta model.

Fining-upwards facies associations of point bar deposits in the Horseshoe Canyon Formation are 10–12 m thick and are composed of lateral accretion packages that range from 1 to 5 m thick and 10s of metres across in the depositional dip direction. Individual lateral accretion packages are characterized by an upwards-fining grain size trend and are bound by intra-point bar erosion surfaces that are commonly associated with bar rotation (Durkin *et al.* 2015). Although these upwards-fining trends are clear in the measured sections from the point bar, the upwards-coarsening trend is not clearly visible in the bayhead delta measured sections and some sections (e.g. TCISH-8 and -10) appear to be fining-upwards overall. Thus the Tibbet Canyon bayhead delta outcrop could be misinterpreted as a fining-upwards succession without careful bed correlation from section to section. Furthermore, with the low-angle surfaces of the bayhead delta, these strata might be interpreted as flat-lying in the subsurface.

The NTG ratio trends within each storey and from section to section (Fig. 16) do not reveal a significant difference between the bayhead delta IHS and the point bar IHS. Most NTG ratio values range from 60 to 80% and do not show any clear spatial or temporal trends. However, the distribution of grain sizes within the IHS intervals is significantly different between the two examples (Fig. 17). Vertical facies proportion curves for the bayhead delta (Fig. 17a) compared with the point bar (Fig. 17b) show inverse trends of upwards-coarsening v. upwards-fining. Furthermore, the mudstone and siltstone fractions that affect flow are distributed differently. As discussed in the bayhead delta characterization, these fine-grained fractions are mainly located at the base of each storey and do not cover more than 80% of the base of the storey-level surface. Conversely, in the point bar vertical facies proportion curve, mudstone deposits are interspersed throughout the vertical section; siltstones at the top of an accretion package cover an average of 50% of the accretion surfaces.

The lateral distributions within a storey of the bayhead delta (Fig. 17a VPC-1 to VPC-5) reveal that the mudstone and siltstone fractions are highest at the base of the storey topset (Fig. 17a VPC-1; *c.* 85% coverage) and lowest at the base of the storey toeset (Fig. 17a VPC-5; *c.* 50%). The mudstone fraction increases from the middle to the bottom (toeset) along the storey (Fig. 17a from VPC-3 to VPC-5). Conversely, the middle of the point bar accretion package contains the highest siltstone proportion (Fig. 17b VPC-3; 100%) at the top of the storey, whereas the highest overall proportion of mudstone is in VPC-2 and decreases towards the topset and toeset. The mudstone is dispersed throughout the vertical section where present in the point bar example.

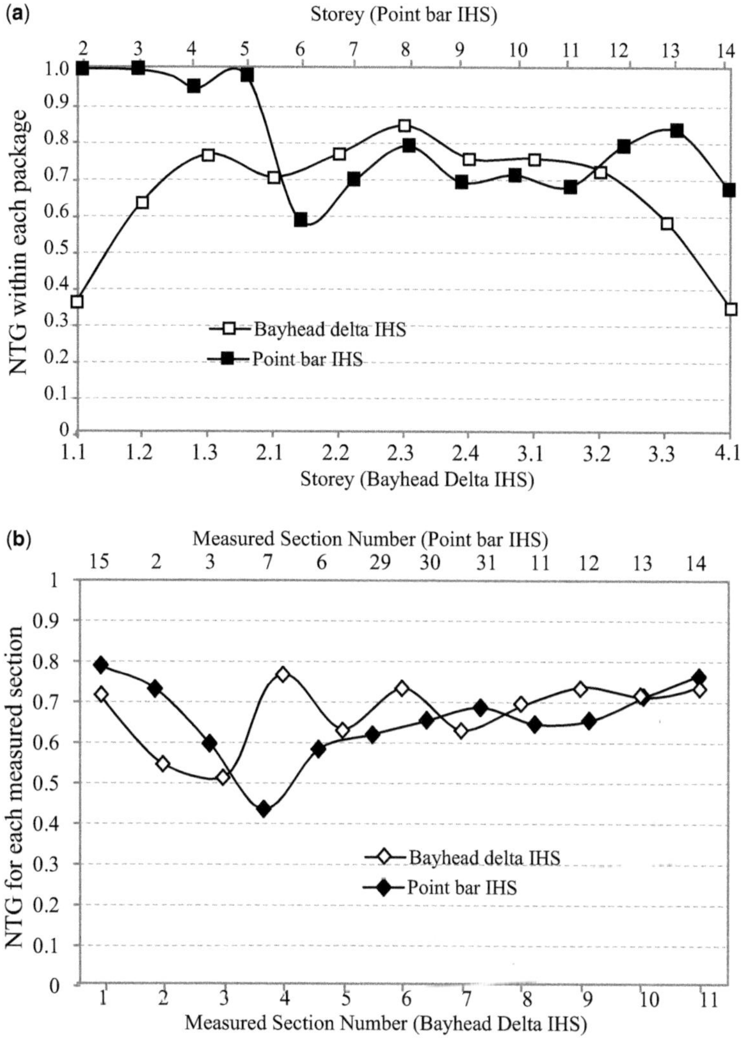

Fig. 16. Comparison of net-to-gross (NTG) ratio trends from (**a**) storey to storey along the inclined heterolithic strata (IHS) of the bayhead delta and point bar from youngest (left) to oldest (right) and (**b**) measured section to measured section. There is no significant difference in the NTG trends in these two different sets of IHS. The variation seen on the youngest and oldest storeys is most probably a result of the occurrence of incomplete storeys that bias the NTG statistic.

The analysis presented here effectively captures the differences in vertical facies proportions between bayhead delta IHS and point bar IHS, and the differences in the distribution of the fine-grained fraction. However, limitations of this comparison are evident in the lateral distributions within a storey as a result of the oblique nature of the bayhead delta outcrop section. The higher sandstone content in the toeset location (Fig. 17a VPC-5) is not intuitive until we consider that the measured sections capturing the toesets are in a more axial position of the delta than the measured sections documenting the topsets, which are in a more off-axis location of the delta (Fig. 3c). In a true depositional dip oriented cross-section, we would expect the proportion of fine-grained material to increase from topset to toeset (Bhattacharya 2003).

However, the documented differences in grain size distribution can significantly affect fluid flow. Models of point bars have shown that fluid preferentially flows along the bottom of the accretion sets, bypassing reserves in the upper portion of the point bar deposit (e.g. Pranter *et al.* 2007). The continuity of mudstone beds plays a critical part

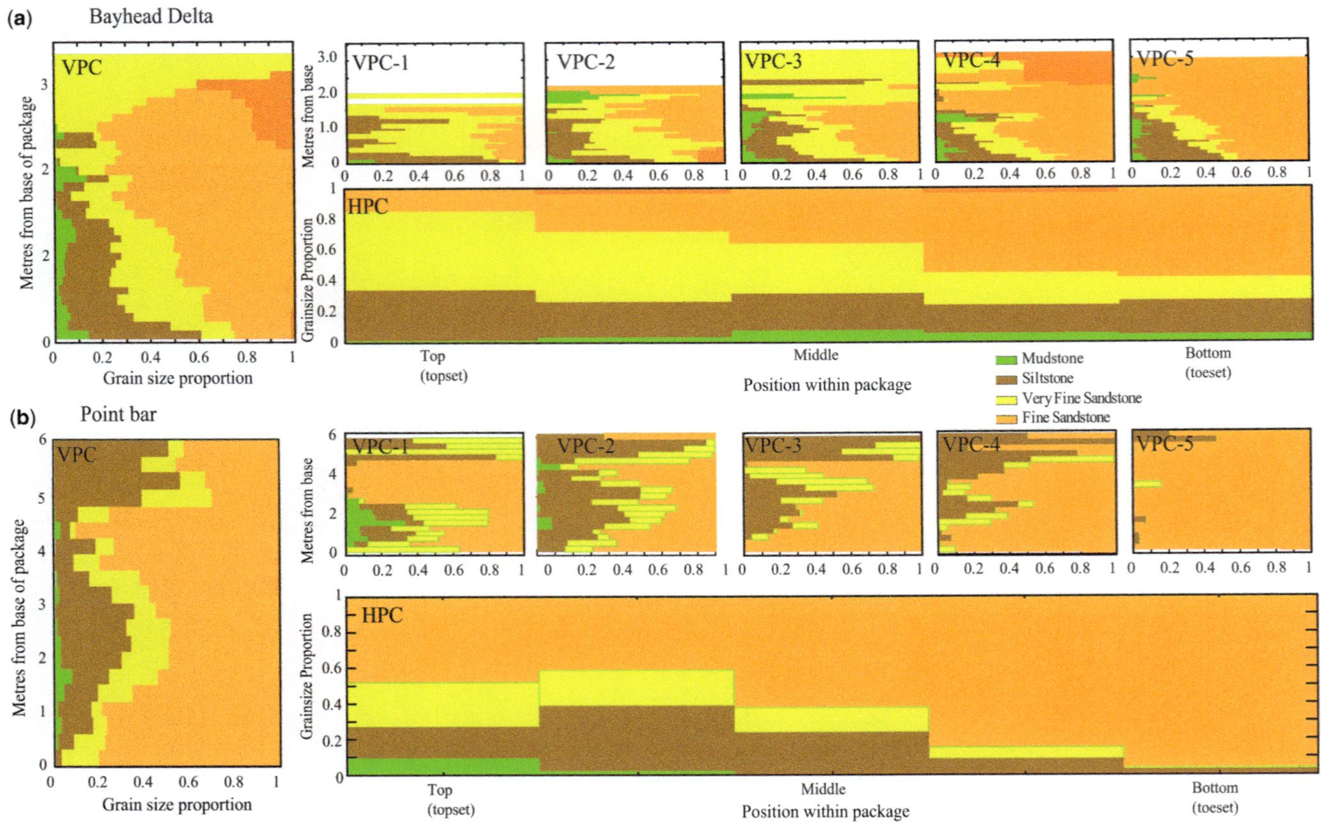

Fig. 17. Vertical and horizontal facies proportion curves for both sets of inclined heterolithic strata. (**a**) The clear upwards-coarsening trend in the vertical facies proportion curve for the bayhead delta is contrasted by (**b**) upwards-fining in the point bar vertical facies proportion curve. Coupled horizontal and vertical facies proportion curves reveal more detail about these vertical trends as a function of the position within the storey. Upwards-coarsening trends remain dominant from topset to toeset in the bayhead delta, with an increase in the proportion of fine-grained sandstone towards the toeset. The upwards-fining trends are also clear from topset to toeset of the point bar inclined heterolithic strata, with limited siltstone and no mudstone present from the middle down to the toeset of each storey. HPC, horizontal facies proportion curve; VPC, vertical facies proportion curve.

in controlling fluid flow in these cases, forcing fluid downwards to the base of the point bar. Where the mudstone units are discontinuous, they act more as baffles to flow by forcing fluids through tortuous flow paths, not necessarily downwards. Conversely, flow in deltaic deposits has been shown to behave in the opposite fashion, with fluid flowing across the top of the delta (e.g. White *et al.* 2004). These two end-member examples illustrate the anticipated differences in flow between these two IHS intervals deposited in different depositional settings, but these examples are deterministically mapped models directly from outcrop (White *et al.* 2004; Pranter *et al.* 2007). Probabilistic characterizations from outcrop, as presented in this study, enable the direct transfer of analogue information into subsurface models to drive spatial grain size distributions. End-member cases of IHS such as these may be used to bound subsurface modelling scenarios, particularly in settings where the interpretation of the depositional environment is uncertain.

Furthermore, proportion curves can provide guidelines for interpreting bayhead delta strata directly from wireline logs in the subsurface and differentiating these deposits from point bar strata. These curves provide insights into whether a section is fining-upwards or coarsening-upwards as a first-order depositional environment interpretation. Surfaces can then be interpreted below mudstone/siltstone beds. The oblique nature of the bayhead delta outcrop section presents shallower dips than a true depositional dip section; however, bayhead delta surfaces probably dip more shallowly than point bar surfaces as a result of the broader depositional area of an estuary v. a channel (Rodriguez *et al.* 2010; Simms & Rodriguez 2014). The risk of misinterpreting the storey-level surfaces and internal storey strata is lower if the initial depositional environment model is interpreted correctly. Quantitative data analysis and core descriptions combined with outcrop analogues can lower the risk associated with correctly correlating wells and potential flow units within bayhead delta reservoirs.

Conclusions

Two facies associations are preserved within the John Henry Member at Tibbet Canyon: estuary fill with tidal deposits and coastal plain. Estuary depositional architectures show a relative decrease in tidal energies over *c.* 65 m of vertical section and this effect is interpreted as a shift from a mixed-energy to a wave-dominated estuarine setting. In addition, Tibbet Canyon strata provide an outcrop-based case study for transgressive to regressive process regime change models. Stratigraphic relationships established in this study suggest that the lower John Henry Member estuarine deposits in Tibbet Canyon correlate up depositional dip to tide-influenced fluvial strata, and down depositional dip to estuarine and related paralic strata, with implications for sequence stratigraphic correlations in the Kaiparowits Plateau.

Statistical analysis of outcrop data quantitatively characterizes a bayhead delta reservoir analogue and provides insight into the development of IHS deposits. The majority of storeys within the bayhead delta section show systematic increases in grain size. Average sandstone grain size trajectories indicate variations in aggradation v. progradation of the delta and probably reflect the sensitivity of this system to small changes in local accommodation. The NTG ratio is highest in the most amalgamated sections near the main axis of the sandstone body. This interval also contains fewer mudstone drapes that could inhibit vertical flow in a reservoir than the off-axis, lower NTG ratio sections.

Comparison between the bayhead delta and a fluvial point bar indicate that they have similar architectures, including accretionary, inclined beds sets with common mudstone drapes. However, deltas prograde and accrete parallel to the flow direction and exhibit an overall upwards-coarsening grain size trend, whereas point bars accrete perpendicular to the flow direction and exhibit an overall upwards-fining grain size trend. The distribution and continuity of mudstone drapes is also distinct between these two examples. In the bayhead delta a higher proportion of mudstone is present from the middle to the toeset of each storey. Conversely, the proportion of mudstone is highest in the top of the point bar accretion sets and absent in the middle to toeset position. In both instances, mudstones are distributed vertically throughout the package and not preferentially towards the base or the top. Siltstones, however, are distributed predominantly towards the base of the bayhead delta storey packages and towards the top of point bar accretion sets. These grain size distributions drive fluid flow either to the top (bayhead delta) or the base of the system (point bar) and, when recognized, can guide the interpretation of the bounding surfaces and the packages within them.

We thank the current and former members of the Rocks2Models research consortium for their support of this research: Chevron, ConocoPhillips, Hess Corporation, Shell and Statoil, as well as many field assistants from the University of Utah. Schlumberger provided an academic licence for Petrel for use in this study. The Grand Staircase Escalante National Monument provided research permits for this research. We also thank David Uličný and an anonymous reviewer for their constructive comments and Gary Hampson for his editorial oversight.

References

Ahmed, S., Bhattacharya, J.P., Garza, D.E. & Li, Y. 2014. Facies architecture and stratigraphic evolution of a river-dominated delta front, Turonian Ferron Sandstone, Utah, U.S.A. *Journal of Sedimentary Research*, **84**, 97–121, https://doi.org/10.2110/jsr.2014.6

Ainsworth, R.B., Flint, S.S. & Howell, J.A. 2008. Predicting coastal depositional style: influence of basin morphology and accommodation to sediment supply ratio within a sequence-stratigraphic framework. *In*: Hampson, G.J., Steel, R.J., Burgess, P.M. & Dalrymple, R.W. (eds) *Recent Advances in Models of Shallow-Marine Stratigraphy*. SEPM Special Publications, **90**, 237–263.

Ainsworth, R.B., Vakarelov, B.K. & Nanson, R.A. 2011. Dynamic spatial and temporal prediction of changes in depositional processes on clastic shorelines: toward improved subsurface uncertainty reduction and management. *American Assocation of Petroleum Geologists Bulletin*, **95**, 267–297, https://doi.org/10.1306/06301010036

Allen, G.P. 1991. Sedimentary processes and facies in the Gironde estuary: a recent model for macrotidal estuarine systems. *In*: Smith, D.G., Reinson, G.E., Zaitlin, B.A., Rainson, G.E. & Rahmani, R.A. (eds) *Clastic Tidal Sedimentology*. Canadian Society of Petroleum Geologists Memoirs, **16**, 29–40.

Allen, G.P. & Posamentier, H.W. 1993. Sequence stratigraphy and facies model of an incised valley fill: the Gironde Estuary, France. *Journal of Sedimentary Research*, **63**, 378–391, https://doi.org/10.1306/D4267B09-2B26-11D7-8648000102C1865D

Allen, G.P. & Posamentier, H.W. 1994. Transgressive facies and sequence architecture in mixed tide- and wave-dominated incised valleys: example from the Gironde Estuary, France. *Journal of Sedimentary Petrology*, **63**, 378–391.

Allen, J.L. & Johnson, C.L. 2010*a*. Facies control on sandstone composition (and influence of statistical methods on interpretations) in the John Henry Member, Straight Cliffs Formation, southern Utah, USA. *Sedimentary Geology*, **230**, 60–76, https://doi.org/10.1016/j.sedgeo.2010.06.023

Allen, J.L. & Johnson, C.L. 2010*b*. Sedimentary facies, paleoenvironments and relative sea level changes in the John Henry Member, Cretaceous Straight Cliffs Formation, Southern Utah, USA. *In*: Carney, S.M., Tabet, D.E. & Johnson, C.L. (eds) *Geology of South-Central Utah*. Utah Geological Association Publications, **39**, 225–247.

Allen, J.L. & Johnson, C.L. 2011. Architecture and formation of transgressive-regressive cycles in marginal marine strata of the John Henry Member, Straight Cliffs Formation, Upper Cretaceous of Southern Utah, USA. *Sedimentology*, **58**, 1486–1513, https://doi.org/10.1111/j.1365-3091.2010.01223.x

Allen, J.R.L. 1964. Studies in fluviatile sedimentation: six cyclothems from the Lower Old Red Sandstone, Anglowelsh Basin. *Sedimentology*, **3**, 163–198.

Armstrong, R.L. 1968. Sevier orogenic belt in Nevada and Utah. *Geological Society of America Bulletin*, **79**, 429–458, https://doi.org/10.1130/0016-7606(1968)79[429:SOBINA]2.0.CO;2

Banerjee, I., Kalkreuth, W. & Davies, E.H. 1996. Coal seam splits and transgressive-regressive coal couplets: a key to stratigraphy of high-frequency sequences. *Geology*, **24**, 1001–1004, https://doi.org/10.1130/0091-7613(1996)024<1001:CSSATR>2.3.CO;2

Bhattacharya, J.P. 2003. Deltas and estuaries. *In*: Middleton, G. (ed.) *Encyclopedia of Sediments and Sedimentary Rocks*. Springer, Dordrecht, 195–202.

Bhattacharya, J.P. 2006. Deltas. *In*: Posamentier, H.W. & Walker, R.G. (eds) *Facies Models Revisited*. SEPM, Tulsa, OK, 237–292, https://doi.org/10.2110/pec.06.84.0237

Bhattacharya, J.P. & Giosan, L. 2003. Wave-influenced deltas: geomorphological implications for facies reconstruction. *Sedimentology*, **50**, 187–210, https://doi.org/10.1046/j.1365-3091.2003.00545.x

Biggs, R.B. 1967. The sediments of Chesapeake Bay. *In*: Lauff, G.H. (ed.) *Estuaries*. American Association for the Advancement of Science Publications, **83**, 239–260.

Billy, J., Chaumillon, E., Féniès, H. & Poirier, C. 2012. Tidal and fluvial controls on the morphological evolution of a lobate estuarine tidal bar: the Plassac Tidal Bar in the Gironde Estuary (France). *Geomorphology*, **169–170**, 86–97, https://doi.org/10.1016/j.geomorph.2012.04.015

Boyd, R., Dalrymple, R.W. & Zaitlin, B.A. 1992. Classification of coastal sedimentary environments. *Sedimentary Geology*, **80**, 139–150.

Boyd, R., Dalrymple, R.W. & Zaitlin, B.A. 2006. Estuarine and incised-valley facies models. *In*: Posamentier, H.W. & Walker, R.G. (eds) *Facies Models Revisited*. SEPM, Tulsa, OK, 171–235, https://doi.org/10.2110/pec.06.84

Bridge, J.S. 2003. *Rivers and Floodplains: Forms, Processes, and the Sedimentary Record*. Wiley-Blackwell, Oxford.

Bridge, P.H. & Leeder, M.R. 1976. Sedimentary model for intertidal mudflat channels, with examples from the Solway Firth, Scotland. *Sedimentology*, **23**, 533–552, https://doi.org/10.1111/j.1365-3091.1976.tb00066.x

Broger, K.E.J., Syhlonyk, G.E. & Zaitlin, B.A. 1997. Glauconite sandstone exploration: a case study from the Lake Newell Project, Southern Alberta. *In*: Pemberton, G.S. & James, D.P. (eds) *Petroleum Geology of the Cretaceous Mannville Group, Western Canada*. Canadian Society of Petroleum Geologists Memoirs, **18**, 140–168.

Burton, D. & Wood, L.J. 2011. Quantitative shale characterization of the tidally influenced Sego Sandstone. *American Assocation of Petroleum Geologists Bulletin*, **95**, 1207–1226, https://doi.org/10.1306/12081010119

Cambell, C.V. 1967. Lamina, laminaset, bed and bedset. *Sedimentology*, **8**, 7–26, https://doi.org/10.1111/j.1365-3091.1967.tb01301.x

Cattaneo, A. & Steel, R.J. 2003. Transgressive deposits: a review of their variability. *Earth-Science Reviews*, **62**, 187–228, https://doi.org/10.1016/S0012-8252(02)00134-4

Chaumillon, E., Féniès, H., Billy, J., Breilh, J.F. & Richetti, H. 2013. Tidal and fluvial controls on the internal architecture and sedimentary facies of a lobate estuarine tidal bar (the Plassac Tidal Bar in the Gironde Estuary, France). *Marine Geology*, **346**, 58–72, https://doi.org/10.1016/j.margeo.2013.07.017

CHENTNIK, B.M., JOHNSON, C.L., MULHERN, J.S. & STRIGHT, L. 2015. Valleys, estuaries, and lagoons: paleoenvironments and regressive-transgressive architecture of the Upper Cretaceous Straight Cliffs Formation. *Journal of Sedimentary Research*, **85**, 1166–1196.

CHOI, K.S., DALRYMPLE, R.W., CHUN, S.S. & KIM, S.P. 2004. Sedimentology of modern, inclined heterolithic stratification (IHS) in the macrotidal Han River Delta, Korea. *Journal of Sedimentary Research*, **74**, 677–689, https://doi.org/10.1306/030804740677

COLEMAN, J.M. & PRIOR, D.B. 1982. Deltaic environments of deposition. *In*: SCHOLLE, P.E. & SPEARING, D. (eds) *Sandstone Depositional Environments*. American Assocation of Petroleum Geologists Memoirs, **31**, 139–178.

COLEMAN, J.M. & WRIGHT, L.D. 1975. Modern river deltas: variability of processes and sand units. *In*: BROUSSARD, M.L. (ed.) *Deltas, Models for Exploration*. Houston Geological Society, Houston, TX, 99–149.

CONEY, P.J. 1972. Cordilleran tectonics and North America plate motion. *American Journal of Science*, **272**, 603–628, https://doi.org/10.2475/ajs.272.7.603

CUMMINGS, D.I., ARNOTT, R.W.C. & HART, B.S. 2006. Tidal signatures in a shelf-margin delta: a product of shelf-edge embayment? *Geology*, **34**, 249–252.

CURRIE, B.S. 2002. Structural configuration of the Early Cretaceous Cordilleran foreland basin system and Sevier thrust belt, Utah and Colorado. *The Journal of Geology*, **110**, 697–718, https://doi.org/10.1086/342626

DALRYMPLE, R.W. 2006. Incised valleys in time and space; an introduction to the volume and an examination of the controls on valley formation and filling. *In*: DALRYMPLE, R.W., LECKIE, D.A. & TILLMAN, R.W. (eds) *Incised Valleys in Time and Space*. SEPM, Special Publications, **85**, 5–12.

DALRYMPLE, R.W. & CHOI, K. 2007. Morphologic and facies trends through the fluvial-marine transition in tide-dominated depositional systems: a schematic framework for environmental and sequence-stratigraphic interpretation. *Earth-Science Reviews*, **81**, 135–174, https://doi.org/10.1016/j.earscirev.2006.10.002

DALRYMPLE, R.W., ZAITLIN, B.A. & BOYD, R. 1992. Estuarine facies models; conceptual basis and stratigraphic implications. *Journal of Sedimentary Research*, **62**, 1130–1146, https://doi.org/10.1306/D4267A69-2B26-11D7-8648000102C1865D

DALRYMPLE, R.W., BAKER, E.K., HARRIS, P.T. & HUGHES, M. 2003. Sedimentology and stratigraphy of a tide-dominated, foreland–basin delta (Fly River, Papua New Guinea). *In*: SIDI, F.H., NUMMEDAL, D., IMBERT, P., DARMAN, H. & POSAMENTIER, H.W. (eds) *Tropical Deltas of Southeast Asia – Sedimentology, Stratigraphy, and Petroleum Geology*. SEPM, Special Publications, **76**, 147–173.

DALRYMPLE, R.W., MACKAY, D.A., ICHASO, A.A. & CHOI, K.S. 2012. Processes, morphodynamics, and facies of tide-dominated estuaries. *In*: DAVIS, R.A. & DALRYMPLE, R.W. (eds) *Principles of Tidal Sedimentology*. Springer, Dordrecht, 79–107.

DECELLES, P.G. & COOGAN, J.C. 2006. Regional structure and kinematic history of the Sevier fold-and-thrust belt, central Utah. *Geological Society of America Bulletin*, **118**, 841–864.

DE MOWBRAY, T. 1983. The genesis of lateral accretion deposits in recent intertidal mudflat channels, Solway Firth. *Sedimentology*, **30**, 425–435, https://doi.org/10.1111/j.1365-3091.1983.tb00681.x

DOLIGEZ, B., BEUCHER, H., GEFFROY, F. & ESCHARD, R. 1999. Integrated reservoir characterization: improvement in heterogeneous stochastic modeling by integration of additional external constraints. *In*: SCHATZINGER, R. & JORDAN, J. (eds) *Reservoir Characterization – Recent Advances*. American Assocation of Petroleum Geologists Memoirs, **71**, 333–342.

DONALDSON, A.C., MARTIN, R.H. & KANES, W.H. 1970. Holocene Guadalupe Delta of the Texas Gulf Coast. *In*: MORGAN, J.P. & SHAVER, R.H. (eds) Deltaic Sedimentation: Modern and Ancient. SEPM Special Publications, **15**, 107–137.

DOOLING, P.R. 2013. *Tidal facies, stratigraphic architecture, and along-strike variability of a high energy, transgressive shoreline, Late Cretaceous, Kaiparowits Plateau, Southern Utah*. MS thesis, University of Utah.

DREYER, T., WHITAKER, M.F., DEXTER, J., FLESCHE, H. & LARSEN, E. 2005. From spit system to tide-dominated delta: integrated reservoir model of the Upper Jurassic Sognefjord Formation on the Troll West Field. *In*: DORÉ, A.G. & VINING, B.A. (eds) *Petroleum Geology: North-West Europe and Global Perspectives – Proceedings of the 6th Petroleum Geology Conference*. Geological Society, London, Petroleum Geology Conference Series, **6**, 423–448, https://doi.org/10.1144/0060423

DRIESE, S.G., LUDVIGSON, G.A. ET AL. 2010. Micromorphology and stable-isotope geochemistry of historical pedogenic siderite formed in PAH-contaminated alluvial clay soils, Tennessee, U.S.A. *Journal of Sedimentary Research*, **80**, 943–954, https://doi.org/10.2110/jsr.2010.087

DURKIN, P.R., HUBBARD, S.M., BOYD, R.L. & LECKIE, D.A. 2015. Stratigraphic expression of intra-point bar erosion and rotation. *Journal of Sedimentary Research*, **85**, 1238–1257, https://doi.org/10.2110/jsr.2015.78

EATON, J.G. 1991. Introduction: tectonic setting along the margin of the Cretaceous Western Interior Seaway, southwestern Utah and northern Arizona. *In*: NATIONS, J.D. & EATON, J.G. (eds) *Stratigraphy, Depositional Environments, and Sedimentary Tectonics of the Western Margin, Cretaceous Western Interior Seaway*. Geological Society of America Special Papers, **260**, 1–8.

ELLIOTT, T. 1974. Interdistributary bay sequences and their genesis. *Sedimentology*, **21**, 611–622, https://doi.org/10.1111/j.1365-3091.1974.tb01793.x

ENGE, H.D., BUCKLEY, S.J., ROTEVATN, A. & HOWELL, J.A. 2007. From outcrop to reservoir simulation model: workflow and procedures. *Geosphere*, **3**, 469–490, https://doi.org/10.1130/GES00099.1

FALIVENE, O., ABRUES, P., GARDINER, A., PICKUP, G., MUNOZ, J.A. & CABRERA, L. 2006. Best practice stochastic facies modeling from a channel-fill turbidite sandstone analog (the Quarry outcrop, Eocene Ainsa basin, northeast Spain). *American Assocation of*

Petroleum Geologists Bulletin, **90**, 1003–1029, https://doi.org/10.1306/02070605112

Feldman, H., Adereti, O. *et al.* 2013. Predicting large-scale reservoir architecture: application of data from shelf edge reservoirs in the eastern Niger Delta. *Nigerian Association of Petroleum Exploration Bulletin*, **25**, 52–80.

Feldman, H.R., Fabijanic, J.M., Faulkner, B.L. & Rudolph, K.W. 2014. Lithofacies, parasequence stacking, and depositional architecture of wave- to tide-dominated shorelines in the Frontier Formation, Western Wyoming, U.S.A. *Journal of Sedimentary Research*, **84**, 694–717, https://doi.org/10.2110/jsr.2014.53

Fenies, H. & Tastet, J. 1998. Facies and architecture of an estuarine tidal bar (the Trompeloup bar, Gironde Estuary, SW France). *Marine Geology*, **150**, 149–169, https://doi.org/10.1016/S0025-3227(98)00059-0

Fenies, H., de Resseguier, A. & Tastet, J. 1999. Intertidal clay-drape couplets (Gironde estuary, France). *Sedimentology*, **46**, 1–15, https://doi.org/10.1046/j.1365-3091.1999.00196.x

Gallin, W.N., Johnson, C.L. & Allen, J.L. 2010. Fluvial and marginal marine architecture of the John Henry Member, Straight Cliffs Formaiton, Kelly Grade of the Kaiparowits Plateau, south-central Utah. *In*: Carney, S.M., Tabet, D.E. & Johnson, C.L. (eds) *Geology of South-Central Utah*. Utah Geological Association Publications, **39**, 248–275.

Galloway, W.D. 1975. Process framework for describing the morphologic and stratigraphic evolution of deltaic depositional systems. *In*: Broussard, M.L. (ed.) *Deltas: Models for Exploration*. Houston Geological Society, Houston, TX, 86–98.

Gingras, M.K., Pemberton, S.G., Saunders, T. & Clifton, H.E. 1999. The ichnology of modern and Pleistocene brackish-water deposits at Willapa Bay, Washington; variability in estuarine settings. *Palaios*, **14**, 352–374.

Gingras, M.K., Räsänen, M. & Ranzi, A. 2002. The significance of bioturbated inclined heterolithic stratification in the southern part of the Miocene Solimoes Formation, Rio Acre, Amazonia Brazil. *Palaios*, **17**, 591–601.

Gooley, J.T., Johnson, C.L. & Pettinga, L. 2016. Spatial and temporal variation of fluvial architecture in a prograding clastic wedge of the Late Cretaceous Western Interior Basin (Kaiparowits Plateau), USA. *Journal of Sedimentary Research*, **86**, 125–147, https://doi.org/10.2110/jsr.2016.11

Gradstein, F.M., Ogg, J.G. & Schmitz, M. 2012. *The Geologic Time Scale 2012*. Elsevier, Amsterdam.

Hancock, J.M. & Kauffman, E.G. 1979. The great transgressions of the Late Cretaceous. *Journal of the Geological Society, London*, **136**, 175–186, https://doi.org/10.1144/gsjgs.136.2.0175

Hassanpour, M.M., Pyrcz, M.J. & Deutsch, C.V. 2013. Improved geostatistical models of inclined heterolithic strata for McMurray Formation, Alberta, Canada. *American Assocation of Petroleum Geologists Bulletin*, **97**, 1209–1224, https://doi.org/10.1306/01021312054

Hettinger, R.D. 1995. *Sedimentological Descriptions and Depositional Interpretation, in Sequence Stratigraphic Context, of Two 300-Meter Cores from the Upper Cretaceous Straight Cliffs Formation, Kaiparowits Plateau, Kane County, Utah*. US Geological Survey Bulletin, **2115-A**.

Holbrook, J.M. 1996. Complex fluvial response to low gradients at maximum regression; a genetic link between smooth sequence-boundary morphology and architecture of overlying sheet sandstone. *Journal of Sedimentary Research*, **66**, 713–722, https://doi.org/10.1306/D42683EC-2B26-11D7-8648000102C1865D

Hooke, J. 2004. Cutoffs galore!: occurrence and causes of multiple cutoffs on a meandering river. *Geomorphology*, **61**, 225–238.

Horne, J.C., Ferm, J.C. & Caru, F.T. 1978. Depositional models in coal exploration and mine planning in Appalachian region. *American Assocation of Petroleum Geologists Bulletin*, **62**, 2379–2411, https://doi.org/10.1306/C1EA5512-16C9-11D7-8645000102C1865D

Hubbard, S.M., Pemberton, S.G., Gingras, M.K. & Thomas, M.B. 2002. Variability in wave-dominated estuary sandstones: implications on subsurface reservoir development. *Bulletin of Canadian Petroleum Geology*, **50**, 118–137, https://doi.org/10.2113/50.1.118

Hubbard, S.M., Smith, D.G. *et al.* 2011. Seismic geomorphology and sedimentology of a tidally influenced river deposit, Lower Cretaceous Athabasca oil sands, Alberta, Canada. *American Assocation of Petroleum Geologists Bulletin*, **95**, 1123–1145, https://doi.org/10.1306/12131010111

Jackson, R.G. 1981. Sedimentology of muddy fine-grained channel deposits in meandering streams of the American Middle West. *Journal of Sedimentary Research*, **51**, 1169–1192, https://doi.org/10.1306/212F7E5A-2B24-11D7-8648000102C1865D

Joeckel, R.M. & Korus, J.T. 2012. Bayhead delta interpretation of an upper Pennsylvanian sheetlike sandbody the broader understanding of transgressive deposits in cyclothems. *Sedimentary Geology*, **274–275**, 22–37, https://doi.org/10.1016/j.sedgeo.2012.07.002

Jordan, T.E. 1981. Thrust loads and foreland basin evolution, Cretaceous, western United States. *American Assocation of Petroleum Geologists Bulletin*, **65**, 2506–2520.

Kauffman, E.G. 1977. Geological and biological overview: Western Interior Cretaceous basin. *Mountain Geologist*, **14**, 75–99.

Kirschbaum, B.M.A. & Hettinger, R.D. 2004. *Facies Analysis and Sequence Stratigraphic Framework of Upper Campanian Strata (Neslen and Mount Garfield Formations, Bluecastle Tongue of the Castlegate Sandstone, and Mancos Shale), Eastern Book Cliffs, Colorado and Utah*. US Geological Survey Digital Data Series, **DDS-69-G**.

Kraus, M.J. & Davies-Vollum, K.S. 2004. Mudrock-dominated fills formed in avulsion splay channels: examples from the Willwood Formation, Wyoming. *Sedimentology*, **51**, 1127–1144, https://doi.org/10.1111/j.1365-3091.2004.00664.x

Labrecque, P.A., Jensen, J.L., Hubbard, S.M. & Nielsen, H. 2011. Sedimentology and stratigraphic architecture of a point bar deposit, Lower Cretaceous McMurray Formation, Alberta, Canada. *Bulletin of Canadian Petroleum Geology*, **59**, 147–171, https://doi.org/10.2113/gscpgbull.59.2.147

LAWTON, T.F., SCHELLENBACH, W.L. & NUGENT, A.E. 2014. Megafan and axial-river systems in the southern Cordilleran foreland basin: Drip Tank Member of Straight Cliffs Formation and adjacent strata, Southern Utah, USA. *Journal of Sedimentary Research*, **84**, 407–434, https://doi.org/10.2110/jsr.2014.33

LI, Y. & BHATTACHARYA, J.P. 2013. Facies-architecture study of a stepped, forced regressive compound incised valley in the Ferron Notom Delta, Southern Central Utah, U.S.A. *Journal of Sedimentary Research*, **83**, 206–225, https://doi.org/10.2110/jsr.2013.19

LIU, S., NUMMEDAL, D. & GURNIS, M. 2014. Dynamic v. flexural controls of Late Cretaceous Western Interior Basin, USA. *Earth and Planetary Science Letters*, **389**, 221–229, https://doi.org/10.1016/j.epsl.2014.01.006

MACEACHERN, J.A., RAYCHAUDHURI, I. & PEMBERTON, S.G.G. 1992. Stratigraphic applications of the *Glossifungites* ichnofacies: delineating discontinuities in the rock record. *In*: PEMBERTON, S.G. (ed.) *Applications of Ichnology to Petroleum Exploration*. Society for Sedimentary Geology Core Workshop, **82**, 169–198.

MACK, G.H., LEEDER, M., PEREZ-ARLUCEA, M. & BAILEY, B.D.J. 2003. Sedimentology, paleontology, and sequence stratigraphy of Early Permian estuarine deposits, south-central New Mexico, USA. *Palaios*, **18**, 403–420, https://doi.org/10.1669/0883-1351(2003)018<0403:SPASSO>2.0.CO;2

MARTINIUS, A.W., KAAS, I., NISS, A., HELGESEN, G., KJREFJORD, J.M. & LEITH, D.A. 2001. Sedimentology of the heterolithic and tide-dominated Tilje Formation (Early Jurassic, Halten Terrace, Offshore Mid-Norway). *In*: MARTINSEN, O.J. & DREYER, T. (eds) *Sedimentary Environments Offshore Norway – Palaeozoic to Recent, Proceedings of the Norwegian Petroleum Society Conference*, 3–5 May 1999, Bergen, Norway. Norwegian Petroleum Society Special Publications, **10**, 103–144, https://doi.org/10.1016/S0928-8937(01)80011-4

MARTINIUS, A.W., RINGROSE, P.S., BROSTROM, C., ELFENBEIN, C., NAESS, A. & RINGAS, J.E. 2005. Reservoir challenges of heterolithic tidal sandstone reservoirs in the Halten Terrace, mid-Norway. *Petroleum Geoscience*, **11**, 3–16, https://doi.org/10.1144/1354-079304-629

MARTINSEN, O.J. & HELLAND-HANSEN, W. 1993. Sequence stratigraphy and facies model of an incised valley fill: the Gironde Estuary, France. *Journal of Sedimentary Research*, **63**, 78–80, https://doi.org/10.1306/D4267B09-2B26-11D7-8648000102C1865D

MCEWEN, M.C. 1969. Sedimentary facies of the modern Trinity Delta. *In*: LANKFORD, R.R. & ROGERS, J.J.W. (eds) *Holocene Geology of the Galveston Bay Area*. Houston Geological Society, Houston, TX, 53–77.

MIALL, A.D. 1985*a*. Architectural-element analysis: a new method of facies analysis applied to fluvial deposits. *Earth-Science Reviews*, **22**, 261–308, https://doi.org/10.1016/0012-8252(85)90001-7

MIALL, A.D. 1985*b*. Sedimentation on an early Proterozoic continental margin under glacial influence: the Gowganda Formation (Huronian), Elliot Lake area, Ontario, Canada. *Sedimentology*, **32**, 763–788.

MIALL, A.D. 2006. *The Geology of Fluvial Deposits*. Springer, Berlin, https://doi.org/10.1007/978-3-662-03237-4

MILLER, K.G., KOMINZ, M.A. *ET AL*. 2005. The Phanerozoic record of global sea-level change. *Science*, **310**, 1293–1298, https://doi.org/10.1126/science.1116412

MILTON, N.J. & BERTRAM, G.T. 1995. Topset play types and their controls. *In*: VAN WAGONER, J.C. & BERTRAM, G.T. (eds) *Sequence Stratigraphy of Foreland Basin Deposits*. American Association of Petroleum Geologists Memoirs, **64**, 1–9.

MORAN, K., HILBERT-WOLF, H.L. *ET AL*. 2010. Attributes of the wood-boring trace fossil *Asthenopodichnium* in the Late Cretaceous Wahweap Formation, Utah, USA. *Palaeogeography, Palaeoclimatology, Palaeoecology*, **297**, 662–669.

MULHERN, J.S. & JOHNSON, C.L. 2016. Time–space variability of paralic strata deposited in a high accommodation, high sediment supply setting: example from the Cretaceous of Utah. *In*: HAMPSON, G.J., REYNOLDS, A.D., KOSTIC, B. & WELLS, M.R. (eds) *Sedimentology of Paralic Reservoirs: Recent Advances*. Geological Society, London, Special Publications, **444**. First published online June 29, 2016, https://doi.org/10.1144/SP444.7

MUSIAL, G., REYNAUD, J.Y., GINGRAS, M.K., FÉNIÈS, H., LABOURDETTE, R. & PARIZE, O. 2012. Subsurface and outcrop characterization of large tidally influenced point bars of the Cretaceous McMurray Formation (Alberta, Canada). *Sedimentary Geology*, **279**, 156–172, https://doi.org/10.1016/j.sedgeo.2011.04.020

NARDIN, T.R., FELDMAN, H.R. & CARTER, B.J. 2013. Stratigraphic architecture of a large-scale point-bar complex in the McMurray Formation: Syncrude's Mildred Lake Mine, Alberta, Canada. *In*: HEIN, F.J., LECKIE, D., LARTER, S. & SUTER, J.R. (eds) *Heavy-Oil and Oil-Sand Petroleum Systems in Alberta and Beyond*. American Assocation of Petroleum Geologists Studies in Geology, **64**, 273–311.

NICHOL, S.L., ZAITLIN, B.A. & THOM, B.G. 1997. The upper Hawkesbury River, New South Wales, Australia: a Holocene example of an estuarine bayhead delta. *Sedimentology*, **44**, 263–286, https://doi.org/10.1111/j.1365-3091.1997.tb01524.x

OLARIU, C. & BHATTACHARYA, J.P. 2006. Terminal distributary channels and delta front architecture of river-dominated delta systems. *Journal of Sedimentary Research*, **76**, 212–233, https://doi.org/10.2110/jsr.2006.026

PAGE, K.J., NANSON, G.C. & FRAZIER, P.S. 2003. Floodplain formation and sediment stratigraphy resulting from oblique accretion on the Murrumbidgee River, Australia. *Journal of Sedimentary Research*, **73**, 5–14, https://doi.org/10.1306/070102730005

PETERSON, F. 1969*a*. *Cretaceous sedimentation and tectonism in the southern Kaiparowits Region, Utah*. PhD thesis, Stanford University.

PETERSON, F. 1969*b*. *Four New Members of the Upper Cretaceous Straight Cliffs Formation in the Southeastern Kaiparowits Region, Kane County, Utah*. Contributions to Stratigraphy. US Geological Survey Bulletin, **1274-J**, J1–J28.

PEIJS-VAN HILTEN, M.T., GOOD, R. & ZAITLIN, B.A. 1998. Heterogeneity modeling and geopseudo upscaling applied to waterflood performance prediction of an incised valley reservoir: Countess YY Pool, Southern Alberta, Canada. *American Assocation of Petroleum Geologists Bulletin*, **82**, 2220–2245.

PLINK-BJÖRKLUND, P. 2008. Wave-to-tide facies change in a Campanian shoreline complex, Chimney Rock Tongue, Wyoming-Utah, U.S.A. *In*: HAMPSON, G.J., STEEL, R.J., BURGESS, P.M. & DALRYMPLE, R.W. (eds) *Recent Advances in Models of Siliciclastic Shallow-Marine Stratigraphy*. Society for Sedimentary Geology, Special Publications, **90**, 265–291.

PRANTER, M.J., ELLISON, A.I., COLE, R.D. & PATTERSON, P.E. 2007. Analysis and modeling of intermediate-scale reservoir heterogeneity based on a fluvial point-bar outcrop analog,Williams Fork Formation, Piceance Basin, Colorado. *American Association of Petroleum Geologists Bulletin*, **91**, 1025–1051.

PRINGLE, J.K., WESTERMAN, A.R., CLARK, J.D., DRINKWATER, N.J. & GARDINER, A.R. 2004. 3D high-resolution digital models of outcrop analogue study sites to constrain reservoir model uncertainty: an example from Alport Castles, Derbyshire, UK. *Petroleum Geoscience*, **10**, 343–352, https://doi.org/10.1144/1354-079303-617

PYRCZ, M.J. & DEUTSCH, C.V. 2014. *Geostatistical Reservoir Modeling*. Oxford University Press, New York.

RAVENNE, C., GALLI, A., DOLIGEZ, B., BEUCHER, H. & ESCHARD, R. 2002. Quantification of facies relationships via proportion curves. *In*: ARMSTRONG, M., BETTINI, C., CHAMPIGNY, N., GALLI, A. & REMACRE, A. (eds) *Geostatistics Rio 2000*. Proceedings of the Geostatistics Sessions of the 31st International Geological Congress, 6–17 August 2000, Rio de Janeiro, Brazil. Quantitative Geology and Geostatistics, **12**. Springer, Berlin, 19–39, https://doi.org/10.1007/978-94-017-1701-4_3

REYNOLDS, T. 1999. Dimensions of paralic sandstone bodies. *American Association of Petroleum Geologists Bulletin*, **83**, 211–229.

RODRIGUEZ, A.B., SIMMS, A.R. & ANDERSON, J.B. 2010. Bay-head deltas across the northern Gulf of Mexico back step in response to the 8.2 ka cooling event. *Quaternary Science Reviews*, **29**, 3983–3993, https://doi.org/10.1016/j.quascirev.2010.10.004

ROY, P.S. 1994. Holocene estuary evolution – stratigraphic studies from southeastern Australia. *In*: DALRYMPLE, R.W., BOYD, R.L. & ZAITLIN, B.A. (eds) *Incised Valley Systems: Origin and Sedimentary Facies*. SEPM, Special Publications, **51**, 241–263.

SHANLEY, K.W. & MCCABE, P.J. 1991. Predicting facies architecture through sequence stratigraphy – an example from the Kaiparowits Plateau, Utah. *Geology*, **19**, 742–745, https://doi.org/10.1130/0091-7613(1991)019<0742:PFATSS>2.3.CO;2

SHANLEY, K.W. & MCCABE, P.J. 1994. Perspectives on the sequence stratigraphy of continental strata. *American Association of Petroleum Geologists Bulletin*, **78**, 544–568.

SHANLEY, K.W., MCCABE, P.J. & HETTINGER, R.D. 1992. Tidal influence in Cretaceous fluvial strata from Utah, USA: a key to sequence stratigraphic interpretation. *Sedimentology*, **39**, 905–930, https://doi.org/10.1111/j.1365-3091.1992.tb02159.x

SIMMS, A.R. & RODRIGUEZ, A.B. 2014. The influence of valley morphology on the rate of bayhead delta progradation. *Journal of Sedimentary Research*, **85**, 38–44, https://doi.org/10.2110/jsr.2015.02

SMITH, D.G., HUBBARD, S.M., LECKIE, D.A. & FUSTIC, M. 2009. Counter point bar deposits: lithofacies and reservoir significance in the meandering modern Peace River and ancient McMurray Formation, Alberta, Canada. *Sedimentology*, **56**, 1655–1669, https://doi.org/10.1111/j.1365-3091.2009.01050.x

STANLEY, K.O. & SURDAM, R.C. 1978. Sedimentation on the front of Eocene Gilbert-type deltas, Washakie Basin, Wyoming. *Journal of Sedimentary Petrology*, **48**, 557–573, https://doi.org/10.1306/212F74D2-2B24-11D7-8648000102C1865D

STEEL, R.J., PLINK-BJORKLUND, P. & ASCHOFF, J. 2012. Tidal deposits of the Campanian Western Interior Seaway, Wyoming, Utah and Colorado, USA. *In*: DAVIS, R.A. & DALRYMPLE, R.W. (eds) *Principles of Tidal Sedimentology*. Springer, Dordrecht, 437–471, https://doi.org/10.1007/978-94-007-0123-6

STRIGHT, L. 2006. Modeling, upscaling, and history matching thin, irregularly-shaped flow barriers: a comprehensive approach for predicting reservoir connectivity. Paper presented at the SPE Annual Technical Conference and Exhibition, 24–27 September 2006, San Antonio, TX, USA, Paper SPE 106528-STU, https://doi.org/10.2118/106528-STU

STROBL, R.S., MUWAIS, W.K., WIGHTMAN, D.M., COTTERILL, D.K. & YUAN, L. 1997. Geological modelling of McMurray Formation reservoirs based on outcrop and subsurface analogues. *In*: PEMBERTON, G.S. & JAMES, D.P. (eds) *Petroleum Geology of the Cretaceous Mannville Group, Western Canada*. Canadian Society of Petroleum Geologists Memoirs, **18**, 292–311.

SUAREZ, M.B., GONZALEZ, L.A. & LUDVIGSON, G.A. 2010. Estimating the oxygen isotopic composition of equatorial precipitation during the Mid-Cretaceous. *Journal of Sedimentary Research*, **80**, 480–491, https://doi.org/10.2110/jsr.2010.048

SZWARC, T.S., JOHNSON, C.L., STRIGHT, L.E. & MCFARLANE, C.M. 2014. Interactions between axial and transverse drainage systems in the Late Cretaceous Cordilleran foreland basin: evidence from detrital zircons in the Straight Cliffs Formation, southern Utah, USA. *Geological Society of America Bulletin*, **127**, 372–392, https://doi.org/10.1130/B31039.1

TÄNAVSUU-MILKEVICIENE, K., PLINK-BJÖRKLUND, P., KIRSIMÄE, K. & AINSAAR, L. 2009. Coeval v. reciprocal mixed carbonate-siliciclastic deposition, Middle Devonian Baltic Basin, Eastern Europe: implications from the regional tectonic development. *Sedimentology*, **56**, 1250–1274, https://doi.org/10.1111/j.1365-3091.2008.01032.x

TERZUOLI, A. & WALKER, R.G. 1997. Estuarine valley fills in the Lower Cretaceous Bluesky Formation, Edson area, Alberta. *Bulletin of Canadian Petroleum Geology*, **45**, 194–217.

THOMAS, L. 2012. *Coal Geology*. 2nd edn. Wiley, Chichester, https://doi.org/10.1002/9781118385685

THOMAS, R.G., SMITH, D.G., WOOD, J.M., VISSER, J., CALVERLEY-RANGE, E.A. & KOSTER, E.H. 1987.

Inclined heterolithic stratification – terminology, description, interpretation and significance. *Sedimentary Geology*, **53**, 123–179, https://doi.org/10.1016/S0037-0738(87)80006-4

Thornton, R.C.N. 1979. *Regional Stratigraphic Analysis of the Gidgealpa Group, Southern Cooper Basin, Australia.* Bulletin of the South Australian Geological Survey, **49**. Department of Mines and Energy, Adelaide.

van Heerden, I. & Roberts, H.H. 1988. Facies development of Atchafalaya Delta, Louisiana: a modern bayhead delta. *American Association of Petroleum Geologists Bulletin*, **72**, 439–453.

Visser, M.J. 1980. Neap-spring cycles reflected in Holocene subtidal large-scale bedform deposits: a preliminary note. *Geology*, **8**, 543, https://doi.org/10.1130/0091-7613(1980)8<543:NCRIHS>2.0.CO;2

White, C.D., Willis, B.J., Dutton, S., Bhattacharya, J.P. & Narayana, K. 2004. Sedimentology, statistics, and flow behavior for a tide-influenced deltaic sandstone, Frontier Formation, Wyoming, United States. *In*: Grammer, G.M., Harris, P.M. & Eberli, G.P. (eds) *Integration of Outcrop and Modern Analogs in Reservoir Modeling*. American Assocation of Petroleum Geologists Memoirs, **80**, 129–152.

Willis, B. 1997. Architecture of fluvial-dominated valley fill deposits in the Cretaceous Fall River Formation. *Sedimentology*, **44**, 735–757, https://doi.org/10.1046/j.1365-3091.1997.d01-48.x

Willis, B.J. 2005. Deposits of tide-influenced deltas. *In*: Ghiosan, L. & Bhattacharya, J.P. (eds) *River Deltas – Concepts, Models, and Examples*. SEPM Special Publications, **83**, 87–129.

Willis, B.J. & Gabel, S.L. 2003. Formation of deep incisions into tide-dominated river deltas: implications for the stratigraphy of the Sego Sandstone, Book Cliffs, Utah, U.S.A. *Journal of Sedimentary Research*, **73**, 246–263, https://doi.org/10.1306/090602730246

Willis, B.J. & White, C.D. 2000. Quantitative outcrop data for flow simulation. *Journal of Sedimentary Research*, **70**, 788–802.

Yang, B.C., Dalrymple, R.W. & Chun, S.S. 2005. Sedimentation on a wave-dominated, open-coast tidal flat, south-western Korea: summer tidal flat – winter shoreface. *Sedimentology*, **52**, 235–252, https://doi.org/10.1111/j.1365-3091.2004.00692.x

Yoshida, S., Steel, R.J. & Dalrymple, R.W. 2007. Changes in depositional processes-an ingredient in a new generation of sequence-stratigraphic models. *Journal of Sedimentary Research*, **77**, 447–460, https://doi.org/10.2110/jsr.2007.048

Recognizing seasonal fluvial influence in ancient tidal deposits

ANNALIZE Q. MCLEAN* & BRENT WILSON

Petroleum Studies Unit, Department of Chemical Engineering, University of the West Indies, Trinidad and Tobago

**Correspondence: annalizemclean@yahoo.com*

Abstract: Seasonal changes in rainfall affect water and sediment discharges in fluvial systems. This is reflected within the rock record by changes in (1) the flow processes and structures created and (2) the composition and volume of sediment. This study demonstrates that high-resolution sedimentological studies of rhythmites in ancient tidal bar deposits reveal a seasonal fluvial influence. The rocks studied belong to the Plio-Pleistocene Erin Formation on Trinidad, deposited under tropical seasonal conditions in the fluvial–tidal transition zone and tidally dominated palaeoenvironments. Time-series and stratigraphic analyses were undertaken of two rock subsections separated by an erosional surface. Spring–neap, monthly and semi-annual cycles were identified within the deposits. Sand-rich and mud-rich couplets showed neap–spring cyclicity, with gradual transitions between flaser, wavy and lenticular bedding. Seasonal changes in fluvial discharge were recognized by minor scours that marked periods of relatively higher discharge. The content and distribution of organic material varied through the seasonally influenced cycle. Rare and dwarfed trace fossils and bioturbated beds also showed a periodic pattern, interpreted as representing changes in palaeoenvironmental conditions due to seasonal freshwater discharge. As a result, reservoirs containing similar deposits may have porosity and permeability variations with seasonal depositional periodicity.

Supplementary material: The full set of measured time-series data is available at https://doi.org/10.6084/m9.figshare.c.3283376

Paralic sequences are deposited in areas with alternating terrestrial and marine deposition. The lateral transition zone between the two sets of environments records spatial and temporal shifts resulting from changes in relative sea-level. The facies within the fluvial–tidal transition zone reflect the alternating conditions caused by both tidal and seasonal fluvial cyclicity. Small-scale heterogeneity in such environments greatly reduces the recovery of hydrocarbons from reservoirs (Choi *et al.* 2011), but the associated bioturbation of laminated layers can increase permeability (Tonkin *et al.* 2010; Gingras *et al.* 2012*a, b*; Hsieh *et al.* 2015). The study of heterogeneity from outcrops allows observation at a scale and resolution not reproducible by three-dimensional seismic, well or core data (Pringle *et al.* 2010). This study highlights the need to consider palaeoenvironmental interpretations with respect to the fluvial–tidal transition zone. This has yet to be applied to Trinidad reservoirs. Qualitative and quantitative methods were used to document relative fluvial influences in ancient tidal deposits from outcrop data. This analysis could be used to describe the expected heterogeneity in similar deposits and potential reservoirs.

The integration of statistical and observational data assists in determining the processes ongoing at the time of deposition and the resulting heterogeneity in the rock record. In areas with mixed influences, high-resolution stratigraphy allows a precise palaeoenvironmental interpretation. This study showed that time-series analysis identifies the tidal cycles in rhythmites, whereas a close look at the corresponding sedimentology confirms cyclicity and reveals trends in depositional processes. Seasonal fluvial variations are recognized in tidally dominated deposits. Also shown are the differences in sedimentary structures and the preservation potential of rocks deposited in palaeoenvironments with varying fluvial influences.

Tidal cycles

Heterolithic rhythmic bedding is a lithofacies interpreted as indicating a strong tidal influence (Tessier *et al.* 1995; Dalrymple 2010*b*; Greb *et al.* 2011). This interpretation is confirmed where tidal periodicity can be identified by the quantitative analysis of time-series data recorded from the deposits (Kvale 1989, 2006, 2012; Dalrymple 2010*b*; Davis Jr & Dalrymple 2012; Longhitano *et al.* 2012). Statistical methods have been used to analyse both ancient (Kvale 1989; Martino & Sanderson 1993; Chan

From: Hampson, G. J., Reynolds, A. D., Kostic, B. & Wells, M. R. (eds) 2017. *Sedimentology of Paralic Reservoirs: Recent Advances*. Geological Society, London, Special Publications, **444**, 287–303.
First published online July 13, 2016, https://doi.org/10.1144/SP444.6

et al. 1994; Choi *et al.* 2001; Couëffé *et al.* 2004; Hovikoski *et al.* 2005; Gent & Sonnenberg 2014) and modern (Choi 2010; Greb *et al.* 2011) tidal rhythmites. Sections with rhythmites are considered to be records of continuous deposition (Kvale 1989; Eisma 1998; Choi & Park 2000; Couëffé *et al.* 2004). Several studies have discussed tidal cycles (e.g. Visser 1980; Chan *et al.* 1994; Tessier *et al.* 1995; Kvale 2006, 2012; Dalrymple 2010*a*, *b*; Davis Jr 2012; Longhitano *et al.* 2012) and a brief description is given here.

Each tidal cycle, from flood to high tide and from ebb to low tide, can deposit one or two couplets, each couplet being a pair of relatively coarser and finer grained layers (Fig. 1a). The resulting laminae indicate deposition due to alternating sediment transport processes. During the ebb and flood tides, sediments are transported and deposited by traction,

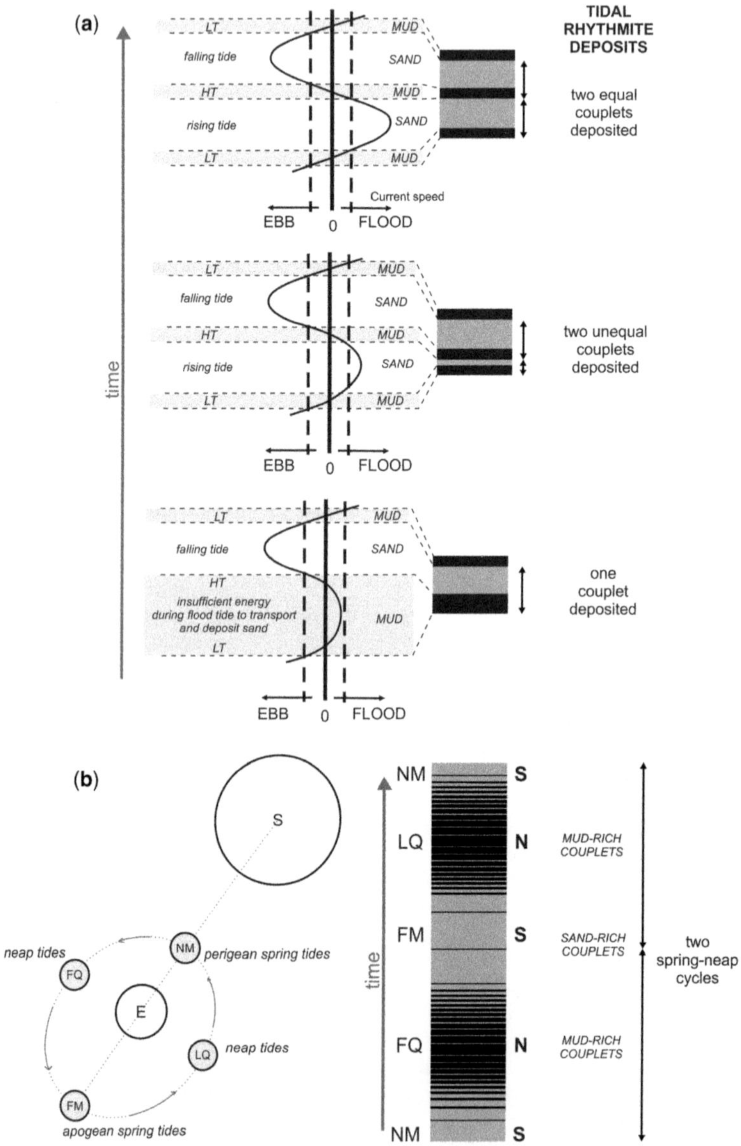

Fig. 1. Tidal rhythmite deposits. (**a**) Couplet deposition during a tidal cycle shown with respect to relative ebb and flood current energies (after Dalrymple 2010*b*). (**b**) Spring–neap cycle deposits during one synodic month with reference to the Moon–Earth–Sun arrangement.

while during slack water conditions at high and low tides deposition is mainly from suspension. The number of couplets per cycle depends on the sediment and water discharge of the area, the inequality of the ebb and flood current energies, inundation v. exposure of the surface during low tide, and the preservation potential of deposits with successive tides (Dalrymple 2010*b*).

The tidal regime of an area determines the duration of each tidal cycle. In a diurnal tidal regime, coastal areas experience one high and one low tide during a lunar day, with a tidal period of 24 hours 50 minutes. Regions with a semi-diurnal tidal regime experience two high and low tides during one lunar day, with a tidal period of 12 hours 25 minutes. This allows for a daily maximum of two couplets deposited in a diurnal regime and four couplets in a semi-diurnal regime. Diurnal inequality refers to two unequal tides in one day, which can lead to alternating couplet thicknesses. A large diurnal inequality can also result in the deposition or preservation of one or no couplets per day (see Table 1).

The period of time required for one revolution of the Moon around the Earth is 29.53 days; this is called a synodic month (Fig. 1b). The range and energy of the tides are greatest at the times of the new and full moons, resulting in spring tides. Neap tides are the conditions of lowest tidal range and energy; they occur during first and third quarters of each synodic month. There are therefore two neap–spring cycles during a synodic month, each having a period of 14.76 days. The periodic increase and decrease in tidal energy results in rhythmic tidal bedding. Spring–neap cycles are recognized in tidal deposits as alternating sand-rich and mud-rich lamina sets. The highest high tides and lowest low tides during spring conditions lead to thicker sand and thinner mud laminae in the preserved couplets. Conversely, during neap conditions, the lowest high tides and highest low tides result in the deposition of thinner sand and thicker mud layers. Successive spring deposits also show inequalities, which can be exaggerated by the forcing effects of lunar perigee (when the Moon is closest to the Earth) and apogee (when the Moon is furthest from the Earth).

Fluvial–tidal transition zones

In tide-dominated environments connected to rivers, the downstream fluvial–tidal transition zone is defined by van den Berg *et al.* (2007, p. 255) as

> ... that part of the river which lies between the landward limit of observable effects of a tidally influenced flow deceleration on fluvial cross-bedding at low river discharge, and the most seaward occurrence of a textural or structural fluvial signature at high river stage.

The variable and combined effects of both fluvial and tidal currents are recognized in the deposits of the transition zone. Fluvial flow decreases in energy seawards, while the tidal current energy first increases due to confinement and then decreases upstream. A tidal prism, or saline wedge, travels upstream from the river's mouth. The resultant hydrodynamic forces lead to current reversals up to the tidal limit and tidal modulation of the river flow just landwards of it in the backwater zone (Dalrymple *et al.* 1992, 2012; van den Berg *et al.* 2007; Dalrymple 2010*b*; Martinius & Gowland 2011). The limits of the transition zone are dynamic, varying with base level changes, fluctuations in seasonal discharge and tidal cycles. High river discharge shifts the landwards limit of tidal influence downstream and low river discharge shifts the downstream limit of fluvial influence upstream. Within a tidal cycle, the flood retards the river's flow and decreases or reverses the velocity, while the ebb draw-down causes an increased velocity downstream. Flood tides during spring conditions affect the fluvial flow further upstream than those during neap conditions. Seasonal variations are recognized in the deposits because a lower river discharge leads to a greater tidal effect further upstream.

Within the transition zone, there is a change in the sedimentological characteristics of the deposits from upstream to downstream. Based on studies of modern and ancient deposits (for example, Dalrymple & Choi 2007; van den Berg *et al.* 2007; Dashtgard *et al.* 2012), sedimentary structures indicative of unidirectional and bidirectional currents are found within the transition zone because the

Table 1. *Periods of tidal cycles and estimated couplets deposited per cycle*

Order	Tidal cycle	Period (solar days)	Couplets per cycle	
			One couplet/cycle	Two couplets/cycle
First	Semi-diurnal	0.5	~1	~2
Second	Synodic neap–spring	14.77	~29	~58
Third	Synodic month	29.53	~58	~116
Fourth	Semi-annual/seasonal	182.6	~365	~730

zone shifts regularly. Lithofacies corresponding to tidally influenced fluvial deposition change downstream to tidal deposits with a decreasing fluvial influence.

In the backwater section, just upstream of the transition zone, Martinius & Gowland (2011) observed that organic particles act as suspended material and are deposited as laminae following spring to neap cyclicity. Tidal modulation causes a reduced, but not reversed, flow that is amplified during spring conditions. The retarded flow allows for the deposition of organic laminae, while the decreasing effect of the flood tide leads to neap deposition that often does not contain organic material.

Bioturbation

Biogenic structures are indicators of environmental conditions, such as the sedimentation rate, the coherency of the substrate, salinity, oxygen level, turbidity, food availability and type, subaerial exposure, water depth, light, temperature and depositional energy during deposition and at the time of surface colonization (McIlroy 2007; MacEachern *et al.* 2007*a*, *b*, 2010; Gingras & MacEachern 2012). The size, diversity and distribution of trace fossils are also indicators of the environmental conditions. The ichnofacies associated with different assemblages of trace fossils are used to infer the palaeoenvironment at the time of their formation (Seilacher 2007).

The distribution of trace fossils in a vertical section can be classified as homogeneous, regularly heterogeneous and sporadically heterogeneous (Gingras & MacEachern 2012). Homogenously distributed biogenic structures indicate a continuous low rate of sedimentation with available food, such as on intertidal flats. In the fluvial–tidal transition zone, a strong seasonal difference in conditions would most probably be reflected by a regularly heterogeneous distribution, with lower sedimentation rates and higher salinity at times of decreased fluvial discharge allowing for colonization by organisms. By contrast, sporadically heterogeneous distributions of trace fossils that show no pattern are more likely to reflect complex and essentially random changes in conditions.

Tidal environments are considered to be stressed as a result of the periodically alternating salinity, water depth, suspended sediment concentration and current direction, duration and intensity. Adaptations to the harsh conditions lead to small sizes, sparse distribution and low diversity of trace fossils (MacEachern *et al.* 2010; Dalrymple *et al.* 2012; Gingras & MacEachern 2012). This brackish-water ichnofacies can then be used to help identify the fluvial–tidal transition zone. Changing conditions upstream through the transition from tidal to fluvial environments lead to a decrease in the size and diversity of biogenic structures (Gingras *et al.* 1999, 2012*b*; MacEachern *et al.* 2007*b*; Gingras & MacEachern 2012). Tidal influence can also be recognized directly if the burrow fill is rhythmic, and indirectly by the feeding patterns, simplicity of burrows, burrow lining and fill material, and rhythmic occurrences and diversities of trace fossils.

Geological setting

The Erin Formation was deposited on the NE South American continental shelf by the palaeo-Orinoco Delta during the Plio-Pleistocene (Kugler 2001). The deposits are found in the Southern Basin of Trinidad, an eastwards extension of the Eastern Venezuelan Basin. Thick sections of the rocks are exposed along the northern and southern coasts of the SW peninsula of the island (Fig. 2) with a structural dip of 3–10° to the NNE. A laterally continuous outcrop extends for more than 300 m at Granville Bay (10° 08′ 17.1″ N, 61° 48′ 04.5″ W). The measurements reported here were taken in 2013. High-energy waves along the coastline have since eroded the outcrop.

According to Kugler (2001), the Erin Formation rests on the Morne L'Enfer Formation and is unconformably overlain by the Pleistocene Cedros Formation. The Erin Formation includes major cycles of lignite and porcellanite, clay, silt and cross-bedded sand. Porcellanites are considered to be clay and shale material that has been baked or fused by the combustion of lignite. The lignitic beds are rich in plant material and often contain petrified tree trunks. The palaeoenvironments of deposition interpreted by Kugler (2001) consist of brackish-water and freshwater estuaries and coastal swamplands.

More recent palaeoenvironmental interpretations of the Erin Formation by the first author include sub-environments in the fluvial–tidal transition zone within a tidally dominated estuary (Fig. 3). The sections reported in this paper were deposited as tidal bar deposits. Heterolithic interlaminated planar beds with sand–mud and silt–mud couplets extend laterally for *c.* 20 m. Trace fossils vary in abundance and distribution and include dwarfed traces of the *Skolithos* ichnofacies and other unidentified burrows, suggesting a brackish water ichnofacies (Pemberton & Wightman 1992; Gingras *et al.* 1999; Buatois *et al.* 2005). Organic fragments sourced from overbank areas and coastal swamps are common among the deposits.

Sediments were deposited in a tropical seasonal palaeoenvironment, with a dry season and a rainy season that lasted for roughly six months each. Similar to present day conditions, a semi-

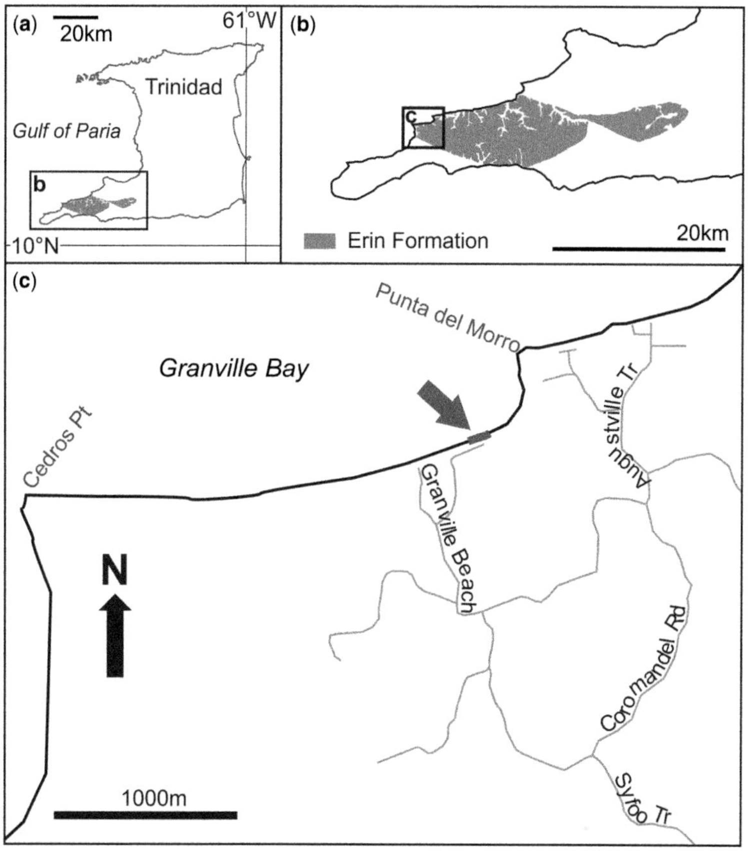

Fig. 2. (**a**) Map of Trinidad showing the SW peninsula of the island. (**b**) SW peninsula of Trinidad showing the Granville Bay location. (**c**) Map of Granville Bay showing the location of the outcrop studied.

diurnal palaeotidal regime can be assumed because the astronomical factors have been constant for several million years (Sonett *et al.* 1996; Choi & Park 2000). The current microtidal range cannot, however, be assumed for the Plio-Pleistocene depositional palaeoenvironments because major differences between geographical settings affect the tidal dynamics of an area (Choi & Dalrymple 2004). The geomorphology of the area includes possible amplification of tidal waves by lateral constriction (Cummings *et al.* 2006) due to the presence of a mountainous terrain and subaqueous and subaerial tectonic ridges on the continental shelf (Pindell & Kennan 2001; Kennan & Pindell 2007; Pindell *et al.* 2009).

Methods

Outcrop data collection

Among the outcrops at Granville Bay, one section contains identifiable tidal rhythmites. An erosive surface divides it into two subsections containing different types and distributions of sedimentary structures. Two subsections were studied: GB1 below the erosive surface and GB2 above it (Fig. 4a). Magnified photographs of the subsections were used to measure the thickness of each couplet of sand–clay, sand–silt or silt–clay laminae. Each photograph included a constant scale in millimetres for measurement of the true stratigraphic thicknesses of the horizontal and wavy parallel-laminated couplets. The thicknesses of discontinuous or pinching couplets were measured at the thickest sections or on the foresets of rippled surfaces so that the maximum deposition for the event could be represented. Where deformation occurred in the section, the beds were traced laterally to continuous laminae that could be measured.

High-resolution stratigraphy

The rhythmic heterolithic bedded facies of the Granville Bay subsections (GB1 and GB2) were

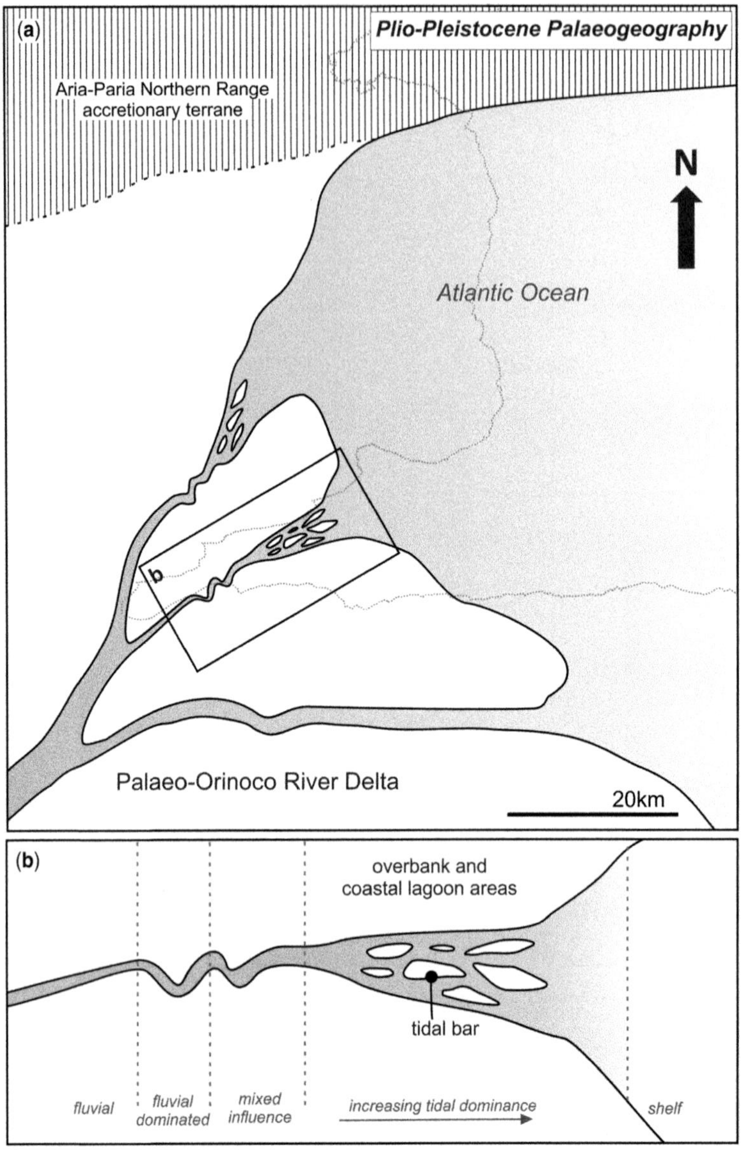

Fig. 3. (**a**) Plio-Pleistocene palaeogeography of the eastward extent of the Eastern Venezuelan Basin during deposition of the Erin Formation. Outline of the current location of Trinidad included. (**b**) Palaeoenvironment of the outcrop deposits studied.

measured and recorded as detailed sedimentary logs. Sedimentary structures and the presence or absence and distribution of organic content were noted. The presence or absence, distribution, intensity and diversity of trace fossils were also recorded using the bioturbation index (BI) of Taylor & Goldring (1993). This is used to distinguish different grades (0–6) based on the percentage of bioturbation, the ability to recognize the primary sedimentary fabric, and the abundance and overlap of burrows. These observations are represented as colour-coded markers at the corresponding couplet numbers. Comparison with the time-series data was then used to identify trends or cycles in the sedimentary features.

Statistical analysis

The thicknesses of 511 couplets from GB1 and 1284 couplets from GB2 were entered into a Microsoft

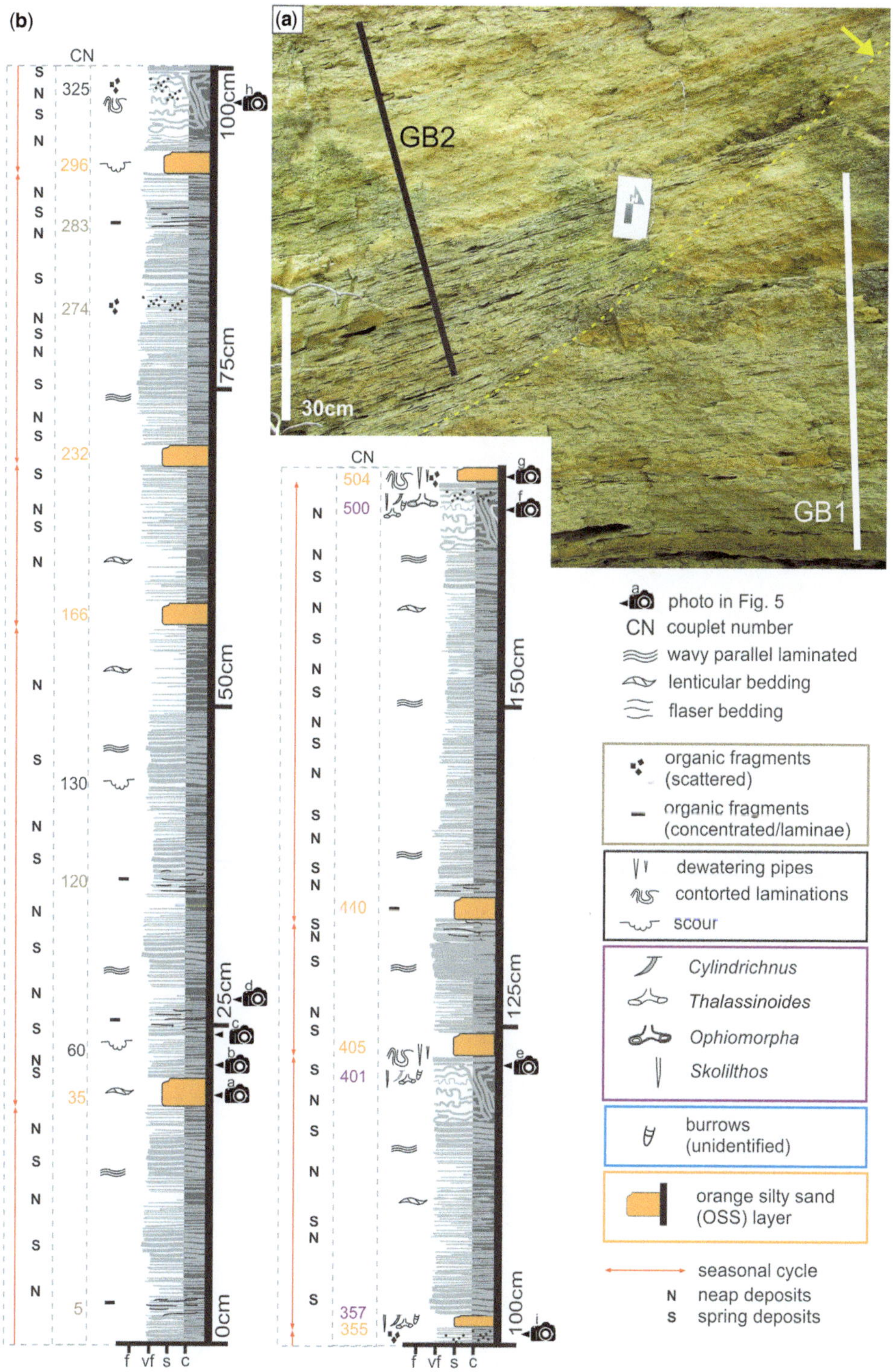

Fig. 4. (**a**) Photograph showing the scoured surface separating subsections GB1 and GB2. (**b**) Sedimentary log of subsection GB1 with interpreted cycles. Couplet numbers corresponding to sedimentary features of interest are listed in the column CN. Locations of photographs through the section are also shown.

(a)
(b)
N
S
N
S
(c)
N
S
N
S
N
S
(d)
N
S
N
S
N
S
(e)
(f)
(g)
Sk
(h)
Sk
Op
(i)
Sk
Sk
Sk
(j)
Th
(k)
N
S
N
S
N
S
N
S
N
Cy
(l)
N
S
N
S
N
S
Sk
0
2
4
6
8
10
12
14

Excel spreadsheet and numbered stratigraphically. Each dataset was then copied into the palaeontological software PAST (Hammer *et al.* 2001; Hammer 2010) for statistical analysis.

Graphs of the couplet thickness against number were generated. Where sinusoidal patterns were recognized, successive peaks and troughs were used to determine the number of couplets in the interpreted cycles. The periods of the cycles were then calculated using:

$$T_n = \frac{(N_c \times t_c)}{24} \text{days} \quad (1)$$

where T_n is the period of the cycle of order n being calculated, N_c is the number of couplets in the cycle and t_c is the time taken for the deposition of each couplet. If two couplets were deposited during each first-order tidal cycle, then $t_c \approx 6.2$ hours. If only one couplet was deposited, then $t_c \approx 12.4$ hours. If deposition was not continuous as a result of erosion or a lack of deposition, the periodicity may be difficult to determine. The statistical methods outlined in the following gave values of N_c or f_n, from which T_n could be calculated using equations (1) and (2).

Autocorrelation. The autocorrelation function was used to test for randomness in the datasets. A 95% confidence interval indicated random, independent values. Histogram plots of the autocorrelation function were interpreted in a similar manner to the thickness plots for values of N_c.

Sinusoidal modelling. The datasets were fitted to sinusoidal models by searching for periodicity in the data. The amplitude, phase and periods (N_c) were output for each of the cycles recognized.

Filtering. The datasets were first transformed by subtracting the mean for each dataset from the thickness values. A bandpass filter was then used to filter out noise and smaller periods. Average values of N_c were recorded from the graph of the filtered response.

Discrete Fourier transform/spectral analysis. Spectral analysis of each time-series dataset produced Lomb and REDFIT periodograms, with power spectra plotted against frequency (cycles per couplet). Successive peaks showed an increase in frequency, which corresponded to a decrease in the order of cycles. The frequency values f_n of major peaks were used to calculate the number of couplets per cycle N_c from:

$$N_c = \frac{1}{f_n} \quad (2)$$

Continuous wavelet transform analysis. Continuous wavelet transform, which is essentially equivalent to fast Fourier transform (Prokoph *et al.* 2000), was used to detect non-stationary cyclicity. Two-dimensional representations of the datasets showed the decomposition of the spectra as image graphs. The Morlet function was used. Warmer colours indicated high-power spectra, representing significant cycle periods in the dataset. Values of y at high-power spectra were used to calculate the number of couplets per cycle using the following steps.

$$\begin{aligned} &\text{From graph axes,} \quad y = \log_2 x \\ &\therefore \quad x = 2^y \\ &\text{In this case,} \quad x = N_c \\ &\therefore \quad N_c = 2^y \quad \text{couplets per cycle} \end{aligned} \quad (3)$$

Description

Sedimentology and ichnology of GB1

The sedimentary log of GB1 is shown in Figure 4b. The recorded sedimentary features are highlighted

Fig. 5. Photographs of sedimentary features observed in subsections GB1 and GB2. (**a**) Lenticular bedding in an orange silty sand (OSS) layer, with organic fragments scattered in sand lenses above and below the OSS. (**b**) Wavy, parallel bedding with interlaminated silt–clay and sand–clay couplets. Thin discontinuous layers of organic material occur at the base of some clay laminae. Clay laminae increase in thickness up-section from spring to neap deposits. (**c**) Alternating sand-rich spring and clay-rich neap deposits, with organic material isolated to neap deposits. (**d**) Single organic fragment in mud-rich layer within interlaminated silt–clay couplets. (**e**) Scour cut into deformed layer and filled with inclined sand–clay couplets. Scour fill includes an OSS layer. (**f**) Contorted layer deforms mud-rich couplets and an OSS layer above it. (**g**) *Skolithos* with rhythmic sediment infill. Bioturbated layer between successive OSS layers. Bioturbation index (BI) 2. (**h**) Bioturbated layer beneath a thin concretion layer (shown by white arrow). Rare *Skolithos* and *Ophiomorpha* occur with unidentified burrows; BI 2. (**i**) *Skolithos* with organic material. Thin concretion layer (shown by white arrow) above bioturbated layer with BI 1. (**j**) *Thalassinoides* in mud-rich layer of mud–silt couplets. (**k**) *Cylindrichnus*. Lenticular beds separate mud-rich couplets. (**l**) *Skolithos* in alternating neap–spring cycles of couplets. Photographs (a–i) from GB1; (j–l) from GB2. Lithology from photographs: light brown or orange colour indicates sand or silty sand, grey colour indicates clay and black colour indicates organic material. Neap (mud-rich) and spring (sand-rich) deposits are interpreted on the photographs as N and S, respectively.

at the corresponding couplet numbers by colour-coded markers.

Sedimentary structures. Parallel and wavy heterolithic bedding structures are common in the subsection, with average couplet thicknesses of 1–2 mm. Lenticular and flaser bedding occur in the sand-rich and mud-rich layers, respectively. The transitions between sand–silt, silt–mud and sand–mud couplets are gradational.

Laterally continuous orange silty sand (OSS) layers are interbedded with layers of heterolithic laminae, with gradual contacts between them. The OSS layers are up to 4 cm thick, poorly sorted and generally massive; some contain lenses of very fine white sand (Fig. 5a).

There is an overall increase in deformation within the beds up-section. Scours, contorted laminations and dewatering pipes occur below some OSS layers near the top of the measured subsection.

Bioturbation. Bioturbated layers with BI 2 (Fig. 5g–i) have a regularly heterogeneous distribution, occurring between successive OSS layers. Thin iron concretions occur at the top of bioturbated layers and are followed in the stratigraphy by continuous parallel and wavy laminated beds. Three such successions occur near the top of the subsection, but are absent in its lower part. Traces include unidentified vertical escape burrows. A few *Skolithos*, *Thalassinoides, Cylindrichnus* and *Ophiomorpha* occur in conjunction with the bioturbated layers.

Organic material. Organic material is not common throughout the subsection, but is concentrated in specific layers and is absent near deformed beds. In layers containing organic material, thin discontinuous laminae of organic fragments occur at the base of clay laminae. The laminae only occur among mud-dominated deposits. Among sandier deposits, organic material is disseminated as fragments. Organic material is found in a few *Chondrites* and *Skolithos* (Fig. 5i), but is absent in the regularly heterogeneous bioturbated layers (Fig. 5j).

Sedimentology and ichnology of GB2

The sedimentary log of GB2 (Fig. 6) shows sedimentary features highlighted by colour-coded markers.

Sedimentary structures. Similar to subsection GB1, wavy parallel interlaminated beds occur throughout the subsection, with lenticular and flaser bedding occurring at sand-rich and mud-rich layers, respectively. Scours and contorted layers are randomly distributed and are found above, below and between OSS layers. The OSS layers in subsection GB2 are massive and are up to 100 mm thick.

Bioturbation. Trace fossils in this subsection consist of rare, dwarfed *Thalassinoides*, *Cylindrichnus* and *Skolithos* (Fig. 5j–l). Regularly heterogeneous bioturbated layers, such as those observed in subsection GB1, are absent in GB2.

Organic material. Disseminated fragments and thin discontinuous laminae of organic material are randomly distributed through subsection GB2.

Statistical results

The statistical analysis of the GB1 and GB2 time-series datasets generated plots (Fig. 7) and values from which N_c and T_n were derived using the appropriate equations. The results are given in Tables 2 and 3.

Analysis

The statistical analysis of each subsection includes interpretations of the tidal cycle, neap–spring, monthly and semi-annual periodicity in deposition. The juxtaposed colour-coded markers and time-series data show trends in depositional conditions (Fig. 8).

GB1

Sedimentary structures. Statistically interpreted cycles of roughly 30 couplets correspond to transitions from mud-rich to silt- or sand-rich layers in the section. This can be interpreted as neap–spring (second-order) cyclicity in a semi-diurnal regime, where only one couplet per tidal cycle is preserved. Alternatively, diurnal inequality may be responsible for the deposition and preservation of only one complete tidal cycle with the stronger tide daily (de Boer *et al.* 1989; Kvale *et al.* 1999; Mazumder & Arima 2005; Dalrymple 2010*b*). From either interpretation, it can be deduced that each couplet represents 12.4 hours in the time-series dataset in this study.

Cycles of *c.* 56 couplets can be interpreted to be monthly (third-order) cycles. Cycles of *c.* 100–150 couplets, corresponding to 2–3 months of deposition, can also be interpreted from the statistical results. These cycles correspond to the OSS layers in the sedimentary log. The massive nature of the OSS layers makes it impossible to statistically determine the time corresponding to the deposition of each layer. Scours, contorted laminations and dewatering pipes are interpreted as deformation due to high-energy flows and rapid deposition (Dzulynski & Smith 1963; Plint 1983; Frey *et al.* 2009).

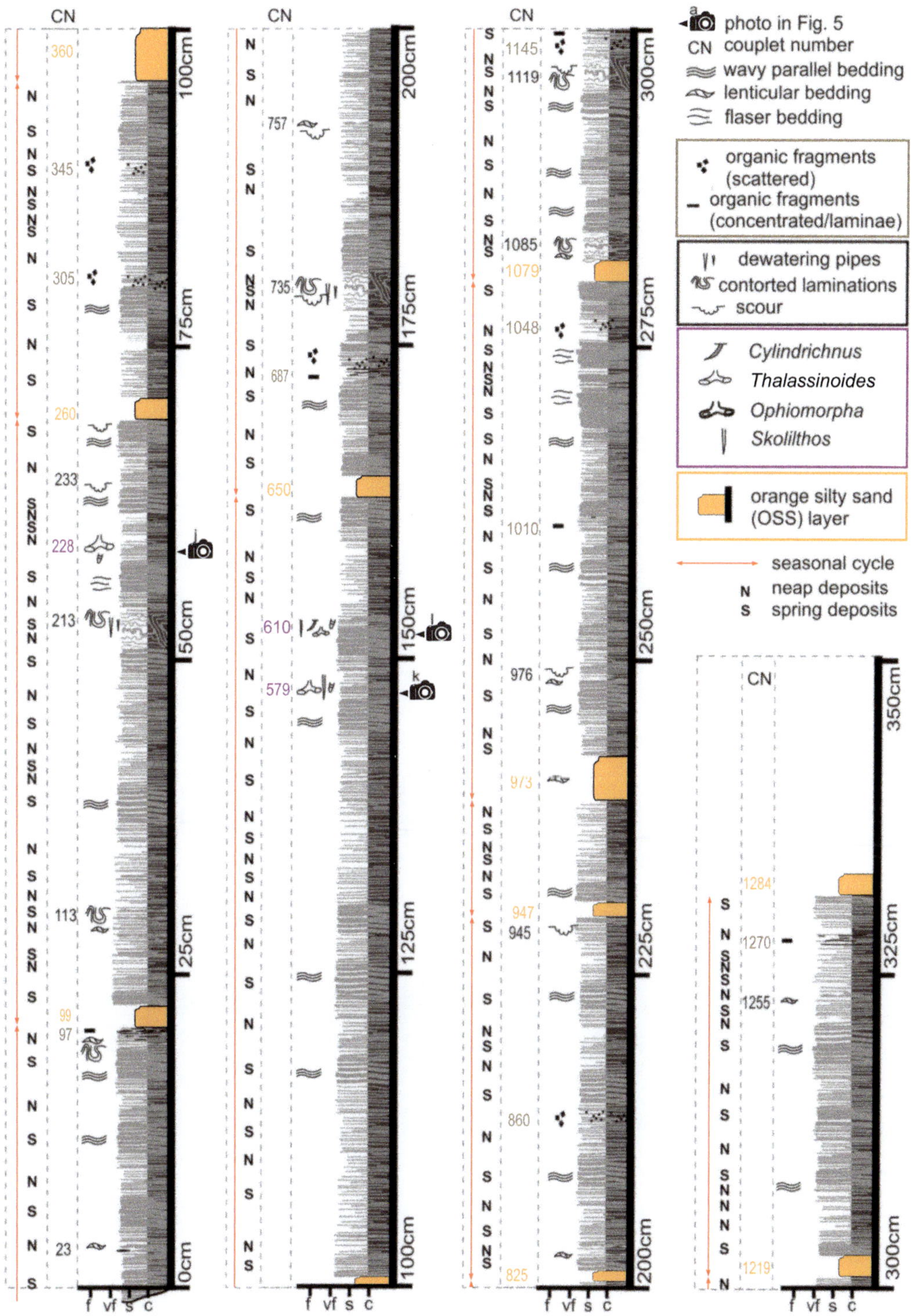

Fig. 6. Sedimentary log of subsection GB2 with interpreted cycles. Couplet numbers corresponding to sedimentary features of interest are listed in the column CN. Locations of photographs through the section are also shown.

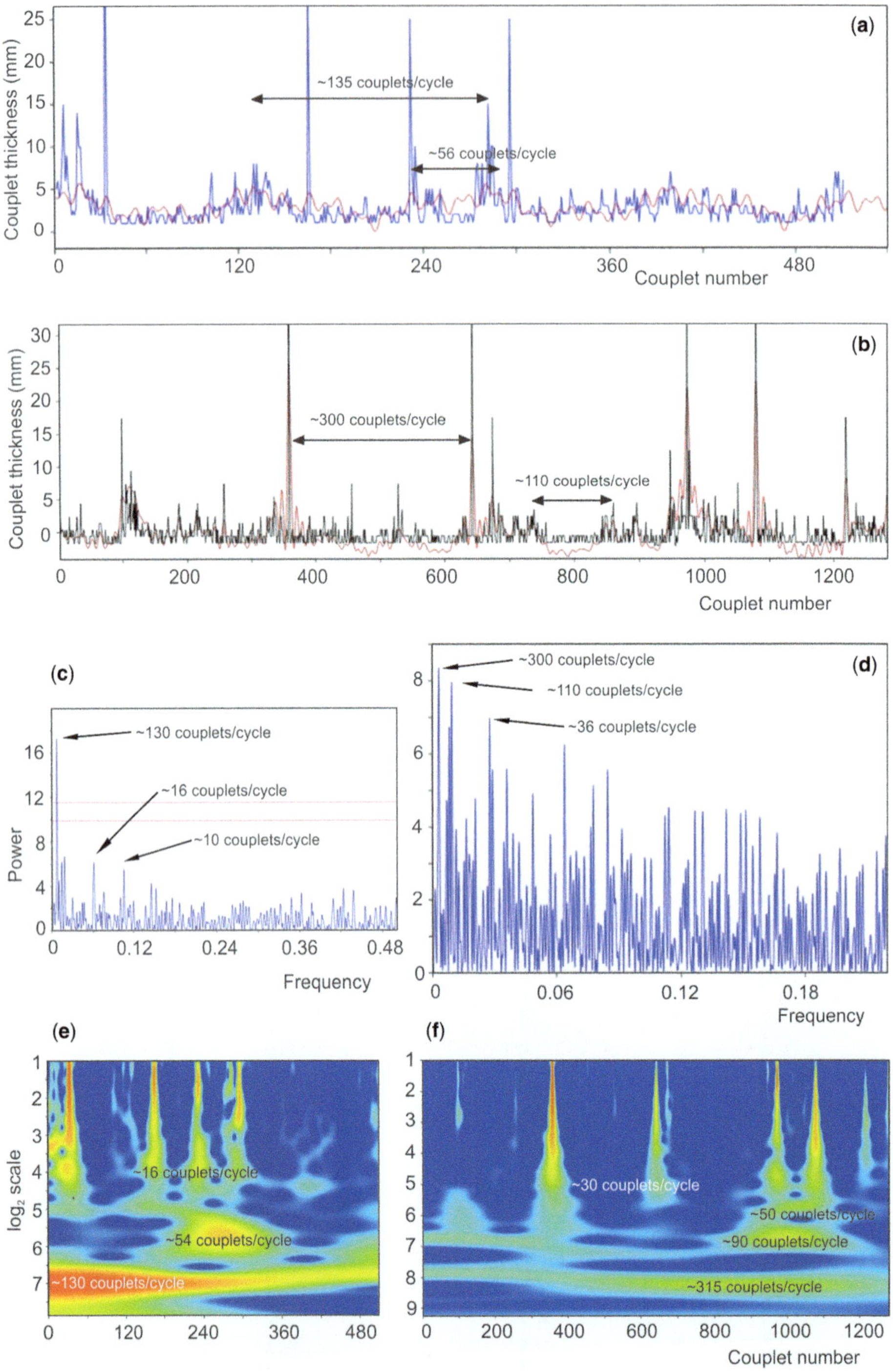

Fig. 7. (**a**) Sinusoidal model fit for time-series dataset of subsection GB1. (**b**) Plot of GB2 original (subtracted mean) and filtered (red line) dataset. (**c**) GB1 Lomb periodogram power spectra plot with noise cut-offs. (**d**) GB2 Lomb periodogram power spectra plot. (**e**) Continuous wavelet transform of GB1 dataset. (**f**) Continuous wavelet transform of GB2 dataset. Annotations in figure give couplet counts, frequencies and periods used in calculations, as interpreted from graphs and output data tables.

Table 2. *Values of N_c from different statistical methods conducted on the GB1 time-series dataset and calculated values of T_n*

Statistical method	N_c (couplets per cycle)	T_n (days)	Order*
Autocorrelation	65	33.58	3
	118	60.97	
	150	77.50	
Sinusoidal modelling	133.8	69.13	
	54.07	27.94	3
	16.51	8.53	2
	9.42	4.87	
Filtering	125	64.58	
Spectral analysis – Lomb periodogram	131.6	67.99	
	16	8.27	2
	9.5	4.91	
Spectral analysis – REDFIT	102	52.70	
	73	37.72	
	9.5	4.91	
	56.8	29.35	3
Continuous wavelet transform	130	67.17	
	54	27.90	3
	16	8.27	

*Cycle orders approximated from theoretical values.

Table 3. *Values of N_c from different statistical methods conducted on GB2 time-series dataset and calculated values of T_n*

Statistical method	N_c (couplets per cycle)	T_n (days)	Order*
Autocorrelation	100	51.67	
	250	129.17	4
	328	169.47	4
	440	227.33	4
	630	325.50	
Sinusoidal modelling	304.8	157.48	4
	107.7	55.65	
	35.97	18.58	
Filtering	300	155.00	4
	110	56.83	
Spectral analysis – Lomb periodogram	303	156.55	
	107.5	55.54	
	35.7	18.45	3
Spectral analysis – REDFIT	322.6	166.68	4
	35.7	18.45	
	107	55.28	
Continuous wavelet transform	315.17	162.84	4
	90.5	46.76	
	32	16.53	2
	52	26.87	3

*Cycle orders interpreted roughly fit theoretical values.

The increase in frequency of contorted laminae up-section can therefore be interpreted as a general increase in energy over time.

Bioturbation. Bioturbated layers with a regularly heterogeneous distribution indicate episodic increased salinity and marine influence (Gingras & MacEachern 2012). The thin iron concretions at the top of the bioturbated layers (Fig. 5h, i) mark periods of lower energy or little to no deposition, during which colonization of the surfaces and the deposits beneath them occurred (MacEachern *et al.* 2007*a*; Gingras *et al.* 2012*b*; La Croix *et al.* 2012). Rare *Skolithos* and *Ophiomorpha* can be interpreted as minor evidence of higher energy conditions (Droser & Bottjer 1989).

Organic material. The deposition of organic fragments is interpreted to have occurred during periods of reduced current energy by flood retardation (Martinius & Gowland 2011) and are scattered among coarser grained spring deposits. Among lower energy neap deposits, the deposition of organic material from suspension possibly preceded the settling of clay particles and this is found as discontinuous laminae. The absence of organic fragments from the regularly heterogeneous bioturbated layers may be due to their removal from the deposits by organisms, leaving feeding traces (fodinichnia) behind (Frey & Howard 1990). Organic material in some *Ophiomorpha*, *Skolithos* and *Chondrites* may be remnants of the actual organisms (Frey & Howard 1990).

GB2

Sedimentary structures. Cyclicity interpreted from the statistical analysis includes spring–neap and monthly cycles corresponding to *c.* 50 and 100 couplets per cycle, respectively. Semi-annual or seasonal (fourth-order) cycles were also interpreted at roughly 300 couplets per cycle. On the logged subsection, the seasonal cyclicity averaged to successive OSS–heterolithic bed couplets. As in subsection GB1, the time corresponding to the deposition of each OSS layer is impossible to determine, again due to the massive texture. The OSS layers can possibly be interpreted as seasonal event beds, deposited at the beginning of the rainy season when a high sediment influx results from river flooding. Heterolithic layers were deposited mainly during the rainy season, with little to no deposition during the dry season. This may have been due to relatively low fluvial discharge and sediment input.

Bioturbation. The rare and dwarfed *Thalassinoides* and *Skolithos* and the absence of bioturbated layers in subsection GB2 indicate relatively low water salinity, with only brief periods of increased salinity. The lack of biogenic structures can be interpreted

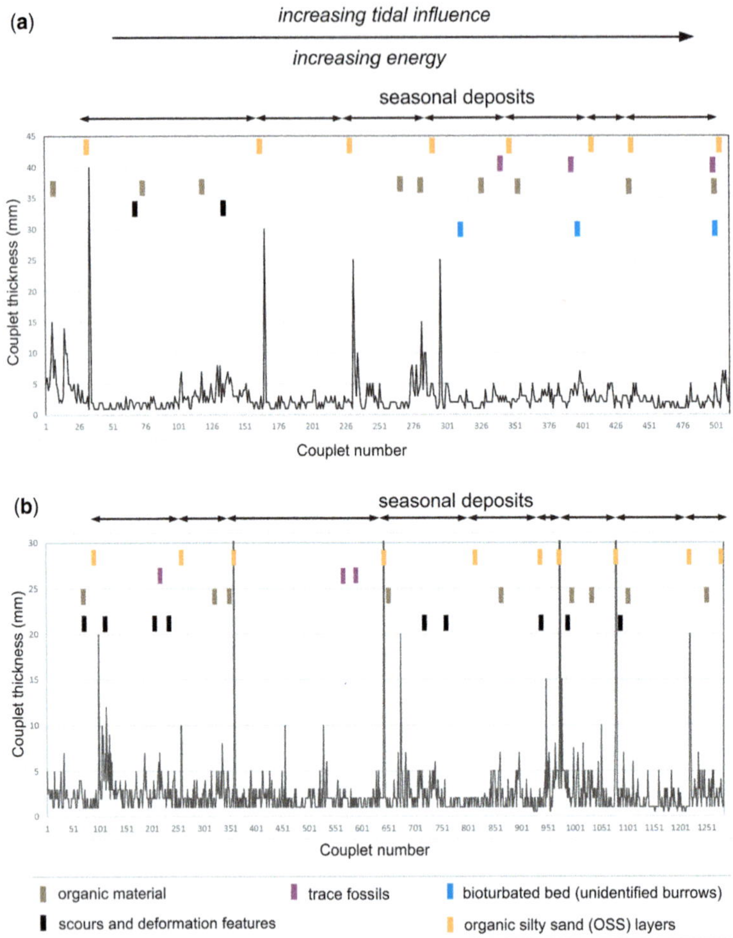

Fig. 8. (**a**) Plot of couplet thickness dataset for subsection GB1. (**b**) Plot of couplet thicknesses dataset for subsection GB2. Colour-coded markers identify major sedimentary features.

as a relatively increased fluvial influence during the deposition of subsection GB2 compared with GB1.

Discussion

The studied section is traversed by an erosion surface. Heterolithic tidal rhythmites and OSS layers are recognized above and below this surface. However, the periodic distribution of sedimentary features differs between the two measured subsections. GB1 includes regularly heterogeneous bioturbation that is absent in GB2. Fluvial influence during the deposition of GB1 is, therefore, interpreted to have been relatively lower than during the deposition of GB2.

The OSS layers, interpreted as being deposited during high river discharge at the beginning of the rainy season, separate successive beds of tidal heterolithic deposits. Semi-annual cyclicity was more easily recognized in the GB2 subsection, possibly due to rapid and continuous deposition during the rainy season. Subsection GB1 was deposited relatively slowly, evidenced by erosion and bioturbated layers (Choi & Park 2000). It is suggested that a relatively lower fluvial influence may result in an incomplete record of tidal deposition. Palaeoenvironmental interpretations may therefore be more precisely determined with respect to fluvial influence.

Sedimentary structures, organic content and bioturbation variations through GB1indicate a general increase in the energy of the depositional system up-section. Changes in sediment supply and fluvial influence in the area may have resulted from tidal channel abandonment. A relative fall in either the local or eustatic sea-level could also have led to the erosion of part of the tidal bar deposits in GB1. The erosional surface is at the base of the subsequent deposition recorded in subsection GB2.

The heterogeneity observed in subsections GB1 and GB2 is periodic, with a low vertical-to-horizontal permeability. Similar reservoir facies (Choi *et al.* 2011) experience reduced fluid recovery. A relatively stronger tidal influence, as interpreted in subsection GB1, can lead to increased vertical permeability in some of the heterolithic beds as a result of the bioturbated layers.

Although tidal signatures and seasonal fluvial influence can be inferred by observing the outcrop, the subsections have a very similar appearance. From the statistical analyses, we were able to identify differences in deposition between GB1 and GB2 relative to the degree of fluvial and tidal influence. Several of the heterolithic beds throughout the outcrop were of comparable thicknesses; however, the fluvial seasonal cyclicity in GB1 and GB2 was interpreted as *c.* 150 and 300 couplets/cycle, respectively. Without quantitative analysis, no significant distinction would have been made between the two subsections and without the comparative qualitative analysis seasonal cyclicity would not be as easily identified in GB1.

Detailed analyses of the kind presented here are unusual in subsurface studies, but the short sections logged and the focus on a detailed analysis of a limited width of outcrop make most of the workflow that we have used directly applicable to the study of continuous conventional core. Such an approach is perhaps increasingly valuable as workers seek to quantify the relative influences of wave, tide and fluvial processes in an attempt to predict lateral facies patterns (Ainsworth *et al.* 2011).

Conclusions

Sedimentological observations of a section of ancient tidal rhythmites influenced the interpretations made from time-series analysis. The relative fluvial influence on tidal deposition can affect the presence, abundance and distribution of sedimentary features, such as contorted beds, bioturbation and the distribution of organic matter. It was found that a greater fluvial influence is evidenced by a more complete depositional record during the rainy season as a result of a relatively higher and continuous sediment input.

Heterogeneity resulting from tidal and seasonal cycles is periodic, affecting the reservoir permeability. Although heterogeneity at this scale is not recognizable from seismic or well data, and may be missing from selective core retrieval, it should not be neglected when considering the estimates and justifications of hydrocarbon recovery using reservoir models. Still, not much is understood of the fluvial–tidal transition zone and the relative fluctuating influence experienced with respect to reservoirs. Additional work needs to be carried out to identify comparable rock sections.

Reservoirs in Trinidad and Tobago tend to be interpreted mainly from facies models, without consideration of the unique properties of each depositional system. Recognizing the fluvial–tidal transition zone in palaeoenvironmental interpretations will, as this study shows, increase and potentially predict the understanding of heterogeneity in similar reservoirs. This can help to account for anomalously low recovery rates and volumes.

This study is part of a PhD project by Annalize McLean. We are grateful for the funding provided by the University of the West Indies. We thank Hasley Vincent and Richard Dawe for guidance throughout this project. We are indebted to Antony Reynolds and two anonymous reviewers for constructive and thoughtful feedback, which helped to improve this manuscript.

References

Ainsworth, R.B., Vakarelov, B.K. & Nanson, R.A. 2011. Dynamic spatial and temporal prediction of changes in depositional processes on clastic shorelines; toward improved subsurface uncertainty reduction and management. *American Association of Petroleum Geologists Bulletin*, **95**, 267–297.

Buatois, L.A., Gringas, M.K. *et al.* 2005. Colonization of brackish-water systems through time: evidence from the trace-fossil record. *PALAIOS*, **20**, 321–347.

Chan, M.A., Kvale, E.P., Archer, A.W. & Sonett, C.P. 1994. Oldest direct evidence of lunar–solar tidal forcing encoded in sedimentary rhythmites, Proterozoic Big Cottonwood Formation, Central Utah. *Geology*, **22**, 791–794.

Choi, K., Jackson, M.D., Hampson, G.J., Jones, A.D.W. & Reynolds, A.D. 2011. Predicting the impact of sedimentological heterogeneity on gas-oil and water-oil displacements: fluvio-deltaic Pereriv Suite Reservoir, Azeri-Chirag-Gunashli Oilfield, South Caspian Basin. *Petroleum Geoscience*, **17**, 143–163, https://doi.org/10.1144/1354-079310-013

Choi, K.S. 2010. Rhythmic climbing-ripple cross-lamination in inclined heterolithic stratification (IHS) of a microtidal estuarine channel, Gomso Bay, West Coast of Korea. *Journal of Sedimentary Research*, **80**, 550–561.

Choi, K.S. & Dalrymple, R.W. 2004. Recurring tide-dominated sedimentation in Kyonggi Bay (west coast of Korea): similarity of tidal deposits in Late Pleistocene and Holocene Sequences. *Marine Geology*, **212**, 81–96.

Choi, K.S. & Park, Y.A. 2000. Late Pleistocene silty tidal rhythmites in the macrotidal flat between Youngjong and Younyou Islands, West Coast of Korea. *Marine Geology*, **167**, 231–241.

CHOI, K.S., KIM, B.O. & PARK, Y.A. 2001. Late Pleistocene tidal rhythmites in Kyunggi Bay, west coast of Korea: a comparison with simulated rhythmites based on modern tides and implications for intertidal positioning. *Journal of Sedimentary Research*, **71**, 680–691.

COUËFFÉ, R., TESSIER, B., GIGOT, P. & BEAUDOIN, B. 2004. Tidal rhythmites as possible indicators of very rapid subsidence in a foreland basin: an example from the Miocene Marine Molasse Formation of the Digne Foreland Basin, SE France. *Journal of Sedimentary Research*, **79**, 749–759.

CUMMINGS, D.I., WILLIAM, R., ARNOTT, C. & HART, B.S. 2006. Tidal signatures in a shelf-margin delta. *Geology*, **34**, 249–252.

DALRYMPLE, R.W. 2010*a*. Interpreting sedimentary successions: facies, facies analysis and facies models. *In*: JAMES, N.P. & DALRYMPLE, R.W. (eds) *Facies Models 4*. Geological Association of Canada, St Johns, 3–18.

DALRYMPLE, R.W. 2010*b*. Tidal depositional systems. *In*: JAMES, N.P. & DALRYMPLE, R.W. (eds) *Facies Models 4*. Geological Association of Canada, St Johns, 201–231.

DALRYMPLE, R.W. & CHOI, K. 2007. Morphological and facies trends through the fluvial-marine transition in tide-dominated depositional systems: a schematic framework for environmental and sequence-stratigraphic interpretation. *Earth-Science Reviews*, **81**, 135–174.

DALRYMPLE, R.W., ZAITLIN, B.A. & BOYD, R. 1992. Estuarine facies models: conceptual basis and stratigraphic implications. *Journal of Sedimentary Petrology*, **62**, 11130–11146.

DALRYMPLE, R.W., MACKAY, D.A., ICHASO, A.A. & CHOI, K. 2012. Processes, morphodynamics, and facies of tide dominated estuaries. *In*: DAVIS, R.A., JR. & DALRYMPLE, R.W. (eds) *Principles of Tidal Sedimentology*. Springer, Dordrecht, 79–108.

DASHTGARD, S.E., VENDITTI, J.G., HILL, P.R., SISULAK, C.F., JOHNSON, S.M. & LA CROIX, A.D. 2012. Sedimentation across the tidal-fluvial transition in the Lower Fraser River. *The Sedimentary Record*, **10**, 4–9.

DAVIS, R.A., JR. 2012. Tidal signatures and their preservation potential in stratigraphic sequences. *In*: DAVIS, R.A., JR. & DALRYMPLE, R.W. (eds) *Principles of Tidal Sedimentology*. Springer, Dordrecht, 35–56.

DAVIS, R.A., JR. & DALRYMPLE, R.W. (eds) 2012. *Principles of Tidal Sedimentology*. Springer, Dordrecht.

DE BOER, P.L., OOST, A.P. & VISSER, M.J. 1989. The diurnal inequality of the tide as a parameter for recognising tidal influence. *Journal of Sedimentary Petrology*, **59**, 912–921.

DROSER, M.L. & BOTTJER, D.J. 1989. Ichnofabric of sandstones deposited in high energy nearshore environments: measurement and utilization. *PALAIOS*, **4**, 598–604.

DZULYNSKI, S. & SMITH, A.J. 1963. Convolute lamination, its origin, preservation, and directional significance. *Journal of Sedimentary Research*, **33**, 612–627.

EISMA, D. 1998. *Intertidal Deposits: River Mouths, Tidal Flats, and Coastal Lagoons*. CRC Press, Boca Raton, FL.

FREY, R.W. & HOWARD, J.D. 1990. Trace fossils and depositional sequences in a clastic shelf setting, Upper Cretaceous of Utah. *Journal of Paleontology*, **64**, 803–820.

FREY, S.E., GRINGAS, M.K. & DASHTGARD, S.E. 2009. Experimental studies of gas-escape and water-escape structures: mechanisms and morphologies. *Journal of Sedimentary Research*, **79**, 808–816.

GENT, V.A. & SONNENBERG, S.A. 2014. Fourier analysis of the laminated facies of the Middle Bakken Member, Sanish-Parshall Field, Mountrail County, North Dakota. Paper presented at the AAPG Rocky Mountain Section Meeting, 20–22 July, 2014, Denver, Colorado.

GINGRAS, M.K. & MACEACHERN, J.A. 2012. Tidal ichnology of shallow-water clastic settings. *In*: DAVIS, R.A., JR. & DALRYMPLE, R.W. (eds) *Principles of Tidal Sedimentology*. Springer, Dordrecht, 57–77.

GINGRAS, M.K., PEMBERTON, S.G., SAUNDERS, T. & CLIFTON, H.E. 1999. The ichnology of modern and Pleistocene brackish-water deposits at Willapa Bay, Washington; variability in estuarine settings. *PALAIOS*, **14**, 352–374.

GINGRAS, M.K., BANIAK, G. ET AL. 2012*a*. Porosity and permeability in bioturbated sediments. *In*: KNAUST, D. & BROMLEY, R. (eds) *Trace Fossils as Indicators of Sedimentary Environments*. Elsevier, Newnes, 835–868.

GINGRAS, M.K., MACEACHERN, J.A. & DASHTGARD, S.E. 2012*b*. The potential of trace fossils as tidal indicators in bays and estuaries. *Sedimentary Geology*, **279**, 97–106.

GREB, S.F., ARCHER, A.W. & DEBOER, D.G. 2011. Apogean-perigean signals encoded in tidal flats at the fluvio-estuarine transition of Glacier Creek, Turnagain Arm, Alaska; implications for ancient tidal rhythmites. *Sedimentology*, **58**, 1434–1452.

HAMMER, Ø. 2010. Time Series Analysis with PAST, http://nhm2.uio.no/norlex/past/TimeseriesPast.pdf

HAMMER, Ø., HARPER, D.A.T. & RYAN, P.D. 2001. PAST: paleontological statistics software package for education and data analysis. *Palaeontologia Electronica*, **4**, 1–9.

HOVIKOSKI, J., RÄSÄNEN, M. ET AL. 2005. Miocene semidiurnal tidal rhythmites in Madre de Dios, Peru. *Geology*, **33**, 177–180.

HSIEH, A.I., ALLEN, D.M. & MACEACHERN, J.A. 2015. Statistical modeling of biogenically enhanced permeability in Tight Reservoir Rock. *Marine and Petroleum Geology*, **65**, 114–125.

KENNAN, L. & PINDELL, J. 2007. *Middle Miocene Orogeny in Trinidad: Record of Caribbean Collision and Template for Younger Deformation*. Geological Society of Trinidad and Tobago, Port of Spain, Trinidad and Tobago.

KUGLER, H.G. (ed.) 2001. Part 4: Paleocene to Holocene Formations. *In*: BOLLI, H.M. & KNAPPERTSBUSCH, M. (eds) *Treatise on the Geology of Trinidad*. Museum of Natural History, Basel, 86–91.

KVALE, E.P. 1989. Daily, monthly, and yearly tidal cycles within laminated siltstones of the Mansfield Formation (Pennsylvanian) of Indiana. *Geology*, **17**, 365–368.

KVALE, E.P. 2006. The origin of neap–spring tidal cycles. *Marine Geology*, **235**, 5–18.

KVALE, E.P. 2012. Tidal constituents of modern and ancient tidal rhythmites: criteria for recognition and analyses. *In*: DAVIS, R.A., JR. & DALRYMPLE, R.W. (eds) *Principles of Tidal Sedimentology*. Springer, Dordrecht, 1–18.

KVALE, E.P., JOHNSON, H.W., SONETT, C.P., ARCHER, A.W. & ZAWISTOSKI, A. 1999. Calculating lunar retreat rates using tidal rhythmites. *Journal of Sedimentary Research*, **69**, 1154–1168.

LA CROIX, A.D., DASHTGARD, S.E. & MACEACHERN, J.A. 2012. Ichnological and sedimentological variability in IHS across the tidal–fluvial transition: McMurray Formation, Southern Athabasca Oil Sands. Paper presented at GeoConvention 2012: Vision, 14–18 May, 2012, Calgary, Canada.

LONGHITANO, S.G., MELLERE, D., STEEL, R.J. & AINSWORTH, R.B. 2012. Tidal depositional systems in the rock record: a review and new insights. *Sedimentary Geology*, **279**, 2–22.

MACEACHERN, J.A., BANN, K.L., PEMBERTON, S.G. & GRINGAS, M.K. 2007*a*. The ichnofacies paradigm: high-resolution palaeoenvironmental interpretation of the rock record. *Applied Ichnology*, **52**, 27–64.

MACEACHERN, J.A., PEMBERTON, S.G., GRINGAS, M.K. & BANN, K.L. 2007*b*. The ichnofacies paradigm: a fifty-year retrospective. *In*: MILLER, W. III (ed.) *Trace Fossils. Concepts, Problems, Prospects*. Elsevier, Amsterdam, 52–77.

MACEACHERN, J.A., PEMBERTON, S.G., GINGRAS, M.K. & BANN, K.L. 2010. Ichnology and facies models. *In*: JAMES, N.P. & DALRYMPLE, R.W. (eds) *Facies Models 4*. Geological Association of Canada, St Johns, 19–58.

MARTINIUS, A.W. & GOWLAND, S. 2011. Tide-influenced fluvial bedforms and tidal bore deposits (Late Jurassic Lourinhã, Lusitanian Basin, Western Portugal). *Sedimentology*, **58**, 285–324.

MARTINO, R.L. & SANDERSON, D.D. 1993. Fourier and autocorrelation analysis of estuarine tidal rhythmites, Lower Breathitt Formation (Pennsylvanian), Eastern Kentucky, USA. *Journal of Sedimentary Petrology*, **63**, 105–119.

MAZUMDER, R. & ARIMA, M. 2005. Tidal rhythmites and their implications. *Earth-Science Reviews*, **69**, 79–95.

MCILROY, D. 2007. Ichnology of a macrotidal tide-dominated deltaic depositional system: Lajas Formation, Neuquén Province, Argentina. *In*: BROMLEY, R.G., BUATOIS, L.A., MÁNGANO, M.G., GENISE, J.F. & MELCHOR, R.N. (eds) *Sediment-Organism Interactions: Multifaceted Ichnology*. SEPM, Tulsa, OK, 195–211.

PEMBERTON, S.G. & WIGHTMAN, D.M. 1992. Ichnological characteristics of brackish water deposits. *In*: PEMBERTON, S.G. (ed.) *Applications of Ichnology to Petroleum Exploration*. SEPM Core Workshop Note, 141–167.

PINDELL, J. & KENNAN, L. 2001. Processes and events in the Terrane Assembly of Trinidad and Eastern Venezuela. *In*: FILLON, R.H., ROSEN, N.C. & WEIMER, P. (eds) *Petroleum Systems of Deep-Water Basins: Global and Gulf of Mexico Experience*. GCSSEPM Foundation 21st Annual Research Conference Proceedings, 159–152, https://doi.org/10.5724/gcs.01.21.0159

PINDELL, J., KENNAN, L., WRIGHT, D. & ERIKSON, J. 2009. Clastic domains of sandstones in Central/Eastern Venezuela, Trinidad and Barbados: heavy mineral and tectonic constraints on provenance and palaeogeography. *In*: JAMES, K.H., LORENTE, M.A. & PINDELL, J.L. (eds) *The Origin and Evolution of the Caribbean Plate*. Geological Society, London, Special Publications, **328**, 743–797, https://doi.org/10.1144/SP328.29

PLINT, A.G. 1983. Sandy fluvial point-bar sediments from the Middle Eocene of Dorset, England. *In*: COLLINSON, J.D. & LEWIN, J. (eds) *Modern and Ancient Fluvial Systems*. International Association of Sedimentologists Special Publications, **6**, 355–368.

PRINGLE, J.K., BRUNT, R.L., HODGSON, D.M. & FLINT, S.S. 2010. Capturing stratigraphic and sedimentological complexity from submarine channel complex outcrops to digital 3D models, Karoo Basin, South Africa. *Petroleum Geoscience*, **16**, 307–330, https://doi.org/10.1144/1354-079309-028

PROKOPH, A., FOWLER, A.D. & PATTERSON, R.T. 2000. Evidence for periodicity and nonlinearity in a high-resolution fossil record of long-term evolution. *Geology*, **28**, 867–870.

SEILACHER, A. 2007. *Trace Fossil Analysis*. Springer, Berlin.

SONETT, C.P., KVALE, E.P., ZAKHARIAN, A., CHAN, M.A. & DEMKO, T.M. 1996. Late Proterozoic and Paleozoic tides, retreat of the moon, and rotation of the earth. *Science*, **273**, 100–104.

TAYLOR, A.M. & GOLDRING, R. 1993. Description and analysis of bioturbation and ichnofabric. *Journal of the Geological Society, London*, **150**, 141–148, https://doi.org/10.1144/gsjgs.150.1.0141

TESSIER, B., ARCHER, A.W., LANIER, W.P. & FELDMAN, H.R. 1995. Comparison of ancient tidal rhythmites (Carboniferous of Kansas and Indiana, USA) with modern analogues (the Bay of Mont-Saint-Michel, France). *In*: FLEMMING, B.W. & BARTHOLOMÄ, A. (eds) *Tidal Signatures and Ancient Sediments*. International Association of Sedimentologists, Special Publications, **24**, 259–271.

TONKIN, N.S., MCILROY, D., MEYER, R. & MOORE-TURPIN, A. 2010. Bioturbation influence on reservoir quality: a case study from the Cretaceous Ben Nevis Formation, Jeanne d'Arc Basin, Offshore Newfoundland, Canada. *American Association of Petroleum Geologists Bulletin*, **94**, 1059–1078.

VAN DEN BERG, J.H., BOERSMA, J.R. & VAN GELDER, A. 2007. Diagnostic sedimentary structures of the fluvial–tidal transition zone – evidence from deposits of the Rhine and Meuse. *Netherlands Journal of Geosciences*, **86**, 287–306.

VISSER, M.J. 1980. Neap–spring cycles reflected in Holocene subtidal large-scale bedform deposits; a preliminary note. *Geology*, **8**, 543–546.

Evolution and architectural styles of a forced-regressive Holocene delta and megafan, Mitchell River, Gulf of Carpentaria, Australia

T. I. LANE[1]*, R. A. NANSON[1], B. K. VAKARELOV[1], R. B. AINSWORTH[1] & S. E. DASHTGARD[2]

[1]*Australian School of Petroleum, University of Adelaide, Adelaide, SA 5005, Australia*

[2]*Department of Earth Sciences, Simon Fraser University, British Columbia, Canada V5A 1S6*

**Correspondence: tessa.lane@adelaide.edu.au*

Abstract: The monsoonal Mitchell River delta and megafan shows minimal anthropogenic disturbance and thus provides a unique opportunity to examine the autogenic and allogenic controls on the evolution of deltas and fluvial megafans. Detailed mapping, vibracoring, trenching and topographic surveying were used to characterize the palaeodistributary channel belts on the megafan and the depositional elements that comprise the delta. Chronological analyses of these data facilitated the reconstruction of the megafan and delta evolution and enabled the identification of discrete periods of delta progradation within the last 6000 years. These results indicate that sediment distribution is controlled primarily by two types of avulsion: (1) delta avulsions, which are frequent ($>16/1000$ years), typically backwater-mediated and associated with local shifts in sedimentation loci; and (2) megafan avulsions, which are less frequent ($>3/1000$ years), but which are also associated with more significant shifts in depositional loci. These links between megafan and delta processes and geomorphology in the Mitchell River region were integrated to develop a new model of channel belt facies associated with fluvial (F), fluvial backwater-affected (F_{BW}), fluvial-dominated, tide-influenced (Ft) and tide-dominated, fluvial-influenced (Tf) channels. This model enables improved predictions of channel belt composition in modern and ancient marginal-marine systems by providing sedimentological and ichnological criteria for distinguishing between channel types.

Supplementary material: Mitchell River region optically stimulated luminescence (OSL) dating methodology and results are available at https://doi.org/10.6084/m9.figshare.c.3280949

Ancient fluvial and deltaic systems are estimated to host a large proportion of the world's remaining hydrocarbon reserves (Tyler & Finley 1991; Keogh *et al.* 2007) and, for this reason, fluvial and deltaic analogues are a necessary component of sedimentological and stratigraphic research. Investigations of modern analogue fluvial and deltaic systems are particularly useful to geologists as they can reveal links between observed modern processes and architecture (e.g. Lang *et al.* 2000; Nanson *et al.* 2012; Martinius *et al.* 2014; Nyberg *et al.* 2015). The Holocene Mitchell River delta and megafan, located in the shallow, intracratonic setting of the Gulf of Carpentaria (GOC), is a well exposed, intact example of a mixed-influence system displaying a complicated network of active and palaeochannels. The Mitchell River deposits thus provide a unique opportunity to examine the link between megafan channel avulsions and coeval deltaic evolution, and the impact of delta evolution on the distribution of channel facies, for direct application to ancient reservoirs.

Fluvial megafans are defined as large radial channel networks characterized by stream-dominated channelized flow, low gradients ($<0.2°$) and large areas (10^3-10^5 km^2) (DeCelles & Cavazza 1999; Horton & DeCelles 2001). They occur in all tectonic settings and climatic zones (e.g. Hartley *et al.* 2010). Fluvial megafans are commonly referred to as distributive fluvial systems, which are fluvial systems that are unconfined downstream of an apex and radiate from that point (e.g. Weissmann *et al.* 2010, 2011, 2013). Some researchers describe a fluvial megafan as a stream-dominated end-member of an alluvial fan (e.g. Galloway & Hobday 1996). However, other researchers restrict the definition of 'alluvial fan' to steep (2–25°) radial depositional systems dominated by fluid and sediment gravity flows (e.g. Blair & McPherson 1994). Sediment distribution on megafans is controlled primarily by autogenic fan avulsion (Schumm *et al.* 1987), which can be affected by a range of allogenic forces. Base level is one such allogenic control which affects megafans that enter a standing body of water (megafan and delta; Fig. 1a). The deltaic portion is defined as an overall regressive depositional system that is built either directly by, or in association with, rivers that enter a standing body of water (Bhattacharya 2010). The boundary between the

From: HAMPSON, G. J., REYNOLDS, A. D., KOSTIC, B. & WELLS, M. R. (eds) 2017. *Sedimentology of Paralic Reservoirs: Recent Advances*. Geological Society, London, Special Publications, **444**, 305–334.
First published online July 7, 2016, https://doi.org/10.1144/SP444.9

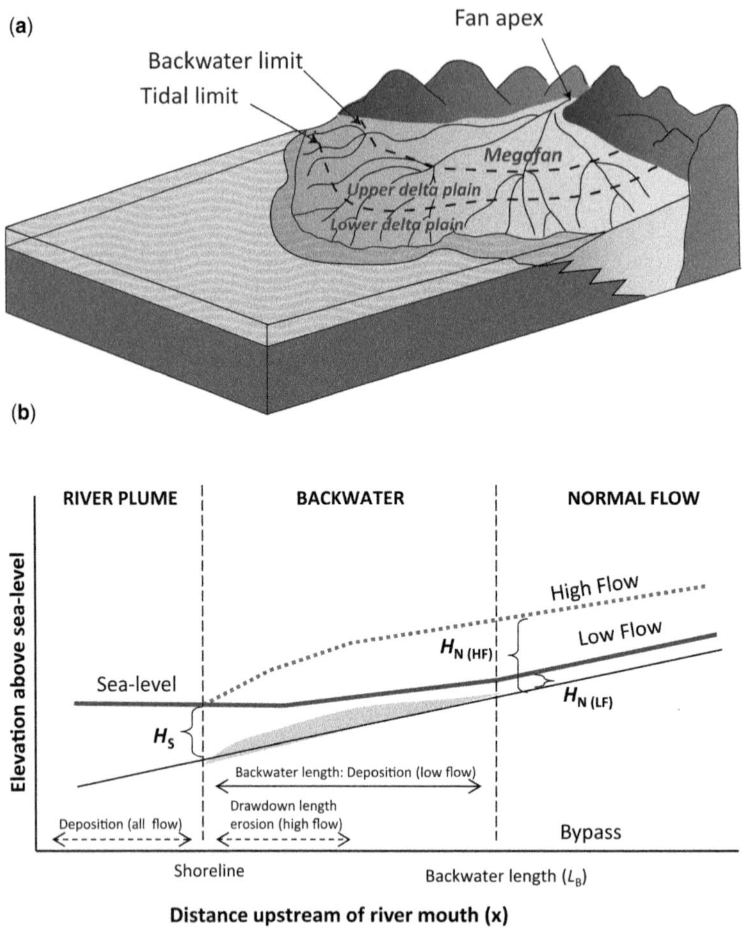

Fig. 1. (**a**) Schematic diagram of a fluvial megafan and associated deltas (after Nemec & Steel 1988; see also Ganti *et al.* 2014). Deltas occur when a fluvial megafan debouches into a standing body of water. The upstream limit of the delta plain is defined by the backwater length (L_B), above which channels cannot feel the hydraulic effects of base level. Megafans and linked deltas may have two distinct avulsion node loci. The first, bed slope mediated avulsions, occur at the apex of the megafan and are related to the break in slope. The second, backwater-mediated avulsions, occur in the vicinity of the landward limit of the backwater length. The division of the delta into the upper and lower delta plain occurs at the tidal limit. (**b**) When a river enters a standing body of water under conditions of low flow, the slope of the bed separates from the slope of the water surface at a distance that is defined as the backwater length (L_B). Upstream of this position is a zone of normal flow. The region between L_B and the shoreline is termed the backwater zone. Seawards of the shoreline, the flow becomes uncontained; this is where the river plume occurs and mouth bars can form. At low flow (solid line), the channel depth at the shore line (H_S) is greater than in the normal flow zone ($H_{N(LF)}$) and the water surface in the backwater zone has a concave water surface profile that leads to spatial deceleration and deposition (see also Lane 1957). Conversely, at high flow (dashed line), when $H_S < H_{N(HF)}$, there is a spatial acceleration in erosion proximal to the mouth within the backwater zone (Lane 1957; Lamb *et al.* 2012). Natural rivers typically have alternating high flow and low flow regimes, so bed aggradation is concentrated in the upstream portion of the backwater zone. As a result of the positive correlation between bed aggradation and avulsion frequency, avulsion nodes are concentrated in the upstream portion of the backwater zone.

purely alluvial portion of the megafan and the deltaic portion is defined by the backwater length (L_B), which is the distance upstream of plume spreading ($x = 0$) to which a river experiences the hydraulic effect of base level (Fig. 1b; Lane 1957). Lamb *et al.* (2012) showed that backwater length scales well to channel depth at the mouth (H_S) and channel bed slope (S_B):

$$L_B = \frac{H_S}{S_B} \quad (1)$$

A model based on the Huanghe River, China has suggested that linked megafan and delta systems are characterized by two types of avulsion: (1) those at the megafan apex, which relate primarily to a break in slope as the fluvial system emerges from relative confinement, are termed bed slope mediated avulsions; and (2) those that relate to backwater processes and are termed backwater-mediated avulsions (Ganti *et al.* 2014). Backwater-mediated avulsions result in more limited shifts in the location of deposition, whereas bed slope mediated avulsions can result in more fundamental shifts in depositional loci (Ganti *et al.* 2014). Although the Huanghe megafan and delta is characterized by a much larger scale and lower depositional slope than the Mitchell system, their processes can be compared and contrasted in a conceptual manner. Linked megafan delta systems are present in both modern (e.g. Jones *et al.* 1993, 2003; Ganti *et al.* 2014) and ancient (e.g. Shiers *et al.* 2014) settings and are under-studied. Identifying the distribution of both types of avulsion, and their effects on delta morphology, is crucial in understanding the distribution of channel and marginal-marine facies and is hence a key focus of this paper.

The deltaic portion of megafans can also be affected by basinal processes such as tides, which often produce a range of mixed-process architectures (e.g. Coleman & Wright 1975; Galloway 1975; Ainsworth *et al.* 2011). For example, tide-influenced channels change their morphology and sedimentary fill in a systematic way as they transition from the continental to the marine zone (Dalrymple & Choi 2007; Davies & Woodroffe 2010), with important implications for facies distributions (Johnson *et al.* 1982; Woodroffe *et al.* 1989; Dalrymple & Choi 2007). The influence of tides on channelized systems is known to introduce significant heterogeneity to channelized fill deposits, typically in the form of inclined heterolithic stratification (Thomas *et al.* 1987), which has implications for petroleum reservoirs composed of these deposits (e.g. Hubbard *et al.* 2011; Fustic *et al.* 2012). Upstream avulsions exert an important control on the balance between fluvial and tidal processes, with important consequences for channel belt fills (e.g. Jones *et al.* 1993, 2003). Understanding the controls on channel avulsion in the context of tidal influence, and other basinal processes, is fundamental to predicting similar facies distributions in the subsurface.

The primary objectives of this paper are:

(1) to link the palaeogeographical evolution of the Holocene Mitchell River megafan and delta to autogenic and allogenic controls; and
(2) to characterize the channel belt style and distribution across the Mitchell River megafan and delta.

Study area

Offshore: Gulf of Carpentaria

The GOC is the world's largest tropical epicontinental seaway, with a total area of *c.* 500 000 km^2 and a catchment area of 1 200 000 km^2 (Reeves *et al.* 2008; Fig. 2). The GOC reaches a maximum depth of 68 m near the eastern margin of the basin (Torgersen *et al.* 1983; Harris *et al.* 2003) and below depths of 50 m the average gradient is 1 : 13 000 (Edgar *et al.* 2003). Despite repeated transgressions and regressions of the shallow Torres Strait (12 m) and Arafura Sill (53 m) that bound the GOC, the associated low and consistent on- and offshore gradients (Fig. 3) have resulted in minimal channel incision during relative sea-level lowstands (Jones *et al.* 2003). The Mitchell River, which is the largest of the channels draining into the GOC, is situated on the eastern margin of this basin (Figs 2 & 4).

Onshore: Mitchell River megafan and delta

The Mitchell River megafan is the largest in Australia, with a drainage basin of 71 630 km^2 (Brooks *et al.* 2009), and flows westwards from the mixed lithology (Raymond *et al.* 2012) Great Dividing Range (Grimes & Doutch 1978) (Figs 2 & 4). The Mitchell River megafan has been operating since the Pliocene and has been divided into five depositional phases: M_1–M_5 (Grimes & Doutch 1978) (Fig. 4a). This study focuses on the M_5 fan, which developed during a forced regression following the maximum Holocene marine transgression (Fig. 4a). The Holocene apex of the megafan has shifted basinwards of the Pleistocene apex (Fig. 4a) as a result of megafan progradation and possibly a reduced sediment supply in the Holocene relative to wetter periods in the Quaternary (Brooks *et al.* 2009).

Holocene sea-level

The last interglacial marine transgression was first felt in the GOC following the breach of the Arafura Sill *c.* 12.2 cal ka BP (Reeves *et al.* 2008). From this time, the sea-level rose at a rate of 12 m/1000 years until *c.* 8 ka and then slowed to 4.5 m/1000 years from 8 ka until the maximum transgression at

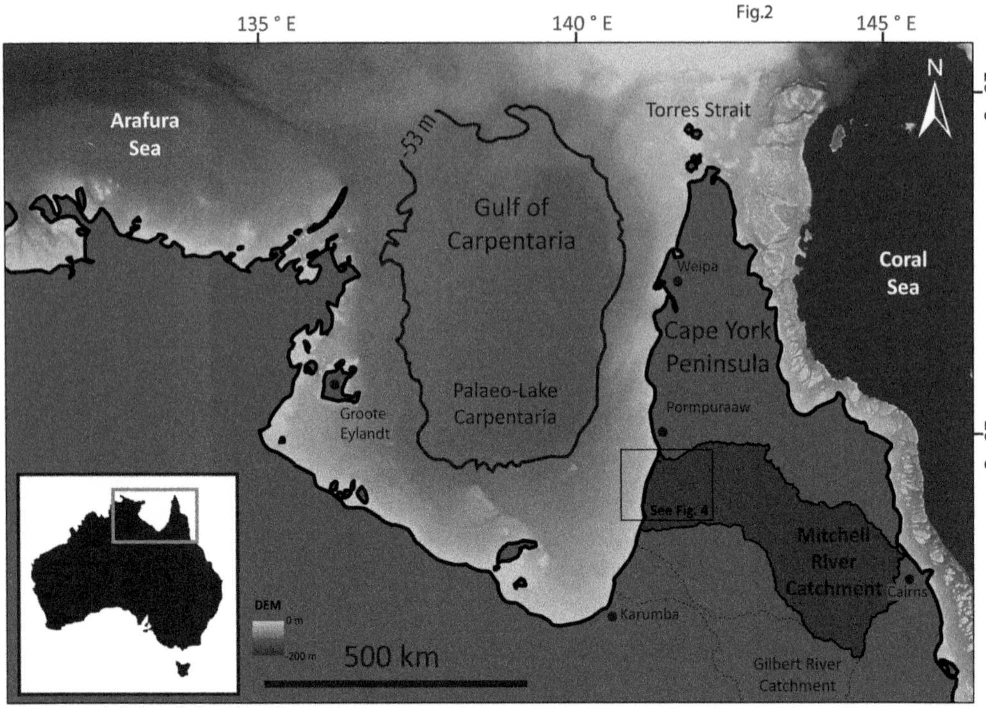

Fig. 2. Location map of the Gulf of Carpentaria showing the Mitchell River catchment (the focus of this study) and the nearby Gilbert River catchment, which is often used to describe similar processes and forms as in the Mitchell River catchment. During periods of low sea-level during the Quaternary, the Gulf of Carpentaria was Lake Carpentaria, with a maximum extent delimited by the −53 m contour (modified from Torgersen *et al.* 1983).

7–6 ka (Thom & Chappell 1975; Thom & Roy 1983). During the maximum transgression, the sea-level reached *c.* 1.0–1.5 m above the present level in northern Queensland. Discussions regarding the details of the sea-level history since the maximum marine transgression are divided (see Lewis *et al.* 2013; Murray-Wallace & Woodroffe 2014). In a study of microatolls in northeastern Queensland, Chappell (1983) inferred a gradual fall of sea-level over the past 6000 years and concluded that there was no evidence for oscillations exceeding the uncertainty range of ± 0.20 m. Other studies indicate oscillations in the rate of fall in sea-level since 6 ka BP, both in the GOC (Woodroffe & Chappell 1993; Jones *et al.* 2003) and along the eastern coast of Queensland (e.g. Larcombe *et al.* 1995; Woodroffe 2009). Whichever scenario best represents the rate of sea-level fall over this period, the resultant forced regression in the southern GOC facilitated the rapid progradation of extensive marginal-marine systems in the SE GOC, including the Macarthur (Woodroffe & Chappell 1993; Jones *et al.* 2003), Gilbert (Jones *et al.* 1993, 2003), Pormpuraaw (Rhodes 1980) and the Mitchell (Nanson *et al.* 2013*b*) river systems. This research provides new evidence for the timing and nature of part of the Holocene sea-level curve in this region.

Holocene climate

Shulmeister (1999) presented a three-stage model for changes in the Holocene effective precipitation (EP) in Australasia, linking the EP to the precessional control of Walker circulation in the Pacific and the intensification of the northern Australian monsoon:

(1) from the beginning of the Holocene until 5 ka BP, there was a gradual increase in EP and temperature;
(2) from 5 to 3.7 ka BP, there was increased EP; and
(3) from 3.75 ka BP to the present, there was a sharp reduction in EP and increased variability in the climate.

Most climatic reconstructions show a similar EP pattern to the model presented by Shulmeister (1999) (e.g. Luly *et al.* 2006; Moss *et al.* 2012). The modern Mitchell River catchment has a monsoonal wet–dry tropical climate (Peel *et al.* 2007)

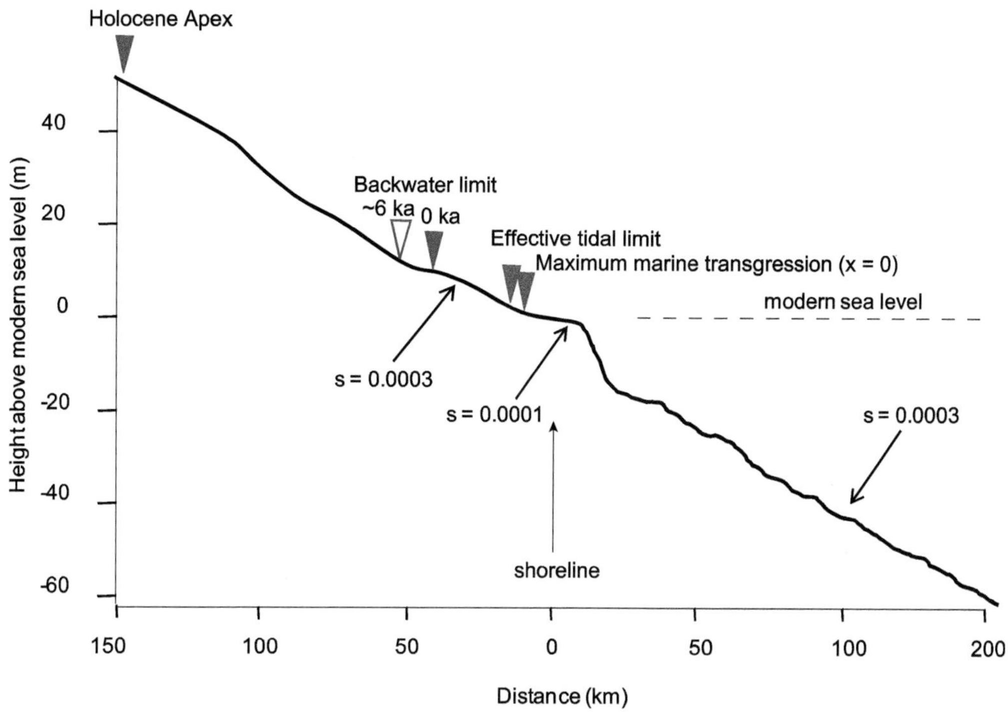

Fig. 3. The longitudinal profile of the lower Mitchell River from the Holocene megafan apex showing the relative position of the backwater limit at *c.* 6 and 0 ka, the maximum Holocene transgression and the effective tidal limit. The river profile is joined from the modern Mitchell River mouth to a shore-perpendicular bathymetric profile (modified from Nanson *et al.* 2013*b*).

as northern Australia has just two seasons: a long dry season (April–November) associated with trade winds that travel from an east to NW direction and a shorter wet season (December–March) associated with a reversal of wind direction (Suppiah 1992).

Wave, tide and fluvial processes

Marginal-marine systems are affected by waves, tides and fluvial processes. The nearest wave height recording station to the Mitchell River delta is located at Weipa, 200 km north of the Mitchell River mouth (Fig. 2). Modelling studies from CSIRO Marine & Atmospheric Research (2010) show a decrease in wave height from this recording station to the Mitchell River delta. Weipa's average wave height is higher during the wet season (0.5 m) than during the dry season (0.35 m). Storms and cyclones, which are a frequent occurrence in the wet season, can drive wave heights up to 3.5 m (Mouhoupt *et al.* 2004).

The tidal range increases towards the south in the GOC. The nearest recording station to the Mitchell River is 250 km to the south at Karumba (Fig. 2), where the average tidal range varies from 0.3 m (neap) to 4 m (spring), with a mean tidal range of 3.3 m (Munro 1984). Wind set up during the wet season, driven by the summer monsoon, causes a sea-level rise of up to 0.5 m in the southeastern GOC (Wolanski 1993) and results in the inundation of the supratidal mud flats that comprise much of the coastal plain (Ridd *et al.* 1988).

Channel flow data for the Mitchell River catchment is limited (Rustomji *et al.* 2010). However, the annual catchment discharge is estimated to be >80 000 000 Ml (see Brooks *et al.* 2009). The flow is highly seasonal and the high fluvial activity corresponds to the summer monsoon from December to March. During floods, the additional discharge is not accommodated by an increase in channel capacity of the Mitchell River downstream, but is partitioned into distributary channels across the megafan (Rustomji *et al.* 2010). This is particularly notable for floods with recurrence intervals >5 years (Rustomji *et al.* 2010).

Methods

Delta mapping and classification

Ainsworth *et al.* (2011) and Vakarelov & Ainsworth (2013) presented a process-based architectural

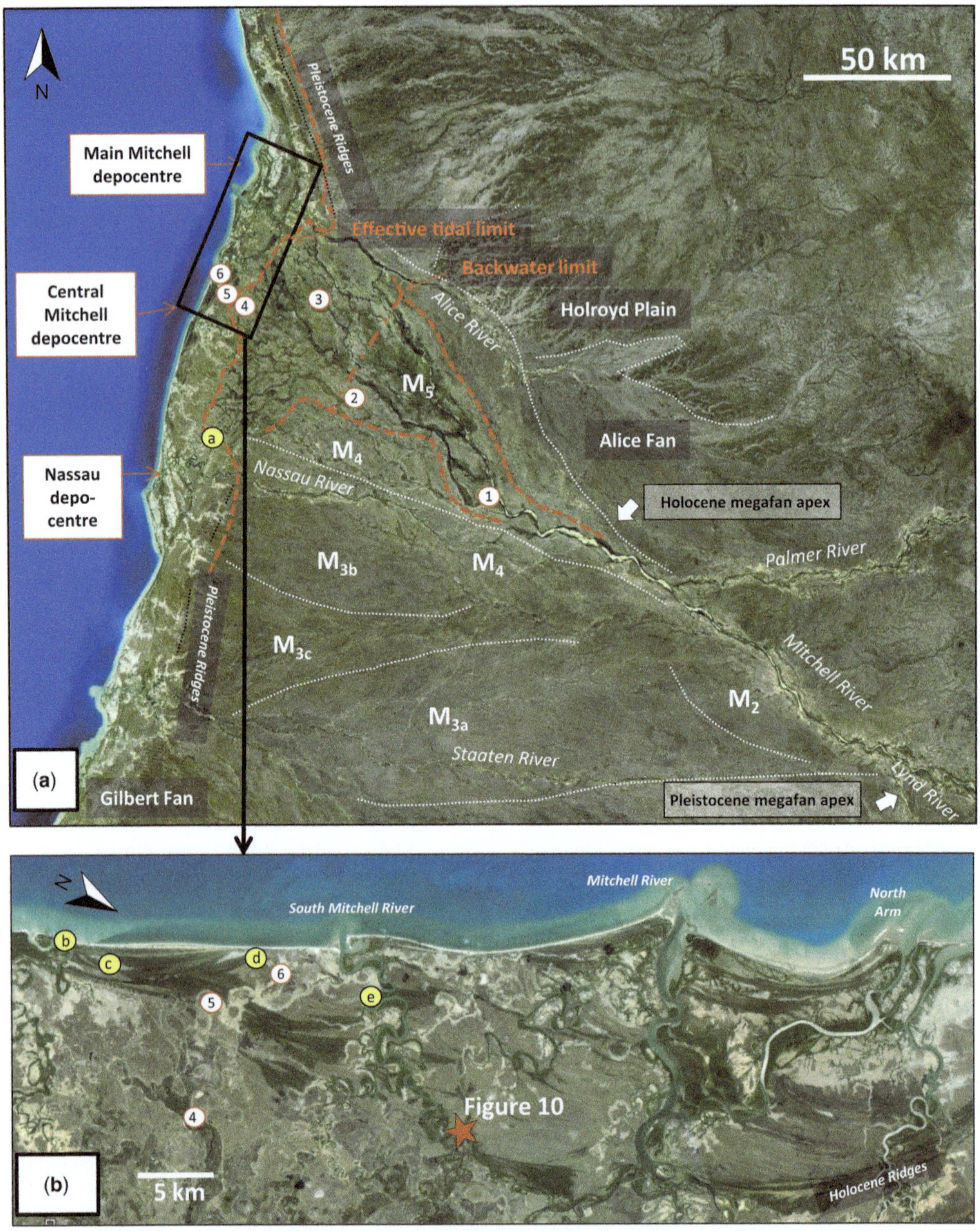

Fig. 4. (**a**) Location map showing evolution of the youngest three Mitchell (*M*) megafans (dashed lines) draining westwards from the Great Dividing Range into the Gulf of Carpentaria (after Grimes & Doutch 1978). Numeric subscripts denote the stages of megafan development (1–5, oldest to youngest). This study focuses on the M_5 megafan and the associated Holocene deltaic units, herein termed the Mitchell River regressive element complex assemblage set. The Mitchell River regressive element complex assemblage set is subdivided into three depocentres, which, from north to south, are the Main Mitchell depocentre, the Central Mitchell depocentre and the Nassau depocentre (red dashed lines). Circled numbers indicate the channel transects presented in Figures 8b and 9; circled letter 'a' indicates the location of stratigraphic site used in the megafan delta reconstruction presented in Figure 8a. (**b**) Close up of the Main Mitchell depocentre. Circled letters indicate the location of other stratigraphic sites used in megafan delta reconstruction presented in Figure 8a and the red star indicates the location of the photograph in Figure 10. Stratigraphic data are summarized in Table 1.

classification for marginal-marine systems. The process classification is a three-tiered extension of the scheme of Galloway (1975) to incorporate all coastal systems. Instead of just a single dominant process being assigned, the scheme allows for primary, secondary and tertiary process components to be described. For example, a depositional unit may be tide-dominated, wave-influenced and fluvial-affected (Tw*f*). Their hierarchical approach provides a method for categorizing marginal-marine systems from the depositional element scale to the parasequence scale, providing a link between sequence stratigraphy and sedimentology. The classification can be applied to both modern (e.g. Nanson *et al.* 2012, 2013*b*) and ancient systems (e.g. Ainsworth *et al.* 2015), enabling an improved comparison between modern and ancient systems. Using this method, depositional elements, which form the building blocks of marginal-marine systems and hence combine to form the units of the subsequent levels of the hierarchy, are grouped to form element complexes (EC) (akin to facies associations). EC units group to form element complex sets (ECS) and element complex assemblages (ECA), which represent depositional environments. ECS and ECA, in turn, form the entire regressive element complex assemblage set (RECAS) (akin to a parasequence), which, in the Mitchell case, represents the Holocene delta. ECS and ECA represent major pulses of shoreline migration and are thus the most useful scale for palaeogeographical reconstruction. A change in process category from one ECS to another results in the placement of an ECA boundary (see Nanson *et al.* 2013*b*). For the purpose of this study, EC, ECS and ECA were mapped and linked to channel avulsions on the megafan.

The Holocene Mitchell River megafan and delta covers a total area of 71 000 km^2, of which the Mitchell RECAS comprises 15 000 km^2. Element-scale mapping was extended from Massey *et al.* (2014) and Nanson *et al.* (2012, 2013*b*) to complete the entire Mitchell River RECAS (Fig. 5a). Mapping of the RECAS and megafan was undertaken using Google Earth imagery and a 30 m Shuttle Radar Topography Mission Digital Elevation Model (Geoscience Australia & CSIRO Land & Water 2011). A total of 13 depositional element types were mapped following the recognition criteria of Nanson *et al.* (2013*b*). Their recognition criteria were based on extensive ground truthing and literature review.

The methodology for assigning process classification uses the percentage component by areal extent of depositional elements that were deposited by primarily fluvial, tide or wave processes (Ainsworth *et al.* 2011; Nanson *et al.* 2013*b*). Unlike most marginal-marine classification systems, this method is semi-quantitative and is more readily translatable to reservoir analogue studies (Massey *et al.* 2014). In some units, the classification has been modified to account for significant post-depositional working (see also Ainsworth *et al.* 2015).

Megafan mapping

Mapping on the megafan was undertaken at the channel belt scale (Fig. 6). The megafan and delta can be subdivided into depositional zones based on the influences of channel hydraulics: the lower delta, upper delta and megafan (Fig. 7). The effective tidal limit was placed at the maximum landward extent of tidal flat deposition, which corresponds to the paralic/fluvial boundary and the limit to which tides have an impact on the channel facies (see also Jones *et al.* 1993, 2003; Queensland Government 2009) and thus defines the boundary between the lower and upper delta plains. The upper delta extends upstream to the backwater length (L_B), the position of which is indicated by the tendency for fluvial channels to avulse and develop more distributive networks (see Chatanantavet *et al.* 2012). The scaling parameter of Lamb *et al.* (2012, equation 1) for the Mitchell River gives an alternative means for estimating an L_B of 20–27 km ($H_S = 6–8$ m, Rustomji *et al.* 2010; and $S_B = 0.0003$, Brooks *et al.* 2009). We set $x = 0$ km (Fig. 7) at the landwards limit of the lower delta plain because during the wet season, when the sea-level rises *c.* 0.5 m higher than dry season levels, there is a considerable amount of channel spreading at this point and the water surface elevation is predicted to be very near sea-level. Despite the complication of tides, bifurcations and the seasonal increase in sea-level, the scaling relationship of Lamb *et al.* (2012) does appear to approximate L_B, as indicated by the concentration of avulsion nodes (Fig. 7), which lends confidence to both approximations.

Stratigraphy and dating

Five beach and chenier ridges were vibracored and dated to determine the thickness of Holocene deposits and the timing of development of the ECS and ECA (Fig. 8a). Six palaeochannel belts were also vibracored and dated to determine downstream changes in channel depth and to link megafan and delta evolution (Fig. 8b). Polyurethane resin peels and X-ray analyses (73 kVp, 5 mA s) were applied to these cores to identify sedimentary and biogenic structures. The elevations of drill tops were surveyed using a Topcon Real Time Kinematic global positioning system and tied to the Australian Height Datum, which approximates the mean sea-level (Roelse *et al.* 1975).

Seventeen samples were collected from these stratigraphic datasets for single-grain optically

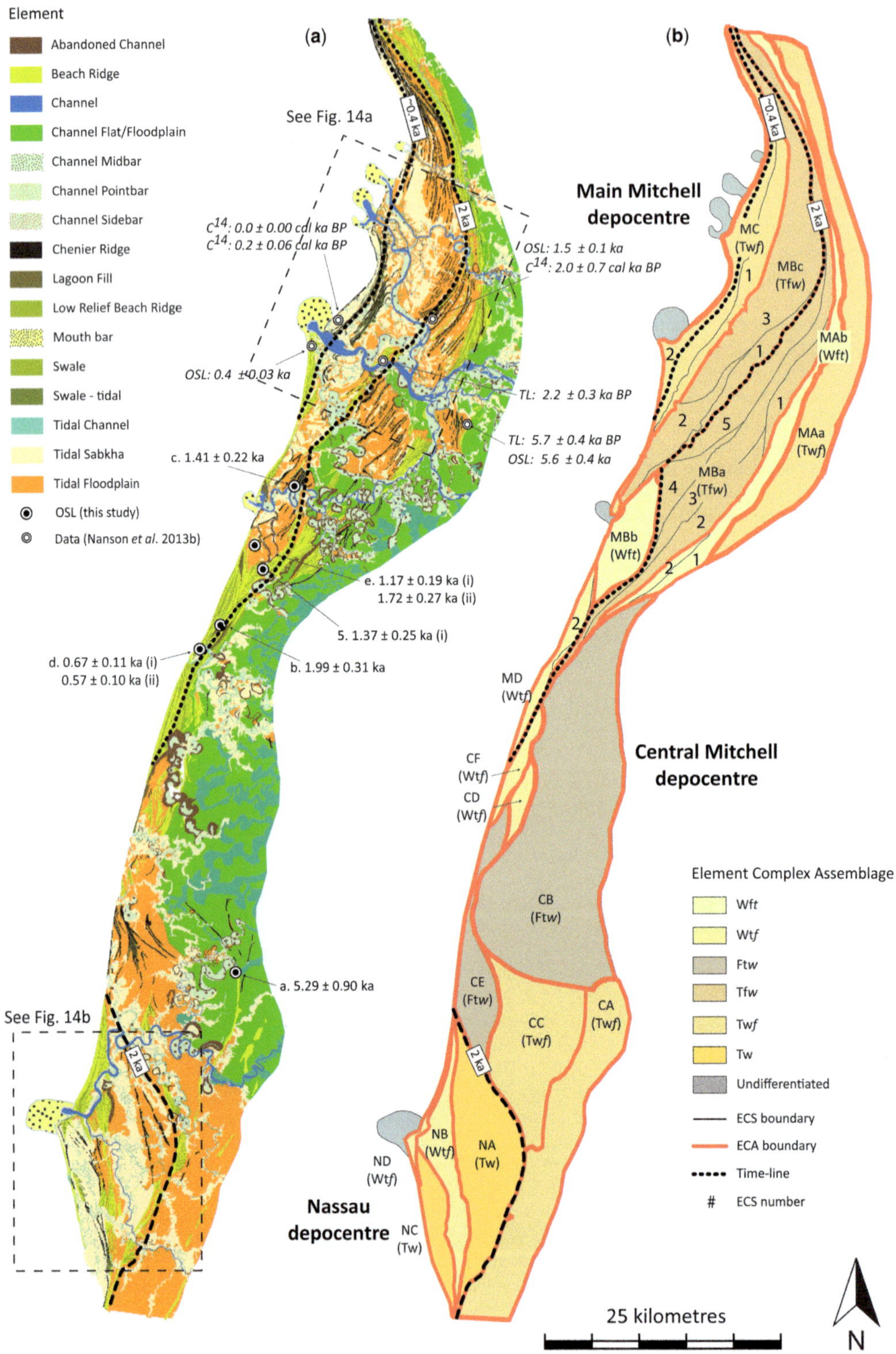
Element
Abandoned Channel
Beach Ridge
Channel
Channel Flat/Floodplain
Channel Midbar
Channel Pointbar
Channel Sidebar
Chenier Ridge
Lagoon Fill
Low Relief Beach Ridge
Mouth bar
Swale
Swale - tidal
Tidal Channel
Tidal Sabkha
Tidal Floodplain
OSL (this study)
Data (Nanson et al. 2013b)
(a)
See Fig. 14a
~0.4 ka
2 ka
C14: 0.0 ± 0.00 cal ka BP
C14: 0.2 ± 0.06 cal ka BP
OSL: 1.5 ± 0.1 ka
C14: 2.0 ± 0.7 cal ka BP
OSL: 0.4 ± 0.03 ka
TL: 2.2 ± 0.3 ka BP
TL: 5.7 ± 0.4 ka BP
OSL: 5.6 ± 0.4 ka
c. 1.41 ± 0.22 ka
e. 1.17 ± 0.19 ka (i)
1.72 ± 0.27 ka (ii)
5. 1.37 ± 0.25 ka (i)
d. 0.67 ± 0.11 ka (i)
0.57 ± 0.10 ka (ii)
b. 1.99 ± 0.31 ka
a. 5.29 ± 0.90 ka
See Fig. 14b
2 ka
(b)
Main Mitchell depocentre
MC (Twf)
MBc (Tfw)
MAb (Wft)
MAa (Twf)
MBa (Tfw)
MBb (Wft)
MD (Wtf)
Central Mitchell depocentre
CF (Wtf)
CD (Wtf)
CB (Ftw)
CE (Ftw)
CC (Twf)
CA (Twf)
NB (Wtf)
NA (Tw)
ND (Wtf)
NC (Tw)
Nassau depocentre
Element Complex Assemblage
Wft
Wtf
Ftw
Tfw
Twf
Tw
Undifferentiated
ECS boundary
ECA boundary
Time-line
ECS number
25 kilometres
N

stimulated luminescence (OSL) age dating (Table 1). A single aliquot regeneration protocol (Wintle & Murray 2006) was used to estimate the equivalent doses. A dose recovery test was conducted on a subset of four samples to test the suitability of the single aliquot regeneration protocol using a subset of five single-grain aliquots per sample and these results supported its suitability for age analysis (Murray & Wintle 2000).

Results

Delta mapping and stratigraphy

The Mitchell River RECAS has prograded up to 17 km (depositional dip) from a shoreline position between the Pleistocene beach ridges (Nanson *et al.* 2013*b*) and the most landward Holocene ridge (Figs 4 & 5). The RECAS extends 130 km along depositional strike and encompasses an area of *c.* 1500 km^2. Three depocentres comprise the RECAS; from north to south these have been named herein as the Main Mitchell, the Central Mitchell and the Nassau depocentres (Fig. 5a). Fieldwork has primarily focused on the Main Mitchell depocentre because it is the largest by area and is currently active. The orientation and position of beach and chenier ridges (shoreline proxies) indicate that the Central Mitchell depocentre is the oldest of the three depocentres and, as such, it is less well preserved and is also largely abandoned. The Nassau depocentre is well exposed and currently active.

Over 11 200 individual elements have been mapped across the delta from which the ECS and ECA units are defined. ECS and ECA level mapping was extended from Nanson *et al.* (2013*b*) to the southern portion of the delta. This extended mapping has necessitated further subdivisions and minor modifications of some previously identified boundaries in the Main Mitchell depocentre. The ECAs in the Main Mitchell depocentre have been named to maintain consistency with previously published work (Nanson *et al.* 2013*b*). For example, ECA B (Nanson *et al.* 2013*b*) is now named ECA MBa, MBb and MBc, where M now stands for the Main Mitchell depocentre and B links to the original ECA B described in Nanson *et al.* (2013*b*). The ECS are labelled alpha-numerically, whereby the letters refer to the depocentre (prefix M, C or N) and then the parent ECA and the numbers refer to their relative timing of deposition. For example, ECS MBa.2 is the second ECS to be deposited in ECA MBa. Seventeen ECA have been mapped across the Mitchell River RECAS (Fig. 5b; Table 2). The Main Mitchell depocentre is composed of seven ECA (Aa, Tf*w*; Ab, Wf*t*; Ba, Tf*w*; Bb, Wf*t*; Bc, Tf*w*; C, Tw*f*; and D, Wt*f*), each representing a pulse of shoreline migration associated with a change in process classification. Major changes of shoreline orientation were observed within five of these ECA in the absence of a change in process category and 14 ECS were defined (ECS Ab.1 and Ab.2, Ba.1–5, Bc.1–3, C.1 and C.2 and D.1 and D.2). The Central Mitchell depocentre consists of six ECA (CA, Tf*w*; CB, Ft*w*; CC, Tf*w*; CD, Fw*t*; CE, Tw*f*; and CF, Wt*f*). However, the lack of surface expression of morphological features prevented mapping at the more detailed ECS level. The Nassau depocentre is defined by four ECA and no subdivision of the ECS units is required.

Channel mapping and stratigraphy

The Mitchell River megafan consists of a radial network of modern and palaeodistributary channels that traverse the megafan surface (Fig. 7). The mapped Holocene portion of the Mitchell River megafan is situated on the northern limb of the greater Mitchell megafan and encompasses a total area of 3000 km^2, of which *c.* 900 km^2 (30%) are composed of channel belt deposits.

Sediment distribution on the upper delta plain and the megafan is characterized by overbank deposition and avulsion. A total of 110 major avulsion nodes were identified from satellite imagery (Fig. 7; Table 3). Full and partial avulsions (*sensu* Slingerland & Smith 2004) have not been distinguished.

Fig. 5. (**a**) Over 11 200 individual elements have been mapped across the Mitchell River regressive element complex assemblage set (RECAS), encompassing an area of 1500 km^2. Elements are the building blocks for subsequent levels of the mapping hierarchy. (**b**) Element complex assemblages (ECA) are discrete pulses of shoreline progradation associated with a change in process classification (coloured according to process category). Major reorganization of the shoreline position without a change in process category defines an element complex set (ECS) within the ECA (black line). A total of 14 ECS and 17 ECA have been mapped across the Mitchell River RECAS. Extended mapping has necessitated further subdivisions of the ECS and ECA units from Nanson *et al.* (2013*b*). However, these units have been named to maintain consistency with previously published work. Mapping and classification shows that there is significant spatial and temporal variation of process category within the Mitchell RECAS. Ft*w*, fluvial-dominated, tide-influenced and wave-affected; Fw*t*, fluvial-dominated, wave-influenced and tide-affected; OSL, optically stimulated luminescence; Tf*w*, tide-dominated, fluvial-influenced and wave-affected; TL, thermoluminescence; Tw, tide-dominated, wave-influenced; Tw*f*, tide-dominated, wave-influenced and fluvial-affected; Wt*f*, wave-dominated, tide-influenced and fluvial-affected.

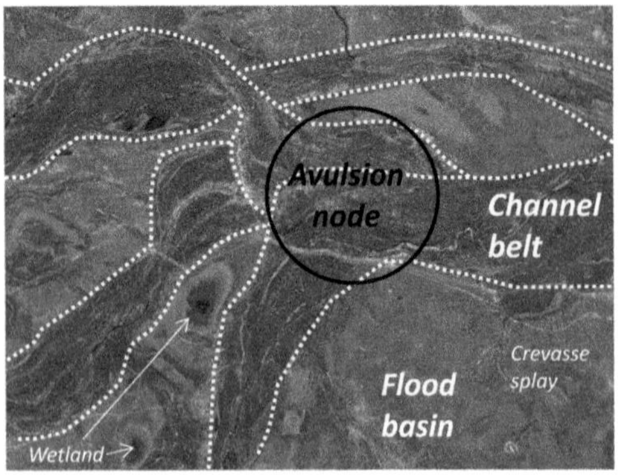

Fig. 6. Channel belts were mapped at the channel belt scale (see Fig. 7 for location) rather than at the more detailed element scale used for the paralic environment. The example shows palaeochannel belts modified by surface scour from ephemeral streams and the tendency for channels to be slowly abandoned or reoccupied by subsequent partial avulsions. The marked avulsion point is nodal because there are two avulsions that have initiated from a single point. Small crevasse splays, wetlands and small channels are features of the megafan, but are not mapped at this resolution.

Smaller scale avulsions associated with anabranching channels on the megafan and delta surface were excluded. Two distinct types of avulsion are recognized: bed slope mediated avulsions on the megafan and backwater-mediated avulsions on the delta plain. The majority of avulsion nodes occur on the upper delta plain, landwards of the Holocene beach ridges, with a further 16 randomly distributed (*sensu* Slingerland & Smith 2004) across the megafan. Sixty delta plain avulsions are related to the formation of the Main Mitchell depocentre, 33 to the Central Mitchell depocentre and one to the Nassau depocentre. Most of the avulsions on the delta also appear to be random; however, four nodal avulsion points are evident (*sensu* Slingerland & Smith 2004). On the lower delta plain, channel bifurcation around mouth bar sediments was identified in three locations (Fig. 7).

Six channel belt sites were selected to reveal downstream changes in channel architecture and morphology from the apex of the megafan to the delta plain (Fig. 4a). The dataset for each location is presented in Table 1; the stratigraphic logs are presented in Figure 8b and cross-sectional diagrams, incorporating surface topography and core datasets, are presented in Figure 9.

Megafan

Major palaeochannel belts on the megafan are typically 1.0–2.5 km wide and of low sinuosity (Fig. 9). Smaller, narrower (<200 m) (unmapped) anabranching channels also traverse the megafan surface (Fig. 6). The channel belt thickness decreases downstream from >11 m (Dunbar core 3) to <9 m (Rutland core 1) (Figs 9.1 & 9.2). Their width to thickness ratios (*sensu* Gibling 2006) are 136–278 and their channel fills fine upwards from gravel to coarse sand and demonstrate a slight decrease in calibre downstream from B-axis (D_B) = 50 mm at Dunbar core 1 (Fig. 8b.1) to D_B = 40 mm at Rutland core 1 (Fig. 8b.2). Cross-lamination was observed in the basal 3–5 m of channel fill and current ripples were present in the top 3 m (Fig. 8b.1 & Fig. 8b.2). Bioturbation and carbonaceous material are common in the top 3 m of sediment. Carbonaceous material consists of mostly roots and most burrows are constructed by reptiles, insects and worms (Fig. 8b.2). The modern Mitchell River megafan channel is straight during the wet season. During the dry season, when discharge is drastically reduced, the channel condenses and occupies a sinuous path, with multiple channels separated by vegetated islands and bars. Historical satellite photographs show that the low flow channel position is dynamic and varies annually. Alluvial sediment on the megafan has been partially indurated (by pedogenic calcrete and Fe–Mn hydroxides/oxides) in places as a result of chemical weathering caused by the monsoonal wet–dry climate (Nanson *et al.* 2005). Indurated sediment is analogous to soft bedrock and restricts channel migration.

Upper delta plain

As fluvial channels reach the delta plain they are affected by backwater dynamics (F_{BW}) and laterally migrate, resulting in scrolled ridge and swale topography (Fig. 9.3). Channel belts on the upper delta plain are typically 0.8–1.5 km wide and 7 m deep, much narrower and slightly shallower than those on the megafan (Figs 9.3 & 9.4). The measured width to thickness ratio is 114–214, slightly less than for the F channels on the megafan. The channel fill consists of fining-upwards medium to coarse sands that are predominantly cross-bedded, with some ripple cross-lamination observed in the top 3 m (Fig. 8b.3). Rare pebbles were observed up to D_B = 30 mm. F_{BW} channel belts are capped with a silty colour-mottled soil and extensive augering across the Racecourse transect (Fig. 9.3) indicated that the silty soil cap varies from 0 to 2 m deep and is deeper in the swales than on the ridges. Bioturbation is comparable with that observed on the megafan and includes roots and a variety of burrows generated by reptiles, insects and worms. The surrounding flood basin is composed of primarily mud with some interbeds of fine to very fine sand (splay deposits).

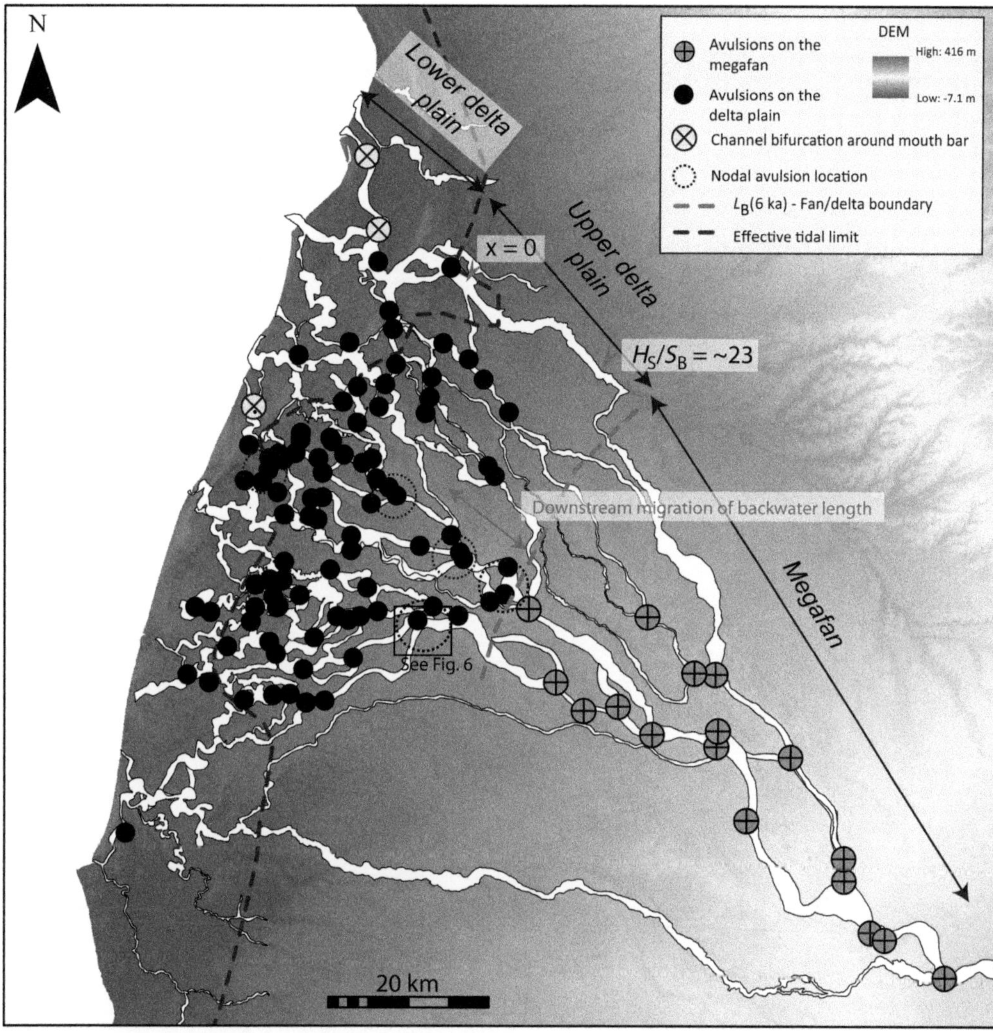

Fig. 7. Avulsion is the primary control on sediment dispersal on the Mitchell River regressive element complex assemblage set. Two distinct types of avulsion are recognized. (1) Avulsions on the megafan, upstream of the backwater limit, are relatively infrequent (>3 ka^{-1}) and are responsible for relatively large-scale changes in deposition at the shoreline. (2) Backwater-mediated avulsions. These are relatively frequent (>16 ka^{-1}) and cause relatively localized shifts in sediment dispersal. The maximum Holocene backwater length is situated where the channels become increasingly distributive. The backwater length has migrated downstream over the last *c.* 6 ka in response to sea-level fall and the progradation of the Mitchell River delta. Channel bifurcation around mouth bars at the shoreline is a feature of sediment dispersal on the Mitchell River, but this process is subordinate to avulsion.

Lower delta plain

The tidal influence increases here as a result of the increased proximity to the shoreline and channel abandonment through avulsion. Lower delta plain channels that are well connected to the catchment are defined as fluvial-dominated, tide-influenced (Ft) and are characterized by scrolled, laterally accreted deposits (e.g. Scroll Site; Fig. 9.5). Ft channel belts are typically 0.6–1.3 km wide and of the order of 6 m deep, scouring into the underlying Holocene offshore muds (Fig. 9.5). The measured width to thickness ratio is 217, which may reflect the increased ability for channel migration through deltaic rather than indurated mega fan sediment (see Nanson *et al.* 2005). The basal 3 m of sediment at core site 5, located 5 km from the modern shoreline (Fig. 4b), consists of

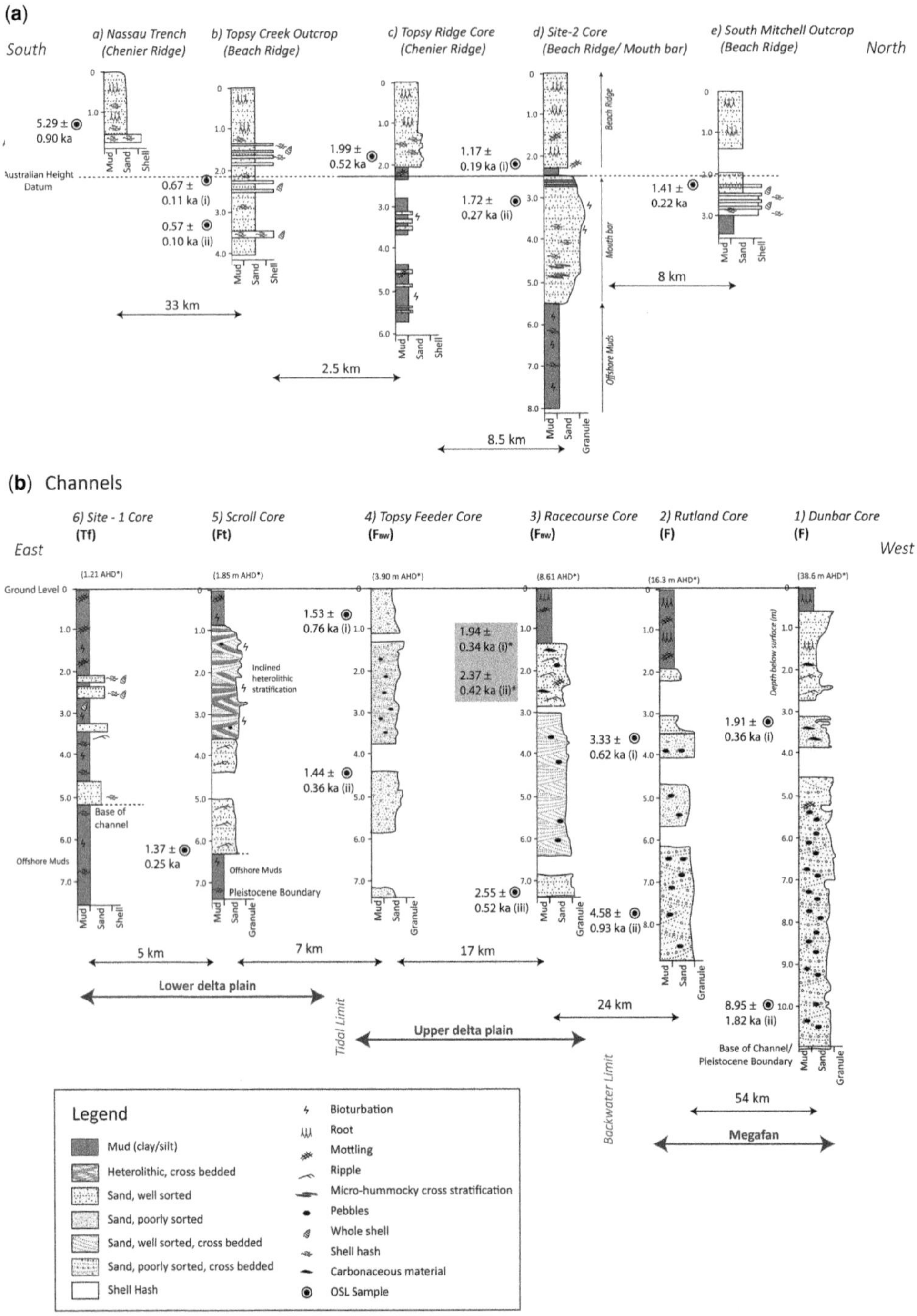

Fig. 8. Stratigraphic logs illustrating the sedimentology at accepted ages of trenched and vibracored transects relating to (**a**) non-channelized units within the lower delta plain and (**b**) channelized environments on the delta and the megafan. *Note that optically stimulated luminescence samples (OSL) Racecourse i and ii are taken from trenches along a transect located *c.* 300 m from the core and are indicated in grey. Channels are presented, from right to left, in a downstream direction. Refer to Figure 4 for locations and Table 1 for summary of OSL results. AHD, Australian Height Datum.

Table 1. *Summary of stratigraphic data*

No.	Site name	Depositional environment	Dataset	Optically stimulated luminescence data				
				Laboratory ID	Field ID	Overburden depth (m)	Age model	Age (ka)
a	Nassau Trench	Chenier ridge (W)	One trench, one OSL	L004/1.1	Nassau Trench	1.10	FMM 4	5.29 ± 0.90
b	Topsy Creek Outcrop	Beach ridge (W)	One outcrop, two OSL	L002/2.15	Topsy Creek Trench (i)	2.15	MAM 3	0.67 ± 0.11
				L002/3.3	Topsy Creek Trench (ii)	3.30	MAM 3	0.57 ± 0.10
c	Topsy Ridge	Beach ridge (W)	One vibracore, two auger hole, one OSL, RTK GPS	L004/1.9	Topsy Ridge	1.9	MAM 3	1.99 ± 0.31
d	Site 2	Beach ridge/mouth bar (W*ft*)	One vibracore, two OSL	L002/2.05	Site 2 (i)	2.05	MAM 3	1.17 ± 0.19
				L002/2.92	Site 2 (ii)	2.92	MAM 3	1.72 ± 0.27
e	South Mitchell Outcrop	Beach ridge (W)	One outcrop, one OSL	L002/2.16	South Mitchell Outcrop	2.16	MAM 3	1.41 ± 0.22
1	Dunbar	Megafan (F)	One vibracore, two OSL, RTK GPS	L002/3.28	Dunbar (i)	3.28	MAM 3	1.91 ± 0.36
				L004/9.9	Dunbar (ii)	9.90	CAM	8.95 ± 1.82
2	Rutland Plains	Megafan (F)	One vibracore, two OSL, RTK GPS	L004/3.65	Rutland Plains (i)	3.65	MAM 3	3.33 ± 0.62
				L004/7.75	Rutland Plains (ii)	7.75	CAM	4.58 ± 0.93
3	Racecourse	Upper delta plain (F)	One vibracore, *c.* 40 auger holes, three OSL, RTK GPS	Adel2036	Racecourse (i)	1.10	MAM 3	1.94 ± 0.34
				Adel2037	Racecourse (ii)	1.60	MAM 3	2.37 ± 0.42
				L004/7.10	Racecourse (ii)	7.10	MAM 3	2.55 ± 0.52
4	Topsy Feeder	Upper delta plain (F)	One vibracore, two OSL, RTK GPS	L004/0.75	Topsy Feeder (i)	0.75	MAM 3	1.53 ± 0.76
				L004/5.3	Topsy Feeder (ii)	5.30	MAM 4	1.44 ± 0.36
5	Scroll Site	Lower delta plain (Ft)	One vibracore, one auger, one OSL, RTK GPS	L004/6.17	Scroll Site	6.17	MAM 3	1.37 ± 0.25
6	Site 1	Lower delta plain (Tf)	One vibracore, RTK GPS					

Refer to Figure 4 for locations.

F, fluvial; Ft, fluvial-dominated, tide-influenced; GPS, global positional system; OSL, optically stimulated luminescence; Tf, tide-dominated, fluvial-influenced; W, wave-dominated; W*ft*, wave-dominated, fluvial-influenced and tide-affected.

Table 2. *Summary of element complex sets and element complex assemblages in the Mitchell River regressive element complex assemblage set*

Depocentre	ECA name	ECS name	Process category		Area (km^2)	Age (ka)
			Element (%)	Modified		
Main Mitchell	Aa		Tf*w*		151	*c.* 5.7
	Ab	Ab.1 & 2	Fw*t*	Wf*t*	79	5.7–3
	Ba	Ba.1–5	Tf*w*		207	3–2
	Bb	–	Tw*f*	Wf*t*	44	2–0
	Bc	Bc. 1–3	Tf*w*		171	2–0
	C	C.1 & 2	Tw*f*		111	2–0
	D	D.1 & 2	Wt*f*		26	2–0
Central	CA		Tf*w*		146	*c.* 5.7
	CB		Ft*w*		330.2	5.7–3
	CC		Tf*w*		111.5	5.7–3
	CD		Fw*t*		12.5	5.7–3
	CE		Tw*f*		53.6	5.7–3
	CF		Wt*f*		14.6	5.7–3
Nassau	NA		Tw*f*	Tw	108.3	2–0
	NB		Tw*f*	Wt*f*	46.2	2–0
	NC		Tw		30.6	2–0
	ND		Wt*f*		8.6	2–0

The methodology for assigning process classification uses the percentage component by areal extent of depositional elements that were deposited by primarily fluvial, tide or wave processes (Element (%)). In some units, the classification has been modified to account for significant post-depositional working or overprinting (Modified).
ECA, element complex assemblage; ECS, element complex set; Ft*w*, fluvial-dominated, tide-influenced and wave-affected; Fw*t*, fluvial-dominated, wave-influenced and tide-affected; Tf*w*, tide-dominated, fluvial-influenced and wave-affected; Tw, tide-dominated, wave-influenced; Tw*f*, tide-dominated, wave-influenced and fluvial-affected; Wt*f*, wave-dominated, tide-influenced and fluvial-affected.

well-sorted, ripple-laminated medium sand with minor shell hash (Fig. 8b.5). The basal sediments are overlain by 2 m of moderately sorted, cross-bedded, sand-rich inclined heterolithic stratified deposits. Granules are common and there are infrequent pebbles (Fig. 8b.5). An exposed outcrop shows that inclined heterolithic stratified deposits in an Ft palaeochannel belt dip at angles between 5 and 20° (Fig. 10). The top metre of sediment consists of bioturbated very fine to fine silty clay (Fig. 8b.5). Trace fossils include threadworm-generated *Skolithos* (low preservation potential), *Cylindrichunus* and *Arenicolites*. A single auger through the abandoned channel indicates that the mud plug is *c.* 1 m thick at this location and is underlain by sand (Fig. 9.5).

Tf channels typically occur as the seawards portion of Ft channels that have been 'switched off' via upstream avulsion. As a result, they inherit their morphology from their fluvial-dominated predecessors, but their fill is predominantly marine. The planform morphology of Tf channels varies from straight to highly sinuous and they are commonly associated with dendritic networks of tidal (*T*) channels that spill out onto extensive tidal flats. Site 1 (Fig. 9.6) is an example of a Tf channel

Table 3. *Avulsion frequency and influences*

Mapped avulsion nodes			Minimum frequency (1000 years^{-1})		Enhanced by	Effect on delta morphology
Megafan		16		3	Monsoonal flow regime, indurated sediment	Large scale (>80 km alongshore)
Delta	Central	33	11	16	Monsoonal flow regime, indurated sediment, backwater (base level, tides, seasonal increase in sea-level)	Smaller scale (10–20 km alongshore)
	Main	60	20			
	Nassau	1	1			
Total mapped nodes = 110			Overall frequency = 18/1000 years			

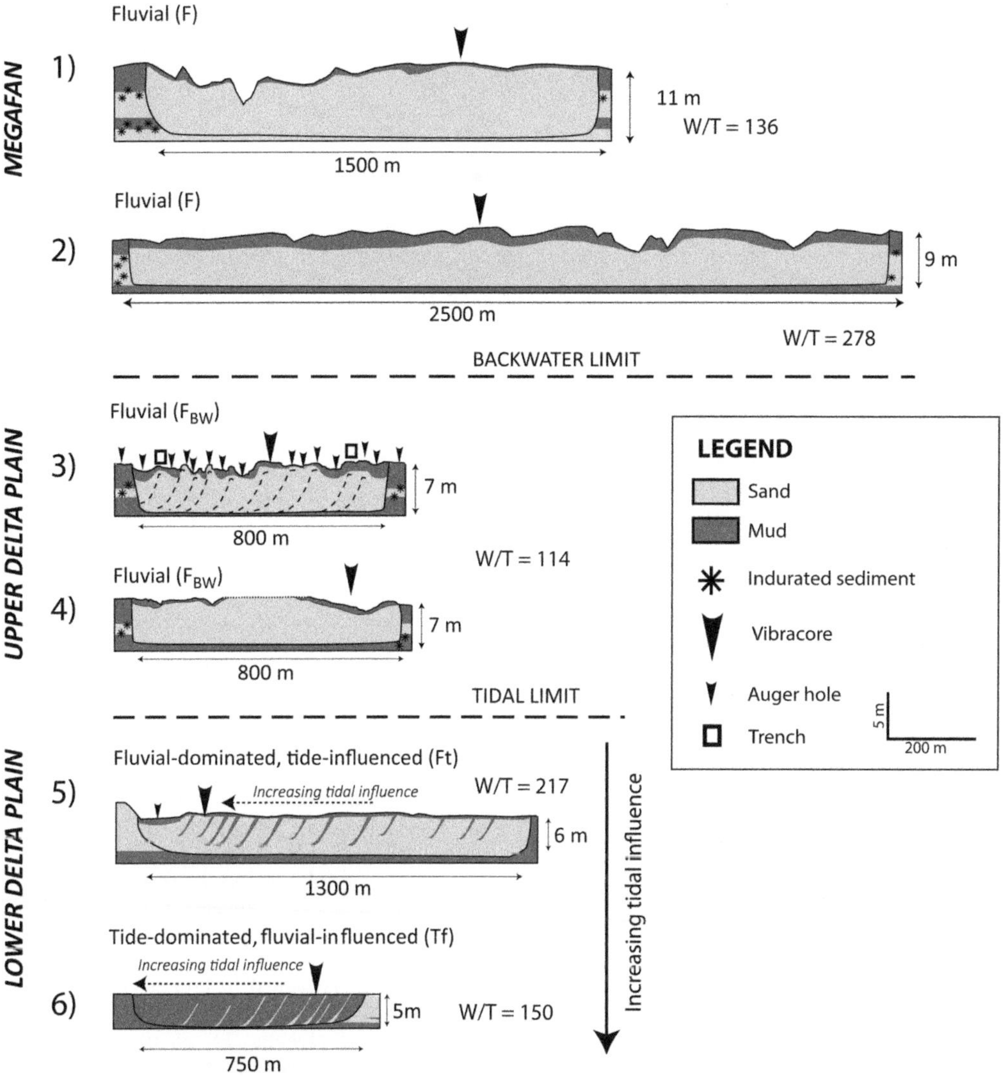

Fig. 9. Cross-sections of palaeochannel belts (flow is towards the reader). Channel belt widths and surface topography were recorded with the RTK global positioning system, whereas channel body depth and fill were obtained from vibracores (summarized in Table 1). The cross-sectional channel facies is interpreted based on the available auger (small arrows) and vibracore (large arrows) data. The presence of subsurface indurated sediment and a sandy layer beneath *c.* 5 m is based on stratigraphic work from the Gilbert River megafan and delta, 150 km south of the Mitchell River (Nanson *et al.* 2005). Numbers 1–6 correspond to the locations indicated in Figure 4. W/T, width to thickness ratio.

and is located on the lower delta plain 4.5 km downstream of Scroll Site (Ft) and 2.5 km inland from the modern coastline (Fig. 4a). The channel belt is 750 m wide and has been largely overprinted by tidal flat (Fig. 9.6). The channel base is *c.* 5 m below the surface and is marked by a 0.2 m channel lag of medium sand, shell hash and rare whole shells (Fig. 8b.6). The width to thickness ratio of this channel belt is 150; however, the width to thickness ratios of Tf channel belts are variable because channels in this zone vary markedly. Their channel fill consists of an overall fining-upwards succession of predominantly blue–grey structureless, probably fluid mud with fine to medium sand containing variable amounts of shell hash and the occasional whole shell (Fig. 8b.6). Sand intervals range in

Fig. 10. An outcrop of a fluvial-dominated, tide-influenced (Ft) palaeochannel on the Mitchell River delta. Lateral accretion forms inclined heterolithic strata of alternating mud, sand and sandy mud. Exposed strata dip at angles of 5–20°. Refer to Figure 4b for location of outcrop.

thickness from 20 cm beds to thin laminations (Fig. 8b.6). The shell content indicates a marine source for the sand (Fig. 8b.6). The overall succession is not heavily bioturbated and most of the trace fossils are *Planolites*, *Skolithos* and uncommon *Arenicolites*. The trace suite is consistent with infaunal responses to a physico-chemically stressed environment, which, in this case, is probably the result of the sedimentation rate and salinity.

A linked megafan delta channel facies model

The stratigraphic results from this dataset are highly correlatable with the channel facies data on the Gilbert River megafan (Nanson *et al.* 1991, 2005) and delta (Jones *et al.* 1993, 2003) (located 150 km south of the Mitchell River; see Fig. 2), which is subject to a similar gradient, climatic regime and a slightly higher tidal range, which lends confidence to the cross-sectional interpretations. From these investigations of channel belt stratigraphy at six field sites, supported by comparable results from similar studies in the GOC, there emerge four discrete channel belt facies:

(1) fluvial (F) channel belts on the megafan;
(2) fluvial (F_{BW}) channel belts in the backwater zone (upper delta plain);
(3) fluvial-dominated, tide-influenced (Ft) channel belts on the lower delta plain; and
(4) tide-dominated, fluvial-influenced (Tf) channel belts on the lower delta plain.

Figure 11 presents a generalized facies model of the downstream trend in channel fill and process. Belts on the megafan are generally formed from low sinuosity channels (Fig. 11a), which are coarse grained and fine upwards. Channel belts on the upper delta plain (F_{BW}) (Fig. 11b) have a scrolled, lateral-accreted morphology and a significant reduction in their cross-sectional area. Indurated sediment outcrops sporadically across the river segments on the Mitchell River megafan and have been recorded on the surface and at depth on the Gilbert River megafan (Nanson *et al.* 1991, 2005). The indurated sediment presented in Figure 11 is inferred from these observations. Fluvial-dominated, tide-influenced channel belts (Ft) on the lower delta plain (Fig. 11c) are characterized by sand-dominated inclined heterolithic strata. Tide-dominated, fluvial-influenced (Tf) channel belts on the lower delta plain (Fig. 11d) are characterized by mud-dominated inclined heterolithic strata.

Discussion

Detailed mapping coupled with coring and OSL dates allowed the palaeogeographical reconstruction of the Mitchell River megafan and delta and an analysis of how the evolution is related to both autogenic and allogenic controls, including avulsion, waves, tides, climate and base level. This, in turn, provides new insights and understanding into the distribution of channel facies and architecture. The following section discusses: (1) the linked evolution of the megafan and delta in response to

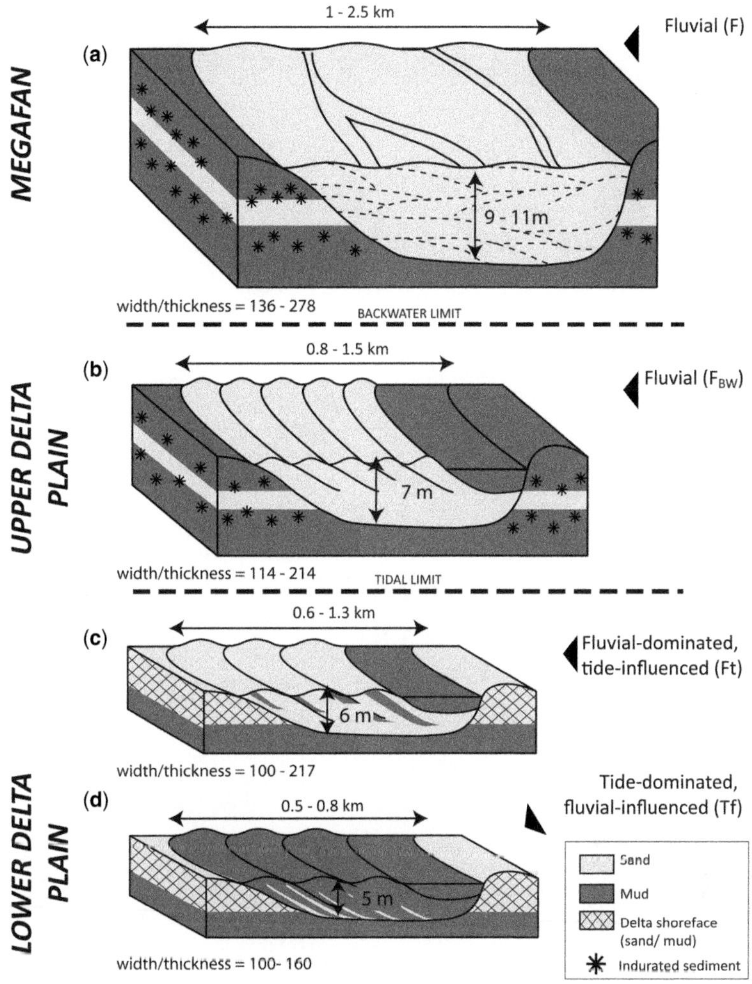

Fig. 11. Schematic diagram showing downstream changes in architecture and sedimentary fill of channel belts on the Mitchell River megafan and delta. (**a**) Fluvial channels on the megafan have deep coarse-grained fill and single thread morphology. (**b**) Fluvial channels on the upper delta plain have a scrolled morphology and a significant reduction in their cross-sectional geometry. (**c**) On the lower delta plain, as the channels become tidally influenced the fill becomes finer grained and more heterolithic (sand and mud). (**d**) Channels nearer the coast that are abandoned by the main fluvial feeder channels are filled with muddier marine-derived material. On the megafan and upper delta plain, the sedimentary facies of the Quaternary alluvium adjacent to the channel belt fill is interpreted based on the stratigraphic results from the Gilbert River megafan located 150 km south of the Mitchell River (Nanson *et al.* 1991, 2005).

autogenic and allogenic controls; (2) avulsion and the factors controlling the distribution of avulsion nodes; and (3) how both the evolutionary history and avulsion pattern control the morphology and sedimentology of channel fills.

Linked megafan and delta evolution

The Mitchell River delta RECAS is composed of three depocentres formed by major avulsions on the megafan (Fig. 5b). The depocentres are composed of children ECA, which vary in process dominance. In fact, there are ECA dominated by each of the three major processes: waves, tides and fluvial energy (Fig. 5b). The change in process category is accompanied by strikingly different morphologies and sedimentary facies. The evolution of the Mitchell River megafan and delta can be summarized in four major phases (Fig. 12) that can be related to major avulsions on the megfan and the ratio of

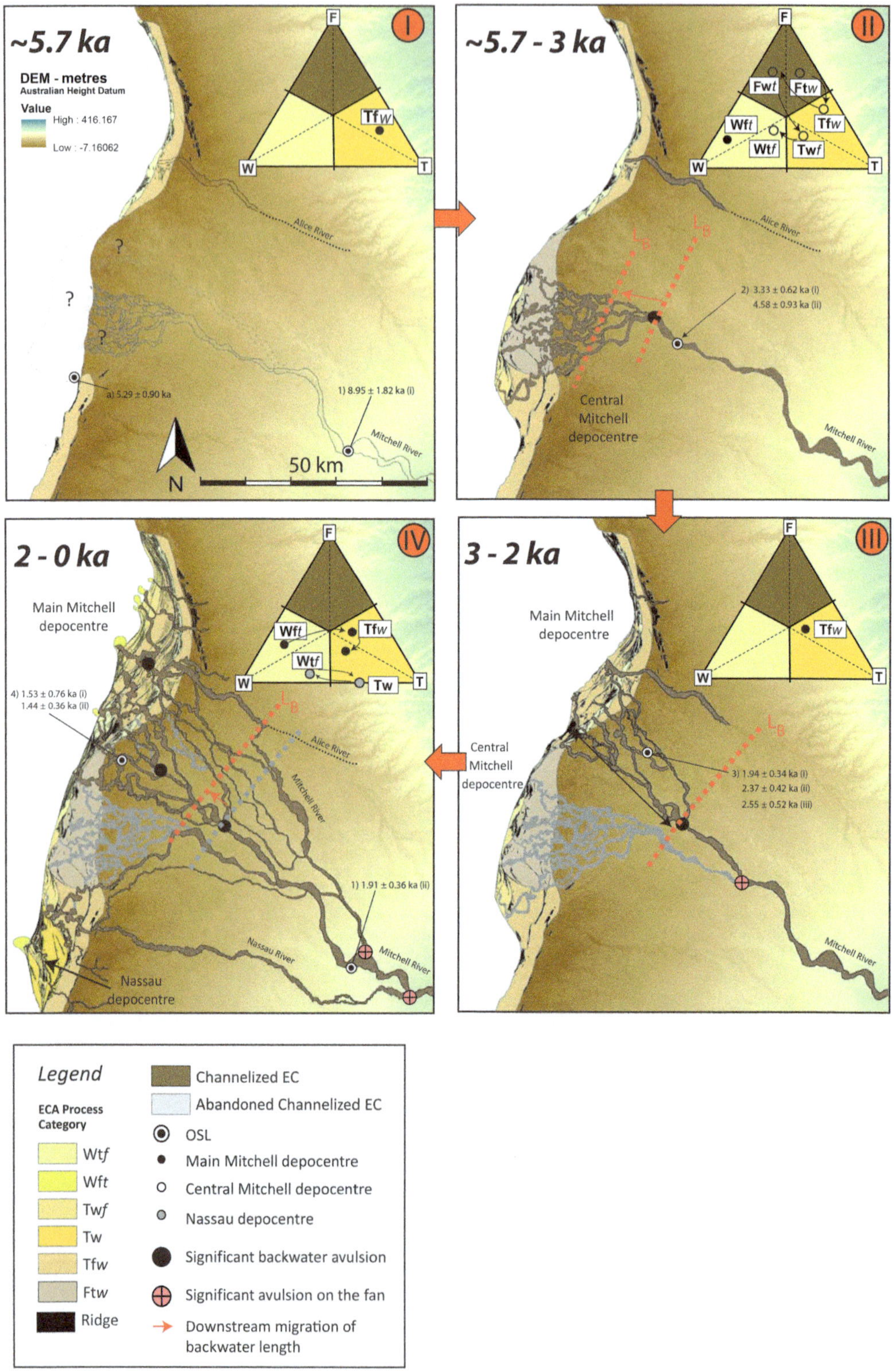
~5.7 ka
I
DEM - metres
Australian Height Datum
Value
High : 416.167
Low : -7.16062
F
W
T
Tfw
Alice River
?
?
?
a) 5.29 ± 0.90 ka
1) 8.95 ± 1.82 ka (i)
Mitchell River
50 km
N
~5.7 - 3 ka
II
Fwt
Ftw
Wft
Tfw
Wtf
Twf
Alice River
L_B
L_B
2) 3.33 ± 0.62 ka (i)
4.58 ± 0.93 ka (ii)
Central
Mitchell
depocentre
Mitchell River
2 - 0 ka
IV
Main Mitchell
depocentre
Wft
Tfw
Wtf
Tw
4) 1.53 ± 0.76 ka (i)
1.44 ± 0.36 ka (ii)
L_B
Alice River
Mitchell River
1) 1.91 ± 0.36 ka (ii)
Nassau River
Mitchell River
Nassau
depocentre
3 - 2 ka
III
Main Mitchell
depocentre
Tfw
Central
Mitchell
depocentre
L_B
3) 1.94 ± 0.34 ka (i)
2.37 ± 0.42 ka (ii)
2.55 ± 0.52 ka (iii)
Mitchell River
Legend
ECA Process
Category
Wtf
Wft
Twf
Tw
Tfw
Ftw
Ridge
Channelized EC
Abandoned Channelized EC
OSL
Main Mitchell depocentre
Central Mitchell depocentre
Nassau depocentre
Significant backwater avulsion
Significant avulsion on the fan
Downstream migration of
backwater length

accommodation to sediment supply (the A/S ratio *sensu* Muto & Steel 1997) (Fig. 13). The regional A/S ratio (Fig. 13, modified from Nanson *et al.* 2013*b*) is qualitative and based on previously published climate and sea-level records. The most conservative sea-level reconstruction for this area is probably that of Chappell & Thom (1986), who suggested a smooth rate of fall from *c.* 6 ka. However, oscillations in the sea-level curve or stalled rates of fall have been inferred from other studies (Murray-Wallace & Woodroffe 2014). Landforms on the Mitchell River delta appear to have responded significantly to changes in sea-level that are less than the resolution of the current dataset. As discussed in the following sections, the results presented in this study advocate a standstill from *c.* 6 to 3 ka followed by a forced regression in the Mitchell River region, with a potential increase in the rate of fall at *c.* 2 ka. The resolution of sea-level curves reconstructed via geomorphological studies is debated and further investigation is necessary to elucidate the rate and timing of sea-level fall in the GOC. Localized variations in the A/S ratio can also result from avulsion processes and localized bathymetric highs as a result of mouth bar deposition. These variations are inferred from the palaeogeographical reconstruction and discussed here.

Phase I of delta development (Fig. 12 I) commenced at the Holocene maximum marine transgression *c.* 5.7 ka (Nanson *et al.* 2013*b*), when the sea-level was 1.5 m higher than the present day (Murray-Wallace & Woodroffe 2014; Fig 13). In the northern and southern portion of the delta, the A/S ratio was relatively high and the relative wave to fluvial influence was high because the fluvial input was sourced only from the Alice River in the north and local catchment drainage in the south (Fig. 12 I). In the Main Mitchell River depocentre, a wave-dominated, fluvial-influenced and tide-affected (Wf*t*) unit (ECA Main Mitchell Ab) acted as a barrier to a tide-dominated, fluvial-influenced and wave-affected (Tf*w*) unit (ECA Main Mitchell Aa; Fig. 5b). On the southernmost portion of the megafan, the ECA Central A extends as a tide-dominated, fluvial-influenced and wave-affected (Tf*w*) unit (Fig. 12 I). During the Holocene transgression, the Mitchell River megafan was operating in the Central Mitchell depocentre region (Fig. 12 I). Evidence for this is seen in nearshore sediment surveys (Rhodes 1980) that indicate high concentrations of sand (80%) in the modern subtidal zone of the Central Mitchell depocentre. The primary process of sedimentation in the Central Mitchell depocentre was markedly different. The Holocene transgression effectively drowned the downstream portion of the M_4 (Grimes & Doutch 1978) megafan, which probably caused a localized bathymetric high. The sediment from the drowned M_4 megafan was probably reworked into M_5 deltaic deposits. The line of maximum marine transgression is evidenced from sparse chenier ridges, which occur 10 km inland from the modern coastline (Fig. 12 I).

Fig. 12. The linked megafan and delta evolution is summarized in four stages. The ternary process diagrams (after Ainsworth *et al.* 2011), located in the upper right and corner of each panel, illustrate the change in process category through time in the Main Mitchell (black dot), Central Mitchell (black circle) and Nassau (grey dot) depocentres. (I) The maximum Holocene marine transgression occurred *c.* 5.7 ka when the sea-level was +1.5 m higher than the present day (Nanson *et al.* 2013*b*) and effective precipitation increased in the catchment area (Shulmeister 1999). Sedimentation is characterized by Wf*t* barrier and associated Tf*w* tidal flat deposition (II). The megafan was operating in the Central Mitchell depocentre. The delta built a series of element complex assemblages in the central portion of the regressive element complex assemblage set. Avulsions on the delta plain cause localized shifts in depositional loci, controlled by the backwater length, which migrated downstream in response to delta progradation. Much of this section is fluvial-dominated, corresponding to the peak in effective precipitation (5–3.7 ka). Aggradation of fluvial sediment suggests a sea-level highstand. (III) A major avulsion on the megafan shifted the sediment supply to the north and the Central Mitchell depocentre was abandoned. A sharp decline in effective precipitation occurred after 3.7 ka, causing a shift from fluvial to tide dominance. The Main Mitchell depocentre built a series of Tf*w* element complex sets in response to a fall in sea-level. Backwater-mediated avulsions on the delta plain caused localized shifts in depositional loci. (IV) From 2 to 0 ka a second pulse of sea-level fall caused the formation of the Nassau depocentre as the Main Mitchell depocentre continued to prograde. During this time, a partial avulsion of the Mitchell River near the apex of the megafan caused an influx of sediment supply to the Nassau depocentre. The subsequent increase in wave to fluvial influence on the Main Mitchell delta may be partially related to the partitioning of flow into the two depocentres. Backwater-mediated avulsions on the lower delta plain caused localized shifts in deposition as the backwater length migrated downstream in response to delta progradation and sea-level fall. Relative changes in effective precipitation, sea-level and process dominance for the Holocene evolution of the Mitchell River regressive element complex assemblage set are shown schematically in Figure 13. EC, element complex; ECA, element complex assemblage; Ft*w*, fluvial-dominated, tide-influenced and wave-affected; OSL, optically stimulated luminescence; Tf*w*, tide-dominated, fluvial-influenced and wave-affected; Tw, tide-dominated, wave-influenced; Tw*f*, tide-dominated, wave-influenced and fluvial-affected; Wf*t*, wave-dominated, fluvial-influenced and tide-affected; Wt*f*, wave-dominated, tide-influenced and fluvial-affected.

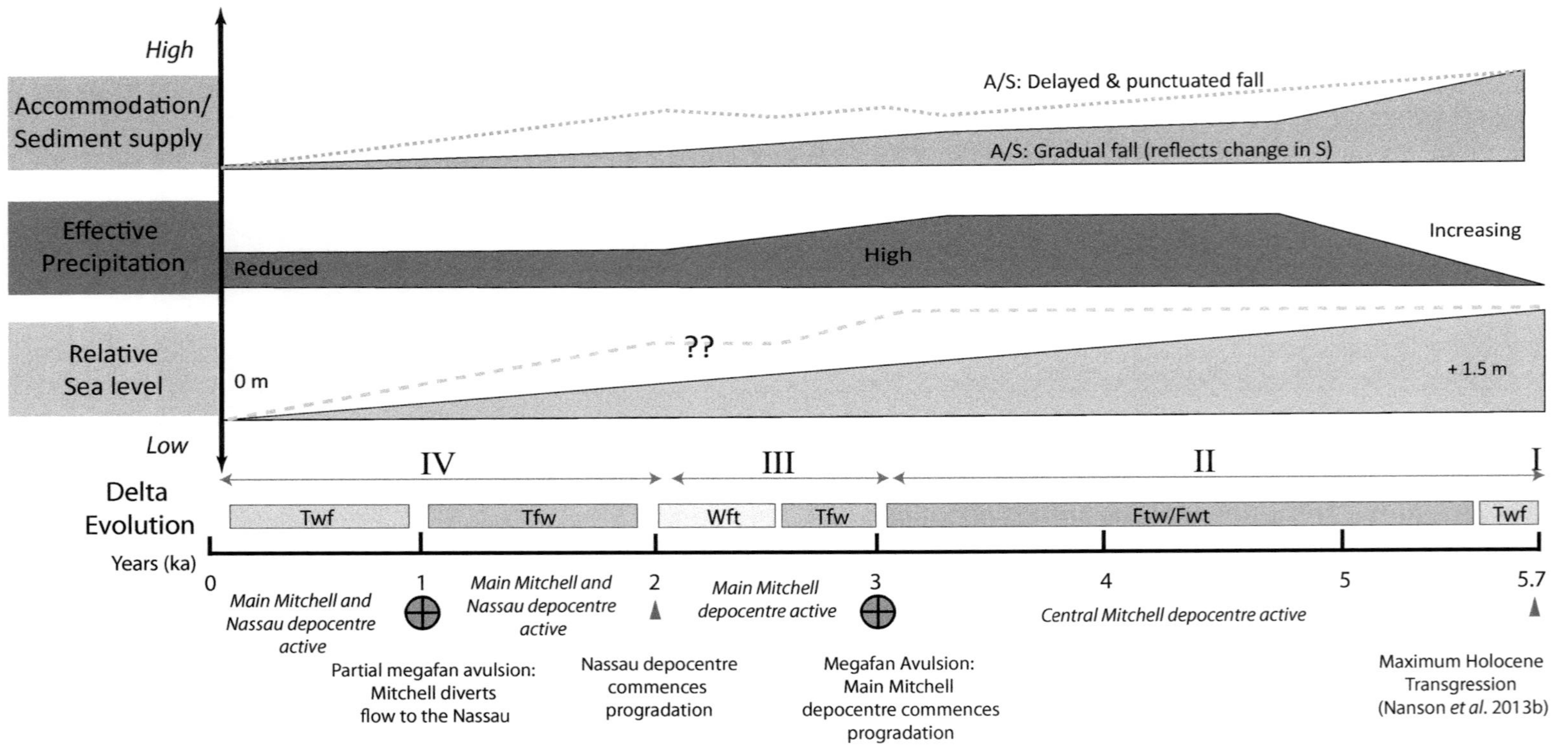

Fig. 13. Schematic summary of the change in process dominance of the Mitchell River delta through the last *c.* 6 ka in response to major avulsions on the megafan, changes in effective precipitation (Shulmeister 1999), sea-level and accommodation to sediment supply (A/S) ratio. The qualitative regional A/S ratio presented here is based on published effective precipitation and sea-level data. Although it is generally accepted that this region has experienced a sea-level fall of *c.* 1.5 m since the maximum Holocene transgression, the details surrounding the sea-level history are unresolved (Murray-Wallace & Woodroffe 2014). The solid polygon represents the most conservative sea-level reconstruction (Chappell & Thom 1986). The dashed lines represent the approximate relative sea-level curve at the Mitchell River delta based on delta morphology and the corresponding A/S ratio. Localized variations in A/S are controlled by avulsions and subtle variations in bathymetry and are not represented in this diagram. Phases I–IV correspond to major phases of linked megafan–delta evolution and are summarized in Figure 12. Modified from Nanson *et al.* (2013*b*). Ft*w*, fluvial-dominated, tide-influenced and wave-affected; OSL, optically stimulated luminescence; Tf*w*, tide-dominated, fluvial-influenced and wave-affected; Tw*f*, tide-dominated, wave-influenced and fluvial-affected; Wf*t*, wave-dominated, fluvial-influenced and tide-affected.

Phase II of megafan delta development (Fig. 12 II) was concentrated in the central portion of the megafan and the associated deltaic units prograded by 10 km to build the 522 km^2 Central Mitchell depocentre. Using the youngest OSL date from Rutland Plains (i) at 3.33 $\pm$ 0.62 ka, the age of this depocentre is estimated as 3–5.7 ka, a period that coincides with the maximum Holocene EP (from 3.7 to *c.* 5 ka; Shulmeister 1999), which was immediately followed by a sharp decline in EP (Fig. 12 II & 13). The A/S ratio in the Central Mitchell depocentre during Phase II was probably lower than the regional A/S ratio (Fig. 13) as a result of the inferred bathymetric high caused by the drowned M_4 fan deposits. In the Central Mitchell depocentre, the ECA Central Mitchell B progrades and is classified as fluvial-dominated, tide-influenced and wave-affected (Ft*w*). This ECA represents the first major pulse of delta lobe formation.

Four subsequent ECA units (ECA CC–F; Fig. 5b) prograded to fill the Central Mitchell depocentre and the process dominance varied throughout its development (Fig. 5b). Beach and chenier ridges increase in density towards the coast within the remaining ECA units (e.g. Central Mitchell E Tf*w*, CF Wt*f*) that comprise the Central Mitchell deopcentre during Phase II (Fig. 5b). Increased A/S ratios caused beach ridges to accrete vertically at the shoreline (cf. Taylor & Stone 1996) and a reduced sediment supply may have been caused by decreased EP (Shulmeister 1999) or partial channel avulsion causing a shift in the major depocentre. The aggradation of fluvial sediment on the Central Mitchell depocentre and the lobate morphology of the ECA units supports the interpretation of a standstill until 3 ka, followed by the downstepping of the Main Mitchell depocentre (Figs 13 & 14).

Phase III of delta progradation was characterized by the enhanced progradation of the Main Mitchell depocentre following a major avulsion on the fan that diverted the flow of the Mitchell River from the Central Mitchell depocentre, where it was captured and joined by the Alice River (Fig. 12 III). The Central Mitchell depocentre was subsequently abandoned and locally eroded. Dated palaeoshorelines (Fig. 5a) indicate that from *c.* 3 to 2 ka the Main Mitchell depocentre prograded as a series of five tide-dominated, fluvial-influenced and wave-affected (Tf*w*) ECS (comprising the ECA Main Mitchell Ba) with a maximum of 6 km of progradation downdip. ECA Main Mitchell Ba shows evidence for delta asymmetry (*sensu* Bhattacharya & Giosan 2003; see also Nanson *et al.* 2013*b*) resulting from longshore drift from south to north (Fig. 5a).

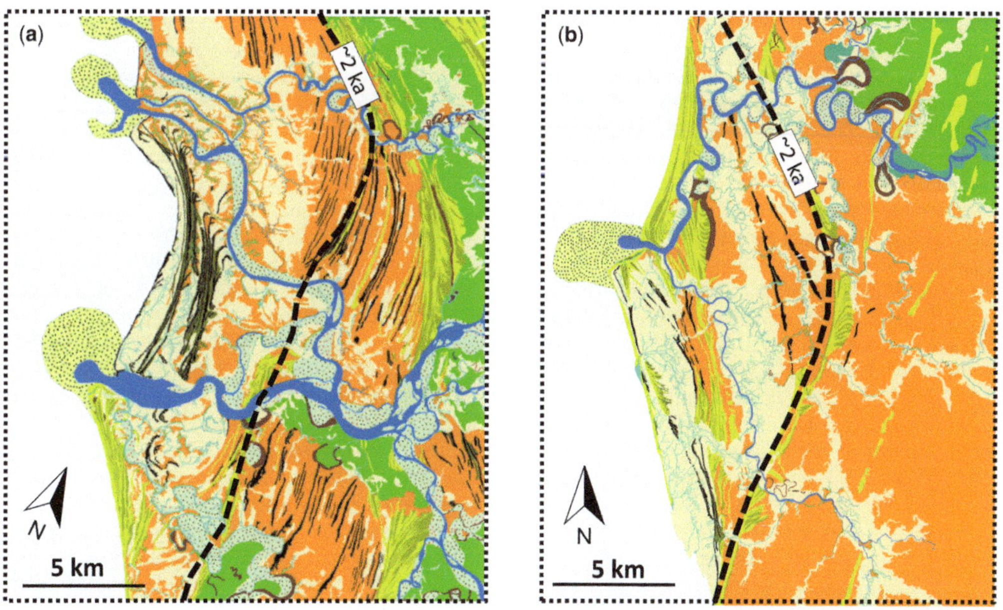

Fig. 14. A detailed comparison of the (**a**) Main Mitchell depocentre and (**b**) Nassau depocentre (see Fig. 4a for location and element key). The presence of tidal sabkha indicates that the Nassau depocentre has a topographic height similar to that of the Main Mitchell depocentre after *c.* 2 ka. In the absence of stratigraphic dating from the region, the low topography has been used to infer a relatively young age for the Nassau depocentre of some time after 2 ka.

The sand accumulated on the southern limbs of these ECS was probably sourced from the eroded portion of the Central Mitchell depocentre.

Phase IV of delta building is ongoing and commenced shortly after 2 ka; it is split between deposition on the northern Main Mitchell depocentre and the Nassau depocentre (Fig. 12 IV). Note that the interpreted timing of Phase IV has shifted slightly from that of Nanson *et al.* (2013*b*) as a result of further stratigraphic dating. The ECA units deposited after 2 ka are characterized by extensive sabkha deposits, which formed as a result of a subtle downstepping during the forced regression (Figs 13 & 14). This could potentially be caused by a change in rate of sea-level fall at about *c.* 2 ka (see also Woodroffe 2009). In the Main Mitchell depocentre, a backwater-mediated avulsion (Fig. 12 IV) initially shifted sediment deposition to the southern limb of the Main Mitchell depocentre. A wave-dominated, fluvial-influenced, tide-affected (Wf*t*) ECA ensued (Main Mitchell Bb; Fig. 5b), which shows some delta asymmetry resulting from a south to north directed longshore current. Another subsequent backwater-mediated avulsion returned the depositional loci to the central portion of the Main Mitchell depocentre and three tide-dominated, fluvial-influenced and wave-affected (Tf*w*) ECS then prograded (comprising ECA Main Mitchell Bc). These ECS units are comparable with those comprising ECA Main Mitchell Ba (Fig. 5b). The youngest ECA on the Main Mitchell depocentre (ECA Main Mitchell C) that is fed by the Mitchell River is tide-dominated, wave-influenced and fluvial-affected (Tw*f*) and was deposited around 0.5 ka (Fig. 5; Nanson *et al.* 2013*b*). The relative increase in wave relative to fluvial influence may be a result of a partial avulsion of the Mitchell River into the Nassau River, or a stalling of relative sea-level fall. Concurrent with the development of ECA Main Mitchell C, a wave-dominated, fluvial-influenced, tide-affected (Wf*t*) ECA (Main Mitchell D) was deposited on the southern flank of the Main Mitchell depocentre (Figs 5b & 12 IV).

The more southerly Nassau depocentre is located on the southern limb of the Mitchell River RECAS and covers an area of *c.* 200 km^2 (Fig. 12 IV). The fluvial activity of the Nassau depocentre is controlled by a partial avulsion at the megafan apex, illustrating the strong effect of megafan avulsion on coastal morphology and sediment supply. In the absence of stratigraphic dating from the Nassau depocentre, we suggest that the low topography can be used to infer a relatively young age for the Nassau depocentre of sometime after 2 ka (Figs 12–14). The onset of progradation is preserved as a tide-dominated, wave-influenced (Tw) ECA (NA; Fig. 5b). ECA NA was deposited in the absence of a direct fluvial sediment supply. An avulsion at the apex of the Holocene megafan (Fig. 12 IV) caused the partial diversion of the Mitchell River into the Nassau depocentre and the initiation of the second ECA (Nassau B) in that area, which is wave-dominated, tide-influenced and fluvial-affected (Wt*f*). This ECA has a cuspate morphology and displays delta asymmetry; the direction of longshore drift was from north to south (Fig. 5a). Following the progradation of ECA Nassau B, the sediment supply was switched off upstream at the avulsion node and a third, tide-dominated, wave-influenced (Tw) ECA (Nassau C) was deposited (Figs 5b & 12 IV). The youngest, and currently active, ECA (Nassau D) is wave-dominated, tide-influenced and fluvial-affected (Wt*f*). This ECA also displays delta asymmetry, with longshore drift from north to south (Fig. 5).

Controls on channel avulsion and depositional loci

Avulsions on the megafan and delta plain have an overriding control on the channel belt distribution and the sediment depositional loci. Two types of avulsion have been identified on the Mitchell River megafan and delta: (1) avulsions on the delta plain; and (2) avulsions on the megafan (Fig. 7). Channel distribution on the delta plain is also controlled to some extent by bifurcation around mouth bars. However, this process plays a subordinate part in determining the delta morphology because waves and tides rework mouth bar sediments into shoreface deposits (cf. Jerolmack & Swenson 2007).

The rate and trajectory of sea-level change influences the frequency of delta channel avulsion (e.g. Stouthamer & Berendsen 2000; Jerolmack 2009). Rapid transgression generally causes more rapid aggradation and increased avulsion frequency than slower transgression (Jerolmack 2009). Conversely, a relative sea-level fall has commonly been inferred to cause fluvial incision and more restricted channel avulsions (e.g. Slingerland & Smith 2004). If a fall in relative sea-level does not cause sufficient channel incision, however, modelling has shown that avulsions can occur as a result of the relative sea-level fall super-elevating the fluvial system (e.g. the Goose River delta; Nijhuis *et al.* 2015). Indeed, the models of Nijhuis *et al.* (2015) have indicated that the avulsion rate may even increase with increased rates of sea-level fall because channels become super-elevated more rapidly.

In the GOC, low and uniform onshore and offshore gradients (e.g. Mitchell River slope = 0.0003; Fig. 3) have minimized fluvial incision during repeated Quaternary regressions (Jones *et al.* 2003). Based on our new chronology and reconstruction of the evolution of the delta (summarized in Fig. 12), we can approximate changes in the rate of channel avulsion and link these to both allo- and autogenic controls.

Avulsions on the Holocene Mitchell River delta are relatively frequent and, based on the total number of mapped avulsion nodes in this region (Fig. 7, Table 3), occur at an estimated rate of 16/1000 years. Considering that channel reoccupation is common (see also Nanson *et al.* 1991; Jones *et al.* 2003), this should be considered a minimum value. Delta avulsions are associated with smaller scale shifts in depositional loci (10–20 km alongshore) than those on the megafan and they are typically the cause of an ECS or ECA boundary (Fig. 12). Hence a more detailed understanding of the frequency of channel avulsion during delta development can provide further insights into delta morphology and facies distribution.

The Central Mitchell depocentre was active between 5.7 and 3 ka during a period of increased EP and a probable sea-level highstand (Fig. 12). Thirty-three avulsion nodes are associated with this earliest depocentre, suggesting a minimum avulsion rate of 11/1000 years (Fig. 7; Table 3). The Main Mitchell depocentre and distributaries subsequently developed after *c.* 3 ka during a period of reduced EP and a relative sea-level fall (Nanson *et al.* 2013*b*). Sixty avulsion nodes are associated with the Main Mitchell depocentre, suggesting an increased avulsion rate of 20/1000 years. Interestingly, such increased avulsion frequency has occurred in spite of decreased EP following *c.* 3 ka and may be attributed to the sea-level fall causing the super-elevation of channels.

Delta avulsions are usually backwater-mediated (Chatanantavet *et al.* 2012) as the base level causes spatial deceleration of the flow (Nittrouer *et al.* 2012), a reduction in stream competence, an increased aggradation rate and, consequently, an increased avulsion frequency relative to the megafan (see Ashworth *et al.* 2004; Jerolmack & Mohrig 2007). In the Mitchell River, the backwater effect is also enhanced by the annual increase in sea-level during the summer monsoon and the corresponding landwards shift of tidal effects (Ali 1995; Jones *et al.* 2003; Nanson *et al.* 2013*b*). Avulsion nodes on the Mitchell River are concentrated on the upper delta plain and those on the lower delta plain occur adjacent to the effective tidal limit (Fig. 7). Avulsion nodes do not typically occur proximal to the shoreline because drawdown during high flows near the mouth of the river (Lamb *et al.* 2012), enhanced by the ebb tide, maintains sediment transport through this reach. Nanson *et al.* (2013*a*) recorded a maximum channel depth of 13 m near the mouth of the modern Mitchell River during the 2012 dry season and showed that the channel depth decreases upstream towards the effective tidal limit.

Seventeen kilometres of delta progradation since 5.7 ka has resulted in a similar length of migration of the backwater length over this period; the most landward avulsion nodes on the delta were probably formed by the oldest avulsions (Fig. 7). However, channel reoccupation is also likely to have occurred such that the landwards–seawards avulsion trend is sometimes more complicated. Four nodal avulsion points are identified on the upper delta plain (Fig. 7) and these are most probably backwater-mediated (Fig. 7). They may be related to either periods of relatively limited net progradation (Ganti *et al.* 2014) or the restricting nature of indurated sediment that has formed sporadically across the upper delta plain and megafan.

The second type of avulsion occurred exclusively on the megafan and these avulsions are significantly less frequent than the backwater-mediated avulsions (>3/1000 years). Avulsions on the megafan have had a dramatic effect on the coastal morphology as they have shifted the depositional loci great distances (>80 km) along the shoreline and are thus associated with depocentre development. Based on observations from the Huanghe River megafan in China, Ganti *et al.* (2014) predicted nodal avulsions initiated from the topographically pinned apex of fan deltas. The currently active Holocene Mitchell (M_5) megafan is entrenched into a Pleistocene megafan and the Pleistocene megafan apex is located upstream of the current Holocene apex (Brooks *et al.* 2009; Fig. 4a). Unlike the Huanghe River, avulsions on the Mitchell megafan appear to occur randomly (*sensu* Slingerland & Smith 2004) on the megafan surface. A similar pattern is observed on the São Lourenço fluvial megafan in Brazil (Assine *et al.* 2014). These avulsions are likely to be an autogenic response to megafan aggradation (see Schumm *et al.* 1987) or to local breaks in slope related to shifting intersection points (Hooke 1967). The intersection point refers to the position where the channel on a fan ceases to be incised, i.e. intersecting the slope of the fan (Hooke 1967). The area downstream of the intersection point is dominated by fluvial deposition, aggradation and the formation of fan lobes (Hooke 1967). The position of avulsion nodes is also related to the indurated Quaternary alluvium that is often associated with rapids and knick-points (Nanson *et al.* 2005). Avulsion can be at least partially attributed to these horizons, which restricts adjustments in channel geometry (Nanson *et al.* 2005).

The complicated network of active, partially abandoned and abandoned channels resulting from the megafan and delta avulsion processes described here combines with purely basinal processes (waves, tides and longshore drift) to result in more complicated delta morphologies than those of non-megafan-fed systems (e.g. the Mississippi River Delta). Although this study does not attempt to provide diagnostic criteria for identifying linked megafan delta systems in ancient settings, it may explain how some ancient delta systems are segmented by frequent lobe switching, particularly when wave or tidal influences might be expected to restrict distributary channel formation via bifurcation around mouth bars. Fluvial megafans are often difficult to resolve in ancient settings (discussed in Miall 2014) and fluvial megafans linked to delta systems are under-represented in the literature.

Controls on channel facies and architecture and implications for reservoir modelling

The new channel facies model (Fig. 11) was applied to the mapped distribution of channels to produce a facies distribution map (Fig. 15). Interpolated stratigraphic sections through the facies zones identified across the megafan and delta (Fig. 16) illustrate the subsurface channel implications of these results. A major reduction of, and increased variability in, cross-sectional area is observed downstream of the backwater limit (Figs 11 & 16). This is probably a result of the partitioning of water and sediment

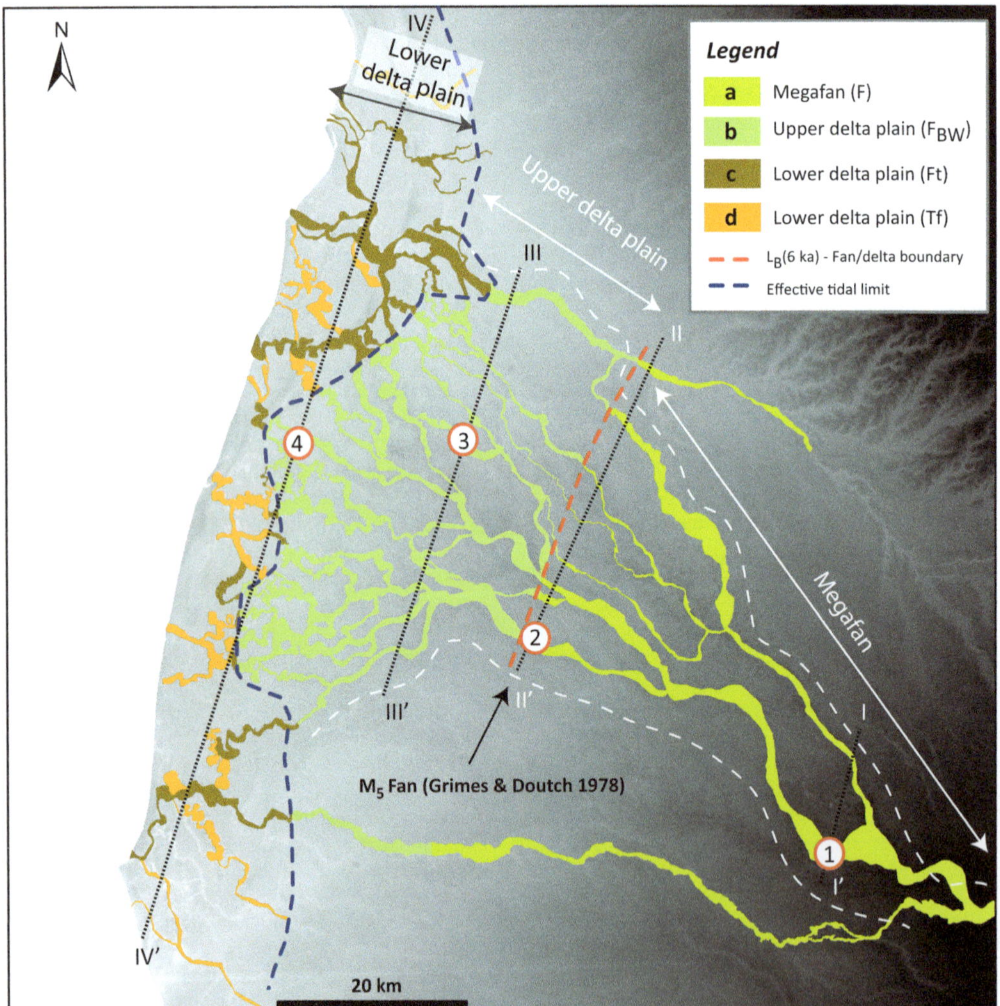

Fig. 15. The spatial distribution of fluvial channels on the Mitchell River megafan (*F*), fluvial channels on the upper delta plain (F_{BW}), fluvial-dominated, tide-influenced (Ft) channels on the lower delta plain and tide-dominated, fluvial-influenced (Tf) channels on the lower delta plain. See Figure 16 for transects.

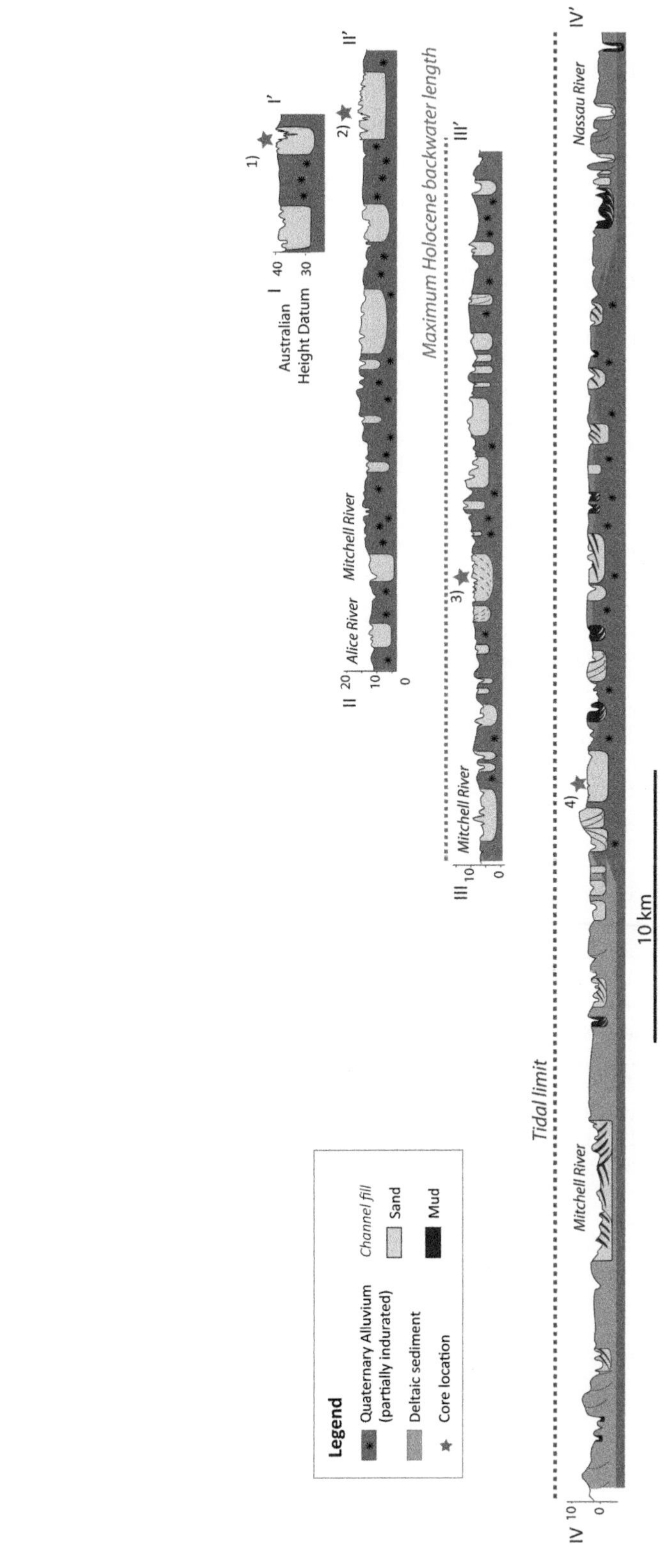

Fig. 16. Interpretive transects through the megafan and delta showing downstream changes in geometry and fill. Refer to Figure 15 for location of transects.

through distributary channels at the upstream limit of the backwater zone. At the inter-channel belt scale, there is an increased potential for the compartmentalization of channel belt deposits (Fig. 16), but at the intra-channel belt scale, heterogeneities that might form baffles and barriers to flow are not extensive (Fig. 11). Both F and F_{BW} channel belts are wide, medium thickness, broad sheets (*sensu* Gibling 2006), which are more typical of meandering rivers or braided low sinuosity rivers than channels on deltas or megafans (Gibling 2006).

On the lower delta plain, the channel geometry is the product of a complicated interplay between the duration of channel belt occupation (and hence the duration of lateral migration) and the relative influence of tides and fluvial flows. Tidal influence, and hence the supply of marine-derived fine material, dissipates away from the shoreline and the distance to which it extends varies seasonally (Nanson *et al.* 2013*b*). The relative tide to fluvial and wave influence is also intimately related to the avulsion history of the megafan and delta. Avulsions can 'switch off', reduce or increase the fluvial supply to channels on the delta, thereby changing the ratio of fluvial to tidal energies along the entire length of the channel.

Tides have a dramatic influence on channel fill style by introducing significant heterogeneity, resulting in the development of muddy inclined heterolithic stratification (Allen 1963; Choi 2011; Johnson & Dashtgard 2014; La Croix & Dashtgard 2014). Inclined heterolithic stratification occurs in channelized and non-channelized environments, but is particularly common in channels with mixed tide and fluvial influences (Thomas *et al.* 1987). In mixed-influence channels, inclined heterolithic stratification is typically attributed to seasonal and fortnightly changes in the turbidity maxima (Hovikoski *et al.* 2008) or flow velocity and suspended sediment concentration (Sisulak & Dashtgard 2012). The sand to mud ratio is related to sediment supply, the degree of tidal influence (e.g. Thomas *et al.* 1987; Dalrymple & Choi 2007) and the degree of flow dampening by tides (Nittrouer *et al.* 2012; La Croix & Dashtgard 2015). On the Mitchell River delta the relative fluvial to tidal influence has a first-order control on the sand to mud ratio, which, in turn, is intimately related to the evolution of the delta (see also Jones *et al.* 1993, 2003). Inclined heterolithic stratification in Ft channel belts has a distinct marine signature (ichnology, shell material and the presence of fluid mud), despite the abundance of clean sand. Sand-rich inclined heterolithic strata are observed as close as 5 km inland from the coast as a result of monsoonal floods, which give the rivers an overwhelming fluvial dominance in the wet season. Where Ft channels are abandoned via an avulsion upstream, they fill with predominantly marine sediment (Tf) at their downstream extent. Tf channels evolve to become muddier as the tidal influence increases. There is a large amount of lateral variability between channel belt fills on the delta plain; along-strike, a muddy channel belt filled with inclined heterolithic strata may occur directly adjacent to a sandy inclined heterolithic strata channel belt fill (Fig. 16 IV).

Conclusions

The Mitchell River and its associated distributaries has constructed a megafan that feeds a mixed-influence delta in a low accommodation basin setting. This analogue study highlights the variability in architecture and fill that can occur within a single deltaic system. The Mitchell River delta has undergone changes in process dominance from overall fluvial dominance to predominantly tide-dominated, in association with avulsion and relatively minor variations in climate and shoreline trajectory. The location of sedimentation on the Mitchell River megafan and delta was driven primarily by avulsion. Two types of avulsion have been described: (1) frequent (>16/1000 years) backwater-mediated avulsions on the delta plain, which have resulted in relatively small-scale shifts in depositional loci of the order of 10–20 km alongshore – these are clustered at the backwater limit and define the boundary between the megafan and delta; and (2) less frequent (>3/1000 years) avulsions on the megafan, which have resulted in shifts in depositional loci in excess of 80 km alongshore and which are also associated with major changes in process dominance. The presence of the Mitchell River megafan has resulted in a complex delta morphology characterized by multiple cycles of lobe building and avulsion. Wave- and tide-dominated features are regularly dissected by channelized features. Avulsion, backwater dynamics and tides have a crucial role in the resulting megafan and delta channel facies.

We have presented a new channel facies model that links changes in channel morphology and fill from near the apex of the megafan to the deltaic shoreline in response to backwater and tidal influences on channel processes:

(1) Fluvial (F) channel belts on the megafan have deep (*c.* 9 m), coarse-grained fill and low sinuosity.

(2) Fluvial (F_{BW}) channel belts on the upper delta plain have a scrolled morphology and a significant reduction in their depth (7 m). On the lower delta plain, channel belt fill is intimately related to the avulsion history of the delta.

(3) Fluvial-dominated, tidally influenced (Ft) channel belts, which can occur as close as

5 km from the coastline, are scrolled and are characterized by sand-rich inclined heterolithic strata.

(4) Ft channel belts that are abandoned by the main fluvial feeder channels via avulsion become progressively more tide-dominated and fluvial-influenced (Tf) and are filled with mud-rich marine-derived inclined heterolithic strata. Along-strike, a mud-rich channel belt filled with inclined heterolithic strata may be directly adjacent to a sand-rich inclined heterolithic strata channel belt fill depending on the avulsion history of the megafan and delta.

We thank the sponsors of Phases I and II of the WAVE Consortium (Apache, Bapetco, BG, BHPBP, BP, Chevron, ConocoPhilips, Nexen, OMV, Shell, Statoil, Todd Energy and Woodside Energy) for partial funding of this research. The Kowanyama Land Management Office, traditional owners of the Kowanyama region, and Viv Sinnamon are thanked for their permission to undertake this research and for providing logistical support in this remote region. We thank Denise Smith, Joanne Lane, Kathryn Amos, Jessica Trainor, Rodney Whitfield, Dugald Smith, Gerald Nanson, Bartosz Cichonski, John Lane and Andrew Fernie for assistance in the field, with special thanks to Brent Peterson from the University of Wollongong for drill rig operations. We also thank Frances Williams and Lee Arnold from the University of Adelaide for OSL contributions and Jorge Garcia Garcia, John Counts and Daniel Lane for assistance in core processing and OSL collection. Thanks are also due to Gary Hampson and an anonymous reviewer for their comments that improved the quality of this work.

References

Ainsworth, R.B., Vakarelov, B.K. & Nanson, R.A. 2011. Dynamic spatial and temporal prediction of changes in depositional processes on clastic shorelines: toward improved subsurface uncertainty reduction and management. *American Association of Petroleum Geologists Bulletin*, **95**, 267–297.

Ainsworth, R.B., Vakarelov, B.K., Lee, C., MacEachern, J.A., Montgomery, A.E., Ricci, L.P. & Dashtgard, S.E. 2015. Architecture and evolution of a regressive, tide-influenced marginal marine succession, Drumheller, Alberta, Canada. *Journal of Sedimentary Research*, **85**, 596–625, https://doi.org/10.2110/jsr.2015.33

Ali, A. 1995. A numerical investigation into the backwater effect on floodwater in the Meghna River in Bangladesh due to the south-west monsoon wind. *Estuarine, Coastal and Shelf Science*, **41**, 689–704.

Allen, J.R.L. 1963. The classification of cross-stratified units with notes on their origin. *Sedimentology*, **2**, 93–114.

Ashworth, P.J., Best, J.L. & Jones, M.A. 2004. Relationship between sediment supply and avulsion frequency in braided rivers. *Geology*, **32**, 21–24.

Assine, M.L., Corradini, F.A., Pupim, F.N. & McGlue, M.M. 2014. Channel arrangements and depositional styles in the São Lourenço fluvial megafan, Brazilian Pantanal wetland. *Sedimentary Geology*, **301**, 172–184.

Bhattacharya, J.P. 2010. Deltas. *In*: Dalrymple, R.G. & James, N.P. (eds) *Facies Models 4*. Geological Association of Canada, GEOtext, **6**, 233–264.

Bhattacharya, J.P. & Giosan, L. 2003. Wave-influenced deltas: geomorphological implications for facies reconstruction. *Sedimentology*, **50**, 187–210.

Blair, T.C. & McPherson, J.G. 1994. Alluvial fans and their natural distinction from rivers based on morphology, hydraulic processes, sedimentary processes and facies assemblages. *Journal of Sedimentary Research*, **A64**, 450–489.

Brooks, A.P., Shellberg, J.G., Knight, J. & Spencer, J. 2009. Alluvial gully erosion: an example from the Mitchell fluvial megafan, Queensland, Australia. *Earth Surface Processes and Landforms*, **34**, 1951–1969.

Chappell, J. 1983. Evidence for smoothly falling sea level relative to north Queensland, Australia, during the past 6,000 yr. *Nature*, **302**, 406–408.

Chappell, J. & Thom, B.G. 1986. Coastal morphodynamics in north Australia: review, and prospect. *Australian Geographic Study*, **24**, 110–127.

Chatanantavet, P., Lamb, M.P. & Nittrouer, J.A. 2012. Backwater controls of avulsion location on deltas. *Geophysical Research Letters*, **39**, L01402.

Choi, K. 2011. External controls on the architecture of inclined heterolithic stratification (IHS) of macrotidal Sukmo Channel: wave v. rainfall. *Marine Geology*, **285**, 17–28.

Coleman, J.M. & Wright, L.D. 1975. Modern river deltas: variability of processes and sand bodies. *In*: Broussard, M.L. (ed.) *Deltas: Models for Exploration*. Houston Geological Society, Houston, TX, 99–149.

CSIRO Marine & Atmospheric Research 2010. http://www.cmar.csiro.au [last accessed 1 August].

Dalrymple, R.W. & Choi, K. 2007. Morphological and facies trends through the fluvial-marine transition in tide dominated depositional systems: a schematic framework for environmental and sequence stratigraphic interpretation. *Earth-Science Reviews*, **81**, 135–174.

Davies, G. & Woodroffe, C.D. 2010. Tidal estuary width convergence: theory and form in north Australian estuaries. *Earth Surface Processes and Landforms*, **35**, 737–749.

DeCelles, P.G. & Cavazza, W. 1999. A comparison of fluvial megafans in the Cordilleran (Upper Cretaceous) and modern Himalayan foreland basin systems. *Geological Society of America Bulletin*, **111**, 1315–1334.

Edgar, N.T., Cecil, C.B., Mattick, R.E., Chivas, A.R., De Deckker, P. & Djajadihardja, Y.S. 2003. A modern analogue for tectonic, eustatic, and climatic processes in cratonic basins: Gulf of Carpentaria, northern Australia. *In*: Cecil, C.B. & Edgar, N.T. (eds) *Climate Controls on Stratigraphy*. SEPM Special Publications, **77**, 193–205.

Fustic, M., Hubbard, S.M., Spencer, R., Smith, D.G., Leckie, D.A., Bennett, B. & Larter, S. 2012. Recognition of down-valley translation in tidally influenced meandering fluvial deposits, Athabasca Oil

Sands (Cretaceous), Alberta, Canada. *Marine and Petroleum Geology*, **29**, 219–232.

GALLOWAY, W.E. 1975. Process framework for describing the morphologic and stratigraphic evolution of deltaic depositional systems. *In*: BROUSSARD, M.L. (ed.) *Deltas: Models for Exploration*. Houston Geological Society, Houston, TX, 87–98.

GALLOWAY, W.E. & HOBDAY, D.K. 1996. *Terrigenous Clastic Depositional Systems: Applications to Fossil Fuel and Groundwater Resources*. 2nd edn. Springer, Heidelberg.

GANTI, V., CHU, Z., LAMB, M.P., NITTROUER, J.A. & PARKER, G. 2014. Testing morphodynamic controls on the location and frequency of river avulsions on fans v. deltas: Huanghe (Yellow River), China. *Geophysical Research Letters*, **41**, 7882–7890.

GEOSCIENCE AUSTRALIA & CSIRO LAND & WATER 2011. *1 Second SRTM Derived Digital Elevation Models User Guide. Version 1.0.4*. Geoscience Australia, Canberra.

GIBLING, M.R. 2006. Width and thickness of fluvial channel bodies and valley fills in the geological record: literature compilation and classification. *Journal of Sedimentary Research*, **76**, 731–770.

GRIMES, K.G. & DOUTCH, H.F. 1978. The late Cainozoic evolution of the Carpentaria Plains, Northern Queensland. *Bureau of Mineral Resources Journal of Geology & Geophysics*, **3**, 101–112.

HARRIS, P.T., HEAP, A. ET AL. 2003. *Geomorphic Features of Continental Margin of Australia*. Record 2003/30. Geoscience Australia, Canberra.

HARTLEY, A.J., WEISSMANN, G.S., NICHOLS, G.J., & WARWICK, G.L. 2010. Large distributive fluvial systems: characteristics, distribution, and controls on development. *Journal of Sedimentary Research*, **80**, 167–183.

HOOKE, R. 1967. Processes on arid-region alluvial fans. *The Journal of Geology*, **75**, 438–460.

HORTON, B.K. & DECELLES, P.G. 2001. Modern and ancient fluvial megafans in the foreland basin system of the central Andes, southern Bolivia: implications for drainage network evolution in fold-thrust belts. *Basin Research*, **13**, 43–61.

HOVIKOSKI, J., RASANEN, M., GINGRAS, M.K., RANZI, A. & MELO, J. 2008. Tidal and seasonal controls in the formation of Late Miocene inclined heterolithic stratification deposits, western Amazonian foreland basin. *Sedimentology*, **55**, 499–530.

HUBBARD, S.M., SMITH, D.G., NIELSEN, H., LECKIE, D.A., FUSTIC, M., SPENCER, R.J. & BLOOM, L. 2011. Seismic geomorphology and sedimentology of a tidally influenced river deposit, Lower Cretaceous Athabasca oil sands, Alberta, Canada. *American Association of Petroleum Geologists Bulletin*, **95**, 1123–1145.

JEROLMACK, D.J. 2009. Conceptual framework for assessing the response of delta channel networks to Holocene sea level rise. *Quaternary Science Reviews*, **28**, 1786–1800.

JEROLMACK, D.J. & MOHRIG, D. 2007. Conditions for branching in depositional rivers. *Geology*, **35**, 463–466.

JEROLMACK, D.J. & SWENSON, J.B. 2007. Scaling relationships and evolution of distributary networks on wave-influenced deltas. *Geophysical Research Letters*, **34**, L23402.

JOHNSON, M.A., KENYON, N.H., BELDERSON, R.H. & STRIDE, A.H. 1982. Sand transport. *In*: STRIDE, A.H. (ed.) *Offshore Tidal Sands, Processes and Deposits*. Chapman & Hall, London, 58–94.

JOHNSON, S.M. & DASHTGARD, S.E. 2014. Inclined heterolithic stratification in a mixed tidal-fluvial channel: differentiating tidal v. fluvial controls on sedimentation. *Sedimentary Geology*, **301**, 41–53.

JONES, B.G., MARTIN, G.R. & SENAPATI, N. 1993. Riverine–tidal interactions in the monsoonal Gilbert River fandelta, northern Australia. *Sedimentary Geology*, **83**, 319–337.

JONES, B.G., WOODROFFE, C.D. & MARTIN, G.R. 2003. Deltas in the Gulf of Carpentaria, Australia: forms, processes and products. *In*: SIDI, F.H., NUMMEDAL, D., IMBERT, P., DARMAN, H. & POSAMENTIER, H.W. (eds) *Tropical Deltas of Southeast Asia – Sedimentology, Stratigraphy, and Petroleum Geology*. SEPM Special Publications, **76**, 21–43.

KEOGH, K.J., MARTINIUS, A.W. & OSLAND, R. 2007. The development of fluvial stochastic modelling in the Norwegian oil industry: a historical review, subsurface implementation and future directions. *Sedimentary Geology*, **202**, 249–268.

LA CROIX, A.D. & DASHTGARD, S.E. 2014. Of sand and mud: sedimentological criteria for identifying the turbidity maximum zone in a tidally influenced river. *Sedimentology*, **61**, 1961–1981.

LA CROIX, A.D. & DASHTGARD, S.E. 2015. A synthesis of depositional trends in intertidal and upper subtidal sediments across the tidal-fluvial-transition in the Fraser River, Canada. *Journal of Sedimentary Research*, **85**, 683–698.

LAMB, M.P., NITTROUER, J.A., MOHRIG, D. & SHAW, J. 2012. Backwater and river plume controls on scour upstream of river mouths: implications for fluvio-deltaic morphodynamics. *Journal of Geophysical Research*, **117**, F1, https://doi.org/10.1029/2011JF002079

LANE, E.W. 1957. *A Study of the Shape of Channels Formed by Natural Streams Flowing in Erodible Material*. US Army Corps of Engineers, Missouri River Division, Omaha, NE.

LANG, S.C., KARSSAN, J., BENSON, J.M., GRASSO, C.A. & AVENELL, L.C. 2000. Applications of modern and ancient geological analogues in characterisation of fluvial and fluvial-lacustrine deltaic reservoirs in the Cooper Basin. *Australian Petroleum Production & Exploration Association Journal 2000*, **40**, 393–415.

LARCOMBE, P., CARTER, R.M., DYE, J., GAGAN, M.K. & JOHNSON, D.P. 1995. New evidence for episodic post-glacial sea-level rise, central Great Barrier Reef, Australia. *Marine Geology*, **127**, 1–44.

LEWIS, S.E., SLOSS, C.R., MURRAY-WALLACE, C.V., WOODROFFE, C.D. & SMITHERS, S.G. 2013. Post-glacial sea-level changes around the Australian margin: a review. *Quaternary Science Reviews*, **74**, 115–138.

LULY, J.G., GRINDROD, J.F. & PENNY, D. 2006. Holocene palaeoenvironments and change at Three-Quarter Mile Lake, Silver Plains Station, Cape York Peninsula, Australia. *The Holocene*, **16**, 1085–1094.

MARTINIUS, A.W., HOWELL, J.A. & GOOD, T.R. (eds) 2014. *Sediment-Body Geometry and Heterogeneity: Analogue Studies for Modelling the Subsurface*.

Geological Society, London, Special Publications, **387**, http://sp.lyellcollection.org/content/387/1.toc

Massey, T.A., Fernie, A.J., Ainsworth, R.B., Nanson, R.A. & Vakarelov, B.K. 2014. Detailed mapping, three-dimensional modelling and upscaling of a mixed-influence delta system, Mitchell River delta, Gulf of Carpentaria, Australia. *In*: Martinius, A.W., Howell, J.A. & Good, T.R. (eds) *Sediment-Body Geometry and Heterogeneity: Analogue Studies for Modelling the Subsurface*. Geological Society, London, Special Publications, **387**, 135–151, https://doi.org/10.1144/SP387.4

Moss, P.T., Cosgrove, R., Ferrier, A. & Haberle, S.G. 2012. Holocene environments of the sclerophyll woodlands of the Wet Tropics of northeastern Australia. *In*: Haberle, S.G. & David, B. (eds) *Peopled Landscapes*. Terra Australis, **34**. ANU ePress, Canberra, 329–341.

Mouhoupt, J., Cunningham, V. *et al.* 2004. *Wave Data Recording Program: Weipa Region, 1978–2004*. Queensland Government Environmental Protection Agency, Coastal Sciences Data Report, **W2004.5**.

Munro, I.S.R. 1984. *Atlas of Operational, Environmental, and Biological Data from the Gulf of Carpentaria Prawn Survey, 1963–65: Part 3 Physical and Chemical Environment*. CSIRO Marine Laboratories, New South Wales (NSW), Australia.

Murray, A.S. & Wintle, A.G. 2000. Luminescence dating of quartz using an improved single-aliquot regenerative dose protocol. *Radiation Measurements*, **32**, 57–73.

Murray-Wallace, C.V. & Woodroffe, C.D. 2014. *Quaternary Sea-level Changes: a Global Perspective*. Cambridge University Press, Cambridge.

Muto, T. & Steel, R.J. 1997. Principles of regression and transgression: the nature of the interplay between accommodation and sediment supply. *Journal of Sedimentary Research*, **67**, 994–1000.

Nanson, G.C., Price, D.M., Short, S.A. & Young, R.W. 1991. Comparative uranium – thorium and thermoluminescence dating of weathered Quaternary alluvium in the tropics of northern Australia. *Quaternary Research*, **35**, 347–366.

Nanson, G.C., Jones, B.G., Price, D.M. & Pietsch, T.J. 2005. Rivers turned to rock: late Quaternary alluvial induration influencing the behaviour and morphology of an anabranching river in the Australian monsoon tropics. *Geomorphology*, **70**, 398–420.

Nanson, R.A., Ainsworth, R.B., Vakarelov, B.K., Fernie, A.J. & Massey, T.A. 2012. Geometric attributes of reservoir elements in a modern low accommodation tide-dominated delta. *Australian Petroleum Production and Exploration Association (APPEA) Journal 2012*, **52**, 483–492.

Nanson, R.A., Ainsworth, R.B., Dashtgard, S.E., Lane, T.I. & Vakarelov, B.K. 2013*a*. Distributary channel geometry and sediment distribution in a modern monsoonal, mixed-Process delta, gulf of carpentaria. Paper presented at the American Association of Petroleum Geologists Annual Convention, Pittsburgh, PA, USA.

Nanson, R.A., Vakarelov, B.K., Ainsworth, R.B., Williams, F.M. & Price, D.M. 2013*b*. Evolution of a Holocene, mixed-process, forced regressive shoreline: the Mitchell river delta, Queensland, Australia. *Marine Geology*, **339**, 22–43.

Nemec, W. & Steel, R.J. 1988. What is a fan delta and how do we recognize it? *In*: Nemec, W. & Steel, R.J. (eds) *Fan Deltas: Sedimentology and Tectonic Settings*. Blackie, Glasgow, 1–13.

Nijhuis, A.G., Edmonds, D.A. *et al.* 2015. Fluvio-deltaic avulsions during relative sea-level rise. *Geology*, **43**, https://doi.org/10.1130/G36788.1

Nittrouer, J.A., Shaw, J., Lamb, M.P. & Mohrig, D. 2012. Spatial and temporal trends for water-flow velocity and bed-material sediment transport in the lower Mississippi River. *Geological Society of America Bulletin*, **124**, 400–414.

Nyberg, B., Buckley, S.J., Howell, J.A. & Nanson, R.A. 2015. Geometric attribute and shape characterization of modern depositional elements: a quantitative GIS method for empirical analysis. *Computers & Geosciences*, **82**, 191–204.

Peel, M.C., Finlayson, B.L. & McMahon, T.A. 2007. Updated world map of the Köppen-Geiger climate classification. *Hydrology and Earth System Sciences Discussions*, **4**, 439–473.

Queensland Government 2009. Mitchell River 726. Map Series Version 3.0, Queensland Wetlands Program, http://www.wetlandinfo.ehp.qld.gov.au/

Raymond, O.L., Liu, S., Gallagher, R., Highet, L.M. & Zhang, W. 2012. *Surface Geology of Australia*, 1:1 000 000 scale, 2012 edn [digital dataset]. Geoscience Australia, Commonwealth of Australia, Canberra, http://www.ga.gov.au

Reeves, J.M., Chivas, A.R. *et al.* 2008. The sedimentary record of palaeoenvironments and sea-level change in the Gulf of Carpentaria, Australia, through the last glacial cycle. *Quaternary International*, **183**, 3–22.

Rhodes, E.G. 1980. *Modes of Holocene coastal progradation, Gulf of Carpentaria*. PhD thesis, Australian National University, Queensland, Australia.

Ridd, P., Sandstrom, M.W. & Wolanski, E. 1988. Outwelling from tropical tidal salt flats. *Estuarine, Coastal and Shelf Science*, **26**, 243–253.

Roelse, A., Granger, H.W. & Graham, J.W. 1975. *The Adjustment of the Australian Levelling Survey 1970–1971*. 2nd edn. Division of National Mapping Technical Report **12**.

Rustomji, P., Shellberg, J.G., Brooks, A.P., Spencer, J. & Caitcheon, G. 2010. *A Catchment Sediment and Nutrient Budget for the Mitchell River, Queensland*. A Report to the Tropical Rivers and Coastal Knowledge (TRaCK) Research Program. Griffith University, Queensland, Australia.

Schumm, S.A., Mosley, M.P. & Weaver, W.E. 1987. *Experimental Fluvial Geomorphology*. Wiley, New York.

Shiers, M.N., Mountney, N.P., Hodgson, D.M. & Cobain, S.L. 2014. Depositional controls on tidally influenced fluvial successions, Neslen Formation, Utah, USA. *Sedimentary Geology*, **311**, 1–16.

Shulmeister, J. 1999. Australasian evidence for mid-Holocene climate change implies precessional control of Walker Circulation in the Pacific. *Quaternary International*, **57/58**, 81–91.

Sisulak, C.F. & Dashtgard, S.E. 2012. Seasonal controls on the development and character of inclined heterolithic stratification in a tide-influenced, fluvially dominated channel: Fraser River, Canada. *Journal of Sedimentary Research*, **82**, 244–257.

Slingerland, R.L. & Smith, N.D. 2004. River avulsions and their deposits. *Annual Review of Earth and Planetary Sciences*, **32**, 257–285.

Stouthamer, E. & Berendsen, H.J.A. 2000. Factors controlling the Holocene avulsion history of the Rhine-Meuse delta (the Netherlands). *Journal of Sedimentary Research*, **70**, 1051–1064.

Suppiah, R. 1992. The Australian summer monsoon: a review. *Progress in Physical Geography*, **16**, 283–218.

Taylor, M. & Stone, G. 1996. Beach-ridges: a review. *Journal of Coastal Research*, **12**, 612–621.

Thom, B.G. & Chappell, J. 1975. Holocene sea levels relative to Australia. *Search*, **6**, 90–93.

Thom, B.G. & Roy, P.S. 1983. Sea level change in New South Wales over the past 15000 years. *In*: Hopley, D. (ed) *Australian Sea Levels in the Last 1500 Years: a Review*. Department of Geography, James Cook University, Townsville, 64–84.

Thomas, R.G., Smith, D.G., Wood, J.M., Visser, J., Caverley-Range, E.A. & Koster, E.H. 1987. Inclined heterolithic stratification – terminology, description, interpretation and significance. *Sedimentary Geology*, **53**, 123–179.

Torgersen, T., Hutchinson, M.F., Searle, D.E. & Nix, A. 1983. General bathymetry of the Gulf of Carpentaria and the Quaternary physiography of Lake Carpentaria. *Palaeogeography, Palaeoclimatology, Palaeoecology*, **41**, 207–225.

Tyler, N. & Finley, R.J. 1991. Architectural controls on the recovery of hydrocarbons from sandstone reservoirs. *In*: Miall, A.D. & Tyler, N. (eds) *The Three-Dimensional Facies Architecture of Terrigenous Clastic Sediments, and its Implications for Hydrocarbon Discovery and Recovery*. SEPM Concepts and Models in Sedimentology and Paleontology, **3**, 1–15.

Vakarelov, B.K. & Ainsworth, R.B. 2013. A hierarchical approach to architectural classification in marginal marine systems: bridging the gap between sedimentology and sequence stratigraphy. *American Association of Petroleum Geologists Bulletin*, **97**, 1121–1161.

Weissmann, G.S., Hartley, A.J., Nichols, G.J., Scuderi, L.A., Olson, M., Buehler, H. & Banteah, R. 2010. Fluvial form in modern continental sedimentary basins: distributive fluvial systems. *Geology*, **38**, 39–42.

Weissmann, G.S., Hartley, A. & Nichols, G. 2011. Alluvial facies distributions in continental sedimentary basins – distributive fluvial systems. *In*: Davidson, S.K., Leleu, S. & North, C. (eds) *From River to Rock Record: the Preservation of Fluvial Sediments and their Subsequent Interpretation*. SEPM, Special Publications, **79**, 327–355.

Weissmann, G.S., Hartley, A.J. *et al.* 2013. Prograding distributive fluvial systems – geomorphic models and ancient examples. *In*: Driese, S.G. & Nordt, L.C. (eds) *New Frontiers in Paleopedology and Terrestrial Paleoclimatology, Paleosols and Soil Surface Analog Systems*. SEPM Special Publications, **104**, 131–147.

Wintle, A.G. & Murray, A.S. 2006. A review of quartz optically stimulated luminescence characteristics and their relevance in single-aliquot regeneration dating protocols. *Radiation Measurements*, **41**, 369–391.

Wolanski, E. 1993. Water circulation in the Gulf of Carpentaria. *Journal of Marine Systems*, **4**, 401–420.

Woodroffe, C.D. & Chappell, J. 1993. Holocene emergence and evolution of the McArthur River Delta, southwestern Gulf of Carpentaria, Australia. *Sedimentary Geology*, **83**, 303–317.

Woodroffe, C.D., Chappell, J., Thom, B.G. & Wallensky, E. 1989. Depositional model of a macrotidal estuary and floodplain, South Alligator River, Northern Australia. *Sedimentology*, **36**, 737–756.

Woodroffe, S.A. 2009. Testing models of mid to late Holocene sea-level change, North Queensland, Australia. *Quaternary Science Reviews*, **28**, 2474–2488.

Transgressive successions of the Mahakam Delta Province, Indonesia

JOSEPH J. LAMBIASE[1]*, RIDHA S. RIADI[1,2], NADIA NIRSAL[1,3] & SALAHUDDIN HUSEIN[4]

[1]*Petroleum Geoscience Programme, Chulalongkorn University, Bangkok, Thailand*

[2]*Present address: VICO Indonesia, Jakarta, Indonesia*

[3]*Present address: Mubadala Petroleum, Jakarta, Indonesia*

[4]*Universitas Gadjah Mada, Yogyakarta, Indonesia*

**Correspondence: joe_lambiase@yahoo.com*

Abstract: Sedimentological and stratigraphic analysis of outcrops and subsurface data in the Mahakam Delta province, combined with concepts of transgressive sedimentary processes derived from a study of the modern Mahakam Delta, indicate that a significant portion of the palaeo-Mahakam Delta succession was deposited during transgressive phases. Some of the transgressive successions resulted from major transgressions and are regionally extensive, but many reflect small, short-lived transgressions within dominantly progradational phases.

Two sandstone facies with significant reservoir potential are common within the transgressive successions. Backfilled distributary sandstones occur in outcrop as 10–20 m thick, fining-upward channel sands that become more marine upward and in the subsurface as elongate, coastline-perpendicular sand bodies with a back-stepping stratigraphic architecture and highly variable thickness. Transgressive shoreline sandstones also fine upward, but they are shoreline-parallel, their lateral extent is controlled by the pre-transgression delta morphology and their thickness depends on the rate of relative sea-level rise.

Both types of transgressive sandstone can be difficult to distinguish from specific progradational sandstones with wireline log data, especially when they occur within predominantly progradational successions. However, their sand body geometries, volumetrics and connectivity are much different from the stratigraphically adjacent sandstones deposited during progradational phases of deltaic deposition.

The Mahakam Delta is located on the east coast of Borneo in Indonesia within the Kutei Basin (Fig. 1). Traditional depositional models for the modern delta characterize it as a mixed river-dominated and tide-dominated delta that is presently prograding based on qualitative assessments of its morphology and facies distribution (Galloway 1975; Allen *et al.* 1976; Gastaldo *et al.* 1995; Allen & Chambers 1998). These models for the modern delta are commonly used as analogues to interpret the outcropping and subsurface successions of the palaeo-Mahakam Delta (e.g. Allen & Chambers 1998; Cibaj & Wiweko 2008), the latter of which constitute the major Kutei Basin reservoir sandstones.

However, a recent quantitative study has determined that the modern delta is presently in a transgressive phase and is depositing a transgressive succession with very high preservation potential (Salahuddin & Lambiase 2013). Recognition that the modern delta is transgressive has three major consequences: (1) it invalidates the use of the modern delta as a viable analogue for progradational subsurface successions; (2) it suggests that transgressive successions may be relatively common in the palaeo-Mahakam Delta; and (3) that the modern delta is a valid analogue for reservoir sandstones within subsurface transgressive successions. This paper summarizes the transgressive character of the modern Mahakam Delta and the succession that is currently being deposited, applies those concepts to selected outcrop and subsurface successions, and discusses the implications for reservoir characterization.

The modern Mahakam Delta

Salahuddin & Lambiase (2013) have presented quantitative hydrodynamic and sedimentological data that demonstrate the transgressive character of the modern delta. The major present day sedimentary processes are (1) backfilling of the distributaries and (2) relatively minor reworking of sediment deposited during the progradational phase that immediately preceded the ongoing

From: Hampson, G. J., Reynolds, A. D., Kostic, B. & Wells, M. R. (eds) 2017. *Sedimentology of Paralic Reservoirs: Recent Advances*. Geological Society, London, Special Publications, **444**, 335–348.
First published online June 29, 2016, https://doi.org/10.1144/SP444.2

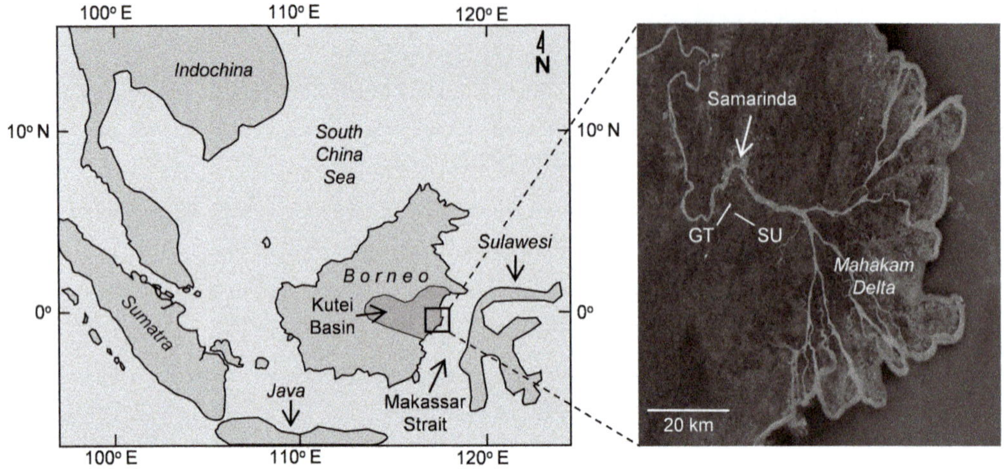

Fig. 1. Location map of the Kutai Basin in eastern Borneo with an inset showing the location of the modern Mahakam Delta and the Stadion Utama Kaltim (SU) and Grand Taman Sari Circuit (GT) outcrops near Samarinda.

transgression. Salahuddin & Lambiase (2013) note that the very low wave energy in the Makassar Straits (Fig. 1), plus rapid subsidence and burial, limits marine reworking to the uppermost progradational strata and preserves the pre-transgression, progradational distributary and interdistributary morphology (Fig. 2).

Sedimentary processes

Ongoing backfilling of the distributaries is generating fining-upwards successions that become increasingly marine upward (Figs 2 & 3; Salahuddin & Lambiase 2013). The current speed and the sediment transport capacity and competence decrease

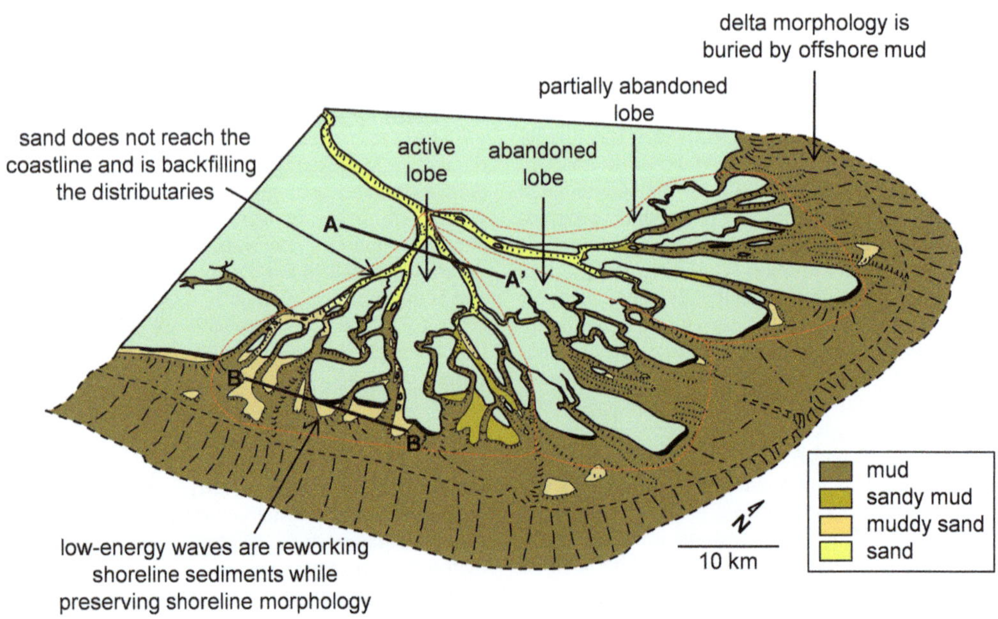

Fig. 2. The morphology, sediment distribution and sedimentary processes on the modern Mahakam Delta (after Husein 2008), plus the active, partially abandoned and abandoned lobes defined by previous workers (e.g. Allen & Chambers 1998). A–A′ and B–B′ are the locations of the schematic cross-sections in Figure 4.

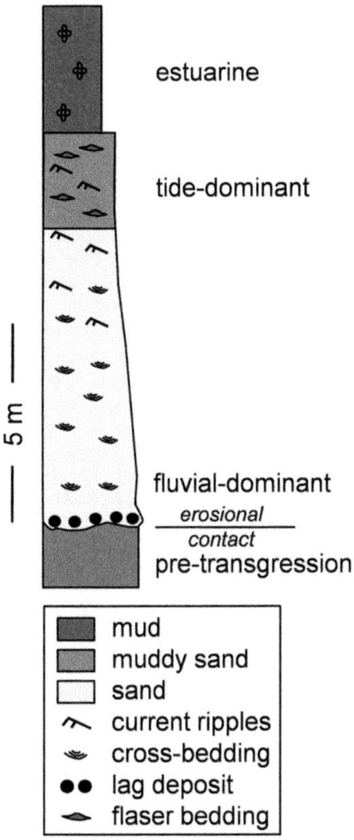

Fig. 3. The stratigraphic column for a backfilled distributary predicted from sedimentary processes on the modern Mahakam Delta (after Husein 2008).

seaward, so that the sediment flooring the distributaries is progressively finer downstream, which generates a fining-upward succession (Salahuddin & Lambiase 2013). These successions also become more marine upward and have excellent preservation potential because of the rapid subsidence rates and minor marine reworking.

Lambiase (2012) demonstrated that a progressively decreasing water surface slope during transgression generates backfilled distributary successions and also often results in avulsion because transgression is more effective at causing channel avulsion than progradation. On the Mahakam Delta, local subsidence rates are a product of eustasy, tectonics and compaction. Consequently, subsidence rates are highly variable, both areally and temporally, so that the amount of backfilling that occurs before avulsion is also likely to be highly variable, as will be the thicknesses of the backfilled successions.

Transgressive avulsion also reconciles the lobe-switching that has been recognized on the modern Mahakam Delta by several researchers (e.g. Allen & Chambers 1998) with ongoing transgression, and suggests that the delta is presently back-stepping (Fig. 2). Relative sea-level rise has been cited as an underlying cause of lobe-switching on the Huang He (Yellow) River and other deltas (Xue 1993; Hori & Saito 2008).

Transgressive sandstones

The ongoing sedimentary processes suggest that two types of transgressive sandstones are currently being deposited. Distributaries are being backfilled with relatively thick, fining-upward sandstones that are expected to be fewer, somewhat thinner and closer together laterally in the upper delta plain than in the lower delta plain (Fig. 4; Salahuddin & Lambiase 2013). As these sands fill the topographically low distributaries, they are laterally adjacent to slightly older, pre-transgression progradational strata (Fig. 4). In contrast, relatively thin sandstones develop directly above pre-transgression progradational strata in interdistributary areas and are separated from them by a transgressive erosional surface generated by marine reworking (Fig. 4).

The three-dimensional geometry of the sandstones within a transgressive succession is expected to be complex and highly dependent on the pre-transgression delta morphology. The backfilled distributary sandstones have the same sinuous shapes as the channels they fill and are orientated quasi-perpendicular to the shoreline, whereas the transgressive shoreline sandstones are shoreline-parallel with a lateral extent that is determined by the distributary spacing (Fig. 5).

The ongoing transgressive lobe-switching means that the backfilled distributary successions are not exactly contemporaneous and that they probably have highly variable thicknesses and lateral extent. It is also likely that some previously abandoned, partially filled distributaries will be re-occupied upstream of the original shoreline and deposit multiple backfilled sandstones with a back-stepping geometry, resulting in a complex three-dimensional sand body distribution within each transgressive succession.

Palaeo-Mahakam Delta outcrops

Transgressive successions are common in the middle Miocene Balikpapan Formation that was deposited by the palaeo-Mahakam Delta and crop out near the town of Samarinda (Fig. 1). In particular, transgressive strata comprise much of the laterally and vertically extensive Stadion Utama Kaltim and Grand Taman Sari Circuit outcrops. The sedimentary structures and trace fossil assemblages in both outcrops are described in detail here and the benthic foram and pollen assemblages were determined from several samples in each outcrop.

Stadion Utama Kaltim outcrop

Nirsal (2010) described four fining-upward units within the 162 m that is exposed in the Stadion Utama Kaltim outcrop (Fig. 6). The lower contact of each unit is an erosional surface at the base of a cross-bedded sandstone that fines upward from coarse to fine sand; the sandstones vary in thickness from 3.5 to 14 m and a change from cross-bedding to ripple cross-lamination accompanies the upward decrease in grain size. Directly overlying each of the basal sandstones are the interbedded marginal marine mudstones, fine sandstones and coals that comprise the rest of each unit (Fig. 6). Important observations are that, in all four stratigraphic units, the sandstones onlap a basal erosional surface and that the onset of marine influence occurs within the basal sandstone and the degree of marine influence increases upward. Upwardly increasing marine influence is marked by flasers, mud drapes and/or marine burrows (most commonly *Ophiomorpha*) that first appear at varying distances from the base of each basal sandstone, become increasingly abundant upward and are transitional into the overlying marginal marine facies (Figs 6 & 7).

Figure 7 illustrates the transgressive character of the stratigraphic succession in a 14 m thick backfilled distributary in the Stadion Utama Kaltim outcrop. There, thick beds of cross-bedded coarse pebbly sandstone fine upward to thin interbedded mudstones; flaser-bedded, mud-draped fine sandstones and sedimentary structures normally associated with tidal processes become more abundant upward, as do marine trace fossils and mangrove and *Nypa* pollen. Most of the succession is channel sandstones that are fluvial at the base and increasingly tide-influenced upward (Fig. 7).

The basal sandstone within each of the four fining-upward units in the Stadion Utama Kaltim outcrop is interpreted as a channel sand that has backfilled a distributary during a transgression, with the overlying marginal marine facies indicating continued transgression. In this interpretation, the basal erosional surfaces represent lowstands that separate four highly aggradational, transgressive stratigraphic units, each of which constitutes a retrogradational parasequence (Fig. 6). The entire outcrop succession may be part of an incised valley fill similar to that recognized in a nearby Miocene

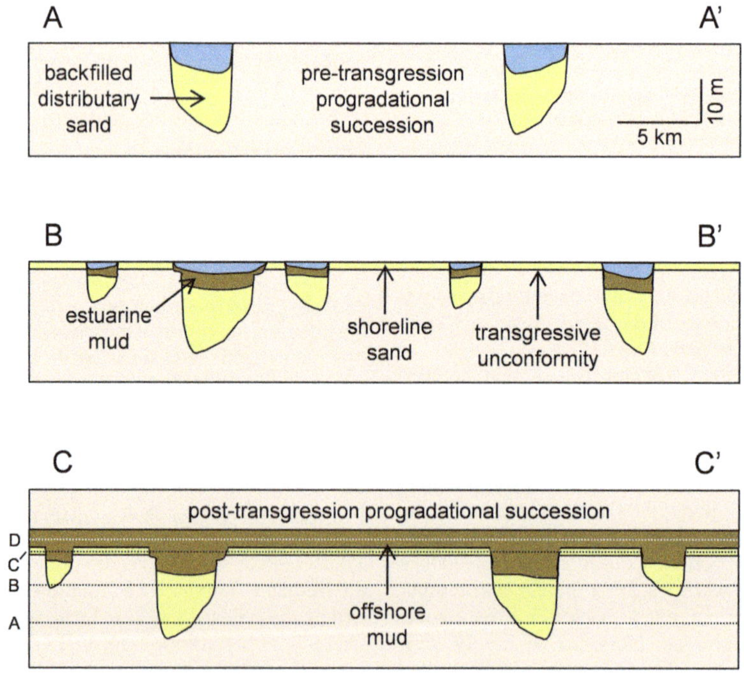

Fig. 4. Schematic cross-sections of the predicted stratigraphy generated by the present day sedimentary processes on the modern Mahakam Delta for the (**a**) upper and (**b**) lower delta plain and (**c**) the lower delta plain assuming that the ongoing transgression continues and is followed by a progradational event. The locations of (a) and (b) are shown on Figure 2; the lithofacies distribution of (c) is shown in Figure 5.

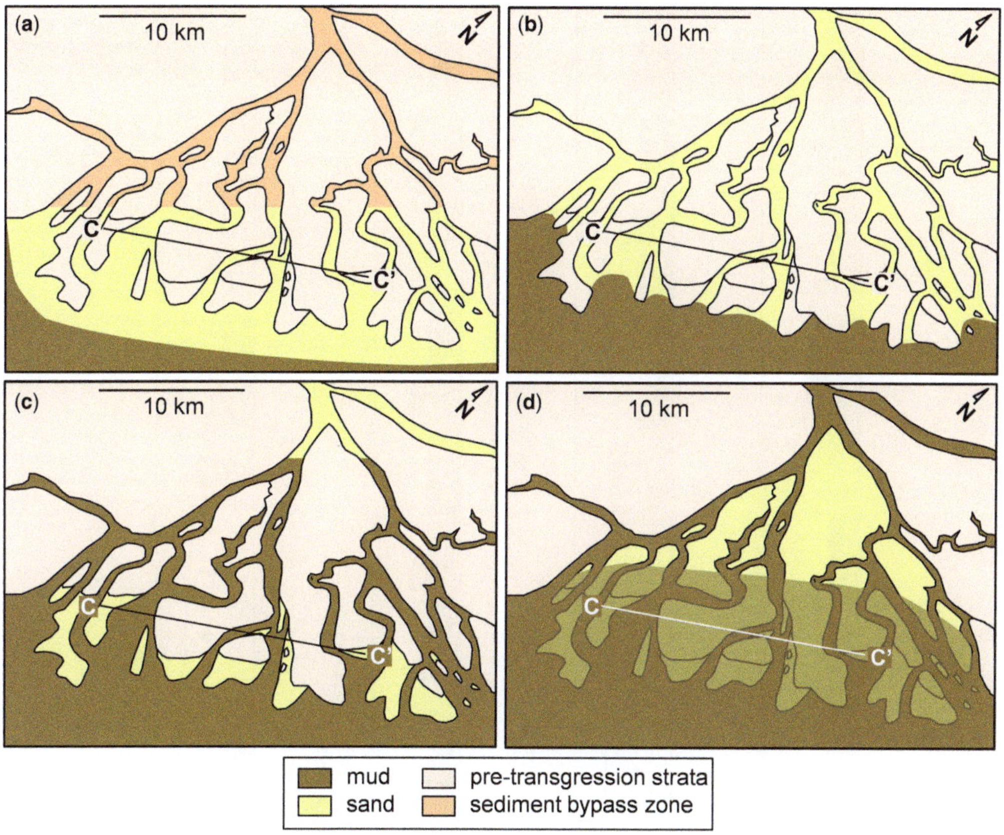

Fig. 5. Schematic lithofacies distribution in the active delta lobe on four stratal slices representing (**a**) earliest transgression, (**b**) distributary backfilling, (**c**) shoreline reworking and estuarine distributaries and (**d**) partial submergence and corresponding to the respective dashed time lines on Figure 4c. Note that the lithofacies are time-transgressive.

subsurface succession (Butterworth *et al.* 2002). A transgressive interpretation is markedly different from traditional, exclusively progradational depositional models for the palaeo-Mahakam Delta (e.g. Allen & Chambers 1998; Mora *et al.* 2000) and previous interpretations of the Stadion Utama Kaltim outcrop considered it to be a progradational fluvial–deltaic succession (e.g. Cibaj & Wiweko 2008).

It is difficult to reconcile the observed significant variation in channel sand thickness over a relatively small stratigraphic interval with a progradational fluvial model because it would require a significant change in channel width (i.e. stream size) in a relatively short period of time. However, the thickness of a backfilled distributary sand is controlled only by the sediment supply rate and the amount of time before the channel avulses as the relative sea-level rises (Lambiase 2012). Subsidence rates were undoubtedly highly variable during deposition of the Miocene palaeo-Mahakam Delta succession because the area was tectonically active so that highly variable channel sand thicknesses are compatible with a transgressive interpretation.

Grand Taman Sari Circuit outcrop

About 1.75 km NW of the Stadion Utama Kaltim outcrop, a *c.* 400 m thick, younger and nearly continuous vertical succession of the Balikpapan Formation is exposed in the Grand Taman Sari Circuit outcrop (Fig. 1). This outcrop extends laterally for more than 600 m across five discrete exposures and consists of 10 marine and non-marine lithofacies. Riadi (2013) and Riadi & Lambiase (2015) have documented that the basal 115 m is a series of fluvial channel and crevasse splay sandstones interbedded with fluvial floodplain mudstones that becomes increasingly sandy and aggradational upwards, suggesting that it is a progradational parasequence set consisting of

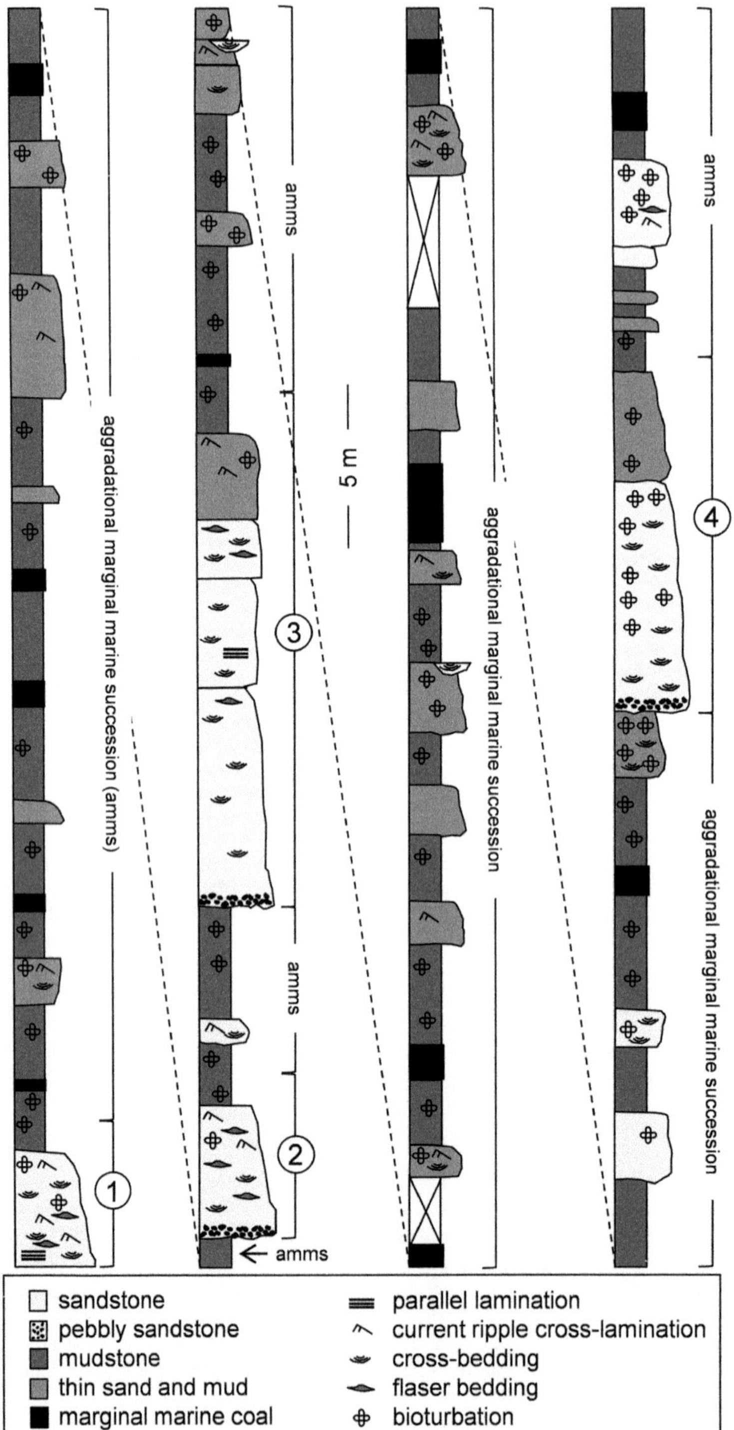

Fig. 6. Stratigraphic column from the Stadion Utama Kaltim outcrop, with the four fining-upward backfilled distributary successions that become more marine upwards numbered 1–4. Figure 7 describes backfilled distributary number 3 in detail (after Nirsal 2010).

Fig. 7. (**a**) A backfilled distributary succession in the Stadion Utama Kaltim outcrop and (**b**) its stratigraphic column. Photos (**c–h**) show that it fines upwards and becomes more marine upwards; the letters adjacent to the stratigraphic column indicate the stratigraphic position of the respective photograph. The facies succession in the photographs is (c) fluvial sandstone, (d) tide-influenced fluvial sandstone, (e) and (f) burrowed and flaser-bedded tide-dominated muddy sandstones, respectively, (g) tide- and wave-influenced marginal marine sandstones and mudstones and (h) bioturbated marine mudstone (after Nirsal 2010).

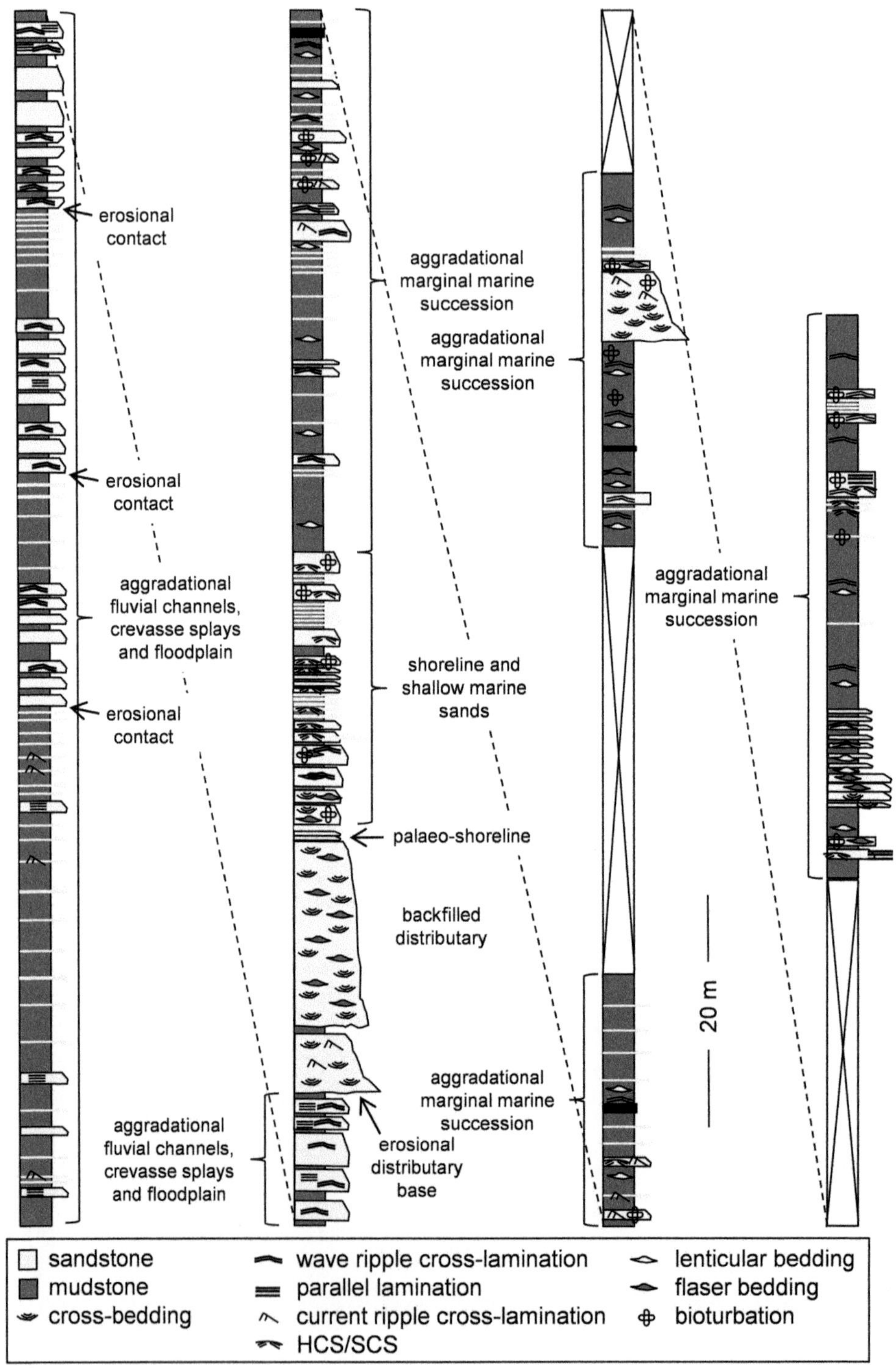

Fig. 8. A composite stratigraphic column for the Grand Taman Sari outcrop illustrating the thicknesses and sedimentary structures in the various units. HCS and SCS indicate hummocky cross-stratification and swaley cross-stratification, respectively. The labelled erosional contacts are interpreted as parasequence boundaries; see the text for an explanation (after Riadi 2013).

transgressive parasequences deposited in an upper delta plain environment (Figs 8 & 9). The top of this non-marine unit is a pronounced erosional surface that extends across the entire outcrop (Fig. 9).

A backfilled distributary, which also extends laterally across the entire outcrop, immediately overlies the erosional surface. The distributary fill succession consists of thick-bedded (>1 m) channel sandstones that vary in total thickness from 5 to 22 m in four exposures that fine upward from pebbly coarse sandstones to fine sandstones; the thickness of the succession increases eastwards, reflecting the lateral position within the distributary and/or the orientation of the distributary relative to the outcrop surface (Riadi 2013; Riadi & Lambiase 2015). There is minor tidal influence in the basal channel sandstones, as indicated by the scattered mud and organic flasers and drapes; tidal influence increases upward where the flasers and drapes become considerably more abundant and the distributary fill unit is directly overlain by marine facies (Figs 8 & 9). As with the Stadion Utama Kaltim succession, the distributary in the Grand Taman Sari Circuit outcrop has been previously interpreted as progradational distributary deposits (e.g. Cibaj & Wiweko 2008), which is inconsistent with strata that become progressively more marine upward.

The distributary fill unit forms the base of the more than 284 m thick transgressive marginal marine, lower delta plain succession that comprises the remainder of the outcrop. It is dominated by relatively thin progradational tidal parasequences that are stacked into retrogradational and aggradational parasequence sets and it also includes two relatively thin, wave-dominant sandstone units that are interpreted as transgressive shoreline sandstones (Fig. 8; Riadi 2013; Riadi & Lambiase 2015). Thus the entire 400 m thick stratigraphic succession in the Grand Taman Sari Circuit outcrop is transgressive and regressive conditions were limited to the time during which the erosional surfaces developed. This is consistent with the outcrop representing transgressive deltaic phases in which sediment was stored onshore and near the shoreline, with the erosional surfaces representing erosion and sediment bypass during intervening progradational deltaic phases that deposited sediment seaward of the outcrop location. The vertical relationship of an upper coastal plain succession underlying a lower coastal plain succession suggests a back-stepping delta (Fig. 9).

Relative abundance of transgressive sandstones

The proportion of transgressive and regressive sandstones in the palaeo-Mahakam Delta succession was estimated using the area of the modern delta (Fig. 1), the areal distribution of the transgressive sandstones

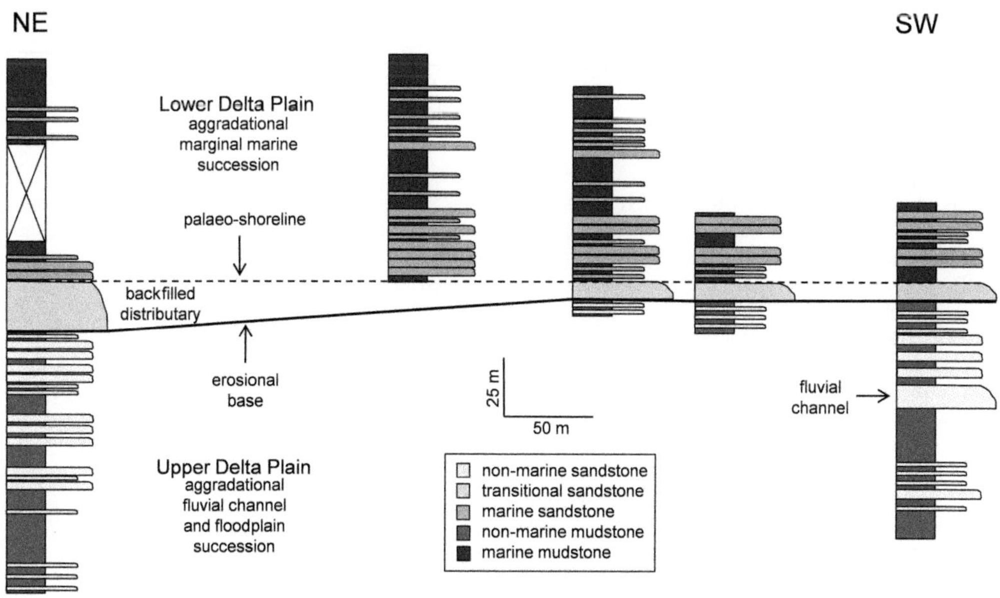

Fig. 9. Lateral and vertical distribution of the upper and lower delta plain successions in the Grand Taman Sari outcrop, plus the lateral extent and thickness variation of the backfilled distributary. Most upper delta plain sandstones are crevasse splay deposits, except for the labelled fluvial channel sandstone (after Riadi 2013).

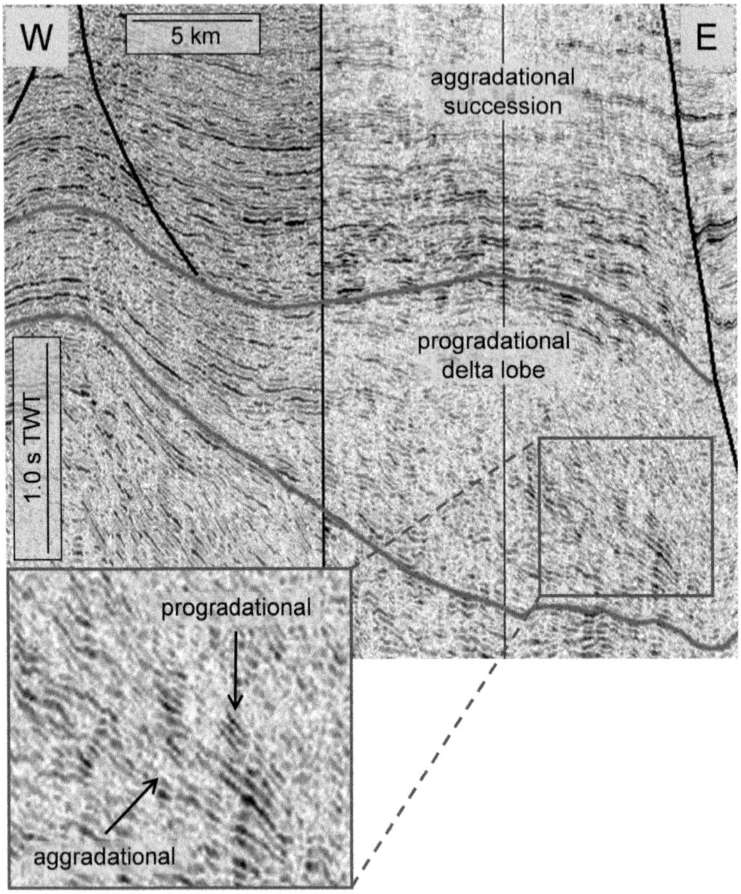

Fig. 10. A seismic line from the Mahakam Delta province illustrating progradational and aggradational stratigraphic signatures, including small-scale aggradational units within a progradational delta lobe.

that are presently accumulating (Fig. 5) and the thickness of the distributary fill and shoreline sandstones in the Stadion Utama Kaltim and Grand Taman Seri Circuit outcrops, respectively (Figs 7 & 9). The modern delta occupies an area of *c.* 2000 km^2, of which *c.* 350 km^2 (17%) is distributaries (Figs 1 & 5). This yields a progradational phase volume of *c.* 40 000 km^3 based on the *c.* 25 m thick clinoform succession deposited by the most recent progradational phase (Salahuddin & Lambiase 2013), compared with the *c.* 5000 km^3 for backfilled distributary sandstones, assuming that the 15 m thick sandstone in the Stadion Utama Kaltim outcrop (Fig. 7) is representative. This suggests that the volume of backfilled distributary sandstones is only about 12.5% as large as that of progradational sandstones after the entire delta is transgressed.

In contrast, the stacked shoreline sandstones in the succession immediately above the distributary fill succession in the Grand Taman Seri Circuit outcrop have an average thickness of about 15 m (Fig. 9), which gives them an approximate volume of 25 000 km^3 or 62.5% of the progradational phase succession volume if the entire delta is transgressed. It is expected that the actual thickness of the transgressive shoreline sandstones is highly variable and it is possible that the Grand Taman Seri Circuit outcrop succession is thicker than most; however, the Grand Taman Seri Circuit example suggests that a significant portion of the total deltaic sandstone succession may be transgressive.

Subsurface successions

Several features common to Kutei Basin seismic and well data reflect the occurrence, distribution and character of transgressive successions and the potential reservoir sandstones within them.

Seismic signature

In general, clastic stratigraphic successions in the Kutei Basin are a composite of progradational and aggradational parasequences. Many unconformity-bound sequences are relatively thick and dominated by progradational parasequences with clinoformal seismic signatures, whereas other sequences are thinner and aggradational, as indicated by parallel reflections (Fig. 10). It is generally recognized that the latter are transgressive successions deposited during regional relative sea-level rises, some of them are almost certainly incised valley fills similar to that described by Butterworth *et al.* (2002). Relatively thin aggradational units are also common within the dominantly progradational sequences (Fig. 10). They are interpreted as the products of relatively small, short-lived and probably local relative sea-level rises and are expected to have facies distributions and stratigraphic architecture similar to the transgressive outcrop successions described here.

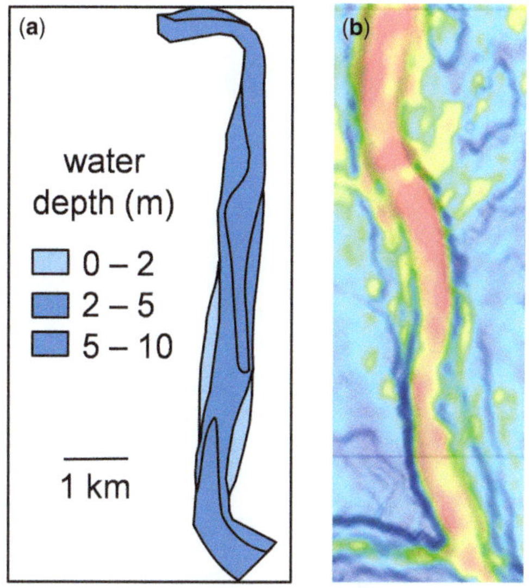

Fig. 11. (**a**) Bathymetry of the Bekapai Baru River, a distributary of the modern Mahakam Delta, plotted at the same scale as a seismic stratal slice (**b**), which depicts a linear, shoreline-perpendicular sandstone (orange) interpreted as a backfilled distributary.

Backfilled distributaries

Sandstone bodies are common on seismic horizon slices extracted from the relatively thin aggradational successions (Fig. 11). Many of those sandstone bodies are shoreline-perpendicular and are tens of kilometres long with estimated thicknesses of 10–20 m, with a few up to 100 m, and widths of several hundreds of metres to 1 km (Prasetyo 2003). This corresponds closely to the orientation and dimensions of the distributaries on the modern delta and the backfilled distributary sandstones exposed in the Stadion Utama Kaltim and Grand Taman Sari Circuit outcrops (Figs 2, 6, 8 & 11). Consequently, these sand bodies are interpreted as backfilled distributary sandstones deposited during relatively short-lived, minor transgressive phases, with the less common >20 m thick sandstones representing stacked backfilled distributary sandstones; they are expected to be common within similar aggradational seismic successions throughout the entire palaeo-Mahakam Delta succession. Salahuddin & Lambiase (2013) noted that the Attaka Field succession described by Trevena *et al.* (2003) corresponds to an aggradational seismic signature and that it can be interpreted as a transgressive succession because it is almost identical in character to the backfilled distributary successions currently being deposited by the modern Mahakam Delta.

Transgressive shorelines

A core from Well X in the Kutei Basin recovered a thin (11 m) wave-dominated transgressive succession where a tidal flat mudstone underlies a wave-dominated, tidally influenced, fine-grained sandstone with an erosional base (Fig. 12). The sandstone is, in turn, overlain by highly bioturbated interbedded thin sandstones and mudstones, with the thickness and proportion of sandstones decreasing upward; high-resolution biostratigraphy confirms that the cored interval is marginal marine and transgressive (Fig. 12). The sandstone is interpreted as a shoreline sand that was deposited by the transgressive reworking of interdistributary sediments; the overlying progressively muddier and more highly bioturbated strata represent progressively deeper shallow marine environments as transgression continued.

Implications for reservoir characterization

The back-stepping transgressive shoreline and backfilled distributary sandstones deposited during transgressive phases have significantly different sand body geometries and connectivity than the fluvial point bar and mouth bar sands that are the prime reservoir targets within the topset beds of progradational deltaic successions in the Kutei Basin. Successful reservoir characterization of the transgressive sandstones depends on recognizing them, as well as understanding their reservoir properties. Recognition is straightforward within relatively thick, laterally extensive transgressive sequences,

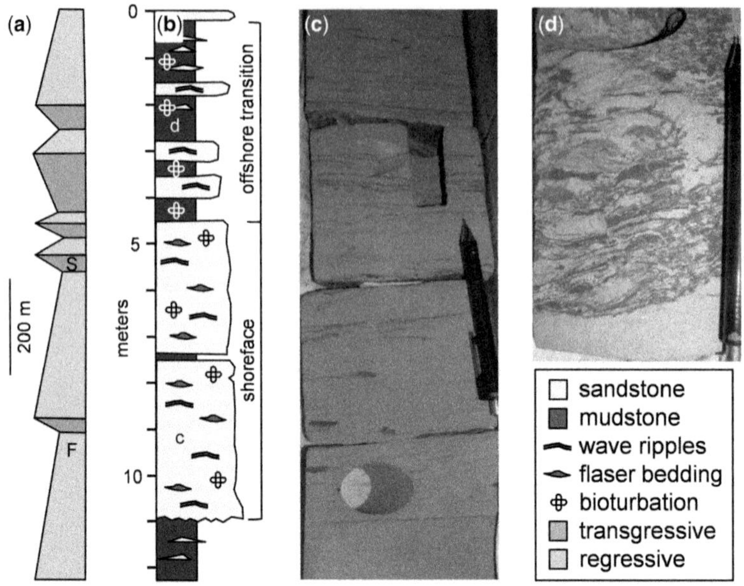

Fig. 12. (**a**) Transgressive and regressive successions in an interval of Well X determined by high-resolution biostratigraphy; S indicates the stratigraphic position of the core in (b), (c) and (d); F indicates the stratigraphic position of the core described in Figure 13. (**b**) Stratigraphic column of a transgressive shoreline succession in a core from Well X. (**c**) Core photograph from a wave-dominant, tide-influenced shoreface sandstone at stratigraphic position C on the stratigraphic column. (**d**) Core photograph from offshore transition heavily bioturbated interbedded sandstones and mudstones at stratigraphic position D on the stratigraphic column. The cores in (c) and (d) are 0.1 m in diameter.

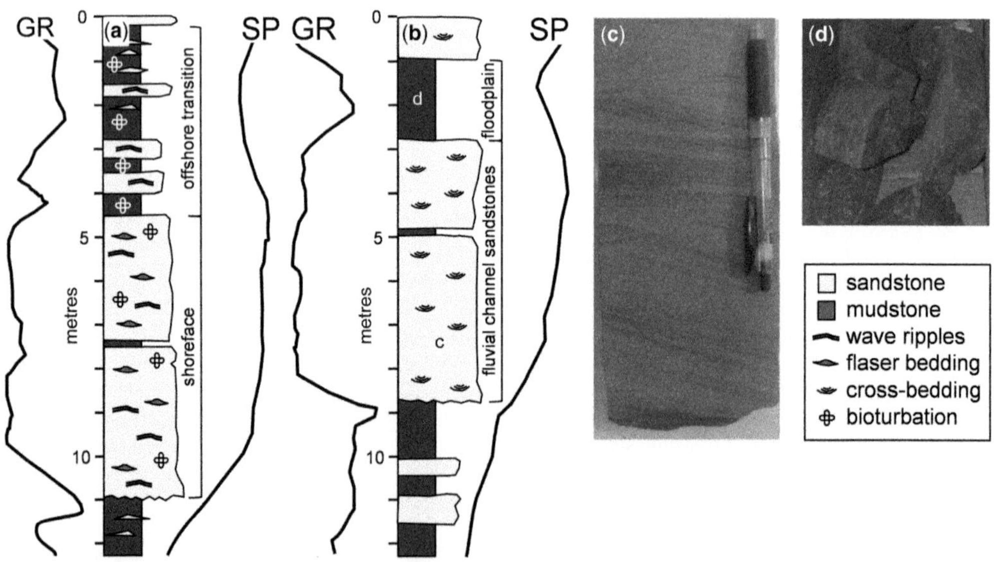

Fig. 13. (**a**) Stratigraphic column of the cored transgressive shoreline succession in Figure 12 with its gamma ray and spontaneous potential log signature. (**b**) Stratigraphic column and gamma ray and spontaneous potential log signature from a fluvial succession, also cored in Well X; see Figure 12 for its stratigraphic position and note the similarity of the log signatures in (a) and (b). (**c**) Core photograph from the fluvial sandstone at stratigraphic position C on the stratigraphic column. (**d**) Core photograph from the microfossil-barren, floodplain mudstones at stratigraphic position D on the stratigraphic column. The core is 0.1 m in diameter.

but it can be problematic where transgressive successions occur within sequences that are strongly progradational (Fig. 10).

Backfilled distributary sandstones

Backfilled distributary sandstones and fluvial point bar deposits have comparable thicknesses and they both fine upwards, which makes it nearly impossible to distinguish them solely with wireline data, plus both facies also can occur adjacent to each other at equivalent depths (Fig. 4). In the absence of core, the key data for identifying backfilled distributary sandstones are high-resolution biostratigraphic analysis of the sandstone and immediately overlying unit, and seismic horizon slices to determine sand body geometry.

The distinction is important because the linear, shoreline-perpendicular backfilled distributary sandstones have a very different geometry to a point bar complex. Lambiase (2012) noted that the presence of backfilled channel sandstones can improve reservoir connectivity relative to purely fluvial systems when the backfilled sandstones are adjacent to multiple point bar complexes. The geometry and total volume of backfilled distributary sandstone reservoirs is also highly variable because it is determined by the number, size and three-dimensional connectivity of the distributaries (Fig. 5). In addition, the sandstone thickness is controlled by channel avulsion so there is no relationship between thickness and channel width in backfilled distributary sandstones, which makes reservoir volume even more difficult to estimate (Lambiase 2012).

Transgressive shoreline sandstones

Similarly, transgressive shoreline sandstones can easily be misinterpreted as fining-upwards fluvial successions when they occur within sequences dominated by progradation. This is clearly demonstrated by the well log signatures and cores from two different stratigraphic intervals in Well X. As described earlier, one interval is a transgressive shoreline sandstone, as evidenced by core and biostratigraphic data (Fig. 12); the other is a fluvial sandstone based on its sedimentary structures and high-resolution biostratigraphic analysis (Figs 12 & 13). However, the two sandstones are nearly identical in thickness and have very similar log signatures (Fig. 13); they would be nearly impossible to distinguish from wireline data alone.

The wave-dominant transgressive shoreline sandstones could also be misidentified as mouth bar deposits if all the sandstones are regarded as progradational. However, they have a different geometry from mouth bar sandstones, and a very different geometry from fluvial channel and point bar sandstones, because their lateral extent is controlled by the pre-transgression delta morphology and their width and thickness are a function of relative sea-level history and wave energy (Fig. 5). As with backfilled distributary sandstones, high-resolution (3 m sample interval) microfossil analysis is the best method of identifying transgressive shoreline sandstones in the absence of core; seismic horizon slicing can also be useful, although many sandstones may be below seismic resolution.

Conclusions

Integrating the sedimentary processes of the modern Mahakam Delta with outcrop examples and subsurface data indicates that transgressive successions are very common and occur in two distinct stratigraphic settings: (1) as laterally extensive sequences and (2) as relatively thin successions within predominantly progradational sequences. Potential reservoir sandstones within those transgressive successions include transgressive shoreline sandstones and, especially, backfilled distributary sandstones, both of which are common.

Backfilled distributary sandstones form relatively linear sand bodies that are quasi-perpendicular to the shoreline; they have a back-stepping stratigraphic architecture with a highly variable geometry and thickness that is controlled by lobe-switching as a result of transgressive avulsion. In contrast, transgressive shoreline sandstones are shoreline-parallel and of variable lateral extent because they are controlled by the pre-transgression delta morphology and tend to be relatively thin as the rate of relative sea-level rise determines their thickness.

Transgressive reservoir sandstones have volumes, sand body geometries and connectivity that are significantly different from stratigraphically adjacent, progradational reservoir sandstones. Both types of transgressive reservoir sandstones can be difficult to recognize in the subsurface, particularly in the thin transgressive successions within predominantly progradational sequences.

Nadia Nirdsal and Ridha Riadi thank PTTEP, Chevron Thailand and the Petroleum Geoscience Programme at Chulalongkorn University for scholarship and research support during their MSc studies. Nadia and Ridha also thank Mike Bidgood, Patrice Brenac and David Shaw for providing biostratigraphic analyses. Salahuddin extends his gratitude to Total E&P Borneo BV and Total E&P Indonésie for sponsoring his PhD scholarship and fieldwork, respectively, and all the Total E&P Indonésie staff technical staff and boat crews who assisted with the fieldwork logistics and data gathering.

References

Allen, G.P. & Chambers, J.L.C. 1998. *Sedimentation in the Modern and Miocene Mahakam Delta*. Indonesian Petroleum Association, Jakarta.

Allen, G.P., Laurier, D. & Thouvenin, J.M. 1976. Sediment distribution patterns in the modern Mahakam Delta. *In*: *Proceedings of the 5th Annual Convention of the Indonesian Petroleum Association. Indonesian Petroleum Association, Jakarta*, 159–178.

Butterworth, P.J., Cook, P. et al. 2002. Reservoir architecture of an incised-valley fill from the Nilam Field, Kutai Basin, Indonesia. *In*: *Proceedings of the 28th Annual Convention of the Indonesian Petroleum Association*. Indonesian Petroleum Association, Jakarta, 537–555.

Cibaj, I. & Wiweko, A. 2008. Recognition of progradational shelf deposits in the middle Miocene of Kutai Basin. *In*: *32nd Annual Convention of the Indonesian Petroleum Association 2008: Meeting Energy Challenges through Cooperation*, 27–29 May 2008, Jakarta, Indonesia. Indonesian Petroleum Association, Jakarta, G-171.

Galloway, W.E. 1975. Process framework for describing the morphologic and stratigraphic evolution of the deltaic depositional systems. *In*: Broussard, M.L. (ed.) *Deltas, Models for Exploration*. Houston Geological Society, Houston, 87–98.

Gastaldo, R.A., Allen, G.P. & Huc, A.Y. 1995. The tidal character of fluvial sediments of the modern Mahakam River Delta, Kalimantan, Indonesia. *In*: Flemming, B.W. & Bartholona, A. (eds) *Tidal Signatures in Modern and Ancient Sediments*. International Association of Sedimentologists, Special Publications, **24**, 171–181.

Hori, K. & Saito, Y. 2008. Classification, architecture and evolution of large-river deltas. *In*: Gupta, A. (ed.) *Large Rivers: Geomorphology and Management*. Wiley, New York, 75–96.

Husein, S. 2008. *Modern sediment dynamics and depositional systems of the Mahakam Delta, Indonesia*. PhD thesis, Universiti Brunei Darussalam.

Lambiase, J.J. 2012. Transgression-driven avulsion of fluvial channels. Abstract presented at the American Association of Petroleum Geologists International Conference and Exhibition, 16–19 September 2012, Singapore. AAPG Search and Discovery Article 90155.

Mora, S., Gardini, M., Kusumanegara, Y. & Wiweko, A. 2000. Modern, ancient deltaic deposits and petroleum system of Mahakam area. *In*: *American Association of Petroleum Geologists–Indonesian Petroleum Association Field Trip Guide Book*. American Association of Petroleum Geologists International Conference and Exhibition, 15–18 October 2000, Bali, Indonesia.

Nirsal, N. 2010. *Facies distribution and stratigraphic development in the palaeo-Mahakam Delta, Indonesia*. MSc thesis, Chulalongkorn University.

Prasetyo, B. 2003. *Facies mapping and reservoir potential of the G58 interval using 3D seismic data in Nilam Field, Sanga-Sanga PSC, Indonesia*. MSc thesis, Universiti Brunei Darussalam.

Riadi, R.S. 2013. *Depositional environments and stratigraphic development of the Grand Taman Sari Circuit outcrop: an analogue for transgressive Mahakam Delta successions*. MSc thesis, Chulalongkorn University.

Riadi, R.S. & Lambiase, J.J. 2015. Outcrop analogues for subsurface sand body geometries in regressive and transgressive Mahakam Delta successions, Indonesia. *In*: *Proceedings of the 39th Annual Convention of the Indonesian Petroleum Association*. Indonesian Petroleum Association, Jakarta, IPA15-G-100.

Salahuddin, & Lambiase, J.J. 2013. Sediment dynamics and depositional systems of the Mahakam Delta, Indonesia: ongoing delta abandonment on a tide-dominated coast. *Journal of Sedimentary Research*, **83**, 503–521.

Trevena, A.S., Partono, Y.J. & Clark, T. 2003. Reservoir heterogeneity of Miocene–Pliocene deltaic sandstones at Attaka and Serang Fields, Kutei Basin, offshore East Kalimantan, Indonesia. *In*: Posamentier, H.W., Sidi, F.H., Darman, H. & Nummedal, D. (eds) *Deltas of the Asia Pacific Region: Modern and Ancient*. SEPM Special Publications, **76**, 235–254.

Xue, C. 1993. Historical changes in the Yellow River delta, China. *Marine Geology*, **113**, 321–329.

Time–space variability of paralic strata deposited in a high accommodation, high sediment supply setting: example from the Cretaceous of Utah

JULIA S. MULHERN & CARI L. JOHNSON*

Department of Geology and Geophysics, University of Utah, 115 S 1460 E, FASB 383, Salt Lake City, UT 84112, USA

**Correspondence: cari.johnson@utah.edu*

Abstract: A previously unstudied section of the John Henry Member (Upper Cretaceous, Straight Cliffs Formation) preserves four stacked regressive–transgressive cycles of paralic strata from the Kaiparowits Plateau in south-central Utah. Meso-scale (10–100 s m thick) shoreface, wave-dominated delta and estuarine depositional environments stack vertically and show the complexity of paralic facies in a single location through time. Correlations with nearby exposures show the palaeogeographical variability updip and along-strike over *c.* 6.5 myr. Such variability highlights the importance of high accommodation settings in preserving transgressive deposits, including landwards-stepping barrier island and lagoon systems. The Buck Hollow section is expanded two to three times compared with correlative successions only 15–40 km away. Tectonics, eustasy and climate contributed to relative shifts in base level, but these regional controls do not explain the dramatic local thickening observed. Local controls on accommodation were quantified through decompaction analysis. The results showed that the expanded thickness of the John Henry Member in Buck Hollow can be explained by differences in decompaction (*c.* 9%), local erosion by fluvial incision (*c.* 5%), early compaction (*c.* 30%) and local structures such as faults (*c.* 100–150%). This outcrop-based study illustrates facies variability within a thick paralic succession and investigates accommodation controls on the preservation of these strata with the goal of improving predictive models for analogous deposits.

Supplementary material: ^{40}Ar/^{39}Ar age data methods and sample location available at https://doi.org/10.6084/m9.figshare.c.3277436.v1

Studies of paralic deposits are generally under-represented in the geological literature, partly as a result of their locations along complex coastal margins with distinct, but dynamic, energy-type controls (wave, tide, rivers; Davies 1964; Galloway 1975; Hayes 1979; Davis & Hayes 1984; Gani & Bhattacharya 2007; Ainsworth *et al.* 2011; Vakarelov & Ainsworth 2013). Paralic strata, particularly because they are linked to transgressive depositional settings, are also broadly perceived to have poor preservation potential (Nichols & Biggs 1985; Davis & Clifton 1987; Demarest & Kraft 1987) and thus there is a lack of robust facies models to guide the recognition and interpretation of these systems. Nevertheless, paralic strata form volumetrically important, but heterogeneous and complex, hydrocarbon reservoirs (Reynolds 1999; Ahlbrandt *et al.* 2005).

In addition to their economic significance, transgressive paralic intervals form an important component of the rock record and are central to sequence stratigraphic interpretations (Arnott 1995). Although progress has been made towards documenting transgressive deposits (Roehler 1988; Devine 1991; Olsen *et al.* 1999; Hubbard *et al.* 2002; Davies *et al.* 2006; Ambrose & Ayers 2007; Plink-Björklund 2008; Sixsmith *et al.* 2008; Allen & Johnson 2011; Kieft *et al.* 2011; Painter *et al.* 2013), models for these settings do not fully encompasses the variability present in both modern and ancient examples, or the interactions of wave, tide, storm and fluvial energy that define coastal settings (Zaitlin & Shultz 1990; Cooper 1993). These models also usually reflect a single snapshot in time and do not consider interactions between the transgressive and regressive components of paralic shorelines through time (Davis & Clifton 1987; Yoshida *et al.* 2004). Sequence stratigraphic interpretations are limited by our understanding of how transgressive deposits stack at the meso-scale (10s to 100s of metres; Yoshida *et al.* 2004; Plink-Björklund 2008; Kieft *et al.* 2011). Similarly, the preservation of transgressive deposits is closely tied to both autogenic and allogenic controls on accommodation (Curray 1964; Schlager 1993; Muto & Steel 1997). In addition to incision and ravinement,

From: Hampson, G. J., Reynolds, A. D., Kostic, B. & Wells, M. R. (eds) 2017. *Sedimentology of Paralic Reservoirs: Recent Advances*. Geological Society, London, Special Publications, **444**, 349–392.
First published online June 29, 2016, https://doi.org/10.1144/SP444.7

tectonics, eustasy, the pre-existing topography and compaction can all play a part in determining the space available for deposition and the likelihood of preservation (Posamentier & Allen 1993), although relatively few studies have investigated the latter two factors (e.g. Reynolds *et al.* 1991).

This study assessed the time–space variability of paralic strata within the Cretaceous Straight Cliffs Formation of southern Utah, USA (Fig. 1). These deposits have received less attention than other Western Interior Seaway successions, yet their interpretation is central to some prominent sequence stratigraphic models (Shanley & McCabe 1991, 1994, 1995). Shoreface, delta and estuary strata of the John Henry Member from a previously unstudied location (Buck Hollow) in the northern Kaiparowits Plateau are documented in detail. The quantitative delineation of facies relationships at the depositional environment scale (10s to 100s of metres vertically) shows the variability possible in a single locality that spans *c.* 7 km along-strike. The temporal variability and evolution are also documented through the 456 m thick succession, spanning *c.* 6.5 myr in duration. Correlations with other locations of John Henry Member deposition in the Kaiparowits Plateau revealed unusual thickening patterns. The possible causes of this local thickening are considered, highlighting some of the unusual characteristics of the Kaiparowits Plateau, with implications for other paralic deposits and foreland basin settings.

Geological background

The late Turonian to early Campanian age Straight Cliffs Formation records fluvial to marginal marine deposition in the Kaiparowits Plateau along the western edge of the Cretaceous Western Interior Seaway (Fig. 1; Peterson 1969*a*, *b*; Eaton & Nations 1991). Subduction of the Farallon plate beneath the North American plate created a broad, asymmetrical foreland basin (Kauffman 1977; DeCelles 2004; DeCelles & Coogan 2006) through flexural loading (Jordan 1981; Pang & Nummedal 1995; Currie 2002; Painter & Carrapa 2013) and dynamic subsidence (Mitrovica *et al.* 1989; Liu *et al.* 2011, 2014). This depression was flooded during the Cretaceous global eustatic highstand (Haq *et al.* 1987; Miller *et al.* 2005) to create an epicratonic seaway (Kauffman 1977; Hancock & Kauffman 1979). The Kaiparowits region was located along the Utah Bight (McGookey *et al.* 1972), a large embayment of the palaeoshoreline in south-central Utah (Fig. 1).

In simple foreland basin models, fold–thrust belts are generally the primary control on both accommodation (through loading) and sediment supply (through erosional exhumation of the orogeny and deposition in the basin; DeCelles & Giles 1996; DeCelles 2012). The preserved geology of the Kaiparowits Plateau shows interactions and conditions more complex than the simplified model. Provenance and sediment dispersal studies indicate that the basin mainly received sediment via distributive fluvial systems fed from the Mogollon Highlands and the Cordilleran Magmatic Arc (Fig. 1a; Lawton *et al.* 2003, 2014; Szwarc *et al.* 2014), with comparatively less input directly from the Sevier fold–thrust belt. These river systems were generally oriented subparallel to the Sevier fold–thrust belt, which generated much of the proximal foredeep accommodation through load-driven subsidence (Jordan 1981; Pang & Nummedal 1995; Currie 2002; Painter & Carrapa 2013). There is also evidence for longshore drift from the north (Allen & Johnson 2010*a*; Szwarc *et al.* 2014). Therefore sediment supply and accommodation were at least partly decoupled during the deposition of the Straight Cliffs Formation. As a result, the basin dynamics and relative sea-level changes within the Kaiparowits region were nuanced and complicated, controlled by processes occurring locally in the basin and in the adjacent Sevier fold–thrust belt, and multiple regional source areas.

Coal resource assessment motivated early investigations of the Straight Cliffs Formation (Peterson 1969*a*, *b*; Vaninetti 1979; Hettinger 2000). Peterson (1969*b*) divided the formation into four members: the Tibbet Canyon Member; the Smoky Hollow Member; the John Henry Member; and the Drip Tank Member (Fig. 2). The lowermost member, the Tibbet Canyon Member, consists of shallow marine facies and overlies the Tropic Shale, recording the transition from offshore marine mudstone deposition to more proximal shoreface and non-marine sedimentation. The Smoky Hollow Member is predominantly terrestrial and consists of carbonaceous floodplain mudstones and thin coals interbedded with isolated fluvial channels. The top of the Smoky Hollow Member is marked by the Calico Bed, an interval of amalgamated, coarse-grained fluvial channels (Bobb 1991). The John Henry Member forms the bulk of the Straight Cliffs Formation and is the focus of this study. The Drip Tank is the uppermost member and is composed of amalgamated fluvial sandstones deposited in distributive fluvial systems (Lawton *et al.* 2014; Gooley *et al.* 2016).

The John Henry Member is the thickest and most laterally variable of the four members of the Straight Cliffs Formation. In the southern and western Kaiparowits Plateau, the John Henry Member preserves coastal plain and fluvial systems flowing NE (Gallin *et al.* 2010; Gooley *et al.* 2016). To the east, the strata transition to offshore, shoreface and tidal

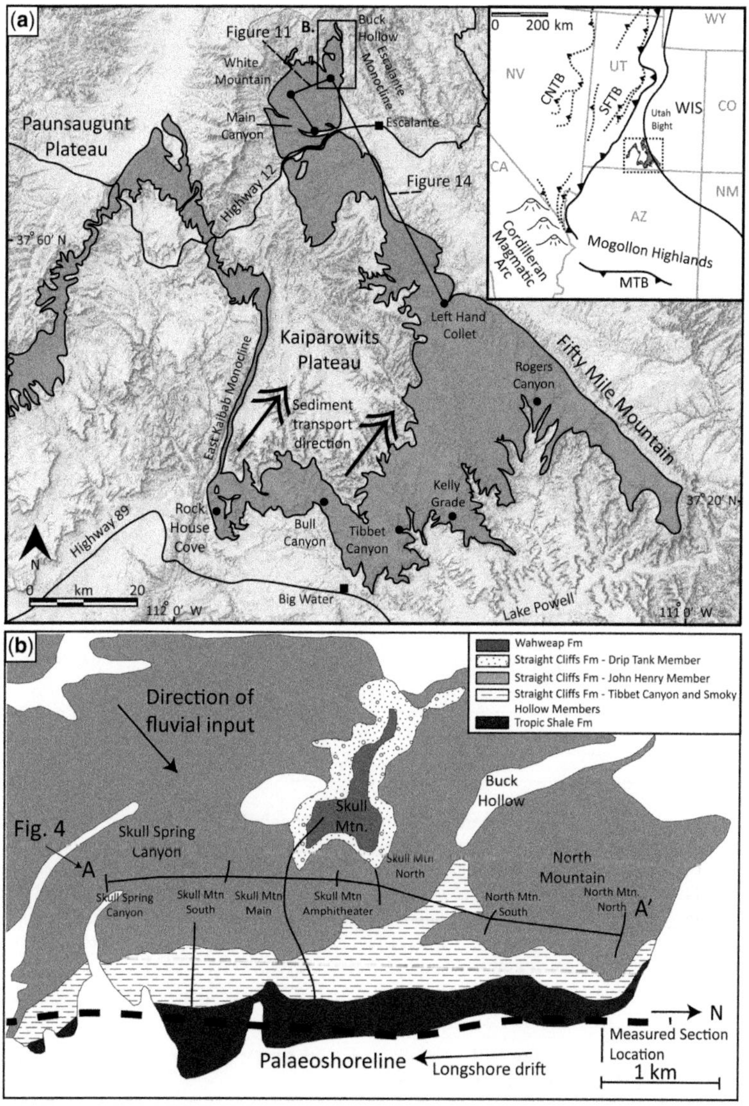

Fig. 1. (**a**) Regional map of the Kaiparowits Plateau of southern Utah. Outcrops of the Straight Cliffs Formation are shown in grey and the locations of previous studies focused on the John Henry Member are labelled. Arrows indicate the general proximal to distal facies relationships in the John Henry Member, ranging from fluvial on the western margin to marine on the eastern margin, with tidal and paralic facies in between. CNTB, Central Nevada Thrust Belt; SFTB, Sevier fold–thrust belt; WIS, Western Interior Seaway; MTB, Maria Thrust Belt. Location of the northern Kaiparowits Plateau geological map shown in (b) and the preliminary stratigraphic correlation lines (Figs 11 & 14) are indicated. Modified from Szwarc *et al.* (2014) and Chentnik *et al.* (2015). (**b**) Simplified geological map of the Buck Hollow field area showing all four members of the Straight Cliffs Formation exposed in an along-strike orientation. Black lines show the locations of measured sections and the location of the correlation line used in Figure 4. Modified from Doelling & Willis (1999).

facies, which are grouped into seven intervals labelled A to G in stratigraphic order of the main shoreface deposits (Fig. 2; Peterson 1969*a*, *b*; Allen & Johnson 2010*a*, *b*, 2011; Dooling 2013). Initial sequence stratigraphic interpretations divided the Straight Cliffs Formation into four unconformity-bound depositional sequences based on a west to east transect through the middle of the Kaiparowits

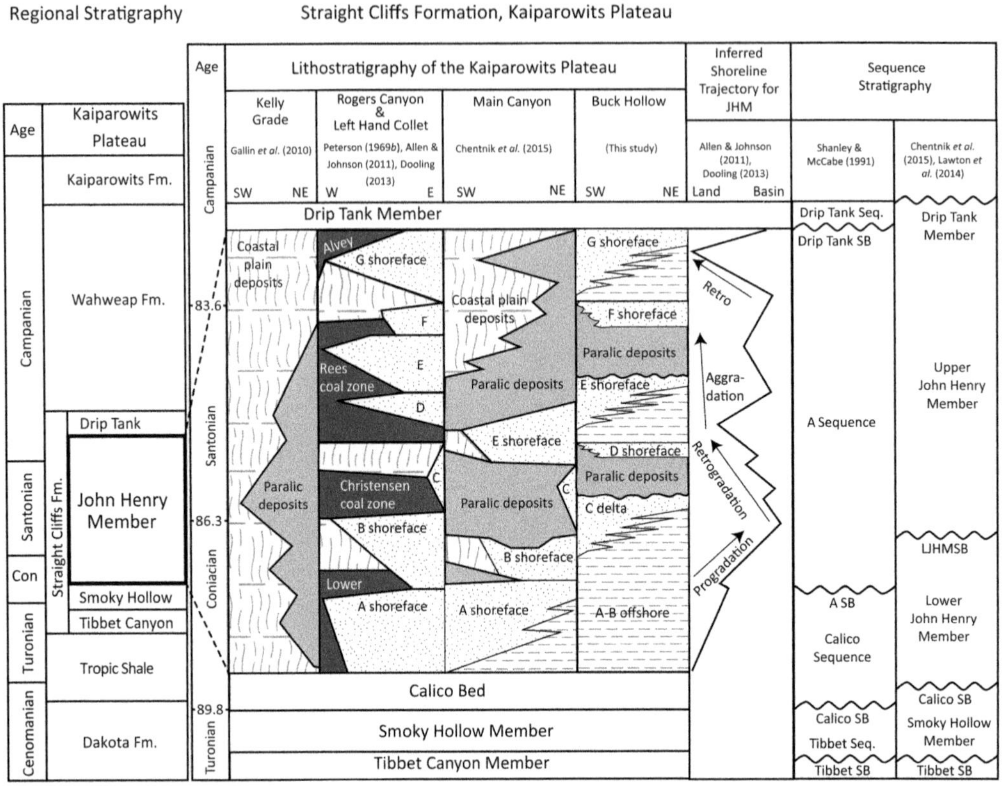

Fig. 2. Stratigraphic summary chart of the Turonian–Campanian Straight Cliffs Formation, including previous lithostratigraphic and sequence stratigraphic interpretations. Seven marine sandstone packages were named A–G by Peterson (1969*a*) and these pinch out landwards into coal zones and coastal plain facies. Relative shoreline movements are based on shoreface pinch-outs and marginal marine facies distributions at Left Hand Collet, Rogers Canyon, Main Canyon and Buck Hollow. Also included are original and updated sequence stratigraphic interpretations for the Straight Cliffs Formation. SB, sequence boundary; LJHMSB, Lower John Henry Member Sequence Boundary; JHM, John Henry Member. Modified from Chentnik *et al.* (2015).

Plateau (Fig. 2; Shanley & McCabe 1991, 1993, 1994, 1995; McCabe & Shanley 1992). Recent stratigraphic assessments from various locations around the plateau have refined this initial model, including the recognition of basin-axial fluvial drainage networks and modification of the sequence boundary interpretations (Allen & Johnson 2011; Szwarc *et al.* 2014; Chentnik *et al.* 2015; Gooley *et al.* 2016).

Notably, the Straight Cliffs succession north of Highway 12 was not included in the early studies (cf. Hettinger *et al.* 1993; Shanley & McCabe 1995). Moving *c.* 14 km to the NE of Main Canyon, this study documents strata in an unstudied portion of the north Kaiparowits Plateau, with the main goals of documenting the time–space variability of a thick, mainly transgressive paralic succession, as well as investigating first-order controls on sedimentation and preservation.

Field analysis

Buck Hollow is located along Posey Lake Road, 15 km north of Escalante, Utah (Fig. 1b). The study area consists of a *c.* 7 km long along-strike outcrop exposure (Figs 1b & 3) that preserves all four members of the Straight Cliffs Formation (Figs 1b & 3). The cliff face has *c.* 1700 m relief and displays the most distal facies of the John Henry Member preserved in the Kaiparowits Plateau. These outcrops are immediately adjacent to the prominent Escalante Monocline, a presumed Laramide-style, basement-involved feature (Sargent & Hansen 1982; Davis 1999). Structural (n = 128) measurements record a local structural dip of 15° W consistently through the Straight Cliffs Formation (Fig. 1b) at Buck Hollow.

Measured section analysis (17 sections, totalling 2008 m) collectively spans from the top of the

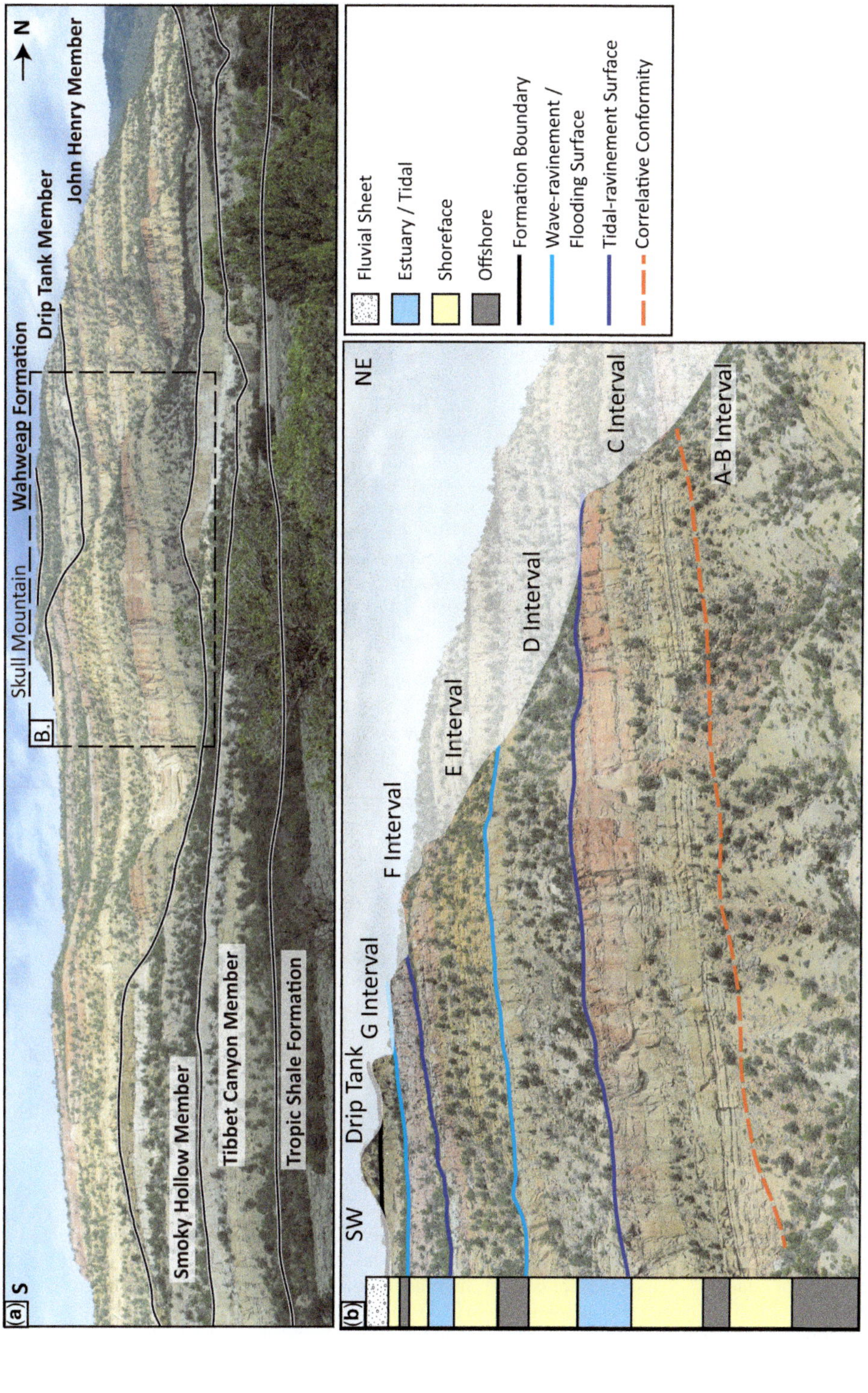

Fig. 3. Outcrop photos of the Buck Hollow field area. (**a**) An oblique photo-pan looking north at the along-strike outcrop exposures in Buck Hollow. (**b**) Photo of the John Henry Member of the Straight Cliffs Formation exposed on the southern limb of Skull Mountain in the Buck Hollow field area. The intervals of the John Henry Member (A–G) are labelled. Major surfaces are marked and the column indicates the broad depositional environments preserved in each interval.

Tropic Shale, through all four members of the Straight Cliffs Formation, into the basal portion of the overlying Wahweap Formation (Figs 1b, 3 & 4). All sections record vertical changes in grain size, texture, sorting, sedimentary structures, bedding, trace and body fossils, and palaeocurrent indicators. Correlations between measured sections were extended by walking along bedding contacts and analysing the aerial photography. Palaeocurrent directions ($n = 1472$) were collected from trough cross-stratification (axes), tabular accretion sets and ripple laminations (Fig. 4; Allen 1966; Miall 1974). Data from Buck Hollow were combined with recent work from White Mountain, Main Canyon (Chentnik *et al.* 2015) and Left Hand Collet (Dooling 2013) and with aerial photography for regional correlation and analysis. Provenance data, specifically sandstone modal compositions and U/Pb detrital zircon geochronology, of the Buck Hollow samples (Szwarc *et al.* 2014) were also considered. Mudstone samples were collected throughout the Straight Cliffs Formation for biostratigraphy (Fig. 5) and examined for palynomorphs (48 samples), nannofossils (18) and forams (18). These data are used here to support depositional environment and general age interpretations; a full assessment of the biostratigraphic data is in progress and will be published at a later date (D. Pocknall pers. comm. 2015). Biotite grains from a bentonite layer in the lower John Henry Member were analysed for $^{40}Ar/^{39}Ar$ geochronology.

Stratigraphy

The paralic strata of Buck Hollow are described using 15 lithofacies (LF) identified based on lithology, primary bedding structures, and trace and body fossils (Table 1). The lithofacies have been combined to describe 15 facies associations (Table 2). The facies associations are grouped to define three depositional environments, interpreted in the following sections. Dinoflagellates and nannofossils help to constrain the timing of deposition and suggest that the John Henry Member at Buck Hollow is late Coniacian to early Campanian in age, generally consistent with other studies (Fig. 5; Eaton 1991; Dyman *et al.* 2002; Szwarc *et al.* 2014).

Facies association 1

Description. Facies associations 1.1–1.5 (Table 2; Fig. 6a) include sandstone and mudstone units that form laterally continuous, coarsening-upwards, stacked cliff faces (17–51 m thick). Facies association 1.1 is composed of planar to wavy-laminated, fissile grey mudstones (LF1, Fig. 6j) and interbedded sandstones (LF2) containing occasional inoceramid and oyster shell fragments. Mudstone intervals contain diverse dinoflagellate assemblages and common nannofossils (Fig. 5). Facies association 1.2 is composed of millimetre-scale laminated mudstones and sandstones (1–10 cm thick; LF2; Fig. 6b) grading vertically into hummocky and swaley cross-stratified sandstone beds (2–50 cm thick; LF3). The sandstone beds (LF3) of facies association 1.2 contain *Arenicolites* (Fig. 6e) and have *Thalassinoides* on the uppermost bedding surfaces. Facies association 1.3 contains hummocky and swaley cross-stratified sandstone beds 0.1–1 m thick (LF3; Fig. 6h, i) grading into planar laminated and trough cross-stratified sandstone beds (LF4; Fig. 6c). These sandstone beds are separated by millimetre- to centimetre-scale, planar laminated mudstone layers (LF1; Fig. 6b). Facies association 1.4 is composed of blocky bedded (0.5–2 m thick), fine- to medium-grained, trough cross-stratified and planar laminated sandstones containing *Skolithos*, *Ophiomorpha* (LF4; Fig. 6d) and *Cylindrichnus concentricus* (Fig. 6g). Beds are laterally continuous and separated by thin (<1 cm thick) interbeds of fissile, laminated, grey mudstones (LF1; Fig. 6k). Facies association 1.5 is a single bed (1–2 m thick) of planar laminated medium-grained sandstone (LF5) with occasional shell fragments and 0.5 cm diameter mudstone clasts.

Interpretation. Facies association 1 is composed of vertically stacked offshore (FA1.1), distal lower shoreface (FA1.2), proximal lower shoreface (FA1.3), upper shoreface (FA1.4) and foreshore (FA1.5) deposits. Facies association 1 is interpreted as a wave-dominated shoreface environment (Table 2; Fig. 6). These sequences were deposited on a wave- and storm-dominated shelf (Plint 2010; Li *et al.* 2011). The packages thicken and shoal vertically, suggesting that the shoreline prograded during deposition (Van Wagoner *et al.* 1988). The preserved trace fossils, nannofossils and the diverse dinoflagellate assemblages are commonly found in marine settings (Figs 5 & 6d–g; MacEachern *et al.* 2010; Ekdale & Harding 2015), suggesting marine deposition for these facies. Large intervals of mudstone (FA1.1; Fig. 3b) represent stacked offshore deposits, the distal expression of a marine shoreface.

Facies association 2

Description. Facies association 2.1 is composed of sharp-based, gently inclined (<5°; Fig. 7j), hummocky and swaley cross-stratified sandstones (LF3) and trough cross-bedded and planar fine- to medium-grained sandstones (LF4). These sandstone units contain *Ophiomorpha*, shell fragments (Fig. 7) and leaf impressions (Fig. 7i). Sandstone beds are

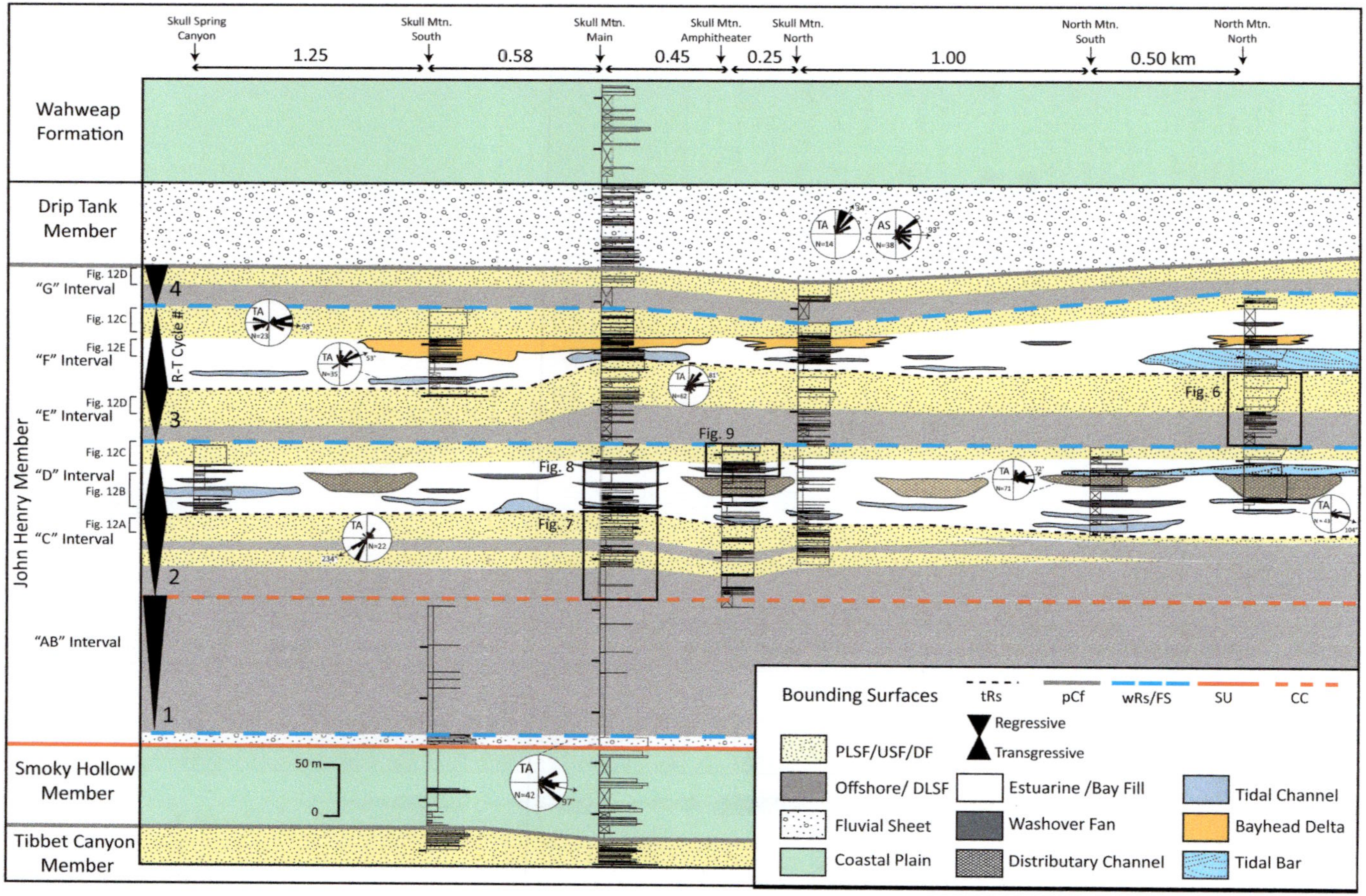

Fig. 4. Stratigraphic correlation of the Buck Hollow field area. Cross-section line and measured section locations are shown in Figure 1b. Black triangles mark the regression–transgression cycles. Also noted are the intervals depicted by the palaeogeographical maps in Figure 12. AS, accretion sets; CC, correlative conformity; DF, delta front; DLSF, distal lower shoreface; PLSF, proximal lower shoreface; pCf, process change from wave- to fluvial-dominated processes; R–T cycle, regressive–transgressive cycle; SU, subaerial unconformity; TA, trough axes; tRs, tidal ravinement surface; USF, upper shoreface; wRs/FS, wave ravinement/flooding surface.

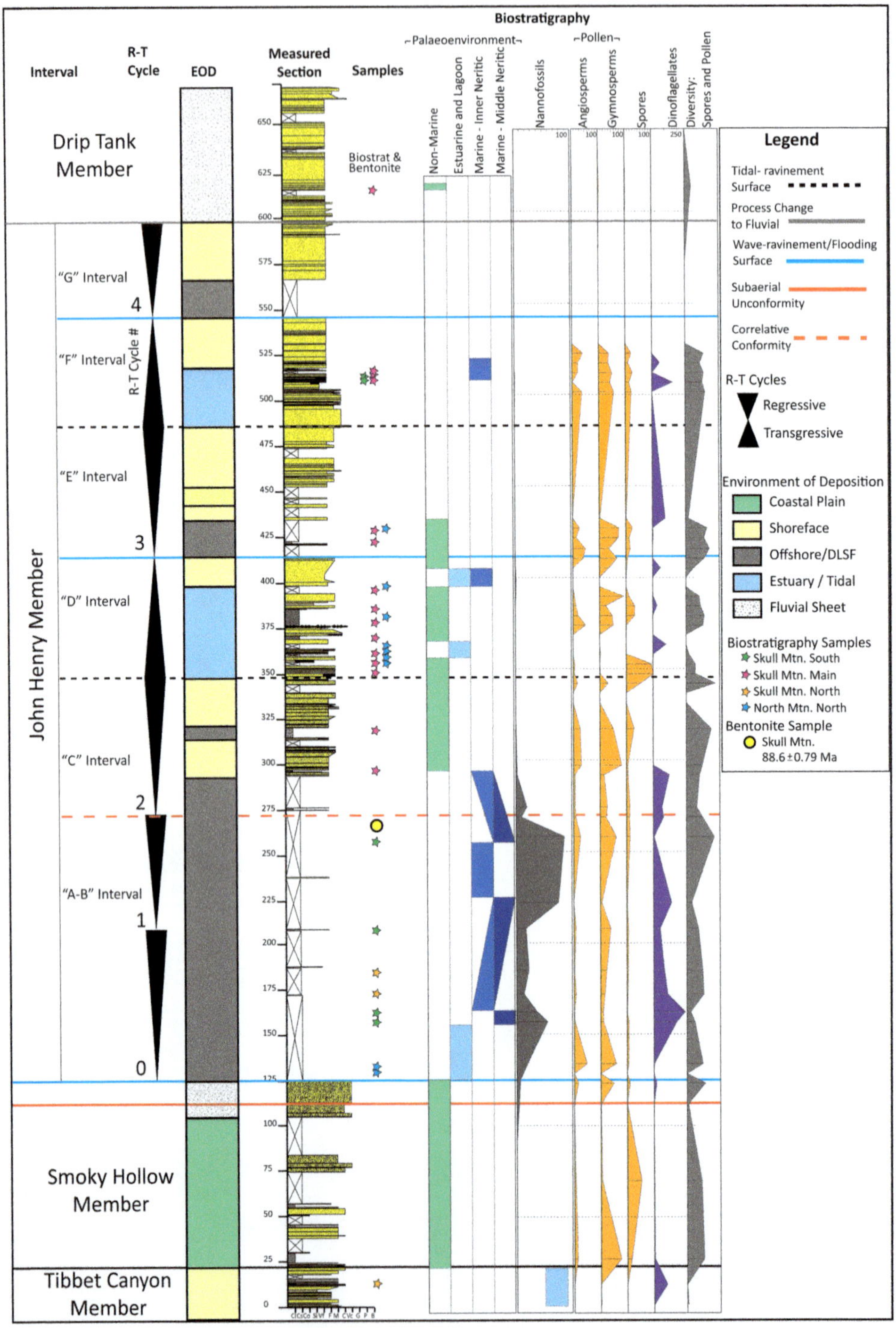

Fig. 5. Biostratigraphic summary data for Buck Hollow plotted beside the Skull Mountain Main measured section (location in Fig. 1b). Sample locations are marked with stars, coloured by location. Interpreted palaeoenvironment track is divided into four sub-environments. The curves of the key fossil groups represent the total count of specimens per sample. Location of $^{40}Ar/^{39}Ar$ dated bentonite sample is marked by a yellow circle. DLSF, distal lower shoreface; EOD, environment of deposition; R–T cycle, regressive–transgressive cycle.

Table 1. *Lithofacies descriptions*

Lithofacies	Lithofacies description and architecture	Trace/body fossils
LF1: grey mudstones	Wavy-laminated mudstones, generally grey-coloured and fissile. Clay-rich with a minor silty component	Small shell fragments, inoceramid and oyster shell fragments. BI = 0–1
LF2: Interbedded light brown–grey mudstones, siltstones, fine-grained sandstones	Planar laminated mudstones and siltstones, 1–10 cm thick, with rare symmetrical wave ripples. Sandstone beds are 0.2–1 m thick, sharp-based and contain abundant HCS. Packages coarsen upwards from silt to very fine-grained sandstone in 0.5–4 m sections	Small shell fragments, *Thalassinoides, Planolites*. BI = 0–3
LF3: Hummocky and swaley cross-stratified sandstones	Outcrop-forming very fine- to fine-grained sandstones, form sharp-based coarsening-upwards packages separated by finer grained sandstones or marine mudstones. Amalgamated sandstone units can be >25 m thick and laterally continuous >10 km. HCS and SCS are abundant. Locally, SCS grades into zones of planar laminations that are gently inclined to the east	Minor inoceramid and oyster shell fragments, *Ophiomorpha, Thalassinoides, Skolithos*. BI = 0–3
LF4: Trough cross-stratified sandstones and planar bedded sandstones	Cliff-forming fine- to medium-grained sandstones. Abundant TCS. Internally, bedding is blocky and consists of thick (>1 m) sandstone beds with occasional mud (*c.* 5 cm) interbeds	Small shell fragments, *Ophiomorpha*. BI = 0–3
LF5: Homogeneous and planar bedded medium-grained sandstones	Planar bedded to massive medium-grained sandstones. Rare TCS	Rare shell fragments. BI = 0–1
LF6: Carbonaceous mudstones and coals	Massive and planar laminated siltstones and mudstones with minor interbeds of very fine-grained ripple-laminated sandstones and coal. The organic content of carbonaceous shales increases vertically at the bed scale, with thin (typically <0.5 m) coals typically occurring as capping units. Siltstones and rippled sandstones are most commonly observed at the base of sandstone bodies	The only fossils observed were root traces. BI = 0–1
LF7: Interlaminated light and dark grey mudstones	Blocky, wavy-interbedded mudstones with millimetre-scale laminations of alternating light and dark grey mudstones. Common (<1 cm thick) coarse-grained sandstone layers	*Skolithos* filled with coarse-grained orange sandstone. BI = 0–2
LF8: Shell hash conglomerate	Small, laterally discontinuous 0.2–2 m beds of shell fragments within a coarse-grained sandstone matrix. Shell fragments range from whole to broken and weathered. Fragment sizes range from 0.5–15 cm long	Oyster shells. BI = 0
LF9: Isolated medium-grained sandstone lenses	Lens-shaped, medium-grained sandstone deposits with internal TCS. Generally, white to yellow coloured. Pockets of convolute bedding and soft sediment deformation. Prevalent oyster shell	Prevalent shell fragments, often large (>15 cm), grouped into lenses and lags. Molluscs and inoceramids, *Teredolites,*

(*Continued*)

Table 1. *Continued*

Lithofacies	Lithofacies description and architecture	Trace/body fossils
	fragments, often large (>15 cm), grouped into lags and lenses. Sporadic large (>20 cm) pockets of *Teredolites*	*Skolithos, Palaeophycus, Ophiomorpha*. BI = 0–3
LF10: Amalgamated trough cross-bedded sandstones with mudstone rip-up clasts	Cliff-forming, amalgamated, TCS, channelized coarse-grained sandstones with rare pebbles. Generally orange-coloured. Channelized beds cut into matrix-supported intervals with highly variably sized mud clasts (0.5–25 cm). Some mud clasts show internal lamination. Matrix is composed of coarse-grained, cemented sandstones	Leaf fossils, *Teredolites*, oyster fragments. BI = 0–2
LF11: Channelized bidirectional sandstones	Laterally confined sigmoidal bars and channel-shaped sandstones. Sandstones typically fine upwards and contain primary sedimentary structures of low-angle planar laminations, wavy-lenticular-flaser bedding, convolute beds, bidirectional, planar and TCS with double mud drapes and abundant reactivation surfaces	Inoceramid and oyster shell fragments; trace fossils are scarce, but include: *Skolithos* and *Planolites*. BI = 0–3
LF12: Tabular, bidirectional cross-stratified sandstones	Medium-grained, well-sorted sandstones contain abundant reactivation surfaces, planar, trough, sigmoidal and herringbone cross-stratification, mud and carbonaceous draped cross-sets, and ball-pillow and flame soft sediment deformation features	Shell fragments of inoceramid and oyster fragments are common, with minor amounts of terrestrial plant debris observed scattered throughout; *Palaeophycus* and *Diplocraterion*. BI = 0–1
LF13: High-angle laminated sandstones	Medium-grained sandstones with high-angle accretion sets and TCS. Beds range in thickness from 0.5 to 1.5 m Some convolute bedding visible towards the top	Occasional shell fragments. BI = 0
LF14: Inclined heterolithic strata	Thinly interbedded, very fine-grained sandstones and mudstones. Inclined beds are between 0.25 and 1.5 m thick and laterally interfinger with medium-grained sandstone deposits, carbonaceous shales; terrestrial material visible within layers. Double mud drapes occur along foresets of low-angle TCS within sandstones	Shell and wood fragments were found with marine trace fossils, ammonite impressions, shark teeth; *Planolites, Cylindrichnus, Thalassinoides* and *Skolithos*. BI = 0–3
LF15: Heavily bioturbated sandstones	Cliff-forming fine-grained sandstone (*c.* 1–3 m thick) intervals with heavy bioturbation	Occasional shell fragments and shark teeth; trace fossils almost exclusively *Ophiomorpha*. BI = 3–6
LF16: Homogenous, tabular sandstones	Homogenous, tabular fine- to medium-grained sandstones	Occasional shell fragments and mudstone clasts concentrated along bedding laminations. BI = 0

Bioturbation Index (BI) follows the scheme of Taylor & Goldring (1993).
HCS, hummocky cross-stratification; SCS, swaley cross-stratification; TCS, trough cross-stratification.

tabular and increase in size from 10 cm to *c.* 2 m thick towards the top of the interval (Fig. 7b, f); these are separated by laterally discontinuous mudstone layers of variable thickness (0.1–1 cm; Fig. 7g). The facies association is stacked into two coarsening-upwards cliff faces (*c.* 21–27 m) to the

Table 2. *Description of facies associations for Buck Hollow*

Lithofacies	Description	Facies association	Environment of deposition	Justification
Facies Association Group 1: wave-dominated shoreface (Fig. 6)				
LF1, LF2, LF3	Recessed and partly covered intervals composed of planar to wavy-bedded, fissile, grey mudstones (LF1) and interbedded mudstones and sandstones (LF2; Fig. 6j) with occasional Inoceramid and oyster shell fragments. Thin sandstones (LF3) display hummocky cross-stratification and are occasionally bioturbated (*Thalassinoides*)	1.1	Offshore	These mudstones are interpreted as offshore strata (Plint 2010) deposited below storm weather wave base (>30; George & Hill 2008), probably transported via hyperpycnal flows (Hill *et al.* 2007; Bhattacharya & MacEachern 2009)
LF1, LF2, LF3	Mudstone deposits (LF1) grading vertically into interbeds of millimetre-scale laminated mudstones and sandstones (LF2) and hummocky and swaley cross-stratified sandstone beds (LF3) 2–50 cm thick	1.2	Distal lower shoreface	These hummocky and swaley cross-stratified interbedded sandstones record shelfal transport and reworking during storms (Duke 1985; Southard *et al.* 1990; Dumas 2005; Dumas & Arnott 2006), suggesting deposition in a distal lower shoreface environment (Clifton 2006)
LF3, LF4, LF1	Hummocky and swaley cross-stratified sandstone beds (LF3) 0.1–1 m thick grading into planar laminated and trough cross-stratified sandstone beds (LF4; Fig. 6h, i). Sandstones separated by millimetre- to centimetre-scale mudstone layers (LF1)	1.3	Proximal lower shoreface	These sandy deposits show evidence of both oscillatory and geostrophic flows, suggesting deposition below the fair weather wave base in the proximal lower shoreface (Reading & Collinson 1996) known to contain *Cylindrichnus concentricus* (Frey 1990)
LF4, LF1	Blocky bedded (0.5–2 m thick) fine- to medium-grained, trough cross-stratified and planar bedded sandstones containing *Skolithos* and *Ophiomorpha* (LF4). Beds are laterally continuous and separated by thin (<1 cm thick) interbeds of fissile, laminated, grey mudstones (LF1)	1.4	Upper shoreface	The higher energy indicators, such as *Ophiomorpha* (Droser & Bottjer 1989), within these deposits suggest deposition in the upper shoreface above the fair weather wave base (Plint 2010)
LF5	Single bed 1–2 m thick bed of planar laminated medium-grained sandstone (LF5) with occasional shell fragments in some locations	1.5	Foreshore	These planar laminated sandstones are foreshore strata deposited in the swash zone (Wright & Short 1984; Masselink & Puleo 2006; Plint 2010)
Facies Association Group Two: wave-dominated delta (Fig. 7)				
LF3, LF4, LF1	Sharp-based, gently inclined (<5°), hummocky and swaley cross-stratified sandstones (LF3) and trough cross-bedded and planar fine- to medium- grained sandstones (LF4). Sands contain *Ophiomorpha*, shell fragments (Fig. 7e) and leaf impressions (Fig. 7i).	2.1	Delta front	The sharp-based nature of the sandstone packages, inclination of bedding and terrigenous material suggest deposition in mouth bars in a delta front with sediment supplied by nearby distributary channels (Walker & Plint 1992; Gingras *et al.*

(Continued)

Table 2. *Continued*

Lithofacies	Description	Facies association	Environment of deposition	Justification
	Beds are tabular and increase in size from 0.1 to 2 m thick (Fig. 7b). Mudstone interbeds are laterally discontinuous and vary in thickness (0.1–1 cm; Fig. 7b)			1998). Dominance of trough cross-bedded sandstones suggests unidirectional current flow in the upper delta front (Charvin *et al.* 2010)
Facies Association Group Three: wave-dominated estuary (Figs 8 & 9)				
LF6, LF1, LF7	Carbonaceous mudstones and siltstones, often capped by 0.1–1 m coal beds (Fig. 8e). Carbonaceous mudstones and siltstones have a high sulphur content, abundant plant material and contain sporadic shell fragments. Interval is predominantly covered and erodes into blocky fragments, suggesting wavy rather than fissile bedding. Tends to create intervals 0.2–5 m thick. Intervals are eroded by higher energy facies	3.1	Proximal carbonaceous estuary fill	These carbonaceous intervals are interpreted as deposition within a low-energy back-barrier estuary or bay with flanking swamps or tidal marsh areas, which provided the terrigenous material (Snedden & Kersey 1981; Holz *et al.* 2000). Carbonaceous shales grade vertically into coals, suggesting the coals are wetting upwards and transgressive (Wadsworth *et al.* 2010)
LF9, LF5, LF8	Lens-shaped medium-grained sandstone deposits with internal trough cross-stratification (LF9). Sandstones have pockets of convolute bedding and soft sediment deformation. Prevalent oyster shell fragments, often large (>15 cm), grouped into lags and lenses (LF8; Fig. 8c). Interval is heavily bioturbated and contains *Ophiomorpha, Palaeophycus* (Fig. 8d) and occasional large (>20 cm) pockets of *Teredolites*	3.2	Washover fans	These deposits are washover fans, which display upper plane bed sedimentation (Barwis & Hayes 1985) with trough and ripple cross-stratification formed by runoff channels (McCubbin 1982; Schwartz 1982). Marine trace fossils suggest deposition within the subaqueous back-barrier
LF14, LF15, LF16, LF7	Sandstone-dominated outcrop (*c.* 3.5 m thick) with a variety of internal facies stacked into 1–50 cm thick beds composed of inclined heterolithic strata (LF14), homogenous sandstone beds (LF16), heavily bioturbated beds (LF15), convolute bedding (Fig. 8g) and laminated mudstones (LF7; Fig. 8i). Bioturbation is prevalent throughout the interval, with burrows both concentrated on bedding planes (*Thalassinoides*; Fig. 8b), within beds (*Cylindrichnus concentricus*; Fig. 8l) and cross-cutting strata (*Skolithos*; Fig. 8f). Occasional shell fragments, shark teeth (Fig. 8k), ammonite impressions (Fig. 8h) and gastropod shells found	3.3	Bayhead delta	These stacked, slightly inclined, heterolithic, upwards-coarsening sands are interpreted as the distal expression of a bayhead delta (Joeckel & Korus 2012). Convolute bedding suggests loading and dewatering during deposition (Dzuynski & Smith 1963). Shark teeth are indicative of an estuarine, marine-influenced setting (Becker *et al.* 2010; Cumbaa *et al.* 2010; Kirkland *et al.* 2013). *Cylindrichnus concentricus* is consistent with shallow marine deposition (Belaústegui & de Gibert 2013; Ekdale & Harding 2015)
LF10, LF6, LF7	Large (30 m thick), amalgamated, lenticular, coarse- to medium-grained, erosive sandstone deposits (LF10;	3.4	Distributary channel	These stacked, amalgamated sand lenses are interpreted as distributary channels entering an

	Fig. 8o) cut through carbonaceous mudstones and coals (LF6). Between the individual channels within the larger amalgamated intervals, large (>10 cm) mudclasts (Fig. 8q) are visible in a coarse-grained matrix. Wood material (Fig. 8p), leaf fossils, *Teredolites* (Fig. 8r) and shell fragments (Fig. 8m) are preserved. Continuous, laminated mudstone layers (LF7) separate the channelized facies			estuary. The matrix and mud rip-up clast layers that create the interbeds between the channel sands represent reworking of estuarine bay fill. Continuous mud layers are interpreted to be fluid mud deposits (Ichaso & Dalrymple 2009). Both terrestrial (wood fragments and plant material) and marine (*Teredolites* and shell fragments) material are visible within these deposits
LF7, LF11	Slope-forming deposits of blocky, wavy-interbedded mudstones with millimetre-scale laminations of alternating light and dark grey mud (LF7). Occasional (*c.* 1 cm) coarse-grained sandstone layers and round burrows, *c.* 3–8 mm in diameter, filled with coarse-grained sandstones. Isolated channelized sandstone lenses (LF11) are confined with these fine-grained deposits. Intervals are non-continuous and isolated	3.5	Tide-dominated central estuary bay fill	These fine-grained, laminated facies represent deposition within the central basin of an estuary (Kvale 2012). Coarse-grained sand interlaminations are the result of storm washover into the central bay. Isolated tidal channels cut through this bay fill
LF11	Lens-shaped outcrops of medium-grained sandstones (LF11) with trough cross-stratification, bidirectional cross-stratification and double mud drapes (Fig. 9b)	3.6	Tidal channels	These tidal channel deposits are fairly common, particularly at the base of the transgressive intervals, probably because they have a relatively high preservation potential (Davis 2012)
LF13, LF12	Medium-grained sandstones with trough cross-stratification and bidirectional cross-stratification (LF13, LF12) and distinct, bed-scale, angled lateral accretion master surfaces (Fig. 9d). Interference ripples visible on master bedding planes. Forms laterally extensive, 3–15 m thick white sheets that can be traced for over 10 km. This facies can have either an erosional base with 1–2 m of relief or conformably overlie shoreface deposits. Occasional trace fossils (*Diplocraterion*, *Palaeophycus* and *Teredolites*) visible (Fig. 9e, f, g, i)	3.7	Tidal bars	The visible lateral accretion master surfaces within these trough and planar cross-bedded sandstones suggest these features are tidal bars (Dalrymple & Choi 2007; Olariu *et al.* 2012) which form in the constricted part of the estuary experiencing a high tidal flow regime (Dalrymple & Rhodes 1995). The clean nature of the sandstone suggests these deposits are either estuarine tidal bars or tidal delta bars (Dalrymple *et al.* 1992) rather than muddier bayhead delta tidal bars (Fenies & Tastet 1998)
LF16, LF7	Stacked, tabular, westwards-inclined fine- to medium-grained sandstone beds (LF16) that vary in thickness laterally. Some beds show 10–30 cm of incision into underlying beds and well-indurated shell fragment lags along the basal incision surface. Some metre-thick beds are homogenous with abundant mudstone clasts. Mudstone interbeds separating sandstones are grey and fissile (LF7)	3.8	Flood tidal delta	The westwards (landwards) stacked, gently dipping beds within this interval suggest flood tidal delta deposition (Fitzgerald *et al.* 2012). Storm action gives the deposits a lobate shape (Boothroyd *et al.* 1985)

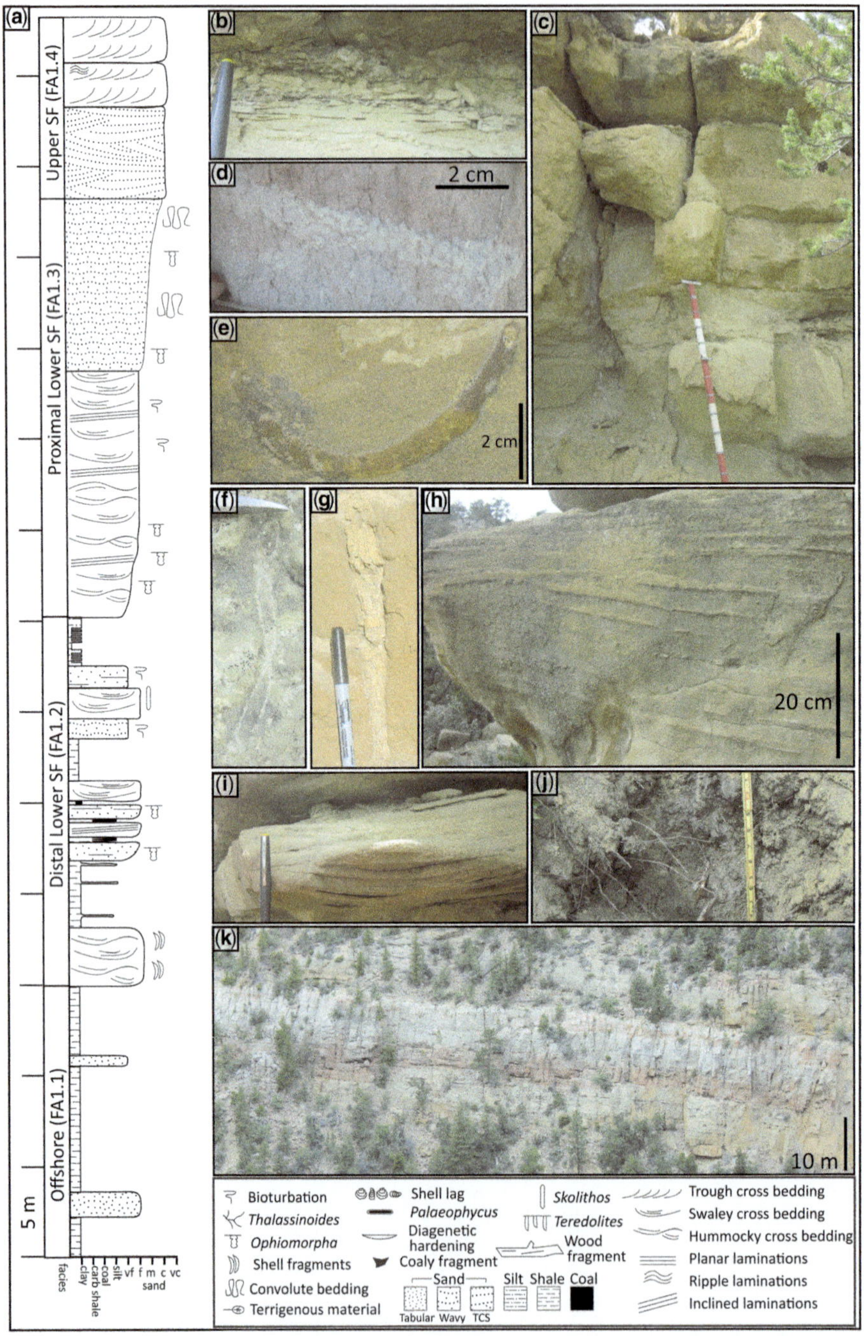

Fig. 6. Representative measured section and field photos of the wave-dominated shoreface facies association (FA1). (**a**) Representative measured section of the shoreface facies association. The lower John Henry Member is characterized by coarsening-upwards parasequences stacked progradationally. Section is the E interval of the North Mountain North section (Fig. 4). (**b**) Interbedded mudstone and sandstone layers (LF2). (**c**) Representative outcrop showing stacked hummocky, swaley (LF3) and trough cross-stratified sandstone beds (LF4). (**d**) *Ophiomorpha nodosa.* (**e**) *Arenicolites.* (**f**) *Palaeophycus.* (**g**) *Cylindrichnus concentricus.* (**h**) Swaley cross-stratification (LF3). (**i**) Hummocky cross-stratification (LF3). (**j**) Grey fissile mudstones (LF1). (**k**) Three progradationally stacked shoreface parasequences from the E interval at Buck Hollow. SF, shoreface; FA, facies association.

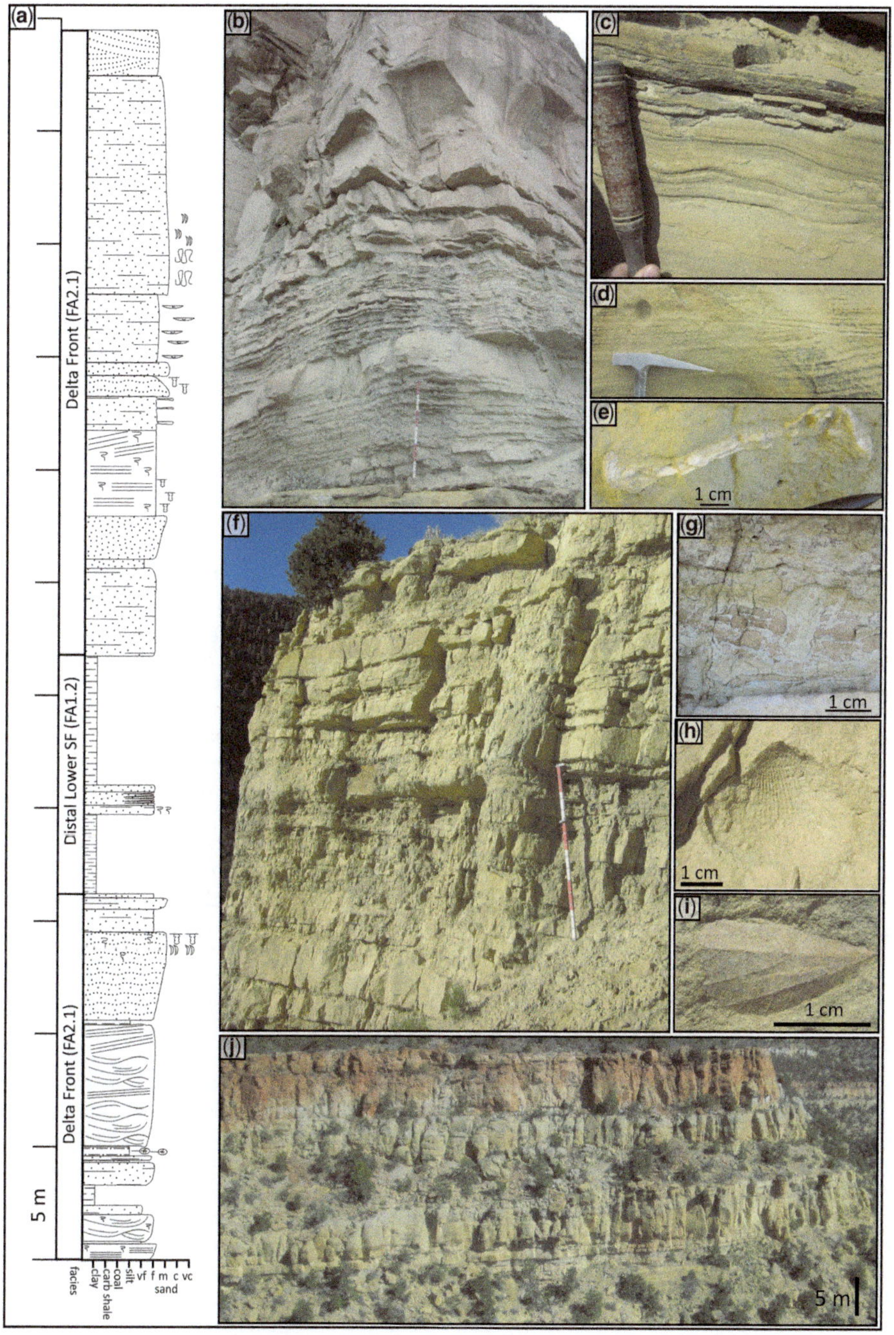

Fig. 7. Field photos and measured section representing the wave-dominated delta facies association (FA2). (**a**) Representative measured section of a delta succession. Section is the C interval from the Skull Mountain Main section (Figs 4, 11 & 14). (**b**) Delta front facies (LF3 & LF4), with wavy bedding (**c**). (**d**) The lower delta front contains swaley cross-stratification (LF3), suggesting wave influence. (**e**) Inoceramid fragment. (**f**) Outcrop photo of the lower delta front showing interbedded mudstones and sandstones (LF2 & LF3). (**g**) Muddy interbedded interval from the delta front showing disturbed beds. (**h**) Shell impression. (**i**) Leaf impression showing proximity to fluvial output. (**j**) Outcrop photo showing progradationally stacked delta successions truncated by tidal ravinement. Legend as Figure 6.

NE (<5°; Fig. 7j). Although tabular, the individual beds pinch out into intervening mudstone intervals over 10 s of metres laterally. However, the cliffs are continuous laterally across the entire study area (*c.* 7 km). The thickness of individual sandstone beds increases vertically (Fig. 7b), showing increased amalgamation towards the top of the cliff faces as the internal sedimentary structures change from swaley cross-stratification to trough and planar cross-stratification (Fig. 7j).

Interpretation. Facies association 2 is composed of vertically stacked offshore (FA1.1), distal lower shoreface (FA1.2) and wave-dominated delta front facies (FA2.1; Table 2; Fig. 7) and is interpreted as wave-dominated delta deposits. The same facies associations were used to interpret the basal portions of the wave-dominated shoreface deposits and the wave-dominated delta deposits because the distal expression of the two environments is indistinguishable in outcrop (Bhattacharya & Walker 1991; MacEachern & Pemberton 1992; Reading & Collinson 1996).

Moving vertically into the upper delta front, the facies show distinct deltaic characteristics. Blocky, gently inclined, trough and planar cross-stratified sandstone beds in the upper portion of the delta front facies are interpreted as storm-reworked, shore-parallel, mouth bar deposits (Fig. 7j; Gingras *et al.* 1998). This sand was output from a fluvial source as mouth bars and then reworked onto the shoreface through wave and storm action (Bhattacharya & Giosan 2003). Individual pulses of sand to the delta front resulted in obliquely exposed clinoforms (0.1–2 m thick and 25–100 m wide; Gani & Bhattacharya 2007; Graham *et al.* 2015). The clinoforms dip (<5°) to the NE, revealing progradation of the delta shoreface (21–27 m thick). Abundant plant material (Fig. 7i), shell fragments (Fig. 7e, h) and wavy bedding (Fig. 7c) within the delta front suggest the proximity of these deposits to the shoreline and rapid deposition. These characteristics indicate proximity to the sediment source, a river influence and rapid sedimentation, distinguishing the delta front from proximal lower and upper shoreface strata. The thinning and pinching out of tabular beds over tens of metres suggest rapid, pulsated deposition, with dewatering contributing to the discontinuous nature of the beds.

Trace fossils are commonly used to distinguish between wave-dominated deltas and wave-dominated shoreface coastlines (MacEachern & Pemberton 1992; Li *et al.* 2011); however, burrowing within both the upper shoreface and delta front at Buck Hollow is not distinct, with the preservation of *Ophiomorpha* and *Skolithos* burrows of similar distribution and intensity in both settings. Pellet-lined *Ophiomorpha nodosa* burrows (Fig. 6d) suggest a high-energy environment (Droser & Bottjer 1989) with water conditions suitable for suspension feeding (MacEachern *et al.* 2010). A low diversity of trace fossils suggests a stressed environment, consistent with rapid deposition (Gingras *et al.* 1998), fitting a wave-dominated delta and/or shoreface experiencing high sedimentation rates.

Facies association 3

Description. Facies association 3 is highly variable (Figs 8 & 9), containing eight sub-environments. Facies association 3.1 consists of carbonaceous shales and coals (Fig. 8e). Commonly capped by 10–100 cm thick coal beds (Fig. 8e), these carbonaceous mudstones and siltstones have significant sulphur residues, abundant plant material (Fig. 8j) and can contain sporadic shell fragments. These fine-grained facies are slope-forming and cut into by isolated bodies of facies associations 3.2, 3.3 and 3.4. Facies association 3.2 consists of lens-shaped, medium-grained sandstone deposits (Fig. 8c) with internal trough cross-stratification (LF9) and pockets of convolute bedding and soft sediment deformation. Throughout the lenses there are prevalent, large (>15 cm) oyster shell fragments (Fig. 8c), which are grouped into lags and lenses (LF8). There is heavy bioturbation, including *Palaeophycus* (Fig. 8d), *Ophiomorpha nodosa* and a few large (>20 cm) *Teredolites* assemblages.

Facies association 3.3 is composed of sandstone-dominated outcrops (*c.* 3.5 m thick; Fig. 8i) subdivided into 1–50 cm thick beds composed of inclined heterolithic strata (LF14), homogenous sandstone beds (LF16), heavily bioturbated beds (LF15), convolute beds (Fig. 8g) and laminated mudstone intervals (LF7). Bioturbation is prevalent throughout the facies association, with burrows both concentrated on bedding planes (*Thalassinoides*; Fig. 8b), cross-cutting strata (*Skolithos*; Fig. 8f) and within beds (*Cylindrichnus concentricus*; Fig. 8l). Minor shell fragments (including gastropods), shark teeth (Fig. 8k) and ammonite impressions (Fig. 8h) occur throughout this interval.

Facies association 3.4 is made up of large (30 m thick), amalgamated, lenticular, coarse- to medium-grained, erosional sandstone deposits (LF10; Fig. 8o), which cut into underlying carbonaceous mudstones and coals (LF6). The amalgamated lenticular sandstones contain wood material (Fig. 8p), leaf fossils, *Teredolites* (Fig. 8r) and shell fragments (Fig. 8m). Between the individual channels, large (>10 cm) mudstone clasts (Fig. 8q) are present in a coarse-grained matrix.

Facies association 3.5 is generally a slope-forming succession, consisting of blocky to wavy-interbedded mudstones with millimetre-scale laminations of alternating light and dark grey mudstone

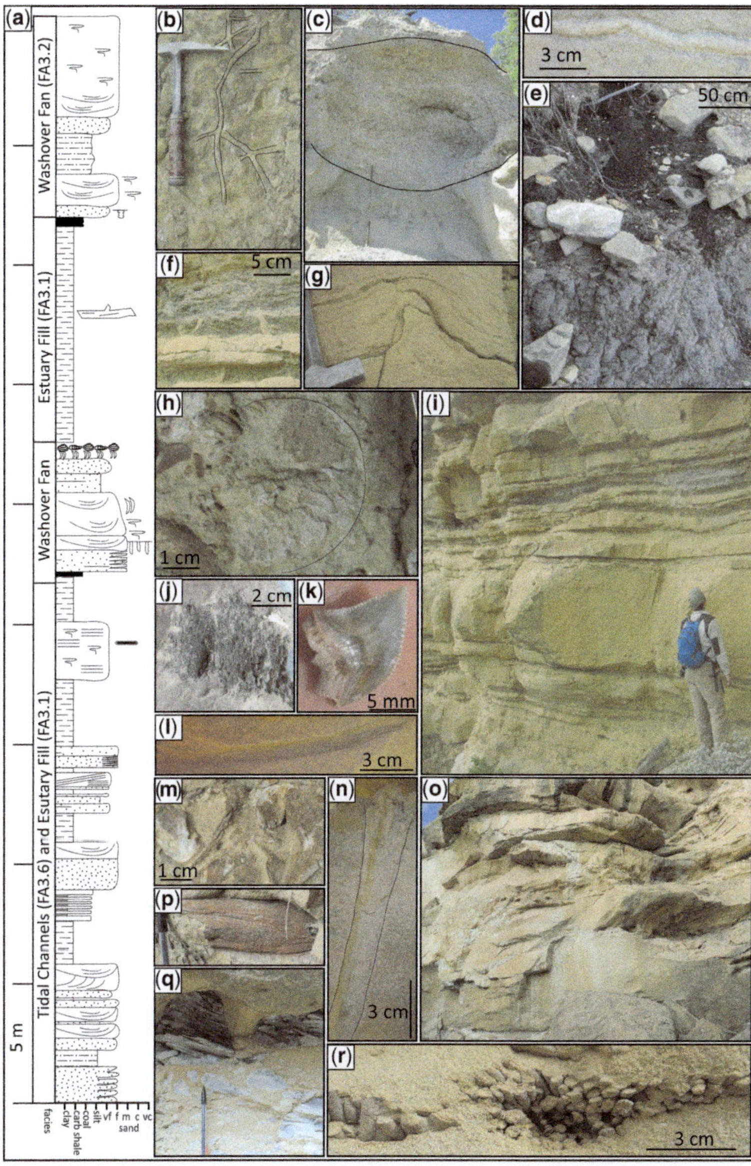

Fig. 8. Field photos illustrating the diversity of estuarine facies (FA3) preserved in Buck Hollow. (**a**) Representative measured section for the estuarine facies association. Section is the lower D interval from the Skull Mountain Main section (Figs 4, 11 & 14). (**b**) *Thalassinoides* burrows on a bedding plane of a bayhead delta sandstone bed. (**c**) Washover fan outcrop (LF9) with an outline around a concentration of shell fragments in the upper part of the fan (LF8). (**d**) *Palaeophycus* fossil from a washover fan outcrop. (**e**) Carbonaceous shale and coal at the base of a washover fan outcrop. (**f**) Vertical trace fossils (*Skolithos*) from the interbedded portion of a bayhead delta outcrop. (**g**) Convolute bedding from a bayhead delta outcrop. (**h**) Ammonite impression from a bayhead delta outcrop. (**i**) Outcrop image of a bayhead showing the wavy and interbedded nature of the beds (LF14). (**j**) Impression of plant material from carbonaceous mudstones (LF6) within the bayhead delta deposits. (**k**) Shark's tooth from a bayhead delta deposit. (**l**) *Cylindrichnus concentricus* from a bayhead delta outcrop. (**m**) Shell fragments from a washover fan outcrop (LF8). (**n**) *Cylindrichnus concentricus* from a distributary channel. (**o**) Distributary channel outcrop (LF10) with metre-scale channels separated by a coarse-grained matrix with abundant mudstone clasts. (**p**) Terrigenous material from a distributary channel outcrop (LF10). (**q**) Large mudstone clasts within a coarse-grained sandstone matrix from the interchannel portion of a distributary channel outcrop (LF10). (**r**) *Teredolites* from an interchannel portion of a distributary channel outcrop. Legend as Figure 6.

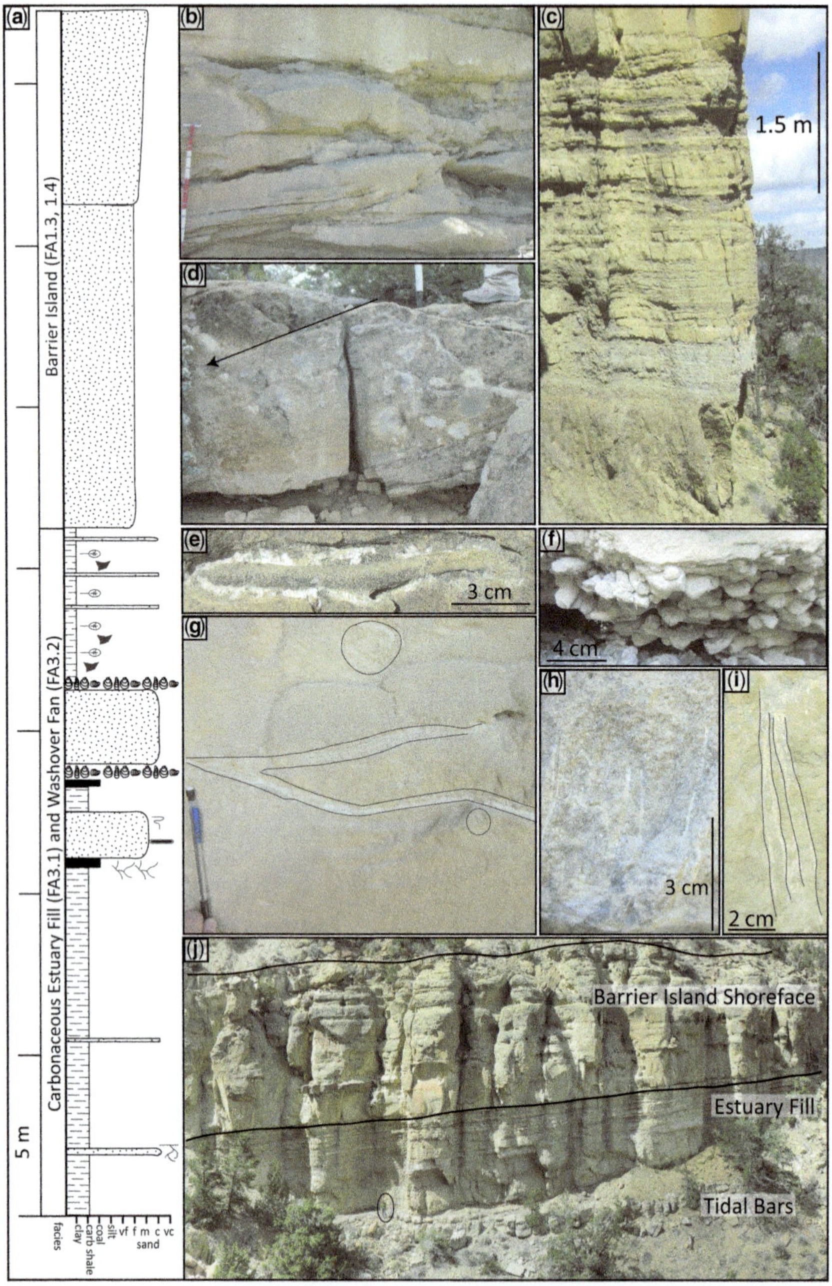

Fig. 9. Field photos illustrating the tidal (FA3) and barrier island facies present in Buck Hollow. (**a**) Representative measured section showing estuarine facies overlain by a barrier island shoreface. Section is the upper D interval from the Skull Mountain Amphitheater section (Fig. 4). (**b**) Tidal channels (LF11). (**c**) Interbedded interval composed of laminated mudstones (LF7) and coarse-grained sandstone lenses (LF16). Mudstones contain abundant leaf and plant impressions. Sandstone lenses are *c.* 1–10 cm thick and contain small coal and mud rip-up fragments. (**d**) Tidal bars with lateral accretion master surfaces (LF13). (**e**) *Palaeophycus* from a tidal bar. (**f**) *Thalassinoides* preserved within a coal bed at the base of a tidal bar. (**g**) Tidal bar showing *Palaeophycus, Cylindrichnus* and *Ophiomorpha*. (**h**) Rhizoliths from a tidal bar. (**i**) *Diplocraterion* from a tidal bar. (**j**) Estuary and barrier island outcrop showing the sharp juxtaposition of barrier island shoreface sandstones on top of interbedded estuary fill. Legend as Figure 6.

(LF7). These deposits include occasional (*c.* 1 cm) coarse-grained sandstone layers and round burrows, *c.* 3–8 mm in diameter, filled with coarse-grained sandstone. Facies association 3.5 is cut into by facies associations 3.6, 3.7 and 3.8. Facies association 3.6 is composed of medium-grained, lens-shaped sandstone bodies (LF11; Fig. 9b) that contain trough cross-stratification, bidirectional cross-stratification and double mud drapes. These sandstones commonly erode into the underlying deposits with 0.5–5 m of relief. Facies association 3.7 is composed of medium-grained sandstones divided into beds by master surfaces with distinct, variably inclined (*c.* 1–20°), lateral accretion generally to the south (Fig. 9d). The beds exhibit trough cross-stratification and bidirectional cross-stratification (LF12, LF13) and form 3–15 m thick sheets. These beds occasionally contain rhizoliths (Fig. 9h) and trace fossils (*Diplocraterion* and *Palaeophycus*; Fig. 9e, g, i). In some locations, facies association 3.7 overlies coal deposits (FA3.1) with distinct *Thalassinoides* at the base (Fig. 9f). Facies association 3.8 displays stacked, tabular, westwards-inclined fine- to medium-grained sandstone beds (LF16), which vary in thickness laterally from 0.2 to 2 m. Some beds have well-indurated shell fragment lags along the basal incision surface, whereas others have abundant mudstone clasts. Fissile grey mudstone interbeds (LF7) separate the sandstones.

Interpretation. Facies association 3 records the juxtaposition of high- and low-energy facies. Areas of fine-grained carbonaceous mudstones and coals (FA3.1; Fig. 8e; Table 2) are interpreted as estuary fill, indicating deposition in a protected back-barrier setting with flanking swamps (Holz *et al.* 2002). These low-energy deposits are cut into by higher energy deposits interpreted as washover fans (FA3.2), bayhead delta deposits (FA3.3) and distributary channels (FA3.4). Overall, facies association 3 is interpreted as a wave-dominated estuary (Table 2). Dinoflagellates from within the estuarine deposits indicate a brackish water setting (Fig. 5).

Washover fans (FA3.2; Fig. 8a, c) record sand entering the estuary during storms (Schwartz 1982; Sedgwick & Davis 2003; Hudock *et al.* 2014) and, with marine shells (Fig. 9c, m), they suggest deposition in a coastal transgressive setting (Roehler 1988; Wadsworth *et al.* 2010). The stacked, inclined, heterolithic, upwards-coarsening sandstone units of FA3.3 (Fig. 8i) are interpreted as the distal portion of a bayhead delta (Joeckel & Korus 2012). These deposits contain thin (1–25 cm) wavy and inclined strata (Fig. 8i) interbedded with finer grained facies, similar to the facies patterns seen in cores from the modern Trinity River bayhead delta in Texas (McEwen 1969). Convolute bedding (Fig. 8g) suggests loading (Dzuynski & Smith 1963) and rapid deposition, known to occur in modern bayhead deltas such as the Atchafalaya delta in Louisiana (van Heerden & Roberts 1988). Shark teeth (Fig. 8k), bioturbation (Fig. 8b, f, l) and ammonite impressions (Fig. 8h) indicate a marine influence (Becker *et al.* 2010; Cumbaa *et al.* 2010; Kirkland *et al.* 2013). Combined, these characteristics help to differentiate bayhead delta deposits from other estuarine sandstone bodies.

Distributary channels (FA3.4; Fig. 8o) are recognized by coarse-grained sandstone lenses containing terrigenous material (Fig. 8p) and fluid mud layers (Ichaso & Dalrymple 2009). These distributary channels pinch laterally into finer grained, central estuary bay fill (FA3.5). Erosional features at the base of the interval are interpreted as tidal channels (FA3.6; Fig. 9b; Hughes 2012). Laterally, the tidal channels (FA3.6) pinch out into fine-grained laminated strata (FA3.5) interpreted as tidally influenced central estuary bay deposits (Kvale 2012). Flood tidal delta (FA3.8) deposits reflect the interaction of both wave and tide processes near a tidal inlet (Fitzgerald *et al.* 2012).

The estuary intervals are capped with a tidal bar complex (FA3.7; Fig. 9d) and back-stepping barrier island strata (FA1.3, FA1.4; Fig. 9a, j). Tidal bar deposits (FA3.7; Fig. 9d) are identified based on lateral accretion master surfaces (Dalrymple & Choi 2007; Olariu *et al.* 2012), creating longitudinal tidal bars (Dalrymple & Rhodes 1995). Shoreface sandstones (FA1.3, FA1.4; Fig. 9j) sharply juxtaposed above estuary fill (Fig. 9c) are interpreted as barrier island deposits. These shoreface deposits are distinguished by a planar erosional surface at the base (Olsen *et al.* 1999) followed by a mix of planar and trough cross-bedded sandstones with abundant *Ophiomorpha* burrows, similar to modern and recent examples (e.g. Carter 1978; Reinson 1992; Willis & Moslow 1994; Tye *et al.* 1999).

Decompaction analysis

One-dimensional basin modelling was performed to extract decompacted thicknesses for representative stratigraphic sections from Buck Hollow and Main Canyon (Fig. 10). Lithologies for the facies at Buck Hollow and Main Canyon were selected to best match the measured section descriptions and field observations. The thicknesses and lithologies of the overburden were compiled and estimated from previously published data (Table 3 and references cited therein). Auto-calculated seawater interface temperature estimates for 37° W latitude (North America) were used (Kauerrauf & Hantschel 2009). Heat flow was held at a constant 60.00 mW m^{-2}.

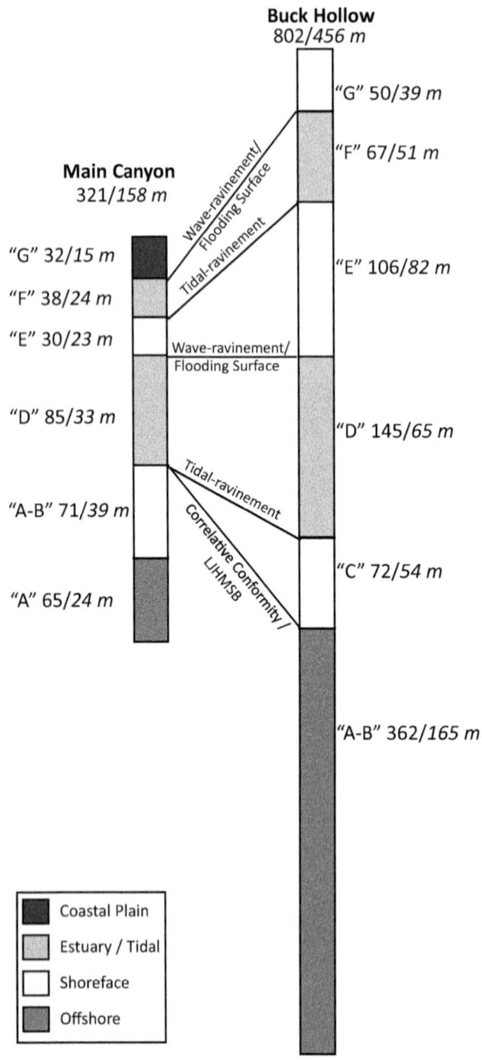

Fig. 10. Schematic stratigraphic columns showing the decompacted and compacted (italics) thicknesses of representative measured sections for Buck Hollow and Main Canyon. Decompaction values were generated through one-dimensional basin modelling; see Table 3 for model inputs.

Decompacted thicknesses were extracted from the thickness overlay feature of the basin history output.

The decompacted thickness of Buck Hollow was 802 m, 1.76× expanded from its present day measured thickness (456 m). The decompacted thickness of Main Canyon was 321 m, 2.03× expanded from its present day measured thickness (158 m). After decompaction, the John Henry Member at Buck Hollow was 481 m (2.5×) thicker than Main Canyon (Table 3). Sensitivity analysis was performed to assess the compaction variability of different coal lithologies (Table 4) and showed that the various coal lithologies tested all expanded three to four times during decompaction.

Discussion

Depositional model

Four regressive–transgressive (R–T) cycles define a depositional model for the John Henry Member at Buck Hollow. The letter-assigned shoreline intervals (A–G of Peterson 1969*a*; Fig. 5) of the John Henry Member were grouped into R–T cycles to describe the sequence stratigraphic architecture. R–T cycles were used (Cattaneo & Steel 2003; Zecchin 2007; Allen & Johnson 2011) rather than parasequence models (Kamola & Van Wagoner 1995; Posamentier & Allen 1999; Hampson & Storms 2003; Catuneanu *et al.* 2009) because R–T cycles are more applicable to successions with significant transgressive deposits (Helland-Hansen & Gjelberg 1994; Arnott 1995; Zecchin 2010). In this study, the cycles begin with regressive surfaces (cf. the T–R cycles of Embry & Johannessen 1993) because major tidal ravinement occurs at the base of the transgressive units, creating irregular and variable boundaries (e.g. Swift 1968; Cattaneo & Steel 2003; Sixsmith *et al.* 2008; Chentnik *et al.* 2015). Conversely, combined wave ravinement/flooding surfaces (Arnott 1995), which mark the base of regressive intervals, are relatively flat across the study area, creating consistent boundaries and datums. The John Henry Member R–T cycles at Buck Hollow (summarized in the following sections in stratigraphic order) are probably third or fourth order, based on their thickness, expression and averaged individual time duration of *c.* 1.5 myr (Vail *et al.* 1977; Vail & Mitchum 1979; Emery & Myers 1996). Within them, numerous higher frequency cycles are present, the expression of which varies by depositional environment (e.g. coarsening-upwards parasequences in shoreface intervals and estuary-infilling coals).

R–T cycle 1 (A–B interval). R–T cycle 1 is composed of a *c.* 150 m thick offshore mudstone interval (FA1.1). The base is a wave ravinement/flooding surface (Fig. 4) characterized by a quartzite pebble conglomerate layer that can be traced regionally into Main Canyon (Chentnik *et al.* 2015) and Left Hand Collet Canyon (Fig. 1; Dooling 2013). This surface is flat and marks the base of the John Henry Member. The conglomerate layer is directly overlain by a thick (*c.* 150–163 m) succession of offshore mudstones consisting of the A and B intervals of Peterson (1969*a*, *b*), henceforth referred to as A–B. The transgressive portion of the cycle cannot

Table 3. *Basin modelling parameters*

Interval	Original thickness (m)	Decompacted thickness (m)	Age (Ma)	Lithology	References
Claron Formation	640		40.0–34.0	Limestone (shaly)	Taylor (1993), Goldstrand (1994), Larsen (2007)
Pine Hollow Formation	137		62.0–55.0	Shale (organic-poor, sandy)	Larsen (2007), Larsen *et al.* (2010)
Cannan Peak Formation	274		70.5–62.0	Conglomerate (typical)	Larsen *et al.* (2010)
Kaiparowits Formation	726		76.1–74.7	Shale (organic-poor, sandy)	Eaton (1991), Lawton & Bradford (2011)
Wahweap Formation	231		81.8–76.1	Sandstone (clay-rich)	Lawton *et al.* (2003, 2014), Jinnah *et al.* (2009), Larsen *et al.* (2010); Jinnah & Roberts (2011)
SCF Drip Tank	65		82.8–81.8	Sandstone (quartzite, typical)	Szwarc *et al.* (2014)
SCF JHM G	39	50	83.5–82.8	Sandstone (typical)	
SCF JHM F	30	39	84.0–83.5	Sandstone (typical)	
SCF JHM F	21	28	84.5–84.0	Sandstone (subarkose, clay-rich)	
SCF JHM E	82	106	85.5–84.5	Sandstone (typical)	
SCF JHM D	16	21	86.0–85.5	Sandstone (typical)	
SCF JHM D	3	12	86.2–86.0	Coal (pure)	
SCF JHM D	46	112	86.8–86.2	Shale (organic-poor, silty)	
SCF JHM C	54	72	88.5–86.8	Sandstone (quartzite, typical)	$^{40}Ar/^{39}Ar$ date (this study)
SCF JHM A–B	165	362	89.1–88.5	Shale (organic-rich, typical)	$^{40}Ar/^{39}Ar$ date (this study)
SCF Smoky Hollow Member	102		93.2–89.1	Sandstone (clay-rich)	(J. Primm pers. comm. 2016)
SCF Tibbet Canyon Member	22		95.0–93.2	Sandstone (typical)	(J. Primm pers. comm. 2016)
Tropic Shale	190		98.0–95.0	Shale (typical)	Eaton (1991)
Dakota and Cedar Mountain formations	57		100–98.0	Sandstone (typical)	Eaton (1991), Dyman *et al.* (2002)

General references: Hintze (1988), Eaton (1991), Doelling & Willis (1999), Doelling *et al.* (2000) and Hettinger *et al.* (2009).
JHM, John Henry Member; SCF, Straight Cliffs Formation.

be distinguished from the regressive portion based on the outcrop exposure or from the biostratigraphic data, therefore the two are not specifically differentiated. This interval is laterally continuous across the entire field area (*c.* 7 km along-strike; Fig. 4).

The upper limit of R–T cycle 1 is a correlative conformity that can be traced updip to the Lower John Henry Member Sequence Boundary in Main Canyon (Fig. 11). This surface exhibits significant (up to 22 m) incision along north–south-trending erosional valleys in Main Canyon (Chentnik *et al.* 2015). However, there is no indication of subaerial exposure in Buck Hollow, only uninterrupted offshore sedimentation, suggesting a correlative conformity (Fig. 4). The position of the correlative conformity surface is estimated to be a few metres

Table 4. *Coal lithology sensitivity analysis results*

Lithology	Decompacted thickness (m)
Coal (pure)	12
Coal (with impurities)	11
Shale (organic-rich, 20% total organic carbon)	9
Shale (organic-rich, 8% total organic carbon)	9
Shale (organic-rich, typical)	9
Shale (black)	9
Coal (silty)	9

below the base of the first well-exposed cliff band of the C interval in R–T cycle 2 (Fig. 3b). This is supported by biostratigraphic data, which imply a relative shallowing (Fig. 5). Within the offshore A–B interval there are several bentonite beds. $^{40}Ar/^{39}Ar$ dating of biotite grains from the uppermost bentonite bed yielded an age of 88.6 ± 0.79 Ma (Fig. 5).

R–T cycle 2 (C and D intervals). R–T cycle 2 is *c.* 140 m thick and records forced regression, resulting in basinwards-stepping delta deposition followed by estuarine deposition during accretionary transgression (Figs 4 & 12). R–T cycle 2 begins above the correlative conformity marking the upper limit of R–T cycle 1. Across that conformable surface, the offshore mudstone deposits of the A–B interval grade into the more proximal siltstone and mudstone units of the basal C interval, with biostratigraphic indicators shifting to more proximal marine environments (Fig. 5). The regressive portion of R–T cycle 2 is composed of a prograding wave-dominated delta (Figs 3b & 12a). The delta interval consists of two *c.* 21–27 m thick, coarsening-upwards, progradationally stacked sequences of shore-parallel mouth bars composed of delta front facies (FA2.1; Figs 8a, b, j, 12a). These two smaller cycles are separated by a *c.* 8 m thick interbedded mudstone and sandstone interval, representing a minor basinwards shift in facies that juxtaposes offshore deposits on top of deltaic upper shoreface deposits. The upper deltaic sequence contains more proximal facies relative to the lower

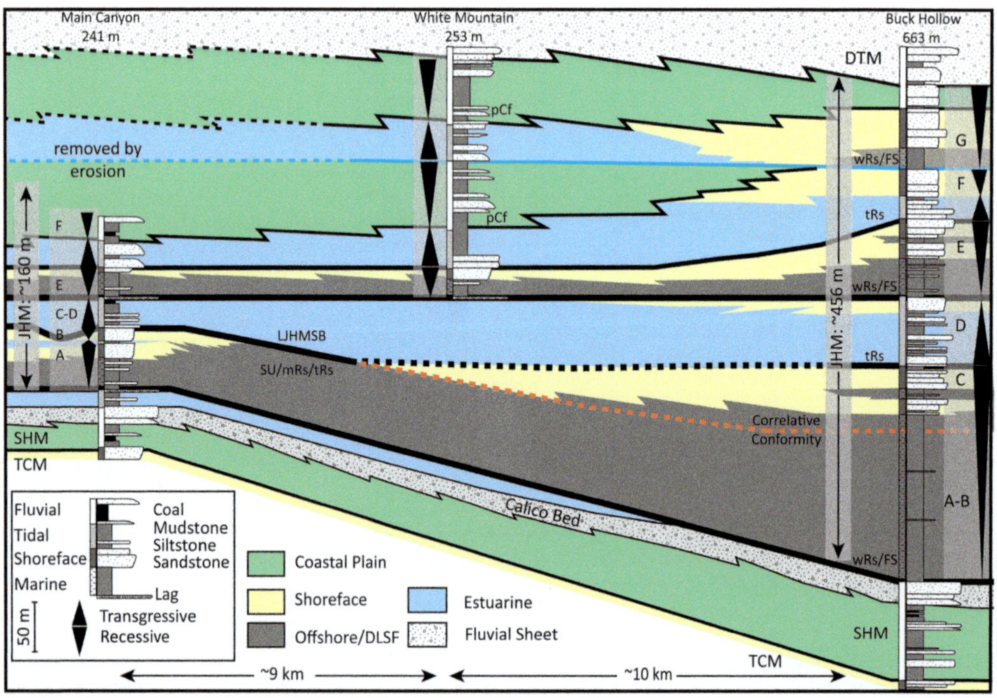

Fig. 11. Regional stratigraphic correlation from SW to NE across the northern Kaiparowits Plateau from Main Canyon and Buck Hollow. Correlation location shown on Figure 1. DLSF, distal lower shoreface; DTM, Drip Tank Member; JHM, John Henry Member; LJHMSB, Lower John Henry Member Sequence Boundary; pCf, process change from tide- to fluvial-dominated processes; SHM, Smoky Hollow Member; SU/mRs/tRs, combined subaerial unconformity, maximum regressive surface and tidal ravinement surface; TCM, Tibbet Canyon Member; tRs, tidal ravinement surface; wRs/FS, wave ravinement/flooding surface. Modified from Chentnik *et al.* (2015).

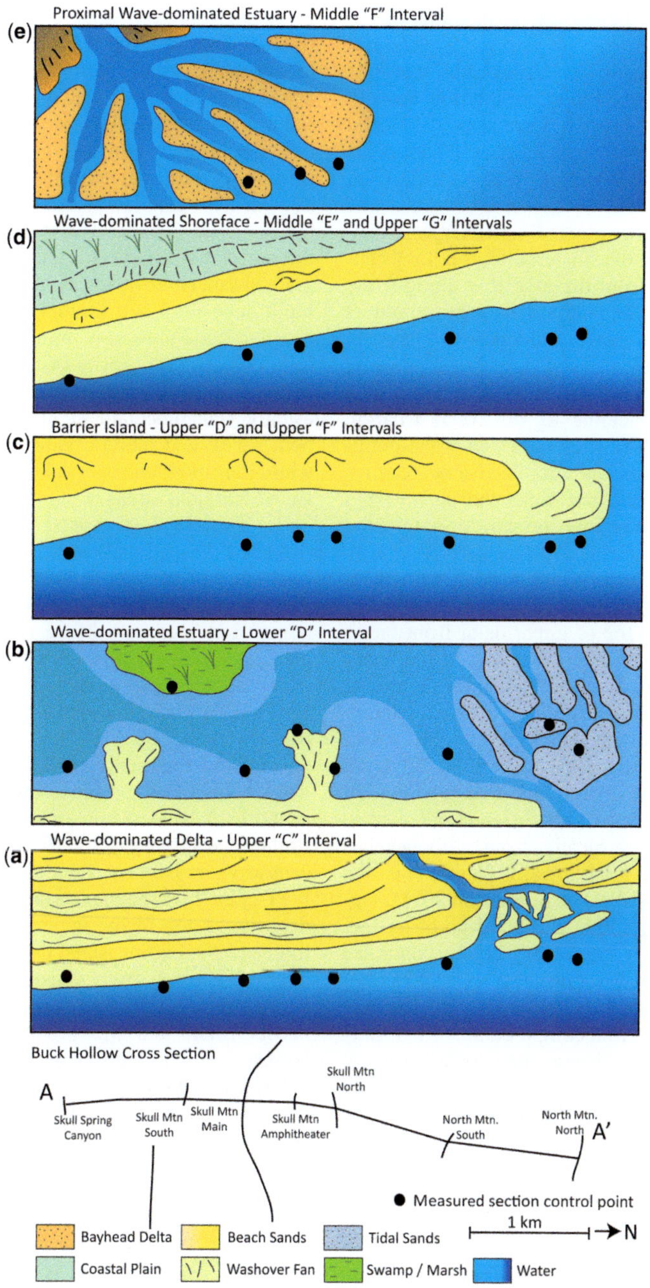

Fig. 12. Schematic palaeogeographical maps for Buck Hollow showing the distribution of depositional environments through time. Black circles mark points controlled by measured section observations. These points are spaced laterally based on the Buck Hollow measured section correlation line shown at the bottom (cf. A–A′ Figs 1b & 4). The points are positioned on the maps to indicate the facies preserved in each measured section. The approximate stratigraphic level of each interval depicted is shown in Figure 4. (**a**) Wave-dominated delta progradation is characterized by direct fluvial output to the shoreline, shore-parallel mouth bars and shorefaces. (**b**) Wave-dominated estuary settings include washover fans, flood tidal deltas, longitudinal tidal bars, central estuary and marsh/swamp environments. (**c**) Barrier island settings are characterized by tidal inlet and shoreface environments that are detached from the mainland. (**d**) Wave-dominated shoreface settings are attached to the mainland. (**e**) Proximal wave-dominated estuaries are characterized by bayhead delta and central estuary environments.

deltaic sequence, suggesting basinwards (NE) progradation.

Strong longshore, storm-driven shelf currents were expected for the Western Interior Seaway (Slingerland & Keen 1999; Hampson 2010) and this was supported by palaeocurrent measurements from trough cross-beds within the delta front (Fig. 5). It is possible that these longshore currents made the delta asymmetrical (Bhattacharya & Giosan 2003); however, the subaerial morphology cannot be determined from the available along-strike outcrop exposures. The mouth bars preserved at Buck Hollow show wave influence and contain abundant terrigenous material (Fig. 7i), suggesting wave-dominated deposition in an area close to a sediment source (Fielding *et al.* 2005; Charvin *et al.* 2010). The delta is classified as wave-dominated because it contains combined-flow features within the delta front, but lacks the channelized and graded beds indicative of a river-dominated delta front (Li *et al.* 2011).

Above the R–T turnaround, a tidal ravinement surface juxtaposes tidal channel facies (FA3.6; Fig. 9b) directly on top of upper delta front facies (FA2.1; Fig. 7a, b), marking the base of the transgressive phase of R–T cycle 2. Laterally variable incision of 1–3 m occurs along this undulatory tidal ravinement surface, which formed the estuary (Fig. 4). The back-barrier was infilled by coals (FA3.1; Fig. 8e) and distributary channels (FA3.4; Fig. 8o) during at least two different fill events. These coals grade vertically from carbonaceous shale to coal in a wetting upwards pattern (Fig. 8e; Wadsworth *et al.* 2010; Allen & Johnson 2011), suggesting the partial infilling and build-up of terrigenous material within the estuary before continued transgression (Allen & Johnson 2011).

Estuary facies of the D interval vary laterally from north to south, reflecting a complex interplay of depositional sub-environments existing contemporaneously (Fig. 12b). Higher energy channelized facies are concentrated in the northern part of the field area (North Mountain; Fig. 1b), whereas the lower energy, finer-grained facies are preserved in the southern portion of the field area (Skull Mountain; Figs 1b & 3b). The facies distribution suggests the estuary did not have a classic funnel shape (Dalrymple *et al.* 1992). Instead, the estuary was probably broad and terraced, similar to the modern day Galveston Bay in Texas (Anderson *et al.* 2008). Galveston Bay shows shallow, protected deposition occurring away from channels and inlets, juxtaposing high- and low-energy deposits in a pattern similar to those preserved in Buck Hollow. Abundant laminated mudstone intervals (FA3.5) and carbonaceous shales and coals (FA3.1; Fig. 8e) suggest that the finer-grained portion of terrestrially derived estuarine sediment was trapped within the estuary.

Washover fans (FA3.2; Fig. 8a, c), tidal channels (FA3.6; Fig. 9b) and tidal bars (FA3.7; Fig. 9d) record the influx of marine sediment into the estuary. Longshore drift delivered clean, winnowed sand to the Buck Hollow shoreline, creating a barrier island and unusually sandy tidal deposits (Fig. 9a, j). The tidal deposits are not as heterolithic as commonly implied by tidal facies models (Longhitano *et al.* 2012). The bimodal grain size of the correlative mudstone and medium- to coarse-grained sandstone facies within the estuary suggests rapid deposition and limited reworking. A similar distribution is present in Winyah Bay, South Carolina, USA, where >50% of the river-derived sediment is trapped within the main body of the estuary (Patchineelam *et al.* 1999). This creates both sandy and muddy bodies within the estuary, while longshore currents deliver sand to the barrier island and tidal deltas.

R–T cycle 2 is capped with a barrier island (FA1.3, FA1.4; Figs 9j & 12c), which records the final stepped transgression during the D interval. The barrier island is bound both above and below by wave ravinement surfaces. The basal wave ravinement surface juxtaposes barrier island shoreface facies above estuary fill (e.g. Demarest & Kraft 1987). This surface is planar and laterally continuous across the Buck Hollow area; however, it is not obvious regionally. Therefore we interpret the wave ravinement surface at the base of the barrier island as a local feature, representing small-scale cyclicity, and the upper surface as the more significant wave ravinement/flooding surface that bounds R–T cycle 2. This upper surface juxtaposes offshore deposits on top of the upper shoreface strata as it truncates the barrier island deposits and caps the D interval. Erosion from wave ravinement removed the subaerial and foreshore components of the barrier island and was probably <10 m, roughly corresponding to the fair weather wave base (Saito 1994; Cattaneo & Steel 2003). These barrier island deposits are similar to other examples of prograding shoreface successions overlying back-barrier facies (e.g. Mellere *et al.* 2005). However, because only back-barrier deposits are present 14 km updip in Main Canyon (rather than time-equivalent shoreface deposits), the D interval shoreface succession section at Buck Hollow is interpreted as barrier islands detached from the mainland. We hypothesize that pulsated transgression (e.g. Rampino & Sanders 1980; Salzmann *et al.* 2013) caused the island to back-step into the accommodation of the correlative estuary and briefly prograde before wave ravinement truncated the island, marking the top of R–T cycle 2.

The thickness of the transgressive portion of R–T cycle 2 (*c.* 65 m) suggests deposition on a shoreline with an accretionary trajectory. Advances in the understanding of shoreline trajectories (Cattaneo & Steel 2003; Bullimore *et al.* 2008; Henriksen *et al.* 2009) and the study of outcrop examples, such as this one, suggest that significant transgressive successions can be preserved in the right circumstances. Specifically, accumulation and preservation should occur on an accretionary shoreline, where the angle of the shoreline trajectory determines the thickness of transgressive deposits (Thorne & Swift 1991; Helland-Hansen & Hampson 2009; Allen & Johnson 2011), allowing for aggradation during transgression. Significant transgressive intervals are not accounted for in most sequence stratigraphic models (Arnott 1995), which predict sediment starvation and condensed deposition, or winnowing and lag deposition, during transgression (Posamentier *et al.* 1988; Galloway 1989; Van Wagoner *et al.* 1990).

Evidence for shoreline progradation in the middle of the transgressive estuary interval (repeated estuary sequences, distributary channel deposits and the development of multiple coals) questions the commonly used end-member distinction between estuaries (Dalrymple *et al.* 1992) and deltas (Bhattacharya 2003, 2010). Based on recent definitions (Dalrymple 2006; Dalrymple *et al.* 2012), an estuary becomes a delta when it is a net exporter of sediment. However, an estuary with ebb dominance can be a net exporter if distributary channels infill with fluvially derived sediment, causing it to act like a delta. Pulsated infilling can occur within a longer duration, overall transgressive period (Roehler 1988; Souza-Filho *et al.* 2009; Hassan *et al.* 2012), defined as an estuary based on current depositional models. The transition from estuary to delta or vice versa is not unusual for the Western Interior Seaway (Roehler 1988; Plink-Björklund 2008; Kieft *et al.* 2011) and is also seen in Quaternary and modern examples (Lessa *et al.* 1998; Gonzalez *et al.* 2001; Milli *et al.* 2013). These observations of pulsated progradation during estuary filling suggests estuaries are capable of net exporting sediment and that a more flexible definition including deltas and other facies may be appropriate for interpreting many ancient successions.

The D interval strata are interpreted as estuarine rather than lagoonal because of the visible river input and evidence for fluvial input to the shoreline in the Buck Hollow area. Both early (Pritchard 1967; Reddering 1980) and recent (Dalrymple *et al.* 1992, 2012; Boyd *et al.* 2006) definitions of estuaries require proximity to a fluvial source. In contrast, early descriptions and definitions of lagoons do not reference fluvial input (Barnes 1980; Kjerfve 1986; Kjerfve & Magill 1989). More recent distinctions state that lagoons have little to no freshwater influx (Davis 1994); however, the term lagoon has been applied to the low-energy areas of back-barrier systems with significant river input (Nichols 1989). The distinction between the two has been the subject of some discussion (Day 1980, 1981; Reddering 1980) because the morphology and characteristics of lagoons and estuaries are determined by a variety of processes, once again reflecting the complexity of paralic settings. The distinction is occasionally avoided by using the term ‘embayment’ to described brackish conditions (MacEachern *et al.* 1998). Improved nomenclature is required, particularly because a variety of classification schemes are currently being used (Hume & Herdendorf 1988; Dalrymple *et al.* 1992; Elliott & McLusky 2002; Potter *et al.* 2010) and some recent facies models (James & Dalrymple 2010) do not distinguish between estuaries and lagoons (Boyd 2010).

R–T cycle 3 (E and F intervals). R–T cycle 3 is *c.* 130 m thick and composed of regressive, prograding shoreface deposits (E interval; Fig. 12d) overlain by transgressive estuary deposits (F interval; Fig. 12e). The base of the R–T cycle 3 is marked by a combined wave ravinement/flooding surface that juxtaposes offshore facies (FA1.1) on the truncated barrier island strata of the D estuary (FA1.3, FA1.4; Figs 3 & 9j), marking the T–R turnaround. The regressive E interval contains three progradationally to aggradationally stacked, upwards-coarsening wave-dominated shoreface successions (Fig. 6a, k) deposited as the shoreline built basinwards. A tidal ravinement surface marks the R–T turnaround within R–T cycle 3, separating the tidal channels of the overlying transgressive F interval from the underlying regressive E upper shoreface deposits. The surface shows highly variable incision (*c.* 0–12 m; Fig. 4). The transgressive F estuary interval preserves tidal channels (FA3.6; Fig. 9b) at the base, deposited during estuary formation. Moving vertically up-section, it contains carbonaceous estuary fill (FA3.1; Fig. 8e) and bayhead delta deposits (FA3.5; Fig. 8i). The bayhead delta deposits in the F interval are laterally discontinuous over *c.* 30–50 m, indicating that a more proximal portion of the estuary is preserved compared with the more distal portion preserved in the D interval. This suggests either more landwards deposition or an estuary that was narrower in the direction perpendicular to the shore. Transgressive barrier island strata (FA1.3, FA1.4; Fig. 9j) overlie the bayhead delta succession. A wave ravinement surface truncates the barrier island and a combined wave ravinement and flooding surface separates R–T cycle 3 from the overlying offshore facies (FA1.1) of R–T cycle 4.

R–T cycle 4 (G interval). R–T cycle 4 is *c.* 60 m thick and is the uppermost cycle of the John Henry

Member. It is composed of the regressive G shoreface interval. The G interval consists of a single prograding wave-dominated shoreface succession (Fig. 12d), which exhibits a normal shoaling shoreface progression (FA1.1–1.5; Fig. 6a; Plint 2010), grading vertically from offshore through foreshore deposits. A process change from marine to fluvial marks the top of the G interval. The facies grade conformably from the foreshore into fluvial deposits. The first clearly trough cross-bedded, medium-grained, channelized sandstones are interpreted as the basal Drip Tank Member based on lithostratigraphic definitions (Peterson 1969*a*). The boundary is gradational, lacking clear evidence for prolonged subaerial exposure. Analysis by Lawton *et al.* (2014) in other parts of the Kaiparowits Plateau moved the location of the sequence boundary associated with the Drip Tank Member to near the top of the interval based on architectural, provenance and palaeocurrent shifts just below the overlying Wahweap Formation (cf. Shanley & McCabe 1991). Accordingly, the top of R–T cycle 4 at Buck Hollow is interpreted as a process change from marine shoreface deposition to fluvial sedimentation, probably below the main sequence boundary.

Energy regime

Energy regime, a classification scheme based on the relative proportion of tide v. wave energy, is used to characterize and compare both modern and ancient coastlines (Hayes 1979; Davis & Hayes 1984; Vakarelov *et al.* 2011; Vakarelov & Ainsworth 2013; Ainsworth *et al.* 2015). The delta, shoreface and estuary facies interpreted for Buck Hollow are consistent with deposition along a wave-dominated coastline (Davis & Hayes 1984; Davis 2013). This is fitting because the Western Interior Seaway was generally microtidal (Slater 1985; Ericksen & Slingerland 1990; Slingerland *et al.* 1996; Slingerland & Keen 1999), allowing wave energy to dominate relative to tidal energy. Energy and process indicators in the strata also support this hypothesis: both the deltaic and shoreface deposits contain abundant hummocky and swaley cross-stratification, grading into trough cross-bedding and planar to wavy laminations (Li *et al.* 2011; Vakarelov *et al.* 2011). The upper shoreface and upper delta front both lack the tidal indicators expected for tidally influenced shoreface deposits, such as ebb-oriented current ripples and oscillation ripples (Dashtgard *et al.* 2009, 2012).

Transgressive estuarine deposits of the John Henry Member display an increased tidal influence (FA2; Figs 8 & 9) relative to the shoreface (FA1; Fig. 6) and deltaic strata (FA2; Fig. 7). This is the result of deposition in a protected back-barrier, sheltered from the wave action of the open coast, with tidal magnification due to estuary funnelling (Dalrymple 2010). The Kaiparowits region was located along the Utah Bight during deposition of the John Henry Member (Fig. 1; McGookey *et al.* 1972; Hampson 2010). The curved embayment of the Utah Bight probably magnified the tides by dampening wave energy, similar to the modern Georgia Bight on the Atlantic coast of the USA. The Georgia Bight (*c.* 1000 km long and *c.* 400 km wide) increases from micro- (0–2 m) to meso-tidal (2–4 m) towards its apex (Davies 1964; Hayes 1994) and is roughly twice as large as the ancient Utah Bight (*c.* 500 km long and *c.* 200 km wide; McGookey *et al.* 1972). The tides heightened by the Utah Bight curvature probably contributed to the distinct tidal facies visible within the transgressive intervals of the John Henry Member. Despite this increased tidal influence, the John Henry Member estuaries are interpreted as wave-dominated. The presence of a barrier island system bounding the estuary aligns the Buck Hollow deposits most closely with the wave-dominated estuary end-member facies model (Dalrymple *et al.* 1992). Process regime changes are known to occur on short (fourth- or fifth-order) timescales (Yoshida *et al.* 2007) and previous work suggests that the Western Interior Seaway strata may record shifts from wave- to tide-dominated deposition (Seidler & Steel 2001; Plink-Björklund 2008). However, this variation in energy regime is not required for the John Henry Member at Buck Hollow. Instead, the coastline remained wave-dominated throughout deposition, leading to wave-influenced shoreline deposits and visible tidal indicators within protected back-barrier settings.

Regional correlation

Regional correlations across the northern Kaiparowits Plateau show the time–space palaeogeographical variability of the foreland basin. The depositional model and relative sea-level curve (Fig. 13) for Buck Hollow are similar, but not identical to, the patterns interpreted across the northern Kaiparowits Plateau at Main Canyon (Fig. 11) and Left Hand Collet (Fig. 14). Analysis of the north Kaiparowits Plateau shows that a regional regression occurred during the A–B–C intervals. The shoreline shifted at least 14 km basinwards between the B and C intervals as the coast prograded towards Buck Hollow (Fig. 13). The shoreface deposits that are preserved at Main Canyon (updip; 14 km SW) and Left Hand Collet (along-strike, 40 km SSE; Fig. 1) during the A and B intervals pinch out into offshore facies in Buck Hollow.

Throughout the deposition of the John Henry Member, raised coal mires in the centre of the plateau (McCabe 1987; McCabe & Shanley 1992;

Hettinger 2000) diverted low gradient, basin-axial fluvial systems NE towards Main Canyon and Buck Hollow (Szwarc *et al.* 2014). The north- and NE-flowing rivers resulted in valley incision at Main Canyon and delta deposition at Buck Hollow during the C interval (Fig. 11). The regression is described as forced because incision occurs at Main Canyon, implying potential down-stepping of the shoreline (Chentnik *et al.* 2015) and corresponding to a significant progradation of the shoreline at Buck Hollow marked by the progradation of delta front facies over offshore deposits. This forced regression led Chentnik *et al.* (2015) to move the most significant hiatus within the lower John Henry Member from the A sequence boundary (as suggested by Shanley & McCabe 1991, 1993; Fig. 2) to the Lower John Henry Member Sequence Boundary (Figs 4 & 11) between the B and C shoreface intervals in the northern Kaiparowits Plateau. This suggestion is supported by the relatively old age for the lower John Henry Member (88.6 ± 0.79 Ma) documented in Buck Hollow (Fig. 5), which implies that the A–B interval is older than previously thought and that a hiatus occurred higher in the section, i.e. during the Lower John Henry Member Sequence Boundary and the C interval.

The forced regression (C interval) is capped by a tidal ravinement surface marking the R–T turnaround in the basin, visible in all three areas (Figs 11 & 14). Stepped, accretionary transgression (D interval) was then preserved across the northern Kaiparowits Plateau. The pulsated rise in relative sea-level resulted in proximal estuarine facies at Main Canyon (Fig. 11), more distal estuarine and barrier island facies at Buck Hollow, and lower energy lagoon and barrier island facies at Left Hand Collet (Fig. 14). Abundant accommodation and sediment supply allowed Buck Hollow to have an accretionary shoreline trajectory (Helland-Hansen & Gjelberg 1994; Helland-Hansen & Martinsen 1996), leading to the deposition and preservation of thick transgressive packages.

The wave ravinement/flooding surface at the base of the E interval was chosen as the datum for regional correlations (Figs 11 & 14) as a result of its consistent expression at the base of easily mappable, extensive exposures of similar facies. The E interval consists of prograding shoreface deposits in all three areas. Shoreface successions at Main Canyon and Left Hand Collet pinch out into offshore deposits at Buck Hollow. Continued progradation led to the deposition of additional shoreface parasequences within the E interval at Buck Hollow (Fig. 4). A tidal ravinement surface truncates the E interval, as the F interval tidal channels cut into underlying upper shoreface strata (Figs 11 & 14). Correlations between Main Canyon and Buck Hollow are less clear for the upper portion of the John Henry Member because coal fires have altered the strata at Main Canyon, creating clinker beds and masking the stratigraphy. Along-strike in Left Hand Collet, a lagoon and barrier island interval (Dooling 2013) correlates to the estuary and barrier island interval at Buck Hollow. The F transgression was followed by a final shoreface progradation (>10 km) during the G interval.

The R–T cycles of the north Kaiparowits Plateau reflect relative sea-level shifts of *c.* 10–50 m over *c.* 1.5 myr duration, interpreted as third- or fourth-order eustatic, tectonic and/or climatic signatures (Vail *et al.* 1977; Vail & Mitchum 1979; Emery & Myers 1996), similar to others interpreted for the John Henry Member (Fig. 1; Allen & Johnson 2010*b*, 2011). The R–T cycles at Buck Hollow do not seem to correspond to the third-order eustatic fluctuations of the most recent global sea-level curves (Fig. 13; Haq *et al.* 1987; Miller *et al.* 2005; Cloetingh & Haq 2015). Along with additional observations outlined in the following sections, this suggests that, in addition to eustasy, other controls influenced relative sea-level and deposition in the Kaiparowits Plateau.

Depositional controls

Regional correlations show that the John Henry Member at Buck Hollow is highly expanded (two to three times) relative to other areas of deposition across the Kaiparowits Plateau. The John Henry Member at Buck Hollow (456 m, compacted) is 289% of the Main Canyon section (158 m thick compacted; Fig. 11) and 179% of the Left Hand Collet section (255 m thick compacted; Fig. 14). These variations in thickness imply local-scale (<15 km) variations in both sediment supply and accommodation, both considered here.

Sediment supply was unusually high at Buck Hollow because the area received sediment from multiple sources. Palaeocurrent data from the southern Kaiparowits region (Fig. 1; Gallin *et al.* 2010; Gooley *et al.* 2016) show that sediment was delivered to the shoreline mainly from the south in large distributive fluvial systems running transverse to the Mogollon Highlands and axial to the Sevier fold–thrust belt (Lawton *et al.* 2014; Szwarc *et al.* 2014). Fluvial sediment supply was further heightened during the C interval through local incision at Main Canyon (Fig. 11). The monsoonal climate of the Western Interior Seaway increased summer runoff (Leier *et al.* 2005; Fricke *et al.* 2010). Monsoons are thought to be responsible for the high volumes of sediment delivered to the Kaiparowits region during deposition of the Drip Tank Member (Lawton *et al.* 2014) and the John Henry Member probably experienced similar effects. Buck Hollow also received sediment from south-directed coastal

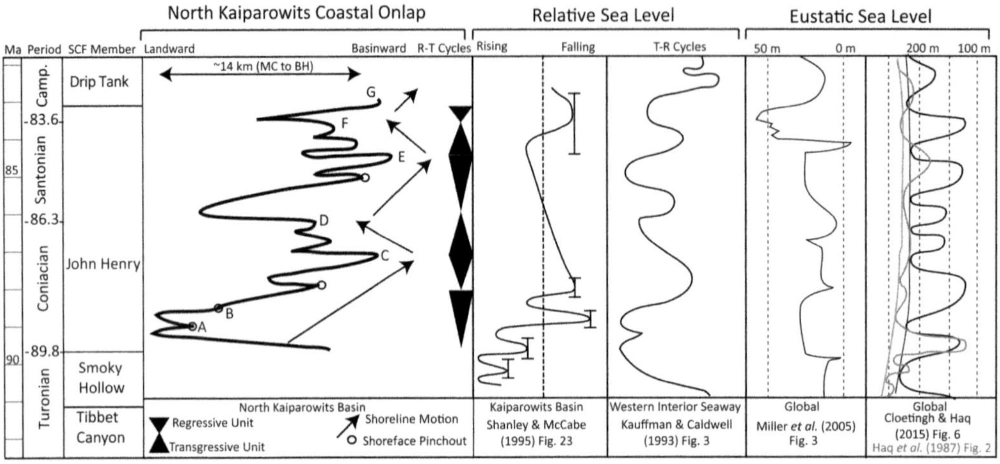

Fig. 13. Coastal onlap, relative sea-level and global eustatic sea-level curves for the Turonian to Campanian. North Kaiparowits coastal onlap curve estimates the relative shifts in the shoreline position based on the stratigraphy at Buck Hollow and Main Canyon and downdip shoreface pinch-outs (into offshore marine) where identified. Vertical bars in the Kaiparowits Plateau curve represent the estimates of uncertainty in the age dates of the sequence boundaries designated by Shanley & McCabe (1995). Camp., Campanian; SCF, Straight Cliffs Formation; MC, Main Canyon; BH, Buck Hollow; R–T, regressive–transgressive.

currents (Slingerland & Keen 1999) via longshore drift and storm reworking. Combined, these three processes supplied copious amounts of sediment to the Buck Hollow shoreline.

Sedimentation rate calculations allow comparisons between localities in the northern Kaiparowits Plateau (Table 5). A lack of precise age control prevents high-resolution calculations of sedimentation rates for each interval within the John Henry Member. Here, estimates were made for the entire John Henry Member (*c.* 4.1–8.5 myr total duration) using youngest maximum depositional age ranges from detrital zircon U/Pb geochronology (Table 5; Szwarc *et al.* 2014). These ages are supported by the new $^{40}Ar/^{39}Ar$ bentonite date and biostratigraphy from this study, which indicate a late Coniacian to early Campanian age range for the John Henry Member (cf. Eaton 1991). Both the biostratigraphic and the zircon data suggest that deposition of the John Henry Member extended into the early Campanian, younger than previously acknowledged (Eaton 1991; Dyman *et al.* 2002). The John Henry Member at Buck Hollow was deposited at a compacted rate of 72.3 m Ma^{-1} and a decompacted rate of 127.3 m Ma^{-1} (Table 5). These rates are higher than the compacted rates calculated for other areas of the plateau (25.1 m Ma^{-1} for Main Canyon, 53.5 m Ma^{-1} for Rogers Canyon, 56.2 m Ma^{-1} for Left Hand Collet; Table 5; locations shown in Fig. 1; Allen & Johnson 2010*b*; Dooling 2013). These rates are also twice the previous subsidence estimates of 25.6 m Ma^{-1} for the John Henry Member (Shanley & McCabe 1995, fig. 21), which was based on a subsidence curve showing 575 m of deposition (compacted) in the Kaiparowits Plateau over 22.5 myr.

Calculated sedimentation rates for all three John Henry Member localities (Table 5) fall within the range of global sedimentation rates calculated on million year timescales (Sadler 1981, 1999; Sommerfield 2006). However, the values are high relative to estimates for other localities around the Western Interior Basin (Table 6). In comparing the strata of Buck Hollow with other formations within the Western Interior Seaway, we also note that some isopach maps, mainly derived from well data (Weimer 1960; Roberts & Kirschbaum 1995; Painter & Carrapa 2013) do not fully account for the preserved outcrop thicknesses in the region (Doelling *et al.* 2000; Hettinger *et al.* 2009).

In addition to a high sediment supply, high rates of accommodation are required to preserve these thick accumulations of strata. Multiple autogenic and allogenic controls could have contributed to the overall regional thickness patterns, as well as the observed two to three times increase in expanded thickness over 14 km along depositional dip in the northern Kaiparowits Plateau. Possible explanations are explored in the following text, including: regional tectonics; eustasy; local structures/tectonics; pre-existing topography; and differential and early compaction (Fig. 15). Decompaction analysis and calculated sedimentation rates help to tease apart the complex and interrelated controls.

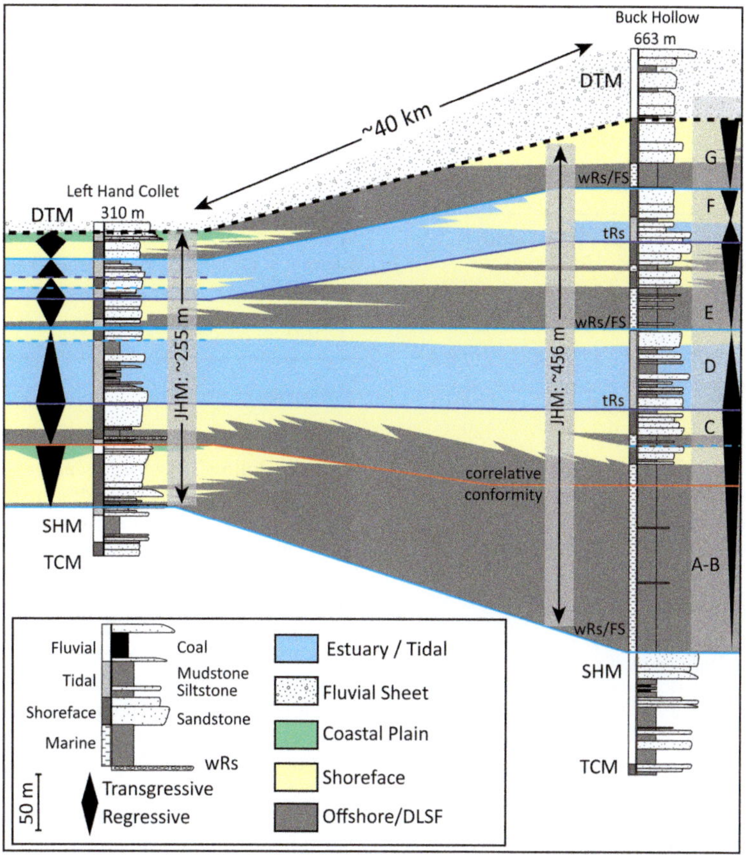

Fig. 14. Stratigraphic correlation between Left Hand Collet (Dooling 2013) and Buck Hollow, correlation location shown on Figure 1. DLSF, distal lower shoreface; DTM, Drip Tank Member; JHM, John Henry Member; SHM, Smoky Hollow Member; TCM, Tibbet Canyon Member; tRs, tidal ravinement surface; wRs/FS, wave ravinement/flooding surface.

High accommodation across the Kaiparowits Plateau was probably the result of eustasy and/or regional tectonics, whereas more local controls are required to explain the dramatic differences in thickness between Buck Hollow and other nearby locations. Global sea-level curves indicate possible short-term *c.* 5–100 m shifts in amplitude during the Late Cretaceous (Fig. 13), interpreted as third-order cycles (Miller *et al.* 2005; Cloetingh & Haq 2015). Eustasy could have contributed to the regional sedimentation patterns. However, we note that new age data and refined sea-level curves indicate a disconnect between regional and global signals and local expressions in the Kaiparowits region (Fig. 13).

Flexural loading by the Sevier fold–thrust belt created tectonic subsidence in the Kaiparowits region (Jordan 1981; Currie 2002; Painter & Carrapa 2013), which was enhanced by dynamic processes (Mitrovica *et al.* 1989; Liu *et al.* 2011, 2014). Detrital zircons from the Sevier fold–thrust belt are more prevalent in the upper portion of the John Henry Member (Szwarc *et al.* 2014), suggesting syn-depositional tectonism and increased sediment supply from uplift-driven, and, possibly, climate-enhanced erosion. Duplex development occurred on the Paxton thrust sheet (DeCelles *et al.* 1995; Ismat & Mitra 2001; DeCelles & Coogan 2006; Szwarc *et al.* 2014) while the Wah Wah and Blue Mountain thrust sheets were also active (Goldstrand 1994). This Sevier tectonism occurred *c.* 200 km west of the Kaiparowits Plateau (DeCelles 2004; DeCelles & Coogan 2006; Tindall *et al.* 2010) and the inferred forebulge was located perhaps as far east as central Colorado (White *et al.* 2002; DeCelles 2004). This places the Kaiparowits Plateau within the medial part of the foredeep. Given the evidence for Mogollon-derived sediment being delivered into the basin (Szwarc *et al.* 2014), the Mogollon Highlands might also be considered

Table 5. *Sedimentation rates for the Straight Cliffs Formation*

	John Henry Member thickness (m)	Duration of deposition (myr)	Sedimentation rate (m Ma^{-1})
Buck Hollow (Szwarc *et al.* 2014; this study)			
Compacted: measured section thicknesses			
Low case	456	8.5	53.6
Middle case	456	6.3	72.4
High case	456	4.1	111.2
Decompacted: calculated thicknesses			
Low case	802	8.5	94.4
Middle case	802	6.3	127.3
High case	802	4.1	195.6
Main Canyon (Chentnik *et al.* 2015)			
Compacted: measured section thicknesses			
Low case	158	8.5	18.6
Middle case	158	6.3	25.1
High case	158	4.1	38.5
Decompacted: calculated thicknesses			
Low case	321	8.5	38.8
Middle case	321	6.3	50.9
High case	321	4.1	78.3
Rogers Canyon (Allen & Johnson 2011)			
Compacted: measured section thicknesses			
Low case	265	5	53.5
High case	265	4.5	58.8
Left Hand Collet (Dooling 2013)			
Compacted: measured section thicknesses			
	281	5	56.2
Kaiparowits Basin (Shanley & McCabe 1995)			
Compacted subsidence			25.56

Table 6. *Sedimentation rates for the Western Interior Seaway*

References	Duration (myr)	Thickness (m)	Rate (m Ma^{-1})
York *et al.* (2011)	2.5	60	24
Sixsmith *et al.* (2008)	0.8	45	56
Valasek (1995)	0.3	10	33
Antia & Fielding (2011)	1	38	38

as a flexural control on subsidence in this part of the foredeep, although only local evidence for Cretaceous-aged thrusting has been clearly documented (Salem 2009).

Regional tectonics, eustasy, climate and the associated variations in sediment supply can explain the regionally high deposition and preservation rates, but they do not easily account for the local differences between Main Canyon and Buck Hollow. Some of the local differences in thickness can be attributed to incision and sediment bypassing. During the C interval, up to 22 m of strata were removed from Main Canyon during incised valley formation (Fig. 15; Chentnik *et al.* 2015). Additional sediment removal could have occurred during other intervals, perhaps up to *c.* 50 m. The erosion and bypassing occurring in Main Canyon relates to delta deposition in Buck Hollow, suggesting an inherent accommodation difference between the two areas at that time. Similarly, pre-existing topography such as a local shelf-edge break could accentuate even relatively small variations in relative sea-level (Dolson *et al.* 1991).

Perhaps the most plausible explanation for the differences in thickness between Main Canyon and Buck Hollow is local structural activity (Fig. 15). Structural deformation in the northern Kaiparowits Plateau is generally thought to have occurred after deposition of the John Henry Member, specifically during early-stage Laramide tectonism, the beginning of which has been extended into the Campanian in this region (Davis & Bump 2009; Tindall *et al.* 2010). A nearby example of syn-tectonic oblique reverse faulting caused 2.8× thickening locally within the Wahweap Formation (80–76 Ma) on the western edge of the Kaiparowits Plateau (Tindall & Davis 1999; Tindall *et al.* 2010). Applying similar fault-driven thickening (*c.* 2.8×) to the compacted thicknesses of Main Canyon and Left Hand Collet yields a range of possible local tectonic accommodation (*c.* 450–700 m) for Buck Hollow (Fig. 15). Located further east, Buck Hollow is in the proximity of local structural features such as the Dutton Monocline (Hettinger 2000), as well as numerous small-scale folds, specifically, a slight fold in Main Canyon that appears to be syn-depositional (Chentnik 2014). To date, there is no reported evidence of Sevier-related blind thrusts propagating this far into the foredeep. Thus, although most of the John Henry Member is Santonian in age, we suggest there could be some overlap with very early, Laramide-style structures, perhaps related to the adjacent Circle Cliffs Uplift (Sargent & Hansen 1982; Davis 1999; Davis & Bump 2009).

Finally, differential compaction is considered as an additional, simple explanation for the observed variation in thickness of the John Henry Member in the northern Kaiparowits Plateau. However, both locations have similar lithologies and overburden histories, and thus preliminary calculations indicate similar expansion for both Buck Hollow (1.76× measured thickness) and Main Canyon (2.03× measured thickness; Fig. 10). Because the decompaction rates of the two areas are fairly similar, the geometries of the regional correlation (Fig. 11) hold, suggesting that the interpreted relationships and surfaces are valid. If 1.76× expansion is applied to the original thickness of Main Canyon (158 m), the decompacted thickness is 278 m. The difference between this and the calculated decompacted thickness (321 m) is 43 m, representing the accommodation accounted for by post-depositional compaction (Fig. 15). This accounts for *c.* 9% of the total (481 m) accommodation difference between the areas. Thickness differences may also be attributed to the presence of coal. Peat to coal compaction rates are not well constrained, with estimates varying from 1.4:1 to 30:1 (Ryer & Langer 1980; Collinson & Scott 1987; McCabe 1987). Sensitivity analysis was carried out to assess whether the variable compaction of coal could be responsible for the differences in thickness in the study areas. Common coal and carbonaceous silt lithologies assessed in PetroMod showed three to four times post-depositional decompaction (Table 4). Therefore the variation in coal thickness (<5 m) between Buck Hollow and Main Canyon is accounted for in the overall calculations and is probably not responsible for the 481 m difference in thickness between the two areas.

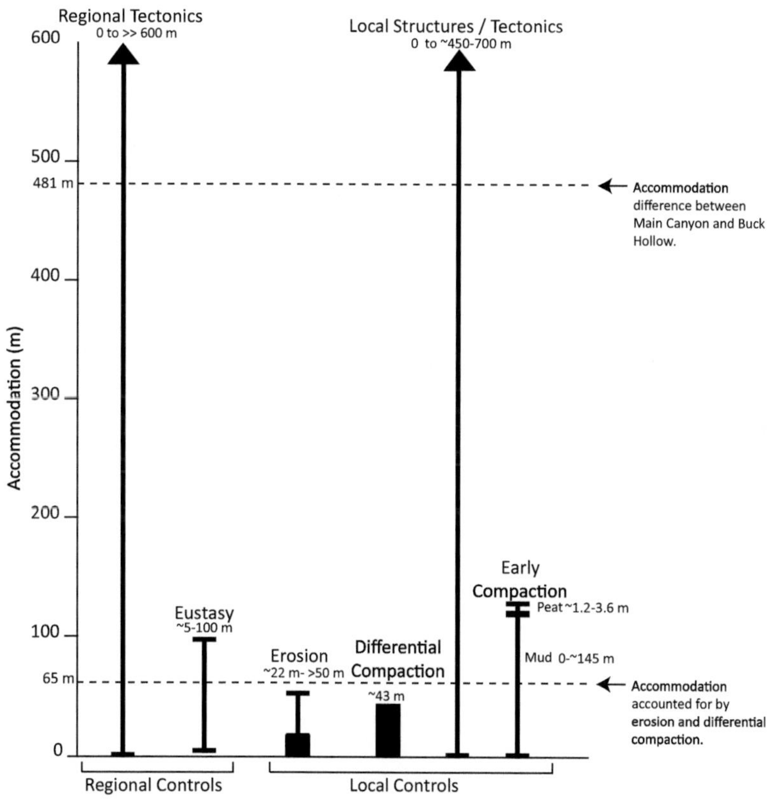

Fig. 15. Schematic plot showing estimates of possible controls on accommodation. The accommodation generated by regional tectonics at the *c.* 10 Ma timescale could range to over 600 m (Currie 2002). Eustatic variation of *c.* 5–100 m occurred during John Henry Member deposition (Haq *et al.* 1987; Miller *et al.* 2005; Cloetingh & Haq 2015). Upper arrow notes the total difference in decompacted thickness between Buck Hollow and Main Canyon (481 m). Lower arrow notes combined accommodation difference (65 m) accounted for by erosion (22 m) measured from outcrop (Chentnik *et al.* 2015) and post-depositional differential compaction (43 m) calculated through back-stripping (black boxes). Estimates for early compaction were made by applying modern rates of early compaction (Table 7) to the thickness values from Buck Hollow (see text). The estimate of the potential accommodation generated by local structures and tectonics (*c.* 450–700 m) was determined by applying expansion values (2.8×) from nearby faulting (Tindall *et al.* 2010) to the John Henry Member at Main Canyon and Left Hand Collet.

Early differential compaction, which creates space from loading and dewatering during deposition, is not often considered as a control on accommodation. Compaction has been studied in detail by soil engineers to understand small-scale compaction (10^0 m) at geologically instantaneous timescales (10^1 years) in modern sediments (Terzaghi 1943; British Geotechnical Society 1975) and it is has been considered at broader timescales for some geological basins (>10^5 years; Athy 1930; Weller 1959). The effects of compaction at the meso-scale (10^1–10^3 years) are poorly understood, yet they can play a key part in deposition and preservation (Guber & Slingerland 1981; Kooi & de Vries 1998; Meckel *et al.* 2006, 2007; Rosati 2009). These meso-scale processes are absent from the mechanical compaction models used in standard basin modelling software.

Early differential compaction could have occurred in both the offshore and estuary facies at Buck Hollow. The *c.* 160 m thick A–B interval may have undergone early compaction and dewatering. Because compaction was taking place, sediment could accumulate without altering the available accommodation. This would have stabilized the relative sea-level locally and allowed prolonged deposition. Recent advances in mudstone depositional processes suggest that hyperpycnal flows deliver abundant sediment to basins (Schieber *et al.* 2007; Schieber & Southard 2009), making accommodation the limiting control on deposition. Compaction is dependent on the overlying load and the duration of deposition, therefore it is difficult to integrate estimates for ancient examples. Mud compaction is estimated to occur at 0.1 mm a^{-1} with >40% compaction possible (Kooi & de Vries 1998). If similar rates are applied to the decompacted thickness of the A–B interval mudstones, a maximum of 145 m of accommodation could have been created over 1.45 myr. If the deposits did not reach their maximum early compaction during deposition, further compaction could have occurred during deposition of the C interval, when 50 m of delta sandstones were rapidly deposited on top of the underlying A–B mudstones.

It is probable that syn-depositional differential compaction also took place during the D and F intervals as barrier island shorefaces stepped landwards, compressing the underlying fine-grained, carbonaceous bay fill and peat (Olsen *et al.* 1999; Rosati *et al.* 2010), which are known to compact during deposition (Nadon 1998). Modern silt and peat compaction rates are highly variable, ranging over three orders of magnitude (Table 7); however, they can be

Table 7. *Early compaction calculations*

Location/study	Substrate	Original thickness (m)	Thickness change (m)	Time interval (years)	Rate (mm a^{-1})	Compaction (%)
Assawoman Island, Virginia (Guber & Slingerland 1981; Gayes 1983; Rosati *et al.* 2010)*	Back-barrier silty mudstones		1.0–1.2	40	25.0	
Theoretical calculations (Terzaghi 1943, p. 265–296)	Silty clay	14	0.31	47	6.6	2.2
Rhine–Muse delta, North Sea (van Asselen 2011)	Peat	10	3	*c.* 100	0.6	30
Romney Marsh, SE England (Long *et al.* 2006)	Peat		3	*c.* 750		
Hammock River Marsh, Connecticut (Bloom 1964)	Sedge peat	10.6		7000		13–44
Bayou Lafourche, Mississippi Delta, Louisiana (Törnqvist *et al.* 2008)	Peat, clay and silt	<40		*c.* 10 000	0.5–5.0	

*Example adjusted for relative sea-level rise of 2 mm a^{-1}.

used to estimate possible early compaction in ancient sediments. If the *c.* 9 m of coaly facies at Buck Hollow compacted by 40% (Bloom 1964; van Asselen 2011), it could have created a maximum of 3.6 m of space in 144 years using a rate of 0.025 m a^{-1} (Rosati *et al.* 2010). This accounts for some, but not all, of the accommodation required to preserve the 10–31 m thick barrier island deposits in the D and F intervals. Early compaction of both the A–B interval mudstones and the coals of the D interval account for a maximum of 148 m of accommodation (Fig. 15), which is 31% of the difference in thickness between Main Canyon and Buck Hollow (Figs 10 & 11).

In summary, although the primary- and more regional-scale controls on accommodation and sediment supply are relatively well known, differentiating the local-scale controls is challenging (Fig. 15). Tectonic and eustatic controls on both accommodation and sediment supply are acknowledged and were probably primary signals in the Cretaceous Kaiparowits region, in addition to climate as a moderating factor on sediment supply. Even at this broad scale, however, questions remain regarding the importance of the Mogollon Highlands, both on sediment input into the basin and the possible additional flexural load, not currently accounted for in most models for the Sevier foredeep. The possible role of dynamic subsidence has also not been fully explored, but probably affected an area much larger than that considered here (Liu *et al.* 2011, 2014). Similarly, at the more local scale, although tectonic or structural controls are the most probable explanation for the dramatic thickening in the northern Kaiparowits Plateau, there is not yet a discrete structure or other feature than can be linked directly to tectonic activity. Initiation of what would become basement-involved Laramide structures seems the best working hypothesis, but this suggests much earlier development of the 'broken foreland' than previously interpreted (McQueen & Beaumont 1989; Dickinson *et al.* 1990). Palaeotopography and differential compaction, particularly the early compaction of mudstone and peat, could be a significant factor of the same order of magnitude as the predicted eustatic excursions for the same timescales (Fig. 15), although these are not typically considered in many analyses of subsidence history.

Conclusions

The John Henry Member at Buck Hollow provides a new example of the time–space variability possible for paralic facies in a single location, providing new data to improve facies, depositional and predictive models. The high accommodation, high sediment supply setting allowed the preservation of multiple thick, stacked R–T cycles, including delta and shoreface successions interbedded with more heterogeneous estuary deposits. Sedimentary structures preserved within the strata suggest that the coastline was wave-dominated throughout the deposition of the John Henry Member. The transgressive estuarine deposits display a clear tidal influence, probably the result of protection from wave reworking provided by barrier islands, as well as tidal magnification due to funnelling and the Utah Bight embayment.

The John Henry Member at Buck Hollow varies from the updip equivalent strata at Main Canyon to the along-strike strata at Left Hand Collet. Within the forced regression of the A–B–C intervals, delta deposition occurred at Buck Hollow, downdip of the incised valleys at Main Canyon. Buck Hollow shows more river influence relative to Left Hand Collet, displaying fluvial input in the form of distributary channels and bayhead delta deposits within the estuary intervals, which were deposited during stepped, accretionary transgressions. Despite these differences, shoreline trends were consistent regionally and major surfaces extend across the basin. Correlations from Buck Hollow both updip and along-strike show the variability possible in a basin with partially disconnected accommodation and controls on sediment supply.

The John Henry Member at Buck Hollow is two to three times expanded relative to other localities across the Kaiparowits Plateau. Although tectonics, eustasy and climate moderated the relative sea-level for the basin as a whole, these regional processes cannot explain the local (<14 km) differences in thickness and unusual geometries preserved in the northern Kaiparowits Plateau. Decompaction calculations show that the unusual geometries are not purely the result of post-depositional differential compaction. Differential compaction does account for *c.* 43 m of accommodation, whereas erosion from incision contributes up to 22 m. However, these two processes only explain *c.* 13.5% of the accommodation differential between Main Canyon and Buck Hollow. Tectonic activity on local structures is the most plausible explanation for the difference in accommodation between the areas and could account for up to *c.* 450–700 m of additional deposition at Buck Hollow. Early compaction, not often considered in foreland basin studies, could have contributed up to *c.* 145 m of accommodation. The quantification of local controls highlights the myriad of processes contributing to the preservation of paralic strata in the Kaiparowits Plateau.

This research is supported by the Rocks2Models research consortium, Cari Johnson and Lisa Stright, principal investigators, with funding from Chevron, ConocoPhillips, Hess Corporation, Shell and Statoil. We are very grateful to

David Pocknall of Hess Corporation for providing biostratigraphic analyses, and Laura Webb and Daniel Jones at the University of Vermont for performing the $^{40}Ar/^{39}Ar$ geochronological analysis cited in this study. Julia Mulhern also received funding from a University of Utah Graduate Research Fellowship, a Rocky Mountain Association of Geologists Foundation Babcock Scholarship and a SEPM Rocky Mountain Section Donald Smith Research Grant. We acknowledge Schlumberger for academic licences for the PetroMod basin modelling software and other software. We also thank the Grand Staircase Escalante National Monument and the Dixie National Forest for their support of field-based research in the Kaiparowits Plateau. We are grateful for the many colleagues and students who assisted in the field and contributed to discussions, particularly B. Chentnik. Finally, we thank Gary Hampson, Massimo Zecchin and an anonymous reviewer for their constructive comments, which improved the original manuscript.

References

Ahlbrandt, T.S., Charpentier, R.R., Klett, J.W., Schmoker, C.J., Schenk, G.F. & Ulmishek, T.S. 2005. *Global Resource Estimates from Total Petroleum Systems*. American Association of Petroleum Geologists Memoirs, **86**.

Ainsworth, R.B., Vakarelov, B.K. & Nanson, R.A. 2011. Dynamic spatial and temporal prediction of changes in depositional processes on clastic shorelines: toward improved subsurface uncertainty reduction and management. *American Association of Petroleum Geologists Bulletin*, **95**, 267–297, https://doi.org/10.1306/06301010036

Ainsworth, R.B., Vakarelov, B.K., Lee, C., MacEachern, J.A., Montgomery, A.E., Ricci, L.P. & Dashtgard, S.E. 2015. Architecture and evolution of a regressive, tide-influenced marginal marine succession, Drumheller, Alberta, Canada. *Journal of Sedimentary Research*, **85**, 596–625, https://doi.org/10.2110/jsr.2015.33

Allen, J.L. & Johnson, C.L. 2010*a*. Facies control on sandstone composition (and influence of statistical methods on interpretations) in the John Henry Member, Straight Cliffs Formation, southern Utah, USA. *Sedimentary Geology*, **230**, 60–76, https://doi.org/10.1016/j.sedgeo.2010.06.023

Allen, J.L. & Johnson, C.L. 2010*b*. Sedimentary facies, paleoenvironments and relative sea level changes in the John Henry Member, Cretaceous Straight Cliffs Formation, southern Utah, USA. *In*: Carney, S.M., Tabet, D.E. & Johnson, C.L. (eds) *Geology of South-Central Utah*. Utah Geological Association, Salt Lake City, 225–247.

Allen, J.L. & Johnson, C.L. 2011. Architecture and formation of transgressive-regressive cycles in marginal marine strata of the John Henry Member, Straight Cliffs Formation, Upper Cretaceous of southern Utah, USA. *Sedimentology*, **58**, 1486–1513, https://doi.org/10.1111/j.1365-3091.2010.01223.x

Allen, J.R.L. 1966. On bed forms and paleocurrents. *Sedimentology*, **6**, 153–190.

Ambrose, W.A. & Ayers, W.B. 2007. Geologic controls on transgressive–regressive cycles in the upper Pictured Cliffs Sandstone and coal geometry in the lower Fruitland Formation, northern San Juan Basin, New Mexico and Colorado. *American Association of Petroleum Geologists Bulletin*, **91**, 1099–1122, https://doi.org/10.1306/03080706040

Anderson, J.B., Rodriguez, A.B., Milliken, K.T. & Taviani, M. 2008. The Holocene evolution of the Galveston estuary complex, Texas: evidence for rapid change in estuarine environments. *In*: Anderson, J.B. & Rodriguez, A.B. (eds) *Response of Upper Gulf Coast Estuaries to Holocene Climate Change and Sea-Level Rise*. Geological Society of America Special Papers, **443**, 89–104, https://doi.org/10.1130/2008.2443(06)

Antia, J. & Fielding, C.R. 2011. Sequence stratigraphy of a condensed low-accommodation succession: lower Upper Cretaceous Dakota Sandstone, Henry Mountains, southeastern Utah. *American Association of Petroleum Geologists Bulletin*, **95**, 413–447, https://doi.org/10.1306/06301009182

Arnott, R.W.C. 1995. The parasequence definition – are transgressive deposits inadequately addressed? *Journal of Sedimentary Research*, **B65**, 1–6, https://doi.org/10.1306/D42681D0-2B26-11D7-8648000102C1865D

Athy, L.F. 1930. Density, porosity, and compaction of sedimentary rocks. *American Association of Petroleum Geologists Bulletin*, **14**, 194–200, https://doi.org/10.1306/3D93289E-16B1-11D7-8645000102C1865D

Barnes, R.S.K. 1980. *Coastal Lagoons: the Natural History of a Neglected Habitat*. Cambridge University Press, Cambridge.

Barwis, J.H. & Hayes, M.O. 1985. Antidunes on modern and ancient washover fans. *Journal of Sedimentary Research*, **55**, 907–916, https://doi.org/10.1306/212F883C-2B24-11D7-8648000102C1865D

Becker, M.A., Wellner, R.W., Mallery, C.S. & Chamberlain, J.A. 2010. Chondrichthyans from the lower Ferron Sandstone Member of the Mancos Shale (Upper Cretaceous: Middle Turonian) of Emery and Carbon Counties Utah, USA. *Journal of Paleontology*, **84**, 248–266, https://doi.org/10.1666/09-053r.1

Belaústegui, Z. & de Gibert, J.M. 2013. Bow-shaped, concentrically laminated polychaete burrows: a *Cylindrichnus concentricus* ichnofabric from the Miocene of Tarragona, NE Spain. *Palaeogeography, Palaeoclimatology, Palaeoecology*, **381–382**, 119–127, https://doi.org/10.1016/j.palaeo.2013.04.021

Bhattacharya, J.P. 2003. Deltas and estuaries. *In*: Middleton, G.V. (ed.) *Encyclopedia of Sedimentology*. Kluwer Academic, Dordrecht, 145–152.

Bhattacharya, J.P. 2010. Deltas. *In*: James, N.P. & Dalrymple, R.W. (eds) *Facies Models 4*. Geological Society of Canada, St John's, 233–264.

Bhattacharya, J.P. & Giosan, L. 2003. Wave-influenced deltas: geomorphological implications for facies reconstruction. *Sedimentology*, **50**, 187–210, https://doi.org/10.1046/j.1365-3091.2003.00545.x

Bhattacharya, J.P. & MacEachern, J.A. 2009. Hyperpycnal rivers and prodeltaic shelves in the Cretaceous Seaway of North America. *Journal of Sedimentary Research*, **79**, 184–209, https://doi.org/10.2110/jsr.2009.026

BHATTACHARYA, J.P. & WALKER, R.G. 1991. River- and wave-dominated systems of the Upper Cretaceous Dunvegan Formation, northwestern Alberta. *Bulletin of Canadian Petroleum Geology*, **39**, 165–191.

BLOOM, A.L. 1964. Peat accumulation and compaction in a Connecticut coastal marsh. *Journal of Sedimentary Petrology*, **34**, 599–603, https://doi.org/10.1306/74D710F5-2B21-11D7-8648000102C1865D

BOBB, M.C. 1991. *The Calico Bed, Upper Cretaceous, southern Utah: a fluvial sheet deposit in the Western Interior Foreland Basin and its relationship to eustasy and tectonics*. MS thesis, University of Colorado.

BOOTHROYD, J.C., FRIEDRICH, N.E. & MCGINN, S.R. 1985. Geology of microtidal coastal lagoons: Rhode Island. *Marine Geology*, **63**, 35–76, https://doi.org/10.1016/0025-3227(85)90079-9

BOYD, R. 2010. Transgressive wave-dominated coasts. *In*: JAMES, N.P. & DALRYMPLE, R.W. (eds) *Facies Models 4*. Geological Association of Canada, St John's, 265–294.

BOYD, R., DALRYMPLE, R.W., ZAITLIN, B.A. & DALRYMPLE, M. 2006. Estuarine and incised-valley facies models. *In*: POSAMENTIER, H.W. & WALKER, R.G. (eds) *Facies Models Revisited*. SEPM Special Publications, **84**, 171–235, https://doi.org/10.2110/pec.06.84

BRITISH GEOTECHNICAL SOCIETY 1975. *Settlement of Structures*. Pentech Press, London.

BULLIMORE, S.A., HELLAND-HANSEN, W., HENRIKSEN, S. & STEEL, R.J. 2008. Shoreline trajectory and its impact on coastal depositional environments: an example from the Upper Cretaceous Mesaverde Group, northwestern Colorado, U.S.A. *In*: HAMPSON, G.J. (ed.) *Recent Advances in Models of Siliciclastic Shallow-Marine Stratigraphy*. SEPM Special Publications, **90**, 209–236.

CARTER, C.H. 1978. A regressive barrier and barrier-protected deposit: depositional environments and geographic setting of the Late Tertiary Cohansey Sand. *Journal of Sedimentary Research*, **48**, 933–949, https://doi.org/10.1306/212F75AE-2B24-11D7-8648000102C1865D

CATTANEO, A. & STEEL, R.J. 2003. Transgressive deposits: a review of their variability. *Earth-Science Reviews*, **62**, 187–228, https://doi.org/10.1016/S0012-8252(02)00134-4

CATUNEANU, O., ABREU, V. *ET AL.* 2009. Towards the standardization of sequence stratigraphy. *Earth-Science Reviews*, **92**, 1–33, https://doi.org/10.1016/j.earscirev.2008.10.003

CHARVIN, K., HAMPSON, G.J., GALLAGHER, K.L. & LABOURDETTE, R. 2010. Intra-parasequence architecture of an interpreted asymmetrical wave-dominated delta. *Sedimentology*, **57**, 760–785, https://doi.org/10.1111/j.1365-3091.2009.01118.x

CHENTNIK, B.M. 2014. *Valleys, estuaries, and lagoons: paleoenvironments and regressive–transgressive architecture of the Upper Cretaceous Straight Cliffs Formation*. MS thesis, University of Utah, Salt Lake City, UT.

CHENTNIK, B.M., JOHNSON, C.L., MULHERN, J.S. & STRIGHT, L.E. 2015. Valleys, estuaries, and lagoons: paleoenvironments and regressive–transgressive architecture of the Upper Cretaceous Straight Cliffs Formation. *Journal of Sedimentary Research*, **85**, 1166–1196.

CLIFTON, H.E. 2006. A reexamination of facies models for clastic shorelines. *In*: POSAMENTIER, H.G. & WALKER, R.G. (eds) *Facies Models Revisited*. SEPM Special Publications, **84**, 293–337.

CLOETINGH, S. & HAQ, B.U. 2015. Inherited landscapes and sea level change. *Science*, **347**, 1258375, https://doi.org/10.1126/science.1258375

COLLINSON, M.E. & SCOTT, A.C. 1987. Implications of vegetational change through the geological record on models for coal-forming environments. *In*: SCOTT, A.C. (ed.) *Coal and Coal-Bearing Strata: Recent Advances*. Geological Society, London, Special Publications, **32**, 67–85, https://doi.org/10.1144/GSL.SP.1987.032.01.06

COOPER, J.A.G. 1993. Sedimentation in a river dominated estuary. *Sedimentology*, **40**, 979–1017, https://doi.org/10.1111/j.1365-3091.1993.tb01372.x

CUMBAA, S.L., SHIMADA, K. & COOK, T.D. 2010. Mid-Cenomanian vertebrate faunas of the Western Interior Seaway of North America and their evolutionary, paleobiogeographical, and paleoecological implications. *Palaeogeography, Palaeoclimatology, Palaeoecology*, **295**, 199–214, https://doi.org/10.1016/j.palaeo.2010.05.038

CURRAY, J.R. 1964. Transgressions and regressions. *In*: MILLER, R.L. (ed.) *Papers in Marine Geology, Shepard Commemorative Volume*. Macmillan, New York, 175–203.

CURRIE, B.S. 2002. Structural configuration of the early Cretaceous Cordilleran foreland-basin system and Sevier thrust system, Utah and Colorado. *The Journal of Geology*, **110**, 697–718.

DALRYMPLE, R.W. 2006. Incised valleys in time and space: an introduction to the volume and an examination of the controls on valley formation and filling. *In*: DALRYMPLE, R.W., LECKIE, D.A. & TILLMAN, R.W. (eds) *Incised Valleys in Time and Space*. SEPM Special Publications, **85**, 5–12.

DALRYMPLE, R.W. 2010. Tidal depositional systems. *In*: JAMES, N.P. & DALRYMPLE, R.W. (eds) *Facies Models 4*. Geological Association of Canada, St John's, 201–232.

DALRYMPLE, R.W. & CHOI, K. 2007. Morphologic and facies trends through the fluvial–marine transition in tide-dominated depositional systems: a schematic framework for environmental and sequence-stratigraphic interpretation. *Earth-Science Reviews*, **81**, 135–174, https://doi.org/10.1016/j.earscirev.2006.10.002

DALRYMPLE, R.W. & RHODES, R.N. 1995. Estuarine Dunes and Bars. *Developments in Sedimentology*, **53**. Elsevier, Amsterdam, 359–422, https://doi.org/10.1016/S0070-4571(05)80033-0

DALRYMPLE, R.W., ZAITLIN, B.A. & BOYD, R. 1992. Estuarine facies models: conceptual basis and stratigraphic implications. *Journal of Sedimentary Research*, **62**, 1130–1146, https://doi.org/10.1306/D4267A69-2B26-11D7-8648000102C1865D

DALRYMPLE, R.W., MACKAY, D.A., ICHASO, A.A. & CHOI, K.S. 2012. Processes, morphodynamics, and facies of tide-dominated estuaries. *In*: DAVIS, R.A., JR. & DALRYMPLE, R.W. (eds) *Principles of Tidal Sedimentology*. Springer, Dordrecht, 79–107.

DASHTGARD, S.E., GINGRAS, M.K. & MACEACHERN, J.A. 2009. Tidally modulated shorefaces. *Journal of*

Sedimentary Research, **79**, 793–807, https://doi.org/10.2110/jsr.2009.084

Dashtgard, S.E., MacEachern, J.A., Frey, S.E. & Gingras, M.K. 2012. Tidal effects on the shoreface: towards a conceptual framework. *Sedimentary Geology*, **279**, 42–61, https://doi.org/10.1016/j.sedgeo.2010.09.006

Davies, J.L. 1964. A morphogenic approach to world shorelines. *Zeitschrift für Geomorphologie*, **8**, 127–142.

Davies, R., Howell, J., Boyd, R., Flint, S. & Diessel, C. 2006. High-resolution sequence-stratigraphic correlation between shallow-marine and terrestrial strata: examples from the Sunnyside Member of the Cretaceous Blackhawk Formation, Book Cliffs, eastern Utah. *American Association of Petroleum Geologists Bulletin*, **90**, 1121–1140, https://doi.org/10.1306/02210604077

Davis, G.H. 1999. *Structural Geology of the Colorado Plateau Region of Southern Utah: with Special Emphasis on Deformation Bands*. Geological Society of America Special Papers, **342**.

Davis, G.H. & Bump, A.P. 2009. Structural geologic evolution of the Colorado Plateau. *In*: Kay, S.M., Ramos, V.A. & Dickinson, W.R. (eds) *Back-Bone of the Americas: Shallow Subduction, Plateau Uplift, and Ridge and Terrane Collision*. Geological Society of America Memoirs, **204**, 99–124.

Davis, R.A., Jr. 1994. Barrier island systems – a geologic overview. *In*: Davis, R.A., Jr. (ed.) *Geology of Holocene Barrier Island Systems*. Springer, Berlin, 1–46.

Davis, R.A., Jr. 2012. Tidal signatures and their preservation potential in stratigraphic sequences. *In*: Davis, R.A., Jr. & Dalrymple, R.W. (eds) *Principles of Tidal Sedimentology*. Springer, Dordrecht, 35–55.

Davis, R.A., Jr. 2013. A new look at barrier-inlet morphodynamics. *Journal of Coastal Research*, **69**, 1–12, https://doi.org/10.2112/SI_69_2

Davis, R.A., Jr. & Clifton, H.E. 1987. Sea-level change and the preservation potential of wave-dominated and tide-dominated coastal sequences. *In*: Nummedal, D., Pilkey, O.H. & Howard, J.D. (eds) *Sea-Level Fluctuation and Coastal Evolution*. SEPM Special Publications, **41**, 167–178.

Davis, R.A., Jr. & Hayes, M.O. 1984. What is a wave-dominated coast? *Marine Geology*, **60**, 313–329, https://doi.org/10.1016/0025-3227(84)90155-5

Day, J.H. 1980. What is an estuary? *South African Journal of Science*, **76**, 198.

Day, J.H. 1981. The nature, origin, and classification of estuaries. *In*: Day, J.H. (ed.) *Estuarine Ecology: With Particular Reference to Southern Africa*. Balkema, Rotterdam, 1–6.

DeCelles, P.G. 2004. Late Jurassic to Eocene evolution of the Cordilleran thrust belt and foreland basin system, western, U.S.A. *American Journal of Science*, **304**, 105–168, https://doi.org/10.2475/ajs.304.2.105

DeCelles, P.G. 2012. Foreland basin systems revisited: variation in response to tectonic settings. *In*: Busby, C. & Azor, A. (eds) *Tectonics of Sedimentary Basins: Recent Advances*. Blackwell, Hoboken, 405–427.

DeCelles, P.G. & Coogan, J.C. 2006. Regional structure and kinematic history of the Sevier fold-and-thrust belt, central Utah. *Geological Society of America Bulletin*, **118**, 841–864, https://doi.org/10.1130/b25759.1

DeCelles, P.G. & Giles, K.A. 1996. Foreland basin systems. *Basin Research*, **8**, 105–123, https://doi.org/10.1046/j.1365-2117.1996.01491.x

DeCelles, P.G., Lawton, T.F. & Mitra, G. 1995. Thrust timing, growth of structural culminations, and synorogenic sedimentation in the type Sevier orogenic belt, western United States. *Geology*, **23**, 699–702, https://doi.org/10.1130/0091-7613(1995)023<0699:TTGOSC>2.3.CO;2

Demarest, J. & Kraft, J.C. 1987. Stratigraphic record of Quaternary sea levels: implications for more ancient strata. *In*: Nummedal, D., Pilkey, O.H. & Howard, J.D. (eds) *Sea-Level Fluctuation and Coastal Evolution*. SEPM Special Publications, **41**, 223–239, https://doi.org/10.2110/pec.87.41.0223

Devine, P.E. 1991. Transgressive origin of channeled estuarine deposits in the Point Lookout Sandstone, northwestern New Mexico: a model for Upper Cretaceous, cyclic regressive parasequences of the U.S. western interior. *American Association of Petroleum Geologists Bulletin*, **75**, 1039–1063.

Dickinson, W.R., Klute, M.A., Hayes, M.J., Janecke, S.U., Lundin, E.R., McKittrick, M.A. & Olivares, M.D. 1990. Paleogeographic and paleotectonic setting of Laramide sedimentary basins in the central Rocky Mountain region: alternative interpretation [by Cather, S. M., Chapin, C. E. *et al.*] and reply. *Geological Society of America Bulletin*, **102**, 256–260, https://doi.org/10.1130/0016-7606(1990)102<0256:PAPSOL>2.3.CO;2

Doelling, H.H. & Willis, G.C. 1999. *Interim Geologic Map of the Escalante and Parts of the Loa and Hite Crossing 30′ X 60′ Quadranges, Garfield and Kane Counties, Utah*. Open File Report **368**, Utah Geological Survey.

Doelling, H.H., Blackett, R.E., Hamblin, A.H., Powell, J.D. & Pollock, G.L. 2000. Geology of Grand Staircase-Escalante National Monument, Utah. *In*: Sprinkel, D.A., Chidsey, T.C. & Anderson, P.B. (eds) *Geology of Utah's Parks and Monuments*. Utah Geological Association Publication, **28**, 189–231.

Dolson, J., Muller, D., Evetts, M.J. & Stein, J.A. 1991. Regional paleotopographic trends and production, Muddy Sandstone (Lower Cretaceous), central and northern Rocky Mountains. *American Association of Petroleum Geologists Bulletin*, **75**, 409–435.

Dooling, P.R. 2013. *Tidal facies, stratigraphic architecture, and along-strike variability of a high energy, transgressive shoreline, Late Cretaceous, Kaiparowits Plateau, southern Utah*. MS thesis, University of Utah, Salt Lake City.

Droser, M.L. & Bottjer, D.J. 1989. Ichnofabric of sandstones deposited in high-energy nearshore environments: measurements and utilization. *Palaios*, **4**, 598–604.

Duke, W.L. 1985. Hummocky cross-stratification, tropical hurricanes and intense winter storms. *Sedimentology*, **32**, 167–194.

Dumas, S. 2005. Experiments on oscillatory-flow and combined-flow bed forms: implications for interpreting parts of the shallow-marine sedimentary record.

Journal of Sedimentary Research, **75**, 501–513, https://doi.org/10.2110/jsr.2005.039

Dumas, S. & Arnott, R.W.C. 2006. Origin of hummocky and swaley cross-stratification – the controlling influence of unidirectional current strength and aggradation rate. *Geology*, **34**, 1073–1076, https://doi.org/10.1130/G22930A.1

Dyman, T.S., Cobban, W.A. *et al*. 2002. *Upper Cretaceous Marine and Brackish Water Strata at Grand Staircase-Escalante National Monument, Utah*. Geological Society of America, Field Trip Road Log, May 2002.

Dzuynski, S. & Smith, A.J. 1963. Convolute lamination, its origin, preservation, and directional significance. *Journal of Sedimentary Research*, **33**, 616–627, https://doi.org/10.1306/74D70ED4-2B21-11D7-8648000102C1865D

Eaton, J.G. 1991. Biostratigraphic framework for the Upper Cretaceous rocks of the Kaiparowits Plateau, southern Utah. *In*: Nations, J.D. & Eaton, J.G. (eds) *Stratigraphy, Depositional Environments, and Sedimentary Tectonics of the Western Margin, Cretaceous Western Interior Seaway*. Geological Society of America Special Papers, **260**, 47–63, https://doi.org/10.1130/SPE260

Eaton, J.G. & Nations, J.D. 1991. Introduction. Tectonic setting along the margin of the Cretaceous Western Interior Seaway, southwestern Utah and northern Arizona. *In*: Nations, J.D. & Eaton, J.G. (eds) *Stratigraphy, Depositional Environments, and Sedimentary Tectonics of the Western Margin, Cretaceous Western Interior Seaway*. Geological Society of America Special Papers, **260**, 1–8, https://doi.org/10.1130/SPE260-p1

Ekdale, A.A. & Harding, S.C. 2015. *Cylindrichnus concentricus* Toots in Howard, 1966 (trace fossil) in its type locality, upper Cretaceous, Wyoming. *Annales Societatis Geologorum Poloniae*, **85**, 427–432.

Elliott, M. & McLusky, D.S. 2002. The need for definitions in understanding estuaries. *Estuarine, Coastal and Shelf Science*, **55**, 815–827, https://doi.org/10.1006/ecss.2002.1031

Embry, A.F. & Johannessen, E.P. 1993. T-R sequence stratigraphy, facies analysis and reservoir distribution in the uppermost Triassic–Lower Jurassic succession, western Sverdrup Basin, Arctic Canada. *In*: Vorren, T.O., Bergsager, E., Dahl-Stamnes, O.A., Holter, E., Johansen, B., Lie, E. & Lund, T.B. (eds) *Arctic Geology and Petroleum Potential*. Norwegian Petroleum Society, Special Publications, **2**, 121–146.

Emery, D. & Myers, K.J. (eds). 1996. *Sequence Stratigraphy*. Blackwell Science, Oxford.

Ericksen, M.C. & Slingerland, R. 1990. Numerical simulations of tidal and wind-driven circulation in the Cretaceous Interior Seaway of North America. *Geological Society of America Bulletin*, **102**, 1499–1516, https://doi.org/10.1130/0016-7606(1990)102<1499:nsotaw>2.3.co;2

Fenies, H. & Tastet, J. 1998. Facies and architecture of an estuarine tidal bar (the Trompeloup bar, Gironde Estuary, SW France). *Marine Geology*, **150**, 149–169, https://doi.org/10.1016/S0025-3227(98)00059-0

Fielding, C.R., Trueman, J.D. & Alexander, J. 2005. Sharp-based, flood-dominated mouth bar sands from the Burdekin River Delta of northeastern Australia: extending the spectrum of mouth-bar facies, geometry, and stacking patterns. *Journal of Sedimentary Research*, **75**, 55–66, https://doi.org/10.2110/jsr.2005.006

Fitzgerald, D.M., Buynevich, I. & Hein, C. 2012. Morphodynamics and facies architecture of tidal inlets and tidal deltas. *In*: Davis, R.A., Jr. & Dalrymple, R.W. (eds) *Principles of Tidal Sedimentology*. Springer, Dordrecht, 301–333, https://doi.org/10.1007/978-94-007-0123-6

Frey, R.W. 1990. Trace fossils and hummocky cross-stratification, Upper Cretaceous of Utah. *Palaios*, **5**, 203–218.

Fricke, H.C., Foreman, B.Z. & Sewall, J.O. 2010. Integrated climate model-oxygen isotope evidence for a North American monsoon during the Late Cretaceous. *Earth and Planetary Science Letters*, **289**, 11–21, https://doi.org/10.1016/j.epsl.2009.10.018

Gallin, W.N., Johnson, C.L. & Allen, J.L. 2010. Fluvial and marginal marine architecture of the John Henry Member, Straight Cliffs Formation, Kelly Grade of the Kaiparowits Plateau, south-central Utah. *In*: Carney, S.M., Tabet, D.E. & Johnson, C.L. (eds) *Geology of South-Central Utah*. Utah Geological Association, Salt Lake City, 248–275.

Galloway, W.E. 1975. Process framework for describing the morphologic and stratigraphic evolution of deltaic depositional systems. *In*: Broussard, M.L. (ed.) *Deltas: Models for Exploration*. Houston Geological Society, Houston, TX, 86–98.

Galloway, W.E. 1989. Genetic stratigraphic sequences in basin analysis I: architecture and genesis of flooding-surface bounded depositional units. *American Association of Petroleum Geologists Bulletin*, **73**, 125–142, https://doi.org/10.1306/703C9AF5-1707-11D7-8645000102C1865D

Gani, M.R. & Bhattacharya, J.P. 2007. Basic building blocks and process variability of a Cretaceous delta: internal facies architecture reveals a more dynamic interaction of river, wave, and tidal processes than is indicated by external shape. *Journal of Sedimentary Research*, **77**, 284–302, https://doi.org/10.2110/jsr.2007.023

Gayes, P.T. 1983. *Primary consolidation and subsidence in transgressive barrier island systems*. MS thesis, Pennsylvania State University, University Park, PA.

George, D.A. & Hill, P.S. 2008. Wave climate, sediment supply and the depth of the sand-mud transition: a global survey. *Marine Geology*, **254**, 121–128, https://doi.org/10.1016/j.margeo.2008.05.005

Gingras, M.K., MacEachern, J.A. & Pemberton, S.G. 1998. A comparative analysis of the ichnology of wave- and river-dominated allomembers of the Upper Cretaceous Dunvegan Formation. *Bulletin of Canadian Petroleum Geology*, **46**, 51–73.

Goldstrand, P.M. 1994. Tectonic development of Upper Cretaceous to Eocene strata of southwestern Utah. *Geological Society of America Bulletin*, **106**, 145–154, https://doi.org/10.1130/0016-7606(1994)106<0145:TDOUCT>2.3.CO;2

Gonzalez, R., Dias, J.A. & Ferreira, Ó. 2001. Recent rapid evolution of the Guadiana Estuary mouth (southwestern Iberian Peninsula). *Journal of Coastal Research*, **34**, 516–527.

GOOLEY, J.T., JOHNSON, C.L. & PETTINGA, L.A. 2016. Spatial and temporal variation of fluvial architecture within a prograding clastic wedge of the Late Cretaceous Western Interior Basin (Kaiparowits Plateau), USA. *Journal of Sedimentary Research*, **86**, 125–147.

GRAHAM, G.H., JACKSON, M.D. & HAMPSON, G.J. 2015. Three-dimensional modeling of clinoforms in shallow-marine reservoirs: part 1. Concepts and application. *American Association of Petroleum Geologists Bulletin*, **99**, 1013–1047, https://doi.org/10.1306/01191513190

GUBER, A.L. & SLINGERLAND, R. 1981. Compaction and lateral flow as processes in barrier island and associated environments. *In*: SLINGERLAND, R., GUBER, A.L. & HANSON, H.W. (eds) *Field Guide to Selected Coastal Geologic Problems on the Central Delmarva Penninsula. Eighth Annual Assateague Shelf and Shore Workshop*. Penn State University, University Park, PA, 1–14.

HAMPSON, G.J. 2010. Sediment dispersal and quantitative stratigraphic architecture across an ancient shelf. *Sedimentology*, **57**, 96–141, https://doi.org/10.1111/j.1365-3091.2009.01093.x

HAMPSON, G.J. & STORMS, J.E.A. 2003. Geomorphological and sequence stratigraphic variability in wave-dominated, shoreface-shelf parasequences. *Sedimentology*, **50**, 667–701, https://doi.org/10.1046/j.1365-3091.2003.00570.x

HANCOCK, J.M. & KAUFFMAN, E.G. 1979. The great transgressions of the Late Cretaceous. *Journal of the Geological Society, London*, **136**, 175–186, https://doi.org/10.1144/gsjgs.136.2.0175

HAQ, B.U., HARDENBOL, J. & VAIL, P.R. 1987. Chronology of fluctuating sea levels since the Triassic. *Science*, **235**, 1156–1167, https://doi.org/10.1126/science.235.4793.1156

HASSAN, S.M., STEEL, R.J., EL BARKOOKY, A., HAMDAN, M., OLARIU, C. & HELPER, M.A. 2012. Stacked, Lower Miocene tide-dominated estuary deposits in a transgressive succession, Western Desert, Egypt. *Sedimentary Geology*, **282**, 241–255, https://doi.org/10.1016/j.sedgeo.2012.09.013

HAYES, M.O. 1979. Barrier island morphology as a function of tidal and wave regime. *In*: LEATHERMAN, S.P. (ed.) *Barrier Islands from the Gulf of Mexico to the Gulf of St. Lawrence*. Academic Press, New York, 1–28.

HAYES, M.O. 1994. The Georgia Bight barrier system. *In*: DAVIS, R.A. (ed.) *Geology of Holocene Barrier Island Systems*. Springer, Berlin, 233–304.

HELLAND-HANSEN, W. & GJELBERG, J.G. 1994. Conceptual basis and variability in sequence stratigraphy: a different perspective. *Sedimentary Geology*, **92**, 31–52, https://doi.org/10.1016/0037-0738(94)90053-1

HELLAND-HANSEN, W. & HAMPSON, G.J. 2009. Trajectory analysis: concepts and applications. *Basin Research*, **21**, 454–483, https://doi.org/10.1111/j.1365-2117.2009.00425.x

HELLAND-HANSEN, W. & MARTINSEN, O.J. 1996. Shoreline trajectories and sequences: description of variable depositional-dip scenarios. *Journal of Sedimentary Research*, **66**, 670–688, https://doi.org/10.1306/D42683DD-2B26-11D7-8648000102C1865D

HENRIKSEN, S., HAMPSON, G.J., HELLAND-HANSEN, W., JOHANNESSEN, E.P. & STEEL, R.J. 2009. Shelf edge and shoreline trajectories, a dynamic approach to stratigraphic analysis. *Basin Research*, **21**, 445–453, https://doi.org/10.1111/j.1365-2117.2009.00432.x

HETTINGER, R.D. 2000. A summary of coal distribution and geology in the Kaiparowits Plateau, Utah. *In*: KRISCHBAUM, M.A., ROBERTS, L.N.R. & BIEWICK, L.R.H. (eds) *Geologic Assessment of Coal in the Colorado Plateau: Arizona, Colorado, New Mexico, and Utah*. US Geological Survey Professional Papers, **1625-B**, J1–J17.

HETTINGER, R.D., MCCABE, P.J. & SHANLEY, K.W. 1993. Detailed facies anatomy of transgressive and highstand systems tracts from the Upper Cretaceous of southern Utah, U.S.A. *In*: WEIMER, R.J. & POSAMENTIER, H.W. (eds) *Siliciclastic Sequence Stratigraphy: Recent Developments and Applications*. American Association of Petroleum Geologists Memoirs, **58**, 235–257.

HETTINGER, R.D., ROBERTS, L.N.R., BIEWICK, L.R.H. & KIRSCHBAUM, M.A. 2009. Geologic overview and resource assessment of coal in the Kaiparowits Plateau, southern Utah. *In*: KIRSCHBAUM, M.A., ROBERTS, L.N.R. & BIEWICK, L.R.H. (eds) *Geologic Assessment of Coal in the Colorado Plateau: Arizona, Colorado, New Mexico, and Utah. Denver, Colorado*. US Geological Survey Professional Papers, **1625-B**, 1–83.

HILL, P.S., FOX, J.M. *ET AL*. 2007. Sediment delivery to the seabed on continental margins. *In*: NITTROUER, C.A., AUSTIN, J.A., FIELD, M.E., KRAVITZ, J.H., SYVITSKI, J.P.M. & WIBERG, P.L. (eds) *Continental Margin Sedimentation: From Sediment Transport to Sequence Stratigraphy*. Blackwell, Oxford, 49–99, https://doi.org/10.1002/9781444304398.ch2

HINTZE, L.F. 1988. *Geologic History of Utah*. Brigham Young University, Provo.

HOLZ, M., VIEIRA, P.E. & KALKREUTH, W. 2000. The early Permian coal-bearing succession of the Paraná Basin in southernmost Brazil: depositional model and sequence stratigraphy. *Revista Brasileira de Geosciencias*, **30**, 424–426.

HOLZ, M., KALKREUTH, W. & BANERJEE, I. 2002. Sequence stratigraphy of paralic coal-bearing strata: an overview. *International Journal of Coal Geology*, **48**, 147–179, https://doi.org/10.1016/S0166-5162(01)00056-8

HUBBARD, S.M., GINGRAS, M.K., PEMBERTON, S.G. & THOMAS, M.B. 2002. Variability in wave-dominated estuary sandstones: implications on subsurface reservoir development. *Bulletin of Canadian Petroleum Geology*, **50**, 118–137, https://doi.org/10.2113/50.1.118

HUDOCK, J.W., FLAIG, P.P. & WOOD, L.J. 2014. Washover fans: a modern geomorphologic analysis and proposed classification scheme to improve reservoir models. *Journal of Sedimentary Research*, **84**, 854–865.

HUGHES, Z.J. 2012. Tidal channels on tidal flats and marshes. *In*: DAVIS, R.A., JR. & DALRYMPLE, R.W. (eds) *Principles of Tidal Sedimentology*. Springer, Dordrecht, 269–300, https://doi.org/10.1007/978-94-007-0123-6_11

HUME, T.M. & HERDENDORF, C.E. 1988. A geomorphic classification of estuaries and its application to coastal resource management – a New Zealand example. *Ocean and Shoreline Management*, **11**, 249–274, https://doi.org/10.1016/0951-8312(88)90022-7

Ichaso, A.A. & Dalrymple, R.W. 2009. Tide- and wave-generated fluid mud deposits in the Tilje Formation (Jurassic), offshore Norway. *Geology*, **37**, 539–542, https://doi.org/10.1130/G25481A.1

Ismat, Z. & Mitra, G. 2001. Folding by cataclastic flow at shallow crustal levels in the Canyon Range, Sevier orogenic belt, west-central Utah. *Journal of Structural Geology*, **23**, 355–378, https://doi.org/10.1016/S0191-8141(00)00101-2

James, N. & Dalrymple, R.W. (eds) . 2010. *Facies Models 4*. Geological Society of Canada, St John's.

Jinnah, Z.A. & Roberts, E.M. 2011. Facies associations, paleoenvironment, and base-level changes in the Upper Cretaceous Wahweap Formation, Utah, U.S.A. *Journal of Sedimentary Research*, **81**, 266–283, https://doi.org/10.2110/jsr.2011.22

Jinnah, Z.A., Roberts, E.M., Deino, A.L., Larsen, J.S., Link, P.K. & Fanning, C.M. 2009. New ^{40}Ar–^{39}Ar and detrital zircon U–Pb ages for the Upper Cretaceous Wahweap and Kaiparowits Formations on the Kaiparowits Plateau, Utah: implications for regional correlation, provenance, and biostratigraphy. *Cretaceous Research*, **30**, 287–299, https://doi.org/10.1016/j.cretres.2008.07.012

Joeckel, R.M. & Korus, J.T. 2012. Bayhead delta interpretation of an Upper Pennsylvanian sheetlike sandbody and the broader understanding of transgressive deposits in cyclothems. *Sedimentary Geology*, **275–276**, 22–37, https://doi.org/10.1016/j.sedgeo.2012.07.002

Jordan, T.E. 1981. Thrust loads and foreland basin evolution, Cretaceous, Western United States. *American Association of Petroleum Geologists Bulletin*, **65**, 2506–2520.

Kamola, D.L. & Van Wagoner, J.C. 1995. Stratigraphy and facies architecture of parasequences with examples from the Spring Canyon Member, Blackhawk Formation, Utah. *In*: Van Wagoner, J.C. & Bertram, G.T. (eds) *Sequence Stratigraphy of Foreland Basin Deposits*. American Association of Petroleum Geologists Memoirs, **64**, 27–54.

Kauerrauf, A.I. & Hantschel, T. 2009. *Fundamentals of Basin and Petroleum Systems Modeling*. Springer, Berlin.

Kauffman, E.G. 1977. Geological and biological overview: Western Interior Cretaceous basin. *Mountain Geologist*, **14**, 75–99.

Kauffman, E.G. & Caldwell, W.G.E. 1993. *The Western Interior Basin in Space and Time*. Geological Association of Canada Special Paper, **39**, 1–30.

Kieft, R.L., Hampson, G.J., Jackson, C.A.L. & Larsen, E. 2011. Stratigraphic architecture of a net-transgressive marginal- to shallow-marine succession: upper Almond Formation, Rock Springs Uplift, Wyoming, U.S.A. *Journal of Sedimentary Research*, **81**, 513–533, https://doi.org/10.2110/jsr.2011.44

Kirkland, J.I., Eaton, J.G. & Brinkman, D.B. 2013. Elasmobranchs from Upper Cretaceous freshwater facies in southern Utah. *In*: Titus, A.L. & Loewen, M.A. (eds) *At the Top of the Grand Staircase: the Late Cretaceous of Southern Utah*. Indiana University Press, Bloomington, 153–194.

Kjerfve, B. 1986. Comparative oceanography of coastal lagoons. *In*: Wolfe, D.A. (ed.) *Estuarine Variability*. Academic Press, San Diego, 63–81, https://doi.org/10.1016/B978-0-12-761890-6.50009-5

Kjerfve, B. & Magill, K.E. 1989. Geographic and hydrodynamic characteristics of shallow coastal lagoons. *Marine Geology*, **88**, 187–199, https://doi.org/10.1016/0025-3227(89)90097-2

Kooi, H. & de Vries, J.J. 1998. Land subsidence and hydrodynamic compaction of sedimentary basins. *Hydrology and Earth System Sciences*, **2**, 159–171, https://doi.org/10.5194/hess-2-159-1998

Kvale, E.P. 2012. Tidal constituents of modern and ancient tidal rhythmites: criteria for recognition and analyses. *In*: Davis, R.A. & Dalrymple, R.W. (eds) *Principles of Tidal Sedimentology*. Springer, Dordrecht, 1–17.

Larsen, J.S. 2007. *Facies and provenance of the Pine Hollow Formation: implications for Sevier foreland basin evolution and the Paleocene climate of southern Utah*. MS thesis, Idaho State University.

Larsen, J.S., Link, P.K., Roberts, E.M., Tapanila, L. & Fanning, C.M. 2010. Cyclic stratigraphy of the Paleogene Pine Hollow Formation and detrital zircon provenance of Campanian to Eocene sandstones of the Kaiparowits and Table Cliffs Basins, south-central Utah. *In*: Carney, S.M., Tabet, D.E. & Johnson, C.L. (eds) *Geology of South-Central Utah*. Utah Geological Association Publications, **39**, 194–214.

Lawton, T.F. & Bradford, B.A. 2011. Correlation and provenance of Upper Cretaceous (Campanian) fluvial strata, Utah, U.S.A. from zircon U–Pb geochronology and petrography. *Journal of Sedimentary Research*, **81**, 495–512, https://doi.org/10.2110/jsr.2011.45

Lawton, T.F., Pollock, S.L. & Robinson, R.A.J. 2003. Integrating sandstone petrology and nonmarine sequence stratigraphy: application to the Late Cretaceous fluvial systems of southwestern Utah, USA. *Journal of Sedimentary Research*, **73**, 389–406, https://doi.org/10.1306/100702730389

Lawton, T.F., Schellenbach, W.L. & Nugent, A.E. 2014. Late Cretaceous fluvial-megafan and axial-river systems in the southern Cordilleran foreland basin: Drip Tank Member of Straight Cliffs Formation and adjacent strata, southern Utah, U.S.A. *Journal of Sedimentary Research*, **84**, 407–434.

Leier, A.L., DeCelles, P.G. & Pelletier, J.D. 2005. Mountains, monsoons, and megafans. *Geology*, **33**, 289–292, https://doi.org/10.1130/G21228.1

Lessa, G.C., Meyers, S.R. & Marone, E. 1998. Holocene stratigraphy in the Paranagua Bay estuary, southern Brazil. *Journal of Sedimentary Research*, **68**, 1060–1076, https://doi.org/10.2110/jsr.68.1060

Li, W., Bhattacharya, J.P., Zhu, Y., Garza, D. & Blankenship, E. 2011. Evaluating delta asymmetry using three-dimensional facies architecture and ichnological analysis, Ferron 'Notom Delta', Capital Reef, Utah, USA. *Sedimentology*, **58**, 478–507, https://doi.org/10.1111/j.1365-3091.2010.01172.x

Liu, S., Nummedal, D. & Liu, L. 2011. Migration of dynamic subsidence across the Late Cretaceous United States Western Interior Basin in response to Farallon Plate subduction. *Geology*, **39**, 555–558, https://doi.org/10.1130/G31692.1

Liu, S., Nummedal, D. & Gurnis, M. 2014. Dynamic v. flexural controls of Late Cretaceous Western

Interior Basin, USA. *Earth and Planetary Science Letters*, **389**, 221–229, https://doi.org/10.1016/j.epsl.2014.01.006

Long, A.J., Waller, M.P. & Stupples, P. 2006. Driving mechanisms of coastal change: peat compaction and the destruction of late Holocene coastal wetlands. *Marine Geology*, **225**, 63–84, https://doi.org/10.1016/j.margeo.2005.09.004

Longhitano, S.G., Mellere, D., Steel, R.J. & Ainsworth, R.B. 2012. Tidal depositional systems in the rock record: a review and new insights. *Sedimentary Geology*, **279**, 2–22, https://doi.org/10.1016/j.sedgeo.2012.03.024

MacEachern, J.A. & Pemberton, S.G.Y. 1992. Ichnological aspects of Cretaceous shoreface successions and shoreface variability in the Western Interior Seaway of North America. *In*: Pemberton, S.G. (ed.) *Applications of Ichnology to Petroleum Exploration*. Society of Sedimentary Geology Core Workshop, **17**, 57–84.

MacEachern, J.A., Zaitlin, B.A., Pemberton, S.G. & Field, J. 1998. High-resolution sequence stratigraphy of early transgressive deposits, Viking Formation, Joffre Field, Alberta, Canada. *American Association of Petroleum Geologists Bulletin*, **82**, 729–756, https://doi.org/10.1306/1D9BC5E3-172D-11D7-8645000102C1865D

MacEachern, J.A., Pemberton, G.S., Gingras, M.K. & Bann, K. 2010. Ichnology and facies models. *In*: James, N.P. & Dalrymple, R.W. (eds) *Facies Models 4*. Geological Association of Canada, St John's, 19–58.

Masselink, G. & Puleo, J.A. 2006. Swash-zone morphodynamics. *Continental Shelf Research*, **26**, 661–680, https://doi.org/10.1016/j.csr.2006.01.015

McCabe, P.J. 1987. Facies studies of coal and coal-bearing strata. *In*: Scott, A.C. (ed.) *Coal and Coal-bearing Strata: Recent Advances*. Geological Society, London, Special Publications, **32**, 51–66, https://doi.org/10.1144/GSL.SP.1987.032.01.05

McCabe, P.J. & Shanley, K.W. 1992. Organic control on shoreface stacking patterns: bogged down in the mire. *Geology*, **20**, 741–744, https://doi.org/10.1130/0091-7613(1992)020<0741:OCOSSP>2.3.CO;2

McCubbin, D.G. 1982. Barrier-island and strand plain facies. *In*: Scholle, P.A. & Spearing, D. (eds) *Sandstone Depositional Environments*. American Association of Petroleum Geologists Memoirs, **31**, 247–279.

McEwen, M.C. 1969. Sedimentary facies of the modern Trinity delta. *In*: Lankford, J.J. & Rogers, J.J.W. (eds) *Holocene Geology of the Galveston Bay Area*. Houston Geological Society, Houston, 53–77.

McGookey, D.P., Haun, J.P., Heele, L.A., Goodell, H.G., McCubbin, D.G., Weimer, R.J. & Wulf, G.R. 1972. Cretaceous systems. *In*: *Geologic Atlas of the Rocky Mountain Region*. Rocky Mountain Association of Geologists, Colorado, 190–233.

McQueen, H.W.S. & Beaumont, C. 1989. Mechanical models of tilted block basins. *In*: Price, R.A. (ed.) *Origin and Evolution of Sedimentary Basins and their Energy and Mineral Resources*. American Geophysical Union Geophysical Monographs, **48**, 65–71, https://doi.org/10.1029/GM048p0065

Meckel, T.A., ten Brink, U.S. & Williams, S.J. 2006. Current subsidence rates due to compaction of Holocene sediments in southern Louisiana. *Geophysical Research Letters*, **33**, 1–5, https://doi.org/10.1029/2006GL026300

Meckel, T.A., ten Brink, U.S. & Williams, S.J. 2007. Sediment compaction rates and subsidence in deltaic plains: numerical constraints and stratigraphic influences. *Basin Research*, **19**, 19–31, https://doi.org/10.1111/j.1365-2117.2006.00310.x

Mellere, D., Zecchin, M. & Perale, C. 2005. Stratigraphy and sedimentology of fault-controlled backstepping shorefaces, middle Pliocene of Crotone Basin, southern Italy. *Sedimentary Geology*, **176**, 281–303, https://doi.org/10.1016/j.sedgeo.2005.01.010

Miall, A.D. 1974. Paleocurrent analysis of alluvial sediments: a discussion of directional variance and vector magnitude. *Journal of Sedimentary Research*, **44**, 1174–1185, https://doi.org/10.1306/212F6C6C-2B24-11D7-8648000102C1865D

Miller, K.G., Kominz, M.A. *et al.* 2005. The Phanerozoic record of global sea-level change. *Science*, **310**, 1293–1298, https://doi.org/10.1126/science.1116412

Milli, S., D'Ambrogi, C. *et al.* 2013. The transition from wave-dominated estuary to wave-dominated delta: the Late Quaternary stratigraphic architecture of Tiber River deltaic succession (Italy). *Sedimentary Geology*, **284–285**, 159–180, https://doi.org/10.1016/j.sedgeo.2012.12.003

Mitrovica, J., Beaumont, C. & Jarvis, G. 1989. Tilting of continental interiors by the dynamical effects of subduction. *Tectonics*, **8**, 1079–1094.

Muto, T. & Steel, R.J. 1997. Principles of regression and transgression: the nature of the interplay between accommodation and sediment supply. *Journal of Sedimentary Research*, **67**, 994–1000, https://doi.org/10.1306/D42686A8-2B26-11D7-8648000102C1865D

Nadon, G.C. 1998. Magnitude and timing of peat-to-coal compaction. *Geology*, **26**, 727–730, https://doi.org/10.1130/0091-7613(1998)026<0727:matopt>2.3.co;2

Nichols, M.M. 1989. Sediment accumulation rates and relative sea-level rise in lagoons. *Marine Geology*, **88**, 201–219.

Nichols, M.M. & Biggs, R.B. 1985. Estuaries. *In*: Davis, R.A. (ed.) *Coastal Sedimentary Environments*. Springer, Berlin, 77–186.

Olariu, C., Steel, R.J., Dalrymple, R.W. & Gingras, M.K. 2012. Tidal dunes v. tidal bars: the sedimentological and architectural characteristics of compound dunes in a tidal seaway, the lower Baronia Sandstone (Lower Eocene), Ager Basin, Spain. *Sedimentary Geology*, **279**, 134–155, https://doi.org/10.1016/j.sedgeo.2012.07.018

Olsen, T.R., Mellere, D. & Olsen, T. 1999. Facies architecture and geometry of landward-stepping shoreface tongues: the Upper Cretaceous Cliff House Sandstone (Mancos Canyon, south-west Colorado). *Sedimentology*, **46**, 603–625, https://doi.org/10.1046/j.1365-3091.1999.00234.x

Painter, C.S. & Carrapa, B. 2013. Flexural v. dynamic processes of subsidence in the North American Cordillera foreland basin. *Geophysical Research Letters*, **40**, 4249–4253, https://doi.org/10.1002/grl.50831

Painter, C.S., York-Sowecke, C.C. & Carrapa, B. 2013. Sequence stratigraphy of the Upper Cretaceous Sego Sandstone Member reveals spatio-temporal changes in depositional processes, Northwest Colorado, U.S.A. *Journal of Sedimentary Research*, **83**, 323–338, https://doi.org/10.2110/jsr.2013.21

Pang, M. & Nummedal, D. 1995. Flexural subsidence and basement tectonics of the Cretaceous Western Interior basin, United States. *Geology*, **23**, 173–176, https://doi.org/10.1130/0091-7613(1995)023<0173:FSABTO>2.3.CO;2

Patchineelam, S.M., Kjerfve, B. & Gardner, L.R. 1999. A preliminary sediment budget for the Winyah Bay estuary, South Carolina, USA. *Marine Geology*, **162**, 133–144, https://doi.org/10.1016/S0025-3227(99)00059-6

Peterson, F. 1969*a*. *Cretaceous Sedimentation and Tectonism in the Southeastern Kaiparowits Region, Utah.*, United States Department of the Interior Geologic Survey, Open File Report **69**.

Peterson, F. 1969*b*. *Four New Members of the Upper Cretaceous Straight Cliffs Formation in the Southeastern Kaiparowits Region, Kane County, Utah. Contributions to Stratigraphy*. US Geological Survey Bulletin, **1274-J**, J1–J28.

Plink-Björklund, P. 2008. Wave-to-tide facies change in a Campanian shoreline complex, Chimney Rock Tongue, Wyoming-Utah, U.S.A. *In*: Hampson, G.J. (ed.) *Recent Advances in Models of Siliciclastic Shallow-Marine Stratigraphy*. SEPM Special Publications, **90**, 265–291.

Plint, A.G. 2010. Wave- and storm-dominated shoreline and shallow-marine systems. *In*: James, N.P. & Dalrymple, R.W. (eds) *Facies Models 4*. Geological Society of Canada, St John's, 167–199.

Posamentier, H.W. & Allen, G.P. 1993. Variability of the sequence stratigraphic model: effects of local basin factors. *Sedimentary Geology*, **86**, 91–109, https://doi.org/10.1016/0037-0738(93)90135-R

Posamentier, H.W. & Allen, G.P. 1999. *Siliciclastic Sequence Stratigraphy: Concepts and Applications*. SEPM Concepts in Sedimentology and Paleontology, **7**.

Posamentier, H.W., Jervey, M.T. & Vail, P.R. 1988. Eustatic controls on clastic deposition I – conceptual framework. *In*: Wilgus, C.K., Hastings, B.S., Kendall, C.G.S.C., Posamentier, H.W., Ross, C.A. & Van Wagoner, J.C. (eds) *Sea-Level Changes: an Integrated Approach*. SEPM Special Publications, **42**, 109–124.

Potter, I.C., Chuwen, B.M., Hoeksema, S.D. & Elliott, M. 2010. The concept of an estuary: a definition that incorporates systems which can become closed to the ocean and hypersaline. *Estuarine, Coastal and Shelf Science*, **87**, 497–500, https://doi.org/10.1016/j.ecss.2010.01.021

Pritchard, D.W. 1967. What is an estuary? Physical viewpoint. *In*: Lauff, G.H. (ed.) *Estuaries*. American Association for the Advancement of Science Publications, **83**, 3–5.

Rampino, M.R. & Sanders, J.E. 1980. Evolution of the barrier: islands of southern Long Island, New York. *Sedimentology*, **28**, 37–47, https://doi.org/10.1111/j.1365-3091.1981.tb01661.x

Reading, H.G. & Collinson, J.D. 1996. *Clastic Coasts*. 3rd edn. Blackwell Science, Oxford.

Reddering, J.S.V. 1980. What is an estuary? *South African Journal of Science*, **76**, 341.

Reinson, G.E. 1992. Transgressive barrier island and estuarine systems. *In*: Walker, R.G. & James, N.P. (eds) *Facies Models: Response to Sea-Level Changes*. Geological Society of Canada, St John's, 179–194.

Reynolds, A.D. 1999. Dimensions of paralic sandstone bodies. *American Association of Petroleum Geologists Bulletin*, **83**, 211–229, https://doi.org/10.1306/00AA9A48-1730-11D7-8645000102C1865D

Reynolds, D.J., Steckler, M.S. & Coakley, B.J. 1991. The role of the sediment load in sequence stratigraphy: the influence of flexural isostasy and compaction. *Journal of Geophysical Research: Solid Earth*, **96**, 6931–6949, https://doi.org/10.1029/90jb01914

Roberts, L.N.R. & Kirschbaum, M.A. 1995. *Paleogeography of the Late Cretaceous of the Western Interior of Middle North America – Coal Distribution and Sediment Accumulation*. US Geological Survey Professional Paper, **1561**.

Roehler, H.W. 1988. *The Pintail Coal Bed and Barrier Bar G – a Model for Coal of Barrier Bar–Lagoon Origin, Upper Cretaceous Almond Formation, Rock Springs Coal Field, Wyoming*. US Geological Survey Professional Paper, 1398.

Rosati, J.D. 2009. *Barrier island migration over a consolidating substrate*. PhD thesis, Louisiana State University, Baton Rouge, LA.

Rosati, J.D., Dean, R.G. & Stone, G.W. 2010. A cross-shore model of barrier island migration over a compressible substrate. *Marine Geology*, **271**, 1–16, https://doi.org/10.1016/j.margeo.2010.01.005

Ryer, T.A. & Langer, A.W. 1980. Thickness change involved in the peat-to-coal transformation for a bituminous coal of Cretaceous age in central Utah. *Journal of Sedimentary Petrology*, **50**, 987–992, https://doi.org/10.1306/212F7B44-2B24-11D7-8648000102C1865D

Sadler, P.M. 1981. Sediment accumulation rates and the completeness of stratigraphic sections. *The Journal of Geology*, **89**, 569–584, http://www.jstor.org/stable/30062397

Sadler, P.M. 1999. The influence of hiatuses on sediment accumulation rates. *GeoResearch Forum*, **5**, 15–40.

Saito, Y. 1994. Shelf sequence and characteristic bounding surfaces in a wave-dominated setting: latest Pleistocene–Holocene examples from northeast Japan. *Marine Geology*, **120**, 105–127, https://doi.org/10.1016/0025-3227(94)90080-9

Salem, A.C. 2009. *Mesozoic tectonics of the Maria fold and thrust belt and McCoy Basin, southeastern California*. PhD thesis, Arizona State University, Tempe, AZ.

Salzmann, L., Green, A. & Cooper, J.A.G. 2013. Submerged barrier shoreline sequences on a high energy, steep and narrow shelf. *Marine Geology*, **346**, 366–374, https://doi.org/10.1016/j.margeo.2013.10.003

Sargent, K.A. & Hansen, D.E. 1982. *Bedrock Geologic Map of the Kaiparowits Coal-basin Area, Utah*. US Geological Survey, Miscellaneous Investigations Map, **I-1033-I**.

Schieber, J. & Southard, J.B. 2009. Bedload transport of mud by floccule ripples – direct observation of ripple migration processes and their implications. *Geology*, **37**, 483–486, https://doi.org/10.1130/g25319a.1

Schieber, J., Southard, J. & Thaisen, K. 2007. Accretion of mudstone beds from migrating floccule ripples. *Science*, **318**, 1760–1763, https://doi.org/10.1126/science.1147001

Schlager, W. 1993. Accommodation and supply – a dual control on stratigraphic sequences. *Sedimentary Geology*, **86**, 111–136, https://doi.org/10.1016/0037-0738(93)90136-S

Schwartz, R.K. 1982. Bedform and stratification characteristics of some modern small-scale washover sand bodies. *Sedimentology*, **29**, 835–849, https://doi.org/10.1111/j.1365-3091.1982.tb00087.x

Sedgwick, P.E. & Davis, R.A., Jr. 2003. Stratigraphy of washover deposits in Florida: implications for recognition in the stratigraphic record. *Marine Geology*, **200**, 31–48, https://doi.org/10.1016/S0025-3227(03)00163-4

Seidler, L. & Steel, R. 2001. Pinch-out style and position of tidally influenced strata in a regressive–transgressive wave-dominated deltaic sandbody, Twentymile Sandstone, Mesaverde Group, NW Colorado. *Sedimentology*, **48**, 399–414, https://doi.org/10.1046/j.1365-3091.2001.00370.x

Shanley, K.W. & McCabe, P.J. 1991. Predicting facies architecture through sequence stratigraphy – an example from the Kaiparowits Plateau, Utah. *Geology*, **19**, 742–745, https://doi.org/10.1130/0091-7613(1991)019<0742:PFATSS>2.3.CO;2

Shanley, K.W. & McCabe, P.J. 1993. *Alluvial Architecture in a Sequence Stratigraphic Framework: a Case History from the Upper Cretaceous of Southern Utah, USA*. International Association of Sedimentologists Special Publications, **15**, 21–56.

Shanley, K.W. & McCabe, P.J. 1994. Perspectives on the sequence stratigraphy of continental strata. *American Association of Petroleum Geologists Bulletin*, **78**, 544–568.

Shanley, K.W. & McCabe, P.J. 1995. Sequence stratigraphy of Turonian–Santonian strata, Kaiparowits Plateau, southern Utah, U.S.A.: implications for regional correlation and foreland basin evolution. *In*: Van Wagoner, J.C. & Bertram, G.T. (eds) *Sequence Stratigraphy of Foreland Basin Deposits*. American Association of Petroleum Geologists Memoir, **64**, 103–136.

Sixsmith, P.J., Hampson, G.J., Gupta, S., Johnson, H.D. & Fofana, J.F. 2008. Facies architecture of a net transgressive sandstone reservoir analog: the Cretaceous Hosta Tongue, New Mexico. *American Association of Petroleum Geologists Bulletin*, **92**, 513–547, https://doi.org/10.1306/01020807017

Slater, R.D. 1985. A numerical model of tides in the Cretaceous Seaway of North America. *The Journal of Geology*, **93**, 333–345, http://www.jstor.org/stable/30085018

Slingerland, R. & Keen, T. 1999. Sediment transport in the Western Interior Seaway of North America: predictions from a climate–ocean–sediment model. *In*: Bergman, K.A. & Snedden, J.W. (eds) *Isolated Shallow Marine Sandbodies: Sequence Stratigraphic Analysis and Sedimentologic Interpretation*. SEPM Special Publications, **64**, 179–190.

Slingerland, R., Kump, L.R., Arthur, M.A., Fawcett, P.J., Sageman, B.B. & Barron, E.J. 1996. Estuarine circulation in the Turonian Western Interior Seaway of North America. *Geological Society of America Bulletin*, **108**, 941–952, https://doi.org/10.1130/0016-7606(1996)108<0941:ecittw>2.3.co;2

Snedden, J.W. & Kersey, D.G. 1981. Origin of the San Miguel lignite deposit and associated lithofacies, Jackson Group, south Texas. *American Association of Petroleum Geologists Bulletin*, **65**, 1099–1109, https://doi.org/10.1306/03B59602-16D1-11D7-8645000102C1865D

Sommerfield, C.K. 2006. On sediment accumulation rates and stratigraphic completeness: lessons from Holocene ocean margins. *Continental Shelf Research*, **26**, 2225–2240, https://doi.org/10.1016/j.csr.2006.07.015

Southard, J.B., Lambie, J.M., Federico, D.C., Pile, H.T. & Weidman, C.R. 1990. Experiments on bed configurations in fine sands under bidirectional purely oscillatory flow, and the origin of hummocky cross-stratification. *Journal of Sedimentary Research*, **60**, 1–17, https://doi.org/10.1306/212F90F7-2B24-11D7-8648000102C1865D

Souza-Filho, P.W.M., Lessa, G.C., Cohen, M.C.L., Costa, F.R. & Lara, R.J. 2009. The subsiding macrotidal barrier estuarine system of the eastern Amazon Coast, northern Brazil. *In*: Dillenburg, S.R. & Hesp, P.A. (eds) *Geology and Geomorphology of Holocene Coastal Barriers of Brazil*. Lecture Notes in Earth Sciences. Springer, Berlin, 347–375, https://doi.org/10.1007/978-3-540-44771-9_11

Swift, D.J.P. 1968. Coastal erosion and transgressive stratigraphy. *The Journal of Geology*, **76**, 444–456, https://doi.org/10.1086/627342

Szwarc, T.S., Johnson, C.L., Stright, L.E. & McFarlane, C.M. 2014. Interactions between axial and transverse drainage systems in the Late Cretaceous Cordilleran foreland basin: evidence from detrital zircons in the Straight Cliffs Formation, southern Utah. *Geological Society of America Bulletin*, **127**, 372–392, https://doi.org/10.1130/B31039.1

Taylor, A.M. & Goldring, R. 1993. Description and analysis of bioturbation and ichnofabric. *Journal of the Geological Society, London*, **150**, 141–148, https://doi.org/10.1144/gsjgs.150.1.0141

Taylor, W.J. 1993. *Stratigraphic and Lithologic Analysis of the Claron Formation in Southwestern Utah*. Miscellaneous Publication Utah Geologic Survey, **93**.

Terzaghi, K. 1943. *Theoretical Soil Mechanics*. Wiley, New York.

Thorne, J.A. & Swift, D.J.P. 1991. Sedimentation on continental margins; VI. A regime model for depositional sequences, their component systems tracts, and bounding surfaces. *In*: Swift, D.J.P., Oertel, G.F., Tillman, R.W. & Thorne, J.A. (eds) *Shelf Sand and Sandstone Bodies: Geometry, Facies and Sequence Stratigraphy*. International Association of Sedimentologists Special Publications, **14**, 189–255.

Tindall, S.E. & Davis, G.H. 1999. Monocline development by oblique-slip fault-propagation folding: the East Kaibab monocline, Colorado Plateau, Utah. *Journal of Structural Geology*, **21**, 1303–1320, https://doi.org/10.1016/S0191-8141(99)00089-9

Tindall, S.E., Storm, L.P., Jenesky, T.A. & Simpson, E.L. 2010. Growth faults in the Kaiparowits Basin,

Utah, pinpoint initial Laramide deformation in the western Colorado Plateau. *Lithosphere*, **2**, 221–231, https://doi.org/10.1130/L79.1

Törnqvist, T.E., Wallace, D.J. *et al.* 2008. Mississippi Delta subsidence primarily caused by compaction of Holocene strata. *Nature Geoscience*, **1**, 173–176, https://doi.org/10.1038/ngeo129

Tye, R.S., Hewlett, J.S., Thompson, P.R. & Goodman, D.K. 1999. Integrated stratigraphic and depositional-facies analysis of parasequences in a transgressive systems tract. *In*: Weimer, R.J. & Posamentier, H.W. (eds) *Siliciclastic Sequence Stratigraphy: Recent Developments and Applications*. American Association of Petroleum Geologists Memoirs, **58**, 99–133.

Vail, P.R. & Mitchum, R.M. 1979. Global cycles of relative changes of sea level from seismic stratigraphy. *In*: Watkins, J.S., Montadert, L. & Dickerson, P.W. (eds) *Geological and Geophysical Investigations of Continental Margins*. American Association of Petroleum Geologists Memoirs, **29**, 469–472.

Vail, P.R., Mitchum, R.M., Jr. & Thompson, S., III 1977. Seismic stratigraphy and global changes of sea level, Part 3: relative changes of sea level from coastal onlap. *In*: Payton, C.E. (ed.) *Seismic Stratigraphy: Applications to Hydrocarbon Exploration*. American Association of Petroleum Geologists Memoirs, **26**, 99–116.

Vakarelov, B.K. & Ainsworth, R.B. 2013. A hierarchical approach to architectural classification in marginal-marine systems: bridging the gap between sedimentology and sequence stratigraphy. *American Association of Petroleum Geologists Bulletin*, **97**, 1121–1161, https://doi.org/10.1306/11011212024

Vakarelov, B.K., Ainsworth, R.B. & MacEachern, J.A. 2011. Recognition of wave-dominated, tide-influenced shoreline systems in the rock record: variations from a microtidal shoreline model. *Sedimentary Geology*, **279**, 23–41, https://doi.org/10.1016/j.sedgeo.2011.03.004

Valasek, D. 1995. The Tocito Sandstone in a sequence stratigraphic framework: an example of landward-stepping small-scale genetic sequences. *In*: Van Wagoner, J.C. & Bertram, G.T. (eds) *Sequence Stratigraphy of Foreland Basin Deposits*. American Association of Petroleum Geologists Memoirs, **64**, 349–369.

van Asselen, S. 2011. The contribution of peat compaction to total basin subsidence: implications for the provision of accommodation space in organic-rich deltas. *Basin Research*, **23**, 239–255, https://doi.org/10.1111/j.1365-2117.2010.00482.x

van Heerden, I. & Roberts, H.H. 1988. Facies development of Atchafalaya Delta, Louisiana: a modern bayhead delta. *American Association of Petroleum Geologists Bulletin*, **72**, 439–453.

Vaninetti, G.E. 1979. *Coal stratigraphy of the John Henry Member of the Straight Cliffs Formation, Kaiparowits Plateau, Utah*. MS thesis, University of Utah.

Van Wagoner, J.C., Posamentier, H.W., Mitchum, R.M., Vail, P.R., Sarg, J.F., Loutit, T.S. & Hardenbol, J. 1988. An overview of the fundamentals of sequence stratigraphy and key definitions. *In*: Wilgus, C.K., Posamentier, H.W., Ross, C.K. & Kendall, C.G. (eds) *Sea-Level Changes: an Integrated Approach*. SEPM Special Publications, **42**, 39–45.

Van Wagoner, J.C., Mitchum, R.M., Campion, K.M. & Rahmanian, V.D. 1990. *Siliciclastic Sequence Stratigraphy in Well Logs, Cores, and Outcrops: Concepts for High-Resolution Correlation of Time and Facies*. American Association of Petroleum Geologists Methods in Exploration Series, **7**.

Wadsworth, J., Diessel, C. & Boyd, R. 2010. The sequence stratigraphic significance of paralic coal and its use as an indicator of accommodation space in terrestrial sediments. *In*: Ratcliffe, K. & Zaitlin, B.A. (eds) *Application of Modern Stratigraphic Techniques: Theory and Case Histories*. SEPM Special Publications, **94**, 201–219.

Walker, R.G. & Plint, A.G. 1992. Wave- and storm-dominated shallow marine systems. *In*: Walker, R.G. & James, N.P. (eds) *Facies Models: a Response to Sea-Level Changes*. Geological Association of Canada, St John's, 219–238.

Weimer, R.J. 1960. Upper Cretaceous stratigraphy, Rocky Mountain area. *American Association of Petroleum Geologists Bulletin*, **44**, 1–20, https://doi.org/10.1306/0BDA5F6F-16BD-11D7-8645000102C1865D

Weller, J.M. 1959. Compaction of sediments. *American Association of Petroleum Geologists Bulletin*, **51**, 468–470, https://doi.org/10.1306/0BDA5C9F-16BD-11D7-8645000102C1865D

White, T., Furlong, K. & Arthur, M. 2002. Forebulge migration in the Cretaceous Western Interior Basin of the central United States. *Basin Research*, **14**, 43–54, https://doi.org/10.1046/j.1365-2117.2002.00165.x

Willis, A.J. & Moslow, T.F. 1994. Stratigraphic setting of transgressive barrier-island reservoirs with an example from the Triassic Halfway Formation, Wembley Field, Alberta, Canada. *American Association of Petroleum Geologists Bulletin*, **78**, 775–791.

Wright, L.D. & Short, A.D. 1984. Morphodynamic variability of surf zones and beaches: a synthesis. *Marine Geology*, **56**, 93–118, https://doi.org/10.1016/0025-3227(84)90008-2

York, C.C., Painter, C.S. & Carrapa, B. 2011. Sedimentological characterization of the Sego Sandstone (NW Colorado, USA): a new scheme to recognize ancient flood-tidal-delta deposits and implications for reservoir potential. *Journal of Sedimentary Research*, **81**, 401–419, https://doi.org/10.2110/jsr.2011.35

Yoshida, S., Johnson, H.D., Pye, K. & Dixon, R.J. 2004. Transgressive changes from tidal estuarine to marine embayment depositional systems: the Lower Cretaceous Woburn Sands of southern England and comparison with Holocene analogs. *American Association of Petroleum Geologists Bulletin*, **88**, 1433–1460, https://doi.org/10.1306/05140403075

Yoshida, S., Steel, R.J. & Dalrymple, R.W. 2007. Changes in depositional processes – an ingredient in a new generation of sequence-stratigraphic models. *Journal of Sedimentary Research*, **77**, 447–460, https://doi.org/10.2110/jsr.2007.048

Zaitlin, B.A. & Shultz, B.C. 1990. Wave-influenced estuarine sand body, Senlac Heavy Oil Pool, Saskatchewan, Canada. *In*: Barwis, J.H., McPherson, J.G. & Studlick, J.R.J. (eds) *Sandstone Petroleum Reservoirs*. Springer, New York, 363–387.

ZECCHIN, M. 2007. The architectural variability of small-scale cycles in shelf and ramp clastic systems: the controlling factors. *Earth Science Reviews*, **84**, 21–55, https://doi.org/10.1016/j.earscirev.2007.05.003

ZECCHIN, M. 2010. Towards the standardization of sequence stratigraphy: is the parasequence concept to be redefined or abandoned? *Earth Science Reviews*, **102**, 117–119, https://doi.org/10.1016/j.earscirev.2010.03.004

Anatomy of a mixed-influence shelf edge delta, Karoo Basin, South Africa

LUZ E. GOMIS-CARTESIO[1]*, MIQUEL POYATOS-MORÉ[1], STEPHEN S. FLINT[1], DAVID M. HODGSON[2], RUFUS L. BRUNT[1] & HENRY DEV. WICKENS[3]

[1]*Stratigraphy Group, School of Earth, Atmospheric and Environmental Sciences, University of Manchester, Manchester M13 9PL, UK*

[2]*Stratigraphy Group, School of Earth and Environment, University of Leeds, Leeds LS2 9JT, UK*

[3]*Geo-Routes Petroleum, 37 Zevendal Road, Kuils River 7580, South Africa*

**Correspondence: luz.gomiscartesio@manchester.ac.uk*

Abstract: The position and process regime of paralic systems relative to the shelf edge rollover is a major control on sediment transfer into deep water. The depositional strike and dip variability of an exhumed Permian shelf edge succession has been studied in the Paardeberg Ridge, Karoo Basin. Siltstone-rich slope turbidites are overlain by 25–75 m-thick prodelta parasequences. These are truncated by a 30 m-thick sandstone-prone unit of tabular or convex-topped sandstones, interpreted as wave-modified mouth bars, cut by multiple irregular concave-upwards erosive surfaces overlain by sandstones, interpreted as distributary channels. The stratigraphic context, lithofacies and architecture are consistent with a mixed-influence shelf edge delta; the erosional base to the unit marks a basinwards shift in facies, consistent with a sequence boundary. Channels become thicker, wider, more erosive and incise into deeper-water facies downdip and correlate with sandstone-rich upper slope turbidites, all of which support the bypass of sand across the rollover. The overall progradational stacking pattern results in a stratigraphic decrease in channel dimensions. The results of this study suggest a predictable relationship between channel geometry, facies and position on the shelf-to-slope profile under a mixed wave and fluvial process regime.

Basin margin progradation and the timing of sediment transport to the oceans is strongly influenced by the position and character of paralic systems relative to the physiography of the shelf edge rollover (e.g. Edwards 1981; Mayall *et al.* 1992; Sydow & Roberts 1994; Morton & Suter 1996; Steel *et al.* 2000, 2003; Muto & Steel 2002; Porębski & Steel 2003; Saller *et al.* 2004; Carvajal & Steel 2009; Covault *et al.* 2009; Dixon *et al.* 2012a). The presence of gullies and channels, which incise into shelf edge deltas and act as conduits for sediment transport, has been widely described and associated with multiple factors, including a lowering of relative sea-level, headward erosion of submarine canyons, gravitational instabilities and slope failure, gravity flows and/or high fluvial discharge, among others (e.g. Pratson *et al.* 1994; Fulthorpe *et al.* 1999, 2000; Muto & Steel 2002; Plink-Björklund & Steel 2002, 2005; Donovan 2003; Porębski & Steel 2003; Posamentier & Kolla 2003; Petter & Steel 2005; Jackson & Johnson 2009; Sanchez *et al.* 2012*b*; Sylvester *et al.* 2012; Prélat *et al.* 2015).

In contrast to reflection seismic data, exhumed shelf margins provide the resolution needed to constrain the interplay of the sedimentary processes responsible for channel and gulley initiation at the shelf edge rollover (Hubbard *et al.* 2010). Typically, however, shelf edge deltas are interpreted to be dominated by a particular process regime (Dixon *et al.* 2012*b*). Mixed-influenced systems are rarely reported (Mellere *et al.* 2003; Pontén & Plink-Björklund 2009; Bowman & Johnson 2014) and the fine-scale downdip changes in channel geometry and infill architecture across the shelf edge rollover have not been documented at outcrop. Most published examples of exhumed rollovers are from high-gradient shelf margins with short slope lengths (Pyles & Slatt 2000; Mellere *et al.* 2002; Plink-Björklund & Steel 2005; Carvajal & Steel 2006; Pyles & Slatt 2007; Pontén & Plink-Björklund 2009; Helland-Hansen 2010). This contrasts with the generally low gradients observed in many seismic datasets (e.g. Pirmez *et al.* 1998; Cattaneo *et al.* 2007; Patruno *et al.* 2015) that are challenging to constrain at outcrop (e.g. Dixon *et al.* 2012*a*, *b*; Jones *et al.* 2013).

The importance of lateral variability in shelf margin physiography, and the distribution of erosional and depositional process regimes, is highlighted in modern (e.g. Olariu & Steel 2009) and

From: HAMPSON, G. J., REYNOLDS, A. D., KOSTIC, B. & WELLS, M. R. (eds) 2017. *Sedimentology of Paralic Reservoirs: Recent Advances*. Geological Society, London, Special Publications, **444**, 393–418.
First published online July 14, 2016, https://doi.org/10.1144/SP444.5

three-dimensional reflection seismic datasets (e.g. Suter & Berryhill 1985; Matteucci & Hine 1987; Poag *et al.* 1990; Tesson *et al.* 1990; Milton & Dyce 1995; Fulthorpe & Austin 1998; Kolla *et al.* 2000; Saller *et al.* 2004; Hadler-Jacobsen *et al.* 2005; Crumeyrolle *et al.* 2007; Ryan *et al.* 2009; Henriksen *et al.* 2011; Moscardelli *et al.* 2012; Sanchez *et al.* 2012*a*, *b*; Bourget *et al.* 2014). Outcrop studies focus on determining clinoform and rollover geometries down depositional dip profiles (e.g. Steel *et al.* 2000; Plink-Björklund & Steel 2002, 2005; Mellere *et al.* 2003; Pyles & Slatt 2007; Uroza & Steel 2008; Dixon *et al.* 2012*a*). Documented outcrop examples with large-scale along-strike control are rare (e.g. Dixon *et al.* 2012*a*; Jones *et al.* 2015).

In this study, a rare example of an exhumed paralic succession in a shelf edge rollover position with extensive depositional strike and dip constraints is presented from the Permian Kookfontein and Waterford formations in the Tanqua depocentre, Karoo Basin, South Africa. The outcrops allow the following objectives to be addressed: (1) to describe the vertical transition from slope to shelf deposits and to identify the shelf edge rollover in a mixed-influence system and (2) to discuss the origin, evolution and infill of channel form features across the slope to shelf transition.

Geological setting

The Late Carboniferous–Triassic Karoo Basin (Fig. 1) has been interpreted as a retro-arc foreland basin with subsidence caused by flexural loading by the Cape Fold Belt (Johnson 1991; Cole 1992; Visser 1993; Veevers *et al.* 1994; Catuneanu *et al.* 1998). However, more recent studies have suggested that the subsidence was due to long wavelength dynamic topography effects driven by the subducting palaeo-Pacific plate (Tankard *et al.* 2009). The fill of the Karoo Basin consists of the >5 km thick Karoo Supergroup (Smith 1990) (Fig. 2a). Glacial deposits of the late Carboniferous Dwyka Group are overlain by Ecca Group post-glacial strata of the Prince Albert, Whitehill and Collingham formations, which show a deepening trend from shallow-water carbonates to basin plain turbidites (De Beer 1992; Turner 1999; Johnson *et al.* 2006). In the Tanqua depocentre, the Collingham Formation is overlain by *c.* 800 m of mudstones of the Tierberg Formation (Fig. 2a). Overlying these mudstones, the Skoorsteenberg Formation consists of sandstone-rich basin floor fans 1–4 (Bouma & Wickens 1991; Wickens 1994; Scott *et al.* 2000; Johnson *et al.* 2001; Andersson *et al.* 2004; Hodgson *et al.* 2006), overlain by a channelized slope wedge of Unit 5 (Wild *et al.* 2005). These deposits are topped by slope to shelf sediments of the Kookfontein and Waterford formations (Wickens 1994; Wild *et al.* 2009; Oliveira *et al.* 2011; Dixon *et al.* 2012*a*, *b*), which were the focus of this study (Fig. 2).

Dataset and methods

The Kookfontein and Waterford formations crop out across the northeastern part of the Tanqua

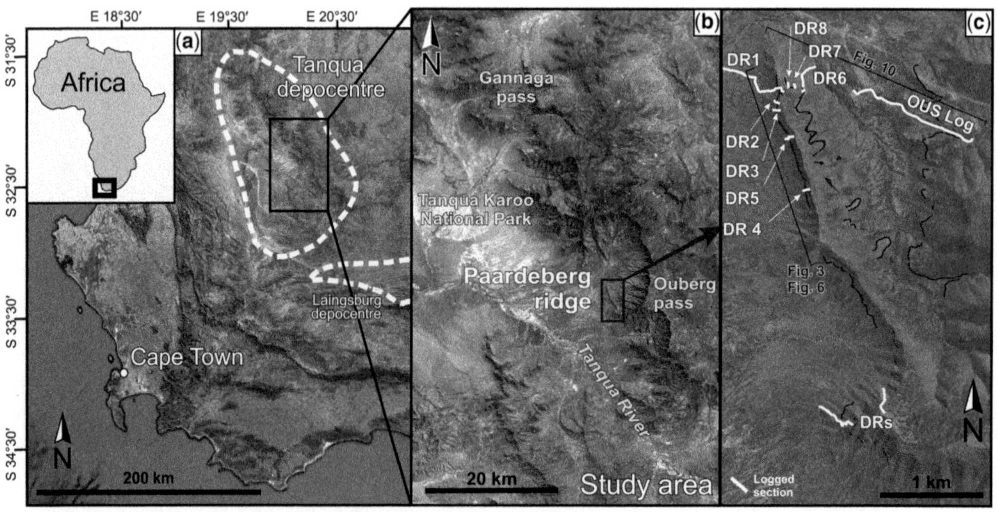

Fig. 1. (**a**) The southwestern Karoo Basin with Tanqua and Laingsburg depocentres outlined. (**b**) The Tanqua depocentre study area. (**c**) Detailed map of the Paardeberg Ridge locality. Log positions are shown in white and black lines indicate mapped erosion surfaces. Images from Google Earth.

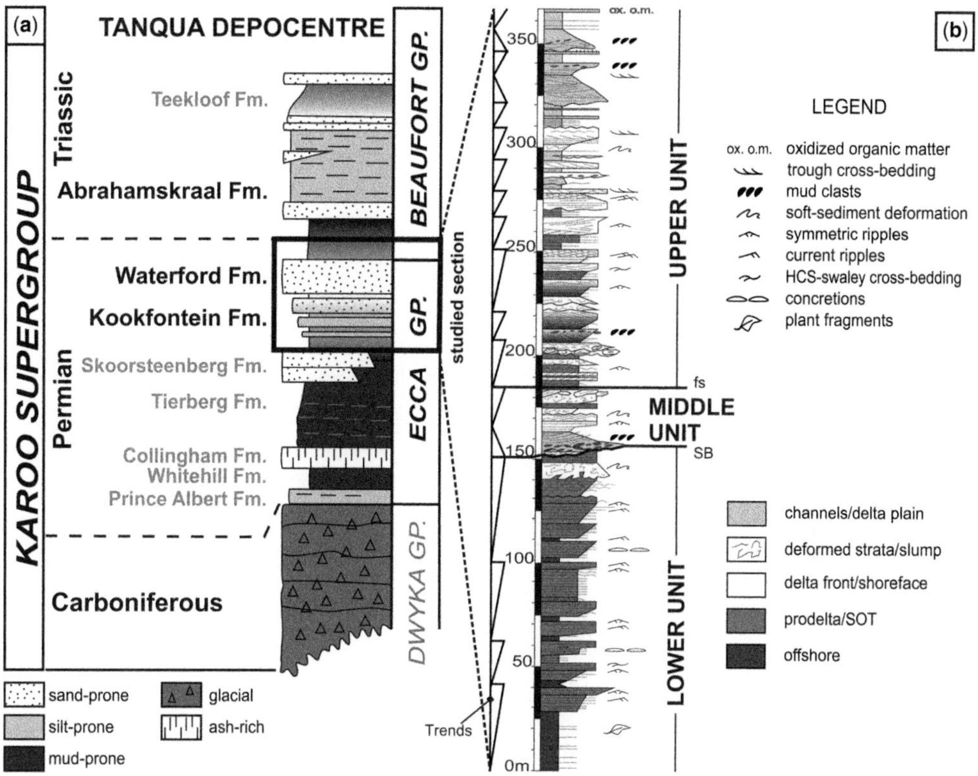

Fig. 2. (**a**) Summary stratigraphy of the Tanqua depocentre adapted from Wickens (1994), Wild *et al.* (2009) and Flint *et al.* (2011). (**b**) Summary log (DRs in Fig. 1c) of the Paardeberg Ridge succession. The middle unit is the focus of this work. HCS, hummocky cross stratification; SOT, shoreface–offshore transition.

depocentre, providing a study area of up to *c.* 1600 km^2. This study is focused on the Paardeberg Ridge locality (Fig. 1b), which exposes the stratigraphic transition from slope to shelf deposits (Fig. 2b) over an area of up to 10 km^2. The outcrop (Fig. 1c) is a 150–400 m-high, 5 km-long NW–SE-trending ridge with a steep SW face and gullied NE face that provides good three-dimensional control (Fig. 1c). The overall tabular and laterally continuous succession has been informally divided into lower, middle and upper units (Figs 2b & 3). The 30–40 m-thick sandstone-rich middle unit displays a complex assemblage of tabular or convex-topped sandstones cut by irregular concave-upwards erosive surfaces overlain by sandstones and is the main focus of this study.

Four regional (1:50 scale) sedimentary logs (DRs, DR1, DR6 and OUS) provide the general stratigraphic context and are combined with 10 detailed logs (1:25 scale), focused on the middle unit, to characterize the different erosional and depositional features in each locality (1090 m of cumulative logged thickness). Physical correlation of the stratigraphic units was constrained by walking out key surfaces between logs, on both faces of the ridge and in intervening areas, to map and capture depositional dip and strike changes in facies and geometries. All observations were recorded on high-resolution photo-panels, satellite imagery and aerial photographs (Figs 1c & 3). Palaeocurrent measurements ($n = 670$) were taken from planar and trough cross-bedding foresets, ripple cross-lamination, primary current lineation, basal tool marks and channel margin orientations (Fig. 4). A stratigraphic hierarchy based on lithofacies, palaeocurrent measurements and key surfaces was developed to capture the stacking patterns across multiple scales and to understand the temporal changes in erosion and deposition.

Facies analysis

Lithofacies were defined based on lithology, grain size and sedimentary structures (Table 1) and are interpreted in terms of depositional processes. Facies associations are described using lateral and vertical

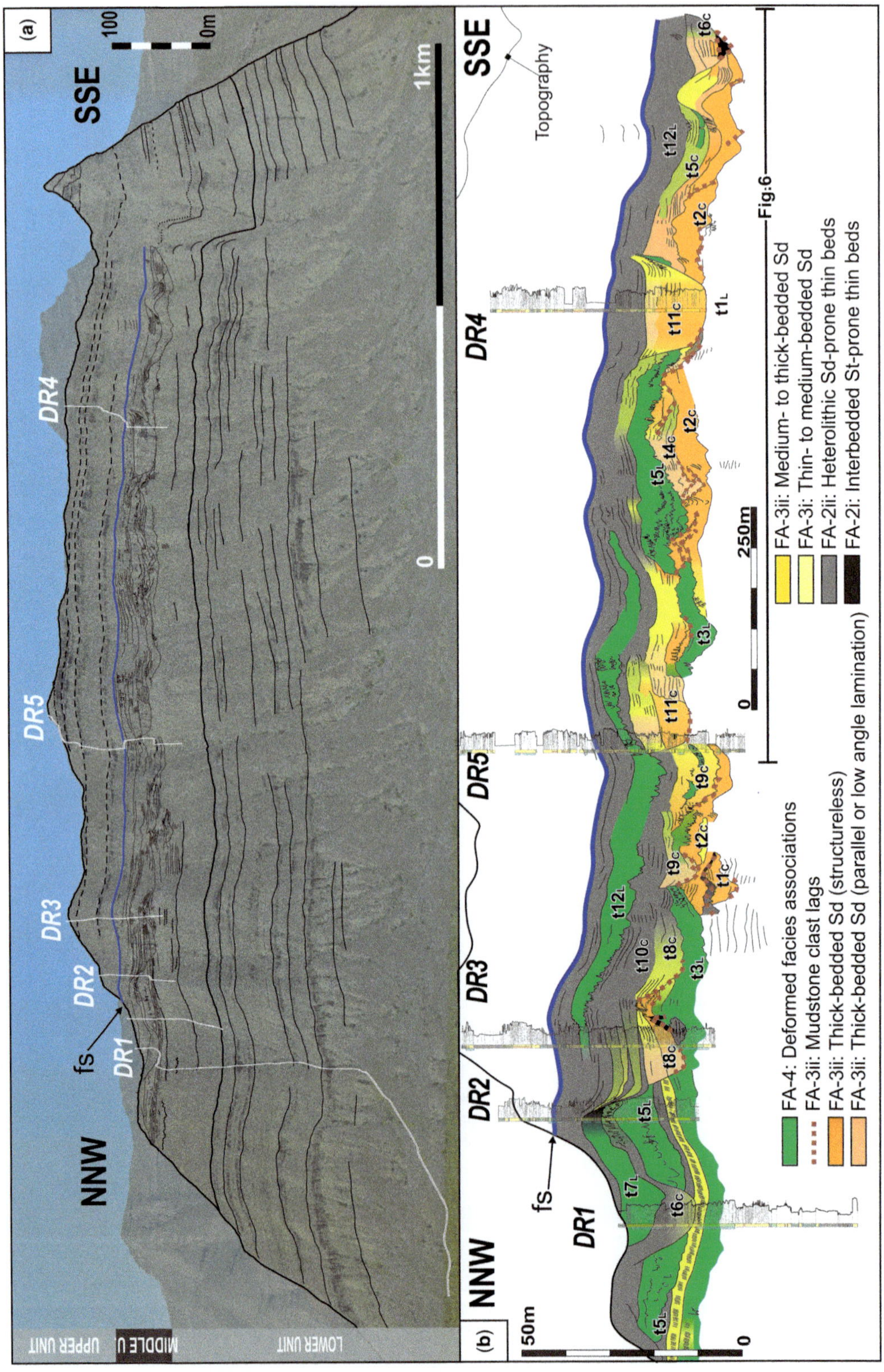
(a)
SSE
NNW
100
0m
1km
0
DR1
DR2
DR3
DR4
DR5
fS
UPPER UNIT
MIDDLE U.
LOWER UNIT
(b)
NNW
SSE
50m
0
250m
Topography
Fig:6
t1C
t1L
t2C
t3L
t4C
t5C
t5L
t6C
t7L
t8C
t9C
t10C
t11C
t12L
FA-4: Deformed facies associations
FA-3ii: Mudstone clast lags
FA-3ii: Thick-bedded Sd (structureless)
FA-3ii: Thick-bedded Sd (parallel or low angle lamination)
FA-3ii: Medium- to thick-bedded Sd
FA-3i: Thin- to medium-bedded Sd
FA-2ii: Heterolithic Sd-prone thin beds
FA-2i: Interbedded St-prone thin beds

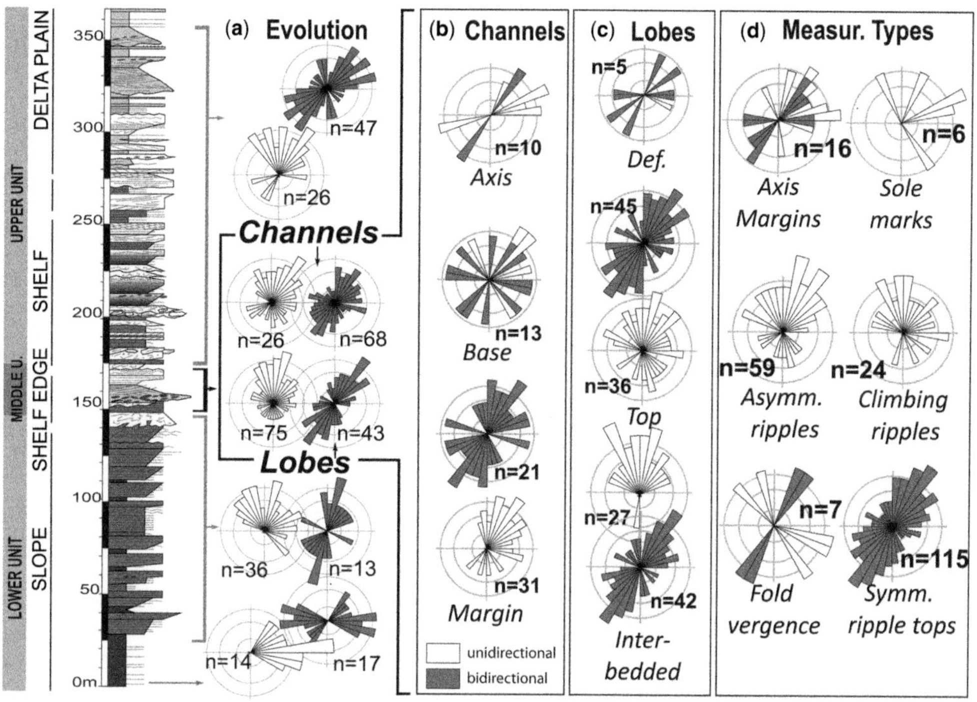

Fig. 4. Rose diagrams showing palaeocurrent distribution for the (**a**) lower, middle and upper units; and comparing (**b**) channelized v. (**c**) lobate bodies, showing measurements within the different parts of the elements; and (**d**) types of measurements. Dark colour represents bidirectional measurements and light colour represents unidirectional measurements. Def, deformation.

relationships of lithofacies and are interpreted in terms of different depositional environments.

Lithofacies

The primary classification is based on three main facies groups according to the maximum grain size observed within single beds (Table 1 and Fig. 5): conglomerate (C), sandstone (S) and mudstone (M); the latter includes the spectrum of grain populations from coarse silt to clay.

Facies associations

FA-1: diffuse laminated to structureless mudstones. Dark grey to green fissile to blocky mudstone bedsets. Commonly structureless fine–medium siltstone (Md, Mo), interbedded with coarse siltstone beds (Ms) with occasional parallel lamination, isolated/starved millimetre-scale ripples or gently undulating laminations. Bioturbation is low to moderate.

These facies are interpreted to represent deposition from suspension fallout, although some of their coarser and laminated components indicate deposition from dilute turbidity currents (Macquaker & Bohacs 2007; Schieber *et al.* 2007) (Table 1). The depositional environment is interpreted to record offshore deposition below storm wave base on the outer shelf or upper slope (Wild *et al.* 2009; Jones *et al.* 2013). This facies association forms the bulk of the strata in the lower part of the regional succession (lower unit).

FA-2i: heterolithic siltstone-prone thin beds. Thin- to very thin-bedded (centimetre- to millimetre-scale)

Fig. 3. (**a**) Vertically exaggerated photo-panorama of the SW face of the Paardeberg Ridge. Log positions are shown in white. (**b**) Architectural detail of the middle unit. Black lines mark the contacts between different elements and red dotted lines show the positions of mudstone clast lags. All depositional elements have been coded – for example t1C or t4L, where 't1' or 't4' is related to the relative time of deposition, inferred from the lateral and vertical relationships between the elements. 'C' or 'L' refers to channelized or lobate body. The oldest erosion surface (t1C) that truncates the distal prodelta facies of the uppermost slope parasequences is interpreted as the basal sequence boundary of the middle unit.

Table 1. *Facies classification, description and process interpretations*

Facies	Lithofacies	Sedimentary structures	Thickness, lithology and textural properties	Process interpretation	Other characteristics	Facies association
Conglomerate facies (C)	Ci	Matrix- to clast-supported intraformational conglomerate, crudely cross-stratified	Centimetre to decimetre scale, poorly sorted and matrix- to clast-supported intraformational conglomerate. Siltstone to sandstone matrix, with mudstone, siltstone or sandstone clasts, 0.5–20 cm in diameter. Irregular contacts, commonly erosive bases, poorly formed cross-bedding and gradational tops (Fig. 5a)	Locally sourced lithic clasts transported as bedload and deposited as basal lag when flow loses energy. Cross-bedding indicates migration of dunes and/or bars. Outsize clast content suggests bank collapse and erosion during high-energy flows or channel lateral migration	Reddening of lags may represent oxidation of iron-rich minerals	FA-3ii
	Cm	Mudstone clast horizon	Horizons of claystone and siltstone rip-up clasts within fine- or medium-grained sandstone beds. Clasts are up to 20 cm in size, rounded to angular and typically poorly sorted. Beds are irregular in thickness and often erosional (Fig. 5b)	Mudstone clasts entrained from erosion of underlying claystone and siltstone material during scouring. Where mudstone clast horizons occur among sandstones, this is interpreted as high-magnitude, low-frequency flows, mobilizing the mudstone clasts	Mud clasts are commonly aligned or imbricated	FA-3ii
Sandstone facies (S)	Ss	Structureless	Centimetre to metre scale, poorly to moderately sorted, very fine to medium-grained sandstone. Sharp base and top, rarely erosive. Few gradational tops (Fig. 5c)	En masse deposition from high-velocity and density sediment gravity flows. Uniform narrow grain size range suggests rapid deposition under upper flow regime conditions supressing bedform development	Internal structures overprinted by intensive bioturbation or dewatering. Locally abundant in plant remains and oxidized organic matter	FA-2, FA-3, FA-4
	Sg	Inverse or normal graded	Centimetre to metre scale, moderately to well-sorted, very fine to fine- or medium-grained sandstone. Base and top can be either sharp or gradational (Fig. 5d)	Normal grading interpreted to reflect evidence of waning flow conditions. Inverse grading reflects waxing flow conditions attributed to river floods	Plant debris and mica; development of composite inverse to normal graded beds	FA-2, FA-3, FA-4

Sp	Parallel bedding	Centimetre to metre scale, moderately to well-sorted very fine to fine-grained sandstone. Sharp top and base, rarely erosive or gradational. Parting lineations are common (Fig. 5e)	Deposition under upper phase plane bed conditions. Parting lineations can be produced by turbulent eddies or microvortices at the bed boundary layer. Also interpreted as representing vertical aggradation under shallow flow conditions	Parting lineation, mud clasts, oxidized organic matter and plant fragments observed in parallel laminae	FA-2, FA-3, FA-4
Sl	Low-angle cross-bedding, swaley or hummocky cross-bedding	Centimetre to metre scale, well-sorted very fine to fine-grained sandstone. Sharp base and top, rarely erosive. Common undulatory bed tops	Deposited under low flow regime conditions within large-scale dunes and barforms. Interpreted as representing deposition in broad bedload sheets during migration downstream and affected by combined or oscillatory flows	Well-sorted rounded grains, mud clasts associated with erosive bases and common symmetrical rippled tops	FA-2, FA-3, FA-4
Sx	High-angle planar or trough cross-bedding	Centimetre to metre scale, moderately sorted very fine to medium-grained sandstone. Sharp base and top, rarely erosive. Few gradational tops (Fig. 5f)	Planar cross-stratification represents migration of two-dimensional subaqueous dunes interpreted to represent deposition within deeper and/or faster parts of a confined/channelized flow. Trough cross-bedding is interpreted to reflect migration of three-dimensional dunes through bedload transportation. Three-dimensional dunes occur under lower flow regime conditions, where deeper scours are most prevalent and are associated with both downstream and laterally accreting barforms	Mud clasts, oxidized organic matter and plant fragments are observed in cross-sets	FA-3, FA-4
Sr	Ripple or climbing ripple lamination	Millimetre to decimetre scale, moderately sorted, very fine to fine-grained sandstone. Generally sharp or gradational bases and asymmetrical rippled tops (Fig. 5g)	Tractional bedforms developed under lower flow regime conditions. Asymmetrical current ripples produced by unidirectional flows. Climbing ripples reflecting higher sedimentation rates	Ripples locally show stoss and lee side preservation	FA-2, FA-3, FA-4

(*Continued*)

Table 1. *Continued*

Facies	Lithofacies	Sedimentary structures	Thickness, lithology and textural properties	Process interpretation	Other characteristics	Facies association
	Sw	Wavy lamination and symmetrical ripple lamination	Millimetre to decimetre scale, well-sorted very fine to fine-grained sandstone. Generally sharp or gradational bases and symmetrical rippled tops (Fig. 5h)	Tractional bedforms developed under lower flow regime conditions with symmetrical crests created by bidirectional currents under orbital wave motion. Secondary ladder-back ripple sets formed between larger ripple troughs	Well-sorted, rounded grains. Common superimposition of interference ripples	FA-2, FA-3, FA-4
Mudstone facies (M)	Ms	Parallel to ripple-laminated, normally and inversely graded fissile siltstones and mudstones	Millimetre to decimetre scale, poorly to moderately sorted coarse siltstone beds. Contacts are generally gradational. Locally sharp based	Deposition from very low-density turbidity/hyperpycnal currents, sometimes associated with river floods or storms. Post-depositional compaction masks primary sedimentary structures	Common appearance of starved and lenticular unidirectional ripples	FA-1, FA-2, FA-4
	Md	Dark grey to black structureless siltstones and mudstones	Millimetre to metre scale, moderately to well-sorted, medium to fine siltstone beds. Generally gradational contacts, locally sharp	Hemipelagic fallout from low current velocities or low suspended sediment concentrations during conditions of low clastic input. Mode of deposition ensures regional coverage of the mudstone deposits in distal settings	Often associated with sideritic concretionary horizons	FA-1, FA-2, (FA-4)
	Mo	Light grey to olive green, structureless to well laminated siltstones and mudstones	Centimetre to metre scale poorly to moderately sorted coarse to fine siltstone beds. Generally gradational contacts, locally sharp	Deposition by direct fallout from suspension, or debris flows leading to a structureless appearance. Green coloration indicates waterlogged environment	Local development of carbonate-rich nodular levels	FA-1, (FA-2)

planar and current–ripple cross-laminated siltstone (Ms), interbedded with minor amounts of ripple-laminated (unidirectional or bidirectional) very fine-grained sandstones (Sr, Sw), which commonly show inverse- to normal-graded composite beds (Sg). These deposits can form laterally extensive tabular packages (several hundred metres) or wedges of slightly inclined strata (a few tens of metres) that drape and/or onlap erosion surfaces. Bioturbation is low to moderate and the presence of organic debris is common.

This facies association is interpreted to record either deposition from low concentration turbulent suspension flows that developed distally or laterally from denser hyperpycnal flows (Plink-Björklund & Steel 2004; Bhattacharya & MacEachern 2009;

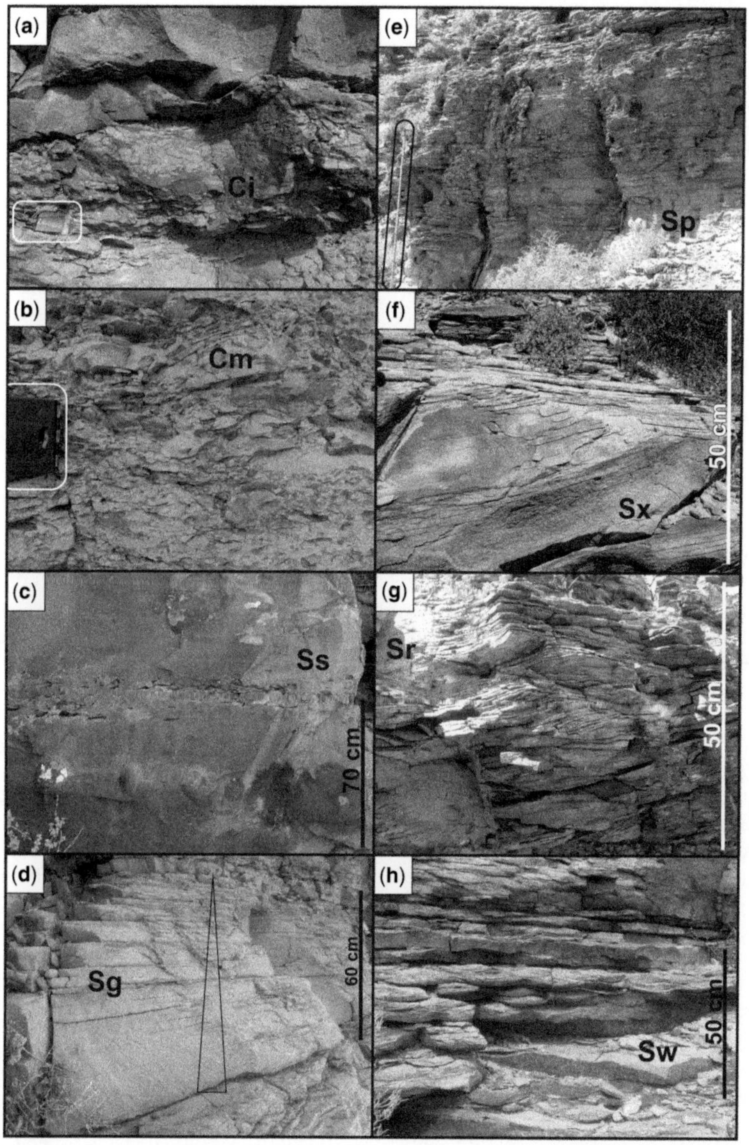

Fig. 5. Representative photographs of lithofacies from the middle unit. Table 1 outlines the lithofacies codes and interpretation of depositional processes. (**a**) Ci, matrix- to clast-supported intraformational conglomerate; (**b**) Cm, mudstone clast horizon; (**c**) Ss, structureless sandstone; (**d**) Sg, inverse to normal-graded sandstone; (**e**) Sp, parallel bedded sandstone; (**f**) Sx, high-angle planar or trough cross-bedded sandstone; (**g**) Sr, ripple or climbing ripple-laminated sandstone; (**h**) Sw, wavy laminated and symmetrical ripple laminated sandstone.

Zavala *et al.* 2011) or deposition from dilute turbidity currents in unconfined settings. This facies association has been interpreted to reflect deposition in a distal prodelta/offshore transition environment relatively far from the sediment feeder system, but still recording the effects of waves.

FA-2ii: heterolithic sandstone-prone thin beds. Laterally extensive (up to several hundreds of metres) 1–15 m-thick packages of thin-bedded (centimetre-scale) parallel or ripple-laminated, inverse- or normal-graded, very fine-grained sandstone (Sl, Sr, Sw) alternating with planar and current–ripple laminated siltstone (Ms). Locally, erosional bed bases and inverse grading are observed in thin sandstone beds and ripple lamination includes asymmetrical or symmetrical forms. In places, this facies association includes medium to thick beds (up to 50 cm) of partially amalgamated, fine-grained sandstone (Ss, Sg, Sl) forming lenticular packages. Bioturbation is moderate to high and the presence of unclassified organic fragments is common.

This facies association is interpreted to record the deposits of moderate to low concentration turbidity currents in both confined (within erosive surfaces) and unconfined settings (tabular and laterally extensive packages). Commonly, the association passes laterally into silt-prone heterolithic deposits (FA-2i) and suggests deposition in a proximal prodelta/shoreface–offshore transition setting (Hampson 2000; Hampson & Howell 2005). The local thicker sandstone beds reflect deposition from rapidly expanding, energetic turbulent suspension flows. This facies association can also be found in the marginal or upper parts of channelized elements (Figs 3b & 6).

FA-3i: thin- to medium-bedded sandstones. Laterally persistent parallel-sided, fine-grained sandstone bedsets intercalated with thinner very fine-grained sandstone bedsets. Beds are up to 20 cm thick. Some beds are normally or inversely graded (Sg), without structures, but most of them show a range of sedimentary structures, including parallel lamination (Sp, Fig. 5e), trough cross-bedding or low-angle or hummocky cross-stratification (Sx, Sl, Fig. 5e, f) and the bed tops exhibit both symmetrical and asymmetrical ripple forms (Sr, Sw, Fig. 5g, h). Climbing ripple cross-lamination is common. Bioturbation is generally from moderate to high and organic fragments are commonly observed in bedding planes.

Normal- and/or inverse-graded beds are interpreted to represent deposition from river-derived waning and/or waxing sediment-laden flows, respectively (e.g. Mulder *et al.* 2003; Plink-Björklund & Steel 2004; Petter & Steel 2006; Olariu *et al.* 2010). The structured beds indicate tractional reworking by high concentration flows and, in the case of climbing ripples, high sedimentation rates possibly associated with an abrupt decrease in flow confinement.

The depositional environment of FA-3i is interpreted depending on the dominance of particular sedimentary structures. The planar, trough and climbing ripple cross-lamination facies association is related to deposition from unidirectional flows in distal mouth bars. Where low-angle, hummocky cross-stratification–swaley cross-stratification in cleaner sandstones predominates, deposition is interpreted to be in a distal wave-influenced delta front or lower shoreface setting (e.g. Hampson 2000; Bhattacharya & Giosan 2003; Hampson & Howell 2005). Thin-bedded sandstones are also observed in the marginal or upper parts of channelized bodies (Figs 3b & 6).

FA-3ii: medium- to thick-bedded sandstones. Laterally extensive or channelized, very fine to fine-grained, predominantly structureless or graded sandstone (Ss, Sg, Fig. 5c, d) bedsets containing amalgamation surfaces, scattered mudstone clasts (Cm, Fig. 5b) and fluid escape structures. Structureless divisions can pass upwards into thick-bedded parallel-laminated (Sp, Fig. 5e) or low-angle cross-laminated sandstones (Sl), which, in turn, pass upwards into asymmetrical or symmetrical ripple-laminated thick-bedded sandstones. Laterally, bedsets either onlap directly onto inclined erosional surfaces or pass laterally into thin-bedded sandstones (FA-3i) or interbedded sandstones and siltstones (FA-2). In channelized examples, the lower part of this facies association usually includes intraformational pebble-sized mudstone clasts in irregular horizons of clast-supported conglomerate to sandy matrix-supported conglomerate in a sandy matrix (Ci). Bioturbation is highly variable, from low to high, and organic fragments can be present in bedding planes.

The nature of the structureless or graded sandstone beds suggests rapid deposition and frictional freezing from high-density sediment gravity flows with a high fallout rate that prevents bedload transport and traction in the base of the flow (e.g. Mutti *et al.* 2003; Tinterri 2007). Basal mudstone clast horizons are residual lag deposits of material eroded and left behind by largely bypassing flows (Stevenson *et al.* 2015). Parallel-laminated beds could be associated with upper flow regime plane bedding or downstream migration of very low amplitude long wavelength bedforms, as has been described in younger fluvial deposits of the Beaufort Group (Turner 1981; Stear 1983, 1985; Wilson *et al.* 2014). Cross- and ripple-laminated beds indicate aggradation and traction processes. This facies association suggests relatively high-energy depositional

(a)

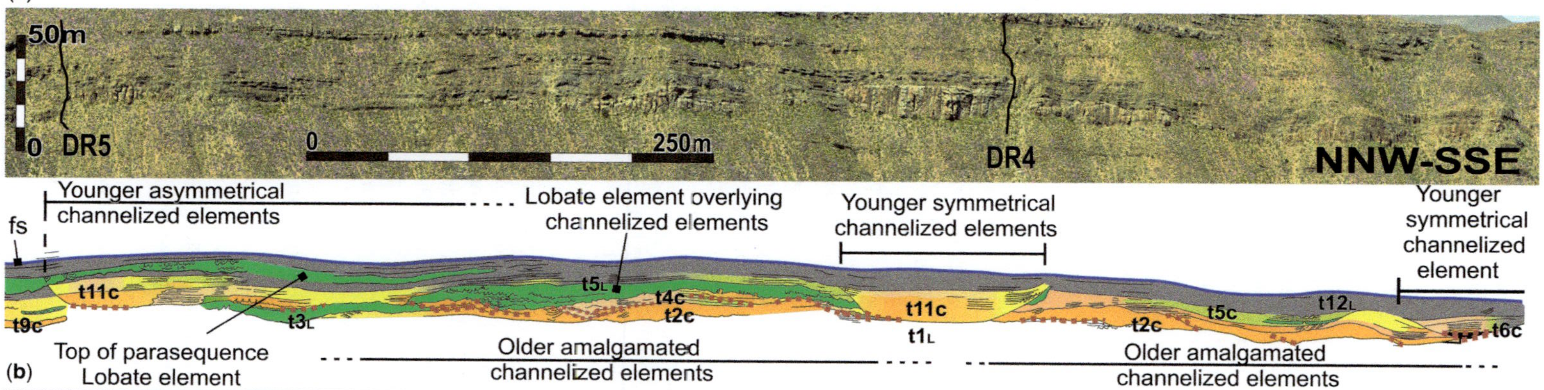

(b)

Dimensions - Channelised Elements													Mean
Time (t)	8	11	2	5	2	8	2	11	9	6	1	6	
WIDTH	>450	>350	>308	>158	>208	133	>133	128	>116	116	>75	75	187.5
THICKNESS	25	>10	>10	>10	>6	20	>7	25	>8	5	>10	15	12.58
LENGTH	-	>317	>362	-	>340	>357	-	>288	>304	>110	>340	-	201.5

Dimensions - Lobate Elements						Mean
Time (t)	12	3	5	5	7	
WIDTH	>667	>625	>408	>325	>168	400.2
THICKNESS	8	>5	>8	>10	>10	8.2

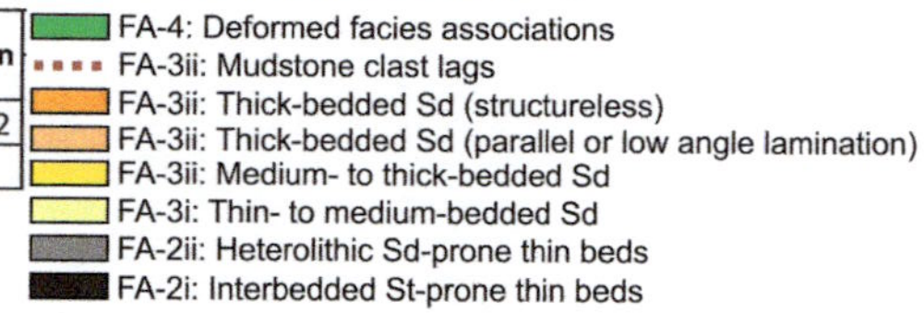

Fig. 6. (**a**) Dimensions, geometries and facies associations observed by depositional element for channelized bodies (symmetrical, asymmetrical or amalgamated fills) and lobate bodies. Sd, Sandstone; St, Siltstone. (**b**) Summary table of width, thickness and length ('>' maximum measured due to outcrop constraints) values for the channelized and lobate elements measured in the middle unit of the Paardeberg Ridge.

environments and is found either filling the central parts of concave-up erosion surfaces (scour- or channel-fills) or in the upper part of lenticular/tabular elements interpreted as mixed-influenced proximal mouth bars (Figs 3b & 6).

FA-4: deformed facies associations. Laterally discontinuous packages of folded and distorted sandy or silty thin- (FA-1, -2) to thick beds (FA-3) that can extend laterally for several hundred metres. They show a wide spectrum of soft sediment deformation types, ranging from the complete destruction of primary sedimentary structures and contorted sandstone clasts floating in a sandy/silty matrix, to units with coherent upwards-directed folds in sandstones. They usually contain large-scale (<5 m high) fluid escape structures and commonly occur associated with erosional and irregular surfaces.

Depending on the original structures and relative stratigraphic position, deformation is interpreted to record a spectrum of processes, from *in situ* foundering to remobilization (Owen 2003; Wild 2005; Wild *et al.* 2009; Oliveira *et al.* 2011; Jones *et al.* 2013), related to failure or gravitational collapse in over-steepened delta front or channel margin settings. These deposits represent *c.* 30% of the middle unit and are also present in the lower part of the upper unit (Figs 2b, 3b & 6).

Stratigraphic evolution

The Kookfontein to lower Waterford formations in the study area are informally divided into lower, middle and upper units (Figs 2 & 3). The lower unit succession starts with a 30–40 m-thick dark grey mudstone (FA-1), with some isolated sandstone bedsets up to 3 m thick (FA-3ii). This is overlain by four 20–75 m-thick heterolithic coarsening- and thickening-upwards packages (FA-1 to FA-2ii), each starting with a regionally extensive mudstone (FA-1). The succession is abruptly overlain by the 30 m-thick middle unit, which consists of tabular or convex-topped deformed sandstones incised by multiple erosive surfaces overlain by sandstones (FA-2 and FA-3). The middle unit is capped by a 2 m-thick regionally extensive mudstone (Fig. 3b), above which the upper unit consists of thinner (10–50 m) sandier and deformed coarsening- and thickening-upwards packages with a progressively lower mudstone content. These packages are overlain by a mudstone-dominated unit, with relatively isolated sandstone beds or bedsets that are either tabular or channelized (FA-3i and FA3-ii). A gradual change in colour from dark grey to green and purple is observed in the uppermost mudstones.

The basal mudstone-prone lower unit with intercalated sandstone beds is interpreted as slope mudstones and turbidites. The overlying four coarsening- and thickening-upwards heterolithic packages of the lower unit are interpreted as dominated by prodelta and offshore deposits. The vertical profile and scale of these stratigraphic packages are consistent with parasequences bounded by flooding surfaces as described by Van Wagoner *et al.* (1990) and their stratigraphic setting is consistent with an upper slope position. This parasequence set is overlain abruptly by erosional channel-fills and deformed lobate bodies (mixed-influence mouth bars) of the middle unit, described and interpreted in detail in the following sections. The regional capping mudstone is interpreted as containing a flooding surface. The sandier, thinner and deformed deposits of the upper unit are interpreted as wave-dominated or mixed-influenced shelf delta/shoreface parasequences (Wf of Ainsworth *et al.* 2011). These are overlain by isolated fluvial sandstone bodies within grey–green and purple floodplain mudstones of the lower Beaufort Group (Wilson *et al.* 2014).

Palaeocurrent analysis and palaeoshoreline orientation

Palaeocurrent analysis in this study, combined with previous publications on the Kookfontein and Waterford formations and underlying submarine fan systems in the Tanqua depocentre, indicate a uniform regional palaeoflow to the NE and north, with a slope and overlying shelf oriented approximately NW–SE (Wickens 1994; Johnson *et al.* 2001; van der Werff & Johnson 2003; Wild *et al.* 2005, 2009; Hodgson *et al.* 2006; Luthi *et al.* 2006; Jones *et al.* 2013). In the mud-prone (FA-1) lower unit, there is a general east to NE unidirectional palaeoflow, with an east–west trend from symmetrical ripples in sandstone beds (Fig. 4a). The overlying parasequences show a north to NE spread in unidirectional current ripples and a NE–SW trend for the bidirectional indicators (Fig. 4a). All this evidence suggests that the NW–SE orientation of the Paardeberg Ridge outcrop is a strike section to the palaeoshoreline and shelf edge during progradation of the basin margin (Fig. 7).

Depositional elements in the middle unit

Architectural descriptions of the Paardeberg Ridge exposure (Fig. 3) are focused on the sandstone-dominated middle unit, which is characterized by two geometries: channelized and lobate.

Channelized bodies

In cross-section (Fig. 3), these bodies have basal concave-up surfaces (Fig. 6) that truncate underlying deposits with up to 20 m of incision. Individual

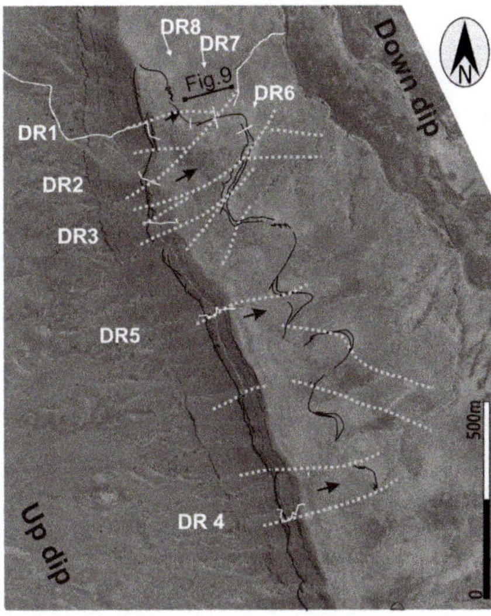

	Channels	
	SW	NE
	up dip	down dip
Thickness	thinner	thicker
Incision/ erosion	less deep	more deeply incised
Mud clast lags	less	more
Soft sediment deformation	less	more
Cutting/ Facies eroded	shallower facies: delta front or other channels	deeper facies: prodelta

Fig. 7. Map of the channelized elements from SW to NE of the Paardeberg Ridge. Black lines represent mapped erosion surfaces. Dotted lines represent correlative surfaces between the two sides of the Paardeberg Ridge. Arrows represent main palaeocurrents. On the right, the table summarizes the differences observed comparing the elements in the updip and downdip positions.

channel bodies (see table in Fig. 6) range from 5 to 20 m thick and 50–400 m wide which, given the strike orientation of the outcrop, are close to the true widths. They can be followed, in planform view (Fig. 7), for up to 350 m downdip and they increase in width and thickness downdip. They are commonly cut by younger erosion surfaces so that only remnant fills are preserved (Fig. 6). Where fully preserved, the cross-sectional geometries range from symmetrical (Fig. 6) to asymmetrical, with one steeper erosive side and the other side showing a low-angle contact and a lateral facies change (Figs 3b & 6). Loaded or slightly deformed bases and abundant mudstone clast conglomerates close to the erosional bases are common (Figs 3b & 6).

Facies associations within channelized bodies stack to form fining- and thinning-upwards packages (Fig. 8), which in most cases also fine and thin laterally (Fig. 6). Packages have basal mudstone clast conglomerates (Fig. 5a, b) overlain by thick-bedded structureless sandstones (FA-3ii, Fig. 5c) that grade vertically and laterally into parallel-laminated (Fig. 5e) or very low-angle laminated thick-bedded sandstones (FA-3ii). These

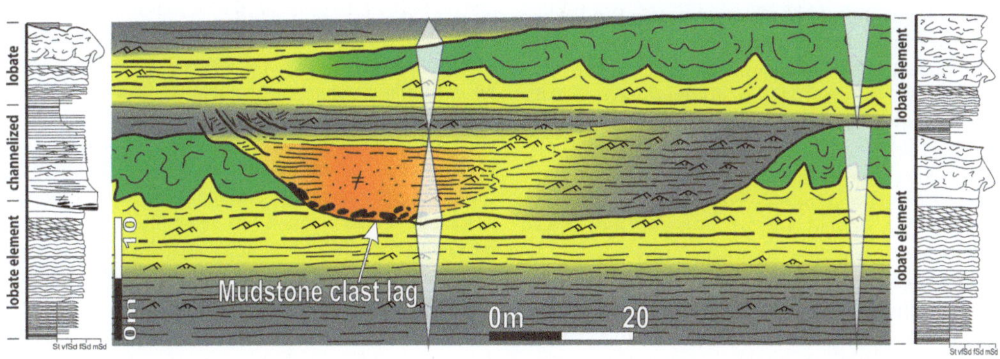

Fig. 8. Sketch showing geometrical relationships observed between channelized and lobate elements in strike section. Colour code is the same as in Figures 3 and 6. Left log shows a lobe element cut by a channelized element and overlain by the off-axis part of another lobate element (triangles in the middle of the figure represent the position of the left log). Right log shows two stacked lobe elements (triangles on the right show position of the right log).

deposits pass gradually upwards into thin-bedded sandstones (<30 cm; FA-3i), which, in turn, may grade vertically and laterally into interbedded sandstones (up to 15 cm thick) and siltstones (FA-2ii and FA-2i), with a gradual decrease in sandstone content (Figs 6 & 8). Sandstones in the thin-bedded and interbedded packages are either structureless, parallel-laminated or with symmetrical and asymmetrical ripple cross-lamination and commonly show symmetrical rippled tops.

These bodies are interpreted to be channel deposits, showing multiple phases of cut and fill. Their facies associations and position are consistent with subaqueous distributary channels in a deltaic setting; their scale, localization and the processes responsible for their origin and fill are discussed in the following sections.

Lobate bodies

These bodies are generally lens-shaped in cross-section with irregular bases (deformed or slightly erosive) and flat to convex-up tops, although the cross-sectional geometry can be modified by subsequent channel erosion (Figs 3 & 6). The axes consist of fine-grained sandstone thinning laterally into finer-grained facies. Lobate bodies (see table in Fig. 6) range from 5 to 10 m thick and from 75 to 600 m wide (true width) and have been mapped for up to 300 m downdip (Fig. 7). Typically, they overlie the thinner, off-axis parts of older lenticular bodies (Fig. 6) or previously deposited channelized elements (Fig. 6), indicating a compensational stacking relationship (Fig. 3b).

Facies associations in lobate bodies stack in coarsening- and thickening-up packages (*c.* 15 m thick) (Figs 6 & 8). A lower section of sand-prone interbedded siltstones and sandstones (FA-2ii) commonly coarsens upwards into thin-bedded sandstone (FA-3i). Sandstone beds are normally graded with parallel lamination and both asymmetrical and symmetrical ripple lamination. Most bed tops show symmetrical ripples. Towards the upper part of the bodies soft sediment deformed packages (FA-4; Fig. 8) show slightly erosive, heavily distorted or loaded bases (Fig. 6). Where the primary fabric is preserved, in most cases it is thick-bedded sandstone (FA-3ii) and/or interbedded sandstone and siltstone (FA-2i and FA-2ii). Flames and other fluid escape structures at loaded bed bases can reach up to 4 m high. Thin- to thick-bedded sandstones with abundant climbing ripple lamination (FA-3, Fig. 5f) are commonly found above, below or between deformed deposits, sometimes eroding the underlying deformed deposits, or in blocks within the deformed packages.

The mixed-process origin with indicators of both river and wave regimes and the widespread soft sediment deformation in these lenticular sandstone bodies makes it challenging to assign a simple interpretation to them. However, the modified coarsening-upwards trends with unidirectional currents, reworked bed tops and the close association with subaqueous distributary channels, lead us to interpret and refer to them as mixed-influence (wave and river) and remobilized mouth bars that were reworked into more strike-elongate lobate geometries.

Architecture and stratigraphic evolution of the middle unit

Cross-cutting relationships between sedimentary bodies allow reconstruction of the relative ages of the channelized and lobate deposits (tx_C or tx_L, respectively in Fig. 3) and therefore also allow the stratigraphic evolution of the middle unit to be interpreted.

The architecture suggests a spatial relationship between the underlying deposits and the positioning of younger elements. The superposition of channels and mixed-influence mouth bars follows a complex temporal distribution (Fig. 3), consistent with lateral compensation processes. Combined with the facies associations, the architecture could also indicate the proximity to an input point at this locality (Olariu & Bhattacharya 2006; Olariu & Steel 2009).

In general, the older channelized elements are sandier, with more structureless sandstone, and wider (Figs 3b & 6) than the younger channels. Commonly, they cut into deeper-water offshore facies and also locally into prodelta and/or deformed mixed-influence mouth bars (Fig. 3b). Most of these older channels show multiple scours with abundant mudstone clast lags (Figs 3b & 6), which may suggest recurrent erosion processes and sediment bypass at the same location over time. Younger channels display a wider spectrum of facies associations and are generally narrower with simpler cuts (Fig. 6). Cleaner basal surfaces, less amalgamation and considerably less mudstone clast lags at their bases are consistent with less erosive and shorter-lived channels. Typically, these younger channels cut shallower-water facies associations (deformed mixed-influence mouth bars) and their fills show a greater degree of wave reworking towards their tops, suggesting deposition in a shallower-water setting.

Lateral and vertical relationships between depositional elements

Typically, a lobate element (or mixed-influence mouth bar) is cut or partially eroded by a channelized element (Figs 6 & 8). The basal incision surfaces of channelized bodies usually cut from the top of the underlying lobate element. The presence of collapsed mixed-influence mouth bar facies in channel-fills has been locally observed (Figs 8 & 9), indicating that the channels were open conduits and suggesting a close relationship between mouth bar deposition and channel-forming processes (Olariu & Bhattacharya 2006). Channelized bodies are, in turn, overlain by younger lobate elements and this pattern is repeated vertically and laterally along the outcrop.

Palaeocurrents in the middle unit

In the channelized bodies, unidirectional measurements from current and climbing ripples, low-angle cross-bedding and basal groove marks show a dominant NE and north trend, with a dispersion of up to 90° (Fig. 4a, b). The orientation of channel margins supports the general SW to NE trend (Fig. 7). Current–ripple laminations from channel margin thin beds show a NE trend and symmetrical ripples are oriented NE–SW (Fig. 4b), consistent with the regional wave reworking measurements (Fig. 4a). Palaeocurrents in the mixed-influence mouth bars or lobate elements show a similar bidirectionality (NE(N) to SW(S)), but the unidirectional measurements show a dominant northwards trend with a spread of 90° (Fig. 4c). Orientations from the upper parts of mixed-influence mouth bars or lobate elements show a wider spread than the lower parts (Fig. 4c). The consistent unidirectional NE trends and the dominant, consistent NE–SW (040–220° to 080–260°) trend of bidirectional measurements from symmetrical ripple crest lines (Fig. 4d) supports the overall NW–SE shoreline orientation.

Discussion

A mixed-influence shelf edge delta

The Paardeberg Ridge locality lies 15 km downdip of, and *c.* 200 m stratigraphically above, the well-documented Tanqua deep-water succession, which consists of basin floor fans 1–4 (Wickens 1994; Hodgson *et al.* 2006) and the Unit 5 lower slope turbidite succession (Wild *et al.* 2005). This regional framework provides an upper slope stratigraphic context for the lower unit of the Paardeberg Ridge (mudstones and 20–75 m thick parasequences dominated by thin-bedded heterolithics). The thinner (10–25 m), sandier parasequences of the upper unit, with the common occurrence of hummocky cross-stratification and symmetrical ripples, are interpreted as wave-dominated and mixed-influenced shelf parasequences and share many characteristics with the Waterford Formation in other parts of the basin-fill (Wild *et al.* 2009; Jones *et al.* 2013, 2015). The sandstone-rich middle unit, which separates the underlying upper slope from the overlying shelf deposits, marks an abrupt increase in the amount of erosion and soft sediment deformation, with proximal facies associations of sandstone-rich mixed-influence mouth bars (FA-3) cut by sandstone-rich distributary channel-fills (FA-2 and FA-3). The stratigraphic context, combined with facies analysis and palaeocurrent data, supports the interpretation of the middle unit as a shelf edge deltaic package that formed during progradation towards the NE. The Karoo palaeoshelf margin is generally considered as a low gradient (Wild *et al.* 2009), stable slope type (*sensu* Ross *et al.* 1994), with limited synsedimentary growth faulting (Jones *et al.* 2013) and widespread amounts of soft sediment deformation and localized slumps (Wild *et al.* 2009; Oliveira *et al.* 2011; Jones *et al.* 2013).

The fill of channelized elements in the middle unit commonly consists of thick-bedded structureless axial sandstones that pass laterally through thin-bedded sandstones to interbedded channel margin sandstones and siltstones (Figs 6 & 8). Bed tops with symmetrical ripples following the regional NE–SW palaeocurrent trend are found in the thin-bedded channel margins and uppermost parts of the channel-fills. Although large storm waves can rework the seafloor at water depths >100 m, the ubiquity of symmetrical ripples and the consistency of the palaeocurrents suggest deposition above the fair weather wave base during channel filling. The thickness and generally structureless character in axial to marginal thick-bedded deposits suggest high-energy conditions, but dune-scale bedforms in channelized elements are rare. The paucity of cross-bedding can be attributed either to unusually high discharges during short, flash flood events (Stear 1983) or to the inhibition of bedform formation due to the narrow grain size range (clay to fine sand) in the whole Karoo Basin (Southard 1971; Turner 1981; Van den Berg & Van Gelder 1993). This limits the spectrum of possible sedimentary structures to lower and upper phase plane bedding and ripples (Rubin & McCulloch 1980; Southard & Boguchwal 1990). Low-angle cross-stratification is locally observed and has been associated with very low amplitude unidirectional bedforms (e.g. Turner 1981; Stear 1983, 1985; Wilson *et al.* 2014), but also to combined flow structures such as hummocky cross-stratification (Nøttvedt & Kreisa 1987; Southard *et al.* 1990). However, its expression is difficult to recognize and interpret in the absence of a wider grain size range. The facies and architectural elements of the studied succession suggest that different process regimes coexisted at the same location during clinothem

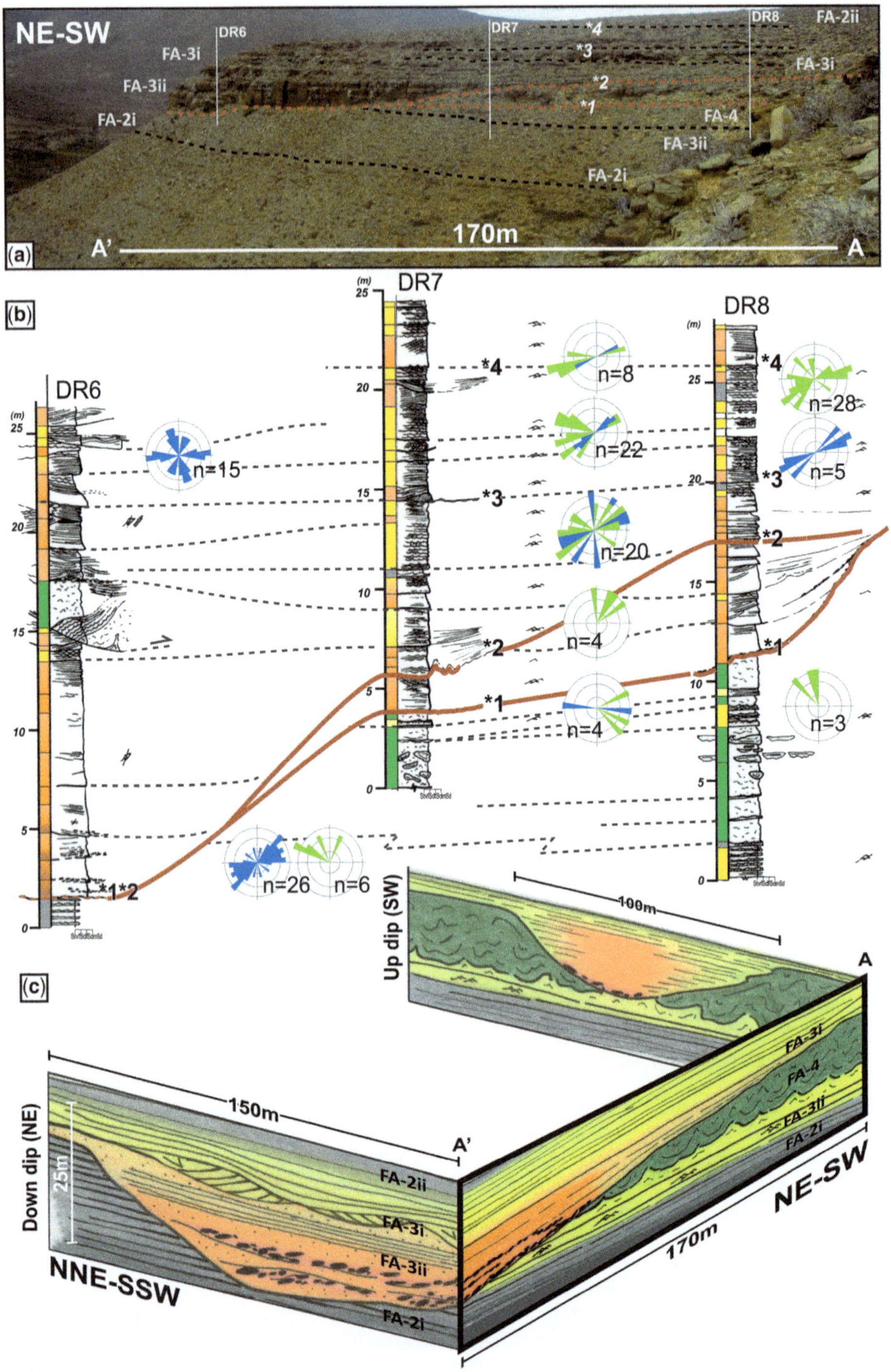

Fig. 9. (**a**) Photograph and (**b**) correlation between DR6, DR7 and DR8 between the SW and NE faces of the Paardeberg Ridge showing downdip deepening and thickening of a channel. Palaeocurrents are parallel with the exposure. *1, *2, *3 and *4 mark key surfaces. Blue represents bidirectional measurements; green represents unidirectional measurements. (**c**) Schematic representation showing the geometrical relationships between channelized and lobate bodies from updip (SW) to downdip (NE). Colour code as for Figures 3 and 6.

progradation and, as such, it is a rare example of an exhumed mixed-influence shelf edge system. Mixed-influence shelf edges have been observed in modern systems (e.g. Ainsworth *et al.* 2011; Olariu 2014) and in a few studies of ancient shelf margins (e.g. Mellere *et al.* 2003; Pontén & Plink-Björklund 2009; Olariu *et al.* 2012; Sanchez *et al.* 2012*b*; Jones *et al.* 2015).

Relative sea-level at the shelf edge

Parasequences of the lower unit show an overall progradational trend (DRs in Fig. 1b and DR1, DR6 and OUS in Fig. 10). The highly tabular character (Fig. 3a) and absence of lateral compensational stacking are interpreted to record relatively high accommodation conditions on the upper slope (Wild *et al.* 2009). The sharp and erosional basal contact of the sand-rich middle unit, with its mixed-influence mouth bars and distributary channels directly truncating upper slope thin beds (Figs 1b, 3b & 10), indicates a basinward shift of the facies belt. Highly erosive elements and evidence of sediment instability, the multi-storey stacking and the limited thickness (*c.* 30 m; Fig. 3b) of the middle unit are consistent with an abrupt change into a more proximal facies succession in a lower accommodation setting.

The abrupt change in depositional environments and stacking patterns in the middle unit does not support a simple progradational trend, which would be expected to display a gradual upwards increase in the sandstone content of parasequences as the delta approached the shelf edge. The basal erosion surface (t1C; Fig. 3b) of the middle unit is therefore interpreted as a sequence boundary (*sensu* Posamentier *et al.* 1988) juxtaposing a paralic succession onto the shelf edge rollover as a response to a relative fall in sea-level. No palaeosol deposit, roots nor desiccation cracks have been found to indicate subaerial exposure of the shelf edge, which suggests that shelf edge accommodation was reduced, but remained subaqueous. A similar situation has been documented in the Laingsburg depocentre by Jones *et al.* (2013), who interpreted this absence of subaerial exposure as consistent with a type 2 sequence boundary (*sensu* Posamentier *et al.* 1988). Correlation of the middle unit 1–3 km downdip to the Ouberg Pass area (Fig. 10) has revealed a succession of sandstone-dominated turbidites, which are interpreted to represent significant sediment bypass through the shelf edge rollover channels. The flooding surface overlying the middle unit (Fig. 3b) marks a relative rise in sea-level. Sufficient new accommodation was subsequently generated on the shelf to allow deposition of shelf parasequences of the upper unit, which are thinner, sandier and display shallower facies associations than those of the lower unit (Figs 1b & 3a).

Shelf edge channel geometry and initiation mechanisms

Additional evidence of a shelf edge rollover setting is derived from mapping (Fig. 7) and correlation between the two sides of the Paardeberg Ridge, which provides a planform control (100–350 m dip section) for the channelized elements of the middle unit (Figs 7 & 9). Channels are thicker, more incised, with a greater amount of mudstone clast lags and soft sediment deformation on the downdip northeastern side of the ridge than on the updip southwestern side (Fig. 9). On the updip side of the ridge, channels cut deformed mixed-influence mouth bar deposits (Figs 3b & 9) or older channel-fills, whereas on the downdip side they also cut down into prodelta and offshore deposits. This basinward increase in channel dimensions and depth of erosion is interpreted to reflect the increased gradient and accommodation across the shelf edge. The steeper gradient enhances the depth of erosion of distributary channels (Olariu & Bhattacharya 2006; Jackson & Johnson 2009) and the onset of gravity-driven density flows and gravitational collapse of channel margins across the rollover (e.g. Bowman & Johnson 2014) (Figs 10 & 11). Residual deposits, amalgamation surfaces and multi-storey stacking in the channels of the middle unit suggest relatively long-lived sediment bypass. This is consistent with the existence of a gullied/scoured slope. Two kilometres further downdip (NE of the Paardeberg Ridge; Figs 1c & 10) the prodelta–offshore succession is abruptly overlain by thick-bedded (0.5–1 m thick) normally graded sandstones that pass from structureless to climbing ripple cross-laminated. The beds are lens-shaped with erosional bases and sole marks oriented NE–SW (parallel to the main palaeocurrent direction). These deposits are interpreted as turbidites and the upper slope expression of the middle unit (Figs 10 & 11).

Although erosional features and sediment conduits (channels and gullies) are commonly observed at the shelf edge rollover (e.g. Porębski & Steel 2003; Sylvester *et al.* 2012; Bowman & Johnson 2014; Prélat *et al.* 2015), the processes involved in their origin require a wide range of possibilities to be considered. Several mechanisms of subaqueous scouring on the outer shelf region are rejected due to the scale and orientation of the incisions or the latitude of the basin at that time (between 40 and 60° S; Faure & Cole 1999). These include glacial scouring or bottom currents from melting (e.g. Ridente *et al.* 2007), longshore currents (e.g. Lewis 1982; Galloway 1998; Mazières *et al.* 2014),

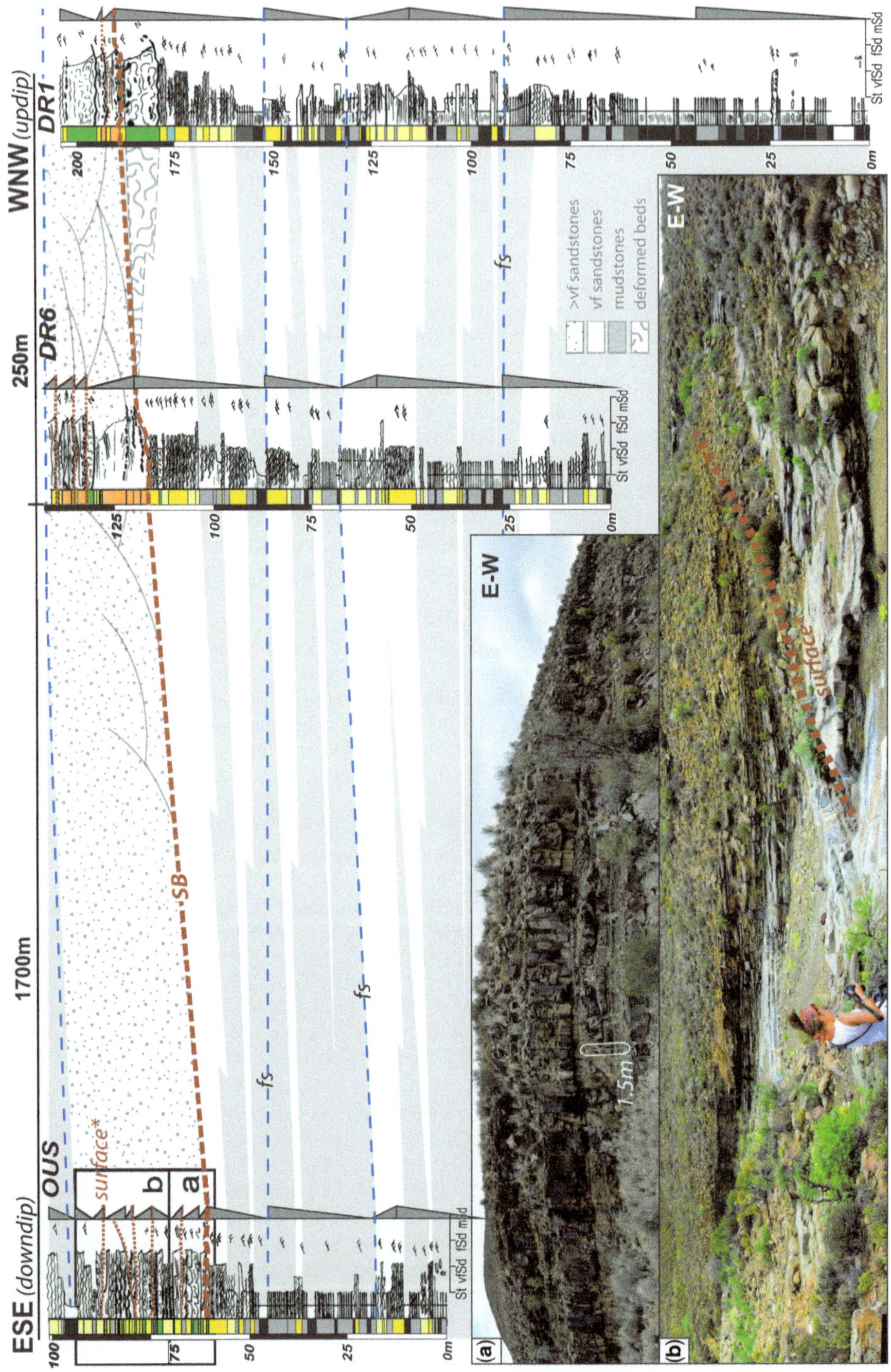

Fig. 10. Correlation between DR1, DR6 and the OUS section, 2 km downdip. The top of the OUS section is interpreted as turbidites deposited from flows that bypassed the channelized shelf edge rollover. Outcrop photographs **(a)** and **(b)** show details of the erosion downdip. Blue lines represent regional flooding surfaces and the red line is the correlation of the basal sequence boundary of the middle unit.

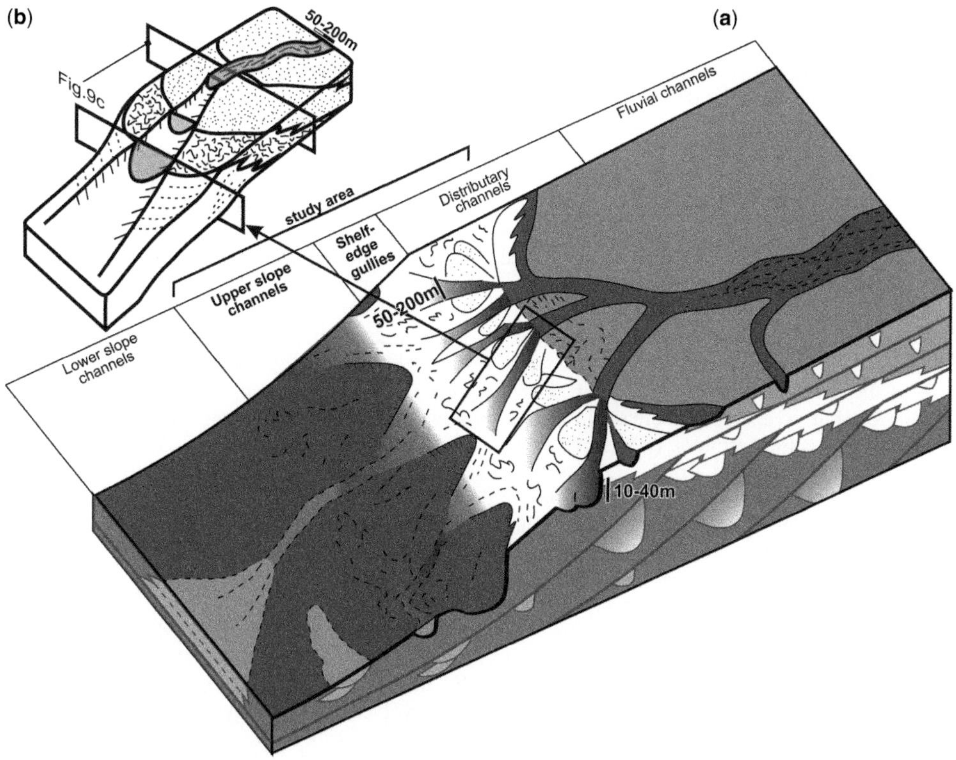

Fig. 11. (**a**) Synthesis of observations and interpretations showing generalized vertical and lateral stacking of depositional elements for a prograding mixed-influence shelf edge rollover. Note the stratigraphic decrease in channel size. (**b**) Detail of a channel morphology in this setting.

shallow-water bottom currents (e.g. Viana *et al.* 2002) and density cascades (e.g. Wilson & Roberts 1995; Shapiro & Hill 1997; Ivanov *et al.* 2004). Erosion from tide- or storm-generated rip currents or surges has been also invoked as a cause for the inception and maintenance of shelf edge conduits (Lewis 1982; Huthnance 1995; Clifton 2006; Normandeau *et al.* 2014); hyperpycnal flows, wave-supported gravity flows or inertial currents can keep conduits open (Huthnance 1995). These mechanisms cannot be discarded as they can develop local defects that potentially evolve into larger channels as the delta progrades and increases sediment supply, leading to loading and destabilization of the area (Lewis 1982).

Steeper slopes at the shelf edge rollover, combined with relatively high sediment loads associated with mouth bar progradation, create conditions for soft sediment deformation to occur (Owen 2003; Oliveira *et al.* 2011). The amount of soft sediment deformation (30%), combined with the observed interaction between channels and mixed-influence mouth bar deposits in the middle unit (Fig. 3b), suggests a close association between channel formation and delta-related instabilities (Oliveira *et al.* 2011). Channel incisions are apparently randomly located, but sometimes overlie the axial location of the underlying mixed-influence mouth bar. The local gradient change around the shelf edge rollover, combined with localized high sediment influx and loading (Wild *et al.* 2009) could have created unstable points around the axial positions of mixed-influence mouth bars. Liquefaction and deformation, particularly in their central and thicker parts, were highly susceptible to local remobilization and gravitational collapse (e.g. Jackson & Johnson 2009), creating discrete bathymetric lows that may have evolved into long-lived conduits through the bypass of gravity flows, generating a highly erosive and channelized shelf edge rollover.

Accretion and progradation of a shelf edge rollover

The deformed and sandier parts of lobate elements are generally 5–10 m thick and 80–600 m wide, whereas the channels are 5–20 m thick and 50–300 m wide (Figs 3b & 6). These geometrical proportions are not typical of terminal distributary systems (e.g. Olariu & Bhattacharya 2006; Olariu *et al.* 2010) because delta-related deposits would be expected to be significantly thicker than the documented channel dimensions. In the Cretaceous Blackhawk Formation of the Book Cliffs (Utah), prodeltaic channels formed by river-derived hyperpycnal flows are just a few metres thick and encased within mudstones (Pattison 2005; Pattison *et al.* 2007).

In shelf-confined mixed fluvial- and wave-influenced delta systems, feeder channels decrease in size as they reach the lower delta plain, where distributaries split, avulse and decrease in dimensions as they become terminal (Bhattacharya & Giosan 2003; Bhattacharya 2006). Observations made in dip and strike sections at both sides of the Paardeberg Ridge outcrop suggest that when distributary channels reached the shelf edge rollover, enhanced erosive and bypass processes created larger incisions and the channels became larger and deeper moving downslope across the shelf edge rollover (Figs 9 & 10). The significant gradient change across the shelf edge rollover may also explain the abrupt facies changes seen in the dip direction. This process might be particularly enhanced when associated with relative sea-level fall (Talling 1998; Muto & Steel 2002; Porębski & Steel 2006) and with the proximity of a feeder system (e.g. Mellere *et al.* 2003; Olariu & Bhattacharya 2006). The cross-cutting relationships between channel elements and lobate bodies suggest that the sudden loss of confinement led to deposition of mixed-influence mouth bars that forced distributary channels to step laterally. The smaller channel size upsection could reflect the start of backstepping as the gradient decreased when approaching the flooding surface above. However, considering that the channels are larger and erode more deeply basinwards (NE; Fig. 10), the upsection reduction in size within the middle unit is proposed to characterize progradation across the shelf edge rollover (Fig. 11).

Conclusions

The Paardeberg Ridge locality allows the spatial and temporal evolution of a NE-prograding shelf margin to be constrained in depositional dip and strike sections. Palaeocurrent data indicate a NW–SE-oriented shoreline with dominant NE–SW wave reworking. Truncation of a shallowing-upwards succession from upper slope turbidites to shelf edge deposits by a 30 m thick sandstone-rich unit of deformed and mixed-influence mouth bars and subaqueous distributary channels represents an abrupt juxtaposition of paralic deposits onto the shelf edge rollover. The basal erosion surface of the unit is correlated downdip to slope turbidite sandstones, interpreted as deposits of flows that bypassed through the channelized shelf edge rollover. This surface is interpreted as a sequence boundary without subaerial shelf exposure.

In strike section, the older channelized elements are wider and more deeply incised, with composite erosion surfaces filled by amalgamated structureless sandstones, and cut into deeper-water facies, suggesting recurrent erosion and sediment bypass. Younger channel-fills are narrower with cleaner cuts that incise into shallower-water facies; their fills are better organized with fewer amalgamation surfaces and basal mudstone clast conglomerates, and more beds with symmetrical rippled tops. This is consistent with a vertical change to less erosive and shorter-lived channels.

A gradient increase at the shelf edge, combined with a high rate of sediment supply associated with deltaic progradation, promoted soft sediment deformation. Liquefied mixed-influence mouth bars were susceptible to local fluidization and remobilization and the resulting irregular surfaces probably evolved into subaqueous sediment conduits. When the distributaries reached an unstable and relatively steep region, enhanced erosion led to larger and deeper incisions across the shelf edge. The interpreted stratigraphic context and the fact that the channels incised into the proximal part of deformed mixed-influence mouth bars suggest that these subaqueous bodies are the shelf edge expression of distributary channels, associated with an abrupt basinwards shift of a mixed-influence deltaic system. This study therefore documents a rare example of the architecture of a progradational mixed-influence paralic succession superimposed on a shelf edge rollover in response to a lowering of relative sea-level.

The work presented here is part of the SLOPE Project, Phase 4. We thank the consortium of sponsors (Anadarko, BHPBilliton, BP, CNOOC-Nexen, ConocoPhillips, E.ON, Engie, Maersk, Murphy, Petrobras, Shell, Statoil, Total, VNG Norge and Woodside) for financial support and strong scientific engagement. The manuscript benefited from insightful reviews by Gary Hampson, Cornel Olariu and an anonymous reviewer. Sarah Cobain, Rhodri Jerrett, Eoin Dunlevy and Sascha Eichenauer are thanked for their collaboration and assistance in the field. We thank the Karoo farmers for access to their land.

References

Ainsworth, R.B., Vakarelov, B.K. & Nanson, R.A. 2011. Dynamic spatial and temporal prediction of changes in depositional processes on clastic shorelines: toward improved subsurface uncertainty reduction and management. *American Association of Petroleum Geologists Bulletin*, **95**, 267–297, https://doi.org/10.1306/06301010036

ANDERSSON, P.O.D., WORDEN, R.H., HODGSON, D.M. & FLINT, S. 2004. Provenance evolution and chemostratigraphy of a Palaeozoic submarine fan-complex: Tanqua Karoo Basin, South Africa. *Marine and Petroleum Geology*, **21**, 555–577, https://doi.org/10.1016/j.marpetgeo.2004.01.004

BHATTACHARYA, J.P. 2006. Deltas. *In*: WALKER, R.G. & POSAMENTIER, H.W. (eds) *Facies Models Revisited*. SEPM Special Publications, **84**, 237–292.

BHATTACHARYA, J.P. & GIOSAN, L. 2003. Wave-influenced deltas: geomorphological implications for facies reconstruction. *Sedimentology*, **50**, 187–210, https://doi.org/10.1046/j.1365-3091.2003.00545.x

BHATTACHARYA, J.P. & MACEACHERN, J.A. 2009. Hyperpycnal rivers and prodeltaic shelves in the Cretaceous seaway of North America. *Journal of Sedimentary Research*, **79**, 184–209.

BOUMA, A.H. & WICKENS, H.D.V. 1991. Permian passive margin submarine fan complex, Karoo Basin, South Africa: possible model to Gulf of Mexico. *Gulf Coast Association of Geological Societies Transactions*, **41**, 30–42.

BOURGET, J., AINSWORTH, R.B. & THOMPSON, S. 2014. Seismic stratigraphy and geomorphology of a tide or wave dominated shelf-edge delta (NW Australia): process-based classification from 3D seismic attributes and implications for the prediction of deep-water sands. *Marine and Petroleum Geology*, **57**, 359–384, https://doi.org/10.1016/j.marpetgeo.2014.05.021

BOWMAN, A.P. & JOHNSON, H.D. 2014. Storm-dominated shelf-edge delta successions in a high accommodation setting: the palaeo-Orinoco Delta (Mayaro Formation), Columbus Basin, South-East Trinidad. *Sedimentology*, **61**, 792–835, https://doi.org/10.1111/sed.12080

CARVAJAL, C. & STEEL, R. 2009. Shelf-edge architecture and bypass of sand to deep water: influence of shelf-edge processes, sea level, and sediment supply. *Journal of Sedimentary Research*, **79**, 652–672, https://doi.org/10.2110/jsr.2009.074

CARVAJAL, C.R. & STEEL, R.J. 2006. Thick turbidite successions from supply-dominated shelves during sea-level highstand. *Geology*, **34**, 665–668, https://doi.org/10.1130/g22505.1

CATTANEO, A., TRINCARDI, F., ASIOLI, A. & CORREGGIARI, A. 2007. The Western Adriatic shelf clinoform: energy-limited bottomset. *Continental Shelf Research*, **27**, 506–525.

CATUNEANU, O., HANCOX, P.J. & RUBIDGE, B.S. 1998. Reciprocal flexural behaviour and contrasting stratigraphies: a new basin development model for the Karoo retroarc foreland system, South Africa. *Basin Research*, **10**, 417–439, https://doi.org/10.1046/j.1365-2117.1998.00078.x

CLIFTON, H.E. 2006. A re-examination of facies models for clastic shorelines. *In*: POSAMENTIER, H.W. & WALKER, R.G. (eds) *Facies Models Revisited*. SEPM Special Publications, **84**, 293–337.

COLE, D.I. 1992. Evolution and development of the Karoo Basin. *In*: DE WIT, M.J. & RANSOME, I.G.D. (eds) *Inversion Tectonics of the Cape Fold Belt, Karoo and Cretaceous Basins of Southern Africa*. A.A. Balkema, Rotterdam, 87–99.

COVAULT, J.A., ROMANS, B.W. & GRAHAM, S.A. 2009. Outcrop expression of a continental-margin-scale shelf-edge delta from the Cretaceous Magallanes Basin, Chile. *Journal of Sedimentary Research*, **79**, 523–539, https://doi.org/10.2110/jsr.2009.053

CRUMEYROLLE, P., RENAUD, I. & SUITER, J. 2007. The use of two- and three-dimensional seismic to understand sediment transfer from fluvial to deepwater via sinuous channels: example from the Mahakam shelf and comparison with outcrop data (south central Pyrenees). *In*: DAVIES, R.J., POSAMENTIER, H.W., WOOD, L.J. & CARTWRIGHT, J.A. (eds) *Seismic Geomorphology: Applications to Hydrocarbon Exploration and Production*. Geological Society, London, Special Publications, **277**, 85–103, https://doi.org/10.1144/gsl.sp.2007.277.01.05

DE BEER, C.H. 1992. Structural evolution of the Cape Fold Belt syntaxis and its influence on syntectonic sedimentation in the SW Karoo Basin. *In*: DE WIT, M.J. & RANSOME, I.G.D. (eds) *Inversion Tectonics of the Cape Fold Belt, Karoo and Cretaceous Basins of Southern Africa*. A.A. Balkema, Rotterdam, 197–206.

DIXON, J.F., STEEL, R.J. & OLARIU, C. 2012*a*. River-dominated, shelf-edge deltas: delivery of sand across the shelf break in the absence of slope incision. *Sedimentology*, **59**, 1133–1157, https://doi.org/10.1111/j.1365-3091.2011.01298.x

DIXON, J.F., STEEL, R.J. & OLARIU, C. 2012*b*. Shelf-edge delta regime as a predictor of deep-water deposition. *Journal of Sedimentary Research*, **82**, 681–687, https://doi.org/10.2110/jsr.2012.59

DONOVAN, A.D. 2003. Depositional topography and sequence development. *In*: ROBERTS, H.H., ROSEN, N.C., FILLON, R.F. & ANDERSON, J.B. (eds) *Shelf Margin Deltas and Linked Down Slope Petroleum Systems: Global Significance and Future Exploration Potential*. GCSSEPM Foundation 23rd Annual Research Conference Proceedings, 493–522, https://doi.org/10.5724/gcs.03.23.0493

EDWARDS, M.B. 1981. Upper Wilcox Rosita delta system of South Texas; growth-faulted shelf-edge deltas. *American Association of Petroleum Geologists Bulletin*, **65**, 54–73.

FAURE, K. & COLE, D. 1999. Geochemical evidence for lacustrine microbial blooms in the vast Permian Main Karoo, Parana, Falkland Islands and Huab basins of southwestern Gondwana. *Palaeogeography, Palaeoclimatology, Palaeoecology*, **152**, 189–213.

FLINT, S.S., HODGSON, D.M. *ET AL*. 2011. Depositional architecture and sequence stratigraphy of the Karoo basin floor to shelf edge succession, Laingsburg depocentre, South Africa. *Marine and Petroleum Geology*, **28**, 658–674.

FULTHORPE, C.S. & AUSTIN, J.A. 1998. Anatomy of rapid margin progradation: three-dimensional geometries of Miocene clinoforms, New Jersey margin. *American Association of Petroleum Geologists Bulletin*, **82**, 251–273.

FULTHORPE, C.S., AUSTIN, J.A. & MOUNTAIN, G.S. 1999. Buried fluvial channels off New Jersey: did sea-level lowstands expose the entire shelf during the

Miocene? *Geology*, **27**, 203–206, https://doi.org/10.1130/0091-7613(1999)027<0203:bfconj>2.3.co;2

Fulthorpe, C.S., Austin, J.A. & Mountain, G.S. 2000. Morphology and distribution of Miocene slope incisions off New Jersey: are they diagnostic of sequence boundaries? *Geological Society of America Bulletin*, **112**, 817–828, https://doi.org/10.1130/0016-7606(2000)112<817:madoms>2.0.co;2

Galloway, W.E. 1998. Siliciclastic slope and base-of-slope depositional systems; component facies, stratigraphic architecture, and classification. *American Association of Petroleum Geologists Bulletin*, **82**, 569–595.

Hadler-Jacobsen, F., Johannessen, E.P., Ashton, N., Henriksen, S., Johnson, S.D. & Kristensen, J.B. 2005. Submarine fan morphology and lithology distribution: a predictable function of sediment delivery, gross shelf-to-basin relief, slope gradient and basin relief/topography. *In*: Doré, A.G. & Vining, B.A. (eds) *Petroleum Geology: North West Europe and Global Perspectives – Proceedings of the 6th Petroleum Geology Conference*. Geological Society, London, 1121–1145, https://doi.org/10.1144/0061121

Hampson, G.J. 2000. Discontinuity surfaces, clinoforms, and facies architecture in a wave-dominated, shoreface-shelf parasequence. *Journal of Sedimentary Research*, **70**, 325–340, https://doi.org/10.1306/2dc40914-0e47-11d7-8643000102c1865d

Hampson, G.J. & Howell, J.A. 2005. Sedimentologic and geomorphic characterization of ancient wave-dominated deltaic shorelines; Upper Cretaceous Blackhawk Formation, Book Cliffs, Utah, U.S.A. *In*: Giosan, L. & Bhattacharhya, J.P. (eds) *River Deltas – Concepts, Models, and Examples*. SEPM Special Publications, **83**, 133–154.

Helland-Hansen, W. 2010. Facies and stacking patterns of shelf-deltas within the Palaeogene Battfjellet Formation, Nordenskiöld Land, Svalbard: implications for subsurface reservoir prediction. *Sedimentology*, **57**, 190–208, https://doi.org/10.1111/j.1365-3091.2009.01102.x

Henriksen, S., Helland-Hansen, W. & Bullimore, S. 2011. Relationships between shelf-edge trajectories and sediment dispersal along depositional dip and strike: a different approach to sequence stratigraphy. *Basin Research*, **23**, 3–21, https://doi.org/10.1111/j.1365-2117.2010.00463.x

Hodgson, D.M., Flint, S.S., Hodgetts, D., Drinkwater, N.J., Johannessen, E.P. & Luthi, S.M. 2006. Stratigraphic evolution of fine-grained submarine fan systems, Tanqua Depocenter, Karoo Basin, South Africa. *Journal of Sedimentary Research*, **76**, 20–40, https://doi.org/10.2110/jsr.2006.03

Hubbard, S.M., Fildani, A., Romans, B.W., Covault, J.A. & McHargue, T.R. 2010. High-relief slope clinoform development: insights from outcrop, Magallanes Basin, Chile. *Journal of Sedimentary Research*, **80**, 357–375, https://doi.org/10.2110/jsr.2010.042

Huthnance, J.M. 1995. Circulation, exchange and water masses at the ocean margin: the role of physical processes at the shelf edge. *Progress in Oceanography*, **35**, 353–431, https://doi.org/10.1016/0079-6611(95)80003-C

Ivanov, V.V., Shapiro, G.I., Huthnance, J.M., Aleynik, D.L. & Golovin, P.N. 2004. Cascades of dense water around the world ocean. *Progress in Oceanography*, **60**, 47–98, https://doi.org/10.1016/j.pocean.2003.12.002

Jackson, C.A.L. & Johnson, H.D. 2009. Sustained turbidity currents and their interaction with debrite-related topography: Labuan Island, offshore NW Borneo, Malaysia. *Sedimentary Geology*, **219**, 77–96, https://doi.org/10.1016/j.sedgeo.2009.04.008

Johnson, M.R. 1991. Sandstone petrography, provenance and plate tectonic setting in Gondwana context of the southeastern Cape-Karoo Basin. *South African Journal of Geology*, **94**, 137–154.

Johnson, M.R., van Vuuren, C.J. *et al.* 2006. Sedimentary rocks of the Karoo Supergroup. *In*: Johnson, M.R., Anhaeusser, C.R. & Thomas, R.J. (eds) *The Geology of South Africa*. Geological Society of South Africa and Council for Geoscience, Pretoria, 461–499.

Johnson, S.D., Flint, S., Hinds, D. & De Ville Wickens, H. 2001. Anatomy, geometry and sequence stratigraphy of basin floor to slope turbidite systems, Tanqua Karoo, South Africa. *Sedimentology*, **48**, 987–1023.

Jones, G.E.D., Hodgson, D.M. & Flint, S.S. 2013. Contrast in the process response of stacked clinothems to the shelf-slope rollover. *Geosphere*, **9**, 299–316, https://doi.org/10.1130/ges00796.1

Jones, G.E.D., Hodgson, D.M. & Flint, S.S. 2015. Lateral variability in clinoform trajectory, process regime, and sediment dispersal patterns beyond the shelf-edge rollover in exhumed basin margin-scale clinothems. *Basin Research*, **27**, 657–680.

Kolla, V., Biondi, P., Long, B. & Fillon, R. 2000. Sequence stratigraphy and architecture of the Late Pleistocene Lagniappe delta complex, northeast Gulf of Mexico. *In*: Hunt, D. & Gawthorpe, R.L. (eds) *Sedimentary Responses to Forced Regressions*. Geological Society, London, Special Publications, **172**, 291–327, https://doi.org/10.1144/gsl.sp.2000.172.01.14

Lewis, D.W. 1982. Channels across continental shelves: corequisites of canyon-fan systems and potential petroleum conduits. *New Zealand Journal of Geology and Geophysics*, **25**, 209–225, https://doi.org/10.1080/00288306.1982.10421410

Luthi, S.M., Hodgson, D.M., Geel, C.R., Flint, S.S., Goedbloed, J.W., Drinkwater, N.J. & Johanessen, E.P. 2006. Contribution of research borehole data to modelling fine-grained turbidite reservoir analogues, Permian Tanqua-Karoo basin floor fans (South Africa). *Petroleum Geology*, **12**, 175–190.

Macquaker, J.H.S. & Bohacs, K.M. 2007. On the accumulation of mud. *Science*, **318**, 1734–1735, https://doi.org/10.1126/science.1151980

Matteucci, T.D. & Hine, A.C. 1987. Evolution of the Cape Fear Terrace: a complex interaction between the Gulf Stream and a paleo-shelf edge delta. *Marine Geology*, **77**, 185–205, https://doi.org/10.1016/0025-3227(87)90111-3

Mayall, M.J., Yeilding, C.A., Oldroyd, J.D., Pulham, A.J. & Sakurai, S. 1992. Facies in a shelf-edge delta; an example from the subsurface of the Gulf of Mexico, middle Pliocene, Mississippi Canyon, Block 109. *American Association of Petroleum Geologists Bulletin*, **76**, 435–448.

Mazières, A., Gillet, H., Castelle, B., Mulder, T., Guyot, C., Garlan, T. & Mallet, C. 2014. High-resolution morphobathymetric analysis and evolution of Capbreton submarine canyon head (Southeast Bay of Biscay—French Atlantic Coast) over the last decade using descriptive and numerical modeling. *Marine Geology*, **351**, 1–12.

Mellere, D., Plink-Björklund, P. & Steel, R.J. 2002. Anatomy of shelf deltas at the edge of a prograding Eocene shelf margin, Spitsbergen. *Sedimentology*, **49**, 1181–1206.

Mellere, D., Breda, A. & Steel, R.J. 2003. Fluvial-incised shelf-edge deltas and linkage to upper-slope channels (Central Tertiary Basin, Spitsbergen). *In*: Roberts, H.H., Rosen, N.C., Fillon, R.F. & Anderson, J.B. (eds) *Shelf Margin Deltas and Linked Down Slope Petroleum Systems: Global Significance and Future Exploration Potential*. GCSSEPM Foundation 23rd Annual Research Conference Proceedings, 231–266, https://doi.org/10.5724/gcs.03.23.0231

Milton, N. & Dyce, M. 1995. Systems tract geometries associated with Early Eocene lowstands, imaged on a 3D seismic dataset from the Bruce area, UK North Sea. *In*: Steel, R.J., Felt, V.L., Johannessen, E.P. & Mathieu, C. (eds) *Sequence Stratigraphy on the Northwest European Margin*. Norwegian Petroleum Society, Special Publications, **5**, 429–449.

Morton, R.A. & Suter, J.R. 1996. Sequence stratigraphy and composition of late Quaternary shelf-margin deltas, northern Gulf of Mexico. *American Association of Petroleum Geologists Bulletin*, **80**, 505–530.

Moscardelli, L., Wood, L.J. & Dunlap, D.B. 2012. Shelf-edge deltas along structurally complex margins: a case study from eastern offshore Trinidad. *American Association of Petroleum Geologists Bulletin*, **96**, 1483–1522, https://doi.org/10.1306/01241211046

Mulder, T., Syvitski, J.P.M., Migeon, S., Faugères, J.C. & Savoye, B. 2003. Hyperpycnal turbidity currents: initiation, behavior and related deposits: a review. *Marine and Petroleum Geology*, **20**, 861–882.

Muto, T. & Steel, R.J. 2002. In defense of shelf edge delta development during falling and lowstand of relative sea level. *The Journal of Geology*, **110**, 421–436, https://doi.org/10.1086/340631

Mutti, E., Tinterri, R., Benevelli, G., DiBiase, D. & Cavanna, G. 2003. Deltaic, mixed and turbidite sedimentation of ancient foreland basins. *Marine and Petroleum Geology*, **20**, 733–755.

Normandeau, A., Lajeunesse, P., St-Onge, G., Bourgault, D., Drouin, S.S.-O., Senneville, S. & Bélanger, S. 2014. Morphodynamics in sediment-starved inner-shelf submarine canyons (Lower St. Lawrence Estuary, Eastern Canada). *Marine Geology*, **357**, 243–255, https://doi.org/10.1016/j.margeo.2014.08.012

Nøttvedt, A. & Kreisa, R.D. 1987. Model for the combined-flow origin of hummocky cross-stratification. *Geology*, **15**, 357–361.

Olariu, C. 2014. Autogenic process change in modern deltas. *In*: Martinius, A.W., Ravnås, R., Howell, J.A., Steel, R.J. & Wonham, J.P. (eds) *From Depositional Systems to Sedimentary Successions on the Norwegian Continental Margin*. Wiley, Chichester, 149–166.

Olariu, C. & Bhattacharya, J.P. 2006. Terminal distributary channels and delta front architecture of river-dominated delta systems. *Journal of Sedimentary Research*, **76**, 212–233, https://doi.org/10.2110/jsr.2006.026

Olariu, C. & Steel, R.J. 2009. Influence of point-source sediment-supply on modern shelf-slope morphology: implications for interpretation of ancient shelf margins. *Basin Research*, **21**, 484–501, https://doi.org/10.1111/j.1365-2117.2009.00420.x

Olariu, C., Steel, R.J. & Petter, A.L. 2010. Delta-front hyperpycnal bed geometry and implications for reservoir modeling: Cretaceous Panther Tongue delta, Book Cliffs, Utah. *American Association of Petroleum Geologists Bulletin*, **94**, 819–845.

Olariu, M.I., Carvajal, C.R., Olariu, C. & Steel, R.J. 2012. Deltaic process and architectural evolution during cross-shelf transits, Maastrichtian Fox Hills Formation, Washakie Basin, Wyoming. *American Association of Petroleum Geologists Bulletin*, **96**, 1931–1956.

Oliveira, C.M.M., Hodgson, D.M. & Flint, S.S. 2011. Distribution of soft-sediment deformation structures in clinoform successions of the Permian Ecca Group, Karoo Basin, South Africa. *Sedimentary Geology*, **235**, 314–330.

Owen, G. 2003. Load structures: gravity-driven sediment mobilization in the shallow subsurface. *In*: van Rensbergen, P., Hillis, R.R., Maltman, A.J. & Morley, C.K. (eds) *Subsurface Sediment Mobilization*. Geological Society, London, Special Publications, **216**, 21–34, https://doi.org/10.1144/gsl.sp.2003.216.01.03

Patruno, S., Hampson, G.J. & Jackson, C.A.L. 2015. Quantitative characterisation of deltaic and subaqueous clinoforms. *Earth-Science Reviews*, **142**, 79–119, https://doi.org/10.1016/j.earscirev.2015.01.004

Pattison, S.A. 2005. Storm-influenced prodelta turbidite complex in the Lower Kenilworth Member at Hatch Mesa, Book Cliffs, Utah, USA: implications for shallow marine facies models. *Journal of Sedimentary Research*, **75**, 420–439.

Pattison, S.A., Ainsworth, R.B. & Hoffman, T.A. 2007. Evidence of across-shelf transport of fine-grained sediments: turbidite-filled shelf channels in the Campanian Aberdeen Member, Book Cliffs, Utah, USA. *Sedimentology*, **54**, 1033–1064.

Petter, A.L. & Steel, R.J. 2005. Deepwater-slope channels and hyperpycnal flows from the Eocene of the central Spitsbergen Basin: predicting basin-floor sands from a shelf edge/upper slope perspective. Paper presented at the AAPG Annual Meeting, 19–22 June 2005, Calgary, Alberta, Canada.

Petter, A.L. & Steel, R.J. 2006. Hyperpycnal flow variability and slope organization on an Eocene shelf margin, Central Basin, Spitsbergen. *American Association of Petroleum Geologists Bulletin*, **90**, 1451–1472.

Pirmez, C., Pratson, L.F. & Steckler, M.S. 1998. Clinoform development by advection-diffusion of suspended sediment: modeling and comparison to natural systems. *Journal of Geophysical Research*, **103**, 24141–24157, https://doi.org/10.1029/98jb01516

Plink-Björklund, P. & Steel, R. 2002. Sea-level fall below the shelf edge, without basin-floor fans. *Geology*, **30**, 115–118.

Plink-Björklund, P. & Steel, R.J. 2004. Initiation of turbidity currents: outcrop evidence for Eocene hyperpycnal flow turbidites. *Sedimentary Geology*, **165**, 29–52.

Plink-Björklund, P. & Steel, R. 2005. Deltas on falling-stage and lowstand shelf margins, the Eocene Central Basin of Spitsbergen: importance of sediment supply. *In*: Giosan, L. & Bhattacharya, J.P. (eds) *River Deltas – Concepts, Models, and Examples*. SEPM Special Publications, **83**, 179–206.

Poag, C.W., Swift, B.A., Schlee, J.S., Ball, M.M. & Sheetz, L.L. 1990. Early Cretaceous shelf-edge deltas of the Baltimore Canyon Trough: principal sources for sediment gravity deposits of the northern Hatteras Basin. *Geology*, **18**, 149–152, https://doi.org/10.1130/0091-7613(1990)018<0149:ecsedo>2.3.co;2

Pontén, A. & Plink-Björklund, P. 2009. Process regime changes across a regressive to transgressive turnaround in a shelf–slope basin, Eocene Central Basin of Spitsbergen. *Journal of Sedimentary Research*, **79**, 2–23, https://doi.org/10.2110/jsr.2009.005

Porębski, S.J. & Steel, R.J. 2003. Shelf-margin deltas: their stratigraphic significance and relation to deepwater sands. *Earth-Science Reviews*, **62**, 283–326.

Porębski, S.J. & Steel, R.J. 2006. Deltas and sea-level change. *Journal of Sedimentary Research*, **76**, 390–403.

Posamentier, H.W. & Kolla, V. 2003. Seismic geomorphology and stratigraphy of depositional elements in deep-water settings. *Journal of Sedimentary Research*, **73**, 367–388, https://doi.org/10.1306/111302730367

Posamentier, H.W., Jervey, M.T. & Vail, P.R. 1988. Eustatic controls on clastic deposition I – conceptual framework. *In*: Wilgus, C.K., Hastings, B.S., Kendall, C.G.S.C., Posamentier, H.W., Ross, C.A. & Van Wagoner, J.C. (eds) *Sea Level Changes – An Integrated Approach*. SEPM Special Publications, **42**, 109–124.

Pratson, L.F., Ryan, W.B.F., Mountain, G.S. & Twichell, D.C. 1994. Submarine canyon initiation by downslope-eroding sediment flows: evidence in late Cenozoic strata on the New Jersey continental slope. *Geological Society of America Bulletin*, **106**, 395–412, https://doi.org/10.1130/0016-7606(1994)106<0395:scibde>2.3.co;2

Prélat, A., Pankhania, S.S., Jackson, C.A.-L. & Hodgson, D.M. 2015. Slope gradient and lithology as controls on the initiation of submarine slope gullies; insights from the North Carnarvon Basin, Offshore NW Australia. *Sedimentary Geology*, **329**, 12–17, https://doi.org/10.1016/j.sedgeo.2015.08.009

Pyles, D.R. & Slatt, R.M. 2000. A high-frequency sequence stratigraphic framework for the shallow through deep-water deposits of the Lewis Shale and Fox Hills Sandstone, Great Divide and Washakie basins, Wyoming. *In*: Weimer, P. (ed.) *Deep-Water Reservoirs of the World*. GCSSEPM Foundation 20th Annual Research Conference Proceedings, 836–861, https://doi.org/10.5724/gcs.00.15.0836

Pyles, D.R. & Slatt, R.M. 2007. Applications to understanding shelf edge to base-of-slope changes in stratigraphic architecture of prograding basin margins: stratigraphy of the Lewis Shale, Wyoming, USA. *In*: Nilsen, T.H., Shew, R.D., Steffens, G.S. & Studlick, J.R.J. (eds) *Atlas of Deepwater Outcrops*. American Association of Petroleum Geologists Studies in Geology, **56**, 485–489.

Ridente, D., Foglini, F., Minisini, D., Trincardi, F. & Verdicchio, G. 2007. Shelf-edge erosion, sediment failure and inception of Bari Canyon on the southwestern Adriatic margin (Central Mediterranean). *Marine Geology*, **246**, 193–207, https://doi.org/10.1016/j.margeo.2007.01.014

Ross, W.C., Halliwell, B.A., May, J.A., Watts, D.E. & Syvitski, J.P.M. 1994. Slope readjustment: a new model for the development of submarine fans and aprons. *Geology*, **22**, 511–514.

Rubin, D.M. & McCulloch, D.S. 1980. Single and superimposed bedforms: a synthesis of San Francisco Bay and flume observations. *Sedimentary Geology*, **26**, 207–231.

Ryan, M.C., Helland-Hansen, W., Johannessen, E.P. & Steel, R.J. 2009. Erosional v. accretionary shelf margins: the influence of margin type on deepwater sedimentation: an example from the Porcupine Basin, offshore western Ireland. *Basin Research*, **21**, 676–703, https://doi.org/10.1111/j.1365-2117.2009.00424.x

Saller, A.H., Noah, J.T., Ruzuar, A.P. & Schneider, R. 2004. Linked lowstand delta to basin-floor fan deposition, offshore Indonesia: an analog for deepwater reservoir systems. *American Association of Petroleum Geologists Bulletin*, **88**, 21–46, https://doi.org/10.1306/09030303003

Sanchez, C.M., Fulthorpe, C.S. & Steel, R.J. 2012*a*. Middle Miocene–Pliocene siliciclastic influx across a carbonate shelf and influence of deltaic sedimentation on shelf construction, northern Carnarvon Basin, northwest shelf of Australia. *Basin Research*, **24**, 664–682, https://doi.org/10.1111/j.1365-2117.2012.00546.x

Sanchez, C.M., Fulthorpe, C.S. & Steel, R.J. 2012*b*. Miocene shelf-edge deltas and their impact on deepwater slope progradation and morphology, northwest shelf of Australia. *Basin Research*, **24**, 683–698, https://doi.org/10.1111/j.1365-2117.2012.00545.x

Schieber, J., Southard, J. & Thaisen, K. 2007. Accretion of mudstone beds from migrating floccule ripples. *Science*, **318**, 1760–1763, https://doi.org/10.1126/science.1147001

Scott, E.D., Bouma, A.H. & Wickens, H.d.V. 2000. Influence of tectonics on submarine fan deposition, Tanqua and Laingsburg subbasins, South Africa. *In*: Bouma, A.H. & Stone, C.G. (eds) *Fine-Grained Turbidite Systems*. American Association of Petroleum Geologists Memoirs, **72**, 47–56.

Shapiro, G.I. & Hill, A.E. 1997. Dynamics of dense water cascades at the shelf edge. *Journal of Physical Oceanography*, **27**, 2381–2394, https://doi.org/10.1175/1520-0485(1997)027<2381:dodwca>2.0.co;2

Smith, R.M.H. 1990. A review of stratigraphy and sedimentary environments of the Karoo Basin of South

Africa. *Journal of African Earth Sciences (and the Middle East)*, **10**, 117–137, https://doi.org/10.1016/0899-5362(90)90050-O

SOUTHARD, J.B. 1971. Representation of bed configurations in depth-velocity-size diagrams. *Journal of Sedimentary Petrology*, **41**, 903–915.

SOUTHARD, J.B. & BOGUCHWAL, L.A. 1990. Bed configurations in steady unidirectional water flows. Part 2. Synthesis of flume data. *Journal of Sedimentary Research*, **60**, 658–679.

SOUTHARD, J.B., LAMBIE, J.M., FEDERICO, D.C., PILE, H.T. & WEIDMAN, C.R. 1990. Experiments on bed configurations in fine sands under bidirectional purely oscillatory flow, and the origin of hummocky cross-stratification. *Journal of Sedimentary Research*, **60**, 1–17, https://doi.org/10.1306/212f90f7-2b24-11d7-8648000102c1865d

STEAR, W.M. 1983. Morphological characteristics of ephemeral stream channel and overbanks splay sandstone bodies in the Permian Lower Beaufort Group, Karoo Basin, South Africa. *In*: COLLINSON, J.D. & LEWIN, J. (eds) *Modern and Ancient Fluvial Systems*: International Association of Sedimentologists, Special Publications, **6**, 405–420.

STEAR, W.M. 1985. Comparison of the bedform distribution and dynamics of modern and ancient sandy ephemeral flood deposits in the southwestern Karoo region, South Africa. *Sedimentary Geology*, **45**, 209–230.

STEEL, R.J., CRABAUGH, J., SCHELPEPPER, M., MELLERE, D., PLINK-BJÖRKLUND, P., DEIBERT, J. & LØSETH, T. 2000. Deltas v. rivers on the shelf edge: their relative contributions to the growth of shelf-margins and basin-floor fans (Barremian and Eocene, Spitsbergen). *In*: WEIMER, P. (ed.) *Deep-Water Reservoirs of the World*. GCSSEPM Foundation 20th Annual Research Conference Proceedings, 981–1009, https://doi.org/10.5724/gcs.00.15.0981

STEEL, R.J., POREBSKI, S.J., PLINK-BJORKLUND, P., MELLERE, D. & SCHELLPEPER, M. 2003. Shelf-edge delta types and their sequence-stratigraphic relationships. *In*: ROBERTS, H.H., ROSEN, N.C., FILLON, R.F. & ANDERSON, J.B. (eds) *Shelf Margin Deltas and Linked Down Slope Petroleum Systems: Global Significance and Future Exploration Potential*. GCSSEPM Foundation 23rd Annual Research Conference Proceedings, 205–230, https://doi.org/10.5724/gcs.03.23.0205

STEVENSON, C.J., JACKSON, C.A.-L., HODGSON, D.M., HUBBARD, S.M. & EGGENHUISEN, J.T. 2015. Deep-water sediment bypass. *Journal of Sedimentary Research*, **85**, 1058–1081.

SUTER, J.R. & BERRYHILL, H.L. 1985. Late Quaternary shelf-margin deltas, northwest Gulf of Mexico. *American Association of Petroleum Geologists Bulletin*, **69**, 77–91.

SYDOW, J. & ROBERTS, H.H. 1994. Stratigraphic framework of a late Pleistocene shelf-edge delta, northeast Gulf of Mexico. *American Association of Petroleum Geologists Bulletin*, **78**, 1276–1312.

SYLVESTER, Z., DEPTUCK, M.E., PRATHER, B.E., PIRMEZ, C. & O'BYRNE, C. 2012. Seismic stratigraphy of a shelf-edge delta and linked submarine channels in the northeastern Gulf of Mexico. *In*: PRATHER, B., DEPTUCK, M.E., MOHRIG, D., VAN HOOR, B. & WYNN, R.B. (eds) *Application of the Principles of Seismic Geomorphology to Continental-Slope and Base-of-Slope Systems: Case Studies from Seafloor and Near-Seafloor Analogues*. SEPM Society for Sedimentary Geology Special Publications, **99**, 31–59.

TALLING, P.J. 1998. How and where do incised valleys form if sea level remains above the shelf edge? *Geology*, **26**, 87–90, https://doi.org/10.1130/0091-7613(1998)026<0087:hawdiv>2.3.co;2

TANKARD, A., WELSINK, H., AUKES, P., NEWTON, R. & STETTLER, E. 2009. Tectonic evolution of the Cape and Karoo basins of South Africa. *Marine and Petroleum Geology*, **26**, 1379–1412, https://doi.org/10.1016/j.marpetgeo.2009.01.022

TESSON, M., GENSOUS, B., ALLEN, G.P. & RAVENNE, C. 1990. Late Quaternary deltaic lowstand wedges on the Rhône continental shelf, France. *Marine Geology*, **91**, 325–332, https://doi.org/10.1016/0025-3227(90)90053-M

TINTERRI, R. 2007. The lower Eocene Roda Sandstone (South-Central Pyrenees): an example of a flood-dominated river-delta system in a tectonically controlled basin. *Rivista Italiana di Paleontologia e Stratigrafia*, **113**, 223–255.

TURNER, B.R. 1981. Possible origin of low angle cross-strata and horizontal lamination in Beaufort Group sandstones of the southern Karoo Basin. *Geological Society of South Africa, Transactions*, **84**, 193–197.

TURNER, B.R. 1999. Tectonostratigraphical development of the Upper Karoo foreland basin: orogenic unloading v. thermally-induced Gondwana rifting. *Journal of African Earth Sciences*, **28**, 215–238.

UROZA, C.A. & STEEL, R.J. 2008. A highstand shelf-margin delta system from the Eocene of West Spitsbergen, Norway. *Sedimentary Geology*, **203**, 229–245.

VAN DEN BERG, J.H. & VAN GELDER, A. 1993. A new bedform stability diagram, with emphasis on the transition of ripples to plane bed in flows over fine sand and silt. *In*: MARZO, M. & PUIGDEFABREGAS, C. (eds) *Alluvial Sedimentation*. International Association of Sedimentologists, Special Publications, **17**, 11–21.

VAN DER WERFF, W. & JOHNSON, S. 2003. High resolution stratigraphic analysis of a turbidite system, Tanqua Karoo Basin, South Africa. *Marine and Petroleum Geology*, **20**, 45–69, https://doi.org/10.1016/S0264-8172(03)00025-4

VAN WAGONER, J.C., MITCHUM, R.M., CAMPION, K.M. & RAHMANIAN, V.D. 1990. *Siliciclastic Sequence Stratigraphy in Well Logs, Cores, and Outcrops*. American Association of Petroleum Geologists Methods in Exploration Series, **7**.

VEEVERS, J.J., COLE, D.I. & COWAN, E.J. 1994. Southern Africa: Karoo Basin and Cape Fold Belt. *In*: VEEVERS, J.J. & POWELL, C.M. (eds) *Permian–Triassic Pangean Basins and Foldbelts Along the Panthalassan Margin of Gondwanaland*. Geological Society of America Memoirs, **184**, 223–279.

VIANA, A.R., HERCOS, C.M., DE ALMEIDA, W., MAGALHÃES, J.L. & DE ANDRADE, S.B. 2002. Evidence of bottom current influence on the Neogene to Quaternary sedimentation along the northern Campos Slope, SW Atlantic Margin. *In*: STOW, D.A.V., FAUGÈRES, J.-C., PUDSEY, C.J., HOWE, J.A. & VIANA, A.R. (eds)

Deep-Water Contourite Systems: Modern Drifts and Ancient Series, Seismic and Sedimentary Characteristics. Geological Society, London, Memoirs, **22**, 249–259, https://doi.org/10.1144/GSL.MEM.2002.022.01.18

Visser, J.N.J. 1993. Sea-level changes in a back-arc-foreland transition: the late Carboniferous–Permian Karoo Basin of South Africa. *Sedimentary Geology*, **83**, 115–131, https://doi.org/10.1016/0037-0738(93)90185-8

Wickens, H.d.V. 1994. *Basin floor fan building turbidites of the southwestern Karoo Basin, Permian Ecca Group, South Africa*. PhD thesis, University of Port Elizabeth.

Wild, R. 2005. *Sedimentological and sequence stratigraphic evolution of a Permian lower slope to shelf succession, Tanqua depocentre, SW Karoo Basin, South Africa*. PhD thesis, University of Liverpool.

Wild, R., Hodgson, D.M. & Flint, S.S. 2005. Architecture and stratigraphic evolution of multiple, vertically-stacked slope channel complexes, Tanqua depocentre, Karoo Basin, South Africa. *In*: Hodgson, D.M. & Flint, S.S. (eds) *Submarine Slope Systems: Processes and Products*. Geological Society, London, Special Publications, **244**, 89–111, https://doi.org/10.1144/GSL.SP.2005.244.01.06

Wild, R., Flint, S.S. & Hodgson, D.M. 2009. Stratigraphic evolution of the upper slope and shelf edge in the Karoo Basin, South Africa. *Basin Research*, **21**, 502–527.

Wilson, P.A. & Roberts, H.H. 1995. Density cascading: off-shelf sediment transport, evidence and implications, Bahama Banks. *Journal of Sedimentary Research*, **65**, 45–56, https://doi.org/10.1306/d426801d-2b26-11d7-8648000102c1865d

Wilson, A., Flint, S., Payenberg, T., Tohver, E. & Lanci, L. 2014. Architectural styles and sedimentology of the fluvial Lower Beaufort Group, Karoo Basin, South Africa. *Journal of Sedimentary Research*, **84**, 326–348, https://doi.org/10.2110/jsr.2014.28

Zavala, C., Arcuri, M., Di Meglio, M., Gamero, H. & Contreras, C. 2011. A genetic facies tract for the analysis of sustained hyperpycnal flow deposits. *In*: Slatt, R.M. & Zavala, C. (eds) *Sediment Transfer from Shelf to Deep Water – Revisiting the Delivery System*. AAPG Studies in Geology, **61**, 1–21.

Index

Figures are shown in *italics* and Tables in **bold**